SYNFUELS HANDBOOK

V. Daniel Hunt
Director, The Energy Institute

Synfuels Handbook

Industrial Press, Inc., 200 Madison Avenue, New York, N.Y. 10157

Notice

This book was prepared as an account of work sponsored by the Industrial Press Inc. Neither the Industrial Press nor The Energy Institute, nor any of its employees, nor any of their contractors, subcontractors, consultants, or their employees, makes any warranty, expressed or implied, or assumes any legal liability or responsibility for the accuracy, completeness, or usefulness of any information, apparatus, product, or process disclosed, or represents that its use would not infringe on privately owned rights.

The views, opinions, and conclusions in this book are those of the author and do not necessarily represent those of the United States Government or the United States Department of Energy.

Public domain information and those documents abstracted or used in full, edited, or otherwise used are noted in the acknowledgments or on specific pages or illustrations.

Library of Congress Cataloging in Publication Data

Hunt, V. Daniel.
Synfuels handbook.

Bibliography: p.
Includes index.
1. Synthetic fuels. I. Title.
TP360.H86 1983 662′.66 82-15751
ISBN 0-8311-1144-5

First Printing

Synfuels Handbook

To Ralph O. Robinson Jr., who raved, rewrote, revised, reviewed, and did all things possible to make me write better. I thank him for his enthusiasm, and interest in a fledgling writer.

Preface

The development of a U.S. synthetic fuel industry has already started to produce domestic energy supplies on a small scale. With further development and demonstration, commercial-sized plants can reduce the country's need for imported petroleum.

It is important to realize the time required to bring a new technology such as synthetic fuels to the marketplace. We must continue the development and production of synthetic fuels now, whether they are derived from oil shale, coal, or biomass.

The *Synfuels Handbook* provides comprehensive coverage of the scientific principles, production processes, environmental constraints, and cost differences of the various synthetic fuel processes. It covers the full range of advantages and disadvantages of each conversion technology and provides comparisons of projects already in operation, as well as those proposed for the near future.

Among other things, the *Synfuels Handbook* is a well-rounded handbook on coal and its use as a basic ingredient in synthetic fuels; on the scientific principles and processes involved in coal liquefaction and coal gasification; on oil shale uses; on the comparison of the significant coal conversion processes; on the utilization of synthetic fuels; on the environmental considerations; and on the comparison of process costs for the different process technologies. In just one volume, it

- Covers all elements of synthetic fuels—from fuel sources to specific process conversion technology
- Provides detailed cost and technical comparisons of process conversion technologies
- Discusses in a balanced manner the environmental impacts attributed to large-scale synthetic fuel development
- Describes the utilization of synthetic fuels and the value of specific products

The *Synfuels Handbook* is based on U.S. Department of Energy and Synthetic Fuels Corporation data; is well illustrated and presented in a clear manner; is supplemented by a bibliography, glossary, abbreviations and acronyms, sources of additional information, and an index for easy reference.

The *Synfuels Handbook* is the most comprehensive and up-to-date publication available describing the principles, processes, and potential of synthetic fuels.

A source of readily accessible information, *Synfuels Handbook* will answer the needs of chemical and mechanical engineers, and be invaluable to professionals, students, and laymen interested in gaining a better understanding about the development of synthetic fuels.

Acknowledgments

The information in the *Synfuels Handbook* has been obtained from a wide variety of authorities who are specialists in their respective field.

I wish to thank the staff of the U.S. Department of Energy's Fossil Energy Office for providing data, reports, photographs, and illustrations, which are used throughout this volume.

This handbook utilizes material of the Engineering Societies Commission on Energy, Inc. (ESCOE) prepared under contract to the U.S. Department of Energy. In particular I acknowledge the work of K. A. Rogers and R. F. Hill in their ESCOE work entitled *Coal Conversion Comparisons*, Report Number FE-2468-51.

I would like to thank those members of the TRW Energy Engineering Division who provided valuable information for this book. They include Mr. Archie Wood for the provision of the draft *Fossil Energy Program Summary for FY 1982*. Dr. P. J. LaRosa, E. T. Coles, Dr. H. L. Fein, J. J. Petros, R. S. Iyer, and R. T. Merritt for the material in the *Synthetic Fuel Cost Estimating Notebook* created by TRW's staff. Appreciation is also extended to TRW staff members who prepared the *U.S. Synthetic Fuels Program*, Spring 1981 Version, for the U.S. Synthetic Fuels Corporation. Portions of Chapter 3 are based on the TRW Report entitled *Achieving a Production Goal of 1 Million B/D of Coal Liquids by 1990*, dated February 8, 1980, Report No. 2623-16-04R1. I appreciate receiving a draft copy of the TRW report entitled *Incentives Relating to Synfuels Objectives and Technologies*, dated November 1980, from Mr. Alex Sherman. I acknowledge the help of Ms. Ann Heywood in providing a copy of TRW's report entitled *Utilization of Synthetic Fuels: An Environmental Perspective* July 1980.

I acknowledge use of portions of the *Gas Energy Review*, produced by the American Gas Association, September 1981, Volume 9, No. 9.

I acknowledge use of portions of Exxon Corporation's report entitled *The Role of Synthetic Fuels in the United States Energy Future*.

The following public domain publications were used as the basic technical resources for this handbook:

- *Coal Data: A Reference*, U.S. Department of Energy, Report No. DOE/EIA-0064, July 1980.
- *Securing America's Energy Future*, U.S. Department of Energy, DOE/S-0008, July 1981.
- *Environmental, Operational, and Economic Aspects of Thirteen Selected Energy Technologies* (U.S. Environmental Protection Agency), EPA-600/7-80-173, September 1980.

- *Synthetic Fuels Development—Earth Science Considerations*, U.S. Department of the Interior/Geological Survey, GPO Stock No. 024-001-03256-2, 1979.
- *Analysis of National Energy Plans*, MITRE Corporation, Report No. MTR-79W00346, November 1979.
- *An Assessment of Oil Shale Technologies*, Office of Technology Assessment, GPO Stock No. 052-003-00759-2, 1980.
- *Oil Shale Projects*, U.S. Department of Energy, April 1981.
- *Coal Conversion*, Technical Report, U.S. Department of Energy, DOE/FE-0010, September 1980.
- *International Coal Technology Summary Document*, U.S. Department of Energy, HCP/P-3885, December 1978.

The preparation of a book of this magnitude is dependent upon an excellent staff, and I have been fortunate in this regard. Judith A. Anderson has effectively served as technical editor of this book. Karen L. Gorbea has carefully proofread and edited major portions of the text. Special thanks are extended to Janet C. Hunt, Anne Potter, Margaret Kubatz, Karen Jackson, and Linda Pines for the word processing and manuscript preparation.

Contents

SYNFUELS HANDBOOK

Chapter 1

Introduction to Synthetic Fuels

This handbook provides a compilation of the significant points of interest to those who want to learn more about synthetic fuels. The term *synthetic fuel* means any solid, liquid, or gas, or combination thereof, that can be used as a substitute for petroleum or natural gas and that is produced by chemical or physical transformation of domestic sources of coal (including lignite and peat), shale, and tar sands. In addition to those fuel resources noted in this definition from the Energy Security Act (P.L. 96-294), there are other synthetic fuels derived from biomass, which include timber and timber waste, animal waste, municipal and industrial waste, sewage, sludge, oceanic and terrestrial plants, and other organic matter. This chapter provides an overview of synthetic fuels (synfuels).

The American energy industry believes that development of synthetic fuel technology and production facilities in the United States is an important national need.

A synthetic fuel industry can start contributing to domestic energy supply by the end of the 1980s, and can grow to a large sustainable volume in the 1990s. Several processes for making synthetic fuels are in advanced stages of development and are ready for application in commercial-sized plants.

Such processes can apply the country's greatest energy strength to its most pressing energy need, by converting abundant domestic resources—oil shale and coal—into liquid and gaseous forms for uses now served by oil and natural gas.

Timely development of a synthetic fuel industry is essential to any realistic expectation that this country's need for imported petroleum can diminish substantially in this century. This development is critically important to national security and to standards of living in the future.

The synthetic fuel industry can represent a substantial strengthening of the country's industrial base; its growth potential offers some of the most significant business and employment opportunities in modern history. But development on a timetable consistent with national security and economic needs will require cooperation, of a degree that may be unprecedented in peacetime, among business, labor, government at all levels, and other interest groups.

The National Need for Synthetic Fuels

The domestic synthetic fuel industry has a potentially important role, but it cannot provide a total solution to America's energy problems: It will not eliminate the need for aggressive efforts in the development of other forms of energy and in energy conservation. The synfuel industry should be viewed in a realistic context—as one of several tools essential to successful management of an inevitable energy transition that already has begun and that will continue for generations.

The U.S. and world economies are shifting from dependence on energy provided principally by oil and gas to a more diversified mix of supplies. The transition will likely lead, during the 21st century, to substantial reliance on energy from renewable and nondepletable sources.

For the United States, the need for a transition to other energy sources began to develop when the country started using oil and natural gas faster than new reserves could be discovered. Domestic petroleum production is declining, and Americans have become dependent on foreign countries for almost half of the oil they use. The rest of the world is now following the pattern—using more petroleum than is being found. Production in some of the Middle Eastern countries with the world's largest oil reserves eventually will plateau and decline—perhaps in the early part of the 21st century.

The transition to nonpetroleum forms of energy will continue for decades, but some of the difficulties involved are already apparent.

U.S. consumers have experienced temporary shortages of oil products on several occasions—including those resulting from the embargo of 1973–1974 and the revolution in Iran. The balance of world oil supply and demand remains precarious. And for years to come, the risk of shortfalls is likely to be greater than the prospect of surpluses.

The rising price of oil imports and the country's increasing dependence on them have contributed to trade balance problems, inflation, declining real discretionary income for individuals, and other economic difficulties. An end to the long-term trend of rising real costs of energy is not in sight.

A crucial question is whether the nation will continue to "muddle along," or will begin to successfully manage its way through the transition—seizing its feasible economic opportunities to save energy and to develop additional domestic energy supply. Unless this happens, succeeding generations of Americans face the prospect of real decline in economic well being, reduced opportunities for employment and advancement, and further loss of independence in international affairs.

There are numerous opportunities to improve the domestic energy balance, but they are not, generally speaking, alternatives to one another. Some can affect energy demand or supply almost immediately. Others can produce benefits within the next decade or two. Still others will have no significant impact until well into the 21st century. Furthermore, not all forms of energy can substitute for the scarce fuels—conventional crude oil and natural gas—in all uses.

Failure to make distinctions concerning the time frames and the uses in which potential sources of energy can help could lead easily to planning based on wishful thinking—with potentially disastrous consequences for U.S. security and long-term economic health.

It is hard to overstate the seriousness of the United States' dependence on imported oil, especially oil from the Middle Eastern members of the Organization of Petroleum Exporting Countries (OPEC). This dependence not only causes a serious economic drain on the national economy, currently about $100 billion per year, but it poses a threat of economic catastrophe in the event of a major disruption of these imports, which could come about from any one of a number of plausible, politically motivated actions by people over whom we have very little influence. This dependence could also lead us into armed conflict, while, at the same time, limiting our ability to conduct the conflict.

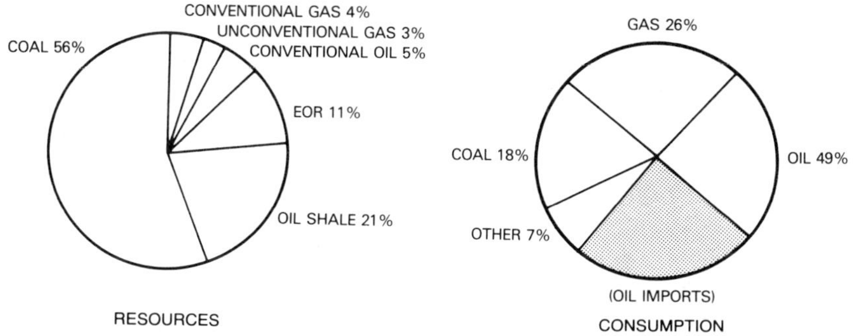

Courtesy: SFC

Fig. 1-1. U.S. energy resource and consumption patterns.

There are a number of actions the United States can undertake to diminish this dependence. One of them is the development of an industry to produce synthetic fuels from domestic resources.

The total domestic fossil energy resources of the United States are sufficient to supply this country for centuries. How much of these resources are ultimately used will depend, of course, on what fraction of them we can extract—which will depend, in turn, on what we are willing to pay for our energy supply. However, there is a disparity between the form of these resources and our current consumption pattern, as shown in Fig. 1-1. Figure 1-1 shows that liquid fuels represent 16% of our resources, but more than 40% of our consumption. Figure 1-2 shows the level of oil imports and domestic production for 1970–1979.

Domestic oil production peaked in the United States about 1970 and then began a slow decline. Demand continued to rise with a consequent increase in oil imports. By

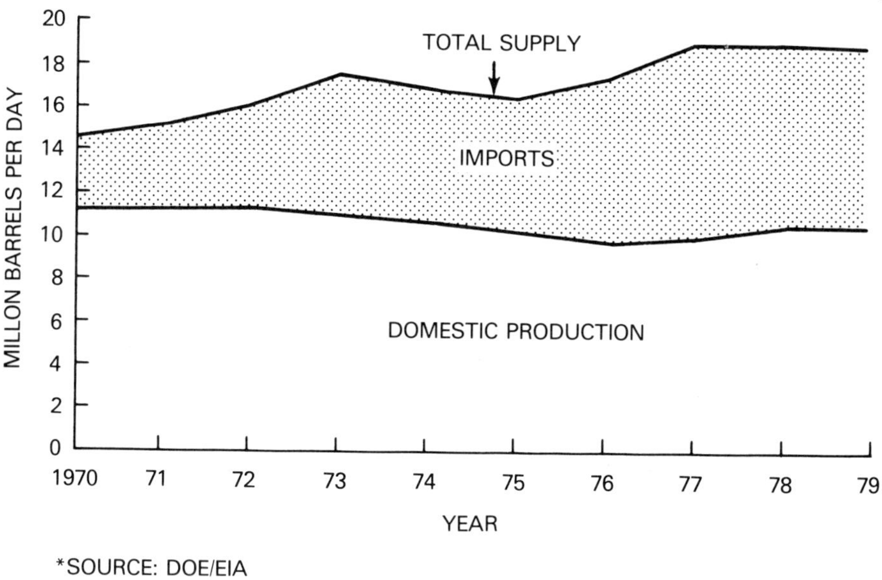

Fig. 1-2. Oil imports and domestic production.

1973, changes in the international oil market led to a situation in which OPEC, theretofore relatively impotent, had the power to raise prices unilaterally. Prices since then have risen in conjunction with crises in the Middle East—the Arab–Israeli War of 1973 and the Iranian crisis of 1978, to cite two of the most recent situations.

Other factors have aggravated our energy problems—inflation, nuclear regulatory problems, ill-advised regulation of oil and natural gas prices, and evident changes in the economies of scale of electricity generation. But, fundamentally, the energy problem in the United States is defined by our dependence on oil from foreign sources, especially the Middle East. Figure 1-3 shows how much oil we import from OPEC and how much comes from the Arab members of OPEC.

- Of about 8.0 million barrels a day imported in 1979, 5.5 million came from OPEC.
- Of the 5.5 million barrels from OPEC, about 3.0 million barrels per day came from Arabian countries in OPEC.

There have been numerous projections of U.S. energy consumption for the remainder of this century. Figure 1-4, typical of these recent forecasts, was developed by the Exxon Corporation (*World Energy Outlook*, 1980).

- There was a high rate of growth of energy demand in the 1960s and early 1970s. Had these rates continued, total energy demand in the United States would

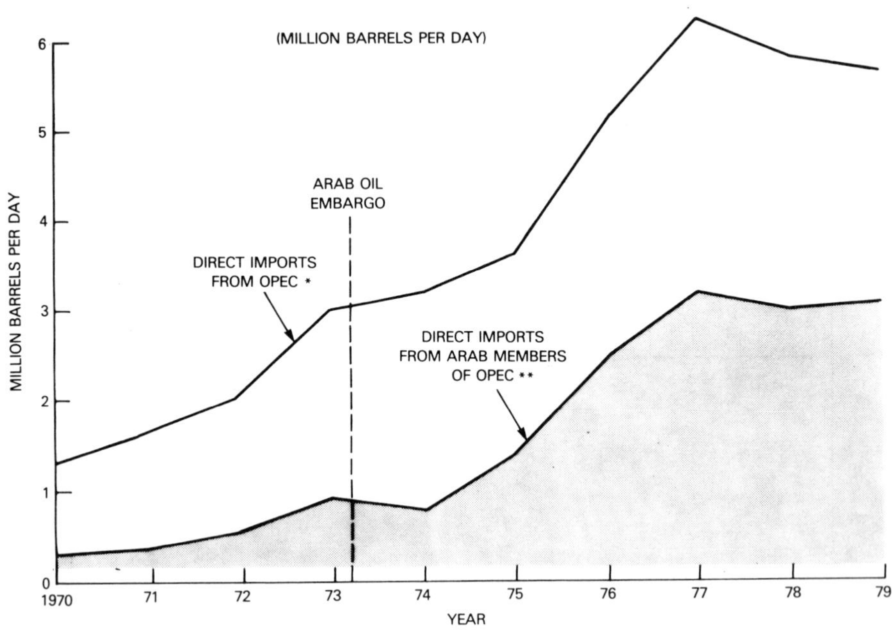

SOURCE: DOE/EIA
*INCLUDES SAUDI ARABIA, IRAQ, QATAR, LIBYA,
**UNITED ARAB EMIRATES, ALGERIA, AND KUWAIT

Courtesy: SFC

Fig. 1-3. U.S. petroleum imports directly from the OPEC countries.

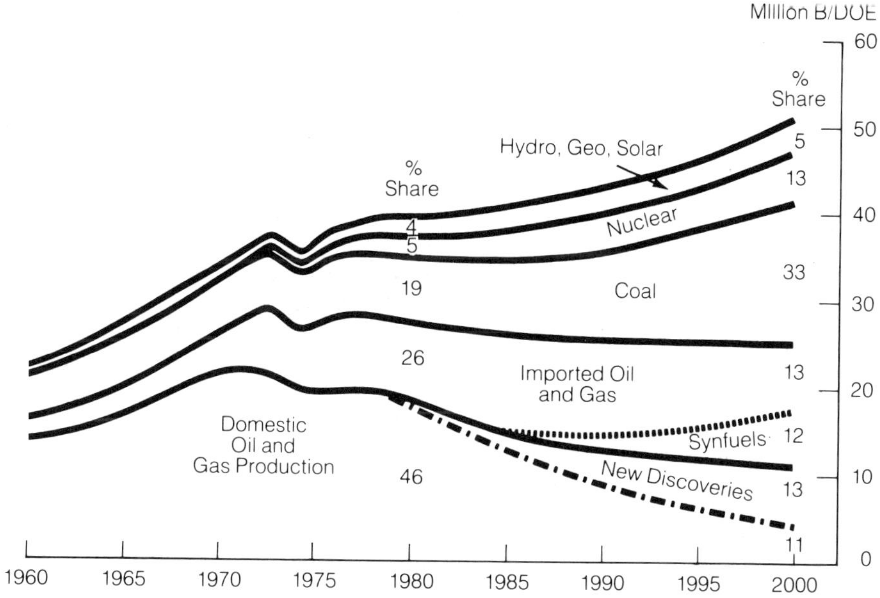

Courtesy: Exxon Corporation

Fig. 1-4. U.S. energy outlook by type of fuel consumed.

have reached 75–80 million barrels of oil equivalent (MMBOE) per day by 2000.

- The United States has experienced a profound change in the 1970s initiated by the 1973–1974 embargo and the rapidly increasing prices of energy. If current trends continue, the United States in 2000 could be using as little as 45 MMBOE per day. Exxon's estimate puts demand at 50 MMBOE.
- With total demand in the 45–50 MMBOE range, oil and gas imports (mainly oil) will be in the range of 15 MMBOE without synfuels. Figure 1-4 highlights the contribution synthetic fuels production can make.
- If the goals set by the Energy Security Act (P.L. 96-294), which established the U.S. Synthetic Fuels Corporation, are met, the United States will be producing 2 MMBOE per day of synthetics by 1992. If the 1987–1992 growth rate continues until 2000, production will reach about 4.5 MMBOE by that time. Under the current administration and the interest rate economic conditions, these goals will not be met.

It may be possible, however, for the United States to significantly reduce its dependence on imports in the 1990s—provided planning and construction of synthetic fuel plants begin now and accelerate rapidly within the next few years. It is projected that these fuels' share of total U.S. energy demand can reach 12% by 2000 and will continue to increase in the 21st century.

Demand for oil and gas is expected to decline gradually as substitution of other fuels, coupled with technological breakthroughs such as improvements in the electric automobile, allows nonpetroleum sources of energy to meet an increasing share of total domestic energy needs. A substantial demand for liquid and gaseous hydrocarbons is expected to continue, however, well into the 21st century.

Potential demand for synthetic fuels—the difference, as shown in Fig. 1-5, between projections of domestic petroleum demand and production—could reach an estimated 15 MMBOE per day early in the 21st century.

Obviously, these projections are subject to variation, but the essential point is that for decades to come there will be a large demand for energy in liquid and gaseous forms that cannot be satisfied with domestic oil and natural gas. In meeting this demand, synthetic fuels are the alternative to continued reliance on imported petroleum through the 1990s and beyond.

If new domestic discoveries of oil and gas should turn out to be greater than expected, and if the nation also has moved rapidly to develop its synthetic fuel potential, the variance from projections would be beneficial—allowing faster progress in reducing the nation's dependence on imported petroleum.

On the other hand, exploration results may fall short of projections. If they do, and if progress in synthetic fuel development also is disappointing, the resulting increase in need of imports would be a serious problem. In view of the present precarious balance of world oil supply and demand and the future outlook, it is unlikely that the needed quantities of foreign petroleum would be available.

Among the potential supplements to domestic oil and gas supplies, synthetic liquids and gases made from oil shale and coal have potential for build-up to large sustainable volumes. A number of processes for making fuels from coal and oil shale have seen successful commercial use in other countries. Other processes have been in research and development stages for years and are approaching commercial readiness.

Shale oil, which is obtained from oil shale (a finely textured sedimentary rock containing an organic substance called *kerogen*), is not quite economically competitive with imported crude oil. The oil shale can be mined, with underground or surface

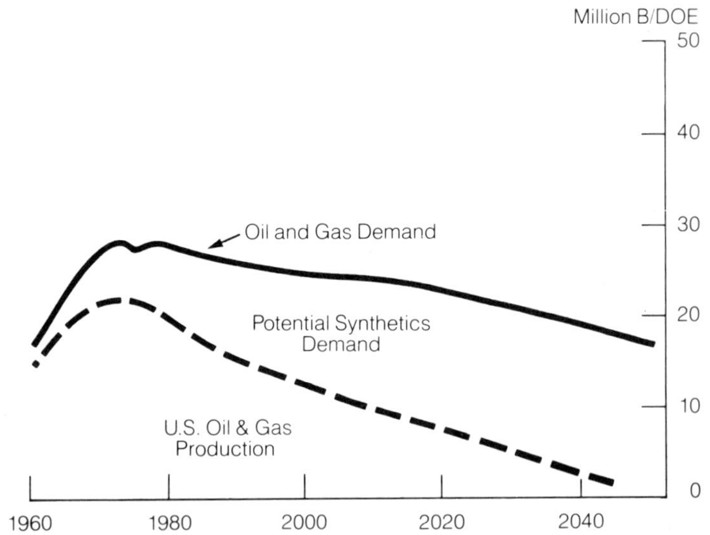

Courtesy: Exxon Corporation

Fig. 1-5. U.S. oil and gas production and demand through 2050.

techniques, then crushed and heated to yield an oil resembling heavy crude. After processing to remove impurities and to lighten the oil, conventional refining methods can turn it into a variety of oil products including gasoline, jet fuel, heating oil, and fuel oil.

Several methods of processing oil shale in large surface retorts have been demonstrated in pilot plants and appear ready for scaling up to commercial size. Efforts to find feasible ways of using *in situ* techniques, which would process the shale underground, are continuing, although early attempts were generally less than successful.

Much research and development work is being done on technologies that convert coal, directly or indirectly, into other liquid fuels. Costs of these liquids are now estimated at 40–60% above shale oil or imported crude oil, but technological advancements are likely to bring costs down. These processes will be needed to allow efficient utilization of a wider range of coals, but are not yet ready for commercial application.

Much synthetic fuel technology, however, is now ready for commercial use, and the domestic resource base is adequate to support a very large industry.

The need for development of a synthetic fuels industry—to ease dependence on imported energy and to meet a coming shortfall of world oil supply—is clearly critical to the national interest. The resources are adequate. The technology is ready for commercial application. And the private sector has the necessary financial, technical, and managerial resources.

There is no need for the federal government to assume the risk of a producer. But governments—at all levels—do have major roles to play.

Coordination by industry and by governments at local and state levels will be essential in adequately planning for population influx and other impacts in the development areas.

Perhaps the greatest responsibility of the federal government, reinforced by state and local governments, is leadership—in building the public awareness of the critical national need to develop domestic synthetic fuels production as rapidly as possible and in creating the economic and regulatory climate to facilitate the new industry's growth.

A second major responsibility of the federal government is to function more effectively in a managerial role—reconciling conflicting priorities and regional interests, and assuring that studies, hearings, and permitting processes proceed without inordinate delays of the sort that have slowed energy projects so often in the past.

Without bold, responsible, bipartisan political leadership, it is unlikely that the promise of synthetic fuels can ever be fully realized. And whether or not the industry's full potential is achieved will depend largely on how well industry, labor, civic leadership, environmentalists, and the governments of cities, counties, states, and the nation can work together.

Vigorous effort must continue if the national interest is to be served. Plans for the country's first synthetic fuels plants are now being made by several firms. But meeting critical national needs will require that work begin now on plans for continued development, beyond these initial plants. What is needed is not just a few new plants, but an entire new industry.

Constraints to Private Sector Development

Known recoverable reserves of coal and oil shale—even after deducting coal to be used conventionally and the energy to be consumed in processing—are capable of providing synthetic fuels equivalent to one trillion (10^{12}) barrels of oil, which is three times as much energy as the U.S. Geological Survey estimates can be provided by the country's remaining proved and undiscovered reserves of oil and gas and which is

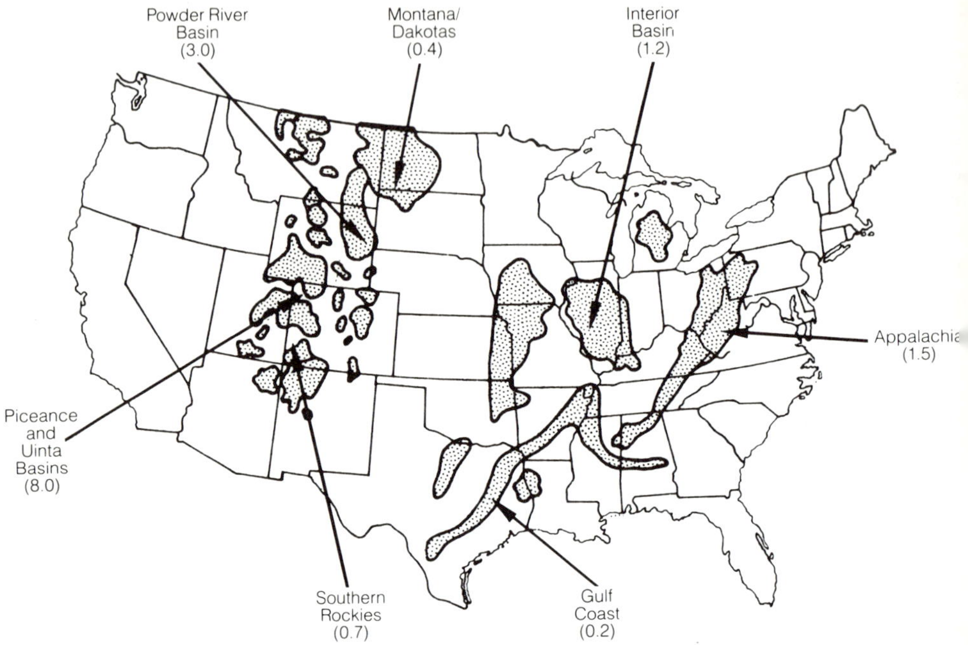

Shale		Coal	
Surface Mines	6.0	Powder River Basin	3.0
Underground Mines	2.0	Other Western	1.1
		Eastern/Gulf	2.9
Total (Piceance & Uinta Basins)	8.0	Total Coal	7.0
		Total Synthetics	15.0

Fig. 1-6. Assumed distribution of synthetic fuel industry (MMBOE per day).

enough to sustain a synthetic fuel industry producing 15 MMBOE per day for 175 years.

The investment required to attain production of 15 MMBOE per day is staggering—almost $80 billion in 1980 dollars (over $3 trillion "as-spent"). This would be spread over 30 years or more. In peak years, this would represent about 1% of the Gross National Product (GNP)—roughly what is now expended for conventional domestic exploration and production of oil and gas. To put it another way, these capital expenditures would constitute about one-tenth of private, nonresidential investment.

Considering the readiness of technology, the size of the resource base, and the financial capabilities of the private sector, a goal of meeting the potential synthetic fuel demand of 15 MMBOE per day by the end of the first decade of the 21st century would be highly ambitious. If the synthetic fuel industry develops at such a speed, it will be possible to begin reducing national dependence on petroleum imports within a decade, reduce them significantly during the 1990s, and ensure adequate supplies of liquid fuel and gas despite an anticipated decline in world oil production.

The task will be massive, however, by any standard of comparison, and will be complicated by the fact that much of the industry will have to be concentrated in arid, sparsely populated parts of the western United States.

The output of a cost-effective 15 MMBOE per day industry, as shown in Fig. 1-6, would include some 8 million barrels per day of shale oil from high-yielding deposits in the Piceance and Uinta Basins in Colorado and Utah. About one-fourth of the shale could be mined underground, but achieving the projected volume would require that the rest be obtained by large-scale surface mining. (Surface mining can recover up to 10 times as much shale as underground room-and-pillar methods.)

The remaining 7 MMBOE per day of synthetic fuels would be derived from coal—3 million from the Powder River Basin of Wyoming and Montana, and another 4 million from other parts of Montana, the Dakotas, the southern Rockies, the Interior Basin (Illinois and parts of adjacent states), Gulf Coast areas, and Appalachia.

SYNTHETIC FUEL PRODUCTION GOAL CONSTRAINTS

To meet the 1992 goal set by the Energy Security Act (P.L. 96-294), about 40 synthetic fuel plants, each with a capacity of 50,000 barrels of oil equivalent per day, would have to be built. Each of these plants would cost about \$2–\$4 billion, \$80–\$160 billion for the 40 plants.

While investments of this size on a national scale are certainly possible—\$200 billion is about one year's defense budget and the interstate highway systems would cost on the order of \$200 billion if constructed now—\$2 billion is a very large outlay indeed for a single company.

Table 1-1 shows, for some large companies with the financial resources to build and operate synfuel plants, the relative measures of the burden of a \$2 billion synthetic fuel investment on overall operation. In other words, to establish a new synfuel industry, industry may need (for the next 20 years) to invest, in untried technologies, amounts equal to industry's annual expenditure on exploration for and production of conventional oil and gas.

Beyond the sheer scale of the task are other important considerations:

- All synthetic fuel technologies are untried at commercial scale in the United States. The effects, even on technologies already applied in South Africa, of operating with U.S. feedstocks and complying with U.S. environmental laws

Table 1-1. Corporate Burden of \$2 Billion Investment

Company	Percentage of Assets	Percentage of Stockholders Equity	Percentage of 1979 Earnings
Exxon	4	9	47
Mobil	7	19	100
Texaco	9	19	114
Standard Oil (California)	11	22	112
Gulf	12	23	151
Du Pont	22	38	213
Dow Chemical	20	51	255
Union Carbide	23	49	360

introduce significant risks to the exploitation of even the most fully developed technologies.

- Foreign oil producers could undercut the prices synfuel producers would have to charge to operate profitably. While such undercutting on a large scale would be a boon to the nation, without protection it would be a disaster to private producers of synthetic fuels.
- The strength of environmental opposition to synthetic fuels in general and to specific projects is as yet unknown. The experience of the nuclear power industry can hardly be encouraging to entrepreneurs considering investments in synthetic fuel plants of billions of dollars.

Figure 1-7 provides a summary of the chances for gains and losses in development of synthetic fuel facilities.

If these barriers of very large capital requirements, uncertainties about the economic and technical performance of synthetic fuel technologies, uncertainties about environmental and socioeconomic effects and opposition stemming from these which might delay synthetic fuel projects are overcome and a large synfuel industry begins to develop, other constraints may come into play which could slow development and increase costs. These constraints relate to such physical factors as materials and equipment availability, coal and water supplies, technical and construction labor, and transportation.

In a recent report, *Achieving a Production Goal of 1 MMBPD of Coal Liquids by 1990*, TRW examines a number of these potential constraints to achieving this production level. The findings, while developed specifically for a coal liquids industry at a specific build-up rate, are generally informative with respect to a synfuels industry. Portions of this report are included in Chapter 3, Coal Liquefaction.

Table 1-2 indicates the potential constraints considered in the comparative analysis.

In the TRW comparison, Bechtel and UOP/Systems Development Corporation concluded that there are significant potential impediments to the 1990 production and

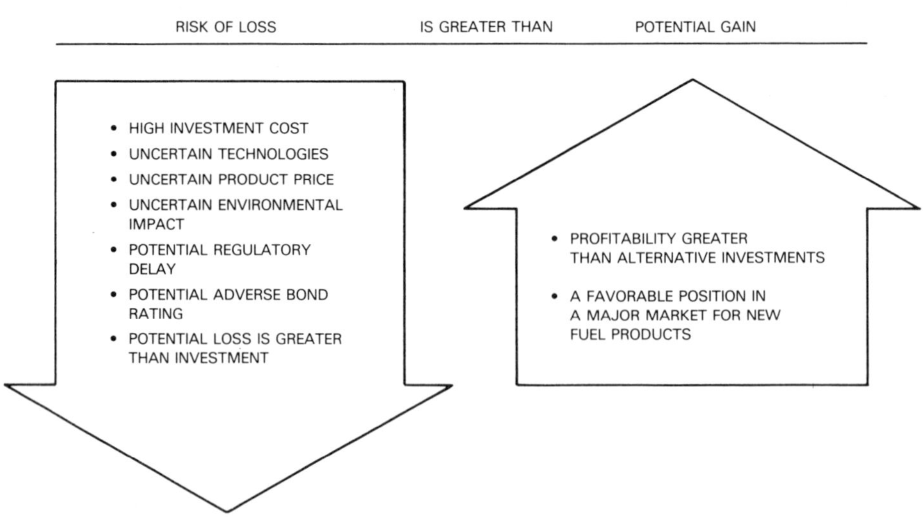

Fig. 1-7. Chances for gains and losses in commercial synthetic fuel developments.

Table 1-2. Summary Findings of *Achieving a Production Goal of 1 MMBPD of Coal Liquids by 1990*

Potential Constraints	Reference Reports			
	Bechtel	MTI	UOP/SDC	Mound
Critical materials availability	P[a]	P	N	
Equipment availability	P	P	P	
Equipment performance		P		
Coal supply	N		N	
Water supply	PQ		PQ	PQ
Technical labor availability	P		P	
Construction labor availability	P		PQ	
Financial/capital availability	N		N	P
Environmental requirements	P		P	P
Permits and licensing	P		P	P
Transportation			X	X
Siting	PQ		PQ	PQ
Technology uncertainties		P		X
Product costs				X
Operating costs			N	X

[a]Legend
X Discussed only—no conclusions.
N No constraint—will be met in normal course.
P A potential constraint that requires action.
PQ A potential constraint in certain locales/regions that requires action.

goal relating to equipment availability/obtainability. Specific items expected to be in short supply are

- Distillation towers
- Heat exchangers
- Compressors
- Large pumps
- Thick-wall pressure vessels

The other studies (MTI and Mound) also concluded that there were significant potential impediments to the 1990 production goal in the areas relating to equipment performance and/or scaleup. Specific items of equipment subject to this risk are

- Gasifiers
- Extractors
- Hydrotreaters
- Oxygen compressors
- Coal slurry heaters

There appears to be no insurmountable problem with the synthetic fuel program in regard to the availability of critical materials and components. As shown in Table 1-3, only nickel would be used in amounts large enough to cause concern—perhaps 14% of available supplies. For the ethanol-from-biomass program, additional nitrogen, potassium, and phosphate fertilizers would be needed, amounting to less than 5% of current availability. For the unconventional gas program, the availability of drill rigs might be a

Table 1-3. Material Requirements for the Synthetic Fuel Program

Material	Total Requirement (10^3 tons)	Domestic U.S. Availability (10^3 tons)	Percentage Used
Cement	1,200	400,000[a]	0.3
Carbon steel	5,000	480,000[a] (All Steel)	1.2
Alloy steel	530		
Stainless steel	260		
Copper	220	8,000[a]	2.7
Aluminum	30	27,500[a]	0.1
Manganese	40	4,800[a]	0.9
Chromium	60	5,200[b]	1.2
Nickel	10	70[c]	14.0

[a] 1972 – 1976 U.S. production.
[b] 1975 stockpile.
[c] Five times 1975 production.

Courtesy: The MITRE Corporation

constraint. The number necessary for the unconventional resources would reach about 700 by 1990, with probably close to 4000 rigs available by that time. Of course, the use of the drill rigs on conventional versus unconventional gas resources will depend upon marketplace decisions by the rig owners.

Components such as pressure vessels, compressors, booms, heat exchangers, valves, and pipes would be needed in amounts less than one-third of U.S. manufacturing capability, according to the Fluor Corporation. In some cases, a rapid buildup in orders might result in delays and price increases.

Trained Manpower

A synthetic fuel industry capable of producing 15 MMBOE per day in the year 2010 would employ an estimated 870,000 people—480,000 in mining and 390,000 in the synthetic fuel processing plants. In peak development years, its growth would also provide 250,000 jobs in construction and 8400 jobs in design engineering. The labor distribution for synthetic fuel projects is shown in Fig. 1-8.

Over the span of 30 years or more, the new industry would directly increase the nation's total number of jobs, over current levels, by 60% in mining, 55% in process industries, 15% in construction, and 35% in the applicable design engineering specialties.

These direct employment effects would be multiplied, of course, by the industry's demand for machinery and other products from suppliers and by growth of new population centers in the development areas.

Careful planning will be required in these areas to develop, almost literally from scratch, transportation systems, housing, school systems, health care facilities, numerous public services, and other essentials needed by thousands of workers and their families.

This will be a formidable challenge, but the task is manageable because the buildup will begin slowly and could be sustained over a long period. Nonetheless, the importance of timely advance planning cannot be overemphasized.

Only the Bechtel and UOP/SCD studies examined the manpower issue in detail for the specific production goal addressed. Both concluded that supply of trained technical

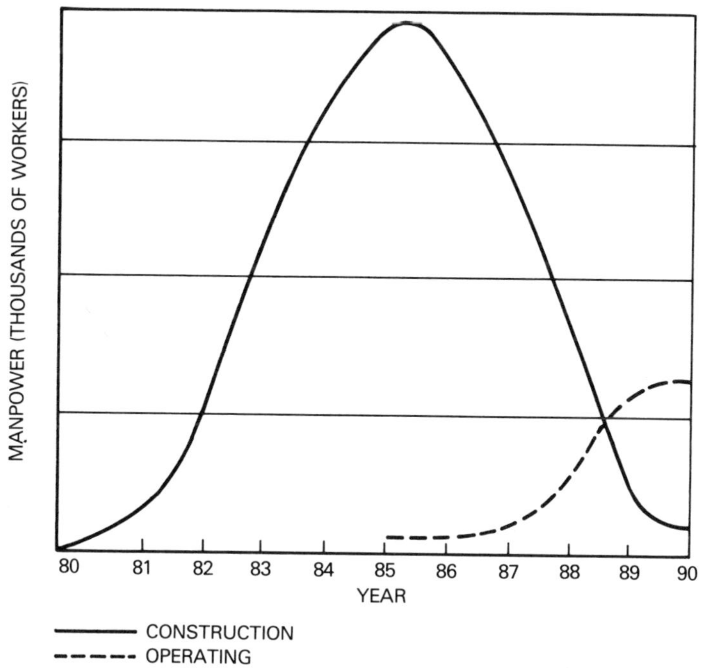

Fig. 1-8. Labor requirement distribution for synthetic fuel development.

manpower (engineers) will be severely taxed, but that, to some extent, this hardship could be eased by attracting trained personnel from other industries. Also, the two studies concluded that construction labor will be in short supply.

Water Availability

Shale oil and coal synthetics require from 2 to 4 barrels of water for each barrel of product. And water supply is a major concern in dry western regions where much of the development would occur.

Unless additional water is transported into these regions, it is unlikely that the industry's production can exceed 7 MMBOE per day—less than one-half the potential demand.

It appears financially and physically feasible to bring the additional needed water to the Piceance and Powder River Basins from the Missouri and several other western rivers—at a resulting increase in costs of about 1%. As a side benefit, the transportation systems also could provide water for agricultural, municipal, and other uses.

Water availability is a western concern. A Colorado Department of Natural Resource report to the U.S. Water Resources Council estimates that a 1.5 million barrels per day synfuel industry could be supported by existing unallocated water supplies in the Upper Colorado River Basin. There would have to be a significant expenditure of funds and effort to build impoundments to retain the water, however. It is reasonable to assume that as second- and third-generation processes evolve, the requirement for water in the Rocky Mountain West will be reduced, at least in part, on a per plant basis.

Three of the four studies (Bechtel, UOP/SCD, and Mound) summarized by TRW concluded that the water supply requirement would not be a deterrent to the coal liquids production goal. However, the Bechtel and UOP/SDC reports conclude that water availability in the western coal regions would not allow siting the total plant capacity for the 1 million barrels per day production goal in this region.

Air and Solid Waste Impacts

The production of synthetic fuels, while involving no unique control problems, does present a problem with the magnitude of emissions, effluents, and solid wastes. Solid waste, in particular, will pose disposal problems of a magnitude not previously encountered. It has been estimated that a 2 MMBOE per day synfuel industry will require mining the equivalent of the Panama Canal every year. Solid waste from oil shale is approximately 80–85% of the feedstock and from coal it is 40–50%. Many of these wastes contain leachable, potential contaminants of the local water resource and must be carefully controlled.

Air emissions are not expected to be a problem in this century. A recent study by the National Commission on Air Quality has suggested that 690,000 barrels per day could be produced in Colorado's Piceance Basin before current air quality limits would be approached. Existing control technology appears to be adequate for all expected emissions, and both developers and states are working to ensure that air quality degradation can be minimized.

Health and Safety Hazards

The development of synthetic fuels is not expected to create any undue hazards to the health and safety of workers. Recent studies by TRW, Tosco, Union Oil, and the U.S. Navy suggest that shale oil products have about the same toxicity as petroleum products. Intermediate process streams from shale oil production, coal liquefaction, and petroleum processing all contain known carcinogens of a group collectively known as polynuclear aromatics (PNAs). In order of potential hazard, coal liquefaction would be highest because of the higher incidence of PNAs, shale would follow, and then petroleum. However, the finished products offered to the marketplace would be virtually indistinguishable from petroleum products and would not reasonably be expected to present any greater potential hazards. The intermediate streams in the processing facilities are expected to be controllable with existing technology, and the hazard potential there too is low. Observance of existing standards of industrial hygiene is expected to mitigate worker exposure to any potential hazards.

Socioeconomic Impacts

The major impact is expected to be the "boom town" effect in the sparsely populated regions of the West. Although mitigation planning is being done and advance payments to local governments are to be made to deal with the rapid influx of population, there is general belief that the allocated funding is too little, especially in the realm of secondary services such as family counseling, drug and alcohol abuse assistance, housing and recreational facilities, and the like. It is believed by many public, private, and state organizations, especially in Colorado, that some mechanism must be found to adequately finance the measures needed to deal with socioeconomic impacts in that interim period between the time population increases begin and projects add significantly to the local tax base.

Synthetic Fuel Technologies

There are a variety of technologies available for producing synthetic fuels from these resources, and they are described in this section in generic terms. For the purpose of this section, synthetic fuel technologies include the following eight major technologies:

- Oil Shale
  - Surface Retorting
  - *In Situ* Retorting
- Coal Liquefaction
  - Direct Hydrogenation
  - Indirect Liquefaction
- Coal Gasification
  - High-Btu Coal Gasification
  - Medium-/Low-Btu Coal Gasification
- Unconventional Gas Recovery
  - Geopressured Methane
  - Methane from Coal Seams (Coalbed Methane)
  - Devonian Shale
  - Tight Gas Sands
- Biomass
  - Liquefaction
  - Gasification
- Tar Sands
  - Surface Recovery
  - *In Situ* Recovery
- Solid Waste
  - Liquefaction
  - Gasification
- Peat

The synthetic fuel resources of interest to the U.S. Synthetic Fuels Corporation are coal (including peat and lignite), oil shale, and tar sands. The major products, markets, and resources are summarized in Table 1-4 for synthetic fuels.

OIL SHALE

Aboveground oil shale retorting technology has achieved a state of near commercial readiness. Several surface retorts have been successfully operated by TOSCO, USBM, Paraho, and Union. Colony Development, Union Oil, and Occidental Oil *(in situ)* had announced plans to begin commercial development. All the planned surface technologies have been demonstrated at pilot scale or larger.

Over the years, many methods for recovering oil from shale have been under development, including surface retorting processes, *in situ* technologies (both modified and true *in situ*), and more esoteric concepts like solvent extraction and biobleaching of oil shales.

Table 1-4. Summary of Synthetic Fuel Products

Resource	Technologies	Major Products	Markets
Shale	Oil shale retorting	Syncrude for refineries to yield gasoline, jet and diesel fuels, etc.	• Commercial/residential • Military • Utility • Industrial
Coal	Direct coal liquefaction	• LPG • Naphtha • Fuel oil	• Commercial/residential • Utility • Industrial • Chemical feedstock
	Indirect coal liquefaction Fischer – Tropsch (Sasol)	• SNG • gasoline • LPG	• Commercial/residential • Utility • Industrial • Chemical feedstock
	M-gasoline	• Gasoline	• Commercial transporation
	Methanol	• Methanol	• Commercial transportation • Chemical feedstock
	High-Btu coal gasification Lurgi slagging Lurgi	• SNG	• Commercial and residential heating
	Medium-Btu coal gasification Texaco Koppers – Totzek	• Medium-Btu gas	• Captive use for heating and as chemical feedstock
Tar sands	Tar sands surface mining and recovery	• Syncrude for refinery use	• Same as shale

Retorting occurs when shale is heated to the temperature at which the organic material within the ore (kerogen) is decomposed into gas, condensable oil, and solid residue. The rate of kerogen decomposition is high at retort temperatures of 900–950°F (480–510°C), and complete decomposition occurs within a few minutes.

The raw shale oil is equivalent to a heavy crude oil that can be hydrotreated to produce a petroleum refinery feedstock low in nitrogen and sulfur with very little residuum, which is well suited for refining into middle distillate fuels. If hydrocracking is chosen for the refinery process, the yield and range of products is particularly desirable: motor gasoline, 17%; jet fuel, 20%; diesel fuel, 54%; and residuum, 9%. In general, shale oil is heavy and viscous, has a high pour point, and is high in nitrogen and oxygen compounds. The hydrocarbons making up shale oil could be classified as being about 20% alkanes, 20% aromatics, 25% aromatic resins, and 35% olefins and naphthenes. By contrast, a typical crude would contain about 15% alkanes, 50% aromatics, and 35% naphthenes with hardly any olefins or resins.

Oil shale is a finely textured sedimentary rock containing the solid, largely insoluble organic material *kerogen*. Oil shale is a marlstone, a fine-grained, thinly layered mixture of dolomite, calcite, quartz, and clays. Kerogen, a solid fossil substance, is embedded in these materials. Kerogen is about 80% carbon, 10% hydrogen, 6% oxygen, 3% nitrogen, and 1% sulfur. High temperatures decompose the kerogen, yielding a raw oil suitable for use as a refinery feedstock. The principal deposits of oil shale are found in Colorado, Utah, and Wyoming. It has been estimated that these deposits contain 2.4 trillion barrels of oil, of which about 600 billion barrels are in rock assaying 25 gallons or more per ton (480 federal, 220 private). The Green River formation contains some of the richest deposits of oil shale in the world—up to 1500 feet thick.

Recovering shale oil from kerogen-bearing rock is basically a simple technology, as shown in Fig. 1-9, although controversy remains as to which of its many variations is the most reliable, the most economic, and the most sound from an environmental viewpoint. Some developers plan to use surface retorting, the most fully developed method of shale oil recovery. Others are pioneering the riskier process of underground retorting, also known as *in situ* retorting.

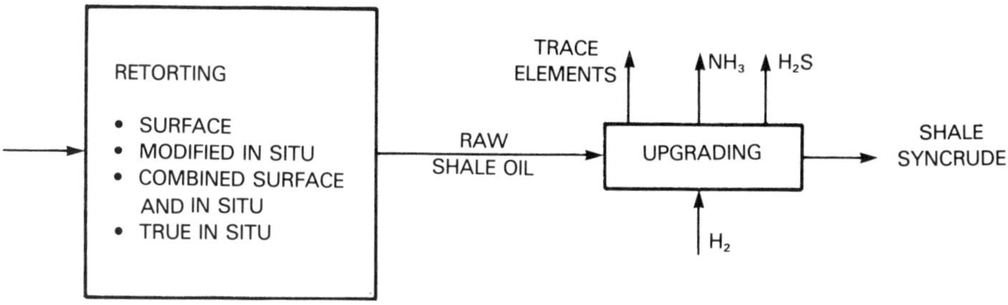

Fig. 1-9. Oil shale processing.

A retort is a kilnlike vessel in which shale fragments are heated to more than 900°F (482°C) to release the kerogen as a vapor, which is then cooled and condensed for recovery. The principles of heating and condensation are the same in all retorting operations. What distinguishes surface from *in situ* processing is the location of the retort. For surface retorting, the rock is mined, brought to the surface, crushed, and heated in free-standing retorts. The *in situ* methods use the earth itself as a retort; shale rubble is blasted into a previously mined columnar underground receptable or fractured in place without prior mining; the rubble is then heated where it lies.

The raw shale oil collected from a retort is too heavy and viscous to flow easily through a pipeline to a refinery. The solution is to upgrade it to a lighter product at the retorting site.

Hydrotreating, a major upgrading step, consists of passing the shale oil over a catalyst in the presence of hydrogen. The oil becomes less viscous and gives up some of its sulfur and nitrogen in the process. These two contaminants are eventually recovered as elemental sulfur and ammonia. The resulting shale syncrude is a light low-sulfur premium-quality refinery feedstock.

Surface Retorting

In surface retorting, as noted previously, the oil shale is mined, crushed, and processed in surface facilities, and the spent shale is disposed of in a landfill or returned to the mine. Examples of surface retorts include the vertical kiln, the circular grate, the rotary kiln, the screw mixer, and the upglow kiln, which uses a rock pump. Surface retorting processes include TOSCO-II, Lurgi–Ruhrgas, and Hytort.

TOSCO-II. The TOSCO-II process is a continuous, indirectly heated retorting process in which preheated, crushed shale, less than ½ in. (13 mm), is contacted with heated ceramic balls in a rotating drum. The ceramic balls are separated from the retort products in an accumulator and recycled through a direct-fired heater to the retort. The spent shale is cooled in a steam generator and moistened for disposal.

Lurgi–Ruhrgas. The Lurgi–Ruhrgas process is a continuous process in which a fine grained heat carrier is circulated from a combustion zone to a retorting zone to supply the retort heat. Heat is transferred from the hot recirculated solid to shale feed in a horizontal-screw-mixed retort.

Hytort. The Hytort aboveground hydroretorting process can produce either substitute natural gas (SNG) or oil from western or eastern shales. A moving-bed retort is used with minus 2 in. (minus 50 mm) shale and overall countercurrent contacting of hydrogen and shale. Hot spent shale preheats the feed hydrogen and hot product gases contact the feed shale.

In Situ Retorting

In situ retorting, as noted previously, is the retorting of oil shale in place, without removing it from the ground by mining. *In situ* processes are often further classified as "true" *in situ* processes involving only the drilling of wells, and "modified" *in situ* processes involving the mining of 20–30% of the shale to develop the underground retort chambers.

Recovery of oil by *in situ* retorting of shale involves fracturing the shale to create enough permeability to permit injection of retorting fluids and recovery of shale oil through wells. The required fracturing is expected to create a large mass of broken shale, ranging in size from dust to several feet thick and varying in assay value from nil to as much as 50 or more gallons/ton. One method of heating is achieved by partial combustion of the fractured shale with injected air or a mixture of air and recirculated gas.

In situ processing has the advantages of

- Avoiding the costs of mining and waste disposal
- Adapting to beds too thick, too badly faulted, or too deep to be readily amenable to conventional mining methods
- Eliminating or reducing need for air pollution controls, which would be required for aboveground processing
- Minimizing problems of possible contamination of surface or subsurface waters by surface disposal of spent-shale wastes

Disadvantages are that *in situ* processing is not adaptable to concurrent recovery of associated minerals, and control of combustion may be very difficult, leading to a low recovery of shale oil. *In situ* processes include oxy-modified and true *in situ*.

The oxy-modified process consists of creating underground retorts of rubblized oil shale by expanding the oil shale into a void utilizing conventional explosives. Processing consists of two basic steps: (1) mining approximately 15–25% of the oil shale deposit and expanding material into this void, and (2) retorting the rubblized oil shale in place.

True *in situ* retorting is suited to shale beds that are thin, deeply buried, or unsuitable for processing by other methods. Underground retorts are prepared from the surface without using mining techniques. The formation is drilled and fractured hydraulically, and explosives are placed in the hydraulic fractures. Blasting induces fracturing for the subsequent retorting step.

COAL LIQUEFACTION

Coal liquefaction is the conversion of coal to liquid fuel. This fuel, which is more versatile and easier to handle than a solid fuel, burns cleanly and has a low ash and sulfur content. To form liquid hydrocarbons from coal, its hydrogen-to-carbon ratio (which is about 0.9:1) must be increased by at least a factor of 2 (crude oil has a hydrogen-to-carbon ratio of about 2:1). Coal liquefaction processes can be divided into four broad categories: direct liquefaction (hydrogenation), solvent extraction, pyrolysis, and indirect liquefaction.

Direct Coal Liquefaction (Hydrogenation)

As illustrated in Fig. 1-10, coal is dissolved, without a prior gasification step as in indirect liquefaction, in a hydrogenating solvent at high temperature [800°F (427°C)] and pressure (1000–2500 psi). The coal is broken down and hydrogenated, yielding a mixture of gases, liquids, solid carbonaceous residues, and ash. The liquid products—coal-derived syncrude—are separated from this mixture, leaving hot [950°F (510°C)] liquids, solid residue, and ash. These "bottoms" can be gasified to produce hydrogen and fuel gas. In coal hydrogenation, hydrogen is added directly. Pulverized feed coal is slurried in a recycled coal-derived solvent mixed with a hydrogen-containing gas and reacted at high temperatures and pressures. The reaction may be noncatalytic, pseudocatalytic, or catalytic. In noncatalytic reactions, certain coals react more readily with hydrogen than other coals do, probably because of the catalytic effect of specific coal ash constituents. In pseudocatalytic reactions, a portion of the reaction product is recycled to the feed slurry to increase the ash content (and thus catalytic activity) in the reactor.

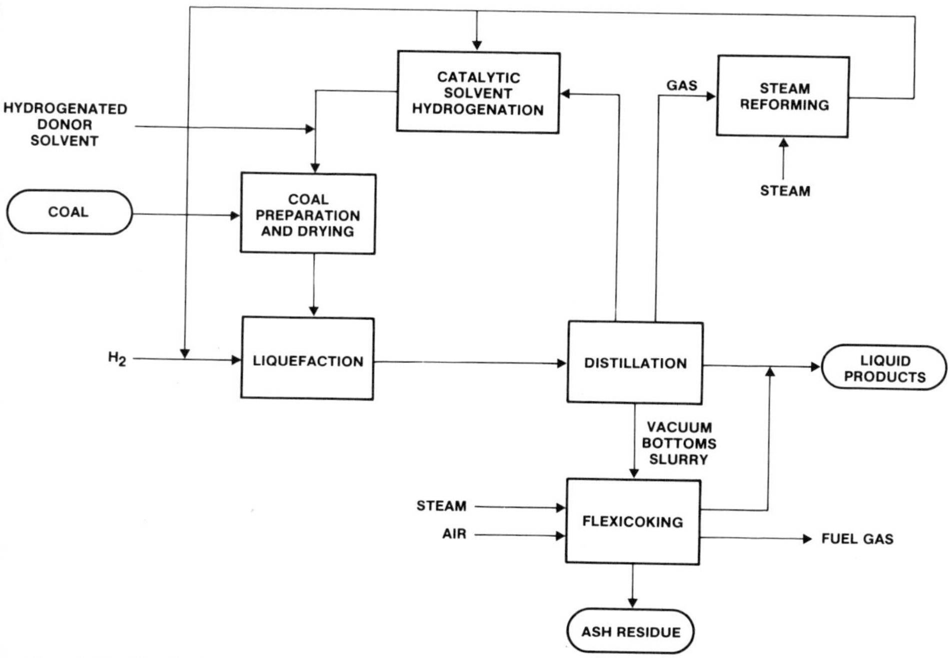

Fig. 1-10. Block diagram of direct coal liquefaction (Exxon Donor Solvent process).

All direct coal liquefaction processes add hydrogen to coal in a solvent to produce liquid products. They differ in the ways they separate "bottoms" from products, generate hydrogen needed for the coal liquefaction process, and catalytically induce hydrogenation of the solvent.

Direct liquefaction is expected to be a more energy efficient route to coal liquids than indirect liquefaction and will thus require less coal per barrel of product produced. Also, plant capital cost is projected to be less for direct liquefaction than indirect. However, direct coal liquids must be upgraded by removing heteroatoms and adding significant quantities of hydrogen before they are suitable for most end-use applications. The cost to upgrade coal liquids is significantly higher than the cost to refine crude oil, because of the severe hydrogen deficiency of coal liquids.

Some of the processes included under direct liquefaction are as follows: H-Coal, Exxon Donor Solvent (EDS), Solvent Refined Coal (SRC-I and SRC-II), and Dow Liquefaction.

H-Coal. The H-Coal process, shown in Fig. 1-11, is a catalytic hydroliquefaction process that converts high-sulfur coal to either a low-sulfur, low-ash boiler fuel or to a refinery syncrude. In this process, coal is dried, crushed, and slurried with recycled oil. Hydrogen is added to the slurry, and the mixture is preheated and charged continuously to the bottom of the ebullated-bed catalytic reactor. The vapor product leaving the top of the reactor is cooled to condense the heavier components to a liquid. Light hydrocarbons, ammonia and hydrogen sulfide are absorbed and separated from the remaining gas, leaving a hydrogen-rich gas which is recompressed and recycled to be combined with the input slurry. The liquid–solid product, containing unconverted coal, ash, and oil, is fed into a flash separator. The material that boils off is passed to an atmospheric distillation

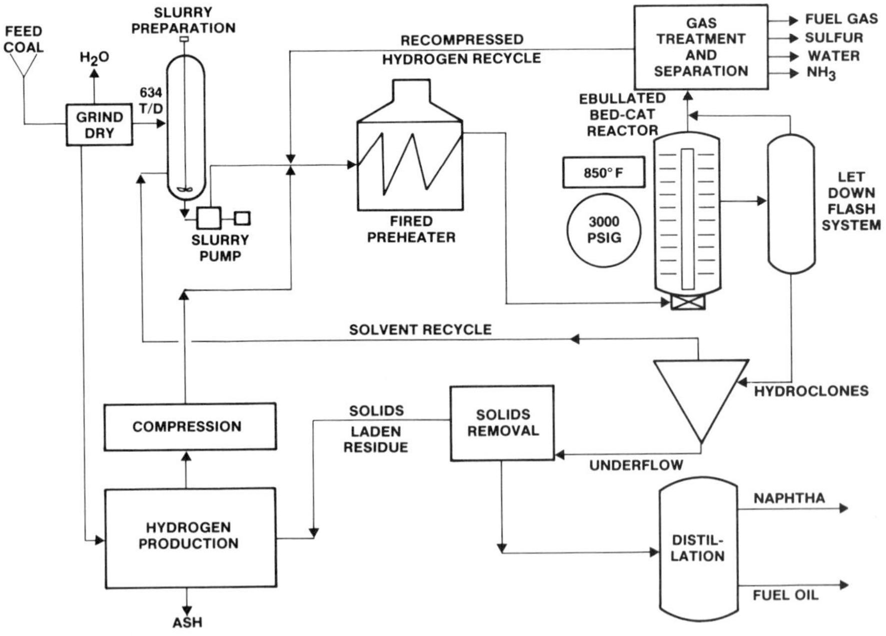

Fig. 1-11. H-Coal process diagram.

with a hydroclone (a liquid–solid separation) and a vacuum still. The gas and liquid products (hydrocarbon gas, hydrogen sulfide, ammonia, light and heavy distillates, and residual fuel) may be further refined as necessary. Heavy distillate is recycled as the slurry medium.

The operating conditions of the H-Coal process can be altered so as to produce various types of primary product. For example, to produce a synthetic crude product, relatively high temperatures and hydrogen partial pressures are used, and the solid–liquid separation can be conducted by vacuum distillation. On the other hand, the production of gas and low-sulfur residual oil requires lower temperatures and pressures.

The H-Coal process is one of several direct liquefaction technologies being developed. Its potential value lies in three areas:

- High-sulfur coals, which cannot be used in conventional coal utilization processes due to environmental considerations, are converted into useful low-sulfur, coal-derived liquids by the H-Coal process.
- A large range of products is available by varying operational modes.
- Operational flexibility, which is the ability to use different types of coal in different modes of operations.

The main characteristics of the pilot plant are:

- *Location*—Catlettsburg, Kentucky
- *Products*—naphtha, distillate fuel oil, boiler fuel oil, depending on mode of operation
- *Production*—1800 barrels per day
- *Feedstock*—200–600 tons bituminous (Kentucky No. 11 and Illinois No. 6) or subbituminous coal (Wyodak) depending on mode of operation

The H-Coal pilot plant has been built. A two-year operating program was planned: shakedown on Kentucky coal; syncrude mode with Illinois No. 6 coal; fuel oil mode with Illinois No. 6 coal; syncrude mode with Wyodak coal. The total project cost was projected to be about $300 million. Committed funding from industrial project participants is about $38 million.

DOE is responsible for management of the Pilot Plant Project. The project has the following non-DOE funding participants: Electric Power Research Institute; Mobil Oil Corporation; Standard Oil Company (Indiana); Ashland Synthetic Fuels, Inc.; Commonwealth of Kentucky; Continental Oil Company. The H-Coal contract provides for participation of these non-DOE organizations through H-Coal Management and Technical Advisory Committees.

Exxon Donor Solvent (EDS). A schematic of the EDS process is presented in Fig. 1-12. In this process, coal is ground and slurried with a process-derived hydrogenated donor solvent. The slurry is heated by a fired heater, and preheated gaseous hydrogen is added. The reaction is carried out in a tubular reactor with no internals. Products from the liquefaction reactor are sent to several stages of separation units for recovery of gas, naphtha, middle distillate, and bottoms, which are comprised primarily of unreacted coal and mineral matter. Distillation is the means for solid–liquid product separation.

The heavy bottoms from distillation are sent to the flexicoker to produce additional liquids and low-Btu gas for in-plant fuel use. Hydrogen for in-plant use is provided by

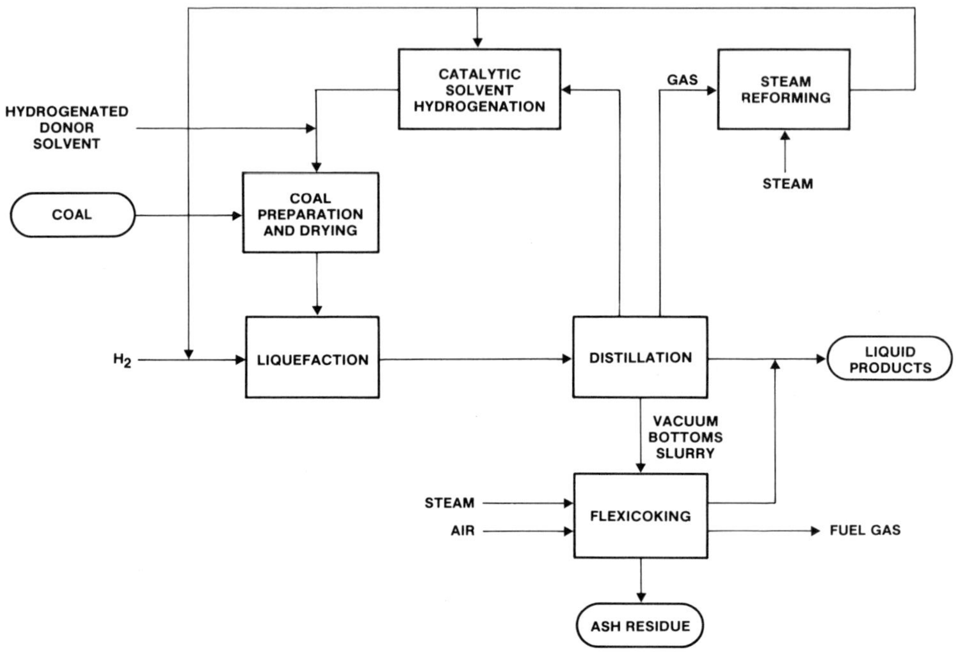

Fig. 1-12. Exxon Donor Solvent process diagram.

steam reforming of gases produced in the process or by partial oxidation of the heavy ends. The hydrogen is recycled to the liquefaction and solvent hydrogenation sections. By varying liquefaction conditions or adjusting solvent properties, the product distribution may be varied over a wide range to respond to market requirements.

EDS is one of four direct liquefaction processes. These processes yield more fuel per ton of coal than indirect processes, and their products are potentially less costly than the products of indirect liquefaction processes.

The main characteristics of the pilot plant are:

- *Location*—Baytown, Texas
- *Products*—naphtha and fuel oil
- *Feedstock*—250 tons per day of Illinois No. 6 coal
- *Process*—EDS with flexicoking operation being added

The pilot plant was started up in March 1980 and completed scheduled shakedown in October. The plant was operated successfully; however, the flexicoker has not yet been incorporated. Final design of the flexicoker is complete.

The cost of the pilot plant and flexicoker program is estimated at $296 million and $64.6 million, respectively. The projects were financed by DOE and industry sponsors as follows: EDS pilot plant ($296 million)—DOE, 50%; Exxon, 22%; EPRI, 12%; JCLD (Japan Coal Liquefaction Co.), 8%; Phillips, 2%; Atlantic, 2%; Ruhrkohle, 2%; AGIP (Italian Nat'l. Oil Co.), 2%. Prototype flexicoker program ($64.6 million)— DOE, 50%; Exxon, 36%; JCLD, 8%; Atlantic, 2%; Ruhrkohle, 2%; AGIP, 2%. Management of the

EDS project was performed by Exxon Research and Engineering Company under the supervision of the project sponsors.

Solvent Refined Coal (SRC-I). The Solvent Refined Coal (SRC-I) process, illustrated in Fig. 1-13, converts high-sulfur, high-ash coals into a substantially ash-free, low-sulfur fuel that is a solid at room temperature. Raw coal is crushed and mixed with a solvent derived from the process. The mixture is pressurized, and hydrogen gas is added. Then the mixture is heated, and undissolved solids are separated out using various solvents. The initial product, when cooled, is a shiny material resembling coal.

The solid SRC-I product can be further processed to yield coke for anodes in the aluminum industry, formed coke for the steel industry, and carbon black. Also, high-quality liquid fuels can be produced by adding more hydrogen in subsequent process steps. SRC-I is one of four direct liquefaction processes.

The main characteristics of the plant are:

- *Location*—Newman, Kentucky
- *Plant Output*—20,000 barrels per day of oil equivalent
- *Product*—solid fuel
- *Feedstock*—6000 tons per day of high-sulfur coal

The SRC-I Demonstration Plant could cost about $1.5 billion: DOE, $1368 million; International Coal Refining Co., $90 million; Commonwealth of Kentucky, $30 million. DOE was the sponsor of the project. Participating industrial partners included

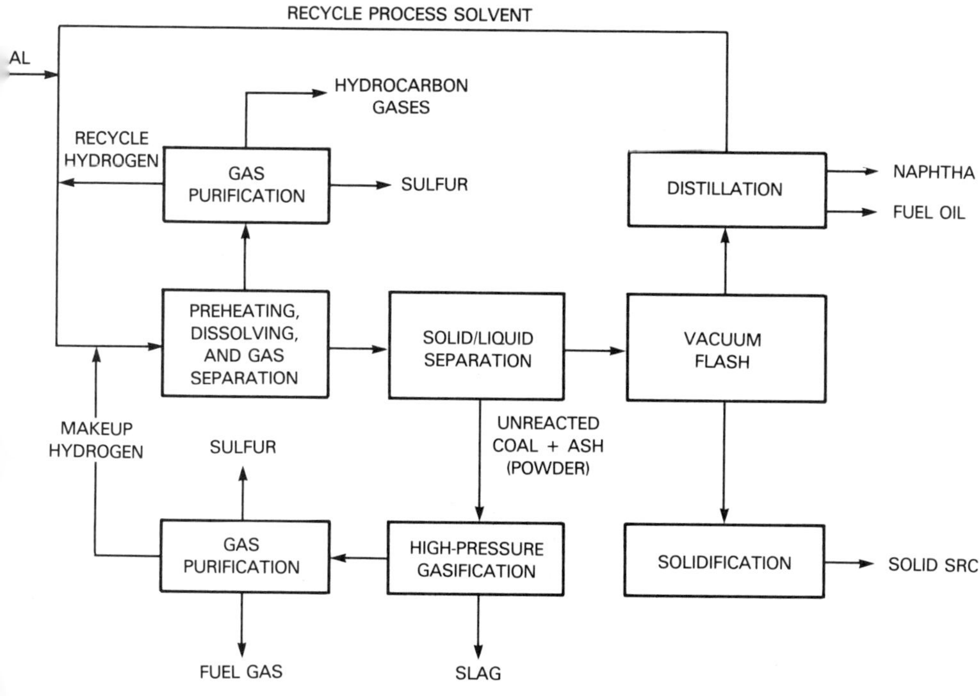

Fig. 1-13. Solvent Refined Coal (SRC-I) solid process diagram.

Southern Company Services, Inc., Air Products and Chemicals, Inc., and Wheelabrator-Frye, Inc.

Solvent Refined Coal (SRC-II). The Solvent Refined Coal (SRC-II) process is illustrated in Fig. 1-14. The primary product is a clean, nonpolluting liquid fuel from high-sulfur coal. Other products are expected to be pipeline-quality synthetic gas, liquefied propane and butane gases, naphtha, and sulfur, ammonia, and tar acid by-products. The process starts by crushing coal, then mixing it with a slurry made by recycling part of the oil from the process. Together with hydrogen, the mixture is heated and passed through a dissolver at high temperature and pressure. Aided by the natural catalysts in the coal, these conditions dissolve the coal. Sulfur is removed as hydrogen sulfide. Ash and undissolved coal are separated from the liquid product and gasified to make the needed hydrogen. SRC-II is one of four direct liquefaction processes.

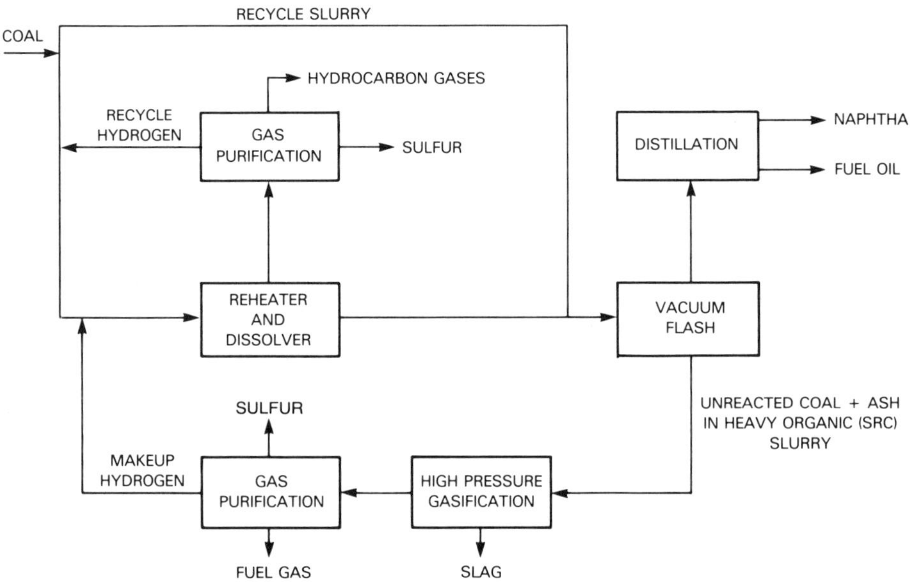

Fig. 1-14. Solvent Refined Coal (SRC-II) liquid process diagram.

The main characteristics of the plant are:

- *Location*—Morgantown, West Virginia
- *Plant Output*—20,000 barrels per day of oil equivalent
- *Products*—synthetic fuel oil, naphtha, and other fuel gases
- *Feedstock*—6000 tons per day of high-sulfur coal

Detailed design of the plant was begun. A site specific environmental impact statement was completed on September 30, 1981. The plant was scheduled to start producing synthetic fuels in 1985.

The demonstration plant is expected to cost about $1.5 billion. Germany and Japan were each to invest 25% of this sum. Gulf Oil Corporation, the industrial partner, would invest $100 million. The remaining $650 million was to be supplied by DOE.

DOE, with the greatest financial interest in the plant, was to be in overall charge of plant design and construction. This responsibility has been delegated to the Oak Ridge Operations Office of DOE, Oak Ridge, Tennessee. Gulf Oil, the developer of the process, through subsidiaries, is managing and coordinating the detailed design of the SRC plant.

Dow Liquefaction. In this process, coal is crushed, dried, and slurried with recycle solvent derived from this process. A catalyst, which consists of an aqueous ammonium molybdate salt emulsified in the slurry, is added. After mixing with process hydrogen, the slurry is preheated to approximately 790°F (421°C) and 2000 psig. Then, the slurry is fed into the liquefaction reactor, where the hydrogenation reaction occurs at approximately 850°F (454°C) and 2000 psig. The reactor product is flashed in two stages and separated as gases, light liquids, and heavy residues (heavy oil, ash, unreacted asphaltenes). Gases are further separated by a cryogenic separator as hydrogen and high-Btu gases. Hydrogen is pumped back to the process. The light liquids are further distilled as liquefied petroleum gas (LPG), naphtha, distillate fuel, and heavy fuel.

The heavy residues are passed into hydroclones. The entire hydroclone overhead stream is recycled as slurry oil. The underflow from the hydroclones goes to a countercurrent solvent deasphalter. The bottoms from the deasphalter are fed to a gasifier where they are contacted with oxygen and steam to convert the residual fuel to carbon monoxide and hydrogen. Off gases from the gasifier are scrubbed and passed to a shift converter to yield the hydrogen needed for the process.

Indirect Liquefaction

Indirect liquefaction processes include a preliminary gasification step which yields a mixture consisting primarily of carbon monoxide and hydrogen. Several types of gasifiers are available, including fixed-bed, fluid-bed, and entrained-bed designs.

The indirect liquefaction process is shown in Fig. 1-15. Coal is heated with steam and oxygen under pressure in a gasifier vessel. The raw gas is cleaned of particulates before a shift conversion step is used to adjust the carbon monoxide to hydrogen ratio for the subsequent synthesis step. The shifted gas is further purified before the synthesis (liquefaction) step.

Indirect processes gasify the coal, clean the intermediate product gases, and then synthesize liquid products such as methanol, gasoline, diesel fuel, and alcohol. Indirect

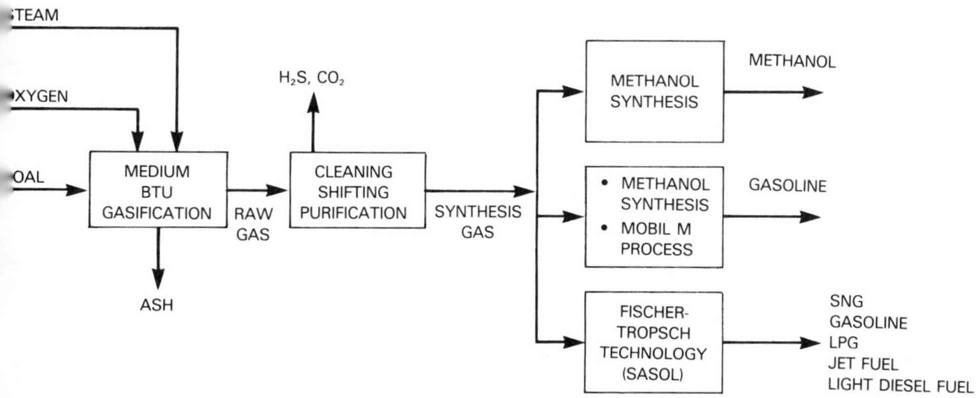

Fig. 1-15. Block diagram of the indirect coal liquefaction process.

processes include: Fischer–Tropsch (Lurgi gasification); Methanol–Lurgi gasification; Mobil M-Gasoline (coal to MeOH to gasoline, Lurgi gasification); Mobil M-Gasoline (gasoline production only).

Fischer–Tropsch (Lurgi Gasification). The Fischer–Tropsch process may be technologically ready for commercialization in the United States using Lurgi pressure gasification with noncaking western coal. Commercial readiness with eastern coal awaits commercial scale demonstration of pressurized gasifiers able to operate with caking coals (e.g., Texaco, Shell–Koppers).

Fischer–Tropsch technology is successfully operating on a commercial scale at the 10,000 barrel per day SASOL I and 60,000 barrel per day SASOL II plants in South Africa. Start-up of an additional 60,000 barrel per day plant (SASOL III) is planned for 1983. These plants produce gasoline, middle distillates (jet fuel, diesel oil), and heavy oil. Active interest in this technology has developed, and plans to license and construct similar plants in the United States are progressing.

The Fischer–Tropsch process is a catalytic conversion system that produces hydrocarbon liquids from coal-derived intermediate-Btu synthesis gas. The process is in commercial operation in South Africa where a Lurgi gasifier is used to produce synthesis gas, which is cleaned of hydrogen sulfide, carbon dioxide, and impurities, and then shift-converted before it enters a catalytic reactor that produces hydrocarbon liquids. Reactors, using different catalysts and temperatures, process gases with different carbon-to-hydrogen ratios into different products. Reactor products include gasoline and diesel/other fuel oils, waxes, alcohols, and ketones.

Methanol–Lurgi Gasification. Sized coal is gasified at pressure with oxygen and steam in a Lurgi gasifier. The raw synthesis gas is then shifted to yield a hydrogen to carbon monoxide ratio of about 2:1, and the shifted gases are quenched with water and fed to an acid gas removal unit to eliminate CO_2 and sulfur compounds. The purified synthesis gas is then fed to the methanol synthesis reactor. This reactor operates at about 1500 psi with a copper catalyst containing zinc and aluminum additives. The raw methanol from the synthesis reactor is dried by distillation.

M-Gasoline. In the Mobil M-Gasoline process, synthesis gas is first converted to methanol using commercially available technology, and the methanol is catalytically converted to high octane gasoline over a molecular-size-specific zeolite catalyst. A 100 barrel per day pilot plant of the methanol conversion step is under construction in West Germany and a 12,500 barrel per day commercial plant operating on reformed natural gas is planned for operation in New Zealand by 1985.

The Mobile M technology is near commercial readiness but has not been demonstrated on a large scale. Results from the West German pilot plant should be assessed before a commercial plant project is launched. Commercial readiness may be achieved soon.

Mobil M-Gasoline (Coal to MeOH to Gasoline, Lurgi Gasification). Crude methanol produced from coal-derived synthesis gas is preheated to 600°F (316°C) at 335 psig and converted in a fixed-bed catalytic reactor to an equilibrium mixture of methanol, dimethyl ether, and water, releasing about 20% of the total heat of reaction. These products are diluted with cooled recycled gas composed mainly of process-derived light hydrocarbons, and are fed to a second catalytic reactor for conversion to gasoline. The inlet temperature of the second reactor is about 680°F (360°C), and the adiabatic temperature rise, which is controlled by the recycle ratio, is about 200°F (93°C). The effluent from the second reactor is condensed and flash evaporated at 100°F (38°C). The gaseous phase is recycled, and the condensed phase—raw gasoline—is separated from the water by-product and sent to storage.

Mobil M-Gasoline (Gasoline Production Only). In the version of the Mobil M process that produces gasoline only, the purge gas (containing some CH_4) is reformed to produce more synthesis gas. The methanol is then used as feed for the Mobil M fuel process (see the previous description).

COAL GASIFICATION

Coal can be gasified to yield four major grades of products: low-Btu gas; medium-Btu gas (for fuel use); medium-Btu gas (for chemical feedstock use); synthetic natural gas (SNG).

High-Btu gasification is ready for commercial development using western coal or lignite with Lurgi gasification. The Lurgi gasifier has undergone 40 years of commercial development and 14 commercial plants are located in Australia, Germany, United Kingdom, India, Pakistan, South Africa, and Korea. There is extensive interest in commercializing the Lurgi technology in the United States. Commercial readiness of high-Btu gasification with eastern coal awaits demonstration of processes which can handle caking coal.

Medium-Btu gas is ready for commercial application to provide process fuel, electric power, or chemical feedstocks. The main uses of the gas are for process heat or power generation and as feedstock for methanol or ammonia synthesis. Tennessee Eastman plans to make acetic anhydride starting with synthesis gas from coal, and there is extensive research in progress to develop other synthesis processes starting with syngas. There is a great deal of commercial interest in all applications for medium-Btu gas.

The major steps in all these processes, illustrated in Fig. 1-16, include gasification of the coal to a mixture of carbon monoxide and hydrogen by heating it severely in the presence of air or oxygen and clean-up of the raw gas. The extent of purification of the product is determined by its intended use.

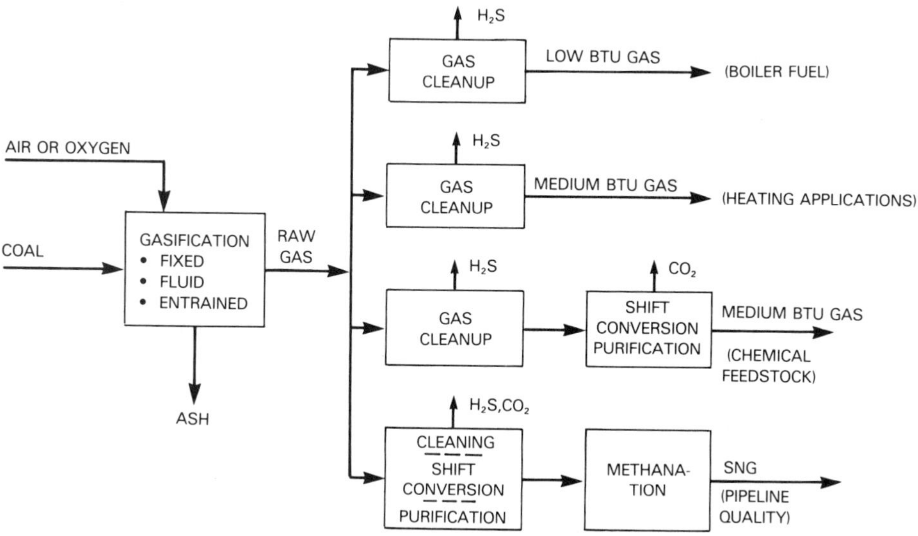

Fig. 1-16. Block diagram of coal gasification process.

Coal can be gasified using a variety of gasifier designs (fixed, fluid, or entrained bed), each of which offers advantages and disadvantages for specific end uses of the product. Air or oxygen can be used in the gasification step; if air is used, the raw gas has a lower heating value. Uses and distinguishing characteristics of the major grades of product gas are described.

Low-Btu Gas (90–130 Btu/scf). Low-Btu gas is used as a fuel gas; consequently, its composition is secondary to its heating value. Since air is used in the gasification step, the product mixture is diluted by nitrogen and, therefore, has a low heating value.

Medium-Btu Gas—Fuel Application (300–400 Btu/scf). The composition of medium-Btu gas destined for use as a fuel gas is relatively unimportant. Oxygen is used in the gasification step, and particulates and sulfur compounds are removed prior to use.

Medium-Btu Gas—Chemical Feedstock Applications (300–400 Btu/scf). When the medium-Btu gas is intended for chemical feedstock (synthesis gas) use, its composition is important. Gasification techniques (entrained bed) which yield greater proportions of carbon monoxide and hydrogen rather than methane are utilized. After the raw gas is cleansed of particulates, the gas is subjected to shift conversion which adjusts the carbon monoxide to hydrogen ratio to that desired for the production of a specific chemical, such as methanol. The shift conversion step produces carbon dioxide as a by-product. Acid gases (the carbon dioxide as well as the hydrogen sulfide contaminant) are removed from the synthesis gas in a purification step to obtain chemical-feedstock-grade gas.

High-Btu Gas—SNG (950+ Btu/scf). Coal is gasified with oxygen using gasification techniques that maximize methane production (fluid or fixed bed). Initial processing proceeds as for the chemical-feedstock-grade gas; however, a methanation step is added to convert the mixture of carbon monoxide and hydrogen to methane. The resulting gas is pipeline-quality SNG.

Coal gasification processes typically involve the reaction of the coal with steam and oxygen to form carbon monoxide and hydrogen. Because this reaction is highly endothermic, some of the coal is reacted in parallel (simultaneously) with oxygen to provide the heat. The latter reaction produces carbon dioxide, or with adequate control of oxygen availability, more carbon monoxide. Thus, the primary fuel gas derived from the coal consists principally of carbon monoxide and hydrogen in varying relative amounts.

High-Btu Coal Gasification

High-Btu gasification technology includes the following processes: British slagging Lurgi; Hydrogasification; Exxon catalytic coal gasification; CONOCO high-Btu gas.

British Slagging Lurgi. Sized coal is fed to the gasifiers, where coal descends through the gasifier and is dried, devolatized, gasified, and combusted. Oxygen and steam are injected into the gasifier from the bottom for combustion and gasification reactions at bed temperatures around 3630°F (2000°C) and 300 psig. At the high-bed temperature, ash is molten as slag, then withdrawn through a central tap in the bottom of the gasifier. The slag is quenched in water to form a glassy frit and removed via a lock hopper.

The raw gas produced in the gasifier is quenched and cooled to condense the tar, oil, phenols, ammonia, and water out of the gas stream. Then, the raw gas passes to a rectisol unit for desulfurization and enters a shift unit to adjust the hydrogen–carbon monoxide ratio to 3, which is required for downstream methanation. The shifted gas passes a CO_2 removal unit for further purification and flows to a methanator, where CO reacts with H_2 catalytically to produce methane.

The quenched liquids and solids from the gasifier are separated as tar and coal fine and liquids. Tar and coal fine are recycled to the gasifier. The liquids are further separated as ammonia, phenol, oil, and water.

Hydrogasification. Pulverized coal and hot hydrogen [2000°F (1094°C)] are fed into a reactor with a rocket injection steam. The reactor is an entrained flow type with short residence time. Raw products from the reactor are separated as gas and oil products as they pass the gas/liquid separation unit. The gas products enter a gas purification unit for desulfurization, and then go to the methanation unit for synthesis of methane from the reaction of CO and H_2. The high-Btu substitute natural gas is separated from hydrogen as the product gases enter the cryogenic separation unit. Hydrogen is recycled to the three reactors.

In this process, hydrogen is generated by the steam reforming process, in which steam reacts with the CO of the syngas. Char produced from the hydropyrolysis of coal in the reactor is withdrawn to a steam and power generation unit for combustion. Power from this unit will supply the oxygen production from air. Oxygen will react with steam supplied from the steam and power generation unit and produce syngas for hydrogen production.

Exxon Catalytic Coal Gasification. Ground coal (minus 8 mesh) is mixed with potassium hydroxide to 15-weight percent K_2CO_3 equivalent and dried. The mixture is fed in the dense phase for multipoint injection at the bottom of the fluidized bed, as it flows through lock hoppers and a high-pressure feeder. Steam and recycle CO and H_2 are preheated to 1550°F (843°C) and injected into the bottom of the fluidized bed as the fluidizing and heat transfer mediums to bring coal to the reaction temperature of 1275°F (691°C). Ninety percent of the feed coal is gasified, and most gases are synthesized as methane. The effluent gas stream flows through cyclones, a venturi scrubber, an acid gas removal system, and a cryogenic distillation unit to recover product methane and a recycle CO/H_2 stream.

Catalyst, char, and ash are withdrawn from the bottom of the reactor and quenched by the recycle gas. The catalyst is recovered in a $Ca(OH)_2$ digestion loop.

CONOCO High-Btu Gas. The high-Btu gasification process being developed by CONOCO is an extension of the conventional Lurgi gasifier originally developed in Germany and England prior to World War II. The product is pipeline-quality gas. As shown in Fig. 1-17, coal is fed into the gasifier at the top through a lock hopper system, and steam and oxygen are introduced near the bottom. As the gasifier incorporates a bottom hearth designed to discharge the ash as a molten slag, no excess steam is used to moderate temperatures in the bottom hearth, and less steam is used than in comparable processes. Raw gas exiting the gasifier is a mix of carbon monoxide and hydrogen, plus impurities which are removed. High-Btu gas is produced by a combined shift/methanation process. After carbon dioxide is removed, the gas is dried and compressed for introduction into a pipeline.

A major gasifier modification in the CONOCO process is the addition of a stirrer which rotates inside the gasifier to break up caking coals. Thus, the plant is a forerunner of facilities designed to overcome the caking tendencies of eastern and midwestern U.S. coals.

The main characteristics of the plant are:

- *Location*—Noble County, Ohio
- *Plant Output*—19 million standard cubic feet per day

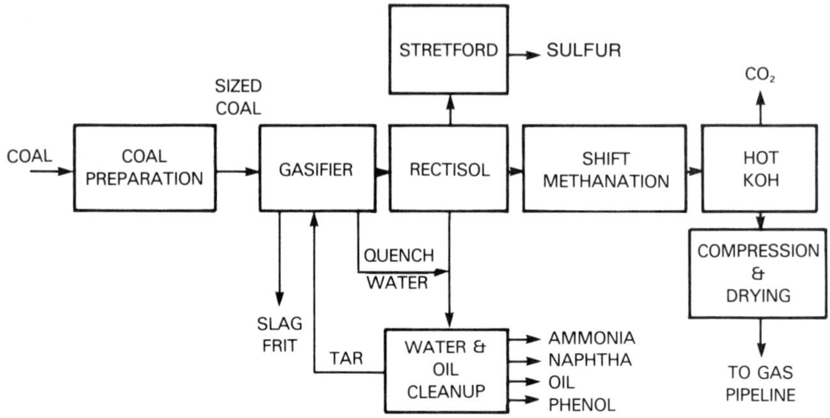

Fig. 1-17. CONOCO high-Btu gas process.

- *Product*—high-Btu gas, pipeline quality
- *Feedstock*—1250 tons per day of high-sulfur coal

Detailed design of the plant was completed in June 1981. If a decision is made to proceed with the demonstration, construction is scheduled for 30 months and operation for 42 months.

The demonstration plant is expected to cost $800 million. DOE has contributed $37 million for the design phase. The remainder of the demonstration plant costs was targeted for 50/50 sharing between CONOCO and DOE.

DOE is the sponsor of the design phase of the project. CONOCO Coal Development Company is the government's industrial partner for the demonstration.

Medium-/Low-Btu Coal Gasification

Medium-/low-Btu gas technology includes the following processes: Koppers–Totzek; Stoic gasification; U-Gas; Texaco coal gasification; Kiln Gas.

Koppers–Totzek. Coal is ground and fed to a gasifier where it reacts with oxygen and steam at 2700°F (1482°C) and 25 psig. Slag falls into a quench tank and is removed for disposal in a water stream. The effluent gas from the gasifier is sent by water to gas scrubbers. Water from the scrubbers and slag tank overflows to a clarifier. Sludge is separated for disposal, and water is recycled to the scrubbers and the slag tank.

The effluent gas from the scrubbers is fed to the acid gas removal system for desulfurization. The clean synthesis gas is produced via the acid gas removal system.

Stoic Gasification. Crushed coal is fed to the upper zone of the gasifier and volatized slowly by upward hot gas. The devolatilized coal then descends into the gasification zone. A portion of the gas flow is removed at the top of the gasification zone as the bottom gas. The rest of the gas from the top of the gasification zone flows upward and devolatilizes the inlet coal, then exits as the top gas. The top gas flows to a precipitator for heavy oil separation, and mixes with solid-free bottom gas as the synthesis gas.

At the bottom of the gasifier, air and steam are injected. Then, air is reacted with coal and forms a fire level to provide enough heat for gasification in the gasification zone. The ash produced is withdrawn from the bottom of the gasifier by grate rotation.

U-Gas. The U-Gas process, shown in Fig. 1-18, was developed by the Institute of Gas Technology. Medium-Btu gas is produced, and solid wastes are removed by an ash agglomeration process.

Steam and oxygen are fed into the bottom of the gasifier where they are mixed with coal in a turbulent manner called "fluidizing." Particles of ash that have lost most of their carbon stick together when they stray into a hot zone in the lower part of the gasifier. As enough ash particles stick together (i.e., agglomerate), they become heavy enough to fall out the inverted-cone-shaped base of the gasifier through a venturi throat. The gas product, a mixture of hydrogen and carbon monoxide with a heating value of 300 Btu/scf, is suitable for industrial purposes within an economical piping range of up to 100 miles.

The Memphis project is intended specifically to develop a way of producing a medium-Btu fuel gas from eastern and midwestern high-sulfur coals. These coals often prove troublesome for conventional gasifiers because of their caking tendencies.

The main characteristics of the plant are:

- *Location*—Memphis, Tennessee
- *Plant Output*—154 million standard cubic feet per year
- *Product*—medium-Btu gas
- *Feedstock*—3100 tons per day of high-sulfur coal

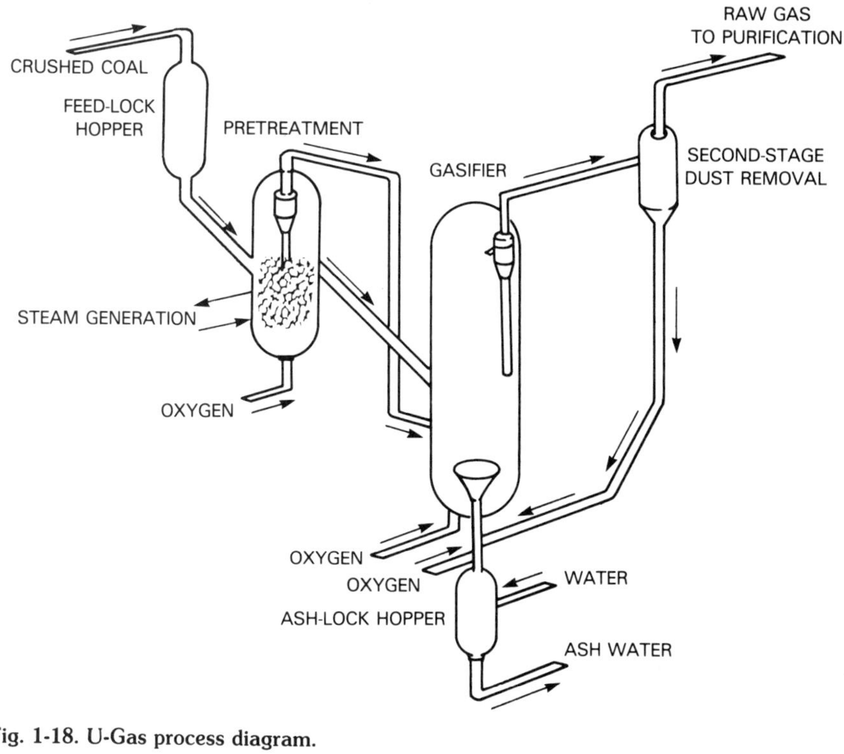

Fig. 1-18. U-Gas process diagram.

The detailed design and environmental impact statement have been completed, and construction can be started. Construction is scheduled to be completed in late 1985.

The total cost of the demonstration project is expected to be $700 million. Memphis initially will fund $100 million, then repay DOE from revenues until a 50/50 sharing is achieved.

The project is managed by DOE and its partner, the City of Memphis Light, Gas, and Water Division.

Texaco Coal Gasification. The coal is ground and mixed with water to form a slurry. The slurry is pumped at high pressure into the gasifier where the slurry reacts with oxygen at 650 psig and 2500°F (1370°C) under the condition of insufficient oxygen for complete combustion. At the gasifier temperature, the ash melts to form slag. Most of the ash in the form of slag falls into a water quench at the bottom of the gasifier. The resultant ash slurry leaves the gasifier and enters the slag dewatering unit. Overflow from the unit is recycled to the process. The remainder of the slag and raw hot gas from the gasifier are cooled in a gas cooling unit to a temperature well below the ash softening point for ash removal. The solid-free gas from the gas scrubber unit is sent to the acid gas removal system to remove hydrogen sulfide and sulfur dioxide. The effluent gas from the acid removal system is the clean fuel.

The various water streams, containing the ash and soot removed from the synthesis gas, are sent to the settler tank where clarified water is recovered for recycling. The settler tank bottoms, containing ash with significant amounts of carbon, can be recycled back and slurried with the coal.

Kiln Gas. Untreated coal is fed to the kiln gasifier, where the rotary gasifier is driven by two motors. The air and steam are injected into the gasifier through ports located in the devolatilization and gasification zones. The effluent gas stream from the rotary gasifier flows through a heat exchanger where 50% of the steam required in the kiln gasifier is generated. The product gas from the heat exchanger enters a gas cooling and tar particulate removal unit and a Stretford sulfur removal system, and finally flows to Unit No. 3 for electricity generation. Ash is removed from the gasifier via wet lock hoppers and sluiced to an existing ash disposal pond. Tars and particulates are separated in a tar particulate separation unit and recycled back to the gasifier. Simultaneously, liquor from the separation unit is blown down to the wastewater pretreatment unit.

UNCONVENTIONAL GAS RECOVERY

U.S. gas production is almost all accounted for by conventional resources for which relatively accurate reserve estimates are readily available. However, these reserves have been declining recently. The 1978 reserve level of 209 trillion cubic feet (Tcf) has declined from a level of 290 Tcf in 1970. Recovery from conventional gas reserves is expected to continue to decline. However, in spite of stringent conservation and fuel-conversion efforts, demand for this clean, versatile fuel will remain high for the foreseeable future. The resulting supply gap for conventional gas has focused increasing attention on unconventional sources of natural gas. Principal unconventional sources under consideration include geopressured methane, methane from coal seams, Devonian shale, and tight gas sands.

Geopressured Methane

The term *geopressured aquifers* characterizes subsurface reservoirs containing methane dissolved in water under high pressures and temperatures. Such water-bearing aquifers have been identified beneath the Gulf of Mexico and coastal regions of Texas and Louisiana.

Economic production requires large amounts of water, which in turn necessitates the drilling of large diameter wells. Currently, the resource is not being produced and may not prove to be economically producible with state-of-the-art technology.

Conceptually, in a total energy recovery system, the geopressured water would first be used to drive a turbine and deliver hydraulic energy to produce electric power. The water would then be flowed through separators, where methane would be extracted. Finally, thermal energy would be recovered by using a binary-cycle converter system to generate electric power. In this technology, the hot water evaporates a secondary fluid, which then drives a turbine generator.

Recovering energy from the geopressured brines will require an extension of the existing technology to handle high flow rates of hot saline water. Rigs are available in the Gulf that can drill the deep large-bore wells that are required. However, working in the geopressured geothermal zones will require special equipment, and delivery times may be long, thereby requiring sufficient lead items for development planning.

The basic technological constraint in developing the potential of geopressured methane is finding reservoirs where the size of the reservoir is large and the rate of production of the brine and the amount of methane dissolved in it are high. This includes finding large, continuous aquifers of high permeability and porosity and determining the state of gas saturation.

Methane from Coal Seams (Coalbed Methane)

Methane, the main constituent of natural gas, is generated during the natural process of coal formation. Coalbed methane gas can be liberated either by mining operations or by operations on virgin coalbeds, whether these are minable or not. A recent coalbed methane resource estimate indicates about 400 Tcf of gas is in place, with about 60 Tcf of that technically recoverable.

In general, there are three modes of gas liberation: during mining operations, after mining operations, and from virgin coalbeds.

During mining operations, gas is released from the coal at the mine face. Gas continues to be released from the roof, walls, and floor of the mine, and from the ribs and pillars of coal left in place. For safety reasons, this gas is carried away in very diluted form by ventilation air. The technology to recover and sell this gas at presently predicted prices does not exist.

After mining operations, gas continues to bleed from any coal remaining or from other adjacent gas-containing strata. With the collapse of the mine roof and the resulting breakage of the immediate roof and superposed strata, additional gas is released from the adjacent reservoir areas to the newly formed gob area. Such gob gas typically exists in a moderately concentrated form (40–90%) and can be collected, processed, and used.

Gas liberation from virgin coalbeds is the principal developing technology, and applies to all virgin coalbeds whether they are to be mined later or are considered to be unminable with today's state of the art. The basic technology is that of drilling into the virgin coal seam; stimulation, if necessary; recovery; and use of the coalbed gas. The techniques include the following:

- *Vertical wells.* This technique consists of drilling vertical holes from the surface to intersect the virgin coal seam. Gas flow experience in such instances has been limited unless the coal seam is highly fractured, permitting the gas to flow readily to the well bore.

- *Slant (directional) wells.* This technique consists of slant drilling a well from the surface with the slant increasing to intersect and remain within a coal seam. The resulting hole will intersect many natural fractures within the seam, permitting greater gas flow into the well bore.
- *Horizontal holes.* This technique consists of drilling holes directly into the face of the coal seam to intersect as many natural fractures as possible. These holes may be drilled from a mine shaft sunk for that purpose, or from mine entries into the coal seam ahead of mining operations, or from a surface outcrop into and following a coal seam.

The greatest constraints on the recovery and use of coalbed methane are the issues of legal ownership of the gas and treatment and disposal of the concomitant coalbed water and gas impurities in accordance with applicable environmental requirements. Significant constraints are also encountered in the stimulation of coal seams to permit gas production.

Devonian Shale

Devonian shale gas production requires some means of stimulation to achieve economic production levels. The primary state-of-the-art stimulation techniques are well-bore shooting and conventional hydraulic fracturing. Currently, Devonian shale gas production is about 0.1 Tcf/year using these stimulation techniques. Most of that production comes from the Big Sandy field of eastern Kentucky and western West Virginia in the Appalachian Basin.

The Devonian shale gas reservoirs range in permeability from 0.001 to 2.0 millidarcies (md) and in porosity from 2 to 14%. Gas in-place estimates for the resource range from 225 to 1900 Tcf with a technically producible portion of about 37 Tcf. The Btu content of the produced gas typically exceeds 1100 Btu/scf, which is considerably greater than the minimum pipeline requirement of 950 Btu/scf.

Well-Bore Shooting. The normal procedure in well-bore shooting is to first fill the entire open hole shale interval with 80% gelled nitroglycerine, then tamp the volume above with sand and/or water, and finally detonate the explosive with a blasting cap. Most Devonian shale wells in favorable geologic areas become on-line producers, even those with only slight gas shows initially. The shooting technique goes back to the early history of Devonian gas play. Disadvantages are that extensive well cleanout is usually required and that production rates after treatment tend to be lower than are possible with hydraulic fracturing and advanced techniques. Experience has further shown that shooting does not necessarily work well in all areas. The best results seem to be obtained in areas containing relatively thick sequences of organic shale that have a predominance of natural fracture systems.

Hydraulic Fracturing Process. The conventional hydraulic fracturing process, sometimes classified as normal hydraulic fracturing, evolved from early fluid injection experience in oil and gas well completion and workover techniques such as squeeze cementing, acidizing, and water injection. This experience led to the concept of fracturing a formation intentionally to improve deliverability. Basically, the process consists of transmitting energy generated by pumping a water-based fluid into a target formation for the purpose of creating a fracture within the formation. Once the formation breaks down and a fracture is created, material such as sand is introduced into the fluid stream to keep the fracture propped open. Theoretically, the established fracture results in improved gas communication from the formation to the well bore and thus increased well deliverability. The standard treatment size for a conventional hydraulic fracture treatment is about 1000 barrels of water and 50,000 pounds of sand. The advantages of hydraulic fracturing over well-bore shooting are increased well productivity and ultimate recovery, with lessened posttreatment well cleanup problems.

Tight Gas Sands

The low permeability western sandstone formations typically yield subcommercial gas flow rates unless artifically stimulated. The leading state-of-the-art stimulation method is massive hydraulic fracturing (MHF). Using appropriate stimulation technology, gas production from the western gas sand sources today amounts to about 1 Tcf per year or 4% of the nation's domestic gas production. That production comes primarily from the less-prevalent blanket-type formations where 600-foot fractures can be reliably produced through stimulation. The more-prevalent lenticular formations have proven to be less favorable targets.

The western tight formations are contained in basins which extend westward from Cotton Valley in northern Louisiana and eastern Texas to the San Juan Basin in New Mexico, and northward through the Rocky Mountain basins to the Northern Great Plains province.

The gas reservoirs range in permeability from near conventional (0.1 md) to very tight (0.001 md) and generally exhibit low gas-filled porosity levels (3–10%). Gas in place for the blanket formations has been estimated to be about 200 Tcf of which half is technically recoverable.

The MHF stimulation technology is designed to overcome the gas production limitations due to low permeability. A large quantity of fluid (up to 1 million gallons) is pumped into a well at pressures greater than the strength of the reservoir rock, thus fracturing the rock outward from the well bore. Propping agents, usually sand, are carried into the reservoir by the fluid and deposited in the fracture to hold it open as stimulation pressure is reduced and the fracturing fluid is expelled. The length of the fracture propped open (generally 500–2000 feet from the well bore) provides a pressure sink with a large surface area for gas flow from formation to well bore. Initial production can range between 500 to several thousand scf/day in a successful well. Total well life typically ranges between 15 and 30 years, although well lives of up to 50 years are not uncommon.

BIOMASS

Biomass represents a renewable, alternative resource base that potentially can be converted into clean liquid fuels which are useful for industrial process heat, electric power generation, and mechanical energy generation. Biomass can be defined as organic matter such as trees, plants, agricultural crops, and grasses that can be grown specifically for the production of fuels.

There are two fundamental methods of converting biomass to fuel: thermochemical conversion and biochemical conversion. *Thermochemical conversion* processes include pyrolysis, gasification, and liquefaction. *Biochemical conversion* processes include anaerobic digestion and fermentation.

Liquefaction

If biomass is to be liquefied, hydrogen must be added at high temperatures and pressures. Liquid fuels can also be obtained by converting synthesis gas (a mixture of carbon monoxide and hydrogen) to methanol or Fischer–Tropsch hydrocarbons. Methanol can then be converted to gasoline by the Mobil M process. Liquefaction essentially includes pyrolysis (tech-air) and catalytic liquefaction processes.

Pyrolysis (Tech-Air). Wood is dried to low (about 7%) moisture content and reduced in size in the wood-preparation section. The dried wood is mixed with recyle process oil and fed to the pyrolyzer. Air enters the pyrolyzer from the bottom and burns part of the wood to provide heat for the pyrolysis reactions at near atmospheric pressure. Char produced from the pyrolysis reactions is withdrawn from the bottom of the reactor and cooled for storage.

The effluent stream containing gases, liquids, and solids is quenched with recycled quench oil and separated in a quench pot as gases and liquids–solids. The liquids–solids stream passes a rotary vacuum filter where the solid solution containing 30–50% solids is recycled to the pyrolyzer as the process oil and the liquid stream as the product oil. A large portion of the oil is recycled and mixed with the effluent stream from the pyrolyzer to quench this stream for separation and dilute ash concentration.

Gases from the quench pot pass a demister and an electrostatic precipitator, thus providing heat for power generation and wood drying.

Catalytic Liquefaction. Wood is dried to 7% moisture content, ground, and slurried with recycle oil, solids, and catalyst. The slurry is compressed, mixed with carbon monoxide from the cryogenic separation unit, and sent to a preheater, where the process fuel provides heat. The effluent slurry at 580–600°F (304–316°C) and 2000 psi is then fed into a dissolver for wood dissolution. Gases leaving the dissolver are cooled and flushed in stages to separate unreacted gases from the oil–solid solution. Part of the oil–solid stream is recycled to the slurry mixing area as the recycle process oil, and the rest of the oil–solid stream is heated and sent to a vacuum flashing unit. The bottoms-containing oil with 30% ash and unreacted wood is recycled to a gasifier for gasification. The oil stream is the catalyst which is fed to the vacuum flashing unit and condensed with the injected steam in the overhead receiver and finally recycled for reuse. The product oil is obtained from the middle of the vacuum flashing unit.

In the gasifier, the recycled oil containing ash and unreacted wood mixes with the dried and ground raw wood, and reacts with steam in a partially oxidized condition. Ash and unconverted carbon are withdrawn, quenched, and fed to the slurrying mix tank. The effluent gases are cooled and directed to an acid gas removal unit where CO_2, hydrogen sulfide, and sulfur dioxide are removed from the gas stream. The clean gas stream is further purifed by removing water in a molecular sieve unit. The product gas stream is separated as hydrogen, methane, and carbon monoxide after the stream passes as a cryogenic separation unit. Carbon monoxide is then mixed with the wood slurry for liquefaction.

Gasification

A number of reactions occur in processes for gasifying biomass. Depending on the reactants added to the biomass feed, the relative amounts of CO, CO_2, H_2, CH_4, and light-hydrocarbon gases in the product fuels gas can be varied. Gasification includes biotherm gas and fluidized-bed gasification.

Biotherm Gas. A peat–water slurry is sent to a digester for the anaerobic digestion process. The raw gases from this digester pass through a gas purification unit and produce methane. The undigested residue from the digester is mechanically dewatered and thermally gasified. Most portions of CO, CO_2, H_2, C_xH_y, and part of ammonia and ash from the thermal gasification process are sent back to the digester. The rest of the gases from the liquid recovery unit will be purified by removing sulfur and carbon dioxide. The clean gases then pass a methanation unit for methane conversion.

Fluidized-Bed Gasification. Woodchip and catalyst are fed at the top of the fluidized-bed gasifier where hot air [1000°F (538°C)] is injected into the bottom as a fluidizing medium. In the bottom of the reactor, woodchip is burned with air to provide heat for gasification at 1 atmosphere. The hot gases produced are quenched by inlet-coal woodchip and exit the gasifier.

The effluent gas stream from the gasifier passes through a cyclone to remove unreacted woodchip and catalyst, both of which are recycled to mix with the raw woodchip. The solid-free gas stream flows to a wet scrubber for desulfurization.

TAR SANDS

Tar sands can be defined as reservoir rock whose interstices are partially to totally filled with a very viscous to semi-solid bitumen (crude oil), which cannot be moved through the rock to a well bore by the inherent reservoir energy. The term is applied to a variety of rock types which are consolidated and unconsolidated and which contain some form of bituminous material. Other terms synonymous with tar sands are bituminous sandstone, oil-impregnated rock, oil sand, and rock asphalt. Tar sands may be distinguished from heavy oil deposits by a difference in viscosity of the contained bitumen or oil. Tar sand bitumen viscosity is so great that production by primary methods is impossible. On the other hand, heavy oil viscosity is low enough to permit production, but probably not at economic rates.

Tar sands are a mixture of bitumen (10–20%), water (5%), and sand clay (84%), and are generally found in the same geographical area as conventional petroleum. Approximately 93% of the U.S. tar sand reserves are in Utah. The total resource is estimated to be the equivalent of about 30 billion barrels of oil.

Tar sand technology was developed primarily in Canada and has been used commercially there for many years. U.S. tar sands are of significantly different constituency from Canadian tar sands, and Canadian process technology may not be directly transferable.

Two methods of processing tar sands, outlined in Fig. 1-19, to produce a syncrude suitable for refinery use are

- *Open Pit Mining for Surface Recovery of the Tar Sands.* The bitumen is extracted from the mined material using a hot water process, and upgraded to a syncrude by a coking, a hydrotreating, or a thermal cracking processing step. This method is commercial in Canada, but produces huge quantities of solid waste.
- In Situ *Recovery.* The bitumen is extracted in-place using hot water, steam, and/or solvents, conveyed to the surface, and then upgraded as above to produce a refinery feedstock. These techniques are not yet commercial, but pilot plant work is being conducted.

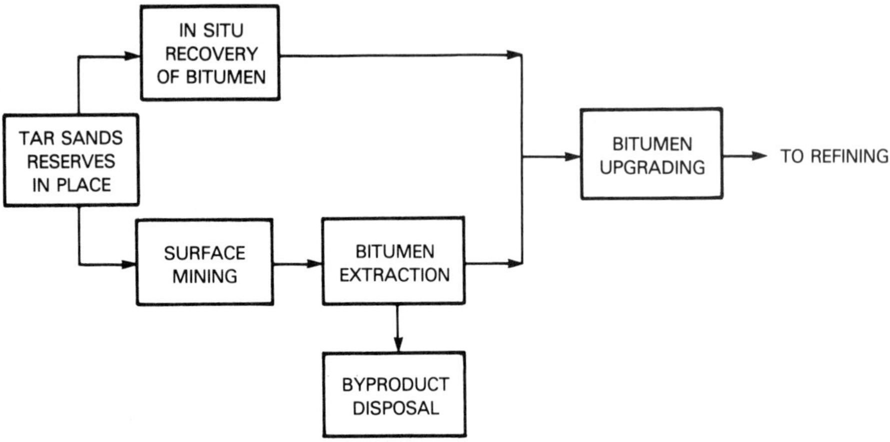

Fig. 1-19. Block diagram of tar sand process.

Surface Recovery

As in recovering coal, surface mining is an economical recovery technique for deposits lying close to the surface with little or no overburden. The following processes are included in surface mining: hot water extraction, solvent extraction, and Lurgi–Ruhrgas.

Hot Water Extraction. With this method, oil sand is fed into the plant and conditioned by agitating the material with hot water at 185°F (85°C) in rotary drums to remove the oil sand lumps. Caustic is added to the conditioning unit for pH control. Then, the conditioned material is screened to separate oversized particles and is sent to a primary separation cell where the hot water and recycled middlings are added for dilution and extraction. As bitumen in the oil sand rises forming a froth on the surface of the cell, the large sand particles settle out. A part of the middlings from the cell are recycled to mix with the conditioned sand oil. The rest of the middlings are directed to conventional flotation cells for scavenging. Bituminous froth is formed as air is injected into the cells and agitated with the middlings. After improving the quality in a settler, the froth mixes with the primary froth from the primary separation cell.

The froth is heated, deaerated, and sent to centrifuges for dehydration and demineralization. Froth treatment tailings from the centrifuges, scavenging cells, and primary separation cell are pumped to a separate area of the tailings plant to recover the floated diluted bitumen.

Solvent Extraction. Through this process, the diatomite feed mixes with a solvent to produce a pumpable slurry in a mix tank. The slurry is then pumped to countercurrent extractor thickeners where the diatomite is washed by progressively clean solvent. The oil-bearing solvent withdrawn from the first thickener is sent to a miscellaneous clarifier where the fines are separated and filtered. The oil-rich solvent is then sent to multiple effect evaporators to move solvent out of the stream. Next, the oil-rich stream is further purified to remove solvent in an oil stripper. The product oil is sent for upgrading, and the solvent is recycled to the countercurrent extractor thickeners.

The diatomite–solvent mixture from the last thickener is directed to a residue stripper where the steam strips the solvent from the diatomite. The stripped solvent flows to a decant tank to separate solvent from water. Solvent is recycled to the thickeners and water is sent back to the residue stripper. The diatomite–water mixture from the residue stripper flows to a residue filter to separate diatomite from water.

Lurgi–Ruhrgas. This process has been discussed in the oil shale section.

In Situ Recovery

A second technique for extracting the bitumen from tar sands is *in situ* recovery, which completely processes the tar sands underground, removing only the valuable bitumen and leaving the sand matrix in place. The *in situ* technique includes forward combustion, reverse combustion, steam stimulation, and steam flood. Only forward combustion and emulsion steam drive will be discussed in this section.

Forward Combustion. Air is introduced through the injection wells and ignited with oil in the tar sands. The combustion provides high thermal energy to distill some of the oil, making it less viscous. During this process, some of the oil is converted to coke which combusts with air and moves the combustion zone forward. The gaseous and oil products are pushed to the production wells, and finally sucked through the production wells to the surface for separation and upgrade.

Emulsion Steam Drive. Steam is injected into the reservoir through the injection wells. The thermal energy from the steam reduces the viscosity of sand oil in the reservoir, and makes the oil more mobile and displaceable. After releasing heat, steam is condensed to water and mixed with sand oil as the emulsion phase. The emulsion mixture is easily driven to the production wells and recovered because of its low viscosity.

SOLID WASTE

If the combustible portion of the nation's waste were used to generate energy, it would yield about 2 quadrillion Btu a year, or 2% of the current U.S. energy use. Furthermore, the inorganics in the waste, if recycled, could save an estimated 1 quad of energy.

The two principal ways of turning municipal solid waste (MSW) into energy are by burning directly in a waterwall incinerator or processing mechanically. The latter method produces refuse-derived fuel (RFD), which can be burned by itself or, more suitably because of its low-sulfur content, can be burned with coal. Solid waste has essentially two technologies: liquefaction and gasification.

Liquefaction

Liquefaction includes the following processes: occidental flash pyrolysis and tech-air pyrolysis.

Occidental Flash Pyrolysis. In this process, municipal refuse is shredded and directed to a magnet unit, air classifier, dryer, and screen for the removal of ferrous metal, aluminum, and glass. The clean stream is shredded again in a secondary shredder and injected to the flash pyrolysis reactor, where the refuse is pyrolyzed at a short residence time to produce char, oil, and medium-Btu gas. The char produced is recycled internally for combustion with air to provide heat for the pyrolysis reactions.

The effluent stream containing char, solids, oil, and gases leaves the reactor at 950°F (510°C) and passes through a cyclone separator to remove char and solids. Then the clean oil and gases are separated in an oil collection system.

Tech-Air Pyrolysis. This process has been discussed under the biomass section.

Gasification

Solid waste gasification includes the following processes: landguard pyrolysis, Resource Sciences pyrolysis, and Andco-Torax.

Landguard Pyrolysis. The solid waste is shredded to minus 4-in. particle size and stored in a surge bin. The shredded particles are fed continuously into a rotary kiln by ram feeders. The solids move downward against a flow of hot gases created by burning fuel oil and air at the low end of the kiln. Pyrolysis takes place at 1800°F (982°C) and forms oil, gases, and char. The char and noncombustible materials fall out of the low end and are quenched. The ferrous metals are mechanically separated.

The hot fuel gases leave the front end at about 1000°F (538°C) and pass through a gas scrubber to produce clean fuel gases.

Resource Sciences Pyrolysis. The solid waste is shredded to 1-in. particle size and fed to a surge bin. The particles are then fed into the top of the reactor, which consists of a vibro-oscillating horizontal ceramic-insulated fluid bed that serves as the furnace hearth and as a conveyor for the refuse through the unit. The top level of the reactor contains radiant-tube heaters that heat the incoming refuse to pyrolysis temperatures [1700°F (927°C)], without oxygen. The fuel gases are processed through a gas-refining subsystem, which includes water condensation and tar and particle removal. Part of the clean gas is pressurized for use in the radiant-tube heaters, and the rest of the clean gas is distributed as a product.

The char produced is withdrawn from the pyrolyzer and sent to a magnetic pulley to remove ferrous metals. The char flow passes through an air classifier to separate char from glass, ash, etc.

Andco-Torax. Unsorted refuse is fed to the gasifier. The refuse is covered with high-temperature air [2000°F (1094°C)] that is preheated by passing it through a blast superheater, heated by the combustion of natural gas or oil. As hot gases pass through the refuse, the refuse slowly settles through the gasifier. Pyrolysis occurs where there is a deficiency of air. The char processed during the pyrolysis gravitates to the bottom and reacts with the preheated air to generate heat for further pyrolysis. The gas leaving the pyrolzer contains 110–150 Btu/scf.

A molten slag is tapped and fritted to produce a black glassy aggregate.

PEAT

Peat, a geologically young coal, consists of partially decomposed plant matter and inorganic minerals that are accumulated in a water-saturated environment over a period of time. Considering the diversity of plant types from which peat may be formed and the extent to which the vegetation may have decomposed, one should not be surprised that there are considerable variations in the properties of the different peat deposits. Peat commonly used for agricultural and horticultural purposes as a soil conditioner is known as sphagnum moss (or fibric) peat. This peat is rich in fiber content and has a high liquid-retention capacity. For energy production, a more decomposed peat such as reed sedge (or hemic) peat is desirable. This peat has a higher heating value and less liquid-retention capacity than moss peat.

Peat, in its naturally occurring state, contains as much as 90% water. To use peat as a source of energy, the bulk of this water must be removed by either mechanical methods or solar energy without the use of fossil energy. This is important because, otherwise, a large fraction of the viable energy in peat will be used in drying.

Conversion of peat to synthetic fuels can be achieved by several processing alternatives that can be grouped together in three major categories: thermal processes, biological processes, and biological–thermal processing combination.

Thermal Processes

Most of the peat conversion work to date is related to thermal processing, primarily pyrolysis and gasification. A number of thermal processes that have been applied to the liquefaction of coal could also be applied to the liquefaction of peat. In thermal processing of peat, the product yields and process thermal efficiencies depend on the type of contacting of the gas–solids and the method of supplying heat for the reactions. It is most important to note that even though the cost of the gasification section of a peat plant is only about 15% of the total plant investment, it is the heart of the plant and controls the efficiency and product distribution of the total plant. In most of the gasification processes, the heat required is generated *in situ* by feeding air or oxygen. A few processes use indirect heat in the form of superheated steam or heated coils in the bed.

The thermal processes could consist of one or more reaction stages, each incorporating one of several types of solids flow, i.e., moving packed bed, fluidized bed, or entrained flow. The gas–solids contacting in the moving packed bed reactors could be either countercurrent or cocurrent (gas and solids both moving down). Gas–solids flow in the entrained flow reactor is cocurrent, while the fluidized-bed reactor most closely corresponds to a completely stirred system.

Biological Processes

Work on the biological conversion of peat is oriented toward the production of methane by microorganisms. Credit for the first significant work in the area of biological conversion of peat to methane goes to the Royal Institute of Technology and dates back to 1926. The mixtures were inoculated with sludge. The results of this work indicated that

- Acidic conditions inhibit digestion.
- Thermally preheated peat is digested more readily than raw peat.
- The addition of calcium acetate increased conversion rates.
- The degeneracy of bacteria occurred gradually.

The thermal and biological processes have both advantages and disadvantages. The primary advantage of thermal processes is faster reaction rates, while the primary advantage of biological conversion processes is that they do not require feed peat dewatering. The primary disadvantage of thermal processing is the requirement of high temperature and sometimes elevated pressure. Two major disadvantages of the biological conversion processes are the necessity of pretreating peat to increase the biologically digestible portion and the dewatering required for the solid by-products.

Biological–Thermal Processing Combination

The Biotherm gas process converts biomass, organic residues, and peat to SNG. This process essentially combines biological and thermal processes for the total conversion of wet peat. The first stage uses a peat–water slurry in an anaerobic digestion system. The undigested residue from the digester is mechanically dewatered and thermally gasified. Gases, ammonia, and ash from the thermal gasifier are fed to the digester. The Biotherm gas process shows potential, but more work needs to be done to obtain design data and to evaluate process economics.

Synthetic Fuel Commercialization

This section focuses on the effectiveness of various economic incentives[1] to encourage synthetic fuel commercialization development. Each incentive serves a different purpose with regard to inherent economics, project risks, marketplace uncertainty, and developer's access to capital. Emphasis is placed on the following seven major technologies:

- Oil Shale
- Coal Liquefaction
- Coal Gasification
  - High-Btu coal gasification
  - Medium-Btu coal gasification
  - Low-Btu coal gasification
- Unconventional Gas Recovery
  - Geopressured methane
  - Methane from coal seams
  - Devonian shale
  - Tight gas sands
- Biomass
- Tar Sands
- Solid Waste

OIL SHALE

Oil shale processing is near commercial readiness; however, the most appropriate process by which the oil will be recovered is still uncertain. Two primary technologies are being considered, each offering different levels of process economics, environmental concerns, and technical readiness; these processes are surface retorting and *in situ* (underground) processing.

Surface retorting technology is considered ready for commercialization. Present estimates of the capital cost of a production-size plant (50,000 barrels per day) are projected to be $1.0–1.3 billion. The estimated production price, $22–40/barrel, is higher than current contract oil prices, but has been comparable to the spot oil market price. A major concern is the potential environmental impact associated with the mining and disposal of shale in pristine environments.

The *in situ* process is still in the developmental stage, and is not expected to be commercially ready until the mid-1980s. Current estimates of capital requirements ($0.9–1.1 billion) and price ($19–26/barrels) are lower than the surface retort process. In addition, the *in situ* process is designed to minimize the amount of surface mining and the subsequent shale disposal problem.

Several financial barriers are encountered by all potential participants, as shown in Table 1-5; the most prominent barriers are a lack of clear economic feasibility and substantial project risks when scaled up from pilot/demonstration size to full-scale com-

---

[1] Incentives under P.L. 96-294 include (i) purchase agreements, price guarantees, and loan guarantees; (ii) loans; (iii) joint ventures; and (iv) corporation construction and contractor operations.

Table 1-5. Oil Shale: Incentives to Overcome Key Barriers

Potential Participants	What They Bring[a]				Major Economic Barriers to Participation			Most Effective Incentives
	Access to Resources	Technical Expertise	Financial Strength	Access to Markets	Economic Feasibility	Project Risk	Access to Capital	
Consortium: • major oil companies • independent oil cos. • technology suppliers • resource owners • others	●	●	●	●	• Production cost \$19 – \$30/bbl higher than current contract oil	• Future price viability uncertain • Technology not demonstrated at full commercial scale • Environmental concerns not resolved	(Not a barrier)	• Price guarantee • Purchase agreement • Loans • Investment tax credit • Accelerated depreciation
Major oil company	●	●	●	●	(No additional)	(No additional)	(Not a barrier)	(Same)
Independent oil company	●	◑/●	◑	●	(No additional)	(No additional)	• Large capital exposure relative to company size	(Same)
Other resource owner	●	○	○/◑	○	(No additional)	All above, plus: • lack technology expertise • lack direct access to markets	• Same as above	All above, plus: • loan guarantees

[a] ● Major contribution
◑ Modest contribution
○ No contribution

Source: Booz-Allen, *Analysis of Economic Incentives to Stimulate a Synthetic Fuels Industry.*

mercial size. In addition, all potential participants face uncertain future world and domestic oil prices and lack of resolution of key environmental issues on use of water, impact of mining in pristine environments, and disposal of shale. Key appropriate incentives may be as follows:

- Loans for projects which face highly uncertain environmental constraints to facilitate the purchase of equipment which meets environmental standards.
- Fixed price guarantee and purchase agreements for protection against uncertain future oil prices to several consortia and major oil companies as an insurance incentive before committing investment funds.
- Investment tax credits and accelerated depreciation to adjust inherent economics to reasonable viability.
- Loan guarantees to overcome financial barriers to entry. Beyond the core group of energy companies and consortia, other resource owners may have interest in owning and operating oil shale plants. These will need aid due to the large financial commitments required.

COAL LIQUEFACTION

The most likely participants in liquefaction projects are project consortia and the major oil companies, both of whom can bring together the major financial and technical resources required to support this technology on a commercial scale. Among the potential participants in liquefaction consortia might be oil companies, coal companies, technology companies, electric utilities, and other resource owners such as the Indian nations. Other possible owner/operators include independent oil companies, coal companies, and electric utilities who can provide either a market for the plant or access to the resource.

The very high level of capital investment required for a commercial-scale facility, coupled with uncertainties about the future economic viability of the facility, currently inhibit aggressive investment by any of the potential participants. Coal liquefaction projects are also faced with important institutional risks:

- Initial plants may lack a market while product characteristics and output levels are not compatible with existing downstream requirements for boiler fuel or refinery feedstock.
- Environmental issues are currently unresolved including plant emissions, solid waste disposal, and coal liquids characteristics.

In addition to the basic problems of economic viability and risk many of the potential owner/operators, with the possible exception of consortia and the major oil companies, are likely to lack the inherent financial strength to support a high-risk multi-billion dollar investment. Electric utilities' participation is further subject to the decisions of their respective regulatory processes.

Coal liquefaction requires additional technical development and the resolution of several institutional issues before major commercial developments are feasible. However, as Table 1-6 shows, a number of economic incentives can assist the development process.

The selection of the most effective set of incentives from the long list of possible ones will be influenced by a number of key factors:

Table 1-6. Coal Liquefaction: Incentives to Overcome Key Barriers

Potential Participants	What They Bring[a]				Major Economic Barriers to Participation			Most Effective Incentives
	Access to Resources	Technical Expertise	Financial Strength	Access to Markets	Economic Feasibility	Project Risk	Access to Capital	
Consortium: • major oil companies • independent oil cos. • coal companies • electric utility companies	●	●	●	●	• Production cost of \$5.00 – \$6.00/MMBtu is higher than prevailing contract prices	• Future price viability is uncertain • Market for initial output is not assured • Environmental concerns not resolved • Direct liquefaction technology not proven at commercial scale	• Production scale projects required \$1.08 to \$3.58 investment	• Price guarantee • Purchase agreement • Investment tax credit • Accelerated depreciation
Major oil company	●	●	◑/●	●	(No additional)	(No additional)	All above, plus: • investment level represents high capital exposure	(Same)
Independent oil company	◑	○	◑	●	(No additional)	All above, plus: • lack experience with leading technologies	All above, plus: • may lack access to sufficient capital at a viable cost	All above, plus: • loan guarantee
Coal company	●	○	○	○	(No additional)	All above, plus: • lack access to captive markets	All above	All above
Electric utility	○	○	○/◑	●	(No additional)	All above, plus: • lack experience with leading technologies	All above, plus: • lack strength to support debt	All above

[a] ● Major contribution
◑ Modest contribution
○ No contribution

Source: Booz-Allen, *Analysis of Economic Incentives to Stimulate a Synthetic Fuels Industry.*

- Negotiated price guarantees appear to address many of the investment barriers to coal liquefaction development; however, they may require both the involvement of the federal government for an extended period of time and a complex system for administration.
- Loan guarantees can broaden the list of potential owner/operators, but may increase the current and future cost of the program to the government.
- Purchase agreements may lessen the high risk of technological or institutional uncertainties.
- Tax credits and accelerated depreciation, which are preferred by the oil industry, can improve project economics and are less complex to administer; however, they require taxable income, which may preclude the involvement of some potential developers.

COAL GASIFICATION

High-Btu Coal Gasification

The technology for the production of pipeline-quality high-Btu gas [900–1000 Btu per standard cubic foot (scf)] is now available even though no commercial-size plants have been built and operated in the United States.

Current estimates of capital requirements for first-generation processes, primarily based on Lurgi and Koppers–Totzek technology, are $1.1–1.8 billion for a 250 million cubic feet per day plant. Projected gas costs range from $5.50 to $8.00 MMBtu. The second-generation technology, i.e., HYGAS, COGAS, and others, which is further from commercial readiness, has similar economics but can use eastern caking coals in addition to western coals.

As in the case of oil shale and coal liquefaction, high-Btu gasification projects require substantial capital and technological expertise. Gas pipeline companies are the most likely developers of high-Btu gas facilities, either alone or through consortia. Other potential participants in gasification consortia include technology companies, coal companies, and other resource owners including Indian nations with large holdings of western coal. Gas distribution companies and coal companies may elect to participate alone, but many lack the financial resources to support projects of this magnitude.

Development of high-Btu coal gasification commercial-scale projects is limited by their high risk level. These risks arise from the combination of (1) the high capital cost of these facilities, particularly when compared to the investment levels of typical gas or coal companies; (2) a high degree of uncertainty about future gas availability and both federal and local regulatory policy; and (3) the current high costs of synthetic gas when compared to conventional gas.

The key incentives appear to be as follows:

- Loan guarantees have been suggested by representatives of the gas industry as key to their participation in high-Btu gas projects. They feel that the high capital investments represented by these projects, coupled with the substantial uncertainty as to gas marketability or price, expose their companies to an unacceptable level of financial risk.
- Purchase agreements may be used separately to stimulate developments by participants without a captive pipeline. Table 1-7 summarizes the effects of incentives on high-Btu coal gasification.

Table 1-7. High-Btu Coal Gasification: Incentives to Overcome Key Barriers

Potential Participants	What They Bring[a]				Major Economic Barriers to Participation			Most Effective Incentives
	Access to Resources	Technical Expertise	Financial Strength	Access to Markets	Economic Feasibility	Project Risk	Access to Capital	
Consortium	●	●	●	●	• Production cost of \$5.00 – \$8.00/MMBtu is higher than natural gas; regional factors dictate feasibility	• Uncertain supplies of competing sources (LNG, Mexican gas, Alaskan gas) • Uncertain regulatory policy will impact project profitability • Environmental constraints may limit western operations	(None)	• Investment tax credits • Accelerated depreciation
Gas pipeline companies	○	○	○/◑	●	(No additional)	All above, plus: • lack experience with coal and gasifier technology	• Financial requirements are large, relative to company sizes	All above, plus: • loan guarantee
Gas distribution companies	○	○	○/◑	●	(No additional)	Same as consortium	Same as above	Same as consortium
Coal companies	●	○/◑	○/◑	○	(No additional)	Same as consortium, plus: • project risk due to noncaptive, regulated market	Same as above	Same as consortium, plus: • price guarantee • purchase agreement

[a]● Major contribution
◑ Modest contribution
○ No contribution

Source: Booz-Allen, *Analysis of Economic Incentives to Stimulate a Synthetic Fuels Industry.*

In addition to the incentive program, the development of high-Btu gas clearly must be coordinated with policies regarding unconventional gas, liquefied natural gas (LNG), Mexican gas, and federal and local gas price regulations. The prices and availability of each of these sources affect the marketability of high-Btu synthetic gas, and, consequently, affect the level of incentives required to make these projects attractive.

Medium-Btu Coal Gasification (MBG)

Process technologies to produce medium-Btu gas (250–350 Btu/scf) from coal are commercially available. Several foreign plants have been built, although none are operating or are in construction in the United States. Medium-Btu gas can be used directly for petrochemical feedstock and can be used efficiently as an industrial and utility fuel. On a technical basis, the industrial market potential has been estimated to exceed 3 quadrillion Btu. However, the fundamental barriers to commercialization are substantial:

- *Economics*. Gas costs are estimated to range from $4.00/MMBtu for large utility-financed multiple-user plants to as high as $6.25/MMBtu for moderate-size industrial plants.
- *Risks*. There is no operating experience in the United States. Total plant costs will be affected by labor and material cost escalation, as well as constraints on siting, safety, environmental controls, and construction of dedicated pipelines.
- *Marketability*. The most economically attractive plants are large multiple-user facilities. Since medium-Btu gas can be economically transported only 150 miles, the market for plant output will be local and limited.

Medium-Btu projects will involve a large number of participants. There is little correlation between coal resource owners, technology suppliers, and end-user markets; for this reason, the natural selection of owner/operators is unclear.

Each potential owner/operator has somewhat unique barriers to entry to the business, as shown in Table 1-8, and often requires different incentives. Consortia face the fundamental barriers to commercialization plus institutional complexities inherent to establishing consortia corporations. The effective incentives are

- Price guarantees
- Investment tax credit
- Accelerated depreciation

Gas companies face an uncertain regulatory climate and large capital exposure relative to typical company size; the combination of issues indicates higher than acceptable risk. To overcome unacceptable risk, the key incentive is loan guarantees.

Coal companies do not have direct access to industrial gas markets. Furthermore, the scale of investment required ($400 million–$1.0 billion) is large, even relative to the size of larger coal companies. For MBG to become a viable alternate fuel, coal companies' financial barriers will have to be mitigated by

- Loan guarantees
- Price guarantees
- Investment tax credit
- Accelerated depreciation

Table 1-8. Medium-Btu Coal Gasification: Incentives to Overcome Key Barriers

Potential Participants	What They Bring[a]				Major Economic Barriers to Participation			Most Effective Incentives
	Access to Resources	Technical Expertise	Financial Strength	Access to Markets	Economic Feasibility	Project Risk	Access to Capital	
Consortium: • fuel users • investors • gas companies • coal companies	○/●	●	●	●	• Production cost \$4.00 – \$6.25/MMBtu nearly competitive with current gas prices	• The multiple user market is complex to penetrate if users are not developers • Availability and price of competing fuels are uncertain	(None)	• Price guarantee • Investment tax credits • Accelerated depreciation
Gas companies	○	○/◑	○/◑	◑	(No additional)	All above, plus: • cost of competing fuels is uncertain • regulatory position on med-Btu gas pricing is uncertain	Large capital exposure for most gas companies	• Investment tax credits • Accelerated depreciation
Coal companies	●	○/◑	◑	○	(No additional)	All above, plus: • uncertain gas marketability	Large capital exposure for most coal companies	• Loan guarantee • Price guarantee • Investment tax credit • Accelerated depreciation
Large fuel users • steel companies • chemical companies • brick companies • others	○	○/◑	○/◑	●	All above, plus: • costs are higher for a single user plant	• Cost of competing fuels is uncertain	None for most major fuel users	• Investment tax credits • Accelerated depreciation

[a] ● Major contribution
◑ Modest contribution
○ No contribution

Source: Booz-Allen, *Analysis of Economic Incentives to Stimulate a Synthetic Fuels Industry.*

Large industrial users can build MBG plants dedicated to individual industrial plants; however, owing to higher per unit costs on a smaller scale, an economic penalty will be paid. The key incentives are

- Investment tax credit
- Accelerated depreciation

Under present conditions and because there is little correlation between potential business partners, it is unclear whether MBG represents an opportunity for the gas business, coal business, user consortia, or financial consortia. For this reason, the selection of incentives beyond investment tax credit and accelerate depreciation will have a major impact on who will be MBG plant owner/operators.

Low-Btu Coal Gasification

Low-Btu coal gasification (LBG) is a user-oriented industrial and utility option. The process technology is mature, and at least 100 foreign installations are known to exist. In the United States, nine LBG plants are in operation, five plants are in operational startup, and at least five plants are in design and/or construction.

Estimates of the technical market opportunities range from 1 to 2 quadrillion Btu. However, market penetration is constrained by several key factors.

Gas costs are somewhat higher than alternate fuels and are estimated to range from $2.7 to $6.50/MMBtu, with likely costs as follows:

- Hot raw gas: $3.50/MMBtu
- Partially cleaned gas: $4.10/MMBtu
- Clean gas: $4.60/MMBtu

In addition, there is substantial uncertainty on site-specific costs, including the extent of refitting boilers, etc., to burn the low-Btu nitrogen-rich gases.

As distinct from other synfuel options, the magnitude of capital investment for LBG plants is not a major barrier. Small LBG plants can be constructed for $1.8–5.8 million, and moderate size multi-unit plants can be built for $10–20 million.

The major participants in LBG development are industrial firms for self-help investment, electric utilities for combined-cycle plant fuel, and third party owner/operators, which could include suppliers of industrial gases and equipment, natural gas distribution companies, and financial ventures.

Also, as distinct from other synfuel options, the key barriers do not vary by type of owner/operator (Table 1-9). The fundamental barriers are unattractive economics and uncertainty about long-term economic benefits. For this reason, the two effective incentives are investment credit and accelerated depreciation.

UNCONVENTIONAL GAS RECOVERY

Geopressured Methane

Geopressured methane represents an immense potential resource base, but its economic recovery under environmentally acceptable conditions remains uncertain. The resource base is concentrated along the Texas and Louisiana Gulf Coast. According to the Department of Energy (DOE), the estimated resource in-place ranges from 1000 to 3000 trillion cubic feet (Tcf).

Table 1-9. Low-Btu Coal Gasification: Incentives to Overcome Key Barriers

Potential Participants	What They Bring[a]				Major Economic Barriers to Participation			Most Effective Incentives
	Access to Resources	Technical Expertise	Financial Strength	Access to Markets	Economic Feasibility	Project Risk	Access to Capital	
Electric utilities	○	◑	●	●	• Projected costs of \$3.40 – \$4.80/MMBtu is not competitive with coal or other fuels	• Uncertain environmental restrictions on direct coal burning	(Not a barrier)	• Accelerated depreciation • Investment tax credit
Industrial fuel users • steel • chemicals • brick • others	○	◑/●	●	●	(No additional)	(No additional)	None except for small users	(No additional)

[a] ● Major contribution
◑ Modest contribution
○ No contribution

Source: Booz-Allen, *Analysis of Economic Incentives to Stimulate a Synthetic Fuels Industry*.

Resource recovery combines conventional drilling and completion technology with water separation and reinjection equipment at the surface. The surface equipment allows the methane to come out of solution at atmospheric pressure, thus separating the gas from the geopressured brine. The energy from the high pressured brine can be used to generate electricity.

Estimated capital requirements and price are highly speculative owing to inadequate information about the methane content of the brine and the rate of fluid flow. DOE has postulated that geopressured methane will cost more than $4/MMBtu and will require investments averaging $4 million per well. The electric power generating facility will represent an additional substantial investment.

The potential participants in geopressured methane include consortia and electric utilities. Though the technology to provide environmentally acceptable recovery methods is still under investigation, potential participants will need to understand the electric and gas markets, as well as being financially strong.

The most likely developers will be consortia, electric utilities, and gas companies. Consortia bring the necessary financial strength through the participation of gas companies, oil companies, technology companies, utilities, and others; electric utilities bring an understanding of the regulated electric generation environment and generation plant technology; and gas companies bring an understanding of the gas markets and extraction technologies.

Although this resource could probably be developed with technologies marginally different from conventional sources, major uncertainties concerning land subsidence and brine disposal have limited its commercial acceptance. The current state of the art of geopressured methane development and land subsidence management has not advanced sufficiently to bring active industry involvement. Further research and testing will be necessary to overcome major technical risk to geopressured development. Research should be directed to:

- *Land subsidence management.* In certain onshore lowland areas, private ownership reverts to the state should subsidence cause the land to submerge.
- *Reinjection/disposal technology.* Brine disposal in the Gulf or reinjection into aquifers raises significant environmental concerns and decreases project economic viability.

Well-defined and properly funded R&D has a greater potential for successful development of geopressured methane than direct funding for commercial development. Technological land environmental uncertainties still need to be resolved or mitigated, financial incentives need to be focused on production, and risk needs to be minimized to stimulate higher levels of industry activity. Table 1-10 provides the potential participants, the barriers to participation, and the most effective incentives. These incentives include one or more of the following:

- Price guarantees
- Investment tax credit
- Accelerated depreciation

Other Unconventional Gas Recovery

The technologies needed to procure gas from the vast resources of tight sands,

Table 1-10. Geopressured Methane: Incentives to Overcome Key Barriers

Potential Participants	What They Bring[a]				Major Economic Barriers to Participation			Most Effective Incentives
	Access to Resources	Technical Expertise	Financial Strength	Access to Markets	Economic Feasibility	Project Risk	Access to Capital	
Consortium	●	●	●	●	• Inherently uneconomic due to high costs	• Technically unproven at commercial scale • Environmental uncertainty may increase investment requirements	(Not a barrier)	• Price guarantee • Investment tax credit • Accelerated depreciation
Electric utilities	○	◑	●	●	(No additional)	(No additional)	• Utility financing may not be available to high risk projects	Same as above
Gas companies	○	◑	●	●	(No additional)	(No additional)	(No additional)	

[a] ● Major contribution
◑ Modest contribution
○ No contribution

Source: Booz-Allen, *Analysis of Economic Incentives to Stimulate a Synthetic Fuels Industry.*

Devonian shales, and methane from coal seams are currently available. Some gas is being produced commercially from each.

Owing to the comparatively low capital requirements, many potential producers are possible. Exploring for these unconventional resources is often similar to conventional exploration; therefore, oil and gas companies have played an important role. However, the relatively low capital requirements associated with these technologies suggest that financial size will not eliminate many interested developers. Recent interest by industrial "self-help" producers and municipal gas systems suggests the diversity of possible participants.

While deregulation of gas produced from Devonian shales and methane from coal seams has improved the economics of some previously marginal resources, most of the resources are still uneconomical at current prices. Gas from tight gas sands has been regulated. Economic thresholds for these resources have been projected by DOE as follows:

- Tight gas sands: $1.5–2.5/MMBtu
- Devonian shales: $1.5–3.00/MMBtu
- Methane from coal seams: $2–3/MMBtu

In addition, the producing industry views investment in unconventional resources as riskier than conventional resources and attaches a higher risk premium to their expected rate of return. As Table 1-11 points out, marginal economics and relative risk are two barriers faced by each potential participant. Investment tax credit and accelerated depreciation are the most effective for overcoming the barriers to entry for these unconventional gas resources, because project profitability is improved and risk is reduced. Price guarantees are also identified as effective in overcoming these barriers but may be administratively more difficult owing to the potential number of wells involved.

Some gas producers have objected to possible incentives for unconventional resources by citing the current availability of cheaper conventional resources. They feel more expensive but subsidized resources would displace the cheaper conventional resources.

BIOMASS

The many processes which yield liquid fuels from biomass are in diverse stages of technical development. Based only on market penetration projections, processes that appear to offer contributions include marine crop production, catalytic liquefaction, and fermentation to produce ethanol. Missions that appear to have greater near-term commercialization potential include:

- Production of ethanol for gasohol blending or unblended use.
- Gasification of wood and low-moisture plants to produce intermediate-Btu gas (IBG), SNG, LBG, and ammonia.
- Pyrolysis of wood and low-moisture plants to produce SNG, fuel oil, and char.
- Combustion of wood and low-moisture plants to produce steam and steam with electricity as a by-product.
- Anaerobic digestion of manure and high-moisture terrestrial crops to produce IBG and SNG.

Table 1-11. Tight Gas Sands, Devonian Shales, and Methane from Coal Sands: Incentives to Overcome Key Barriers

Potential Participants	What They Bring[a]				Major Economic Barriers to Participation			Most Effective Incentives
	Access to Resources	Technical Expertise	Financial Strength	Access to Markets	Economic Feasibility	Project Risk	Access to Capital	
• Consortium	●	●	●	●	(Not a barrier)	• Industry regards these resources as high risk as compared to conventional oil and gas resources	(Not a barrier)	• Investment tax credit • Accelerated depreciation • Price guarantee
• Gas companies	◑	◑	●	●	(No additional)	(No additional)	(No additional)	(Same)
• Large companies					(No additional)	(No additional)	(No additional)	(Same)
oil	●	●	●	●				
coal	●	◑	●	○				
• Small developers	◑	○	◑	○	(No additional)	(No additional)	(No additional)	(Same)
• Municipalities	◑	○	●	●	(No additional)	(No additional)	(No additional)	(Same)

[a] ● Major contribution
◑ Modest contribution
○ No contribution

Source: Booz-Allen, *Analysis of Economic Incentives to Stimulate a Synthetic Fuels Industry.*

Overall biomass program success (as measured by the levels of future quad [one quadrillion Btu] production) can best be achieved by emphasizing production methods, procedures, and policies designed to increase feedstock availability. Loan guarantees are needed to expand gasohol production facilities. The market presentation of biomass fuel products is sensitive to feedstock scenario and costs.

Gasohol produced from agricultural feedstocks has become our most promising near-term option for producing synthetic fuels. Gasohol is the only synthetic fuel currently produced and distributed by major fuel corporations.

Financial incentive systems of various types, including excise tax reductions, loan guarantees, and investment tax credits, should be continued to increase the use of available crop and timber lands for the production of energy crops and combination energy–food/fiber crops, and encourage construction and operation of biomass fuel and chemical product production facilities.

TAR SANDS

Tar sands in the United States represent a source of hydrocarbon material that could help to satisfy the growing energy demands. Tar sand activities have increased over the past several years, and a growing interest is being shown in the development of a tar sand industry. However, these activities are not now, nor have they ever been, great enough to ensure a substantial development in the near term. Unless substantial capital is diverted to tar sand exploitation or a breakthrough occurs in the development of an economically viable and generally applicable oil recovery technology, significant oil production from U.S. tar sand will develop slowly and occur only over the long term.

Several important constraints to commercial development of tar sands, technical and nontechnical, have been identified. The technical problems include:

- Low permeability in reservoirs with relatively high (greater than 60%) bitumen saturation. Sufficient natural or induced permeability to allow injectivity to steam at sufficient rates must exist for successful application of the process. This problem could restrict application of the process to reservoirs with relatively low bitumen saturation.
- Process control has not been sufficiently demonstrated to take maximum advantage of the steam drive process. No theoretical or practical data base exists in the public-need area for process control in a fractured media. Optimization of control will enable greater productivity to the tar sand resource.
- Appropriate well completion techniques and hardware for relatively shallow and low productivity tar sand wells are not readily available. Techniques and hardware, developed by trial and error, should be standardized and made widely available to the industry.

The nontechnical constraints include:

- Questionable availability of water for oil recovery processes
- Undeclared moratorium (since 1965) on leasing of federally owned tar sand properties in Utah
- Lead time required for the environmental assessment
- Locations and magnitudes of markets for tar and bitumen

Cooperative agreements (i.e., joint venture under P.L. 96-294) appear to be the necessary incentive to encourage the commercial development of U.S. tar sand resources. The contractors under such an incentive should be required to share at least 50% of the cost of the project.

SOLID WASTE

If the combustible portion of this nation's waste were used to generate energy, it would yield about 2 quadrillion Btu/year, or 2% of the current U.S. energy use. While the recovery of energy and materials from waste grew enormously during the 1970s, from almost nothing to 30 plants capable of processing almost 8 million tons of municipal solid waste (MSW) annually, the trend has not been without some problems.

Solid waste is as difficult to apply technologically as it is to dispose of environmentally. The development of systems and processes to recover energy and recyclables takes longer and is more expensive than that foreseen in the early 1970s.

There are two principal (and rival) uses of turning MSW into energy: It can either be burned indirectly in a device called a "waterwall incinerator" to generate steam and electricity, or it can be technically processed. The latter technique produces a shredded product known as refuse-derived fuel (RDF).

RDF, say many resource recovery experts, has several clear advantages. It can be conveniently stored and transported and can be produced to a constant, specified standard to meet the handling and burning requirements of a boiler. Generally, RDF has been burned with coal in existing utility boilers as a way of reducing costs and because most utility boilers are large enough to accept all the RDF a community can produce. There is no technical reason, however, why a boiler could not be built specifically to handle RDF. However, RDF has run into some problems. Fuel users have complained about its high-ash content and the risk of long-term damage to their boilers. Attempts are now being made to sieve the waste from which it is made more thoroughly to reduce the noncombustible content. Air classifiers also are being used to winnow the waste, but considerably more R&D is needed to improve the quality of the product.

The waterwall incinerator is, by contrast, generally considered a proven technology, because most American installations are copies of plants operating successfully in Europe or elsewhere. Plant capital and operating costs are high, but waste with a wide range of properties can be burned in this way without prior processing.

Most waste-to-energy plants, however, still have a poor reputation, which is the result of the technology not being developed along conventional, maturing lines but rather "continuing, as it began—as an art." Also, the field was not profitable enough to attract private sector R&D funds. The implementors perceived technical and economic risks, and were fearful that the plant they built would not work and the recovered materials and energy products could not be sold on a consistent basis. Until a proven plant is operational, communities will not support the technology.

Oddly enough, utilities, which might be considered prime customers for the RDF produced by resource recovery plants, have been loath to be involved. There are a variety of reasons for this:

- Lack of capital
- Reluctance to risk using equipment by utilizing unfamiliar fuel
- Sheer inconsequentiality of RDF in the average utility fuel mix

The potential financial, regulatory, technical, and political problems of burning solid waste, therefore, are frequently considered not worth the effort by utility executives. The key likely incentives necessary to encourage prudent innovation in waste-to-energy plants are load guarantees and price guarantees. Such incentives would encourage the construction of many new plants, because owners would be assured that the plants ultimately would work and customers would know that the new fuels would not result in large maintenance bills from the long-term effects of using RDF. In addition to their production of energy, the development of RDF systems will make a substantial contribution to reducing our nation's waste disposal problem.

Chapter 2

Coal: An Overview

Introduction

Coal is an important energy source in the United States. Although its contribution to the nation's total energy supply has declined as compared with earlier years, coal continues as a mainstay of the economy. During the 1970s, annual coal production fluctuated from 561 million to about 776 million tons; in 1979, its value exceeded $18 billion. Most of the coal was consumed within the United States, chiefly to generate electricity. Collected in one place, all the coal mined during a single year would make a pile about 1000 feet high, covering 1 square mile. Yet, even after more than 2 centuries of coal production, the United States has at least 200 billion tons of coal in reserve for future use.

Coal is a heterogeneous material containing organic and mineral matter. The organic matter is composed primarily of carbon and small amounts of hydrogen, nitrogen, oxygen, and sulfur. The mineral matter consists chiefly of clay minerals, mineral forms of sulfur (mostly pyrite), calcite, and smaller amounts of other minerals. Moisture is also considered a constituent of coal in the bed or seam and a part of the coal when subjected to analytical testing. The percentage content of organic matter is used to distinguish coal from peat, graphite, and other carbon-bearing rocks. To be classified as coal, the "rock" must contain more than 50% by weight carbonaceous material. Although peat also is a carbonaceous sediment, by definition it is not a coal.

The appearance of as-mined coal varies in color from dark brown to black, while the natural seam surface may be dull, silky, or glossy. Coal textures range from earthy or fibrous lignite, to tough, brittle, and hard anthracite. Lignite is the least friable of all coals, but it, as well as most subbituminous coals, will "slack" (crack, split, and crumble) when exposed to the atmosphere and weather.

Coal had its beginning as ancient plants that lived in swamps millions of years ago. Geological processes working over vast spans of time compressed and altered the plant remains, increasing the percentage of carbon present, thus producing the different ranks of coal: lignite, subbituminous, bituminous, and anthracite. The resulting coalbeds, sometimes called seams, are interlayered between beds of sandstone, shale, and limestone. The coalbeds range in thickness from less than 1 inch to more than 100 feet. According to some estimates, an accumulation of 5–10 feet of relatively compacted plant material was needed for each foot of coal.

Coal is mined from more than 300 beds, but more than one-half of the annual production comes from only about 20 beds. Most of the total tonnage is from beds 2–6 feet thick.

When coal is burned, its fossil energy—the energy stored in the ancient plant material—is released. It ranges from about 14 million British thermal units (Btu) per ton for lignite to about 29 million Btu per ton for some bituminous coals. Using an average of 22 million Btu per ton, a ton of coal has the same heating value as 22,000 cubic feet of natural gas, 159 gallons of distillate fuel oil, or 1 cord of seasoned hardwood.

The most extensive coal seams of North America are found in rocks of the Mississippian and Pennsylvanian age, 270–350 million years old. Every major system of rocks formed since then contains coal. Most of the world's commercially important coals are no older than the Mississippian, but Devonian coals, 350–400 million years old, have been mined in Europe. Although some Permian coal (225–270 million years old) is mined in this country, this coal is much more abundant in Australia and the Soviet Union. The largest American reserves, other than those of the Mississippian and Pennsylvanian periods, are the western subbituminous coals and lignites of the Cretaceous and Tertiary ages (1–135 million years old). Coal-forming conditions probably existed several hundred times, and, as the map shown in Fig. 2-1 suggests, some of the regions involved must have covered tens of thousands of square miles.

ORIGIN AND FORMATION OF COAL

Geologists generally agree that specific conditions must have existed to permit the accumulation of plant material and, later, the formation of coal. These conditions were

- A swampy or marshy environment and a climate favorable to extensive plant growth
- Some subsidence in the area during accumulation of debris and compaction of the deposited plant material to permit further accumulation

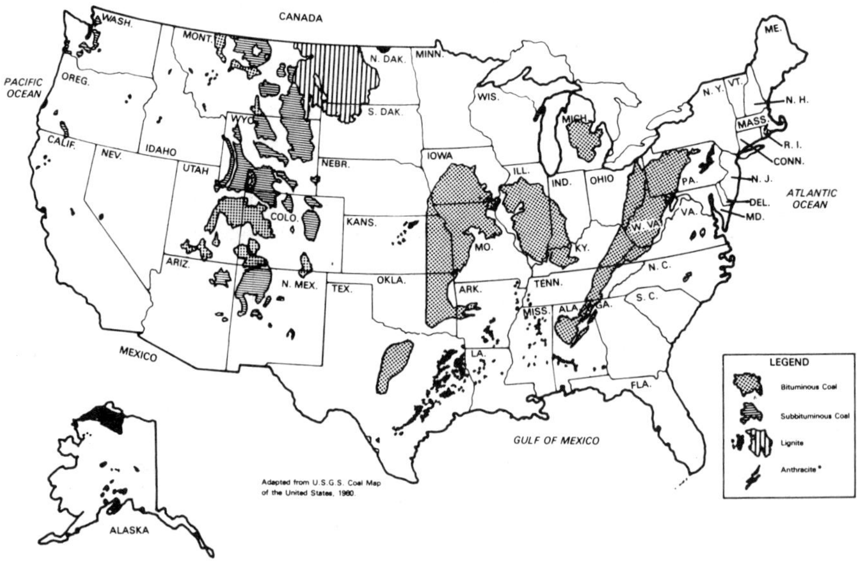

*Principal deposits in Pennsylvania; small deposits in Arkansas, Colorado, New Mexico, and Virginia. Anthracite - meta-anthracite deposits in Massachusetts and Rhode Island

Courtesy: DOE

Fig. 2-1. Coalfields of the United States (coal rank not distinguished in Alaska).

- Sufficient water to exclude air and thus cause decay
- Proximity to the sea or to other vital areas so that the material eventually could be buried by sediments
- A site of accumulation where erosion of either the plant material or the forming coal could not occur

The initial chemical and bacterial action and the effect of a changing environment eventually produced lignite, the least metamorphasized of all coals. With deeper burial and, over a long period of time, exposure of the coal to moderate heat and greater pressure, further decomposition occurred and produced coals of a higher rank.

Long-term environmental effects in areas where original plant material was deposited had much to do with the properties of coals. In the Midwest, for example, Illinois coals were inundated by marine water soon after formation, imparting a high concentration of sulfur. By contrast, the low sulfur coals of the Fort Union Basin of North Dakota developed under primarily freshwater conditions. The fact that the lignites of the Great Plains are much younger and had their own special set of environmental conditions when formed helps to explain why these coals contain more moisture and oxygen than Appalachian coals.

Classification of Coals

RANKING

Although various coal ranking systems are used in the United States, the one generally accepted is the American Society for Testing and Materials (ASTM) rank classification based on the degree of metamorphism (coalification) of the parent plant material.

Calculated on a moist, mineral-matter-free (MMF) basis, fixed carbon content and calorific value are the standard criteria for ranking by the ASTM method (Table 2-1), although agglomerating (caking) characteristics and percentage volatility are also considered. The mineral matter is excluded from the analysis because it does not reflect the degree of metamorphism of the coal; however, bed (equilibrium) moisture is included.

Coals that have heating values of 14,000 Btu/lb or more on a moist basis or contain 69% or more fixed carbon on a dry basis are all ranked by fixed carbon, regardless of their heating values. All other coals, those with less than 14,000 Btu/lb and containing less than 69% fixed carbon, are ranked by calorific value on a moist basis. "Moist" in this context means the coal's inherent (bed or seam) moisture, not visible water on the coal's surface.

The important properties of coals arranged by increasing rank are shown graphically in Figs. 2-2 and 2-3. In both figures, rank increases from left to right. Figure 2-2 shows that the heating value, sometimes expressed as "Rank Btu," tends to increase with rank, reaching a peak in the low-volatile bituminous coals. Figure 2-3 illustrates the fact that, as rank increases, moisture and volatile matter percentages tend to decrease, with a corresponding general increase in fixed carbon content.

PETROGRAPHY

Another method of measuring and evaluating the properties of bituminous coals is petrography, which uses a "type" classification based on the various coal components. From the viewpoint of petrography, bituminous coals consist of two basic materials—

Table 2-1. Classification of Coals by ASTM Ranking

Class	Group	Fixed Carbon Limits (%) (Dry, mineral-matter-free basis)		Volatile Matter Limits (%) (Dry, mineral-matter-free basis)		Calorific Value Limits (Btu/lb) (Moist, mineral-matter-free basis)		Agglomerating Character
		Equal or Greater Than	Less Than	Greater Than	Equal or Less Than	Equal or Greater Than	Less Than	
I Anthracite	1. Meta-anthracite	98	—	—	2	—	—	
	2. Anthracite	92	98	2	8	—	—	Nonagglomerating
	3. Semianthracite	86	92	8	14	—	—	
II Bituminous	1. Low volatile bituminous coal	78	86	14	22	—	—	
	2. Medium volatile bituminous coal	69	78	22	31	—	—	
	3. High volatile A bituminous coal	—	69	31	—	14,000	—	Commonly Agglomerating
	4. High volatile B bituminous coal	—	—	—	—	13,000	14,000	
	5. High volatile C bituminous coal	—	—	—	—	11,500	13,000	
		—	—	—	—	10,500	11,500	Agglomerating
III Subbituminous	1. Subbituminous A Coal	—	—	—	—	10,500	11,500	
	2. Subbituminous B Coal	—	—	—	—	9,500	10,500	
	3. Subbituminous C Coal	—	—	—	—	8,300	9,500	Nonagglomerating
IV Lignite	1. Lignite A	—	—	—	—	6,300	8,300	
	2. Lignite B	—	—	—	—	—	6,300	

Source: ASTM, Standard Specification D388.

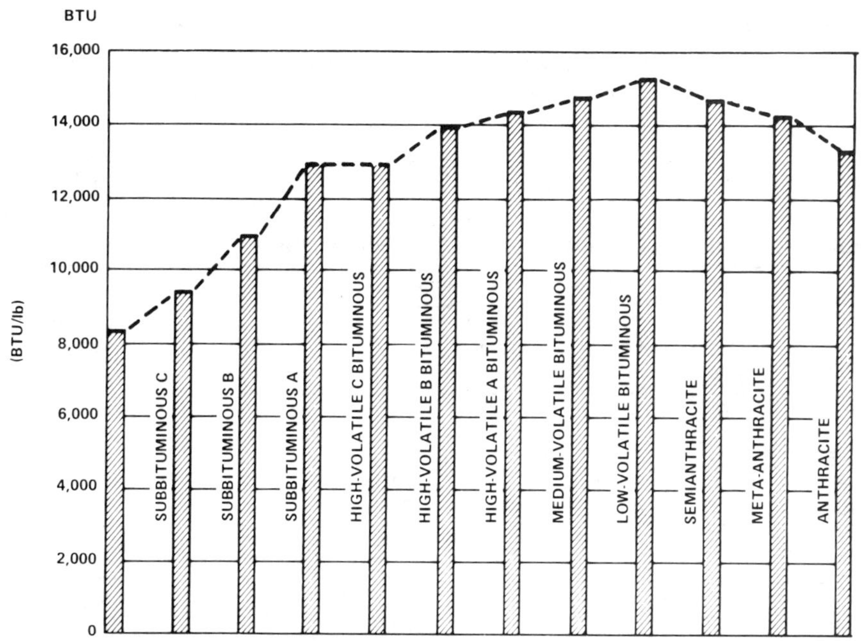

Fig. 2-2. Calorific value by rank (moist, MMF basis).

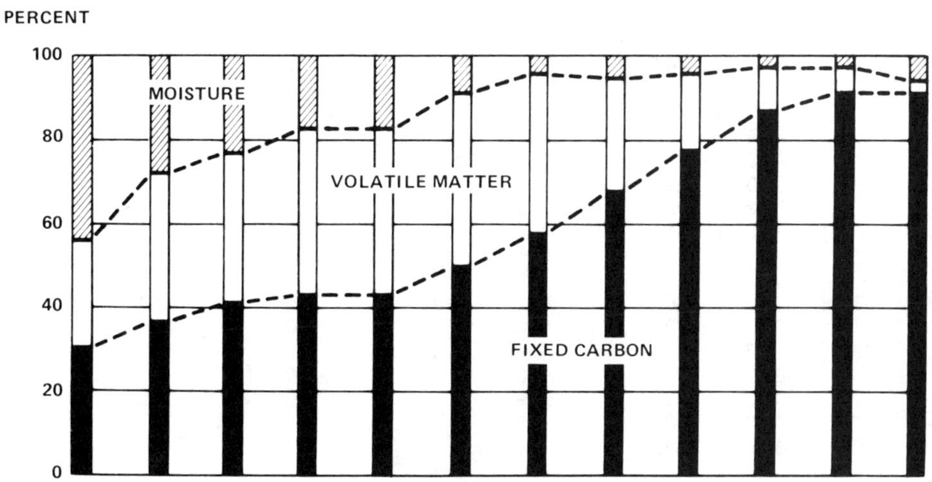

Fig. 2-3. Moisture, volatile matter, and fixed carbon by rank given in respective columns of Fig. 2-2 (moist, MMF basis).

minerals and organic "macerals." The latter form the carbonaceous, combustible fraction of the coal, and both occur as grains, particles, or fragments.

Petrography is used primarily in the study of coals utilized for metallurgical purposes and to date has found little application in the analysis of steam coals.

Coal Consumption

In the early 1900s, coal was the nation's major fuel, supplying almost 90% of its energy needs. Later, coal's importance declined, mainly because of competition from petroleum and natural gas. In 1950, coal accounted for about 38% of the energy consumed in the United States. In recent years, coal consumption has totaled more than 600 million tons annually, as shown in Table 2-2, but its share of the total energy

Table 2-2. Coal Consumption by End-Use Sector, 1949 – 1979 (Million Short Tons)

Year	Electric Utilities	Industry and Miscellaneous			Transportation	Residential and Commercial	Total
		Coke Plants	Other Industry and Miscellaneous	Total			
1949	84.0	91.4	121.2	212.6	70.2	116.5	483.2
1950	91.9	104.0	120.6	224.6	63.0	114.6	494.1
1951	105.8	113.7	128.7	242.4	56.2	101.5	505.9
1952	107.1	97.8	117.1	214.9	39.8	92.3	454.1
1953	115.9	113.1	117.0	230.1	29.6	79.2	454.8
1954	118.4	85.6	98.2	183.9	18.6	69.1	389.9
1955	143.8	107.7	110.1	217.8	17.0	68.4	447.0
1956	158.3	106.3	114.3	220.6	13.8	64.2	456.9
1957	160.8	108.4	106.5	214.9	9.8	49.0	434.5
1958	155.7	76.8	100.5	177.4	4.7	47.9	385.7
1959	168.4	79.6	92.7	172.3	3.6	40.8	385.1
1960	176.6	81.4	96.0	177.4	3.0	40.9	398.0
1961	182.1	74.2	95.9	170.1	0.8	37.3	390.3
1962	193.2	74.7	97.1	171.7	0.7	36.5	402.2
1963	211.3	78.1	101.9	180.0	0.7	31.5	423.5
1964	225.4	89.2	103.1	192.4	0.7	27.2	445.7
1965	244.8	95.3	105.6	200.8	0.7	25.7	472.0
1966	266.5	96.4	108.7	205.1	0.6	25.6	497.7
1967	274.2	92.8	101.8	194.6	0.5	22.1	491.4
1968	297.8	91.3	100.4	191.6	0.4	20.0	509.8
1969	310.6	93.4	93.1	186.6	0.3	18.9	516.4
1970	320.2	96.5	90.2	186.6	0.3	16.1	523.2
1971	327.3	83.2	75.6	158.9	0.2	15.2	501.6
1972	351.8	87.7	72.9	160.6	0.2	11.7	524.3
1973	389.2	94.1	68.0	162.1	0.1	11.1	562.6
1974	391.8	90.2	64.9	155.1	0.1	11.4	558.4
1975	406.0	83.6	63.6	147.2	[a]	9.4	562.6
1976	448.4	84.7	61.8	146.5	[a]	8.9	603.8
1977	477.1	77.7	61.5	139.2	[a]	9.0	625.3
1978	481.2	71.4	63.1	134.5	[a]	9.5	625.2
1979[b]	528.8	77.1	65.9	143.0	[a]	9.1	680.9

[a] Less than 0.05 million short tons.
[b] Preliminary.
Note: Sum of components may not equal total due to independent rounding.
Note: After 1977, small amounts of bituminous coal and lignite consumed by the transportation sector are includ in the Other Industry and Miscellaneous category.
Source: U.S. Department of Energy, Energy Information Administration, *Annual Report to Congress*, Volume Tw 1979.

consumed in the country has been less than 20%. On a per capita basis, coal consumption in 1978 amounted to about 16 pounds per person each day.

About 75% of the coal consumed in recent years, averaging more than 450 million tons annually during the late 1970s, has been used to generate electricity, as shown in Fig. 2-4. This coal, often termed steam coal, is burned to produce steam that drives electric generators. Each ton of coal consumed in an electric power plant generates about 2000 kilowatt-hours of electricity. A pound of coal supplies enough electricity to light ten 100-W bulbs for about 1 hour. Coal-fired power plants account for about 45% of the total electricity generated in the United States.

The second largest use of coal, totaling more than 70 million tons annually, is to produce coke for the steel industry. One ton of metallurgical-grade coal (a premium coal that meets certain specifications) can be converted in special high-temperature ovens into about 1400 pounds (0.7 ton) of coke as shown in Fig. 2-5. The coke is used in smelting iron ore in blast furnaces. About 1.2 tons of cokes are consumed per ton of pig iron produced. The process of converting coal into coke yields useful byproducts: benzene, coal tars, naphtha, and similar coal chemicals that are used to produce a number of familiar products such as solvents, varnishes, perfumes, medicines, dyes, and plastics.

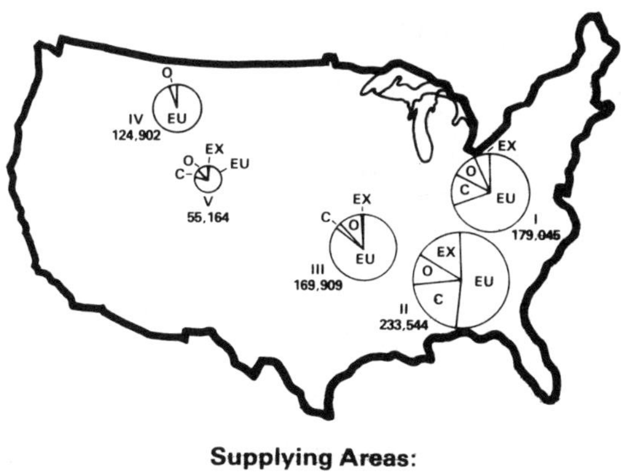

Supplying Areas:

I. Northern Appalachian Area Districts 1,2,3,4,6.

II. Southern Appalachian Area Districts 7,8,13.

III. Interior & Gulf Area Districts 9,10,11,12,14,15.

IV. Northern Great Plains, Pacific Districts 19,21,22,23.

V. Rocky Mountain Area Districts 16,17,18,20.

End-Use Code:

EU—Electric Utilities
C—Coke Plants
O—All Others[1]
EX—Exports (Except Canada)

[1] Includes all other industrial, retail sales, railroad fuel, Great Lakes vessel fuel, Great Lakes commercial docks, coal used at mines and sales to employees, and destinations and consumer uses not revealable.

Source: Form EIA 6

Source: Form EIA 6

Fig. 2-4. End-use distribution by supplying area, 1979 (thousand short tons).

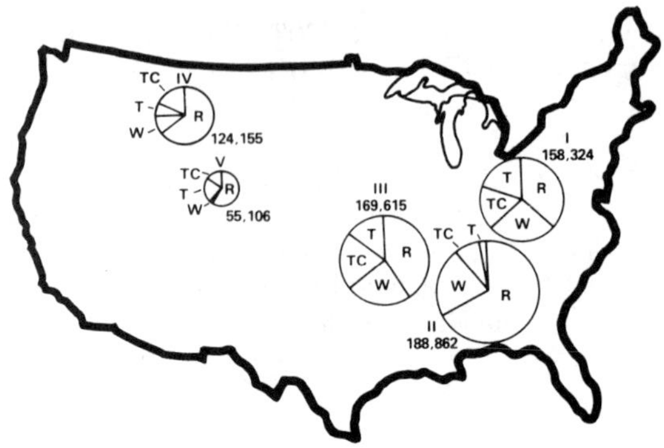

Supplying Areas:

I. Northern Appalachian Area Districts 1,2,3,4,6.

II. Southern Appalachian Area Districts 7,8,13.

III. Interior & Gulf Area Districts 9,10,11,12,14,15.

IV. Northern Great Plains, Pacific Districts 19,21,22,23.

V. Rocky Mountain Area Districts 16,17,18,20.

Transportation Mode Codes:

R—All-rail

W—Water transport (river, Great Lakes, tidewater)

T—Truck

TC—Tramway, Conveyor and private railroad

[1] Excludes railroad fuel, Great Lakes vessel fuel, Great Lakes commercial docks (consumer use not available), coal used at mines and sales to employees, and destinations and consumer uses not revealable.

Source: Form EIA 6

Source: Form EIA 6

Fig. 2-5. Shipments of coal for U.S. consumption, 1979, by producing area and transportation mode (thousand short tons).

Smaller amounts of coal are used for heating in foundries; in various industrial, manufacturing, and commercial establishments; and in homes.

Coal Resources

The United States has coal resources totaling more than 3.9 trillion tons: approximately 2.2 trillion tons are estimated undiscovered resources and 1.7 trillion tons are identified resources. Included in the latter are about 438 billion tons that comprise the demonstrated reserve base, as shown in Fig. 2-6. The reserve base tonnage consists of coal that was considered to be technically and economically minable as of January 1, 1976. Nearly 297 billion tons was estimated to be minable by underground methods and about 141 billion tons by surface methods. However, the amount of coal that can be recovered from the reserve base varies from area to area. It ranges from less than 50% in some underground mines to more than 90% at some surface mines. In underground mines, some coal must be left untouched to form pillars needed to prevent the mine from caving and the surface from subsiding. At both underground and surface mines, natural features such as folded, faulted, and interlayered rock strata reduce the amount of coal that can be recovered. Some coal deposits underlie towns and cities and, consequently, will not be mined. Environmental and man-made restrictions, as well as the lack of suitable technology, could limit the mining of some coal deposits. For the nation as a whole, about one-half of the reserve base, or 219 billion tons, is estimated to be recoverable. According to the World Energy Conference, the estimated recoverable coal reserves in the United States rank first in the world, accounting for about 31% of the total.

Coal production began in the eastern United States during the mid-1700s. A century later, coal was being mined in about a dozen states. Before the turn of the 20th century, coal mining was reported in most of the nation's coal fields. Today, coal is mined in 26 states; the leading coal producers are Kentucky, Pennsylvania, West Virginia, Wyoming, Illinois, Ohio, and Virginia, as shown in Fig. 2-7.

Annual coal production reached more than 200 million tons during the late 1800s. Since 1900 it has ranged from less than 300 million tons to more than 700 million tons, reflecting economic conditions, market demand, coal miners' strikes, environmental restrictions, and world wars. The cumulative coal production since 1890 totals nearly 45 billion tons, as shown in Fig. 2-8. Worldwide, the United States is a major coal producer. In 1978 the nation's coal output accounted for 17.1% of the world total of 3.9 billion tons, ranking third after the U.S.S.R. (20.5%) and the People's Republic of China (17.4%).

The United States coal industry is comprised of more than 6000 mines employing more than 200,000 miners. About 80% of the annual production is from about 1200

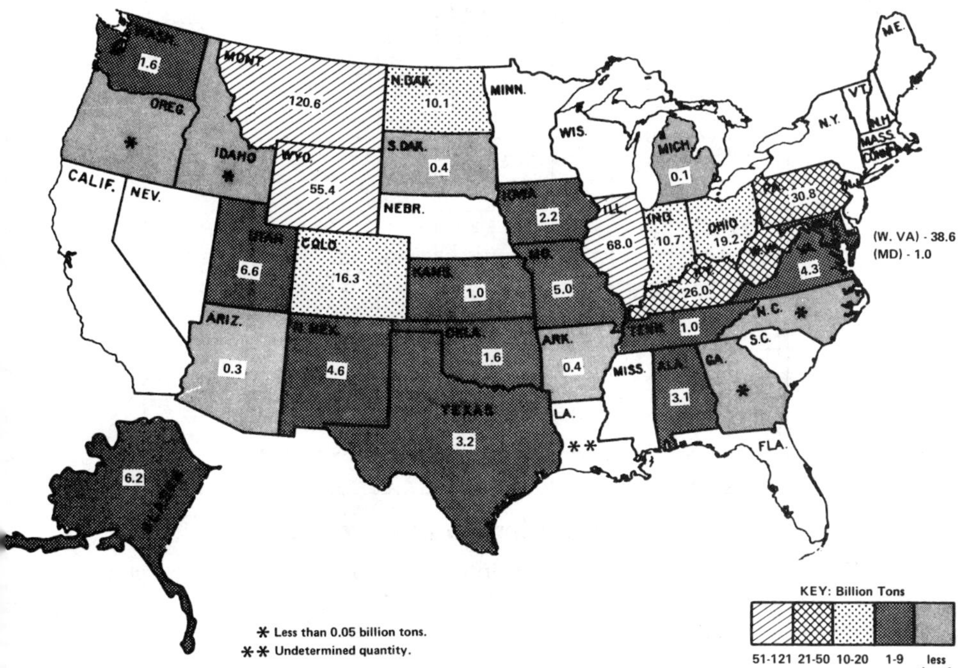

Source: Bureau of Mines, U.S. Department of the Interior

Note: The recoverability ranges from 40 to 90 percent for individual deposits. Fifty percent or more of the total demonstrated coal reserve base is estimated to be recoverable.

Source: Bureau of Mines, U.S. Department of the Interior

Fig. 2-6. Demonstrated coal reserve base, January 1, 1976 (U.S. total: 438.3 billion tons).

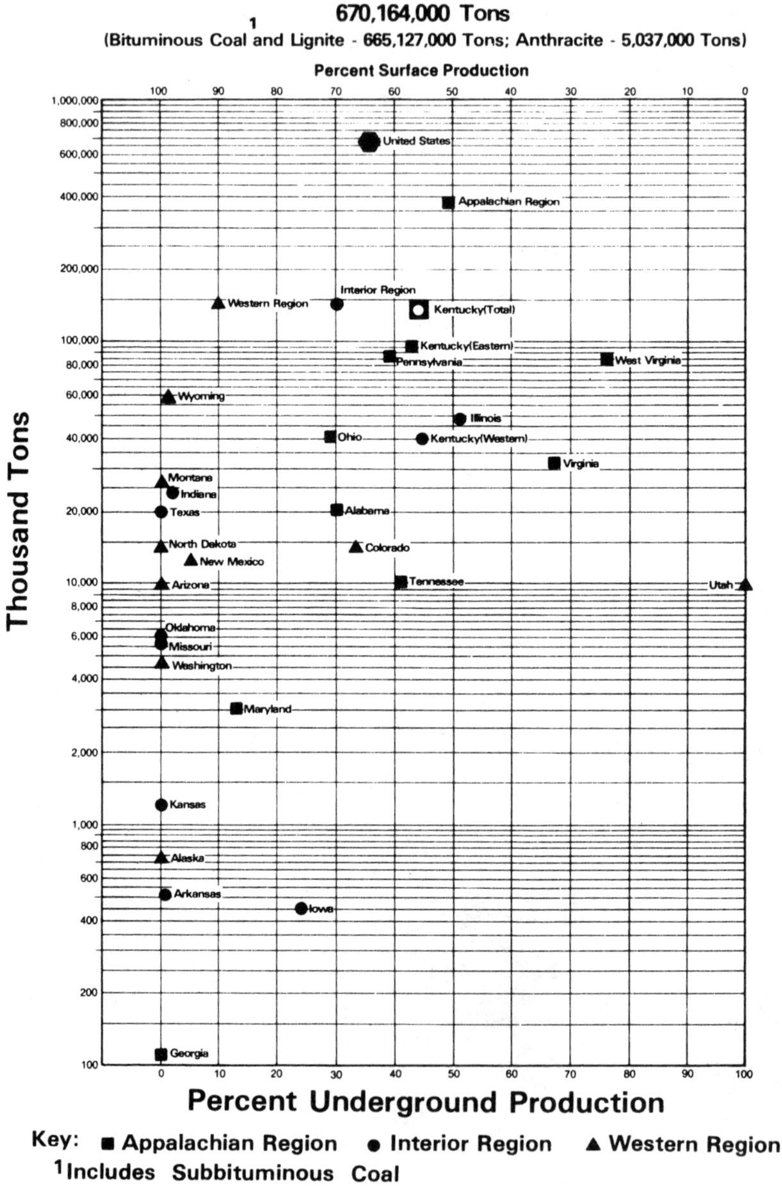

Fig. 2-7. U.S. coal production in 1978.

mines, each producing over 100,000 tons annually. Historically, underground mines have provided most of the production. Recently, however, surface mines have accounted for more than one-half of the total coal produced. The mining method used, whether underground or surface, depends on the terrain and the depth of the coal.

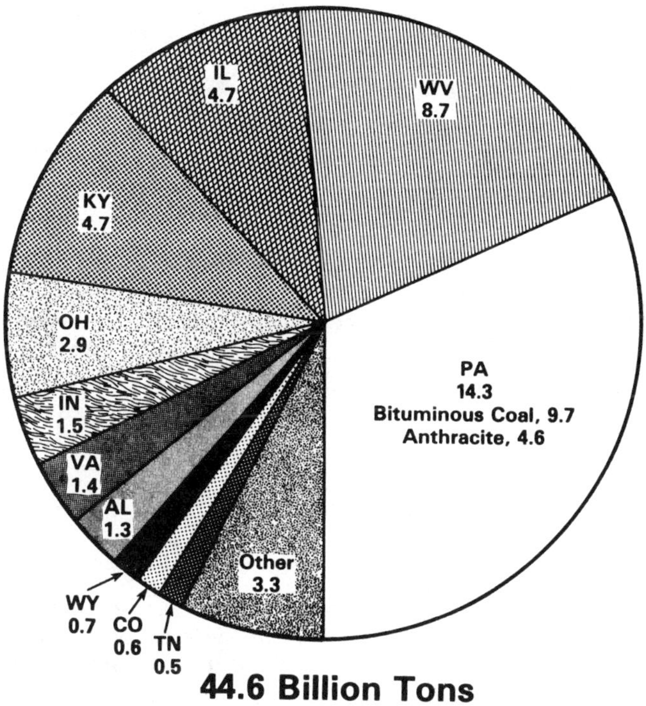

Fig. 2-8. Cumulative coal production by states 1890 – 1978 (billion tons).

Coal Analysis

PROXIMATE ANALYSIS

Proximate analysis is a series of laboratory tests providing information on the behavior of coal when heated under specified conditions. As defined in ASTM D3172, this analysis yields the total moisture of the coal sample "as-delivered" to the laboratory, the percentage content of volatile matter, and the percentage of ash. Fixed carbon is arrived at by difference.

MOISTURE

Coal in its raw state contains varying amounts of moisture. Accurate determination of this moisture is complicated by the fact that as soon as the coal is exposed to the atmosphere it begins to dry. For this reason, ASTM recommends letting the coal "air dry" before grinding for laboratory analysis. The "air-loss" moisture represents the surface moisture of the coal. The inherent moisture or water trapped within the coal is determined by drying the sample in an oven while following rigid specifications for temperature, time, and equipment, and then measuring the weight loss. ASTM D3173 provides the following formula for calculating moisture loss during oven drying:

$$\text{Inherent moisture (\%)} = (A - B)/A \times 100$$

where

A = weight of sample used
B = weight of sample left after heating

When this value is added to the water lost during air drying, as determined by ASTM method D2013, D2961, or D346, the sum represents the total moisture of the coal on an as-received basis.

ASH

Ash is the noncombustible residue left after complete combustion of a coal sample. The ash is derived from the noncombustible minerals in the coal and usually weighs slightly less than the mineral matter. For coals high in calcium, the ash weight can be higher due to retention of sulfur oxides. ASTM D3174 gives the following formula for ash analysis:

$$\text{Ash (\%)} = (A - B)/C \times 100$$

where

A = weight of capsule (porcelain container) and ash residue
B = weight of empty capsule
C = weight of analysis sample used

Ash differs in composition from the inorganic constituents present in the coal. Water in clays and calcium sulfate are driven off; carbonates produce carbon dioxide, and pyrite is converted to ferric oxide. Because each of these reactions involves a loss in weight, formulas (ASTM D388) have been developed to convert percentage ash to a mineral matter basis.

VOLATILE MATTER

Volatile matter is the portion of coal driven off as a gas or vapor when the sample is subjected to a standardized heating test (ASTM D3175); it consists of hydrocarbons and other gases resulting from distillation and decomposition. Corrected for moisture, the measured weight loss serves to calculate volatile matter content:

$$\text{Weight loss (\%)} = (A - B)/A \times 100$$

where

A = weight of sample used
B = weight of sample after heating

$$\text{Volatile matter (\%)} = C - D$$

where

C = weight loss (%)
D = moisture (%)

FIXED CARBON

As defined in ASTM D3172, the percentage of fixed carbon in a sample is the difference between the sum of percentage moisture, ash, and volatile matter, subtracted from 100, or

$$\text{Fixed carbon (\%)} = 100 - [\text{moisture (\%)} + \text{ash (\%)} + \text{volatile matter (\%)}]$$

ULTIMATE ANALYSIS

As the name implies, ultimate analysis is a laboratory procedure for determining the elemental composition of a coal. In ASTM D3176, percentage carbon and percentage hydrogen refer to the content of these elements in the gaseous products driven off by complete combustion of the sample. Total sulfur, nitrogen, and ash are for the sample as a whole, while percentage oxygen is by difference. Sometimes phosphorus and chlorine are reported in ultimate analyses. Ultimate analysis is highly important in the design of power generating plants in which the data are used to compute combustion air requirements, weight of products of combustion, and flue gas losses.

Table 2-3 summarizes the proximate and ultimate analyses in the form used in U.S. Bureau of Mines publications. It permits comparison of the three commonly used methods for reporting coals. Preparation engineers prefer to work with coal on an air-dried, surface-moisture-free basis, because calorific value and other parameters can be calculated for any moisture content. The drop in the hydrogen value and other parameters can be calculated for any moisture content. The drop in the hydrogen and oxygen values going from the as-received to the dry basis occurs because the hydrogen and oxygen contents of the free water (surface moisture) are included with the hydrogen and oxygen values of the coal. Some laboratories do not assign the water to hydrogen and oxygen but report moisture separately.

CALORIC (HEATING) VALUE

Calorific value refers to that part of coal that is combustible, i.e., carbon and hydrogen in the organic matter and both organic and pyritic sulfur. It is measured by combustion of the coal in a calorimeter or "bomb" submerged in water whose temperature is measured before and after combustion.

For anthracite and bituminous coals, the gross heating value (Btu content including latent heat of vaporization) can be calculated approximately, with data taken from the ultimate analysis, by Dulong's formula:

$$\text{Heating value (Btu/lb)} = 14{,}600\ C + 62{,}000\ H - \tfrac{1}{8}\ O + 4{,}050\ S$$

where C = carbon, H = hydrogen, O = oxygen, and S = sulfur—expressed in fractions of a pound. For low-rank coals, the heating value obtained by this formula is generally less than its actual value.

TRACE ELEMENTS

Minute quantities of virtually all metallic and nonmetallic elements were introduced into coals in any of the following ways:

- As part of the inert material washed into the vegetation at the time of accumulation
- As a biochemical precipitate from water flooding the plant material or young coal
- As a minor constituent of the original plant cells
- As a later addition, perhaps millions of years later, carried by groundwater moving downward and laterally

Most of the trace elements occur in about the same concentration as their estimated average in the earth's crust, with 25–30 elements usually in greater concentration. The more common elements include vanadium, manganese, barium, gold, silver, copper,

Table 2-3. Proximate and Ultimate Analyses for Coal from the Upper Kittanning Seam, Pennsylvania

Condition or Basis	Proximate (%)				Ultimate (%)[a]					Heating Value (Btu/lb)
	Moisture	Volatile Matter	Fixed Carbon	Ash	Carbon	Hydrogen	Oxygen	Nitrogen	Sulfur	
As-Received	3.39	20.61	66.53	9.47	77.29	4.59	5.61	1.73	1.31	13,619
Dry		21.33	68.87	9.80	80.00	4.36	2.69	1.79	1.36	14,098
Moisture and Ash-Free		23.65	76.35		88.69	4.83	2.99	1.98	1.51	15,629

[a] In the ultimate analysis, moisture on an as-received basis is included in the hydrogen and oxygen.
Source: Fryling, *Combustion Engineering*, Comb. Engr., Inc., 1966.

lead, zinc, tin, arsenic, antimony, uranium, germanium, beryllium, boron, scandium, yttrium, platinum, cobalt, and cadmium.

Research on trace elements may eventually permit segregation or concentration of certain undesirable elements in the coal-washing plant refuse to minimize their presence in the cleaned coal. At present there is no commercial technique for altering the trace element content of coal.

The listing of trace elements in Table 2-4 is in order of decreasing affinity for clean coal fractions. The sequence may be divided into the following groups:

- Elements with high organic affinity—germanium, beryllium, boron—and closely associated with the clean coal
- Elements with low organic affinity—mercury, zirconium, zinc, arsenic, cadmium—and closely associated with the ash
- Elements of varying organic affinity—the remaining elements—and associated to varying degrees with the clean coal and ash

Although an element may be listed among those with high organic affinities, its occurrence in inorganic combination is not precluded. Among those found in high concentration in clean coal, boron also is found in illite (a clay mineral) and probably would appear in the feed of a preparation plant handling an Illinois coal. Similarly, a

Table 2-4. Affinity of Elements for Pure Coal and Ash as Determined from Float – Sink Data

	Illinois Coal Seams			
	Davis	Dekoven	Colchester No. 2	Herrin No. 6
Clean coal—lightest specific gravity fraction (elements in organic combination)	B	Ge	Ge	Ge
	Ge	Ga	B	B
	Be	Be	P	Be
	Ti	Ti	Be	Sb
	Ga	Sb	Sb	V
	P	Co	Ti	Mo
	V	P	Co	Ga
	Cr	Ni	Se	P
	Sb	Cu	Ga	Se
	Se	Se	V	Ni
	Co	Cr	Ni	Cr
	Cu	Mn	Ph	Co
	Ni	Zn	Cu	Cu
	Mn	Zr	Hg	Ti
	Zr	V	Zr	Zr
	Mo	Mo	Cr	Pb
	Cd	Pb	Mn	Mn
Mineral ash—specific gravity greater than 1.60 (elements in inorganic combination)	Hg	Hg	As	As
	Pb	As	Mo	Cd
	Zn		Cd	Zn
	As		Zn	Hg

Source: Harold J. Gluskoter, "Mineral Matter and Trace Elements in Coal," *American Chemical Society Advances in Chemistry Series, No. 141* (1975), 22 pp.

number of elements usually concentrated most heavily in high-specific-gravity fractions also may be in organic combination.

All trace elements in Table 2-4 are volatile and tend to be emitted to the atmosphere when coal is burned. When commercial preparation processes for adjusting trace elements in coals are developed, the preparation engineer will be able to alter the composition of the clean coal by removing elements of low organic affinity, thus making it possible to control partially the composition of elements leaving the stack.

Sulfur in Coal

The partial removal of sulfur from coal to permit the coal's combustion within governmental guidelines regulating sulfur dioxide emissions has become a major objective of coal preparation. To help visualize the magnitude of the problem, consider that in 1975 United States electric utilities purchased 431 million tons of coal with average sulfur contents ranging from 0.5 to 10.0% by weight.

FORMS OF SULFUR IN COAL

Sulfur in coal occurs in three forms: organic, sulfate, and pyritic. Organic sulfur, which is an integral part of the coal matrix and which generally cannot be removed by direct physical separation, comprises from 30 to 70% of the coal's total sulfur content.

The sulfate content normally is a water-soluble oxidation product that can be removed readily during coal washing. Sulfate sulfur contents usually are less than 0.05% in fresh coal and thus are negligible.

Pyritic sulfur is the mineral pyrite that occurs in coal as discrete particles, although often of microscopic size. A dense mineral, it has a specific gravity of about 5.0, compared with coal which has a specific gravity ranging between 1.3 and 1.7.

SULFUR REMOVAL TECHNOLOGIES

Available methods of controlling sulfur oxide emissions from stationary combustion sources are divided into the following major categories:

- The physical removal of pyritic sulfur through coal washing (beneficiation) prior to combustion
- The removal of sulfur oxides from the combustion flue gas (FGD)
- The conversion of coal to a clean fuel by such processes as gasification, liquefaction, and chemical extraction

Of the three methods, physical removal of pyritic sulfur is the most developed method technologically and, potentially, the lowest in cost. However, the amount of sulfur reduction obtainable is limited because only pyritic sulfur responds to processing. Removal of sulfur oxides from stack gas by use of scrubbers is being adopted on an increasing scale, although both capital and operating costs are substantial and the disposal of scrubber sludge may present major problems. Conversion to clean fuels and chemical cleaning appear costly, both in money and in energy consumed during the conversion process, but this situation is likely to change in the near future owing to intensive research and development projects in the public and private sectors.

SULFUR REDUCTION POTENTIAL OF U.S. COALS

The sulfur reduction potential of bituminous coals through beneficiation varies among mines and coal seams and can be determined reliably only by washability analyses of each coal. These tests help select the beneficiation circuits and equipment best suited to improve the specific coal. Washability analyses evaluate those coal properties that determine a coal's amenability to improvement by physical methods. This includes stage crushing to release impurities and specific gravity fractionation to demonstrate the quality and quantity of the product. A washability study is made by testing the coal sample at preselected, carefully controlled specific gravities. This is called a "float–sink" analysis or specific gravity separation. Chemical analyses of the various specific gravity fractions of the coal are used to compile the washability data that indicate how well the coal can be beneficiated.

In its 1976 Report of Investigations, RI-8118 (2-5), the U.S. Bureau of Mines evaluated the washability of 455 raw channel samples from currently producing U.S. mines. It is a useful reference work for assessing sulfur reduction in regional coal beds. One must understand, however, that the float–sink data obtained from channel samples cannot be considered representative of the quality of the product loaded at the mine where the sample was taken, but only as indicative of the quality of the coal bed in that particular region. Furthermore, float–sink data produced under laboratory conditions represent nearly perfect specific gravity separations that can only be approached but not equalled in commercial practices. The difference between laboratory and practice increases as the specific gravity of the separating fluid decreases.

Important results of RI-8118 (2-5) can be summarized as follows:

- The 227 northern Appalachian region coal bed samples evaluated averaged 2.01% pyritic sulfur and 3.01% total sulfur. Only 4% of the raw coal samples would meet the pre-1977 EPA sulfur dioxide emission standard for new sources. Coals from the Upper and Lower Freeport and Upper Kittanning coal beds could be upgraded to meet the 1976 but not the 1977 New Source Performance Standard (NSPS) utilizing coal preparation techniques. The coals of this region are especially amenable to pyritic sulfur liberation.
- The 35 southern Appalachian region coal bed samples evaluated contained, on the average, 0.37% pyritic sulfur and 1.04% total sulfur.
- The 10 Alabama region coal bed samples evaluated contained, on the average, 0.69% pyritic sulfur and 1.33% total sulfur. The pyrite does not readily liberate from these coals. Thirty percent of the raw coal samples would meet the 1976 emission standard with no preparation.
- The 95 eastern Midwest region coal samples contained, on the average, 2.29% pyritic sulfur and 3.92% total sulfur. These coals contained an average of 1.63% organic sulfur and, consequently, could not be upgraded to meet the 1976 EPA sulfur dioxide emission standard. On the average, none of the principal coal beds could be upgraded sufficiently to meet the 1976 emission standard or the more restrictive 1977 NSPS.
- The 44 western Midwest region coal samples evaluated contained an average of 3.58% pyritic sulfur and 5.25% total sulfur. These coals contained, on the average, 1.67% organic sulfur and thus generally could not be upgraded to meet the 1976 emission standard. Less than 3% of the raw coal samples would meet that standard without preparation. These coals contained, on the average, more than 1% higher pyritic and total sulfur content than those of the eastern Midwest region.

- The 44 western region coal bed samples evaluated contained an average of 0.68% total sulfur which would produce 1.1 pounds SO_2/MMBtu and would thus meet the 1976 EPA SO_2 emission standard as mined. Seventy percent of the raw coal samples would meet that standard without preparation. These are generally subbituminous coals of high inherent moisture and low-Btu content.
- The 455 U.S. coal samples evaluated contained, on the average, 1.91% pyritic sulfur and 3.02% total sulfur. Only 14% of the raw coal samples could meet the pre-1977 EPA SO_2 emission standard of 1.2 pounds SO_2/MMBtu without preparation.

In reviewing the previous information, note that all values are on a moisture-free basis and, therefore, are not directly comparable with data based on a moist, as-received basis, especially in the case of subbituminous and lignitic coals, in which moisture contents of 30% or more are not uncommon.

Coal Mining

Coal mining methods fall into one of two general categories: underground or surface mining (see Fig. 2-9). In the United States in 1977, 272 million tons were produced from underground and 417 million tons from surface mines. More than 67% of the nation's coal reserve base, using current technology, can best be mined by underground methods; however, economic and technological changes may make more of this base accessible to surface mining techniques (see Table 2-5).

The production output of the mining system is the product of the performance of the continuous miner subsystem and other related interfaces. Although the miner has a capability for 10–12 tons a minute, the lesser capacity of the other subsystems and the

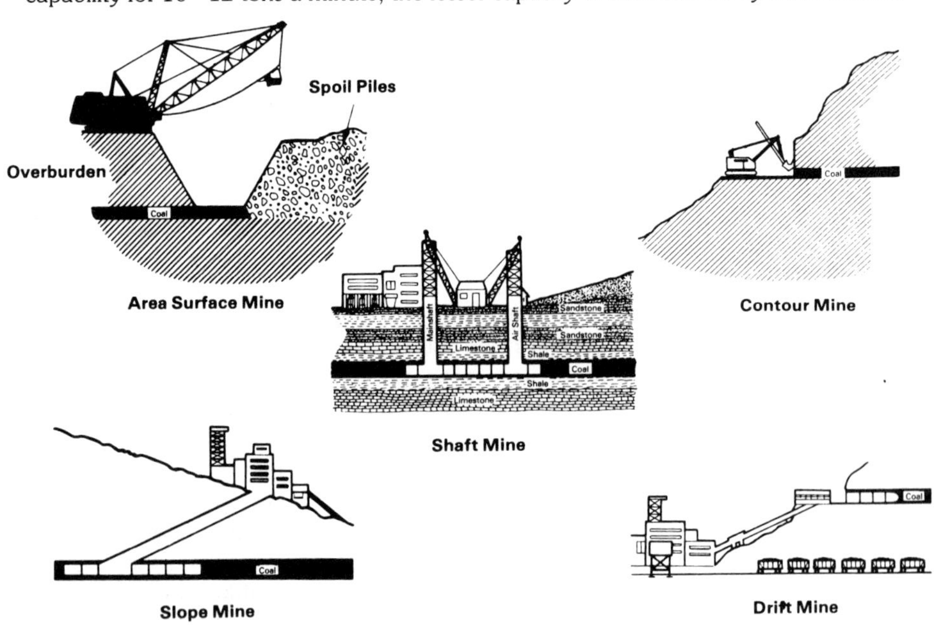

Fig. 2-9. Coal mining methods.

fact that the miner is in a production mode only 24% of an 8-hour shift limit the available output to less than 300 tons per shift.

UNDERGROUND COAL MINING

The two major underground coal mining methods currently used in the United States and Europe are room and pillar and longwall. About 95% of underground coal is mined in the United States by the room and pillar method, and most of the remainder by longwall as shown in Fig. 2-10.

Other underground coal mining methods are shortwall and hydraulic mining. In the United States, shortwall mining is practiced only to a limited extent. Hydraulic mining is not currently used, but it may be adopted for a few extra-thick seams in the West before 1990. The particular advantage of hydraulic mining is in sloping seams.

The room and pillar system of underground coal mining, as shown in Fig. 2-11, leaves undisturbed blocks of coal in a regular grid to prevent collapse of overhead rock strata. Approximately one-half of the coal is normally left behind in pillars to prevent subsidence. In some cases, room and pillar mining may be followed by pillar robbing on the retreat, allowing the roof to collapse. This technique leads to subsidence and cannot be practiced in areas where the resulting surface damage cannot be tolerated.

Room and pillar mining can be conducted with varying degrees of mechanization. Up to the early 1930s, the chambers were excavated by drilling boreholes, blasting, manually loading shattered coal into short-haul carts, and propping the roof. The cycle would then be repeated. Mechanization has almost eliminated this type of mining. This mechanization includes the continuous mining system, which led to peak productivity in 1969 at an average of 15.6 tons per man day. Since then, productivity has dropped; in 1976, it had declined to 8.6 tons. The decline in productivity has been caused by more stringent safety regulations and a variety of labor relations problems in the bituminous coal industry. This trend in declining productivity has been widely recognized in the United States, and current technology programs are designed

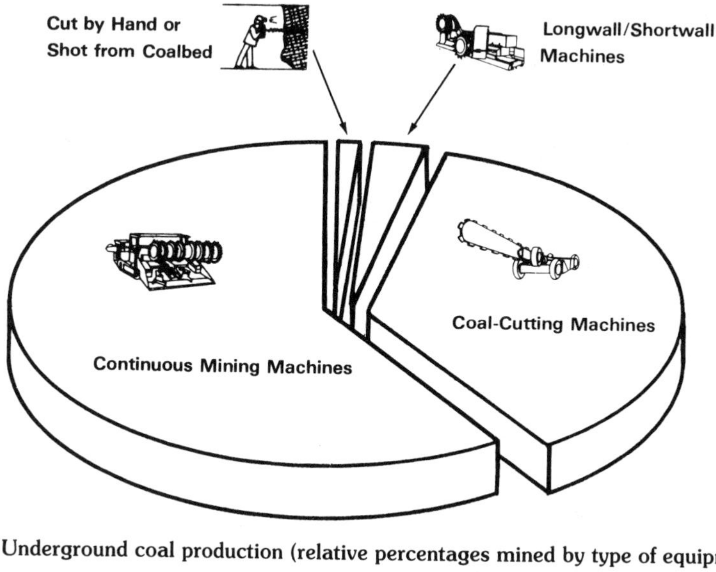

Fig. 2-10. Underground coal production (relative percentages mined by type of equipment).

Table 2-5. Demonstrated Coal Reserve Base of the United States, January 1, 1976, by Method of Mining

State	Potentially Minable by Underground Methods (million short tons)				Potentially Minable by Surface Methods (million short tons)					Grand Total
	Anthracite	Bituminous	Sub-bituminous	Total	Anthracite	Bituminous	Sub-bituminous	Lignite	Total	
Alabama	—	1,724.2	—	1,724.2	—	284.4	—	1,083.0	1,367.4	3,091.6
Alaska	—	617.0	4,805.9	5,423.0	—	80.5	640.7	14.0	735.2	6,158.2
Arizona	—	—	—	—	—	325.5	—	—	325.5	325.5
Arkansas	88.6	163.1	—	251.7	7.8	107.0	—	25.7	140.5	392.2
Colorado	25.5	8,467.9	3,972.1	12,465.4	—	676.2	149.2	2,965.7	3,791.0	16,256.4
Georgia	—	0.5	—	0.5	—	0.4	—	—	0.4	0.9
Illinois	—	53,128.1	—	53,128.1	—	14,841.2	—	—	14,841.2	67,969.3
Indiana	—	8,939.8	—	8,939.8	—	1,774.5	—	—	1,774.5	10,714.4
Idaho	—	4.4	—	4.4	—	—	—	—	—	4.4
Iowa	—	1,736.8	—	1,736.8	—	465.4	—	—	465.4	2,202.2
Kansas	—	—	—	—	—	998.2	—	—	465.4	998.2
Kentucky, East	—	9,072.5	—	9,072.5	—	4,467.6	—	—	4,467.6	13,540.1
Kentucky, West	—	8,510.4	—	8,510.4	—	3,900.4	—	—	3,950.4	12,460.8
Louisiana	—	—	—	—	—	—	—	a	a	a
Maryland	—	913.8	—	913.8	—	134.5	—	—	134.5	1,048.3
Michigan	—	125.2	—	125.2	—	1.6	—	—	1.6	126.8
Missouri	—	1,418.0	—	1,418.0	—	3,596.0	—	—	3,596.0	5,014.0
Montana	—	1,385.4	69,573.5	70,958.9	—	—	33,843.2	15,766.8	49,610.1	120,568.9
New Mexico	2.3	1,258.8	889.0	2,150.1	—	601.1	1,846.8	—	2,447.9	4,598.0
North Carolina	—	31.3	—	31.3	—	0.4	—	—	0.4	31.7
North Dakota	—	—	—	—	—	—	—	10,145.3	10,145.3	10,145.3
Ohio	—	13,090.5	—	13,090.5	—	6,139.8	—	—	6,139.8	19,230.2
Oklahoma	—	1,192.9	—	1,192.9	—	425.2	—	—	425.2	1,618.0
Oregon	—	0.3	14.5	14.5	—	—	2.9	—	2.9	17.5
Pennsylvania	6,966.8	22,335.9	—	29,302.7	142.7	1,391.8	—	—	1,534.4	30,837.1

Table 2-5. *(continued)*

State	Potentially Minable by Underground Methods (million short tons)				Potentially Minable by Surface Methods (million short tons)					Grand Total
	Anthracite	Bituminous	Sub-bituminous	Total	Anthracite	Bituminous	Sub-bituminous	Lignite	Total	
South Dakota	—	—	—	—	—	—	—	426.1	426.1	426.1
Tennessee	—	627.2	—	627.2	—	337.9	—	—	337.9	965.1
Texas	—	—	—	—	—	—	—	3,181.9	3,181.9	3,181.9
Utah	—	6,283.8	1.1	6,284.9	—	267.9	—	—	267.9	6,552.8
Virginia	137.5	3,277.0	—	3,414.5	—	888.5	—	—	888.5	4,302.9
Washington	—	255.3	835.3	1,090.6	—	—	481.5	8.1	489.5	1,580.1
West Virginia	—	33,457.4	—	33,457.4	—	5,149.1	—	—	5,149.1	38,606.5
Wyoming	—	4,002.5	27,644.8	31,647.2	—	—	23,724.7	—	23,724.7	55,371.9
Total[b]	7,220.7	182,019.6	107,736.1	296,976.3	150.5	46,905.0	60,688.9	33,616.6	141,361.0	438,337.3

[a]Quantity undetermined (basic resource data do not provide the detail required for delineation of reserve base).
[b]Data may not add to totals shown due to rounding.
Source: *Coal Facts*. National Coal Association—1978, 1979.

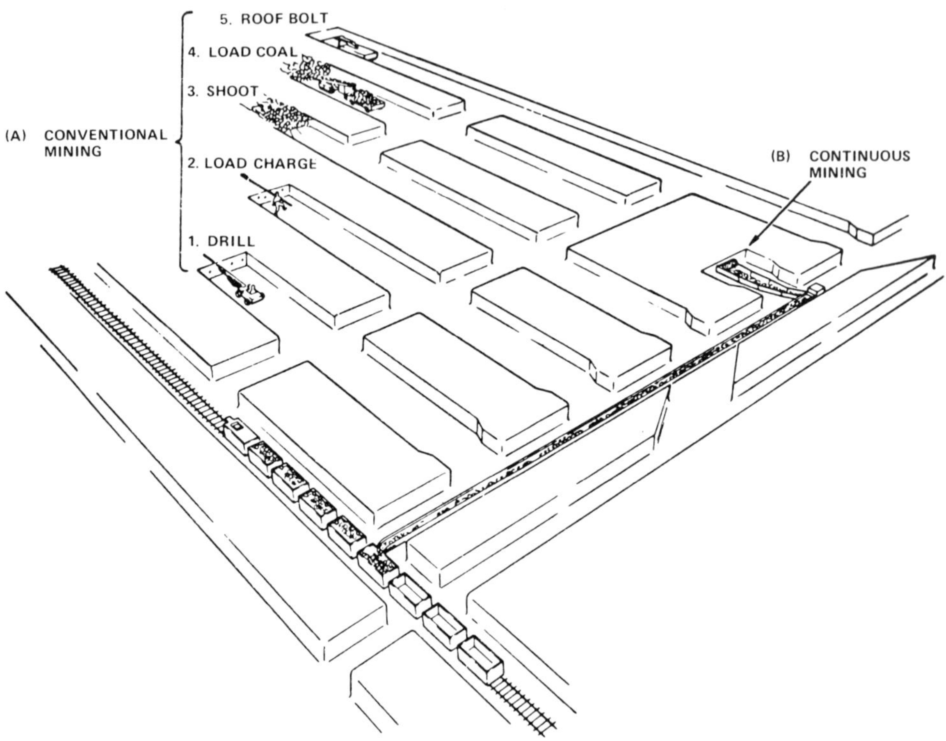

Fig. 2-11. Illustration of room and pillar using (A) conventional (blasting) and (B) continuous mining technique.

to reverse this trend without compromising either the safety of miners or the environment.

Longwall mining of underground coal utilizes conventional room and pillar continuous mining equipment to develop panels of coal which may be 1 mile long, 600 feet wide, and 4 or more feet high. Figure 2-12 illustrates the type and location of equipment used in this mining technique. A high-horsepower mining machine removes the coal by shearing or plowing a slice of coal up to 30 inches thick from the face of the panel. The broken coal is moved to the headpiece side of the panel by the face conveyor and out of the mine by the main haulage belt conveyor. Self-advancing hydraulic roof supports, lined up parallel to the face, move up one by one as the shear or plow travels across the face. The overburden (gob) collapses behind the roof supports, thus relieving the pressure on the coal face.

The principals of longwall mining are not new. The technique was used in England toward the end of the 17th century and has matured through advances in equipment design to the point where it now accounts for approximately 95% of the coal mined in England. The technique was also widely used throughout other European mines. Geological conditions and increased pressure at greater depths in European mines result in roof and floor conditions that do not permit large areas to remain unsupported or supported by roof bolts, thus precluding the use of the room and pillar mining method.

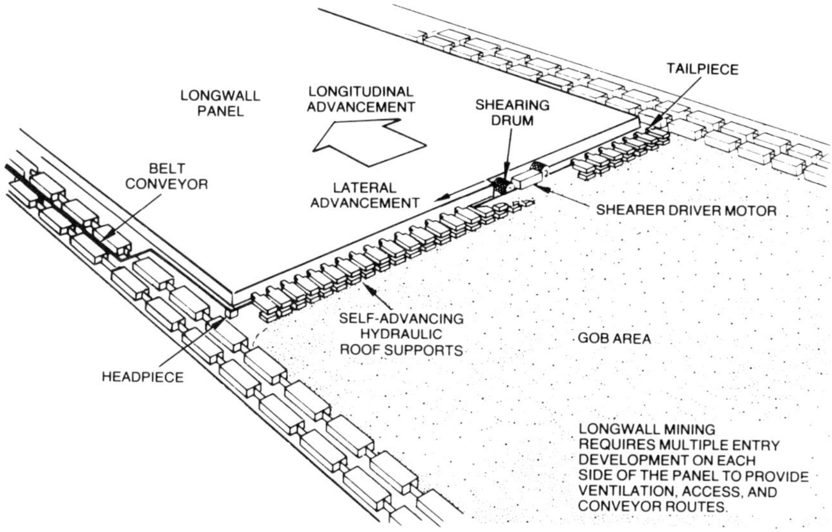

Fig. 2-12. Illustration of longwall mining techniques.

The efficiency of the longwall technique has progressively improved through introduction of self-advancing supports and the flexible-armored face conveyors. This fully mechanized version, usually imported equipment, was introduced to the United States in 1960. During the past several years, equipment manufacturers, singly and in conjunction with European manufacturers, have made several advancements toward improved productivity and safety.

The increasing interest in longwall mining is attributable to the following factors:

- Improved economics (less cost per ton)
- The system's inherent high productivity (700–900 tons per shift with new faces exceeding 1000 tons)
- Health and safety considerations (compliance with laws and regulations; roof support, ventilation, and dust requirements)
- Increased recovery of coal (present overall recovery closely approximates the room and pillar rate with a greater potential as new techniques and equipment are developed)
- Predictable subsidence
- Adaptability to thick and multiple seams
- Capability to mine at greater depths

There are several constraints that continue to limit application of this technology:

- The nonfracturing and noncaving characteristics of rock strata above the abundantly available shallow depth seams do not permit mining by the longwall method.
- Reluctance of mine operators to commit the large initial capital investment,

which can run as high as $10 million to initiate operations. Mine size must be larger than 1 million tons per year to justify the capital investment.

- Surface subsidence, although more predictable and uniform, is of concern in urban areas.

Several government studies are currently underway to identify and eliminate items that limit the productivity and to increase the economic attractiveness of longwall mining. Studies and research projects are also directed toward automation techniques and toward field trials and demonstrations of improved longwall equipment and techniques. Successful completion of these and industrial sector advancements could increase longwall output from the current 5% to as much as 60% of underground coal production in the United States.

Shortwall mining is similar to longwall mining. The difference is the face length, typically 150 feet, and the face equipment, which are usually the continuous miners and shuttle cars. This method is somewhat less capital intensive than longwall mining because the same coal extraction and conveying equipment can be used in mine development and production mining operations. Coal recovery is comparable to the longwall method. Shortwall mining, like longwall mining, can be used only in areas in which subsidence is acceptable.

The underground hydraulic mining of coal involves the use of high-pressure (2000 psi) water jets to break the coal loose and convey it from the face as a slurry, either in open flumes, in steel pipes, or in flexible hose pipelines. Hydraulic mining is especially applicable to steeply pitched or unusually thick seams of coal because of the following:

- Conventional coal mining machinery cannot handle seams thicker than about 10–12 feet.
- Hydraulic mining moves the operator safely away from rockfalls in the vicinity of the working face.
- Respirable dust generation levels are lower.

In Canada, Kaiser Resources is presently operating a 4000-ton-per-day hydraulic mine in British Columbia; in the Soviet Union, 10 hydraulic mines are in operation in the Donets and Kuznetsk coal basins. Hydraulic mining is also practiced in the Federal Republic of Germany. Demonstrations of hydraulic mining are being planned in the United States.

SURFACE COAL MINING

The production of coal from surface mines has increased during the last decade because production costs are lower than those for underground mines. This trend is expected to continue.

The two general approaches used in the surface mining of coal are area mining and contour mining. In 1977, the production of surface-mined bituminous coal amounted to 417 million tons, with 48 million tons from contour mines, 275 million tons from area mines, and 3 million tons from a supplemental approach known as auger mining.[1]

[1] Auger mining is a drilling process used in conjunction with other surface mining approaches. It uses very large bits, often in twin- and triple-drill gangs, to bore into the coal seam. Individual holes may be 30–95 inches in diameter and 150 feet or more in depth. The only mineral which is auger mined is coal. Although primarily a surface technique, it may be used underground where poor roof conditions or other factors preclude the use of more standard approaches.

Area mining is practiced on relatively flat terrain, typified by the land from Ohio to the Rockies. A box cut or trench is first made through the overburden to expose the coal seam. Coal is extracted from the seam, and succeeding cuts are made to expose a fresh seam for mining. The overburden, or spoil, from each succeeding cut is used as fill for the preceding cut as shown in Fig. 2-13. Large draglines or shovels (Fig. 2-13) are used in most surface area mines.

Dragline casting has improved through economies of scale. Equipment with bucket capacities of 75 cubic yards or more and booms of 300 feet are becoming common. The major limitation to dragline casting is the dumping radius limit. Large stripping ratios[2] necessitate the rehandling of approximately 10–40% of the overburden with draglines, shovels, and trucks or bulldozers, and add to the net production cost of surface-mined coal. A system being developed to overcome this limitation would use continuous conveyors to transport overburden to spoil banks. Testing and evaluation of hardware for this concept are planned for the early 1980s. Thus, techniques for improving productivity of area mining could be instituted before 1990.

Contour mining is used in mountainous or hilly terrain. The operation starts at a coal seam or outcrop of coal on the side of a mountain and follows the seam along an approximation of a contour line. The topsoil is removed by shovel, bulldozer, scraper or dragline, as dictated by the terrain, and is held aside for the reclamation operation. A bench is created from which coal is extracted and loaded into haulage vehicles for removal from the site. Succeeding cuts are made, and the overburden is put into the pit

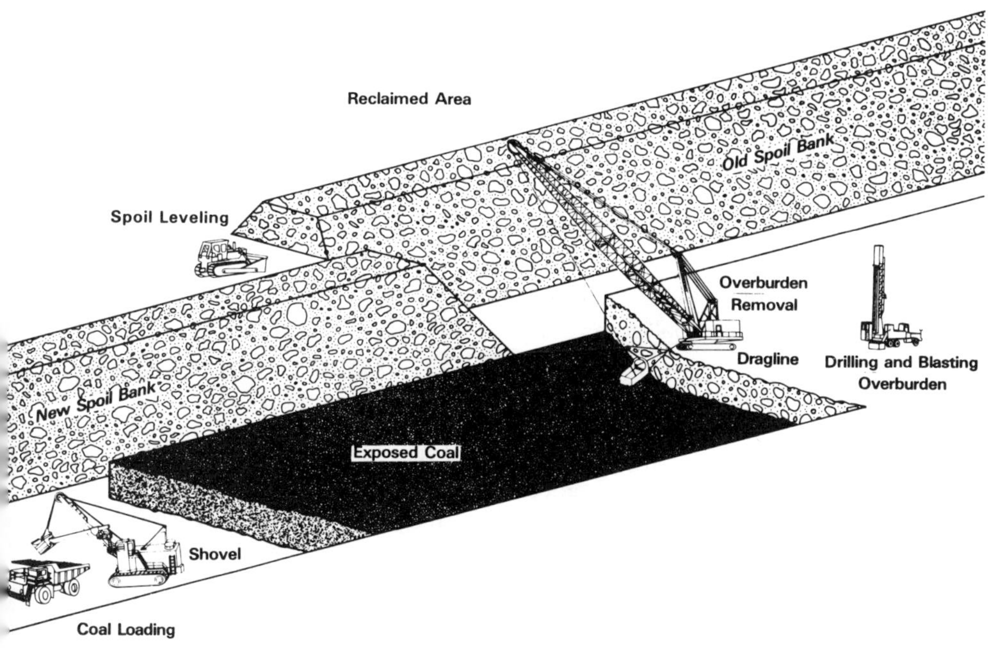

Fig. 2-13. Area surface mining with dragline and shovel.

[2] Stripping ratio is the cubic yards of overburden moved per ton of coal recovered.

from which coal was extracted. The technique of contour mining is illustrated in Fig. 2-14. When cuts at a given contour bench involve an overburden ratio that becomes uneconomical, the area is restored and the mine is abandoned, converted to an underground operation, or auger mined to extract some additional coal.

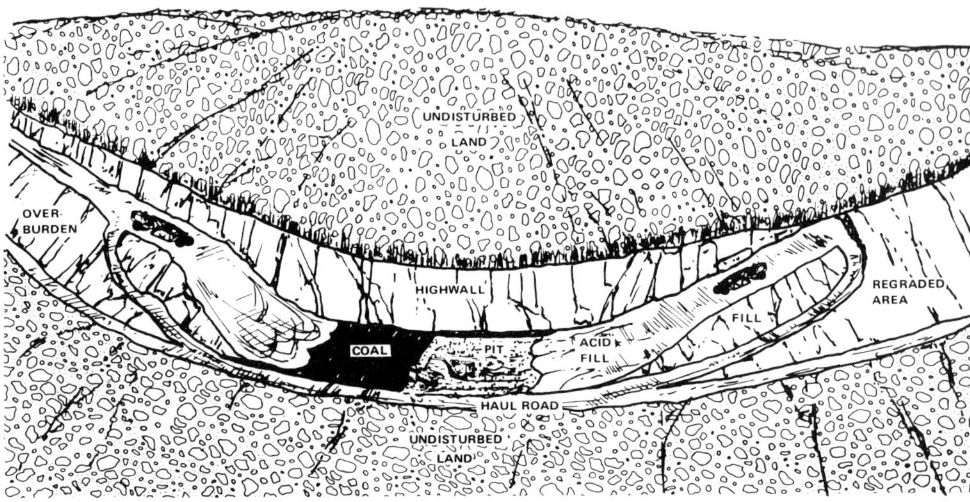

Fig. 2-14. A representative contour-mining operation.

Conventional contour mining involves considerable hauling and back hauling of overburden, usually by truck. It is a labor-intensive process that would benefit from the application of conveyors or other automated hauling methods which would reduce traffic and decrease labor requirements. Improved hauling systems would yield the following:

- Decreased production cost
- Increased productivity
- Improved land reclamation in steeply sloped areas
- Improved safety conditions

Bucket Wheel Excavators. Many of the western surface mines will be very large, producing between 5 and 20 million tons per year. Large bucket wheel excavators working in conjunction with improved belt conveyors and stackers are being employed by some coal companies in lieu of draglines or shovels for large surface mines. Such excavator systems have already proved successful in Germany and other European mining sites, and there are approximately 10 such units operating in the United States—most of them in Illinois.

Bucket wheel excavators, under favorable circumstances, are the most productive excavating machines used in surface mining. They effectively remove unconsolidated overburden that does not require blasting. They are unique among excavating machines because they are designed to operate continuously, in contrast to stripping shovels and draglines which operate intermittently. This type of system may allow continuous mining and transporting of coal and overburden in high-volume applications in the West.

All forms of surface mining depend on the availability of machinery to move large quantities of overburden and coal. Earth-moving machinery that have been developed since the mid-1920s have grown to impressive sizes. In 1925, the largest shovels had a volume of 10 cubic yards; by 1964, however, this figure had reached 170 cubic yards. Dragline buckets as large as 220 cubic yards are in use, and 300-yard buckets are possible. Today, all of the larger electric draglines (typically 6000 horsepower or more) are walker-mounted.[3] Future advances will not be a result of increases in equipment size, but will be due to improved haulage systems that support or replace the draglines and shovels.

Coal Preparation

Coal preparation (beneficiation) is the means by which coal producers are able to deliver a finished product, tailored to meet the needs of each end-user. Figure 2-15 illustrates the basic flow of coal through a preparation plant. Removal of ash-producing minerals results in cleaner burning fuels with enhanced heating values (Btu/lb). Beneficiation produces the feedstock of highly uniform size required by modern power plants and combustion facilities. Finally, through removal of mineral and pyritic sulfur (sulfur not chemically bound to the carbon in coal), beneficiation reduces the cost of compliance with environmental standards. Beneficiation techniques will play a key role in the effort to substitute coal for petroleum and natural gas in our economy, especially as modern extraction techniques designed to increase productivity and recovery produce a run-of-mine (ROM) coal with increased impurities.

Coal preparation plans are tailored to the properties of a specific mine and to the requirements of the consumers that the mine serves. The intensity of treatment can vary widely. ROM steam coal may be only crushed, and the coarse rock removed at a cost of 50¢ per ton. High-grade metallurgical coking coal may be crushed and treated by flotation, at a cost of \$3–4 per ton, not counting a loss of 20% of the carbon content in the waste discarded.

The basic preparation plant operations are as follows:

- Size reduction (breaking, crushing, and grinding)
- Screening (size classification)
- Cleaning
- Dewatering and drying

SIZE REDUCTION AND CLASSIFICATION

Size reduction and classification have been stable technologies for many decades, and no major changes are anticipated in the near future. Coal processors may use a wide variety of jaw crushers, roll crushers, hammer mills, rotary mills and breakers, and ring-roll mills. Similarly, in the area of size classification, practically every conceivable variation of sieve has been tested.

---

[3] Walkers use large footlike pads that can be lifted by the machine and advanced in unison. When the pads supporting the machine have been advanced, the body of the machine can then be moved to its original position in relation to the pads. This results in the machine "walking" along the ground.

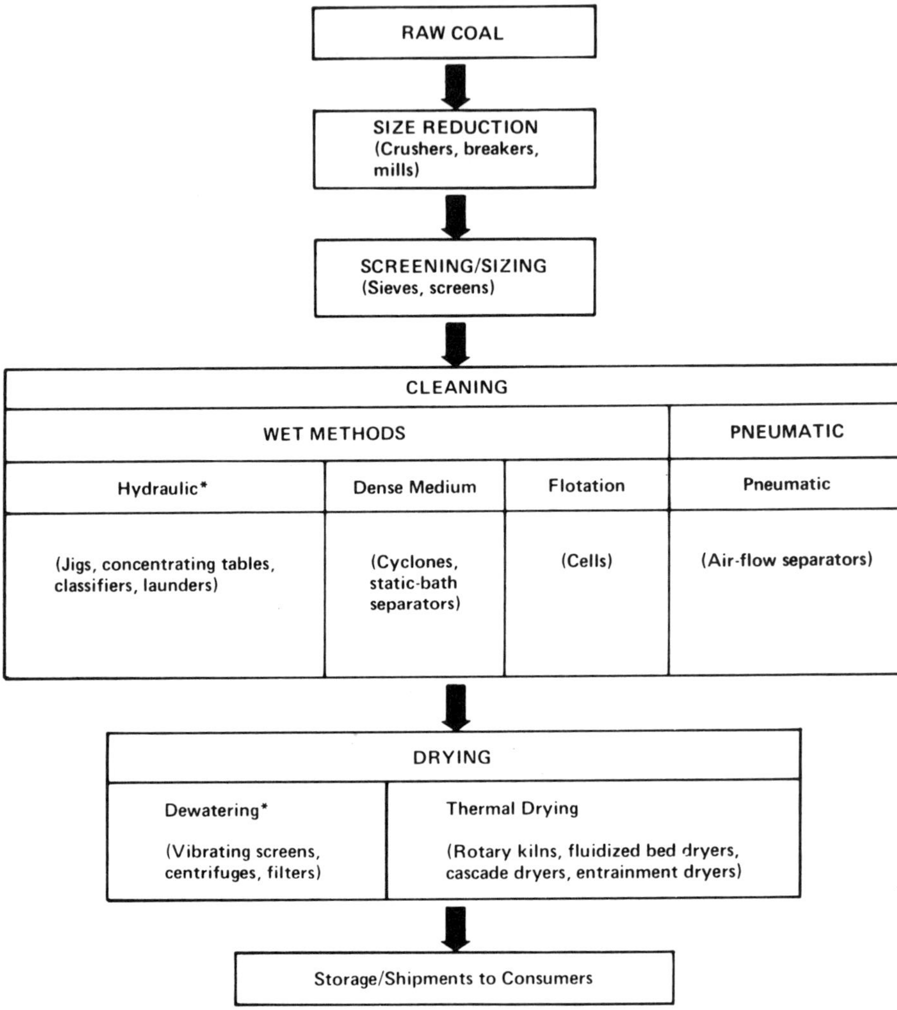

*Principal method

Fig. 2-15. Basic flow of coal through a preparation plant.

COAL CLEANING

Cleaning consists of the removal of all extraneous matter mixed in with the coal as it is delivered from the mine—shale, clay, pyrite, and blasting wire, for example. About 60% of underground-mined coal and 25% of surface-mined coal are subjected to mechanical cleaning.

There are two distinct classes of coal cleaning technology. The first includes techniques that were developed generally prior to World War II. They tend to be well established and have undergone only slight refinement. They are usually mechanical in nature, are not specifically directed toward sulfur removal, and result in a comparatively modest increase in the cost of coal. The second class of coal cleaning technologies is a result of the recent requirements for sulfur reduction. This class is expensive and more sophisticated, and is undergoing intensive development.

Currently, wet methods of cleaning dominate the coal beneficiation field, with only 2.5% of coal being upgraded by pneumatic methods. Most wet methods take advantage of the difference in specific gravity between coal and the common mine waste minerals. The specific gravity of "pure" coal may vary between 1.23 and 1.72, depending on rank and ash content. Most of the mineral impurities are roughly twice as dense as coal.

A demand for more thorough coal cleaning processes has arisen recently. The competitive economic climate has also changed so that more sophisticated technology (which usually means more expensive technology) can be introduced. For example, there are new federal limits to the amount of sulfur dioxide that a coal-burning installation may emit. Therefore, more complete sulfur removal is needed than mechanical coal cleaning provides. The incremental cost of improved coal cleaning that will be tolerated in the marketplace is directly related to the equivalent cost of flue-gas scrubbing to remove SO_2 after combustion.

The presence of sodium in coal causes fly ash to foul boiler tubes and gas passages, leading to loss of performance. Deposits on turbine blades lead to corrosion or imbalance. Thus, another requirement for coal cleaning is the removal of sodium and other alkali metal salts and oxides.

Wet methods of coal cleaning can be grouped into three major categories:

- Dense media
- Hydraulic (including jigs, tables, launders, and water-only cyclones)
- Flotation

Both fine coals and coarse coals are treated by dense media and hydraulic methods, although the approach differs for the two sizes. Flotation can be applied only to fine coal.

Dense media methods use a slurried suspending fluid with a superficial specific gravity between that of coal and mine waste. In the United States, the preferred suspending medium is an aqueous slurry of fine (90% minus 325 mesh) magnetite. This "heavy media separation" process was introduced as the Tromp Process in 1938 and a variation was developed by American Cyanamid in 1940. Magnetite can be contained within the cleaning system because it can be washed from the coal and waste, and recovered from the aqueous effluent by magnetic separation.

Cyclone separators are often used to enhance dense media processes. They are superior to static bath separators in sharpness of density discrimination, because cyclones can expose the slurry mixture to a centrifugal acceleration force field much greater than the normal force of gravity. The advances of dense media cyclones were demonstrated at Dutch State Mines during World War II; but they were not adopted in the United States until 1961.

Hydraulic coal cleaning includes the use of jigs, tables, and launders. Jigs are by far the most important devices in the coal preparation field today because of their versatility and high capacity. A typical coal cleaning jig is shown in Fig. 2-16.

Jigs stratify a slurry mixture of coal and waste by alternately expanding and contracting the particulate bed by means of vertical pulses of water. This results in layers of

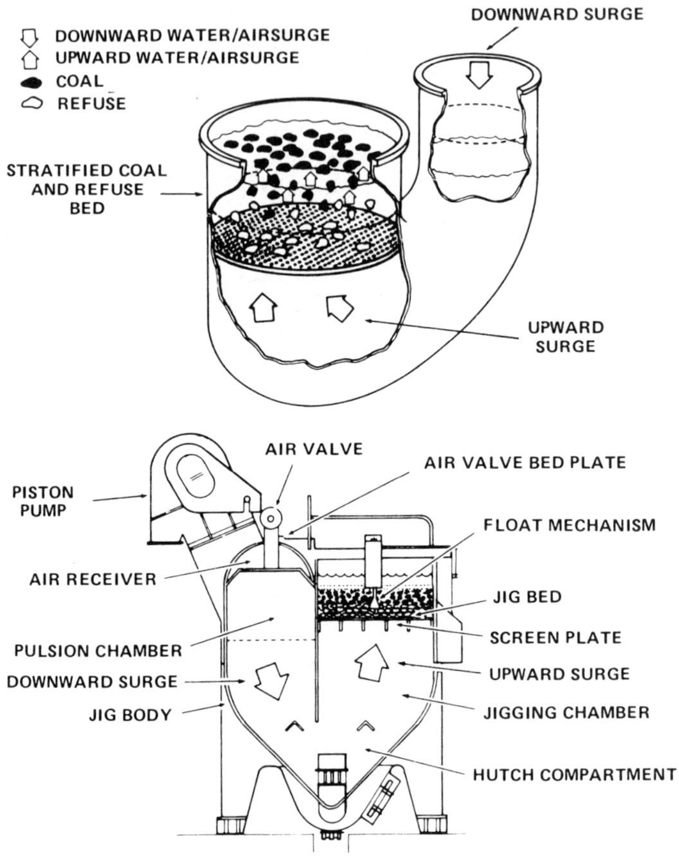

Fig. 2-16. Jig washing action principle (two views).

particles arranged in order of increasing specific gravity from top to bottom of the bed. The jig is a flat rectangular box, with an open top and perforated bottom, that contains the slurry bed being stratified. Water is pulsed laterally up and down through the false bottom to accomplish the separation.

Because they classify material by specific gravity rather than by particle size, jigs can handle both large and small coal lumps and particles, often as a single feed mixture. Slightly less than one-half the coal washing in the United States takes place in jigs (124 million tons out of 276 million tons in 1975). The German Baum jig, introduced into the United States in 1928, uses compressed air to produce the vibratory water impulses and is the most popular in the West. In recent years, however, the Tazub jig has become more popular in the Federal Republic of Germany.

Normally, a jig discharges cleaned coal, middlings, and refuse. The middlings may be recycled, or they may be ground fine to be classified in another jig or in other coal beneficiation devices. Jigs treating fine coal (minus ½ inch) may have a bed of coarse granulated feldspar laid over the perforated plate below the active zone. Waste from these jigs percolates down through the feldspar bed and the support plate for discharge below.

Deister tables[4] are used to concentrate fine coals (typically 3/8 inch to 200 mesh). This equipment was first introduced in 1917 for anthracite coal; it was adopted for use for bituminous coal into the 1920s, and wide application occurred in the 1940s. In 1975, Deister tables were the third most popular coal beneficiation devices, yielding 29 million tons.

Launders are relatively crude devices compared to modern jigs and Deister tables. A launder is essentially a sloping trough. It often has a riffled bottom, like an old-fashioned washboard. Fine ROM coal slurry (1/2 inch to 60 mesh) is fed to the upper end, and purified coal slurry is discharged from the lower end. Waste is trapped as a bottom sediment along the length of the trough. Laundering accounted for 2.7 million tons of cleaned coal in 1975.

Like many other minerals, fine coal can be beneficiated by flotation. Bubbles of air admitted to the flotation cell attach themselves selectively to the coal particles and lift them out of the slurry as a froth, leaving the waste behind in the slurry.

Flotation machines, called cells, are simply mechanically agitated tanks containing a slurry of fine coal and water, with an air supply to bubble up through the turbulent slurry contents. Three types of chemicals are added in small quantities to modify the surface-active characteristics of the system:

- *Frothers*, which act like soaps to reduce drastically the surface tension of waste and generate a stable foam. Amyl alcohol, butyl alcohol, and certain cresols are frothers.
- *Collectors*, or *promoters*, which are reagents to enhance the ability of coal to capture air bubbles without producing a similar effect on waste. An example of a collector is kerosene. Methyl isobutyl carbinol is a combination reagent, acting as both frother and collector.
- *Modifying reagents*, which is a catch-all term for all other types of flotation modifiers. The most important are depressing agents (e.g., NaCN), activating agents (e.g., Na_2S), pH regulators (e.g., Na_2CO_3), dispersing agents (e.g., $NaO_2 \times SiO_2$), and protective colloids (e.g., glue).

Large-scale demonstrations of processes for upgrading coal finer than 1/2 inch are planned for the early 1980s, and could be commercialized before 1990. One such process is high-gradient magnetic separation (HGMS), a new means for separating small, weakly magnetic particles of pyrite from coal. Both wet and dry HGMS units will be evaluated. Oil agglomeration, a technique for separating ash minerals from coal by selective coating with a thick oil film, will be demonstrated; a combined process of oil agglomeration and wet HGMS will also be developed. Other new or improved coal cleaning techniques are in various stages of development, including fine-particle dense-medium cyclones, two-stage flotation, and low-cost sodium removal for western coals.

The removal of additional sulfur is the goal of several proprietary processes now being developed. Ten processes are listed in Table 2-6. Laboratory tests suggest that sub-

[4]The Deister table is simply a solid broad flat plate, sloping slightly from the horizontal, with an array of shallow parallel ridges forming a riffle pattern across the natural waterflow direction. A rapid reciprocating motion (about 250 strokes per minute) is imposed parallel to the ridges. This motion propels the waste trapped in the valleys of the riffle toward one end of the table. Coal is washed over the ridges of the riffle and goes over the downhill beam side of the table.

Table 2-6. Status of Chemical Coal Cleaning Processes

Process	Percent Sulfur Removal Pyritic	Organic	Development Status
1. Magnex	0–90+	0	1 TPD Pilot Units
2. KVB	0–90+	0–40	Preliminary Bench Scale
3. Battelle/Columbus	0–90+	0–70	Small Pilot Unit
4. Meyers	0–90+	0	8 TPD Pilot Plant
5. Ledgemont	0–90+	0	Bench Scale
	0–90+	0–25	Bench Scale
6. Bureau of Mines	0–90+	0–40	1–4 lb/hr Pilot Unit
7. Ames	0–90+	0–25+	Bench Scale
8. Chlorinolysis	0–90+	20–70	Bench Scale
9. Mixed Leachant (Soviet Union)	0–90+	0	5–10 TPD Pilot Plant
10. General Electric	0–90+	0–70	Preliminary Bench Scale

stantially complete removal of pyritic sulfur is possible; some of the processes may also remove much of the organic sulfur (sulfur chemically bound to carbon atoms). The sponsors estimate an incremental cost of $14–22 per ton for the tested coal.

While several of these processes that involve the oxydesulfurization of coal show promise, chemical cleaning of coal requires substantially more development and may not be widely practiced before 1990.

DEWATERING AND DRYING

After wet-process cleaning (97.5% of coal cleaning is wet) or slurry pipeline transportation, coal must be dewatered and dried because wet coal is heavier to transport and may cause problems in cold weather.

It is easy to dewater coarse coal: Vibrating screens or solid bowl centrifuges can drain stoker coal (1¼ to ¼ inch) to 5% or less moisture content. For finer coals (¼ inch to 28 mesh), screen-type centrifuges are used. Pressure filters and vacuum disk filters are used for preliminary dewatering, for example, to convert a slurry containing 80% water to a compact filter cake containing perhaps 30%. Solid-bowl centrifuges are used for dewatering the finer-sized coals (minus 28 mesh). For final drying, thermal ovens are needed, either direct or indirect fired. The fluidized-bed dryer is a good example of a direct-fired unit for fine coals.

Coal Transportation

Coal is moved from the mine or beneficiation plant to the point of use by several different modes of transportation. River barges, railroads, and slurry pipelines provide long distance transportation; short hauls may be accomplished by either highway truck or conveyor belt. It is not always necessary to move the coal very far. In some cases, it can be converted to another form of energy at the mine mouth or beneficiation plant site. Except for very short distances, electric power transmission is more expensive than coal haulage by railroad barge. Liquid and gaseous products of coal conversion may, however, be cheaper to transport than coal, particularly if they are transported by pipeline.

The cost comparisons for various modes of coal transportation are shown in Table 2-7.

Table 2-7. Cost Comparison for Various Modes of Coal Transportation

Modes of Transportation	Energy Cost (Btu per Ton – Mile)	Economic Cost[a] (Mils per Ton – Mile)
1. River Barge	400 – 500	6
2. Slurry Pipeline	410	7 – 19
3. Railroad	340 – 390	6 – 10
4. Highway Truck	2,600 – 2,800	34
5. Conveyor Belt	3,000	38

[a]Exclusive of transportation energy costs.

RIVER BARGE

The least expensive (and, in many cases, also the most energy-efficient) mode of bulk cargo shipment is a barge load of coal towed down a navigable river or canal. The minimum energy cost in still water (neither bucking the current nor drifting downstream) is about 400–500 Btu per ton-mile. A typical economic cost, exclusive of the transportation energy costs, is about 6.4 mils per ton-mile.[5]

Unfortunately, there are limitations to shipping by barge. Navigable waterways do not extend everywhere, and they reach very few coal mine sites; thus, most coal that moves by water is transshipped at least once en route from mine to consumer. In the North, freezing temperatures periodically close waterways, and drought or floods interrupt water traffic. Moreover, it is impractical to handle spot shipments or even scheduled, repetitive shipments of less than 1000 tons by barge. For all these reasons, barge hauling must be supplemented by other modes of shipment.

RAILROAD

Full trainload shipments of 9000 tons can be made with reasonable economy, but the cost per ton-mile increases rapidly as the lot size decreases. Coal shipments in conventional freight trains are even more expensive because hopper cars have extended periods of idleness. No coal is moved in less than carload lots.

A comparatively recent innovation in coal transportation by rail is the concept of a unit train, dedicated to moving coal from mine to consumer. The concept was used to transport petroleum products from Texas to New Jersey during World War II in an effort to protect petroleum shipments from submarine attack. In 1960, the first application to coal was made in Alabama. By 1974, unit trains were hauling 135 million tons of coal annually out of a total of 388 million tons hauled by all railroad means.

Railroad unit trains offer a compromise between cost and energy efficiency, on the one hand, and convenience, on the other hand. The energy cost is about 340–390 Btu per ton-mile with diesel locomotives and is slightly higher on electrified railroads. The monetary cost of moving by railroad has been estimated at 6–10 mills per ton-mile. The cost estimate includes the following:

[5]This cost estimate is for a 360-mile transfer of 22,500 tons (15 barges of 1500 tons each) in dedicated tows. Shoreside terminal costs at both ends of the haul are included.

- Provision of track[6] 4.7 mills per ton-mile
- Unit train operating costs[7] 1.2 mills per ton-mile
- Total minimum transportation cost[7] 5.9 mills per ton-mile

This estimate excludes the cost of fuel, taxes, and land rentals; thus, it understates the true cost of moving coal in unit trains.

SLURRY PIPELINES

Coal slurry pipelines are a relatively new technology. The coal is crushed and then mixed with water to form a fluidlike mixture, which is pumped through a tubular pipeline. At the destination, the coal and water are separated.

Coal slurry pipelines should be comparatively efficient. The actual pumping energy consumption can be as low as 410 Btu per ton-mile, but the energy cost for grinding the coal and preparing the slurry must be added. This latter energy charge could be acceptable when the pipeline is long, but it becomes more important for short pipelines. For example, the energy cost of slurry preparation for the proposed Wyoming–Arkansas pipeline, which will be 1040 miles long, has been estimated at 150 Btu per ton-mile. The same slurry-preparation energy debit, applied to the relatively short (273 miles) Black Mesa pipeline, is 590 Btu per ton-mile. Since finely ground feedstock is most often used, the charge for grinding is absorbed by the consumer who would be required to pay for it prior to use anyway. These energy cost figures exclude pumping power for returning "ink" from the coal destination to the slurry source should this prove to be necessary.

"Ink" is the black slime liquor that is drained from coal at the end of the pipeline, which typically contains 20% coal fines that are smaller than 40 microns, resulting from coal particles rubbing together during pipeline transit. Disposal of ink is a very serious problem, because it may not be dumped into surface-water supplies or pumped into a disposal well, unless the disposal site is effectively quarantined from groundwater resources. The Mohave plant serviced by the Black Mesa pipeline referenced above has tried to recycle ink through the coal dewatering centrifuges, but this resulted in filter cake so moist that too much fuel gas was required to dry it. Currently, ink at the Mohave station is impounded in a 35-acre waterproof basin where surface evaporation will dry it up.

There is a conceptually simple solution to the complementary problem of insufficient water supply at the slurry pipeline origin and ink disposal at destination, i.e., pipe the ink back to the coal mine for reuse as slurry medium. This solution has never been tried. There may be technical problems (such as an intolerable rise in slurry viscosity) and the costs of pipeline transportation of coal would certainly be increased. The ink return pipeline would not cost as much as the original slurry pipeline; its diameter would be about 60% as large to carry 36% of the slurry volumetric flow. When right-of-way and most field construction expenses are allocated to the slurry pipeline, the ink return pipeline would increase capital costs by 35–40%. These additional costs would adversely affect the ability of slurry pipelines to compete economically with railroads.

[6] Provision of track includes only two cost items: annual maintenance and 15% per year of roadbed capital cost at current replacement prices totaling $328,000 per mile. Government taxes on privately owned right-of-way, land rentals, or other forms of real estate expenses are not included. Coal traffic assumed in these estimates is 70 million tons per year.

[7] Excluding transportation energy costs.

It has also been suggested that ink be burned using wet air oxidation technology, but it is too early to forecast the potential of such systems.

Coal slurry pipelines less than about 12 inches in diameter are unlikely to transport coal cheaper than railroads. A 12-inch pipeline could carry about 2.6 million tons of coal per year, enough to support about 800 MW of baseload generating capacity. Such a line would have to be operated at not less than 60% of capacity, delivering at a minimum enough coal to support about 500 MW of baseload generating capacity. The maximum size of the pipeline is limited essentially by the quantity of coal consumed at the destination.

Transportation costs of moving coal by pipeline are still uncertain. Only the Black Mesa pipeline in Arizona is in operation today. The transportation cost estimates range from 7 to 19 mills per ton-mile. These estimates do not include costs for water supply at slurry source, ink disposal costs at the destination, or pumping energy costs. An obstacle to further expansion of the slurry pipeline is acquisition of the necessary right of way. The solution to this problem will require legislative action on the right of eminent domain.

HIGHWAY

Trucks are the most convenient and flexible mode of coal transportation over short distances and in small lots of 5–20 tons. In such operations, there are no practical alternatives to trucks. On the other hand, trucks cannot compete with railroads over long distances. Trucks consume about 2600 Btu per ton-mile. The shipping cost of moving a truckload (25 tons) of coal over a short haul (50 miles) on a new highway in level country is at least 34 mills per ton-mile. This cost is based on truck operating and capital costs only; transportation fuel costs are excluded and highway costs are included only in the form of taxes.

CONVEYOR BELTS

Conveyor belts or cable belts several miles long are successfully in use today; however, very few are longer than 3000 feet. Belts are useful primarily in manufacturing plants or underground mines where haulage distances are not great. No coal is moved via long-distance belt series. In very difficult terrain and with distances of up to 5 miles, however, belt transport might be less expensive than modes requiring roads or railroads.

The energy cost of moving 200 tons of coal per hour on a series of level 30-inch belts is approximately 3000 Btu per ton-mile. The minimum dollar cost is 38 mills per ton-mile.

Environmental Impact of Mining, Preparation, and Transportation of Coal

UNDERGROUND MINING

The three principal environmental concerns associated with the underground mining of coal are mine drainage, solid waste disposal, and subsidence. In addition, underground mining is an especially hazardous occupation.

Mine drainage results from the infiltration of surface water through the surrounding strata to the mine. This drainage must be pumped back to the surface, where it is potentially a pollutant, to avoid mine flooding. The composition of mine drainage varies with the specific geologic area but is often acidic, with high sulfate content, and high levels of aluminum, calcium, magnesium, and manganese. Treatment of mine drainage is site-specific and can be expensive. Treatment may require one or more of the following:

- Neutralization
- Aeration
- Chemical treatment to precipitate dissolved solids
- Removal of suspended solids

The requirement for treatment can be reduced in some cases by preventing infiltration through water diversion, mine sealing, and surface restoration to promote runoff rather than penetration.

Problems associated with solid mine waste disposal include discharge of particulates, noxious gases, acid mine runoff, erosion, and landslides. The accepted practice is to lay down refuse in compacted layers and to cover the pile with earth. This technique protects the refuse from leaching by rainwater and allows vegetation cover to be established. In a small number of cases, profitable employment of the solid waste has been achieved through backfilling of abandoned mines, the manufacture of bricks and cinder blocks, the construction of secondary roads, and combustion as a low-grade fuel for generation of electricity.

Subsidence is controlled by leaving pillars of coal in mined areas to support the overburden. Pillars cannot be removed in areas where the resulting surface subsidence will damage buildings, farmlands, lakes, or streams. Subsidence is of less concern in rugged areas with few buildings or little farming; it is in such areas that fuller coal recovery can be accomplished.

The safety of underground coal mining is likely to be enhanced by two developments. The first of these is the fielding of the "miner–bolter," a continuous mining machine with an integrated roof bolter. This machine will aid the control of the roof in the immediate area of the working face—the most hazardous area. The second development is the implementation of coalbed methane degasification practices. Pilot projects to drain methane from coalbeds in advance of mining operations are now underway. Should such practices become widespread, they will reduce the probability of catastrophic gas field mine explosions.

SURFACE MINING

The chief environmental concerns associated with surface mining of coal are related to overburden removal and reclamation.

Overburden removal entails disruption of large surface areas, and stabilization of the exposed banks and spoil piles for contour mining necessary to prevent erosion. The leaching action of rainwater on loose shale and soil can cause high levels of dissolved solids and high acidity. A commonly employed technique in hilly areas is to compact the spoil in layers in order to enhance resistance to leaching and erosion.

Reclamation of surface-mined terrain starts with proper storage of the top soil. Once the coal is removed, the overburden and top soil are replaced to restore the land to an approximation of its original contour or, in some approved cases, to an alternate, more useful configuration. Revegetation is location-specific; in the northeastern United States and similar climates, about 90% of the restored areas will support vegetation. Under the Surface Mine Act, mining will be precluded in areas where it cannot be demonstrated that vegetation can be restored. The techniques of revegetation include acceleration of the natural succession through the use of native species, combination planting, and soil enrichment.

Reclamation of areas disturbed by surface mining operations is a controversial subject that has received increasing attention during the past several years. Research and development is being conducted in the areas of overburden and topsoil distribution, and in seeding and planting for a wide variety of topographic and climatic conditions. These conditions are especially difficult in the arid western regions of the United States. Thus, reclamation is more likely to be an obstacle to expanded coal production there than in the East and Midwest where water is abundant.

COAL PREPARATION FACILITIES

Coal preparation facilities pose environmental concerns for air, land, and water.

Dust abatement is the chief consideration in maintaining air quality. Normal mechanical sizing and classification processes generate a quantity of coal and waste fines that must be controlled. In addition, preparation facilities must handle an increasing quantity of fine coal as a result of mechanization. The techniques for particulate abatement are well developed, and the complexity of the facility will be determined by economics and the existing ambient air quality standards. Control technologies include electrostatic precipitation, cloth bag filters, and cyclones.

Landfill disposal of refuse from coal preparation facilities involves many of the same concerns as those for mine-waste disposal. The content of carbonaceous material in weathered reject solids is such that refuse pile fires can start spontaneously and pose a serious air pollution and safety problem. Permanent burial of weathered refuse must follow strict landfill procedures to prevent leaching and erosion from surface water. As in the disposal of mine waste, prevention is the key to successful antipollution operations.

As more coal is subjected to wet cleaning, the problem of treating "black water" containing suspended dust also increases. Because of water quality standards, all new facilities will be required to recycle process water unless the discharged water can meet all of the water quality standards. Standard methods for removing solids prior to recycling water include thickening, vacuum filtration, settlement in ponds, and pressure filtration.

The introduction of chemical cleaning techniques could pose an environmental hazard from the release of process chemicals. Several of the proposed processes incorporate caustic or acidic leaching solutions, and require the treatment of effluents.

TRANSPORTATION

Some problems encountered during conventional transportation are noise from unit trains and dust from trucks and trains. Those related to dust can be controlled by available techniques such as covering of open gondolas or using nonfreezing chemical wetting agents. Noise from unit trains through small western towns is a serious problem, and track rerouting may be the only solution. However, rerouting creates additional problems of cost, land right of way, and other institutional problems. Coal slurry pipelines possess a unique set of environmental problems, including "ink" disposal and a potential for spills similar to petroleum pipelines.

STORAGE OF COAL

Coal storage requires techniques for controlling the emission of dust, preventing acidic runoff as a result of leaching by rainwater, and avoiding spontaneous coal-pile fires. The means of accomplishing these ends are well developed and include the same site-management techniques used in mining and preparation.

Chapter 3

Coal Liquefaction

Introduction

Liquid fuels play a vital role in the U.S. economy; in many cases, they fill needs for which other energy forms cannot substitute efficiently or economically—in transportation, for example. Liquid hydrocarbons are also the basis of this nation's chemical industry. Many commonplace items—plastics, pesticides, drugs, and paints, to name just a few—have a common heritage in oil. Without it, each would be either uneconomical or impossible to manufacture.

Although demand for liquid fuels has increased steadily in the United States, petroleum production peaked in 1970 and began to decline in 1979, when almost one-half the nation's oil was supplied by foreign countries, at a price tag of more than $80 billion. But even at elevated prices, world reserves, depleted by rising consumption, may not be able to supply world needs beginning in the 1990s and reserves may decline rapidly after the turn of the century.

One way to supplement diminishing oil supplies in the future could be to manufacture synthetic liquids from the nation's coal reserves. The coal beneath the United States—some 430 billion tons are considered mineable—makes up one-half the known reserves of the free world and has five times the energy value of domestic recoverable oil and gas.

From 1 to 3 barrels of oil can be made from each ton of coal. Coal-derived synthetic fuels could substantially prolong the usefulness of that portion of the nation's transportation, manufacturing, and electric utility industry that must rely on a steady flow of liquid hydrocarbons.

HISTORY

Coal liquefaction is not new. Prior to World War II, a number of coal-to-liquid plants were built in Germany in anticipation that oil supplies might be cut off. Complex, bulky, and expensive as they were, these processes helped fuel the German war machine.

Small-scale feasibility studies of German technology were conducted in the United States during the 1930s, but the effort was largely moot because of the discovery of cheaper East Texas oil in 1930. In 1944, U.S. interest revived somewhat, and Congress passed the Synthetic Liquid Fuels Act. The act provided $85 million in research funds, and two experimental plants were built and operated until the program expired in 1955; then, just as in 1930, another large oil discovery killed interest in coal liquids—this time the find was in the Middle East.

What appeared to be a cheap, abundant energy source in the 1950s has today become the focal point of America's efforts to maintain its economic and national security. The near monopoly of foreign suppliers caused the price of world oil to increase almost eightfold in the decade of the 1970s. Even at today's prices—roughly $35 a barrel—there is no assurance that portions of foreign supplies will continue to flow uninterrupted in the future, or that prices will not continue their upward spiral.

In view of this, a domestic synthetic fuels industry, historically hindered by unfavorable economics, is now being viewed as vital to an energy-secure marketplace in the late 1980s and beyond.

HOW IS IT DONE?

Solid coal consists mainly of carbon and ash, and contains only small amounts of hydrogen, along with other elements such as sulfur, oxygen, and nitrogen. Oil, on the other hand, is made up of molecules with a much higher proportion of hydrogen linked to its carbon atoms.

Therefore, the basic way to turn coal into oil is to increase the amount of hydrogen. Techniques for doing this vary widely. Costs and difficulties depend largely on what is required to reach the desired ratio between carbon and hydrogen atoms in the synthetic oil molecule. In general, the more hydrogen added, the "lighter" are the resulting liquids.

There are three basic ways to change coal into liquid fuels: (1) break coal into two fractions: one that is rich in carbon (char) and one that retains most of the hydrogen (oil); (2) break coal apart with steam and oxygen to form a gas, then rebuild the gaseous molecules into liquids; or (3) add hydrogen directly to the coal.

The first approach is called "pyrolysis" or "carbonization." It may hold some potential, particularly since it is relatively simple, works at moderate temperatures and pressures, and produces two fuels: a heavy synthetic crude oil and char (which can be converted to gas). The drawback to date, however, has been the low yields of liquids, roughly a barrel or less per ton of coal.

The major efforts in the United States to produce liquids from coal have focused on the other two methods, commonly referred to as "indirect" and "direct" liquefaction techniques, respectively.

INDIRECT LIQUEFACTION

The indirect process is so-named because coal is first broken apart with steam and oxygen to form a gas (called "synthesis gas"), which is primarily a mixture of carbon monoxide and hydrogen as shown in Fig. 3-1. Then the gas is chemically recombined to form liquids, the exact nature of which depend on the individual process.

Although this two-step route is less efficient and perhaps more costly than direct techniques, it relies on technology used commercially in the world today.

A large number of plants are operating worldwide producing methanol (or "methyl alcohol") in commercial quantities, some using coal as the feedstock. While these plants are small compared to those that would be used in a major synthetic fuel industry, the risks of scaling up the technologies are less since the basic processes remain the same.

The key to indirect liquefaction is the first step—gasification. Up to 80% of the costs of liquids produced this way depend on the cost of producing the synthesis gas.

Several gasifiers are currently being used, primarily in Europe and Asia. Most, however, work only with coals similar to those found in the western United States. Move eastward into the Midwestern and Appalachian regions and the coals tend to cake when subjected to the heat and pressures required for gasification. Because of this, the first American coal-to-liquid plants may rely on noncaking western coals, with later plants moving to the East as new techniques are developed to overcome the caking tendencies.

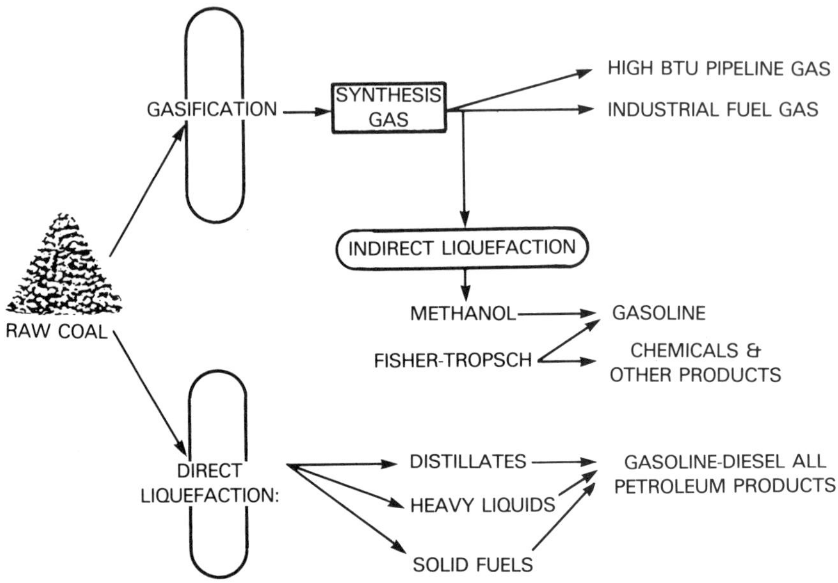

Fig. 3-1. Indirect liquefaction process diagram.

Once the synthesis gas leaves the gasifiers and is cleaned, it can be converted to liquids in either of two ways. In one technique, methanol is produced; in the other, a process known as "Fischer–Tropsch" is used to produce a variety of liquid fuels and chemicals.

Methanol Production

Methanol is currently made commercially in this country by "splitting" natural gas into carbon monoxide and hydrogen—the same gases produced when coal is gasified. Since subsequent steps in the "indirect" liquefaction process cannot distinguish clean synthesis gas produced from coal from that produced using natural gas, the conversion of coal to methanol requires no major development other than improving the gasification step. It is a commercially available technology. The actual technique for recombining carbon monoxide and hydrogen to form methanol involves passing the gases over a zinc or copper catalyst, which speeds the reaction between the two gases but does not enter into the reaction itself. Aided by the catalyst, the gases combined to form methanol.

Until a few years ago, this conversion was a high-pressure step operating at 8000 pounds per square inch. Imperial Chemical Industries, however, developed a low-pressure process that operates at about 1500 pounds per square inch and uses a copper catalyst with zinc and aluminum additives. The success of Imperial's low-pressure technique led other companies to develop similar processes, and today at least four other firms offer licensed low-pressure methanol synthesis processes.

Methanol can be used in a variety of applications. It is a stable liquid at room temperature and is less explosive than gasoline, although somewhat more toxic. It can be used in turbines to provide electricity for peak demands or it may be added to gasoline as a fuel extender, much like ethanol is used today to make "gasohol." Methanol, however, contains only half as much energy as an equivalent amount of gasoline. An automobile using only methanol would have to carry twice as much fuel to travel the same distance if it were gasoline powered.

Unfortunately, when methanol and gasoline are mixed and the mixture comes into contact with a small amount of water, the methanol and water settle to the bottom of the tank, leaving the gasoline floating on top. This "phase separation" inhibits engine performance. In addition, methanol will attack fuel tank liners and plastic components in the fuel system.

Mobil M-Gasoline Process

Converting methanol to gasoline would solve many of the problems of blending methanol into gasoline. A jointly funded program between the Department of Energy (DOE) and the Mobil R&D Corp. has successfully developed a process for doing this.

The key to Mobil's M-Gasoline process is a synthetic "zeolite" catalyst, one of a family of specially developed catalysts used in petroleum refining. When the catalyst contacts methanol under the proper conditions, the carbon and hydrogen molecules are recombined to form high-octane, unleaded gasoline. This unique example of "molecular engineering" has been tested at a 4 barrel per day plant at Mobil's Paulsboro, New Jersey, research laboratories, and in March 1980, DOE and the Federal Republic of Germany agreed to scale-up the process in a $33 million, 100 barrel per day experimental plant near a German methanol production facility. The pilot plant will test a new "fluid bed" version, in which catalyst particles are suspended and vigorously agitated by a mixture of methanol vapor, recycled product gases, and unreacting nitrogen. Earlier processes used the "fixed bed" approach, in which catalyst particles stay essentially immobile while the methanol passes through them.

The government of New Zealand announced in November 1979 that it would purchase the fixed bed version of Mobil's process to convert methanol made from offshore natural gas into 13,000 barrels per day of gasoline. Also, DOE announced in February 1980 that it would begin contract talks with W.R. Grace & Co. of New York to fund the conceptual design to a full-size plant that would convert coal to methanol, then use the Mobil process to produce gasoline. The plant proposed for Baskett, Kentucky, would turn 29,000 tons per day of high-sulfur coal into 16,000 tons of methanol first, which would then be converted to 50,000 barrels per day of gasoline.

Fischer–Tropsch Process

The best-known "indirect" liquefaction method is a process developed in 1925 by the German chemist team of Franz Fischer and Hans Tropsch. It was used in several German plants at the end of World War II to provide nearly 15,000 barrels per day of military fuels.

Instead of recombining the carbon monoxide and hydrogen from the gasifier to form methanol, the Fischer–Tropsch process uses a different catalyst to produce not only gasoline, but jet fuel, diesel oil, middle distillates, heavy oil, waxes, and a variety of chemical products.

In the 1950s, the South African government began construction of a small Fischer–Tropsch plant, which, although completed in 1956, did not begin full operations until 5 years later.

In the South African plant, known as SASOL-1 (for South African Coal, Oil and Gas Ltd.), synthesis gas is made using 16 German-developed Lurgi gasifiers. The gas is fed to two different processes: a German process known as ARGE and a more advanced American (Kellogg) process called SYNTHOL. The basic difference between the two is that the synthesis gas and catalyst move rapidly through the SYNTHOL reactor whereas the catalyst in the ARGE vessel remains essentially fixed. In both, however, the carbon monoxide and hydrogen from the gasifiers contact an iron catalyst and are recombined into a wide variety of liquids which must be separated before being used.

The SASOL-1 plant manufactures about 10,000 barrels per day of liquids—about 1 barrel for each ton of coal. A new, larger plant, termed SASOL-2, began operations early in 1980. With a projected cost of $2.8 billion, it will use 36 large Lurgi gasifiers and produce about 40,000 barrels per day of liquids using solely the SYNTHOL liquefaction step. There are now plans to double the capacity of the new plant in the early 1980s.

Although commercially proven, the Fischer–Tropsch process has some drawbacks. Thermal efficiencies are relatively low, around 45–50%, and even improvements made during the last two decades have succeeded only in pushing liquid yields to around 1.6 barrels per ton of coal. This may make Fischer–Tropsch products more expensive than other coal-derived synthetic fuels, although it is difficult to compare the real costs of SASOL products with those that might be produced in the United States.

Converting coal to gasoline can produce a varied slate of by-products, as shown in Fig. 3-1, depending on the process. Two conceptual coal-to-gasoline plants located in the West are compared in Table 3-1, one using the M-Gasoline process and the other using the Fischer–Tropsch technique.

Table 3-1. M-Gasoline and Fischer–Tropsch By-products

By-products	M-Gasoline	Fischer–Tropsch
Input		
Coal (millions tons/day)	27.3	27.8
Water (gallons/minute)	6,300	6,600
Product		
Gasoline (barrels/day)	22,045	13,580
Synthetic Natural Gas (million cubic feet/day)	148.5	173.3
LPG (C_3) (barrels/day)	2,555	1,107
Butanes (barrels/day)	2,205	146
Diesel Fuel (barrels/day)	—	2,307
Fuel Oil (barrels/day)	—	622
Alcohol (thousand pounds/day)	—	510
FOE[a] (barrels/day)	45,550	44,950
Other		
Power (MW electrical)	5.31	3.31
Coal Fines (thousand tons/day)	1.2	0
Ammonia (tons/day)	103	103
Sulfur (tons/day)	61	61
Efficiency (higher heating value) (%)	62	58

[a] All products converted to fuel oil equivalent—6.0 million Btu per barrel.

DIRECT LIQUEFACTION

In theory, "direct" liquefaction processes—where hydrogen is added directly to coal without the gas step—offer better economics than their indirect counterparts. This is primarily because efficiencies approach 70%, and yields of up to 3 barrels of oil per ton of coal can be obtained. Direct liquefaction techniques can be tailored more easily to produce a more exact slate of higher-value products.

Like indirect liquefaction, the roots of direct hydrogenation began in Germany in the early 1900s. Even before World War I, Frederich Bergius had started work on "destructive hydrogenation" of coal. Improved by the I.G. Farben chemists and others between wars, the direct liquefaction process ultimately formed the basis for German plants producing 75,000–100,000 barrels per day of synthetic liquids in the latter days of World War II.

The bulk of federally sponsored research in the United States has focused on direct liquefaction. DOE's program was begun in 1962 by the Office of Coal Research in the Department of the Interior (the office was the predecessor to the fossil energy program now underway in the Department of Energy).

Three major variations of the "direct" liquefaction route have since progressed to large-scale development: the Solvent-Refined Coal (SRC), Exxon Donor Solvent, and H-Coal processes.

Solvent-Refined Coal (SRC)

Oldest of the liquefaction processes in DOE's development program was the SRC program, which dates back to 1962 when Spencer Chemical Company was awarded a research contract by the Office of Coal Research. The goal was to study the feasibility of removing ash from coal, making it more economical to transport. Spencer completed the contract in1965 with successful demonstration of the process in a 50 pound per hour laboratory unit.

During the early research work, Gulf Oil Corporation acquired Spencer and, after a reorganization, assigned the contract to the research arm of the Pittsburgh & Midway Coal Mining Co.

With the success of the small-scale unit, the government awarded a contract to Pittsburgh & Midway in 1966 to scale-up the process to a pilot plant capable of processing up to 50 tons of coal per day. In 1969, Stearns-Roger Corp. completed design of the unit, but construction funds were not available until late 1971. In June 1972, the contract with Pittsburgh & Midway was extended to include construction and operation of the pilot plant.

Rust Engineering Company, now a subsidiary of Wheelabrator-Frye, began building the pilot plant in July 1972 at the U.S. Army's Ft. Lewis, near Tacoma, Washington. The plant was sited in a relatively remote area near Lake Sequalitchew to take advantage of existing water supplies, sewer systems, and topography.

In October 1974, the plant began to operate. Total construction costs were $32 million, and the average yearly operating costs were approximately $15 million. Some 250 local tradesmen were needed to build the plant and its operations employ about 175.

As the plant was originally designed, raw coal is pulverized and mixed with oil recycled from the process. The mixture is subjected to a temperature of about 850°F (454°C) and a pressure of almost 2000 pounds per square inch. At these conditions, the coal dissolves and combines with hydrogen, while the ash along with the sulfur is separated out.

The hot liquid solution is then cooled to form a solid material which can be transported and handled in facilities designed for coal. This version of the process is called SRC-I.

Test runs at the Ft. Lewis plant in 1975 produced 3000 tons of the solid SRC-I fuel. Much of this fuel was transported to the Georgia Power Company's utility power plant in June 1977 and burned in a successful demonstration. The test showed the fuel to be a clean-burning alternative to high-sulfur coal, which would require an expensive scrubber to be installed.

Not all of the early test runs at Ft. Lewis were successful, however. Engineers frequently encountered problems with the techniques designed to separate the liquid products from the undissolved ash moving through the system. Faced with largely ineffective filtering systems, plant operators revamped the facility to recycle back through the liquefaction vessel the stream of material which they could not otherwise separate.

The result was a liquid product that remained a fluid at room temperature. Not only had the solids separation problem been eliminated, but the new variation of the process created a fuel that could substitute directly for petroleum. The process was termed SRC-II.

(Subsequently, engineers developed deashing methods that use various solvents to separate the undissolved materials from the hot slurry before it is cooled into a solid SRC-II fuel. The once-serious problem of solid–liquid separation is no longer considered a technical obstacle.)

In 1977–1978, almost 6000 barrels of liquid SRC fuel were produced. A portion of this was burned in a successful test conducted by Consolidated Edison in New York during the summer of 1978.

Concurrently with the federal effort on SRC, the Edison Electric Institute teamed with the Southern Company Services, a subsidiary of the Atlanta-based Southern Company, in 1972 to study key steps of the SRC process. Catalytic, Inc., a subsidiary of Air Products & Chemicals, was contracted to design, build, and operate a 6 ton per day pilot plant at Wilsonville, Alabama. The facility, designed solely for the SRC-I process, was completed in August 1973. The Electric Power Research Institute assumed the responsibilities of the Edison Electric Institute in April 1973, and the federal government began funding work at the Wilsonville plant in 1976. With subsequent improvements, the plant cost about $8.4 million and runs at an average annual cost of about $7 million.

In all, eight different coals, representing various regions of the United States, have been tested at Ft. Lewis and Wilsonville.

Successful operation at both plants gave DOE the confidence in 1978 to take the next and final step in preparing the SRC technology for commercial use. In July 1978, the department awarded contracts to Pittsburgh & Midway, representing Gulf Oil, and to Southern Company Services, representing a joint venture of Air Products & Chemicals and Wheelabrator-Frye. The contracts called for designing, building, and operating 6000-ton per day "modules" of a commercial SRC plant.

Pittsburgh & Midway began designing a facility for a site outside Morgantown, West Virginia, that would use the SRC-II (liquids) process. Southern Company's team was working on a design of a plant to use the SRC-I (solids) process and proposes to site the facility near Newman, Kentucky.

Preliminary designs were completed in July 1979, and DOE authorized both firms to move into the detailed engineering phase. Following completion of detailed environmental impact studies by the end of 1980, site preparation was scheduled to begin. Operations at both plants were planned to begin in 1984 with the output from each facility projected to be the equivalent of 20,000 barrels of oil.

Full-size, commercial SRC facilities would replicate the components of these first demonstration modules. Financed by the private sector, these future synthetic fuel plants could each produce the equivalent of up to 100,000 barrels of oil per day beginning in the late 1980s.

Federal funding for these projects in the 1982 budget has been eliminated. Project development is uncertain at this time.

Exxon Donor Solvent

Another approach to "direct" liquefaction is the Exxon Donor Solvent process which originated in the Exxon Research and Engineering Company's laboratories in 1966.

Crushed and dried coal is fed into a reactor where temperatures are maintained at 800–880°F and pressures are from 1500 to 2000 pounds per square inch. Hydrogen is supplied in two forms with roughly one-half coming from a coal-derived liquid that has circulated through a separate vessel where it picked up hydrogen with the help of a special catalyst. Additional hydrogen is fed as a gas.

By keeping the catalyst separate from the harsh environment of the liquefaction vessel—a technique employed routinely in petroleum processing—catalyst life is prolonged. The action of the solvent carrying hydrogen from one reactor to "donate" it to the coal in a separate vessel accounts for the name "Donor Solvent."

By 1971, the sequence of steps had been developed at Exxon's laboratories in Baytown, Texas, and Florham Park, New Jersey. Through 1975, Exxon has spent about $32 million to develop the process and had constructed and operated small test units ranging from ¼ to 1 ton per day of coal input.

In July 1977, the federal government signed an agreement with Exxon to scale-up the technology to a 250 ton per day pilot plant. The cooperative agreement called for DOE to fund half of the program, with the remainder provided by Carter Oil Company (now called Exxon Coal USA), the Electric Power Research Institute, and the Phillips Petroleum Company. Later, these industrial partners were joined by Atlantic Richfield Co. and by firms from the Federal Republic of Germany and Japan.

Construction of the plant began in May 1978 adjacent to an Exxon refinery at Baytown, Texas. Following a formal dedication of the facility in February 1980, engineers began preparing for the first coal to be fed into the system.

The process produces liquids suitable for motor gasoline blending stocks, low-sulfur fuel oil, and utility fuel. From 2.5 to 3 barrels of liquid can be produced from each ton of coal.

H-Coal Process

Unlike the Exxon Donor Solvent process, the H-Coal method adds hydrogen to a coal slurry with the help of a catalyst introduced directly into the liquefaction vessel.

The H-Coal process was first developed by Hydrocarbon Research, Inc., Morristown, New Jersey, a subsidiary of Dynalectron Corp. It is a variation of the H-Oil process used in refineries to convert heavy oil residues into lighter products.

Early development of the H-Coal process beginning in 1964 involved research with a benchscale unit processing 25 pounds of coal per day and a larger 3 ton per day process development unit. Both small-scale experimental units are located at Hydrocarbon Research's test site at Trenton, New Jersey.

A variety of coals have been tested in a small-scale research facility, including eastern, midwestern and western bituminous coals, western subbituminous coals, lignites from Texas and North Dakota, and Australian brown coal.

Based on these smaller tests, a decision was made in1973 to move to a 200–600 ton per day pilot plant. Ashland Synthetic Fuels, Inc., a subsidiary of Ashland Oil, was awarded the primary construction and operating contracts by DOE. Subsequently, the construction portion of the contract was transferred to Badger Plants, Inc. Hydrocarbon Research, Inc. supplied the engineering and research information.

In addition to these firms, other participants were the Commonwealth of Kentucky, Conoco Coal Development Company, the Electric Power Research Institute, Mobil Oil, and Standard Oil of Indiana.

In early 1980, construction of the H-Coal pilot plant was completed at a site adjacent to the Ashland Oil refinery near Catlettsburg, Kentucky. Following pressure tests, the plant started processing oil in the spring of 1980, while engineers prepared for gradually feeding in portions of coal in early summer.

Initially the plant can process 200 tons of coal per day, producing roughly 600 barrels of synthetic crude oil suitable for refining. The amount of coal can be increased to 600 tons per day and operating conditions changed to produce 1800 barrels per day of low-sulfur fuel oil.

Coal Liquefaction Process Descriptions

This section provides descriptions of various coal liquefaction processes. The commercial and industrial importance of converting coal to liquid fuels and chemicals was first recognized in post-World War I Germany. Development of coal liquefaction proceeded along three basic paths: indirect conversion followed by Fischer–Tropsch synthesis, direct liquefaction, and pyrolysis.

In indirect conversion, coal is first gasified to produce synthesis gas, which is then purified and, in some cases, undergoes a shift reaction to increase hydrogen content. The resulting synthesis gas can then be chemically reacted to produce methanol or a slate of chemical intermediates that can be further upgraded to gasoline. The subsequent conversion of methanol to gasoline can be performed with the use of an additional process.

In indirect liquefaction, coal is first dissolved in an appropriate solvent, often a product fraction of the process. Liquid fuels are then produced by reacting the solution of coal and solvent with hydrogen.

Pyrolysis involves exposing the coal to very high temperatures in an inert or oxygen-deficient atmosphere. The coal is converted into coal (char) and yields tar, liquids, and fuel gases as by-products. Flash pyrolysis processes are currently at benchscale and thus are not expected to be available for commercial-scale operation until after 1990.

The products derived from coal liquefaction can be used as

- Low-ash, low-sulfur boiler fuels suitable for electric power generation and for steam generation for industrial use
- High-grade fuels such as gasoline, heating oil, and chemical feedstocks

The various liquefaction processes result in significantly different products: Fischer–Tropsch yields a wide range of hydrocarbons, including many alcohols and other oxygen-bearing materials; H-Coal gives heavy and light hydrocarbon fuels still deficient in hydrogen and requiring further treatment if used for any purpose other than boiler fuel; Exxon's Donor Solvent process also provides fuels that, without further processing, are primarily useful as boiler fuel, but there is some marketable LPG (liquefied petroleum gas) product; and M-Gasoline produces premium gasoline and some LPG product.

The solid product of SRC-I can be used in a variety of market applications, including existing coal-fired boilers, where the product can be burned with only minor modifications to the boiler, and in some boilers that burn heavy fuel oil. Fuel by-products can be burned in existing oil-fired boilers. A modified process will result in liquid products.

The principal application of the SRC-II process is as new fuel oil in industrial and utility boilers to substitute for petroleum. SRC liquid fuels in which essentially all sulfur and ash are removed can be used directly in compliance with emission standards (probably) without further stack-gas cleanup or scrubbing.

Recently, the production of liquid hydrocarbons from coal has been demonstrated by several ventures. The SASOL plant in South Africa is the most modern example. The SASOL plants use indirect conversion of coal to make gasoline and other liquid products.

Direct conversion by a liquefaction process to produce syncrude or a heavy industrial fuel on a commercial scale is not currently being done. However, Germany used direct liquefaction to make about 90% of its aviation fuel in World War II, and a commercial-scale liquefaction process was employed in England until about 1958.

The area, labor force, and resource requirements for a synfuels plant will vary depending on the specific project and local conditions. However, to provide a rough idea as to the scale of a synfuel plant, Table 3-2 was prepared and presents the estimated requirements for two types of coal liquefaction plants, each yielding 60,000 barrels per day of liquids.

Process descriptions for indirect and direct coal liquefaction processes are described in the following paragraphs.

Table 3-2. Synfuel Plant Resource Requirements

Characteristic	Coal Liquefaction Process	
	Indirect	Direct
Coal feed (tons per day)	35,500	26,700
Water requirements (gallons/minute)	11,800	7,840
Products (barrels/day)	60,000 Hydrotreated liquids, C_3-C_5 hydrocarbons, heavy alcohols	60,000 Low-sulfur fuel oil, naphtha, LPG
Operations and maintenance personnel (persons)	3,570	2,750
Land (acres)	1,860	770

Source: *Production of Synthetic Liquids from Coal*, Bechtel National, Inc., DOE Report 03137-1.

INDIRECT COAL LIQUEFACTION

A particular advantage of indirect liquefaction is that essentially all of the sulfur and nitrogen present in the coal can be separated in the gaseous phase and thus eliminated from the liquid products. These materials are difficult and expensive to remove to a very low concentration with direct processes.

The two indirect liquefaction processes receiving significant attention are the Fischer–Tropsch and the Mobil-M (methanol) process. A modification of the Fischer–Tropsch process is in commercial use in South Africa. A range of hydrocarbon products are obtainable with this process. The Mobil process is in the development stage. The principal product of this process is gasoline.

Environmental problems common to fossil energy facilities will also apply to coal liquefaction facilities. Liquefaction facilities do present some unique problems due to incomplete combustion resulting in a wide variety of organic compounds, reducing conditions resulting in H_2S and other reduced sulfur compounds, and catalytic processes producing spent catalyst with associated environmental concerns. These problems are generally common to all liquefaction processes. Since no large scale plants are in operation in the United States, the only available data on emissions and effluents are estimated from pilot plant operations and cannot be quantified for a commercial operation.

Projected thermal efficiencies for indirect coal plants (i.e., Fischer–Tropsch and Mobil) producing pipeline quality synthetic natural gas (SNG) and gasoline and/or diesel fuel are in the 50–60% range, the Mobil process being the more efficient. Liquid fuels from indirect liquefaction plants are projected as being more costly than from direct processes.

Although there remain unanswered questions relating to coal liquefaction (e.g., commercial demonstration, environmental impacts, costs), the successful development of the technology would provide a valuable energy alternative and allow greater utilization of our nation's coal reserves. Additionally, liquid fuels are generally easier to store, transport, and utilize than solid fuels, and during liquefaction, impurities (e.g., sulfur) can be removed, making it possible to produce an environmentally acceptable liquid fuel from various ranks of coal.

The basic objective of coal liquefaction is to convert coal to liquid fuels with minimal production of gases, liquids, and organic solid residues. All ranks of coal can be liquefied, although some are more attractive than others. The liquid products vary both with the type of coal used and the particular process applied.

There are several methods for producing a liquid fuel from coal. As with gasification, either hydrogen has to be added or carbon removed from the compounds in the coal. In bituminous coal, for example, the carbon-to-hydrogen ratio by weight is about 16 to 1; in fuel oil the ratio is about 6 to 1. Although liquefaction is a complex process, it can be viewed as a change in the carbon-to-hydrogen ratio that can be accomplished by one of several processes (e.g., indirect liquefaction). The chemical structure of the coal influences the type of chemical reactions that will take place during liquefaction. This structure varies with rank of coal. Coal liquefaction processes can be grouped into four distinct categories: direct hydrogenation (e.g., H-Coal), solvent extraction (e.g., Solvent-Refined Coal), pyrolysis (e.g., clean coke), and indirect liquefaction (e.g., Fischer–Tropsch).

In direct hydrogenation, hydrogen is added catalytically to coal in a reactor under high pressure and temperature resulting in vapor and liquid phases which are cooled to separate the products, refined to remove by-products, and, depending on the fuel product desired, processed further. The process conditions (temperature, pressure, and amount of hydrogen added) determine the fuel produced. Processes and products in this category are *H-Coal*, which produces boiler fuel or synthetic crude, and *Synthoil*, which produces synthetic crude or fuel oil.

The solvent extraction process liquefies coal through indirect transfer of hydrogen to the coal using a process-derived solvent and a hydrogen atmosphere. Processes and products in this category are *hydrocarbonization*, which produces fuel oil; *clean coke*, which produces coke and liquid fuels; and *flash pyrolysis*, which produces fuel oil, coke, and gas.

Indirect liquefaction involves the initial gasification of coal to produce a mixture of CO and H_2 (synthesis gas), which is purified and converted to liquid fuels by reaction over appropriate catalysts to produce alcohols (methanol) or paraffinic hydrocarbons. A particular advantage of indirect liquefaction is that essentially all of the sulfur and nitrogen present in the coal can be separated in the gaseous phase and thus eliminated from the liquid products. These materials are often difficult and expensive to remove to a very low concentration with direct and indirect liquefaction in that direct processes produce a significant amount of potentially carcinogenic aromatic organic compounds. The indirect category processes and products are the *Mobil Process* which produces gasoline and *Fischer–Tropsch* which produces liquid fuels and chemical products. The Fischer–Tropsch process is significant in that it is the only large commercial coal liquefaction plant in operation. (It is located in South Africa.) Discussions on the above two indirect liquefaction processes follow.

Mobil Methanol Technology

The Mobil Oil Corporation is developing an improved process for production of motor fuels by indirect liquefaction. The product gasoline has a high octane rating and is free of heavy ends so that product upgrading is not necessary.

The Mobil process involves the conversion of methanol to gasoline. When starting with coal, coal is first converted to a synthetic gas with subsequent conversion to methanol by proven commercial technology. The methanol is then converted to gasoline by means of the Mobil process. Figure 3-2 indicates the basic concept.

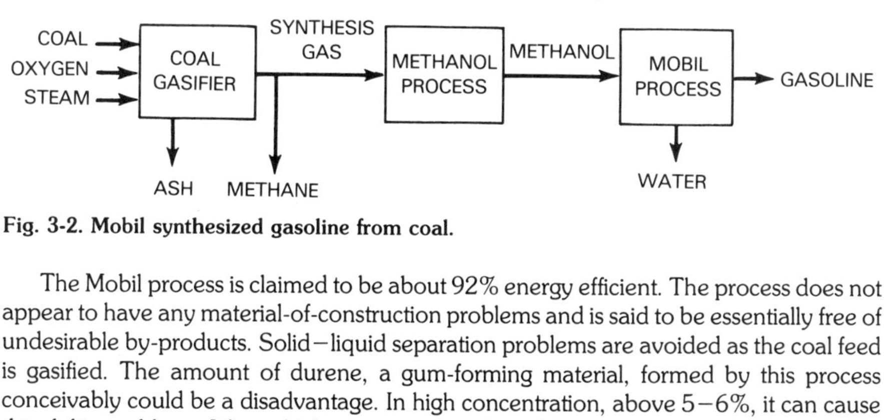

Fig. 3-2. Mobil synthesized gasoline from coal.

The Mobil process is claimed to be about 92% energy efficient. The process does not appear to have any material-of-construction problems and is said to be essentially free of undesirable by-products. Solid–liquid separation problems are avoided as the coal feed is gasified. The amount of durene, a gum-forming material, formed by this process conceivably could be a disadvantage. In high concentration, above 5–6%, it can cause drivability problems. It has a high octane and is a desirable constituent if the concentration can be kept down. Mobil claims that knowledge to date indicates they can control the durene content to less than 4%, a value within desired limits.

In the overall Mobil indirect liquefaction process (Fig. 3-3), synthetic gas is produced from coal via any of the existing medium-Btu coal gasification processes (e.g., Lurgi, Koppers–Totzek, Texaco). When the synthesis gas is shift converted to the proper carbon-to-hydrogen ratio, methanol becomes an alternative product from any process that can produce high-Btu gas. In practice, the synthesis gas is converted to methanol by any of a number of proven processes (e.g., the low-pressure methanol process). The yield of methanol is maximized by a combination of optimum reactor conditions and catalyst and the recycle of unreacted gases. High catalyst selectivity limits the production of ethers, ketones, and higher alcohols. Carbon dioxide is removed prior

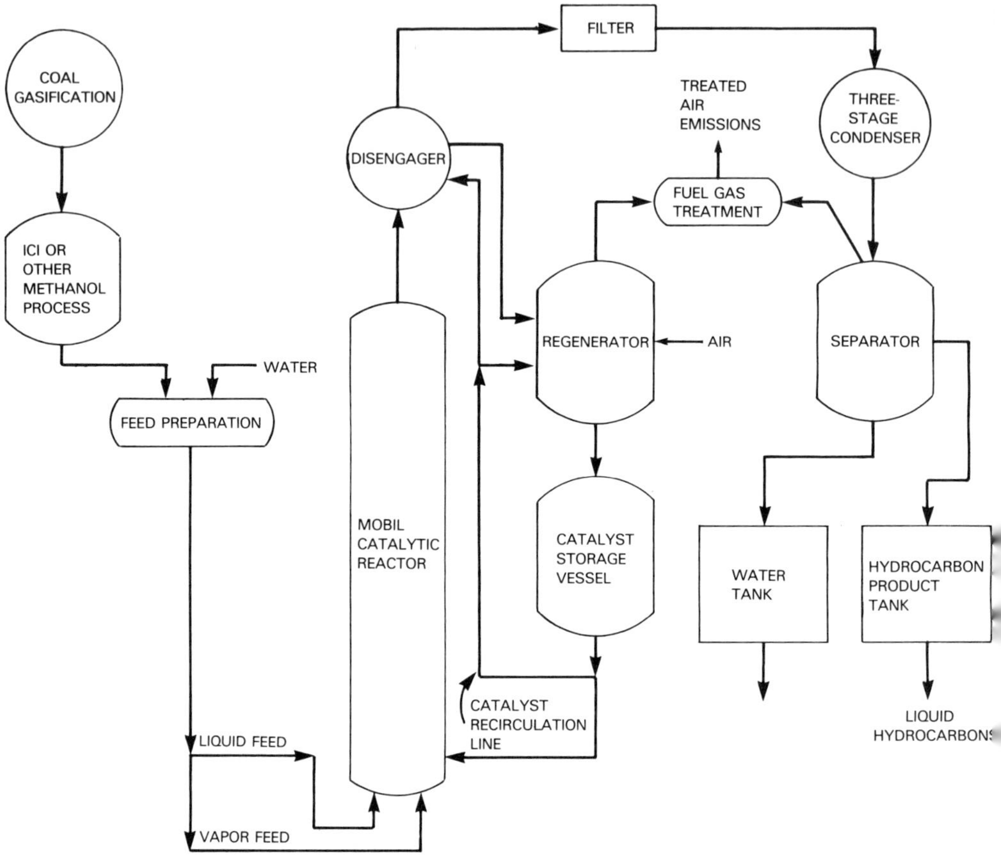

Fig. 3-3. Mobil catalytic process.

to converting the synthesis gas to methanol, thereby qualifying the product as a feed for the Mobil process.

The Mobil process converts methanol into a high-octane gasoline by dehydration over a shape-selective zeolite catalyst. The secret of the process is really the catalyst, a unique zeolite identified a few years ago. Mobil's initial development efforts are directed at the fixed-bed reactor configuration. Even so, considerations are also being given to tubular and fluid-bed catalytic reactor units.

In the process, methanol is blended with water and charged as a vapor into the reactor. The product is separated from the catalyst, filtered, and condensed, and the water is separated from the hydrocarbon components. The principal output components are premium gasoline (approximately 90% of Btu output) and LPG (approximately 10% of Btu output).

Fischer–Tropsch Technology

The Fischer–Tropsch (FT) indirect liquefaction process is based on the FT catalytic reactions that were discovered in 1923. The FT reactions can cover a range of combinations using different metallic oxide catalysts to react hydrogen and carbon monoxide to form a mixture of olefins, paraffins, and alcohols. These reactions are exothermic and are not specific to the formation of any single compound. The range of products from the FT process depends on the reaction and temperature conditions as well as the type of catalyst used and the composition of (input) reactants.

In the complete scheme of the FT process, the coal is initially gasified (e.g., using a Lurgi gasifier), then cleaned of hydrogen sulfide, carbon monoxide, and impurities, and finally shift converted before it enters a catalytic reactor.

The major reason for interest in the FT process is that a commercial plant using a modification of this method is currently operating in South Africa (SASOL I). In the commercialized process, two reactors, using different catalysts and temperatures and pressures, process gases with different carbon and hydrogen ratios into different products. Its main drawbacks are that a great deal of reaction heat is produced (i.e., efficiency suffers) and the process is apparently expensive. As previously indicated, the process can produce a wide variety of hydrocarbons ranging from methane to light oils. The different products are produced with varying degrees of overall thermal efficiency.

The gasification can be accomplished with any of a number of different gasifiers. SASOL I, as shown in Fig. 3-4, employs 13 Lurgi high-pressure, steam–oxygen coal gasifiers to produce a product gas containing carbon monoxide, tars, and oils as the main components. The crude product gas is cleaned of carbon monoxide, hydrogen sulfide, organic sulfur, ammonia, and phenol. The cleaned gas is partitioned into two streams. One stream is adjusted to a hydrogen-to-carbon monoxide ratio of 2 to 1 and fed to a fixed-bed catalytic reactor (ARGE) that is operating at 450°F (232°C) and 360 psi. The

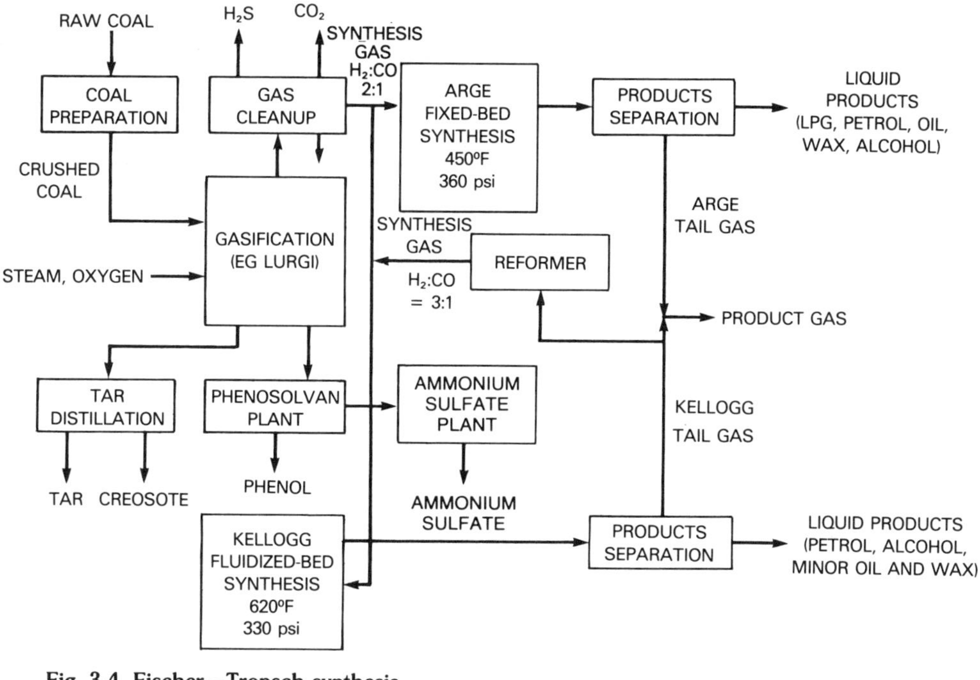

Fig. 3-4. Fischer – Tropsch synthesis.

products from this reactor are mainly straight chain and medium boiling oils, diesel oil, LPG, and some alcohols.

The other stream of purified gas (i.e., gas from the Lurgi gasifiers) is combined with reformed product gas (to increase the hydrogen-to-carbon monoxide ratio) and sent to a fluidized-bed reactor (i.e., SYNTHOL: the U.S. developed Kellogg synthesis). The operating conditions are 620°F (327°C) and 330 psi. The products from this reactor are mainly gasoline, fuel oil fractions, and various chemical products. The gasoline has a lower octane rating than the natural petroleum-based gasoline that is currently marketed in the United States.

Overviews on SASOL I and SASOL II follow:

	SASOL I	**SASOL II**
Location:	Sasolburg, South Africa	Sedunda, South Africa
Description:	Gasification in Lurgi gasifiers Two Fischer–Tropsch synthesis units; (1) ARGE fixed-bed unit; temperature, 450°F; pressure, 360 psi; catalyst, pelleted precipitated iron (2) Kellogg SYNTHOL process, high-velocity entrained-flow reaction using a double promoted iron catalyst	Gasification in Lurgi gasifiers, Fischer–Tropsch synthesis unit using the Kellogg SYNTHOL process
Size:	10,000 barrels per day	Nominal 40,000 barrels per day
Status:	In commercial production since 1956	Commenced operation in 1981
Years in operation:	24	
Coal type:	Subbituminous	Subbituminous
Major products:	Liquid fuels, chemicals, and fuel gas	Liquid fuels (gasoline is the major product)

Process variabilities for indirect liquefaction processes are such that any efficiency value must be used with caution. Owing to the proprietary nature of indirect liquefaction technology (e.g., SASOL), a detailed heat balance was not obtainable. Even so, a specific comparison between the FT and the Mobil processes was obtained. A performance comparison between the Mobil methanol-to-gasoline technology and the commercially available FT technology for the production of the motor gasoline meeting U.S. quality standards is described in the following discussion. The description covers complete conceptual plant complexes using the Lurgi dry-ash pressure technology to gasify subbituminous coal.[1] Except for the Mobil process, processes used are commercially available. Coproduction of products, namely SNG, LPG, and gasoline, is practiced. Efficiency values are not based on detailed mass balances and associated heat balances, but on the known performance of specific processes. A thermal efficiency comparison between the commercial FT technology and the Mobil methanol-to-gasoline technology for the production of motor gasoline (with other output products) from U.S. subbituminous coal is provided by Table 3-3.

Table 3-3. Thermal Efficiencies

	Methanol-to-Gasoline		Fischer–Tropsch	
	Btu/hour (10^6 Btu)	Percentage of Input	Btu/hour (10^6 Btu)	Percentage of Input
Input				
Coal	19,383		19,708	
Coal fines (excess)	(872)		—	
Methanol	—		3	
Total input	18,511		19,711	
Output				
SNG	6,067	32.8	7,243	36.8
C_3 LPG	247	1.3	176	0.9
C_4 LPG	385	2.1	26	0.1
10 RVP gasoline	4,689	25.3	2,842	14.4
Diesel fuel			514	2.6
Heavy fuel oil			147	0.7
Subtotal	11,388	61.5	10,948	55.5
Alcohols			290	1.5
Sulfur	19	0.1	19	0.1
Ammonia	83	0.5	83	0.4
Power[a]	18	0.1	11	0.1
Total output	11,508	62.2	11,351	57.6

[a] Direct thermal equivalent value (thermal efficiencies are highly dependent on product mix).

Applications

The conversion of coal to liquids has never been accomplished commercially in the United States. In the past, coal-to-oil plants were constructed in a number of countries. Currently there is only one plant, in South Africa, producing liquids from coal. This plant, SASOL I, employs an indirect liquefaction process.

The successful development and commercial demonstration of a coal liquefaction technology would provide a valuable energy alternative and would allow greater utilization of the nation's coal resources. The development and future commercial use of coal liquefaction technology are dependent on many factors. Some of the more important include

[1] M. Schreiner, *Research Guidance Studies to Assess Gasoline from Coal by Methanol-to-Gasoline and SASOL-Type Fischer–Tropsch Technologies*, Mobil Research and Development Corporation for DOE, Report FE-2447-13, August 1978.

- The demonstration (large scale) of a viable technology
- The ability to satisfy environmental concerns
- The ability to produce a commercially usable liquid fuel with an acceptable overall efficiency rating and cost

Regulatory Impacts

Each liquefaction technology will have to be evaluated separately as to regulatory impact. Because of the difference in technologies and also varying state and local regulations, siting of a major coal liquefaction facility must be approached on a case-by-case basis.

A coal conversion facility must comply with regulations and standards including requirements of the Clean Air Act, the Clean Water Act, Safe Drinking Water Act, the Resource Conservation and Recovery Act, the Toxic Substances Control Act, the Federal Nonnuclear Research and Development Act, and the National Environmental Policy Act, as well as applicable state laws. Failure to comply has the potential of halting all progress toward commercialization.

Current standards for hazardous air pollutants limit mercury, beryllium, and lead emissions. These standards conceivably could put a limit on coal types that can be utilized in future demonstration plants. Since effluent guidelines have not been developed for most fossil energy technologies, permit requirements are determined on a case-by-case basis by states or by EPA.

Disposal of specific materials used in coal liquefaction may be regulated in the future. Currently, solid waste disposal must comply with stringent standards. Monitoring is required and state or EPA permits for all landfills will be required by April 1, 1988.

The Resource Conservation and Recovery Act (RCRA) has guidelines for the land disposal of solid wastes (40 CFR 241). These standards set minimum levels of performance for any solid waste land disposal site. Additional standards have been proposed for disposal of solid wastes that contain hazardous pollutants. All future coal liquefaction facilities will have to abide by these solid waste standards.

Undoubtedly, the coal conversion industry would benefit from the experience of the petroleum industry in dealing with complex organic substances and new processes while complying with governing statutes.

Performance

Currently, there are no commercial coal liquefaction plants in the United States; therefore, all projections are based on available data associated with SASOL I and assessments relating to SASOL II and efforts in the United States still in the developmental stage.

The overall thermal efficiency of the FT process is dependent on the output product mix. The projected overall thermal efficiency of a modern FT plant that would predominately produce pipeline quality SNG with a lesser amount of liquid fuels (32% on a Btu basis) has been estimated to be approximately 55% (see Table 3-3). The Mobil process producing pipeline quality SNG with almost as much gasoline (i.e., on a Btu basis) would have an overall thermal efficiency of approximately 60%. Noteworthy is the fact that a greater percentage of the Mobil process output is in the form of gasoline and that these efficiency values are based on giving credit to all output products. When these processes are optimized for gasoline yield and credit is not given for product gas, the efficiency values are significantly less (e.g., 32% for FT).

The general expectation, based on substantial development efforts and overseas experience, e.g., SASOL I, is that efficiencies of indirect liquefaction processes will not measurably improve (over currently anticipated values) in the foreseeable future. System losses are well understood and substantial efforts have already been directed at utilizing all system available energy.

DIRECT COAL LIQUEFACTION

Direct hydrogenation, solvent extraction, and pyrolysis are classified as direct liquefaction processes. Projected efficiencies for direct coal liquefaction facilities are in the 60–70% range. Exact values for coal conversion efficiencies are difficult to estimate and thus an exact value cannot be given until commercial demonstration takes place. In 1977, DOE estimated the cost to produce a synthetic crude to be $4–6 per million Btu.

In the United States, the direct liquefaction processes receiving DOE support that are closest to near-term commercial demonstration are H-Coal, Solvent-Refined Coal, and Donor Solvent. Detailed descriptions of these processes follow.

H-Coal

The H-Coal pilot plant is located in Catlettsburg, Kentucky. This 600 ton of coal per day test facility recently commenced operation. The process is a catalytic hydroliquefaction process that converts high-sulfur-content coal to boiler fuels and to syncrude. A schematic of the process is provided in Fig. 3-5. Coal is crushed to minus 60 mesh, dried, and then slurried with recycled oil and pumped to a pressure of about 200 atmospheres. Compressed hydrogen is added to the slurry, and the mixture is preheated and charged continuously to the bottom of the ebullient-bed catalytic reactor. The upward passage of the internally recycled reaction mixture maintains the catalyst in a fluidized state. (Catalyst activity is maintained by the semicontinuous addition of fresh catalyst and the withdrawal of spent catalyst.) The temperature of the ebullient-bed catalytic reactor is controlled by adjusting the temperature of the reactants entering from the preheater. Typically, the temperature of the mixture entering the reactor is 650–700°F (343–371°C).

Vapor product leaving the top of the reactor is cooled to separate the heavier components as a liquid. Light hydrocarbons, ammonia, and hydrogen sulfide are absorbed from the gas stream and sent to a separator and a sulfur recovery unit. The remaining hydrogen-rich gas is recompressed and combined with the input slurry. The liquid from the condenser is fed to an atmospheric distillation unit. The liquid–solid product from the reactor, containing unconverted coal, ash, and oil, is fed into a flash separator. The material that boils off is passed to the atmospheric distillation unit that yields light and heavy distillate products. The bottoms product from the flash separator (solids and heavy oil) is further separated with a hydrocyclone, with a liquid–solid separator, and by vacuum distillation.

The gas and liquid products, composed of hydrocarbon gas, hydrogen sulfide, ammonia, light distillate, heavy distillate, and residual fuel, may be further refined as necessary. A portion of the heavy distillate is recycled as the slurry medium. The stream containing the unreacted carbon and some liquid will eventually be processed in a commercial installation to produce additional hydrogen needed for the process.

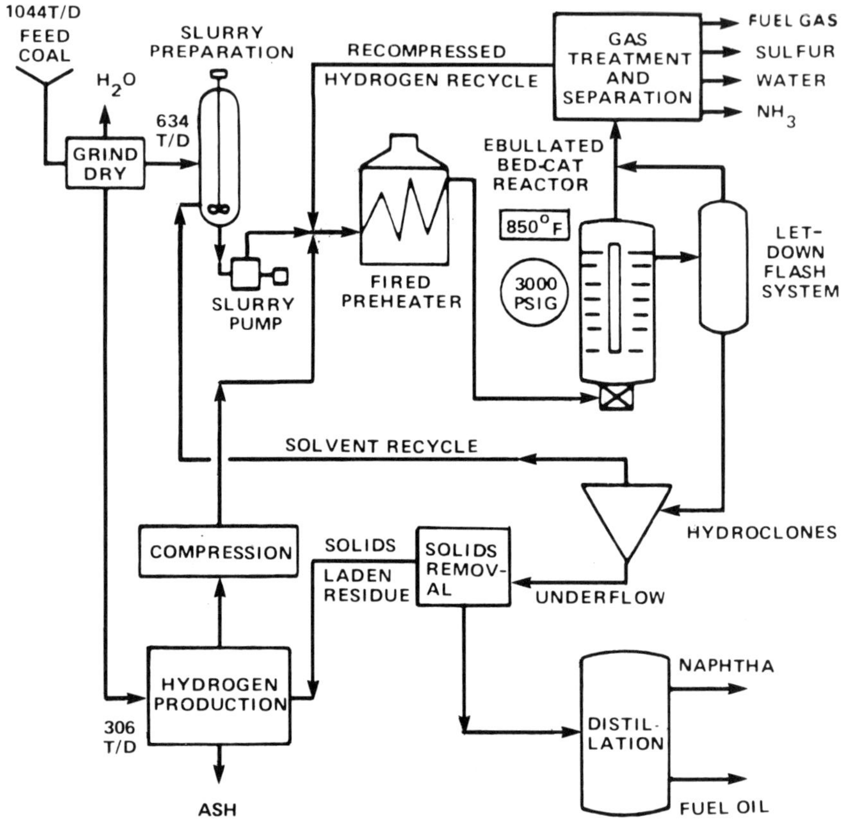

Fig. 3-5. H-Coal process.

The specific operating conditions of the H-Coal process affect the type of fuel produced. For example, to produce synthetic crude, more hydrogen is required, and there is a lower yield of residual fuel oil. In this mode, the separation of the solids (unconverted carbon and ash) from the liquid can be accomplished by vacuum distillation, and a special liquid–solid separation unit is not required. To produce clean fuel gas and low-sulfur residual oil as major products, the temperature and pressure in the ebullient-bed reactor are lowered, and less hydrogen is required.

Solvent-Refined Coal

Despite the growing energy shortfall in the United States, coal production in the United States increased an average of only 2% each year during the 1970s. Among the reasons for slow growth is the fact that coal, in its natural form, has several drawbacks:

- As a solid material, coal is more difficult and costly to handle, transport, and use than are liquids or gases.
- In addition to heat, coal produces ash, which can present serious disposal problems, particularly to urban area power plants.

- Some coal, particularly that mined from the bituminous seams found in the eastern United States, has a high sulfur content. Direct burning of high-sulfur coal presents air pollution problems.

All of these drawbacks can be overcome by converting coal to liquid and gaseous synthetic fuels. In addition, synthetic fuels from coal can replace petroleum and natural gas in many applications that coal itself could never fill. Changing coal into a liquid or gas is achieved by increasing the hydrogen content of pulverized coal in processes involving heat and pressure.

Coal liquefaction is not new. A coal hydrogenation process was described in 1913 by Friedrich Bergius, a German chemist who later received a Nobel Prize in chemistry for his work. Research was conducted in many countries in the 1920s and 1930s—including the United States. But only in countries which lacked large domestic petroleum deposits did the laboratory technology develop into production plants. Germany was the world leader. A coal hydrogenation plant was in operation there in 1927. By 1944, when a fuel-short German war machine depended upon it, 120,000 barrels of synthetic fuels per day were provided by coal hydrogenation plants operating in Germany. This fuel was extremely expensive, and environmental protection was not a serious concern.

The only coal liquefaction plants in the world today have been built by the South African Coal, Oil, and Gas Corp. (SASOL). The first SASOL plant produces 10,000 barrels per day, and an even larger plant began operation in March 1980.

Meanwhile, a number of other coal liquefaction processes are being developed, principally in the United States. Some of them show promise of being more efficient and cost-effective than those used in the early German and present South African liquefaction plants.

One leading U.S. technology is the Solvent-Refined Coal (SRC) process, which is particularly effective in converting the high-sulfur, high-ash coal found in abundance in the United States into clean-burning synthetic fuels.

The SRC process, like most coal liquefaction technologies, has its roots in theories developed in German synthetic fuel laboratories in the 1920s. The process derives its name from the fact that an oily solvent is used to carry pulverized coal through the liquefaction plant.

Extensive research and development programs needed to turn the early solvent-refined coal experiments into a large-scale technology have taken place in the United States in the 1960s and 1970s.

Systematic SRC research began in 1962 when the Pittsburgh & Midway (P&M) Coal Mining Co., which the next year became a subsidiary of Gulf Oil Corporation, was awarded a contract by the Office of Coal Research of the U.S. Department of the Interior. Its objective was to evaluate the technical feasibility of the solvent-refined-coal process as applied to U.S. coals.

Initially, the principal objective of the SRC program was to investigate how efficiently an ashless solid boiler fuel could be produced. But during the early laboratory studies it was found that as ash was removed, so was most of the sulfur—a primary constituent of urban pollution.

During the early research, Gulf scientists envisioned the potential for a clean-burning solid substitute for coal. Used instead of coal, the solid SRC product would not only minimize the ash disposal problem at coal-burning power plants, but air pollution would also be substantially reduced.

By 1965, a 50 pound per hour laboratory facility at Merriam, Kansas, had successfully tested the principles of the SRC process.

In 1966, the Office of Coal Research contracted with P&M to design and build a pilot plant capable of processing 50 tons of coal per day. The government's pilot plant was to provide engineering data needed for the design of an even larger demonstration plant—as well as samples of solvent-refined-coal products for performance combustion tests and environmental and health studies.

A site for the pilot plant was selected at Fort Lewis, Washington, near Tacoma. P&M began building the SRC plant in 1972, and began operating it for the government in September 1974.

The SRC pilot plant has used primarily high-sulfur bituminous coals from eastern United States seams. Coal from Pennsylvania, Kentucky, West Virginia, and Ohio has been processed at Fort Lewis.

Initially, the plant operated only in a version of the process, which became known as SRC-I, that produced the solid product. It is a liquefaction process, because the product—when first separated from the ash, sulfur, and solvent—is in a liquid form. But when heat and pressure are removed, SRC-I turns to a glassy solid which, without further refining, can be burned like coal in electric power generating plants.

The SRC-I solid fuel contains considerably less ash or sulfur, and pound for pound has one-third more Btu's (heating value) than the bituminous coal from which it was produced.

Meanwhile, ongoing laboratory work indicated early in the 1970s that modifications to the SRC-I process could result in the production of a liquid rather than solid product. The liquid is easy to handle and transport, and its characteristics are much like petroleum-derived boiler fuel used in oil-fired electric power plants. Gulf scientists believed the liquid product of SRC could also be refined into other petroleumlike products.

Moreover, it was found that the solvent used to dissolve the pulverized coal could be produced in this new version of the SRC process, thus eliminating the need to constantly add outside solvents. This new version of the SRC process was named SRC-II.

In 1976 the Fort Lewis plant began intermittent SRC-II production. Using about 30 tons a day of high-sulfur eastern U.S. coal, the plant can produce about 70 barrels of clean-burning liquid fuel each day. The initial SRC-II runs at Fort Lewis confirmed the results of the laboratory work, and in January 1978, P&M submitted a proposal for construction and operation of an SRC-II demonstration plant to the then newly formed U.S. Department of Energy (DOE).

DOE awarded P&M a study contract in July 1978, and in September 1979 the go-ahead was given for the engineering of an SRC-II demonstration plant using commercial-scale equipment, and designed to process about 6000 tons of high-sulfur coal daily into the equivalent of about 20,000 barrels per day of SRC-II fuels; it is currently planned to be constructed in Morgantown, West Virginia, with startup in 1984.

At the same time, another prime contractor was selected to design, build, and operate a demonstration plant to investigate the commercial feasibility of the SRC-I process.

The product from the original SRC mode (SRC-I) is a nearly ash-free low-sulfur solid fuel at ambient conditions. In a modification of the process (SRC-II) a distilled liquid product results. The SRC-I product is projected to serve as a clean fuel substitute for high-sulfur coal. The SRC-II product can be used directly as a boiler fuel or, with further hydrogenation, as a refinery feedstock for conversion to conventional fuels.

Figure 3-6 is a schematic of the SRC-I process. The coal is first pulverized and mixed with a coal-derived solvent in a slurry mix tank. The slurry is mixed with hydrogen, which is produced by other steps in the process, and is then pumped through a fired preheater and passed into a dissolver where about 90% of the moisture- and ash-free coal is dissolved. Several other reactions also occur in the dissolver: the coal is depolymerized and hydrogenated, which results in an overall decrease in product molecular weight; the solvent is hydrocracked to form lower-molecular-weight hydrocarbons that range from light oil to methane; and much of the organic sulfur is removed by hydrogenation in the form of hydrogen sulfide.

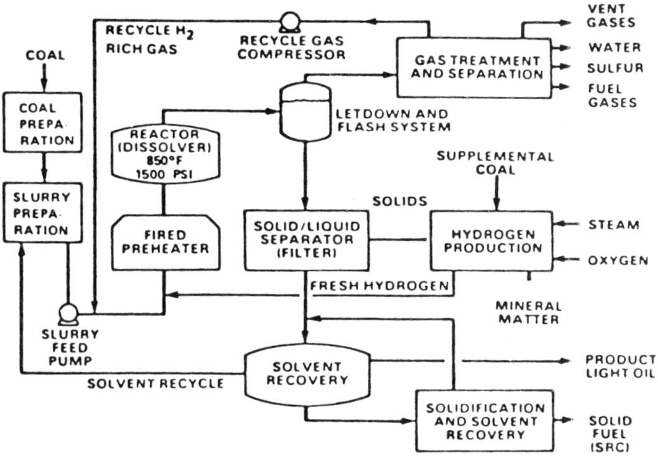

Fig. 3-6. SRC-I process.

From the dissolver, the mixture passes to a separator where the gases are separated from the slurry of undissolved solids and coal solution. Raw gas is sent to a hydrogen recovery and gas desulfurization unit. Hydrogen recovered is recycled with the slurry coming from the slurry mix tank. Hydrocarbon gases are given off, and hydrogen sulfide is converted to elemental sulfur.

The slurry of undissolved solids and the coal solution pass to a filtration unit where undissolved solids are separated from the coal solution. In the commercial-scale process, the solids are sent to a gasifier–converter where they react with supplemental coal, steam, and oxygen to produce hydrogen for use in process. The coal solution passes to the solvent-recovery unit, and the final liquid product, solvent-refined coal, is produced. The solvent-refined coal has a solidification point of 350–400°F (177–204°C) and a heating value of about 16,000 Btu/lb.

As previously indicated, modification of the SRC process (SRC-II) produces an all-distillate liquid instead of a solid residual fuel as the principal product. Figure 3-7 is a schematic of the SRC-II process. In the modification, part of the product slurry is recycled as solvent for the pulverized coal feed instead of 450°F-plus boiling-range distillate. As a result of the increased severity of reaction conditions, the hydrogen reaction is greater and

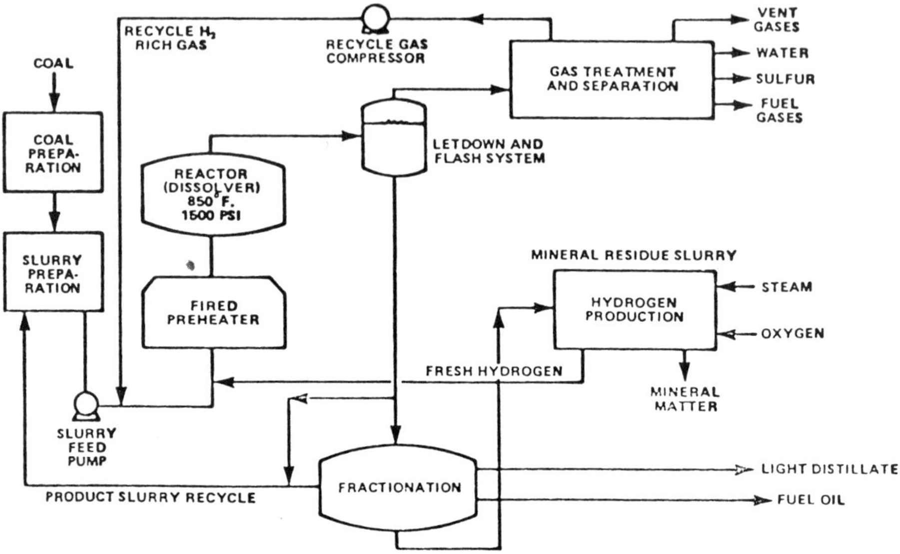

Fig. 3-7. SRC-II process.

a major part of the coal is converted to a liquid distillate product. The quantity of unconverted coal and vacuum residue is controlled so it is in balance with the requirements for gasifier feed to produce the process hydrogen requirements. This eliminates the solid/liquid separation step (filtration) required for production of fuel in solid form.

SRC-II Products

While the SRC-II products have similar petroleum counterparts, they are chemically different in some respects. Accordingly, most of the production of the demonstration plant will be used as materials for laboratory and large-scale field tests by the industries which are counting heavily on the output of future SRC-II refineries.

The first large-scale commercial use of the SRC-II products is expected to be in fuel oil applications, such as public utility power generation and industrial plant heating. Later, as refining technology advances, the majority of the liquid products of SRC-II plants might be turned into transportation fuels, such as jet aircraft and diesel fuels. It is expected that gaseous products will be used principally as feedstocks for petrochemicals. In the 21st century, the coal refining industry is expected to be as fully diversified as the petroleum industry is today.

Distillate Fuel Oil. The major product of the SRC-II process—a distillate fuel oil containing less than 0.3% sulfur—initially would be used as a nonpolluting fuel for production of electrical power and steam in eastern United States areas where public utilities and industry currently use petroleum fuel oils.

In an SRC-II commercial plant equivalent to a 100,000 barrel-per-day petroleum refinery, approximately 60,000 barrels of distillate fuel oil would be produced—enough for the electric generation needs of a city of 3 million people.

Many industrial plants that currently use fuel oil and gas to fire steam plants are faced with costly energy supply interruptions—or conversion to coal. The availability of SRC-II fuel oil will be an extremely attractive alternative to coal conversion in all but the largest industrial applications.

Naphtha. The SRC-II process produces considerable amounts of naphtha—a light, low-boiling-range liquid. Naphtha can be upgraded to produce a high-octane unleaded gasoline blending stock that can be used to supplement gasoline supplies derived from petroleum. An SRC-II commercial plant's naphtha production could be converted into 800,000 gallons of gasoline per day, enough for the needs of a city of 250,000 people. However, the demands of the marketplace might make naphtha even more valuable as a feedstock for chemicals used in manufacturing. For example, naphtha could be a valuable source of aromatics, used in the production of nylon for such products as fabrics, high-strength manufacturing components, and fishing line.

Methane and Other Product Gases. Typically, a commercial SRC-II plant would produce about 50 million standard cubic feet of methane pipeline gas during a 24-hour period. This would supply the needs of about 100,000 gas-heated homes.

By the time a commercial plant could be in service in the late 1980s, the methane quantity could vary, upward or downward, from the 50 million standard cubic feet because of processing options which could be implemented to increase the volume of other products. Also, varying amounts of the methane will be used for fuel in the SRC-II process, and thus would not reach the market.

Ethane and propane also would be useful as fuels, but more likely they will be converted into ethylene and propylene, feedstocks for a wide variety of materials involved in the manufacture of textiles and durable goods. Ethylene is used in the production of polyester fabrics and hard, durable plastics; propylene is used to produce acrylics.

Donor Solvent

The chief features of the Exxon Donor Solvent (EDS) process are illustrated in the process schematic, Fig. 3-8. A pilot plant (250 tons of coal per day) is located in Baytown, Texas, next to an Exxon refinery. Crushed coal is liquefied in a noncatalytic reactor in the presence of molecular hydrogen and the hydrogen-donor solvent. The liquefaction reactor operates at 800–850°F and 1500–2000 psi.

The hydrogen-donor solvent is a 400–850°F boiling-range material. The solvent is a catalytically hydrogenated recycle stream fractionated from the middle boiling range of the liquid product. After hydrogenation, the solvent is mixed with fresh coal feed and pumped through a preheat furnace into the liquefaction reactor. Slurry leaving the liquefaction reactor is separated by distillation into gas, naphtha, distillates, and a vacuum bottoms slurry. The vacuum slurry is coked to produce additional liquids.

The plant is "balanced" in that it is self-sufficient in both process fuel and hydrogen requirements. Process fuel and hydrogen are produced by gasifying the coke and by reforming C_1-C_2 gases from the liquefaction process.

The process is simple, and critical processing steps are adaptations of Exxon's base in petroleum refinery technology. Distinguishing features are the decoupled configuration of the liquefaction and catalytic hydrogenation sections and the use of vacuum distillation for solids/liquid separation. The catalyst does not contact coal minerals or high-boiling liquids, thereby leading to longer catalyst life at high activity. Use of hydrogenated rather than unhydrogenated recycle solvent produces a very significant improvement in process operability, particularly in downstream processing vessels. Also, hydrogenated solvent produces higher distillate product yields than unhydroge-

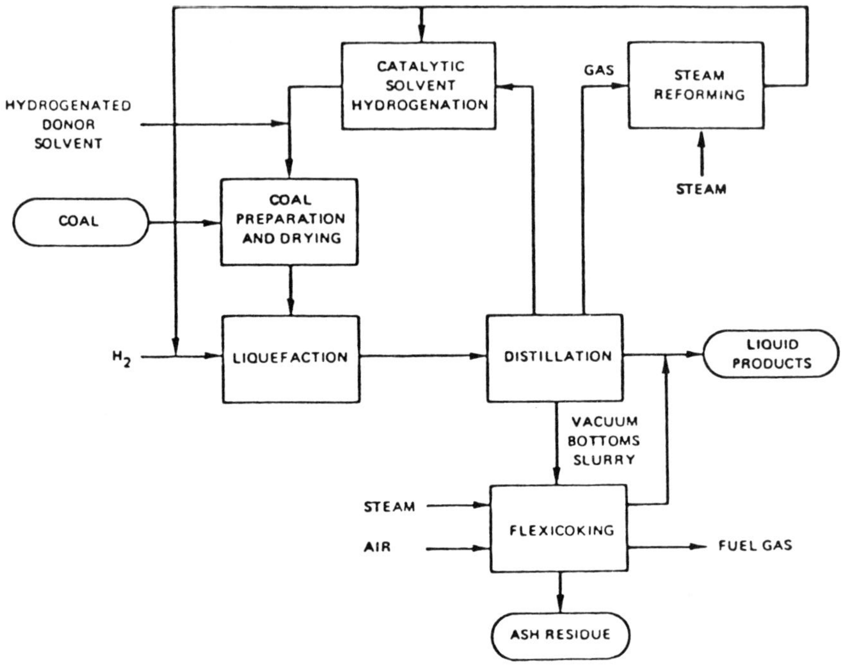

Fig. 3-8. Donor solvent liquefaction process.

nated solvent. The use of mechanical separation devices for solids/liquids separation is avoided.

The process gives high yields of low-sulfur liquids from bituminous or subbituminous coals or lignites. For Illinois bituminous coal, the liquid yield is 2.6 barrels of C_4+ liquid per ton of dry coal. Ammonia and elemental sulfur are the only by-products of significance. By varying liquefaction conditions or adjusting solvent properties, product distribution may be varied over a wide range.

An estimated heat balance is given by Table 3-4. Diagrammatically, this is illustrated by the heat flow diagram, Fig. 3-9.

Performance

Currently, there are no coal liquefaction facilities operating in the United States. Therefore, all projections are based on technology still in the development stage. An estimated efficiency range of 60–70% for direct liquefaction processes has been calculated.

It is difficult to provide realistic and confident estimates for coal conversion efficiencies. Also, efficiencies reported often do not indicate whether the value is for a plant that purchases all, part, or none of the supplemental energy needed in the conversion. (These purchases include electricity, steam, and other utilities.)

An efficiency range of 60–70% for direct hydrogenation (H-Coal) has been calculated by H. Perry. Overall energy efficiencies for various coal liquefaction processes are in

Table 3-4. Estimated Heat Balance for a Commercial-Scale EDS Plant

	Btu/day (10^6 Btu)	Percentage of Total Energy Input
System Products		
Liquids	323,071	61.72
Sulfur, ammonia	8,309	1.59
System Losses		
Ash, combustible and sensible heat	26,882	5.13
Stack losses	20,039	3.83
Energy losses via water and air	136,853	26.14
Liquefaction and solvent hydrogenation (9.80%)		
Flexicoking (6.44%)		
Hydrogenation and recovery (6.72%)		
By-product recovery, offsites, and miscellaneous (3.18%)		
Other miscellaneous	8,309	1.59
Energy Input		
Coal (cleaned)[a]	488,761	96.37
Electrical power[b]	34,702	6.63

[a] Coal—Illinois No. 6; 10,574 Btu/lb as-received prior to cleaning.
[b] Power based on 8500 Btu/kWh to generate.

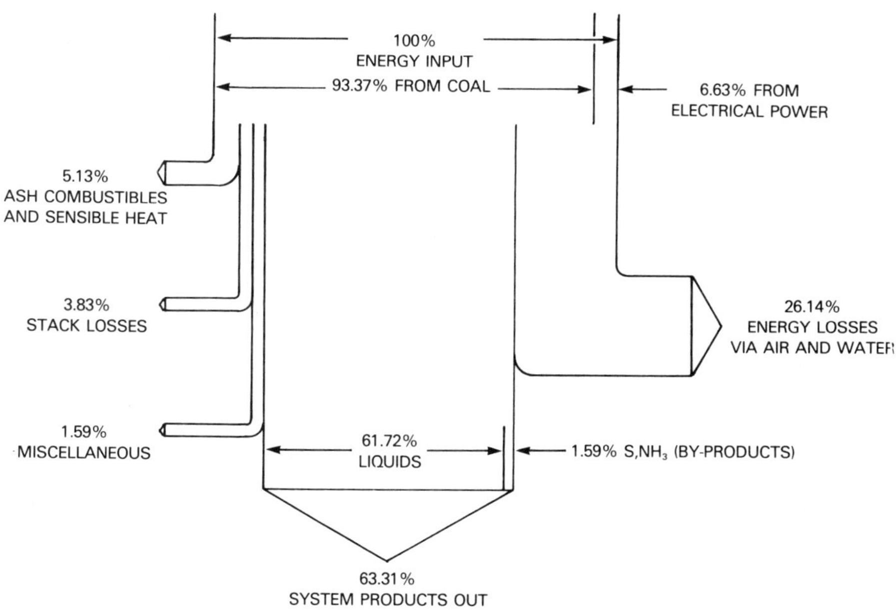

Fig. 3-9. Energy flow diagram for a commercial-scale EDS plant.

the 62–69% range. Limiting factors that can reduce efficiencies significantly are specific to each process and output mix and must be separately computed.

Liquefaction Production Potential

The United States, because of economic and strategic considerations, must significantly increase the utilization of coal, our largest energy resource. To this end, President Carter, in July 1979, called for the development of facilities for the production by 1990 of 2.5 million barrels per day (B/D) of oil substitutes from a variety of sources. Of this total, the range of 1.0–1.5 million B/D was suggested as a goal for the production of liquids or equivalent from coal.

In line with President Carter's energy message, the Department of Energy undertook four separate and independent studies to identify the critical elements in a liquefaction operation and to assess the risks in using commercially available equipment in the construction of coal liquefaction plants which would achieve the 1990 production goal. These studies were documented by the following reports:

- *Production of Synthetic Liquids from Coal: 1980–2000: A Preliminary Study of Potential Impediments, Final Report*, Bechtel International Inc., San Francisco, California, Contract No. ET-78-C-01-313/M001.
- *Project C-Clear, An Assessment of Commercial Coal Liquefaction Process Equipment Performance and Supply*, Mechanical Technologies, Inc., Latham, New York, Contract No. DE-AT21-77MC0-2165, Task 11.
- *Synthetic Transportation Fuels from Coal*, Mound Facility, Monsanto Research Corporation, Miamisburgh, Ohio.
- *Feasibility Assessment—Production of Synthetic Fuels from Direct and Indirect Liquefaction Processes*, UOP Inc./Systems Development Corporation, McLean, Virginia, Contract No. ET-78-C-01-3117.

Since the studies which produced these reports were performed under somewhat different scenarios, DOE undertook the task of compiling a summary of each study report and correlating and developing a consensus of the findings in terms of what constitutes the impediments to achieving the 1990 coal liquids production goal. This section presents the TRW results of that effort.

Table 3-5 indicates the impediments considered in the reference reports and summarizes the conclusions reached by each study. Bechtel and UOP/SDC, because of their contacts with architect/engineering firms, provided broad coverage of the potential problems with a construction program orientation. MTI, because of their expertise in equipment performance and availability assessments, covered these aspects in their analysis. The Mound study supplements the MTI study to yield a "third" entity to be compared with the Bechtel and UOP/SDC findings.

To simplify development of consensus conclusions, the potential impediments listed in Table 3-5 were grouped into the following categories:

- Equipment
- Coal supply
- Manpower
- Water supply

Table 3-5. Summary Findings

Potential Impediment	Reference Reports			
	Bechtel	MTI	UOP/SDC	Mound
Critical materials availability	P		N	
Equipment availability	P	P	P	
Equipment performance		P		
Coal supply	N		N	
Water supply	PQ		PQ	PQ
Technical labor availability	P		P	
Construction labor availability	P		PQ	
Financial/capital availability	N		N	P
Environmental requirements	P		P	P
Permits and licensing	P		P	P
Transportation			X	X
Siting	PQ		PQ	PQ
Technology uncertainties		P		X
Product costs				X
Operating costs			N	X

Legend

X Discussed only—no conclusions
N No impediment—will be met in normal course
P Potential impediment—requires action
PQ Potential impediment in certain locales/regions

- Material
- Financial
- Environmental
- Permits and licensing

The extent of the categories has been summarized in the following generalized statements.

EQUIPMENT

Two of the four studies (Bechtel and UOP/SDC) concluded that there are significant potential impediments to the 1990 production goal relating to equipment availability/obtainability. Specific items expected to be in short supply are distillation, heat exchangers, compressors, large pumps, and thick-wall pressure vessels. However, the two studies point out that the anticipated supply shortage of this equipment could be overcome by giving necessary timely incentives to the industry to increase its production capacity.

The other studies (MTI and Mound) concluded that there are significant potential impediments to the 1990 production goal in the areas relating to equipment performance and/or scaleup. Specific items of equipment subject to this risk are gasifiers, extractors, hydrotreaters, oxygen compressors, and coal slurry heaters.

COAL SUPPLY

All studies concluded that the supply of coal, from both reserves and mining capacity, will not be an impediment to the U.S. 1 million B/D production goal. The mining industry has ample expansion capability to meet the necessary mine production to feed coal liquefaction plants.

Comparison Matrix

Studies	Study Emphasis and General Conclusion	Equipment Considerations
Bechtel Study	• Requirement Aspects of One and Three MMBPD Coal Liquids Industry by the Year 2000 Was Studied • No Major Impediments Seen for 1 MMBPD Scenario • Potential Impediments for the 3 MMBPD Scenario. Also in the Areas of Materials and Equipment Manpower Plant Permits and Licenses. (The 3 MMBPD Scenario Assumes Production Level of 1 MMBPD by 1990)	• Potentially Critical Equipment Whose Availability is of Concern for the 3 MMBPD Program Are Drag Lines, Heat Exchangers, Pressure Vessels, Centrifugal Compressors, Alloy and Stainless Steel Valves
UOP Study	• Constraints to Producing 1 MMBPD by 1990 Was Examined • The Limitation on the Availability of Air Separation Plants, Compressors, and Technical Manpower are the Major Constraints Identified	• Availability of Air Separation Plant Fabrication Capability Is the Most Severe Limitation. Limited Number of Vendors Serve the Market. Recommend Government Funding of New Capacity or Provide Incentives for Existing Fabricators to Expand Capacity • U.S. Capacity for the Production of Compressors Will be Strained • Need for Pressure Vessels Will Tax Fabrication Capacity in Certain Wall Thickness and Weight Categories But Given Sufficient Incentives Industry Capacity can be Expanded
MTI Study	• Use Scenario of 1 MMBPD by 1990 • Considers Performance Risk of Each Piece of Equipment and Supply Risk Based on Scenario • Concludes Substantial Risk Exists in Both Commercial Equipment Performance and Supply	• 4% of the Identified Items are of High Performance Risk. Included in this Category are Gasifiers, Extractors, Extract Hydrotreaters, Oxygen, Compressors, and Coal Slurry Heaters • 15% of the Identified Items are of Medium Performance Risk • Types of Equipment with High Performance Risk are also Those With High Supply Risk • Supply Risks for Oxygen Plants and Large Reciprocating Pumps are also High
METC Mound Study	• General Discussion of the Production of Synthetic Transportation Fuels from Coal • Report Consists of Seven Sections: Market, Technology Status and Exploitation, Technological Problems and Uncertainties, Economics, Financial Barriers, Socioeconomic Impact, Environmental • No Particular Scenario Used	• Recognizes Problems Relating to Scaleup. Dry Coal Feeding to Pressurized Vessels. Separation of Solid and Liquid Products From the Reaction Mixture

Coal Supply Considerations	Manpower Considerations	Water Supply Considerations
• Coal Requirements for the 3 MMBPD Program would be Approximately One-Fourth of the Total Coal Production at that Time	• For the 3 MMBPD Case Chemical Engineers Working in this Area must be Increased by One-Third by 1985 • Pipefitters, Welders, Boiler-makers, Electricians will be in Short Supply in Certain Regions. 30 to 80 Percent of the Available Manpower will be Required by the Liquids Program	• Water Supply should not be an Impediment Since Only 5 Percent of Current Consumption would be Required by this Coal Liquids Program
• No Major Problems Foreseen	• The Availability of Qualified Degree and Nondegree Personnel will be a Major Constraint	• Water Requirements and the Need to Establish Local Infrastructure Limit the Number of Economically Operated Coal Liquid Plants in Western U.S. • Water for Coal Conversion Plants in the Western Coal Regions will be a Problem and an Expense but not a Deterrent • Smaller Liquid Plants (10,000 T/D) Should be Located in Appalachia. Larger Plants (30,000 T/D) could be Located in the Illinois Basin
• Not Addressed in this Study	• Not Addressed in this Study	• Not Addressed in this Study
• Recognizes the Need for Increased Mining Capacity	• Recognizes Potential Manpower Shortage	• Appears to Indicate that Water Supply is not a Major Impediment. Recognizes that Problems are Highly Site Specific

Comparison Matrix (*continued*)

Studies	Material Considerations	Financial Considerations
Bechtel Study	• Chromium Supply for use in Equipment could cause Concern as all the National Requirement is Currently Imported. The 3 MMBPD Scenario would Utilize About 3 Percent of the National Chromium Consumption	• In the 1980 – 2000 Period for the 3 MMBPD Scenario the Capital Required will be About $201 Billion or One-Eighth of that Required by the Energy Industry. This Requirement is Within Manageable Limits
UOP Study	• Critical Material Analysis for only 12 Fischer – Tropsch Plants Supplying 1 MMBPD of Coal Products Conducted • Concludes that the 1985 Demand for Cobalt and Molybdenum will be Less Than 1% of the Total National Demand • 1985 Demand for Nickel and Chromium will be 1.5% and 7% Respectively of the Total National Demand	• Requires Large Capital Commitment Irrespective of the Liquefaction Process Chosen. Capital Cost is not Sensitive to Process Used
MTI Study	• Not Addressed in this Study	• Not Addressed in this Study
METC Mound Study	• Recognizes that High Temperature and Pressure Conditions Under which Synfuels Reactions occur Result in Equipment and Materials Problems Need Continued Work to Improve the Reliability and Stability of Special Materials for this Technology	• Discusses Various Types of Incentives Such as Price Guarantee, Tax Credits, and Accelerated Depreciation Loan Guarantees • Does not Favor Direct Grants, However Recognizes Small Technology Companies may Desire Grants

Environmental Considerations	Permits and Licensing Considerations
• Recognizes the Need to Comply with Various Environmental Considerations but does not Consider them to be Major Impediments	• Reporting Requirements must be Reduced, Review Periods Shortened and Standards Relaxed
• Environmental Acceptances of Coal Conversion Plant is Uncertain Because of Lack of Emission Data	• Environmental and/or Siting Considerations may Ultimately Prove to be the most Significant Barrier to Meeting Coal Liquid Production Goal As Federal Requirements are Expedited State Imposed Requirements may Impose Delays in Permitting Procedures
• Not Addressed in this Study	• Not Addressed in this Study
• Socioeconomic and Environmental Impact must be Taken into Consideration While Locating Liquefaction Plants • Recommends Continued Study of the Potential Hazards Due to Exposure and Emissions of Chemicals from Liquefaction Plants	• Has Presented a Table Listing Relevant Federal Environmental Regulations and a Brief Analysis of the Impact which they will have Upon all Liquefaction Facilities Regardless of Location • Recognizes that State and Local Regulations have to be Reviewed on a Site-by-Site Consideration

MANPOWER CONSIDERATIONS

Only Bechtel and UOP/SDC studies examined this issue in detail for the specific production goal addressed. Both concluded that supply of trained technical manpower (engineers) will be severely taxed, but that to some extent, this hardship could be eased by attracting trained personnel from other industries. Also, the two studies concluded that construction labor will be in short supply, but could be somewhat alleviated by reducing the training period of the craftsmen and by extending the work week.

WATER SUPPLY

Three of the four studies (Bechtel, UOP/SDC, and Mound) concluded that the water supply requirement would not be a deterrent to the coal liquids production goal. However, the Bechtel and the UOP/SDC reports conclude that water availability in the western coal regions would not allow siting the total plant capacity for the 1 million B/D production goal in this region.

MATERIAL CONSIDERATIONS

Bechtel and UOP/SDC reports concluded that the availability of critical materials, i.e., cobalt, nickel, molybdenum, will not be a barrier to achieving the coal liquids production goal. The availability of chromium may be a constraint because this material is currently imported and the anticipated demand may reach 7% of the total national trend.

FINANCIAL CONSIDERATIONS

Two of the four reports (Bechtel, UOP/SDC) concluded that the availability of capital will not be a barrier to the coal liquefaction production goal.

ENVIRONMENTAL CONSIDERATIONS

The lack of emissions data relating to the coal liquids production was viewed by three studies (Bechtel, UOP/SDC, and Mound) as creating uncertainties and possible delays in environmental acceptance by the regulating bodies.

PERMITS AND LICENSING CONSIDERATIONS

Bechtel, UOP/SDC, and Mound reported that the permitting and licensing group's requirements will impose time and schedule constraints on construction planning. UOP/SDC and Mound indicate that state rather than federal requirements may impose greater hindrance to site development.

Bechtel International Inc. Assessment

Bechtel analyzed the resource requirements and reviewed major obstacles to the daily production of several million barrels of synthetic coal liquids. Furthermore, the study *Production of Synthetic Liquids from Coal, 1980–2000: A Preliminary Study of Potential Impediments, Final Report*, Bechtel International Inc., San Francisco, California, December 1979, sought to identify the industry infrastructure needed to support the commercial readiness of the coal liquefaction process. A selected list of critical resource items and their domestic/international availability was developed and examined, and the impact of their supply on the various synthetic coal liquids programs was evaluated.

The study approach was to develop representative, or generic, direct and indirect coal liquefaction conceptual designs from available technology and costs data. The generic processes were to employ technology that would be considered "commercial" by the mid or late-1980s. The size of the generic construction mobilization was considered reasonable at the outset of the program. The product slate was directed toward unrefined liquid fuels rather than diesel oil or gasoline. The generic processes were to use a wide range of coals to permit siting in most coal-producing regions across the country.

Because of the dearth of conceptual design data in the literature, Bechtel developed generic plant designs by using in-house design expertise. Bechtel assumed that because it is first-generation technology, the indirect process will be used at the outset of the liquids program, and the direct process will be introduced 2–4 years later as a second-generation technology. The products of either of these processes will be limited to boiler fuels and/or other liquid products that require further upgrading. Both processes use all by-product gases within the plant; consequently, these facilities are energy independent. Cost estimates were developed from equipment lists, as well as material and labor estimates, which enable the determination of an order-of-magnitude cost estimate and target plant construction schedule for both processes.

TYPE OF PLANT CONSIDERED

The generic plant was assumed to be of 60,000 B/D size and considered to be modular for multiple plant expansion, contingent on coal and water supply and environmental conditions. Either the process-type plant was to be located at the coal mine mouth or in close proximity to coal mines, minimizing transportation requirements.

Indirect Coal Liquefaction

Indirect coal liquefaction utilizes two major processing steps. First the coal is gasified to produce a synthesis gas consisting mostly of hydrogen (H_2) and carbon monoxide (CO). The synthesis gas is then reacted in the presence of a catalyst to form liquid products. The product slate for this process is approximately 58% hydrotreated naphtha, 33% C_3-C_5 hydrocarbons, and 9% heavy alcohols. Figures 3-10 and 3-11 present schematic illustrations of this process mode.

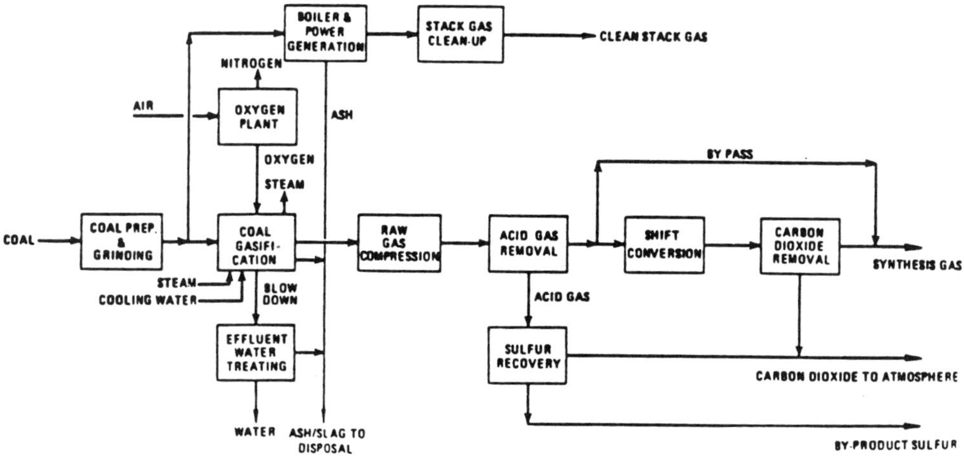

Fig. 3-10. Indirect process—coal gasification.

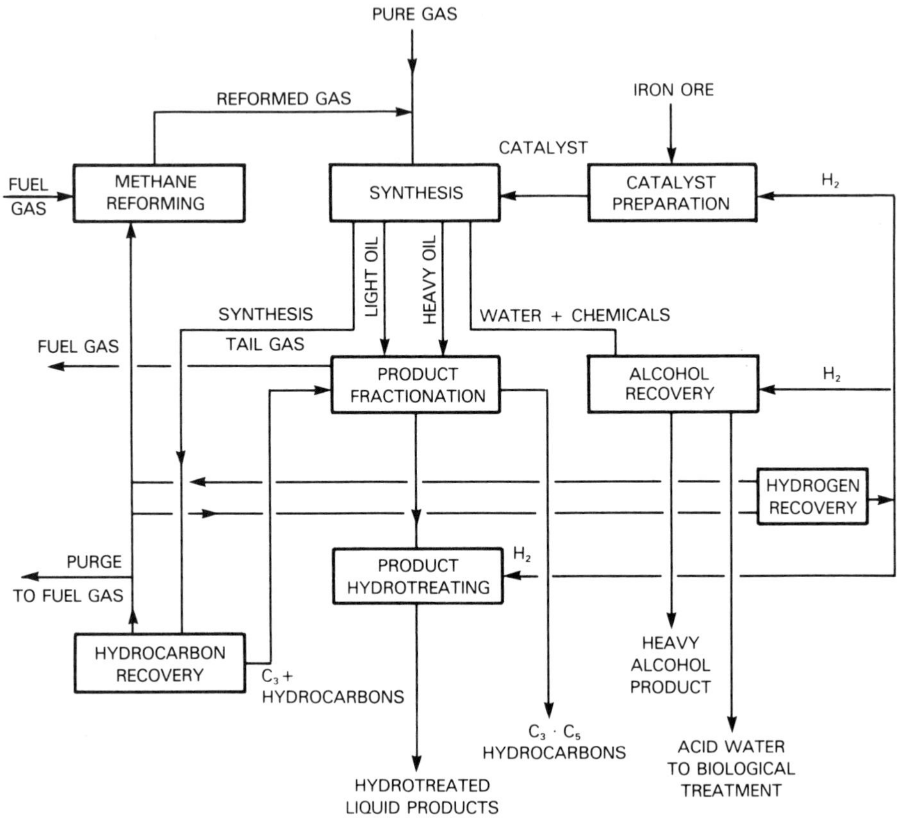

Fig. 3-11. Indirect process—liquid hydrocarbon synthesis.

Direct Coal Liquefaction

Direct coal liquefaction employs a solvent oil to dissolve and hydrogenate coal to produce liquid products. The product slate for the direct process is approximately 60% low-sulfur oil, 33% C_5−400F naphtha, and 7% C_3-C_4 LPG. Figure 3-12 is a schematic illustration of this process mode.

SCENARIO EMPLOYED

The construction schedules were developed in two phases by Bechtel. In the first phase, national production scenarios were set. DOE selected intermediate production goals of 1 million B/D by 1990 for the 3 million B/D scenario for the year 2000, and 350,000 B/D for 1990 for the 1 million B/D scenario goal for 2000. In constructing the schedules it was assumed that no commercial-scale plant could be completed before the end of 1985. For the less aggressive program implied by the 1 million B/D scenario, it was assumed that no plant would be completed before 1986. Figure 3-13 shows the respective production levels of the two scenarios over the 20-year period through the year 2000.

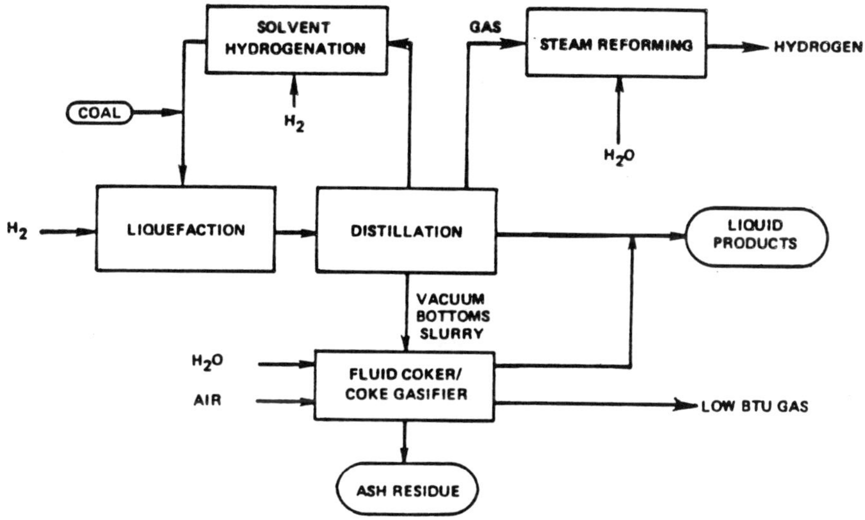

Fig. 3-12. Direct coal liquefaction plant.

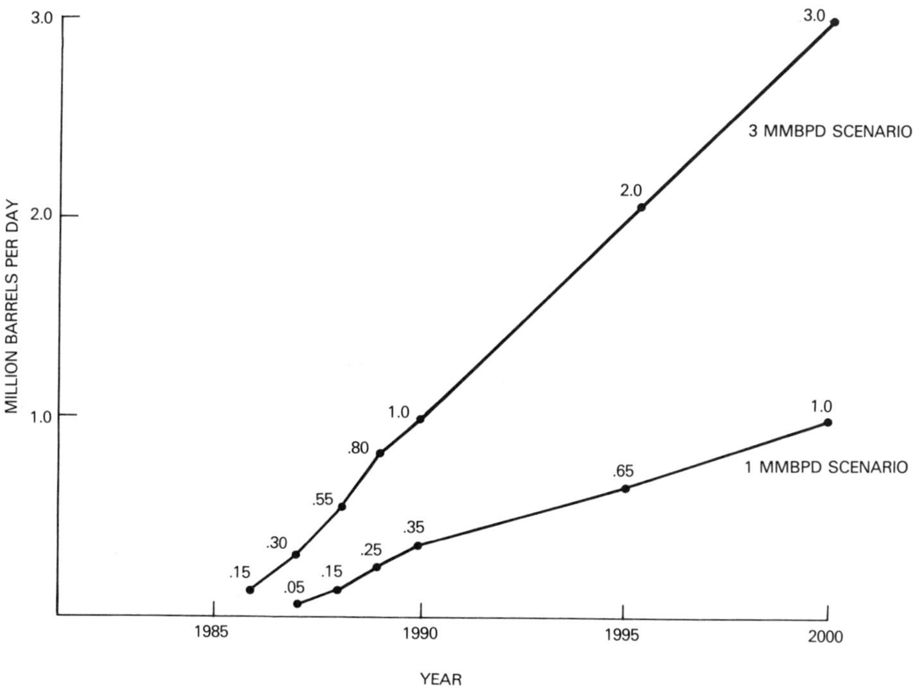

Fig. 3-13. Coal liquids production (at the start of the year).

The only initial commercially available technology is the first-generation indirect process. With DOE concurrence, the first direct liquefaction process plants were assumed to be introduced in 1987. For the 3 million B/D scenario, the 1 million B/D production in 1990 was to consist of 600,000 B/D from indirect plants and 400,000 B/D from direct process plants.

Table 3-6 shows the facility completion schedule for the two scenarios. As shown, these particular scenarios imply a rapid growth in construction in the first decade of the 1980 to 2000 program, followed by a period of stagnant or even declining activity. Bechtel states that the schedule(s) does not represent an assessment of "best" or "most likely" production levels.

The second phase or scenario development required that the national scenarios be disaggregated to regions within the United States. These regional scenarios (see Table 3-7) are an attempt to define a reasonable distribution of liquefaction plants, as there exists no consensus about specific plant location. To ensure reasonableness of the scenarios, a general rule was used that coal liquids production in any region would not be assumed to be more than had been postulated in at least one other study (Stanford Research Institute, *Impacts of Synthetic Liquid Fuels Development: Assessment of Critical Factors*, 1977).

Key features of the scenario are:

- Overall energy consumption grows at a 1.6% annual rate during the 1980–2000 period.
- Oil consumption peaks in the early 1980s and then declines at a 1.1% average annual rate to 14.7 million B/D by 2000. Domestic oil production (other than from coal liquids) is maintained at about present levels, as increases in enhanced oil recovery and shale oil compensate for declines in conventional oil production.
- Coal production rises at a 6 percent annual rate from about 700 million tons per year (MMT/Y) in 1979 to 2460 MMT/Y in 2000 (including requirements of 493 MMT/Y for the coal liquefaction of the 3 MMBPD scenario). Of the year 2000 production, 1200 MMT/Y is assumed to be from eastern mines and 1260 MMT/Y from western mines.
- Electric power generating capacity increases at a 2.4% annual rate to 950 GWe in 2000. Coal-fired power increases from 213 GWe in 1978 to 450 GWe in 2000.

IMPEDIMENTS CONSIDERED

Potential impediments to the coal liquids program are described in the following paragraphs.

Engineering Manpower

Engineering Requirements. The estimated engineering design manhour requirements for the direct and indirect liquefaction processes are, respectively, 1.5 and 3 times greater than refinery requirements. Contributing to the greatly increased engineering effort are process complexity; handling of solid, mixed, and liquid phases; high temperatures and pressure; and ash and waste disposal.

Table 3-6. Facility Completion Schedules

	Number of Plants by Year															Total
	1985	86	87	88	89	90	91	92	93	94	95	96	97	98	99	85–99
3 MMBPD Scenario	3	3	5	5	4	4	4	4	4	4	4	4	4	4	5	61
1 MMBPD Scenario		1	2	2	2	1	2	1	1	2	1	1	2	1	2	21

Table 3-7. Regional Scenarios

Region	Year															Total
	1985	86	87	88	89	90	91	92	93	94	95	96	97	98	99	
3 MMBPD Scenario																
3 South Atlantic			1	1		1		1	1		1		1			7
4 East North Central	1	1	2		2	1	2	1	1	1	2	1	2	1	3	21
5 East South Central		1		1		1		1		1		1		1		7
6 West North Central	1		1	2		1	1		1	1		1		1	1	11
7 West South Central									1			1			1	3
8 Northern Mountain	1	1	1	1	2		1	1		1	1		1	1		12
Total	3	3	5	5	4	4	4	4	4	4	4	4	4	4	5	61
1 MMBPD Scenario																
3 South Atlantic							1				1					2
4 East North Central			1	1		1				2			1		2	8
5 East South Central		1						1				1				3
6 West North Central			1		1		1									3
7 West South Central														1		1
8 Northern Mountain				1	1				1				1			4
Total	0	1	2	2	2	1	2	1	1	2	1	1	2	1	2	21

Greater engineering manpower is required for the indirect process since it is, in effect, two processes in series, gasification and synthesis, requiring considerably more equipment, piping, and instrumentation engineering.

Analysis of Engineering Manpower Requirements. The 3 million B/D production scenario could require up to 1300 additional chemical engineers for coal liquefaction process work by 1984; this represents a 65% increase in the major engineering force in 6 years. It appears that the supply of other engineering disciplines should be adequate.

The 3 million B/D scenario by 2000 will tax the supply of chemical engineers for process and project work and will be further constrained by the projected growth in the chemical, petrochemical, and nuclear-fuel-processing industries.

Manual Labor Considerations

Peak employment of craftsmen occurs during the period 1980–1990 for the coal liquids construction period. As a consequence of the current shortage of skilled construction manpower, especially pipefitters, pipefitter–welders, boilermakers, boilermaker–welders, and electricians, a shortage of manpower for a coal liquids construction program could develop rapidly in the areas with the greatest possible concentration of plants. The regions most affected will be the East North Central/East South Central (Ohio River Valley), the West North Central, and the Northern Mountain Regions.

During the past 10 years, a sharp decline in the willingness of craftsmen to follow the work has drastically altered the historical manpower–mobility pattern in the construction industry. Major factors that influence decisions on where to work include higher hourly wages, better benefit packages, the proximity of the job to home or to a favorite recreational area, and the housing and recreational facilities that are available at the project.

Impact of the Coal Liquids Program. The 3 million B/D coal liquids construction program builds rapidly from the zero level in 1980 to the employment of 73,000 manual employees nationally by 1986, the peak year, accounting for approximately 2% of the manual construction employment of 1985.

- Management could make better use of the labor force through a stabilization agreement that eliminates wasteful labor practices.
- Use of training programs could be accomplished by the upgrading of existing craftsmen through use of on site training programs and expansion of current apprenticeship programs. The primary skill shortage will be welding.
- Expanded use of "nonjourneymen" or "helpers" to perform the less skilled work now done by fully qualified journeymen.
- Initiation of a comprehensive recruitment program to attract minorities and females to the construction industry.
- A program of constant coordination with vocational and trade schools, as well as government training programs to ensure optimal use of existing training facilities.

Materials and Equipment

The majority of the new, major energy-related expansions have been foreign with domestic activity flat or generally declining. The following factors have contributed to slowing the expansion rate: controls of return on investment; environmental restraints and their associated delays of approach; lack of definitive energy policy; and foreign competition.

Potentially Critical Items

Chromium—used to form alloys with iron, nickel, or cobalt. The United States imports over 92% of its chromium, as domestic costs currently make it economically infeasible to utilize U.S. resources. Availability of chromium is dependent on the world market.

Alloy and stainless-steel valves—made of cast, forged or fabricated materials. Restraints that valve manufacturers must overcome to meet the demand for a specialized valve are

- Availability of quality castings and forgings
- Availability of chrome, molybdenum, and cobalt
- Availability of qualified machinists or time to train new ones
- Sufficient time to plan physical expansion, if required

Draglines—in large-size excavators capable of stripping or excavating overburden and/or ones of various types including coal. Draglines are basically manufactured from iron and steel castings, forgings, and fabricated mill shapes. Draglines for the coal liquids program in 1987 could be accommodated with a lead time for ordering of 2–2½ years.

Pumps and drivers (less than 1000 hp)—pumps are generally manufactured from ferrous castings, forgings, and machined posts. In this power range the drivers are almost exclusively electric motors. There is more than adequate capacity. Three shift operations roughly double single-shift production. European and Japanese manufacturing capacity is availability.

Centrifugal compressors and drivers (less than 1000 hp)—compressors are manufactured from iron and steel castings, forgings, and fabricated posts. Suppliers expressed confidence that U.S. industry could accommodate the project peak requirements in the year 1984.

Heat exchangers—devices used to transfer heat from one medium to another. The heat exchanger industry has recently experienced significant declines in production. Two major constraints are (1) manpower to operate the machines and to weld and (2) availability of heat-treated plate for the heat-exchanger plates. Requirements for the 3 MMBPD coal liquids program could strain domestic capacity. Mitigating factors are

- Foreign manufacturing capacity is high.
- The industry can rapidly expand production as demand develops.
- The potential exists for reducing heat-exchanger requirements through the use of alternative processes.

Pressure vessels—a tank or cylinder or closed shell that is subjected to use in fluctuating pressures. This industry caters to petrochemical needs and is somewhat reactive to demand; consequently, the heavier the tangible demand, the greater the expansion to meet the challenge.

- Availability of chromium is dependent upon world market; therefore, political changes or economic sanctions could disrupt the U.S. supply.
- Foreign suppliers may have to be called upon to meet some of the requirements in peak demand years.

- Consensus of energy-related manufacturers is that they can satisfy the requirements of the coal liquids programs, provided firm program commitments are issued on a timely basis.

Capital Requirements

For the combined 3 MMBPD and gasoline scenario, cumulative capital requirements (in constant 1978 dollars) are projected to be $1530 billion for the energy sector in the 1980–2000 period. Of this, $201.3 billion are related to the development of coal liquids.

Annually, total capital requirements increase at an average rate of 2.7% in the 3 MMBPD combined scenario from about $52.5 billion in 1980 to $884 billion in 2000, of which $12.2 billion are related to coal liquids. In the 1 MMBPD combined scenario, annual capital requirements increase at a 2.3% rate to a $82.2 billion level, of which $4.6 billion are related to coal liquids.

These estimates include the capital required for the design, preparation, construction, and startup of both new and replacement plants.

Assessment of Capital Availability. Capital expenditures have increased in real terms, at an 8.7% annual rate (adjusted by the implicit price deflator for nonresidential fixed investment—NRFI), during the 1970s. Even though capital requirements are projected to increase at a faster rate than energy consumption, the rate of growth in energy consumption is such that project capital requirements for energy will probably decline relative to other economic sectors, given conditions of sustained economic growth.

Environmental Reports and Permit Acquisition

The preparation and acquisition of major permits indicate that a minimum of 3 years lead time is necessary prior to initiation of construction. Major environmental permits typically required include air permits through the state and/or local air quality management district, water pollution control permits, mining and extraction permits, and Corps of Engineers permits.

- *Licensing Procedures and Problems:* Assumes all air, water quality, and waste management as well as environmental reporting regulations will apply to the siting and construction of each 60,000 B/D coal liquid plant. Relaxed environmental regulations will be in the form of reduced reporting requirements; shortened review periods for reports; and relaxed standards, with best available control technology and lowest possible emissions.
- *Site Selection:* At present there is no federal law or regulation with respect to overall energy facility siting. At least 26 states have power plant siting laws. Selection of the most feasible site for the facility is based upon engineering, economic, environmental, and resource limitations. Areas which include national or state parks or grassland, areas of critical or sensitive habitat, cities and towns, airports and transportation facilities, and prime agricultural lands are either eliminated or considered poor for location of industrial facilities.
- *Air Quality:* Air quality is managed by state and local air quality management districts under the Clean Air Act and amendments. Standards established by EPA include five major pollutants: carbon monoxide (CO), oxidants (OX), total suspended particulates (TSP), sulfur dioxide (SO_2), and nitrogen oxide (NO_x).

- *Water Quality:* The Clean Water Act requires that industrial plants achieve zero discharge to public waters without a permit, and imposes stringent pollution control requirements on all discharges.
- *Solid Waste Management:* Depending on the coal source, the amount of material residue remaining from the liquefaction process will vary. The type of waste expected from these operations consists of digested biological sludge from a biological oxidation unit, oily sludge from the API separator and dissolved air flotation unit, ash from the fluid coker/coke gasifier, and solids removed from the boiler feedwater cold-lime treating unit blowdown. There will be wide variation from one state to the next in licensing requirements for solid waste disposal facilities.
- *Environmental Impact Statement (EIS):* The EIS preparation program may be accompanied by an environmental baseline monitoring/study program as well as a permit acquisition program. Results in identification of required local, state, and federal permits. Major environmental issues for synthetic liquids processing are worker health and safety; potential public health risks from process materials, by-products, and emissions; and management of solid wastes and leachates.
- *Environmental Impacts of Coal Liquids Development:* The following generic impacts have been identified: Coal mining impacts and potential for reclamation; potential public health risks from process materials; management of solid wastes and leachates; air quality with respect to attainment and nonattainment areas and the ambient levels of particulates, sulfur oxides, PSD limits, fugitive dust, oxides of nitrogen, and hydrocarbons; water quality, both surface and ground supplies; worker health and safety; and socioeconomic impacts such as changes in population and labor force characteristics, community and regional economic characteristics, and the boom-town effects.

Water Resources

Availability of water for coal liquids plants can serve as a limit or constraint with regard to development within various regions. In some areas of the West, water shortages are already apparent, and the water development issue is one of extreme political sensitivity.

Water Requirements. Water requirements for typical direct coal liquefaction plants are estimated at approximately 9 million gallons per day (MGD) (10,100 acre feet per year). The indirect coal liquefaction process requires 1.7 times as much water as the direct process or 15.3 MGD (17,250 acre feet per year).

Water Availability. A surplus of 14,000 MGD through the year 2000 exists in the West Virginia, Ohio, and Kentucky (Ohio River Valley) region. Illinois in the Upper Mississippi Water Resource Region projects a surplus of 500 MGD. For either scenario, adequate surface water is available in these regions without reducing in-stream flows for maintaining optimal fish and wildlife conditions.

Total demand by coal liquids facilities in North Dakota, Wyoming, and Montana is 259 MGD, just over 10% of the apparent surplus. Under the prevailing system of purchased water rights, most of the available surface water supply in this area has already been allocated.

In Texas, water consumption requirements for coal liquids facilities are estimated at 28 MGD in the year 2000, which is much less than the apparent surplus.

Water requirements for members of the labor force, their families, and towns must also be evaluated.

Other Considerations

Process Materials Requirements. The coal liquefaction process has highly specific requirements for process catalysts and chemicals. Although there may be difficulties in obtaining specific process materials for specific processes, process material availability is unlikely to be a constraint to the overall coal liquids programs.

Transportation. Coal transport requirements will be minimal, since it is assumed that coal liquids plants would be mine-mouth or in close proximity to mines.

Annual Operation and Maintenance Costs. Costs include those for labor, materials, equipment, utilities, and various miscellaneous items, but excluding fuel costs and debt service. These costs are too small to have any significant macroeconomic impact and could not be considered constraining from an overall economic viewpoint; these costs are a small portion of total annual costs.

BECHTEL CONCLUSIONS

The overall conclusion, summarized in Table 3-8, is that either the 1 or 3 MMBPD scenario could be implemented without major dislocation to the economy as a whole. However, action must be planned and taken to overcome potential labor, material, and equipment impediments. Licensing and permitting procedures for liquefaction plants must be expedited. These actions are particularly necessary if the 3 MMBPD program is to be achieved:

- *Engineering Manpower:* For the 3 MMBPD coal liquids scenario it is concluded that the supply of chemical engineers will be taxed. The supply of other engineering disciplines should be adequate, although this scenario will increase the competition for qualified engineers of all types.
- *Manual Labor:* To meet the labor requirements in the West North Central and Northern Mountain regions, there will need to be a maximization of the present labor force, enhanced management–labor–government cooperation, expanded use of nonjourneymen, expanded apprenticeship and training programs, and programs and coordination with vocational schools.
- *Equipment and Materials:* Coal availability will not present a direct constraint to the coal liquids program. Sufficient water supplies can be made available to supply the requirements of either scenario.
- *Capital Requirements:* Capital requirements for the coal liquids program are expected to be manageable.
- *Environmental Issues and Permit Acquisition:* Although the environmental impacts have not yet been definitively determined, it is expected that coal liquids plants can be designed and constructed to meet environmental conditions. Unless the permitting lead time can be reduced, the 3 MMBPD scenario schedule will be difficult to achieve.

Mechanical Technologies Inc. Assessment

The MTI report *Project C-Clear: An Assessment of Commercial Coal Liquefaction Process Equipment Performance and Supply*, Mechanical Technologies, Inc., Latham,

Potential Impediments Matrix (Case #1)

	Engineering Manpower	Manual Labor	Materials and Equipment	Plant Permits and Licenses
1 million B/D—Year 1990	Resources Adequate—No Foreseeable Impediments—21 Plants Required			
3 million B/D—Year 2000	Chemical Engineers increased by one-third by year 1985. Supply of chemical engineers for process design and project work will be taxed.	30 – 80% of available pipefitters, pipefitters – welders, boilermakers, boilermaker – welders and electricians required (union craftsmen only)	• 490 million tons of coal annually by year 2000 or one-quarter of total coal production at that time. 1980 – 2000 cumulative capital $201 billion or one-third of capital required by total energy industry. • 61 plants required, coal supply availability not a direct constraint to program. • Draglines—88% of present mfg. cap. • Heat exchangers—75% of current mfg. cap. • Pressure Vessels—13% of natl. mfg. cap. • Centrifugal compressors (10,000 hp ea) 13% of mfg. cap. • Alloy & stainless steel valves—8% of valve mfg. cap. • Chromium—3% of natl. consumption (chromium imported) • Suff. water supplies	Reduce reporting requirements, shorten review periods and relax standards while maintaining best available control technology. *Environmental* • Coal mining impacts • Public health risks from process materials • Effects of disposal of solid wastes and leachates • Worker health and safety • Air quality permits (state and/or local) • Mining and extraction permits • Corps of Engineers permits. 3 years lead time.

New York, January 1980, presents an assessment of the risks involved in utilizing commercially available equipment to support a coal liquefaction commercialization effort. The selected approach was to synthesize a generic coal liquefaction process plant configuration including lists and duty specifications of all major equipment items. Each item was then categorized with respect to two areas:

- *Performance risk:* the risk that the commercially available equipment item will not perform satisfactorily in its required service.
- *Supply risk:* the risk that the required equipment item will not be commercially available in the quantities required to support the commercialization effort.

For the categorization effort, the services of an eight-member ad hoc team of experts in process plant equipment were utilized, and their consensus used to assign each equipment item to a low-, medium-, or high-performance risk category. Supply risk categories were determined by comparing the required number of each equipment item to the available foreign and domestic production capacity.

The equipment assessment was carried out only for the process steps and associated equipment specific to the coal liquefaction process. The process steps analyzed were coal receiving, storage, and reclaim; coal preparation; slurry preparation; extraction; hydrogen compression; solids separation; and gasification. Process areas not evaluated were gas processing, solvent hydrogenation, and distillation. Also, consideration was not given to instrumentation, inert gas purge or blanketing, solvent purge and flushing, health and safety, or environmental controls.

TYPE OF PLANT CONSIDERED

The plant considered was a generalized coal liquefaction process synthesized by MTI from published data on commercially viable coal liquefaction processes and is shown in a block flow diagram by Fig. 3-14. The process steps included

- Extraction of coal in the presence of molecular hydrogen, donor solvent, and recycled extract slurry
- Solids separation by settling distillation
- Extract hydrogenation
- Gasification of unextracted coal residue using a generalized process operating with a granular solid feed under slagging conditions with discharge of water-quenched slag
- Gas processing to remove C_1-C_3, CO_2, S, NH_3, acid
- Solvent hydrogenation
- Distillation to separate solvent from light and heavy oil products

The overall plant material balance was as follows:

Feeds	Balance
Dry coal	6,000 T/D
Coal moisture	312 T/D
Oxygen	3,876 T/D
Water	472 G/M

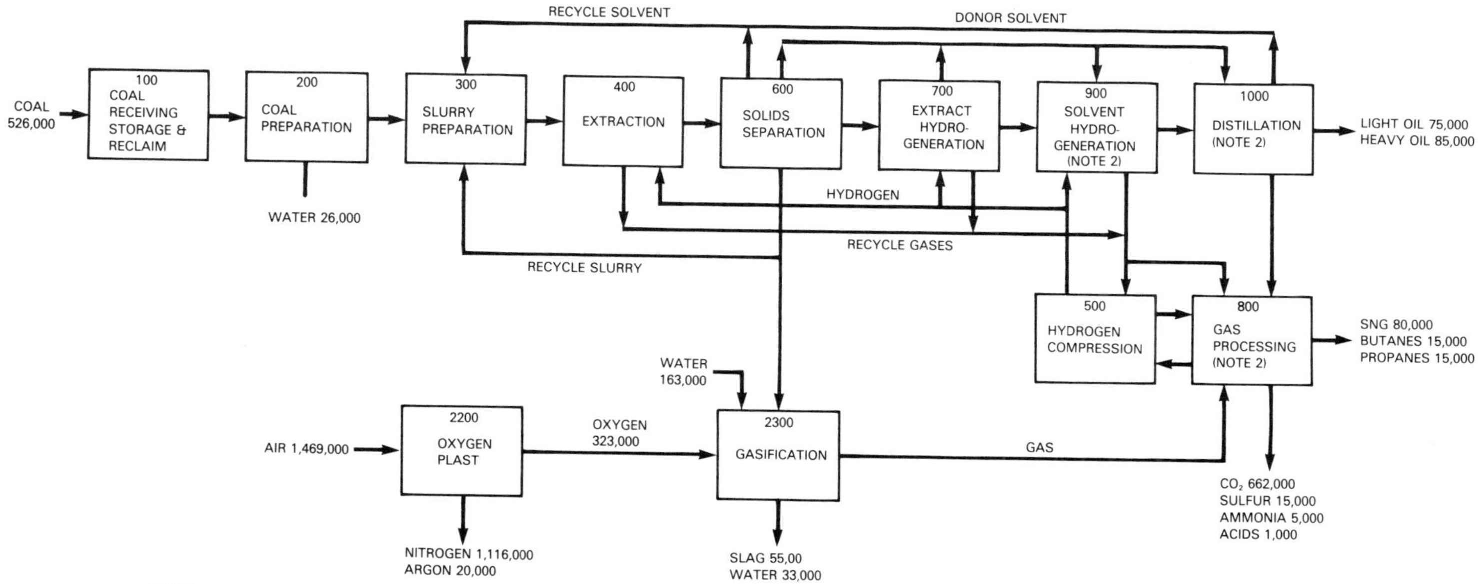

NOTES:

1. FLOW RATES IN POUNDS PER HOUR, BASED ON 6,000 TPD DRY COAL.

2. SECTIONS 800, 900, & 1000 ARE NOT CONSIDERED FURTHER IN THIS PRELIMINARY BROAD-BRUSHED ANALYSIS.

Fig. 3-14. Coal liquefaction plant block flow diagram.

Products	
Heavy oil	5,800 B/D
Light oil	5,200 B/D
Butanes	1,700 B/D
Propanes	1,700 B/D
Substitute natural gas	38 MMscfD
By-products	
Carbon dioxide	7,940 T/D
Water	118 G/M
Slag	660 T/D
Sulfur	164 T/D
Ammonia	60 T/D
Tar acids	12 T/D

The commercialization scenario considered in this study entails the construction and operation of 10 full-scale plants by 1990, with each plant producing 100,000 B/D FOE. Each plant consisted of five identical process trains, each operating on 600 T/D coal feed and producing 20,000 B/D FOE with assorted by-products, e.g., CO_2, slag, H_2O, S, NH_3, and acids. Designs, procurement, construction, and shakedown of each plant were assumed to take 6 years, and the plants were located west of the Mississippi River and were served by overland and river transportation facilities.

IMPEDIMENTS CONSIDERED

The impediments to commercialization considered in this study were (1) the capability of existing commercially available equipment to perform satisfactorily in a coal liquefaction plant, and (2) the availability of the equipment in quantities necessary to support a large-scale commercialization effort. To evaluate these barriers, assessments were made of equipment performance risk and equipment supply risk.

The equipment performance risk represents the risk of equipment failing to perform to specification at operating conditions. The equipment supply risk covers the risk of industry being unable to provide the quantity of equipment necessary to meet coal liquefaction plant construction schedules.

The equipment lists obtained from the generalized process flow diagram were evaluated independently by members of the risk evaluation team. Their inputs were used to assign each of the identified equipment items to low-, moderate-, and high-risk category. The criteria used for selecting the performance risk categories are

- *Low-Risk Items:* The equipment has adequate application data from services under comparable operating conditions. The service record indicates that little development effort is required for confident use of the equipment in a full-scale plant. Sufficient data exist to conduct equipment reliability/availability assessments in order to identify the exact development needs.
- *Moderate-Risk Items:* The equipment operability data are limited by or vary slightly from coal liquefaction process requirements. However, sound engineering judgment indicates that a major development effort is not required to use the equipment, with confidence, in a full-scale plant, and that a reasonable amount of data are available for equipment reliability/availability assessment in order to define the development effort.

- *High-Risk Items:* The equipment operability data are for process environments which vary significantly from those found in coal conversion services, and a substantial development effort will be required to modify and/or scale up existing commercial equipment. Equipment reliability/availability assessment can only be estimated due to scarce operability data.

The risks associated with failure to supply a sufficient number of the specific types of process equipment necessary for the scheduled full-scale coal liquefaction plants are represented by two categories. The criteria for the determination of the two risk types are

- *Low Supply Risk Equipment:* The number of items of a given type of equipment required to support the proposed coal liquefaction plants over the next 11 years is less than 25% of the total domestic and foreign production capacity for those items, assuming that present production level remains constant.
- *High Supply Risk Equipment:* The number of items of a given type of equipment required to support the proposed coal liquefaction plants over the next 11 years is greater than 25% of the total domestic and foreign production capacity for those items, assuming that present production level remains constant.

To determine the present production capacity, such factors as present domestic and foreign production capacities and production lead times were considered. The foreign capacity was limited to sources in western European countries and Japan.

MTI'S CONCLUSIONS

The principal conclusion reached in this study is that substantial risks exist in both the performance and supply of equipment needed for commercialization of direct coal liquefaction.

Four percent (19 items) of identified equipment items are of high-performance risk, for which significant development efforts are required for confident use of the presently available commercial equipment and technology in full-scale plants.

The equipment items in the high-performance risk category are slurry heater; oxygen compressors; extractor, stages 1, 2, and 3; extract hydrotreater guard chamber; extract hydrotreater; and gasifier.

Fifteen percent (68 items) of identified equipment items are in the moderate-performance risk category for which further development effort is also required in order to use the presently available commercial equipment with confidence in full-scale plants. The development effort for moderate-risk items is small in comparison to that for high-risk items. Representative items in this category include feed slurry circulating pump, feed slurry injection pump, extractor overhead control valve, extractor slurry pressure letdown valve, extract hydrotreater separator, flash still heater, slurry flash still heater, and extractor overhead waste heat boiler.

Thirty percent of equipment items in the slurry preparation, extraction, and extract hydrogenation sections of the plant fall in either the high- or moderate-performance risk categories.

Virtually all of the equipment items comprising the solids handling equipment, separation, and miscellaneous types are in the low-performance risk category. Typical equipment items in this category are steam condensate high-pressure injection pump, steam condensate low-pressure injection pump, makeup hydrogen compressor, primary still reboiler, primary still condenser, magnetic separator, screen, and thermal dryer package.

Equipment supply risks for various types of equipment are given below:

Equipment	Number of Equipment Per Plant	Equipment Supply Risk
Pumps	420	High
Compressors	60	High
Valves	80	High
Heat exchangers—shell and tubes	340	High
Fired heater	150	High
Separators	30	Low
Vessels/reactors	675	High
General solid handling equipment	365	Low
Miscellaneous	165	Low

Also, the equipment for large reciprocating pumps and oxygen plants over the next 10 years is between 50 and 100% of the presently available domestic and foreign production capacity. Since the demand for these equipment items from the noncoal-conversion industrial users is expected to persist, the risk of equipment supply to support the assumed commercialization scenario is very high.

Monsanto Research Corporation Assessment

The Mound Facility Report *Synthetic Transportation Fuels from Coal, Mound Facility*, Monsanto Research Corporation, Miamisburg, Ohio, November 1979, presents a general discussion of the production of synthetic transportation fuels from coal. Seven topics were covered in the report:

- *Market Study.* This section presents a price perspective of fossil fuels together with historical and projected future consumption patterns. A potential timetable for the commercial development of coal liquefaction is described.
- *Technology Status and Exploitation.* This section provides technological descriptions and the developmental status of the major direct and indirect coal liquefaction processes.
- *Technological Problems and Uncertainties.* This section describes the technological problems and uncertainties associated with coal liquefaction in general.
- *Economic Potential.* This section compares the economics of the various coal liquefaction processes. It is based on ESCOE's July 1979 report.
- *Financial Barriers.* This section analyzes the advantages and disadvantages of a variety of economic incentives for synfuels development. The incentives discussed are price, tax, risk reduction, capital access, surcharges, mandatory production, and removal of regulatory disincentives.
- *Socioeconomic Impact.* This section describes potential socioeconomic impacts of a commercial-size coal liquefaction plant and potential measures for their mitigation. Impacts discussed relate to plant size and labor requirements and impact on local and regional communities.

- *Environmental.* This section discusses environmental impacts with respect to the region surrounding a proposed liquefaction plant and possible mitigative measures that might be used.

TYPE OF PLANT CONSIDERED

No particular type of plant is considered. Descriptive material is presented for a variety of plants while the Financial Barriers, Socioeconomic Impact, the Environmental sections relate merely to a representative commercial-size coal gasification plant.

SCENARIO EMPLOYED

No particular commercialization scenario was analyzed. The potential problems were discussed in relation to a single, commercial-size coal liquefaction plant.

IMPEDIMENTS CONSIDERED

The report provides descriptions of technical problems, financial incentives that might promote commercialization, socioeconomic impacts, and environmental impacts. However, there is no discussion of specific impacts that these problems might have on coal liquefaction commercialization other than that they will constitute impediments that should be mitigated.

MONSANTO CONCLUSIONS

The major conclusions reached in this study are as follows:

- Coal liquefaction is a way to ease the shortage of liquid fuels.
- Technical problems and uncertainties associated with coal liquefaction relate to separation problems, environmental emissions, volume mining of coal, materials problems, scaleup problems, product testing, dry coal feeding to pressurized vessels, and other technological uncertainties.
- A barrier to commercialization arises from indecision over whether to commercialize first- or second-generation technology. Issues relating to this choice are uncompetitiveness of synthetic fuels, limitations on federal budget, thermal efficiency, product mix, and replacement of first-generation components with advanced systems.
- A crash program in coal liquefaction would be severely impacted by material and manpower shortages within the construction industry.
- The cost of a coal liquefaction plant is fairly constant regardless of the technology used.
- A highly capital intensive industry requires incentives that minimize exposure to capital losses and encourage investment; and industry sensitive to the cost of coal requires protection from unexpected price increases. Actions that the government can take to encourage the production of synthetic fuels are remove regulatory and institutional barriers; provide tax incentives such as accelerated depreciation and higher investment credits to reduce risk to investors; design programs to fit the particular designs and characteristics of the technology or process; and design programs that consider the industrial organization of the producers, i.e., utilities versus private companies.

- Regardless of the precautions taken during the development of the coal liquefaction and support facilities (mines, disposal area, etc.), the regions surrounding these facilities will experience some form of boom-town growth. Conflicts will arise between the rural social order which currently exists in the region and the new urbanized society which will accompany growth. Early planning is required to handle these impacts.
- Regardless of the precautions taken during the development and operation of the coal liquefaction facilities and support facilities (mines, disposal areas, etc.), the regions surrounding these facilities will experience some form of adverse environmental impact. To handle these impacts, the developer of each liquefaction facility will be required to conduct an in-depth analysis of the various environmental and resource-related requirements.

UOP/Systems Development Corporation Assessment

In the study *Feasibility Assessment: Production of Synthetic Fuels from Direct and Indirect Liquefaction Processes*, UOP/SDC, McLean, Virginia, January 1980, two direct and two indirect liquefaction technologies were evaluated from the standpoint of which technologies, or mixes thereof, offer the best opportunity for achieving the goal of producing 1 million B/D of coal liquids by 1990. The specific thrust was to identify those factors that might be limiting in the effort to commercialize such technologies. The general premise was that identification of impediments in the program is essential to minimize their effects, thereby facilitating as rapid a construction schedule as is feasible.

No engineering work was performed. The availability of nonconfidential information was reviewed and utilized as the basic source of process information. A representative or generic technology for each type of liquefaction process was characterized for the analysis. The characterizations, owing to the limitation of available data, contain elements of several different processes, including the following: energy raw material requirements (coal, water, air); product yields and efficiencies; catalyst and chemical requirements; plant requirements (construction materials, major equipment, engineering, and construction manpower); siting considerations; environmental considerations; capital costs; operating costs; and construction and equipment delivery schedule considerations.

Industries providing the necessary manpower and equipment were surveyed regarding their capacity to produce the equipment and provide the manpower. Plant requirements were compared with industry capacities to identify the areas where bottlenecks might arise. Limitations that might be imposed by siting and environmental considerations were investigated. Economic considerations and financial constraints were addressed from the standpoint of comparison of projected capital requirements with past annual financial activity of the hydroprocessing industry.

TYPE OF PLANT CONSIDERED

This study was not performed on a single type of liquefaction technology. Several mixes of four liquefaction technologies were studied. Two generic or representative direct liquefaction processes, i.e., a fuel oil mode (SRC-I) and a syncrude/SNG mode (SRC-II), were considered along with two indirect liquefaction modes, i.e., Fischer–Tropsch and methanol-to-gasoline. Table 3-9 shows the overall energy balance for each mode.

Table 3-9. Energy Balance

	Direct				Indirect			
	Fuel Oil		Syncrude/SNG		Fischer – Tropsch		M-Gasoline	
	10^9 Btu/SD	Percentage of Coal	10^9 Btu/SD	Percentage of Coal	10^9 Btu/SD	Percentage of Coal	10^9 Btu/SD	Percentage of Coal
Inputs								
Coal	765.00	100.00	857.6	100.00	753.00	100.00	444.3	100.0
Power	63.61	8.3	2.4	0.27	—	—	—	—
	828.6		860.0					
Products (Net)								
SNG			48.0	5.60	268.82	35.70	145.6	32.8
LPG			127.0	14.81	36.31	4.82	5.9	1.3
Mixed C_4's			12.6	1.47			9.2	2.1
Naphthas	53.5	7.0	95.4	11.12	108.07	14.35		
Gasoline							112.5	25.3
Diesel fuel					51.24	6.80		
Diesel fuel + Heavy fuel oil (bracketed)	95.5	12.5	333.1	38.84				
Heavy fuel oil					18.46	2.45		
Boiler fuel	361.5	47.3						
Alcohols					12.21	1.62		
Coke	104.0	13.5						
Subtotal	614.5	80.3	616.1	71.84	495.11	65.75	273.2	61.5
Sulfur[a]	8.10	1.1	9.4	1.09	8.03	1.07	0.5	0.1
NH_3			3.6	0.41			2.0	0.5
Power[b]					15.56	2.07	1.2	0.3
Tar			1.6	0.18				
Total	622.6	81.4	630.7	73.52	518.70	68.89	276.9	62.4

[a] At 6.0 MMBtu/bbl fuel oil equivalent.
[b] At 6.0 MMBtu/bbl fuel oil equivalent and power at 9500 Btu/kWh.

A nominal plant size of 30,000 tons/day of coal feed was selected as the mean of the range of likely plant sizes that would be utilized to produce the overall production goal of 1 million barrels per day by 1990.

SCENARIO EMPLOYED

The analysis focuses on the goal of producing 1 million barrels per day of coal liquids by 1990 by a multiplant program. Plants were situated in the three major coal-producing regions of the country, i.e., the western region, the central region, and the eastern region.

The estimated investment for the multiplant program is about $26 billion. Spread evenly over 7 years, the average yearly expenditure of about $3.7 billion results. This amounts to about 50% of the estimated 1980 expenditures of the U.S. hydrocarbon processing industry of about $7 billion. Favorable conditions of expansion of industrial capacity for critical resources, e.g., air separation plants, compressors, and technical manpower, were assumed in the scenario.

IMPEDIMENTS CONSIDERED

The following impediments and/or constraints to achieving the stated coal liquids production goal are identified and analyzed in this analysis: air separation plants, technical manpower, compressors, thick-wall vessels, siting considerations, environmental considerations, interdependence of projects, and capital and operating costs considerations.

The salient points of the analyses on each identified impediment are summarized in Table 3-10.

Table 3-11 summarizes the required resources or specific component in terms of peak requirements as a percentage of the availability in the time frame considered.

These figures reflect the equipment and personnel requirements for each process investigated as a percentage of the national production capacity or availability that would have to be utilized to meet the schedule for the designated scenario. Five cases were developed. In each of the first four, the entire capacity of 1 million B/D FOE of coal products was considered to consist of multiple plants of just one of each of the four processes (i.e., 1 million B/D of direct liquefaction in the fuel oil mode, 1 million B/D of direct liquefaction in the syncrude/SNG mode, etc.). In the fifth case, labeled 4 × 25% combined, each of the four process modes was represented to the extent of about 250,000 B/D.

As a basic premise, the study assumed that if the requirement for any item exceeded 20% of the availability or national production capacity of that item, dislocations in the supply and demand relationship for that item would recur.

The analysis was refined to reflect a more-detailed project schedule to account for the time required for major item engineering, purchasing, manufacturing, and delivery. Then year-by-year schedules of the average and peak requirements for the three requirement categories considered most difficult to remedy were compared to the national production capacity or availability of the item. Figure 3-15 shows the results of this more-detailed analysis for a case of continued current capacity and a case for a 10% annual growth in current capacities and availability.

UOP/SYSTEMS DEVELOPMENT CONCLUSIONS

The following conclusions were developed by the study:

- Of the four liquefaction process routes and various multiplant mixes, none stood out as being significantly freer of constraints.

Table 3-10. Impediment Identification

Constraint	Process	Causes/Specific Requirements	Potential Solution Path	Notes
Air-separation plants Aluminum distillation tower Brazed aluminum heat exchanger	All	Limited fabrication capacity Shop fabricated to ensure quality Shipping dimensions limit size Peak demand is 48% to 122% of national output	Increase fabrication capacity Dedicate fabrication capacity Field fabrication Develop processes which use only enriched air	
Technical manpower	All	Limited number of persons employed in the industry Personnel dispersed into small offices Lead time to graduate new engineers Peak demand is 9% to 30% of national technical base	Entice personnel from other industries Expand role of nondegreed personnel	Causes major dislocations in other sections of the economy
Thick-wall vessels and piping	All	Limited U.S. fabrication capacity Peak demand is driven by size 1″ = 6% to 25% 1″ – 4″ = 5% to 188% 4″ – 8″ = 0 to 34% 8″ – 14″=0 to 400% Percentage of U.S. fabrication capacity	Use smaller – thinner wall units Field fabricate—multi-layer Use prestressed concrete Use cast iron elements	None of the solutions are done commercially at this time Syncrude/SNG is most impacted

Table 3-10 *(continued)*

Constraint	Process	Causes/Specific Requirements	Potential Solution Path	Notes
Compressors	All	Limited U.S. fabrication capacity Peak demand is 11% to 22% of national fabrication capacity	Increase U.S. fabrication capacity Use more foreign units	Methanol route is most impacted
Valves	All process routes	Severe service environment Long lead time Limited availability for supply in larger sizes (6″). Valves larger than 10″ are custom designed and not stocked		
Western region	All process routes	Water requirements Absence of infrastructure Manpower availability Limited transportation Competition for area's coal High construction cost	Transbasin diversion Dry cooling towers Reservoirs to regulate flow Design to minimize water requirements	Sites located near coal source to minimize cost of hauling
Eastern region	All process routes	Higher cost coal Rugged terrain Air pollution cost	Use smaller plants	Site earlier plants in the East
Interdependence of projects	All process routes	Competition for components, resources, funds, manpower, etc., in U.S. and world	Area not investigated in study	

Environmental Plant emissions to air to water waste disposal noise use of land Raw coal liquids transportation and storage Raw coal liquid refining Raw coal liquid combustion Exposure to hazardous chemicals	All process routes	Lack of accepted criteria for siting the synthetic fuels industry lack of emissions data Specific state requirements may prevail if federal requirements are relaxed	Expedite preconstruction procedures, i.e., energy mobilization board Careful site planning	Six years may be required from start of siting to start of construction
Capital requirements	All process routes	Costs are large regardless of process route Cost not sensitive to technology		
Operating costs		Operating costs are not a major criteria for selection of process mix		
Strategic metals/ minerals for catalysts and alloy steels	Fischer–Tropsch plants producing 1 MMB/D FOE	Based on total estimated 1985 demand: cobalt 1% molybdenum 1% nickel 1.5% chromium 7%		Data not available for other process routes

Table 3-11. Peak Requirements vs Availability of Resources (at 1 MMBPD Capacity for Each Process Mode)

Resource or Component	Capacity or Availability	Peak Requirements as Percentage of Availability				
		Fuel Oil Mode	Syncrude/SNG Mode	Fischer–Tropsch	Methanol/ Gasoline	4 × 25% Combined[a]
Compressors	540	21	18	18	40	25
Vessels						
CS 1″	10,000	8	7	6	25	9
Alloy 1″	6,000	0	0	0	3	1
1–4″	500	18	5	126	188	89
4–8″	100	0	0	34	31	18
8–14″ <100 tons	30	0	0	0	0	0
100–800 tons	20	140	260	0	0	95
>800 tons	5	0	400	0	0	100
Air separation (oxygen) plant (T/D/Y)	18,000	125	144	312	162	184
Construction labor (people)	5,000,000	<1	<1	<1	1.2	1
Contractor's technical personnel (people)	45,000	24	13	10	48	25

[a] Approximately 250,000 B/D capacity each of the four process modes.

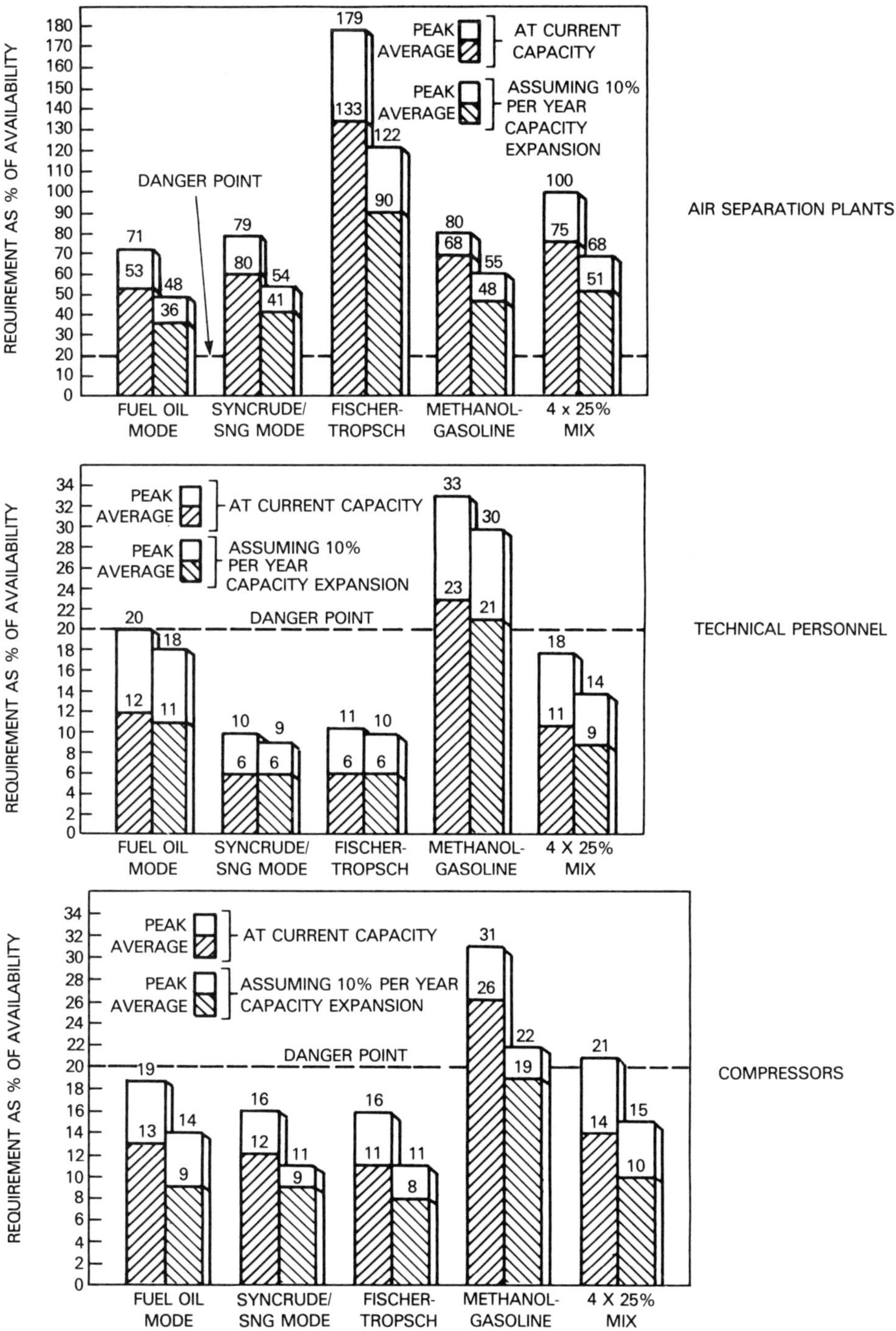

Fig. 3-15. Peak requirements vs availability.

- The single severest constraint to achieving the stated production goal for coal liquids is the industrial capacity required to produce the necessary number of air separation plants utilized in the processes.
- Achieving the 1 million B/D production goal by 1990 will require a significant increase in fabrication capacity of air separation plants or the dedication of approximately 50% of current fabrication capacity to the project.
- Environmental and/or siting considerations may ultimately prove to be the most immovable barrier to meeting the coal liquids production goal. As federal requirements are expedited, state-imposed requirements may impose delays in permitting procedures.
- The following conclusions relate to siting considerations:

  Western Region

  Each of the five western regions has sufficient coal reserves to support from 6 to 110 plants of the 30,000 T/D size for a 30-year plant life.

  The mining industry has sufficient production slack and/or capability to expand production capacity to meet projected requirements.

  In the western regions, water availability will not be a deterrent, but will be expensive.

  Transportation is limited and will increase costs.

  Central States Region

  The Illinois Basin appears most favorable for location of liquefaction plants. The area has adequate coal, water, transportation, and manpower resources to satisfy projected requirements.

  Eastern States Region

  This region appears best for the small, initial liquefaction plants. Coal reserves, transportation, water, and manpower are adequate to meet projected requirements.

DOE Coal Liquefaction Program

The Department of Energy has continued a broad-based research and development program for coal liquefaction.

The development of commercial coal liquefaction processes will provide a route for increasing domestic production of clean fossil fuels and feedstocks within environmental guidelines. Synthetic fuels derived from coal liquefaction processes can replace petroleum products most effectively in two distinct markets. One market uses low-ash, low-sulfur boiler fuels suitable for clean generation of electric power and industrial steam. The other market uses high-grade fuels such as jet and diesel fuel, gasoline, heating oil, and chemical feedstocks.

DOE's liquefaction activity is designed to conduct long-term, high-risk, high-payoff R&D to facilitate the establishment of a synthetic liquid fuels industry. With the recent creation of the Synthetic Fuels Corporation, the DOE coal liquefaction technology program is to be reconfigured to support a number of processes in tandem from the laboratory stage through process development units. Responsibility for demonstration activities—the precursor to commercialization—will be shifted to the corporation. This is consistent with the administration's policy to focus the federal role on long-range research and development while relying on private market forces to set the pace for the introduction of synfuel technologies into the marketplace.

Specifically, the activity is planned to accomplish the following objectives:

1. To develop improved indirect liquefaction processes to produce liquid fuels from synthesis gas made from coal by the late 1980s.
2. To promote improvements designed to reduce costs of the major direct liquefaction processes by evolutionary integration of advanced liquefaction concepts into these processes.
3. To continue design efforts in FY 1981 for large-scale plants using the major direct liquefaction processes for continuity, while project participants determine their interest and capability for pursuing such plants on their own with or without the assistance of the Synthetic Fuels Corporation.
4. To finish early process testing at large pilot plants, using major direct liquefaction processes.

The general strategy is to support several liquefaction processes at the same time from laboratory scale through process development unit (PDU) operations.

This approach is based on several considerations. First, as is evident in today's crude oil-based liquid fuels industry, market needs for a wide range of fuels produced from a variety of diverse coal types, coupled with the need for many plants to produce significant quantities, may require several liquefaction processes. By the same token, the future synthetic liquid fuels industry, which will replace a large share of this fuel market, cannot rely on any single process configuration. Second, the processes under development support each other in many ways. Third, all processes under development require supporting research to improve overall operability, increase efficiency, and reduce costs while meeting environmental, health and safety, and socioeconomic requirements.

Four direct liquefaction processes (SRC-I, SRC-II, H-Coal, and EDS) are in advanced development stages:

- Solvent-refined coal processes (SRC) I and II are being tested in 6 ton per day and 50 ton per day coal pilot plants currently in operation. Technical demonstration plants of 6000 ton per day size (one for each process) are in design stages.
- H-Coal is undergoing continued development in a 600 ton per day pilot plant that began startup in FY 1980. Continued support work is provided by bench-scale units and a 3 ton per day PDU.
- Exxon Donor Solvent is undergoing development in a 250 ton per day pilot plant that began startup during FY 1980. Continued laboratory and engineering research support backs up the operation.

Third-generation direct liquefaction processes include disposable catalyst hydrogenation, advanced two-stage liquefaction, and flash liquefaction (pyrolysis). To define the potential of these newer processes for scaled-up engineering studies, bench-scale and PDU studies continue.

Indirect liquefaction processes include development of technology to process liquid fuels via synthesis gas from coal. An accelerated effort is underway in the United States to develop new, improved indirect processes tailored to U.S. coals and markets. Design and construction of a small (100 barrel per day) methanol-to-gasoline pilot plant based on a Mobil fluidized-bed process is underway, and a 35 barrel per day liquid/slurry phase synthesis gas-to-methanol plant is being prepared for operation.

Research and development activities provide backup to these processes with technological improvements that, in turn, can significantly improve the process economics. Two major problem areas common to most direct liquefaction processes are under study: (1) utilization of solids-containing bottoms for hydrogen production and (2) durability of auxiliary equipment. Significant effort is devoted to engineering studies and analyses required to guide and evaluate process development work.

A major effort is directed toward environmental concerns associated with coal liquefaction processes: the presence of potentially toxic and carcinogenic organic compounds in coal liquid products and residues, and toxic trace minerals in the products and by-products.

The environmental impact of coal liquefaction processes is being investigated under an effort designed to analyze and characterize hazardous materials appearing in various process streams or being emitted from operating installations. Studies are also being conducted at the pilot plant level to develop adequate occupational safety and health procedures.

DIRECT HYDROGENATION

One direct hydrogenation project, the H-Coal process, will be active in FY 1982, to be completed during that period.

Ebullated-Bed (H-Coal) Pilot Plant

The H-Coal process is a catalytic hydroliquefaction process that converts high-sulfur coal either to a boiler fuel meeting sulfur emission regulations or to a refinery syncrude (Fig. 3-16). Coal is dried and crushed to minus 40 mesh, then slurried with recycled oil and pumped to a pressure of 200 atmospheres. Compressed hydrogen is added to the slurry, and the mixture is preheated and charged continuously to the bottom of the ebullated-bed catalytic reactor. Upward passage of the internally recycled reaction mixture maintains the catalyst in a fluidized state, with catalyst activity maintained by the semicontinuous addition of fresh catalyst and the withdrawal of spent catalyst. The temperature of the ebullated-bed catalytic reactor is controlled by adjusting the temperature of reactants entering from the preheater. Typically, the preheater outlet temperature is 700–750°F (371–399°C).

The vapor product leaving the top of the separator is cooled to condense the heavier components as a liquid. Ammonia, hydrogen sulfide, and light hydrocarbon gases are absorbed and separated from the remaining gas, leaving a hydrogen-rich gas that is recompressed and recycled to be combined with the input slurry along with makeup hydrogen. The remaining liquid–solids, containing unconverted coal, ash, and oil, are fed into a flash separator. The separator overhead is sent to an atmospheric distillation unit. The bottoms, containing solids and heavy oil, are further separated with a hydroclone, and the hydroclone overflow is recycled back to the reactor. The hydroclone bottoms undergo further separation to provide the liquefaction products.

The process may be operated to produce refinery syncrude or a low-sulfur, low-ash boiler fuel suitable for use in industrial boilers. To produce refinery syncrude, the coal residence time is low, more hydrogen is required, and the products are separated by

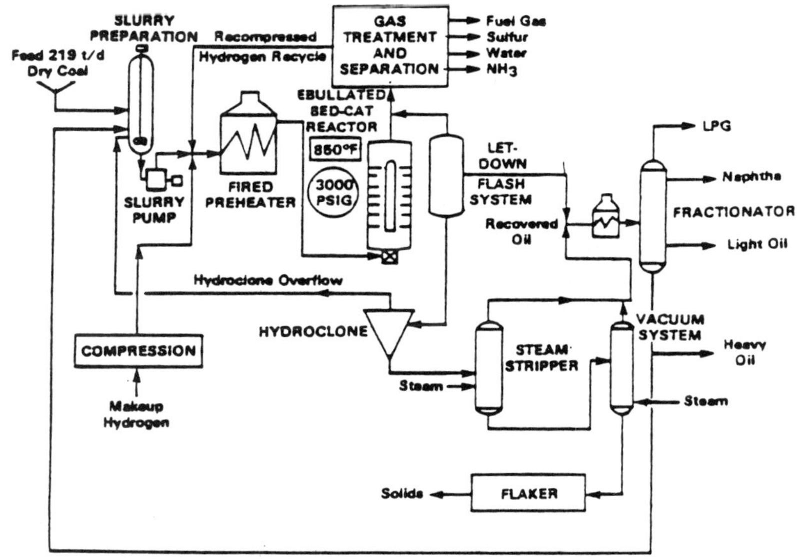

Fig. 3-16. H-Coal process, syncrude mode.

distillation. To produce a low-sulfur, low-ash liquid boiler fuel, the throughput of the ebullated-bed reactor is increased, hydrogen consumption is decreased, and the solids are separated from the liquid product by nondistillation methods. In the H-Coal pilot plant, the separation is done with a Lummus Anti-Solvent De-ashing System.

The ash and liquid products (hydrocarbon gas, light and heavy distillates, and residual fuel) may be further refined as necessary. In the commercial configuration, the vacuum bottoms streams containing unreacted carbon and some liquid will eventually be processed on site to produce the hydrogen needed for the process. In the H-Coal pilot plant, hydrogen is supplied by the adjacent oil refinery.

The H-Coal process was developed by Hydrocarbon Research, Inc. (HRI) as a further application of the ebullated-bed processing technology originally used to convert heavy oil petroleum residues into lighter fractions (H-Oil process). Early development of the H-Coal process was under the mixed sponsorship of DOE and an industry consortium. An independent evaluation in 1976 supported the technical feasibility of the H-Coal process.

Based on the experimental results obtained from a bench-scale and a 3 tons per day PDU operating under both syncrude and boiler fuel modes, final detailed designs of a 200 and a 600 tons per day pilot plant were completed in December 1977.

The pilot plant site in Catlettsburg, Kentucky, was selected by the Office of Coal Research, Department of the Interior, at the recommendation of HRI, from several sites offered by the private partners. Badger Plants, Inc., was responsible for construction of the pilot plant, which was completed in FY 1980. Pilot plant operations are the responsibility of Ashland Synthetic Fuels, Inc., with technical support through a subcontract with HRI. The pilot plant is designed to convert up to 600 tons of coal per day into as much as 1800 barrels of synthetic oil per day.

The industry consortium sponsoring the H-Coal pilot project is composed of the Electric Power Research Institute, Ashland Oil, Inc., Standard Oil of Indiana, CONOCO Coal Development Company, Mobil Oil Company, the Commonwealth of Kentucky, and Ruhrkohle AG. Startup of the pilot plant in Catlettsburg, Kentucky, began in FY 1980; operation of the syncrude mode was completed in FY 1980, and operation of the boiler fuel mode began in early FY 1981. Material balance data obtained are being used to establish process yield data as a function of coal type and operating conditions.

In FY 1982, operation of the plant in the boiler fuel mode will be completed. Within the guidelines of new policy, DOE will complete ongoing operations in FY 1982. Since the summer of 1981, many of the equipment problems encountered during startup were overcome, and on February 17, 1982, the first long-term operation of the plant began. The plant is currently converting up to 200 tons per day of Illinois No. 6 coal into synthetic crude oil suitable for refining. Early this fall, the facility will be switched to the production of boiler fuel and will process up to 600 tons per day of Illinois No. 6 coal. The second series of test runs will be completed in FY 1982, and the plant will be shut down. If appropriate, DOE will undertake discussions with the project's industrial participants or possibly other interested firms should they wish to continue privately funded operations to support their own commercialization plans.

SOLVENT EXTRACTION

Solvent extraction projects involve the use of a solvent to obtain a liquefied product from coal. The solvent acts as the agent transferring hydrogen to the coal to promote liquefaction. (Solvent-refined-coal projects are described in this section.)

The Exxon Donor Solvent (EDS) process is a noncatalytic process that liquefies coal by the use of a hydrogen-donor solvent obtained from coal-derived distillate. The donor solvent transfers hydrogen to the coal, thus promoting liquefaction of the coal. The process utilizes engineering and design technology similar to that used in the petroleum industry. It is sufficiently flexible to allow for different varieties of coal feed, and it produces a wide spectrum of liquid products depending on market demands.

Coal is ground and slurried with the recycle donor solvent. The slurry is heated, and preheated gaseous hydrogen is added. A tubular reactor with no internals is used. Products from the liquefaction reactor are sent to several stages of separation units for recovery of gas, naphtha, middle distillate, and bottoms comprised primarily of unreacted coal and mineral matter. Distillation is the means for solid–liquid product separation.

The heavy bottoms from distillation go to a Flexicoker that produces additional liquids and low-Btu gas for in-plant fuel use. Hydrogen for in-plant use is provided by steam reforming of C_1-C_2 gases produced in the process or by partial oxidation. The hydrogen is recycled to the liquefaction and solvent hydrogenation sections.

A portion of the middle distillate product is sent to the solvent hydrogenation step, using a catalytic fixed-bed reactor to produce donor solvent to be recycled to the slurry preparation step. Depending on the ultimate product utilization, the primary liquid products may be refined further.

The EDS pilot plant is balanced: it is self-sufficient in both process fuel and hydrogen requirements. The process (Fig. 3-17) gives high yields of low-sulfur liquids from either bituminous or subbituminous coals or lignites. For Illinois bituminous coal, the liquid yield is determined to be 2.8 barrels of C_4/1000°F (538°C) liquids per ton of dry coal feed.

The project was started in 1966 with Exxon Research and Engineering Company (ER&E) funding. The developmental program is structured into five phases: predevelop-

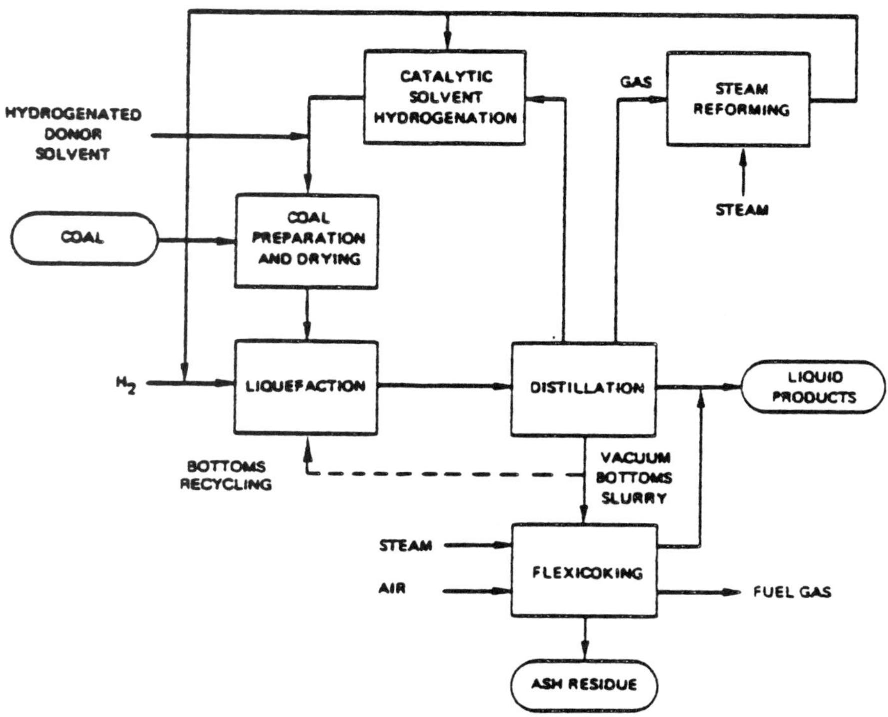

Fig. 3-17. Donor solvent process.

ment, planning and design, detailed engineering, construction, and operation. Through 1975, cumulative ER&E funds of $32 million were expended on phases 1 and 2, resulting in the development and demonstration of the process in laboratory-scale reactors of up to 0.5 ton per day.

In August 1975, ER&E submitted an unsolicited proposal to the Energy Research and Development Administration (ERDA) to cost-share an estimated $268 million future-development cost on a 50/50 basis. Because the project cost was substantial, ER&E also sought financial participation from other industrial partners. Because industrial cost-sharing participants were not firmed up at the time ER&E could start phase 3, that phase was divided into two separate phases to permit discrete funding: phase 3-A, laboratory and engineering research and development, and phase 3-B, detailed engineering.

Phase 3-A spanned the period between January 1976 and mid-1977. Total cost of this phase was $13 million, shared by ERDA and ER&E on a 50/50 basis.

Phase 3-B spans the period between mid-1977 and FY 1982. In FY 1982, the testing program will continue using a limited range of coals. The 250-ton per day plant near Baytown, Texas, has been undergoing full testing operations on Illinois No. 6 coal since January, following a successful startup that began last summer. Since the FY 1982 budget request represents a decrease in the government's financial commitment, the funding level will be presented to the Sponsors Management Committee (in accordance with the cooperative agreement) for their decision regarding further use or disposition of the facility in view of the government's indication of limited funding.

The 1982 milestones are subject to consideration by the Management Steering Committee.

INDIRECT LIQUEFACTION

This broad-based catalyst and process R&D project is designed to develop substantially improved indirect liquefaction technology. Specific goals include the following:

- Improve the integration of liquefaction with coal gasification, especially from second-generation pressurized gasifiers for eastern bituminous caking coals, by developing technology that can convert low-ratio hydrogen/carbon monoxide producer gas directly to liquid fuels.
- Improve process selectivity for premium liquid fuels, such as gasoline, by use of shape-selective catalysts.
- Achieve higher process efficiency by using more thermally efficient ebullating-bed or liquid/slurry phase reactors.
- Develop a simpler method of upgrading conventional Fischer–Tropsch liquids.

To achieve a substantial improvement over existing indirect liquefaction methods, technology will be developed that integrates efficiently with second-generation coal gasification systems. Existing technology requires hydrogen-rich feed gas and results in a low thermal efficiency. Catalysts and reactor systems that allow the direct use of the clean, that is, low-hydrogen-content, synthesis gas will be evaluated at bench scale, and the most promising candidates tested at pilot scale. Three candidate fuels that could be produced are methanol, fuel-grade alcohols, and hydrocarbons.

The preferred route would be hydrocarbons since this requires the lowest hydrogen content in the synthesis gas per Btu of fuel produced. However, technology for use of fuel-grade alcohols is developed further and will also be evaluated. This program was expanded from earlier Advanced Research and Technology Development activity in FY 1981 to complement the development of second-generation coal gasifiers in the mid-to-late 1980s.

A limitation of existing technology is the inability to produce a quality product in a selective fashion. The use of modern zeolite-based catalysts provides a potential solution to this problem. This effort will build on the pioneering work done under prior DOE funding. Data will also be obtained on the ability of zeolite catalysts to upgrade Fischer–Tropsch liquids to high-octane gasoline in a simple, efficient process.

A major engineering difficulty in synthesis gas conversion is the high exotherm that occurs when the synthesis gas is converted to a liquid. Ebullating-bed and liquid/slurry phase reactors have a potential to provide good temperature control and allow efficient recovery of the heat produced. These reaction systems will be evaluated at both the bench and pilot scales, and the data will be used to develop detailed design models.

Construction of the 100 barrel per day fluidized-bed methanol-to-gasoline pilot plant is currently underway in collaboration with West Germany. One of the participants is Mobil Oil Research and Development Corporation. It is expected that the fluidized-bed mode of operation will permit a higher thermal efficiency for the process and a greater overall selectivity to gasoline production.

During FY 1981, a 35 barrel per day liquid-phase methanol pilot plant was installed at La Porte, Texas. Testing of catalysts will be initiated for conversion of synthesis gas to methanol at higher thermal efficiencies than state-of-the-art technology. The prime contractor for this work is Air Products & Chemicals, Inc. (APCI). This pilot plant may subsequently be used to examine selective slurry-phase catalysts for the conversion of low-ratio H_2/CO producer gas to hydrocarbons.

Catalyst development work is being carried out at the Pittsburgh Energy Technology Center, Union Carbide, APCI, and Battelle National Laboratory (BNL). The catalysts include shape-selective dual-function metal-cluster catalysts and bimetallic catalysts.

In addition, Mobil is investigating a two-step process for the conversion of low-ratio H_2/CO gas initially to low-molecular-weight hydrocarbons over modified Fischer–Tropsch catalysts, followed by conversion of the light hydrocarbons over a zeolite catalyst to synthetic gasoline. This process would offer significant economic advantages over state-of-the-art technology.

During FY 1982, the requested funding will be utilized for the following activities:

- Complete revamp of the liquid phase methanol pilot plant.
- Complete construction and initiate operation of the M-Gasoline pilot plant in Concept I.
- Evaluate ongoing catalysts and process development projects to select candidates for acceleration to large-scale testing in FY 1983.

Third-Generation Processes

The third-generation processes activity includes research and technology development for process improvements that offer potentially significant advantages over processes now in development. These advanced processes are in laboratory- and bench-scale development or use larger continuous process units, including PDUs. Processes selected for support offer the potential for significant improvements in one or more of the following areas: hydrogen utilization, operability, product quality and yield, plant investment and operating costs, and types of coal utilized.

Disposable Catalyst Hydrogenation

This process (Fig. 3-18) uses inexpensive catalysts to reduce liquefaction severity, thereby substantially reducing hydrogen consumption and cost while increasing liquid product yield in coal liquefaction. This can be achieved in several ways: (1) by using certain mineral constituents in coal such as pyrite and pyrhotite as catalysts, (2) by using inexpensive materials and naturally occurring ores as catalysts, (3) by slurrying coal with inexpensive synthetic catalysts, and (4) by removing mineral constituents in coal that negatively affect coal dissolution or poison the other catalyst constituents or both. The program involves initial testing of the mineral matter and inexpensive ores in autoclaves, followed by bench-scale and PDU testing for the most promising catalysts. The program also involves a detailed characterization of the coal mineral matter, and includes attempts to improve the catalytic activity of the mineral matter. In the future, the most promising candidate will be tested in an existing DOE facility. Thus it is possible that an effective disposable mineral catalyst obtained from coal could be readily incorporated into the SRC-II or EDS process.

Briefly, coal is dried and mixed with a process-derived recycle slurry and the disposable catalyst. The coal slurry is compressed to 2000–3000 psi with the addition of recycle hydrogen and the "in-balance" hydrogen produced by gasifying the solids-containing bottoms, and preheated. The preheated gas/coal slurry goes to a slurry-phase pressure reactor. Gaseous and coal/oil/catalyst slurry discharged from the reactor is sent to a gas

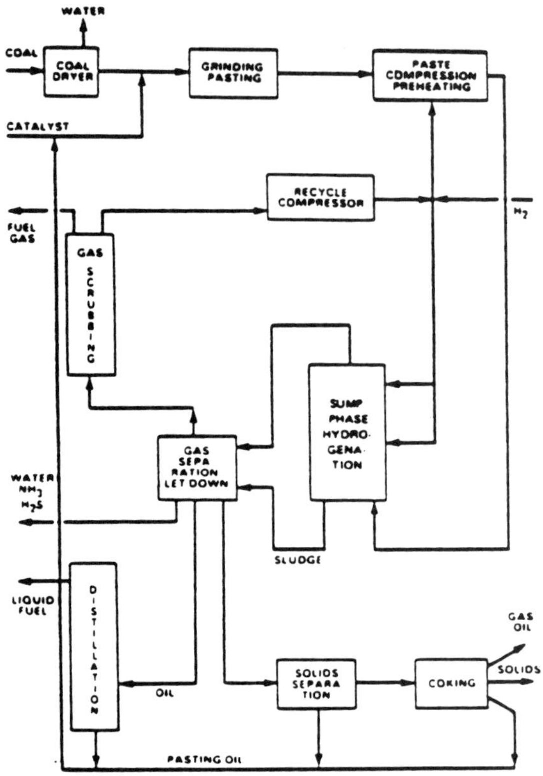

Fig. 3-18. Disposable catalyst hydrogenation process.

separator and letdown system. The gas is recycled. Part of the slurry is recycled, and part is distilled to obtain distillate fuel oil products. The distillation bottoms are gasified to produce sufficient synthesis gas that, after conversion to hydrogen, joins the recycled gas.

The most promising catalysts identified in the bench-scale testing program were tested in the 1200 pounds of coal per day unit at the Pittsburgh Energy Technology Center. The candidates include activated mineral matter derived from coal, inexpensive ores, and metallic wastes. The role of some components of the coal mineral matter in liquefaction has been established. New coal cleaning techniques that leave the catalytically active components in the coal while removing the inert matter were explored. Coals cleaned by these new techniques were tested in a continuous bench-scale unit. Liquefaction of West Virginia and Kentucky coals, in particular, were markedly improved by the addition of active disposable catalysts.

Preliminary correlation between hydrocracking and hydrogenation functions of the catalytic mineral components in coal and the production of highly efficient recycle process solvent has been identified.

In FY 1982, high-surface-area high-activity synthetic expendable catalysts will also be prepared and evaluated. In addition, the effect of disposable catalysts on short-contact-time (low-severity) coal liquefaction will be determined.

Pyrolysis and Hydropyrolysis (Flash Liquefaction)

Flash liquefaction (Fig. 3-19) involves very rapid heat-up and pyrolysis of the coal, generally under entrained-flow, under high-pressure hydrogen (flash hydropyrolysis), or in an inert or reducing gas at atmospheric pressure (flash pyrolysis) to produce some liquid, as well as gas and char. This subprogram involves at present two principal contractors: Rockwell International (RI), for the flash hydropyrolysis work, and Occidental Research (OR), for the flash pyrolysis work.

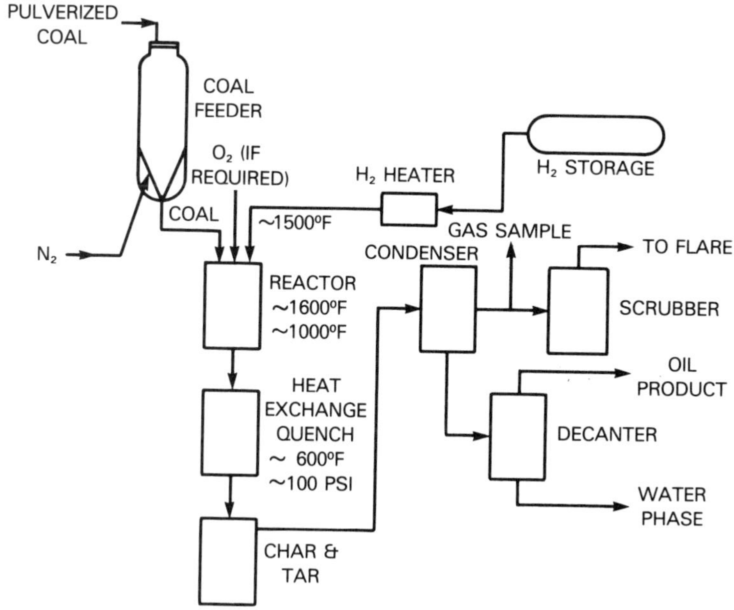

Fig. 3-19. Partial liquefaction by direct hydrogenation (flash hydropyrolysis) PDU.

The RI approach utilizes a unique coal injector and mixing device to enable rapid heat-up and feeding of the coal under hydrogen pressure without plugging of the injector. Heat is provided by the preheated hydrogen. At 1600°F (871°C), 1000-psig hydrogen, and a 100-millisecond contact time, 35% of the carbon from a western Kentucky coal is reported as liquids and 17% as methane. The development work is currently being conducted using a 1 ton per hour PDU in conjunction with RI's reactor. Numerous very short runs have successfully been completed.

In a series of short duration runs with a 1 ton per hour PDU, RI has defined the optimum reaction conditions and method of product quenching for their system. In addition, various design modifications have been determined for preheating the hydrogen and scaleup of the feeder for injecting and mixing hydrogen with coal.

During FY 1981, extensive mechanical modifications of the 1 ton per hour PDU will have been completed to enable operation for extended periods of time to assess process operability, performance, and equipment reliability. Materials and component testing are being continued to define the practical limits of the process operating conditions.

During FY 1982, long duration runs on the 1 to 4 ton per day continuous PDU will be conducted to confirm data on process parameters and component design. In addition, engineering studies will also be made for a commercial plant to determine the design uncertainties and process economics, including the cost of refining the raw products to salable products.

The primary goal of the OR work is to define a set of conditions that will enable the use of eastern bituminous caking coals to produce acceptable yields of liquid with improved quality. For this purpose, the effect of carrier gases of different composition, and of product quenching with a hydrogen donor, are being investigated. If proven successful, this technology can be adapted to the pyrolysis step of the COGAS process now under consideration for demonstration plant development.

Occidental Research completed the investigation on the effects of carrier gas/heat carrier on pyrolysis yield and initiated the study of product quenching by hydrogen donor. During FY 1982, this study and the study of subsequent hydrogen-donor regeneration will continue.

Process Research/Technology Development

The objective of this effort is to increase the technology database that can be immediately applied to coal liquefaction processes under development. The following projects are in progress:

- Development of improved H-Coal catalysts. Efforts are continuing at American Oil Company (AMOCO) and Pittsburgh Energy Technology Center involving catalyst screening, aging tests on the more promising catalysts in a bench-scale ebullated-bed unit, catalyst functionality studies, and characterization of the spent catalyst.
- Determination of reactor fluid dynamics relevant to existing processes. Cold-flow models developed by Air Products and Chemicals, Inc. and AMOCO are being investigated to predict the SRC-I and H-Coal reactor performance, respectively. The models are then refined to reflect actual process performance.
- Identification of properties of coal/recycle solvent slurries and coal liquids critical to the design of preheaters and other unit operations. Oak Ridge National Laboratory, Pittsburgh Energy Technology Center, and Brookhaven National Laboratory are continuing to obtain data on coal liquids viscosity, thermal conductivity, heats of reactions, and other physical and chemical data under conditions in the preheater and reactor.
- Development of coal liquids refining R&D aimed at tailoring technology to coal liquids from the different liquefaction processes. Universal Oil Products (UOP) and Chevron are carrying out bench-scale pilot plant refinery tests and making engineering evaluations to develop preferred coal liquids refining schemes, product slates, and estimated refining costs. Catalyst development studies will be completed at the Filtrol Corporation in FY 1981.

The superior performance of an improved H-Coal catalyst has been demonstrated and will be tested in the H-Coal PDU. A fluid dynamics model for the SRC-I and H-Coal reactors has been developed based on cold-flow model studies, and will be correlated with SRC and H-Coal pilot plant performance data. Various properties of coal liquid slurries, such as viscosity, density, and thermal conductivity, are being determined under preheater and reactor conditions. Data correlations for a detailed process variable study of lignite liquefaction in the SRC mode have now been completed. The refining studies at UOP and Chevron have been completed. The results indicate that the cost of refining a distillate syncrude such as the product from the SRC-II process is about the same as that for refining a medium-gravity sour petroleum crude.

Two-stage liquefaction (TSL) is a third-generation process in which low-severity (short-contact-time) coal liquefaction extract is produced in the first stage; the deashed product, topped to remove light oils, is then upgraded in a second-stage expanded-bed hydrocracker to "clean" distillate hydrocarbons. This process has the potential of a marked cost reduction over current second-generation processes as a result of substantially reduced hydrogen consumption and increased liquid product yields. The objective of current work is to demonstrate in an integrated PDU the feasibility of the TSL process and to assess its technical and economic advantages over second-generation processes.

In brief (see Fig. 3-20), coal is dried and mixed with a process-derived oil and charged to a fired heater under hydrogen pressure. The slurry is then rapidly heated to 800–850°F (427–454°C). The slurry may then pass to an adiabatic reactor, depending upon

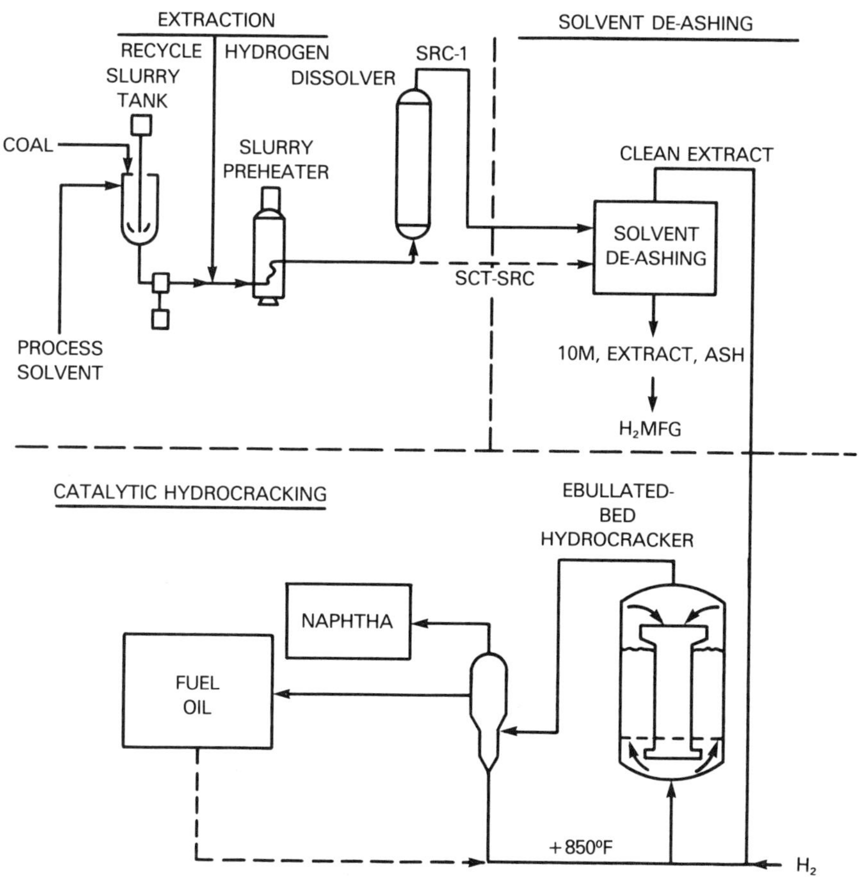

Fig. 3-20. Two-stage liquefaction process.

the residence time required to dissolve about 90% of the coal with minimum hydrogen consumption. Depending upon the coal, solvent characteristics, temperature, pressure, and the effectiveness of any disposable catalyst added, this residence time may range from 5 to 20 minutes. For reactive coals, the reaction occurring in the preheater may suffice without requiring an additional reactor. The effluent slurry is then topped to remove 500–650°F (260–343°C) distillate and then deashed using the Lummus anti-solvent process. The overflow is pumped to the Lummus-refining unit for catalytic hydrocracking into naphtha and fuel oil. The underflow is transferred to a gasification unit for production of hydrogen and fuel gas.

This project, which evolved from the process research/technology development program, will address the following items:

- Systems integration of a relatively complex process sequence.
- Optimum process conditions and process solvent to facilitate short-residence-time (low-severity) coal liquefaction.
- Constraints on and conditions required for solvent deashing of high-viscosity coal extracts.
- Upgrading high-viscosity deashed coal extracts in ebullated-bed catalytic reactors.

SUPPORT STUDIES/ENGINEERING EVALUATIONS

The FY 1982 effort supports generic studies applicable to liquefaction processes and new project starts. These studies include:

- Engineering assessments and independent technical status reviews.
- Environmental and toxicological assessment, including coordination of the environmental data for the different processes.
- Health effect studies related to upgrading of coal liquids.
- Ancillary processing studies, such as hydrogen generation.
- Standby and maintenance support for unused liquefaction facilities.

Experienced architect/engineering firms are engaged in this program to conduct reviews and to assist in appraising the technical status, commercial feasibility, and projected economics of various liquefaction processes.

Evaluation of the Texaco partial oxidation process for generating hydrogen from SRC residue has continued, and similar tests on H-Coal and EDS residues are underway. Commercial alternatives to the Texaco process for bottoms processing are also being examined.

Toxicity studies of coal liquids and process streams have continued and work has now begun on upgraded coal liquids.

Standby and maintenance support has been provided for the Cresap, West Virginia, facility and for the synthoil PDU facility. It has been determined that there are no further program requirements for these facilities, and action is being initiated to dispose of them. Technical support is provided by Pittsburgh Energy Technology Center to the major liquefaction pilot plant programs and to process development projects.

LIQUEFACTION DEMONSTRATION PLANTS

Solvent-Refined-Coal Demonstration Plant (SRC-I)

The major change in DOE's responsibility under new policy is reflected in the liquefaction demonstration plants category. In1978, DOE began a program to design, construct, and operate single modules of a full-size commercial plant that would demonstrate two variations of the Solvent-Refined Coal process. Preliminary designs have been completed for both plants, and detailed process designs are now underway.

The administration now recommends that the Synthetic Fuels Corporation consider providing support for continuing the designs and beginning construction of the plants in the event the private participants request such support. Authority to conduct such less-than-full-scale technical demonstrations is contained within the Energy Security Act, 1980 (P.L 96-294), creating the independent corporation.

Process design will continue on the SRC-I demonstration plant proposed for Newman, Kentucky, until Congress acts on the recently submitted recision request. Should Congress concur in the decision, the project would be terminated on or before the end of FY 1981. This, however, does not preclude the industrial partners from pursuing the demonstration and commercialization of this technology with funding solely from the private sector, or applying for financial assistance from the Synthetic Fuels Corporation.

The SRC-I process was selected for demonstration in FY 1978 because (1) the production of the SRC-I product is expected to be responsive to the need for fuels for utility boilers and other industrial markets; (2) the technology is the most mature of the liquefaction processes by virtue of the successful operation of the Wilsonville and Fort Lewis pilot plants, which have produced clean-burning solid fuel from high-sulfur coals; and (3) preliminary economic evaluations indicate that this process is cost competitive with other liquefaction processes.

The SRC-I demonstration program was initiated in July 1978 with the award of a contract to Southern Company Services, Inc. (SCS) for performance of preliminary plant design. The contractor completed work on phase zero, preliminary design, on July 31, 1979.

The contractor prepared a preliminary demonstration plant design and cost estimate, long-lead procurement plans, market and economic assessments, a detailed management plant for subsequent phases, a cost-sharing proposal, and environmental analyses for the project. As part of phase zero, a joint venture composed of Air Products & Chemicals, Inc. (APCI) and Wheelabrator-Frye, Inc. submitted a proposal to share with the government in the costs of proceeding through phases 1, 2, and 3.

Solvent-Refined-Coal Pilot Plants

The administration's request for funds will support the continued operation of the SRC pilot plant at Ft. Lewis, Washington, if the SRC-II demonstration project moves forward under the aegis of the Synthetic Fuels Corporation. In that case, the operation of the pilot plant is expected to be jointly funded with the corporation in accordance with the present international agreement wherein R&D funding is provided for the project. In the absence of joint support associated with the demonstration plant, the funds will be used to redirect the SRC pilot-plant efforts to long-term research and development.

The SRC pilot-plant activities are designed to facilitate the successful scaleup and precommercial demonstration of the SRC-I and SRC-II processes. This involves, therefore, verification of the designs for the demonstration plants through operation of pilot plants at the proposed conditions with the desired coals; evaluation of the components and specific design elements for scaleup such as preheaters, slurry mixing, and solid/liquid separation; erosion and corrosion tests; and evaluation of coals at pilot-plant scale before scaleup to demonstration plant operation.

To attain these objectives the process and equipment development have been conducted by means of the 30−50 ton per day SRC-I/II pilot plant at Fort Lewis, Washington, operated by Pittsburgh and Midway Coal Mining Company; the 6 ton per day SRC-I pilot plant at Wilsonville,Alabama, cofunded with the Electric Power Research Institute and operated by Catalytic, Inc. under subcontract to Southern Services Company, Inc.; and the 1 ton per day SRC-II PDU at Harmarville, Pennsylvania, operated by Pittsburgh and Midway Coal Mining Company under subcontract to Gulf Research and Development Co.

Fort Lewis Pilot Plant

The pilot plant at Fort Lewis, Washington, was originally designed to develop the SRC-I process. In the SRC-I mode the plant is capable of processing 50 tons per day of coal into a low-sulfur deashed solid. The plant was later converted to operate in the SRC-II mode. In this mode it can process 30 tons per day of coal into low-sulfur liquid fuels.

Wilsonville Pilot Plant

Operation of the 6 ton per day SRC-I pilot plant at Wilsonville, Alabama, was originally financed by industry through EPRI. Since 1976 it has been supported jointly by DOE and EPRI to continue process development work in SRC-I. Wilsonville provides a flexible test unit for investigating process performance on a variety of coals and for evaluating various component and process improvement options. To date, process performance data have been obtained on nine coals. Significant progress has been made in developing solid−liquid separation options. Activities will be terminated during FY 1981 and a recision of $1 million has been proposed to Congress.

The Future

Both the Exxon and H-Coal pilot plants are large enough so that with the experience gained from the SRC demonstration modules, both processes could be scaled to commercial size, leapfrogging the demonstration-plant stage. All three techniques could be in commercial practice by the 1990s, and would join the indirect liquefaction plants expected to be built in the 1980s.

The key hurdle confronting a synthetic fuels industry today, regardless of the process, is cost—both of the fuel itself and of the plant that produces it. Today, projected costs of coal liquids produced commercially range from $30 to $45 a barrel, the higher figure representing perhaps the products from "indirect" processes and the lower figure reflecting an optimistic projection of products from the more-advanced "direct" techniques. Until the first plant is built in the United States, though, the actual prices will remain highly uncertain. The continued high interest rate is a serious limitation on investment in synthetic fuel facilities.

Likewise, the true costs of the plants themselves can only be estimated today. Each SRC demonstration module is expected to cost more than $1 billion, although subsequent modules will probably cost somewhat less. Full-size plants have been estimated to cost from $2 billion to $4 billion each, and the actual costs by the time construction is completed could be more.

This approach—moving ahead with technologies that are commercially ready while maintaining a research and development effort to improve these techniques—will lay the foundation for a national synthetic fuels industry that can help America make the transition to a more reliable energy economy.

Chapter 4

Coal Gasification

Introduction

The science of turning coal into gas is hardly new; it has been practiced and abandoned periodically during the past two centuries. In 1813, the original London Bridge was lit by gas made from coal. Lighting for many European and American streets in the 19th and early 20th centuries was provided by "town gas," or as it was sometimes known, "water gas." The fuel was made by reacting coal with steam as shown in Fig. 4-1. The basic principles of coal gasification involve heating the coal under pressure and in the presence of steam and air or oxygen. The chemical bonds of coal break apart under these conditions, and a synthetic gas is formed. The process was expensive and troublesome to maintain, the efficiencies were low, and the operation was dirty.

Despite its drawbacks, however, gas made from coal played a major role in the United States until well into the 1900s, particularly in supplying markets far from natural gas fields but near coal deposits. As late as 1932, the residential gas market in the eastern United States was supplied mainly with synthetic gas made from coal.

When long-distance pipelines were constructed during World War II, cheaper natural gas became more convenient nationwide, and the use of coal gas in the United States declined. By the end of the war, synthetic gas was supplying less than 10% of the heating value of all gas sold in the United States.

Today interest in synthetic gas from coal is resurging, both in the United States and overseas. Several small-scale gasifiers suited for manufacturing fuel gas on-site (at the user's plant) are either operating or are on the drawing boards of industrial firms. And many groups, in both government and industry, are looking again at the possibility of using coal gasification as the first step in producing vitally needed liquids.

In many cases, the technology is available—and has been for several years. Obstacles have been the high cost and risk of first-of-a-kind ventures.

Concurrent with this effort is a joint government/industry research program that is attempting to improve existing processes by increasing the variety of usable coals, reducing costs, and producing gases tailored for different uses.

WHY COAL GASIFICATION?

The prime reason for coal gasification is an increasing interest in finding substitutes for oil, particularly that being imported from foreign suppliers. America spent almost $90 billion in 1980 for foreign oil—more than $10 million an hour. With that payment flow jobs and the ability to maintain our economic and national security. Moreover, there is no assurance that further interruptions in the supply of foreign crude will not occur as prices continue upward.

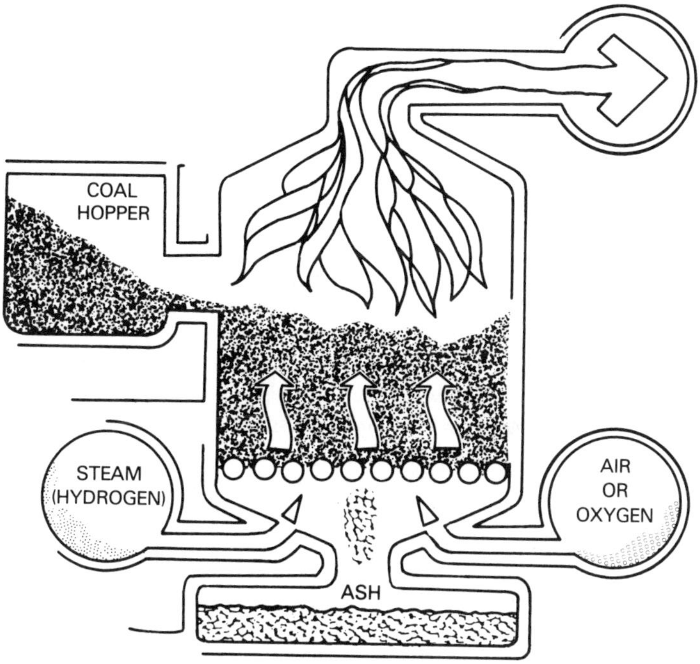

Fig. 4-1. Basic principles of coal gasification.

In many applications, gas can easily substitute for oil—in utility and industrial boilers, for example. But the demand for gas could eventually outrun easily attainable natural supplies. Although constituting only 4% of the nation's conventional energy reserves, natural gas supplies roughly 25% of consumption. America's proven producible gas reserves add up to roughly a 10-year supply at present demand rates, and, even today, some users have been forced to convert to other, high-cost energy supplies, including foreign oil.

Increased prices, brought on by the Natural Gas Policy Act of 1978, have opened new domestic supplies of natural gas for the near future, but over the long term the extent of the nation's gas reserves is at best uncertain.

WHY USE COAL?

The driving reason to use coal is the amount of coal available. Perhaps as much as 430 billion recoverable tons lie underneath the United States, almost one-half the known free world reserves. As the Persian Gulf is to oil, America is to coal.

But coal is not without its disadvantages. It is bulky, difficult to transport, and often laden with sulfur. As a solid, it is in the wrong form for 55% of American homeowners who rely on gas for heating and cooking.

Similarly, many industries, particularly the smaller ones, are geared to use gas. Changing to another fuel could mean expensive replacement of combustion equipment. And any decision to burn coal directly in many of these installations might mean a sizable and additional investment for capital, land, technical talent, and environmental cleanup equipment.

A more acceptable alternative may be to convert the coal into clean-burning, easily transportable gas that can be used in equipment now burning natural gas or oil.

Changing coal to gas could also open up new supplies for the chemical industry, much of which depends on a steady flow of hydrocarbons to manufacture thousands of products that have become commonplace in today's life. By using well-known chemical techniques, gases from coal can be rebuilt into a spectrum of products—from fertilizers to gasoline.

HOW IS IT DONE?

The basic chemical process is to react coal with water (as steam). The technique was discovered in the alchemists' laboratories as far back as 1780.

While the sizes and shapes of the equipment or the way heat is supplied to break the coal apart may change, the chemical transformations that take place in today's gasifiers are based on this 200-year-old discovery.

As shown in Fig. 4-2, coal is first crushed and usually dried, then fed into a large vessel where it reacts with steam and either air or oxygen. The reaction usually occurs at high temperatures [from 1500 to 3500°F (816 to 1927°C), depending on the process] and high pressures [up to 1600 pounds per square inch (psi)]. Under these conditions, coal begins to break down, and hydrogen atoms from the steam, along with oxygen atoms, begin to link up with the carbon atoms of coal to produce a mixture of gases. This mixture is usually made up of carbon monoxide (CO), hydrogen (H_2), and varying amounts of methane (CH_4).

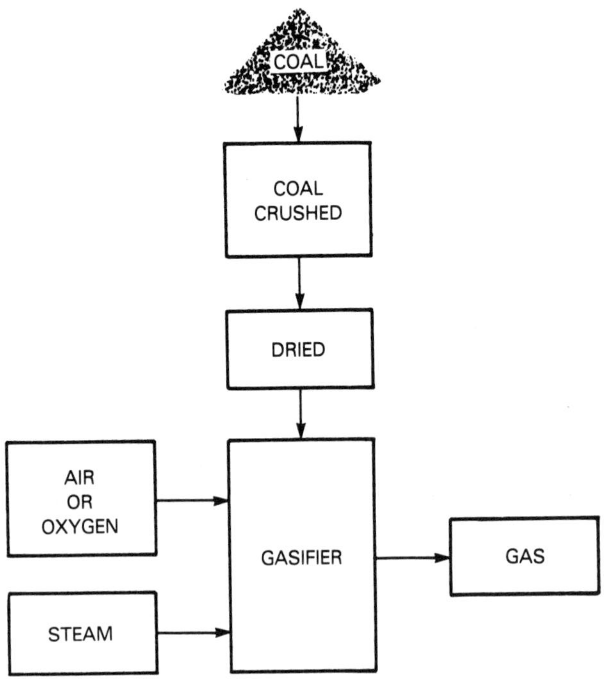

Fig. 4-2. Coal gasification block diagram.

If air has been used, the gases also contain a high percentage of nitrogen and have a heating value roughly one-tenth that of natural gas—or about 100 Btu per cubic foot. This relatively low heating value accounts for the name "low-Btu" gas. In the 1930s, it was called "town gas."

Injecting oxygen rather than air into the gasifier reduces the nitrogen content and raises the heating value of the gas to the "medium-Btu" range—or roughly one-third that of natural gas, or about 300 Btu per cubic foot.

Because of these heating values, neither low- nor medium-Btu gas can be piped economically the long distances of natural gas. The gases, however, are ideally suited for industrial purposes, for utility boilers or turbines, or for use as chemical feedstocks near the point of production.

If the goal is a high-quality, or "high-Btu," gas (also called substitute natural gas, or SNG), the percentage of methane in the gases must be boosted (raising the heat content to 950–1000 Btu per cubic foot). Methane is the chief constituent of natural gas. To make it from coal gases, two additional steps are required. First, some of the carbon monoxide made in the gasifier must be reacted with steam to form additional hydrogen. This step, called *shift conversion*, sets up the proper ratio of gases for the next step, called *methanation*.

In methanation, the carbon monoxide and hydrogen flow over a nickel-based alloy, as shown in Fig. 4-3. This nickel acts as a catalyst, causing the gases to recombine to form methane. After purification, the methane can be fed directly into a natural gas pipeline.

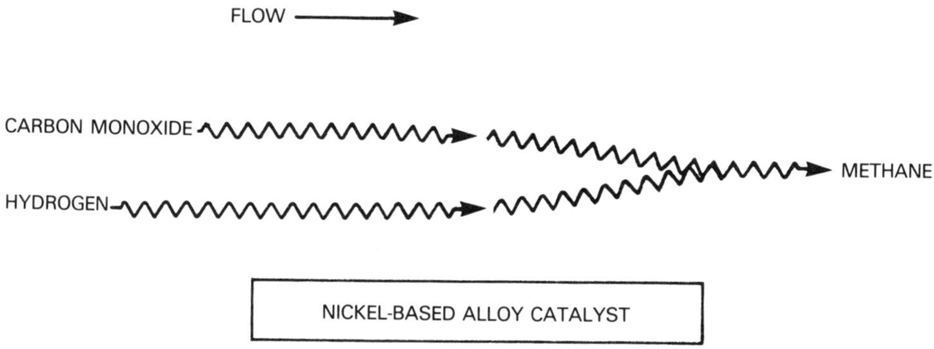

Fig. 4-3. Methanation schematic diagram.

WHAT ARE THE OBSTACLES?

The foremost obstacle is the high cost of the gas. None of the synthetic gas processes—even those using the most advanced technology—are yet able to compete in the United States with natural gas or petroleum at today's prices.

Best estimates place the cost of pipeline-quality, high-Btu gas in the range of $4.00–$5.50 per thousand cubic feet (1982 dollars), roughly double the current national average for natural gas. (Lower-quality synthetic gas may cost somewhat less.)

The synthetic gas plants themselves are also expensive. Even if companies could be assured that the price of synthetic gas could compete with other gas sources in the 1980s, a commercial 30,000-ton-per-day coal gasification project represents such a massive capital investment—as much as $1.5–$3 billion—that even the largest companies to date have balked at plant development.

Another obstacle is the coal itself. Unlike many areas of the world, the United States contains coals that vary widely with geographic region. Each type of coal displays different characteristics when subjected to the conditions required for gasification.

For example, most high-sulfur bituminous coals found in the eastern United States tend to swell and stick together when heated. The caked coal often clogs the gasifier, shutting down the process.

Western subbituminous coals and lignites, on the other hand, gasify much more easily without caking. These coals behave similarly to European coals. Therefore, many of the gasifiers sold today overseas could find application in the western United States.

But large-scale use of western coal may compete in some areas with agriculture, particularly in the use of sometimes scarce water. It may be possible, however, to use deep aquifers or water supplies unsuited for other purposes or perhaps to use various water-conserving techniques to meet the demands of a western gasification project.

Still, if western coal use is to increase, transportation networks must be extended to ship either the coal or the synthetic gas to the more populous, industrial sectors of the nation. And special attention must be paid to the impacts coal gasification plants will have on communities unaccustomed to the rapid influx of the thousands of workers that will be needed.

Environmental implications of making gas from coal must also be a major concern. While gasifiers remove much of the sulfur and other potential pollutants from the gaseous product, they also produce residual wastes that will have to be controlled. The amount and type of these wastes vary from process to process, although studies indicate that the quantities could be substantially less than from an equivalently sized power plant burning coal and using scrubbers.

Testing is now underway to determine if the solid wastes from coal gasification plants could leach into waterways and groundwater. Initial results have shown that the ash from several gasification processes has leaching properties well below the levels proposed in recent legislation. Conventional landfill and ash disposal techniques commonly used by coal-fired power plants are expected to be sufficient for disposing of this material.

Toxic elements from synthetic gas plants, unless adequately controlled, could adversely affect human health. Tars and aromatic hydrocarbons resulting from synthetic fuel production are believed to contain cancer-causing agents and other harmful substances that will require careful handling. Technology does exist to alleviate many of these problems, although the uncertain and changing nature of environmental regulations has been a key factor in stalling large commercial commitments to synthetic fuel plants.

HOW FAR HAVE WE COME?

Several processes have been developed to convert coal into gas. Many are used commercially around the world today, primarily in European countries and in South Africa. The larger units generally produce a medium-Btu gas for manufacturing chemicals. Some of these processes have been on the drawing boards of many American industries for years, but no large-scale plant has ever been built in the United States.

Many techniques being considered for pioneer U.S. ventures trace their heritage back to the work of German engineers in the 1920s and 1930s. One such benchmark technology, the Lurgi gasifier, as shown in Fig. 4-4, was first developed in 1936. Today,

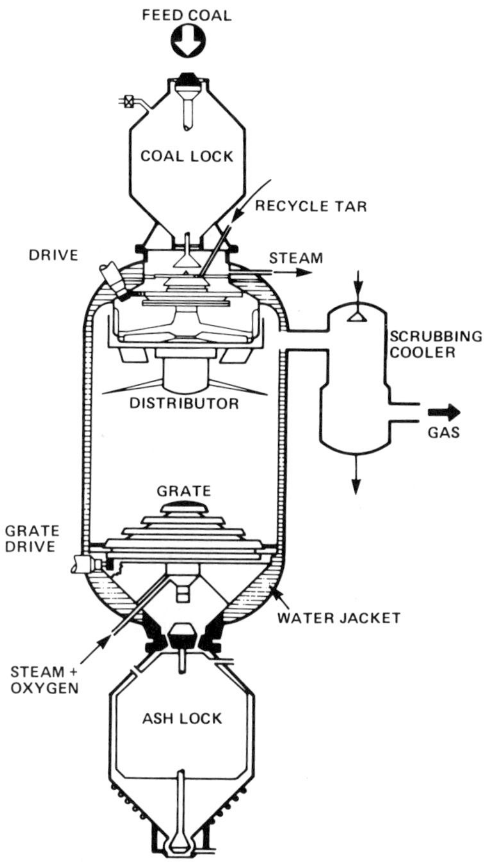

Fig. 4-4. Lurgi gasifier diagram.

much-improved versions of it have been the choice for most of the commercial synthetic pipeline gas projects now being planned.

The Lurgi gasifier will form the core of a project now planned for Beulah, North Dakota, in Mercer County. This effort, the most active of any major coal gasification project in the United States, has been in the planning stages since the early 1970s. Designed to produce 138 million cubic feet per day of high-Btu gas, the project is projected to cost close to $1.5 billion, including an adjacent mine and related piping equipment.

In August 1978, the Department of Energy (DOE) testified before the Federal Energy Regulatory Commission, which sets pricing regulations for interstate transport of gas. DOE voiced its support for regulations to permit the five gas companies involved—a consortium headed by American Natural Resources of Detroit, Michigan—to pass a portion of the plant's cost on to their customers should it fail for nontechnological reasons.

DOE also supported "rolling in," or averaging, the price of the synthetic gas with that of the companies' conventional sources. On November 21, 1979, the FERC agreed with most of the requests.

Industrial firms are reassessing the economics and markets for plants using any one of a family of "first-generation" gasifiers, including not only the Lurgi but also the German-developed Koppers–Totzek and Winkler gasifiers. Recently, the American-developed Texaco and the Shell (Dutch) Koppers gasifiers have joined the ranks as potential candidates for commercial plants producing high- or medium-Btu gas either for direct use or as a source of chemicals for making liquids.

For certain firms, particularly those located near readily available coal and needing a source of process heat, smaller-scale low-Btu gasification is becoming more attractive. Although only a few companies operate low-Btu gasifiers today, estimates show that today's energy situation may cause as many as 400 or more firms to install on-site gasifiers by the year 2000. In this category, such units as the STOIC, Wellman–Galusha, or Riley–Morgan gasifiers are considered "off-the-shelf" technology.

WHAT'S IN THE FUTURE?

Even though commercially available gasifiers are being used today in various parts of the world and interest is growing in the United States, they have some drawbacks.

The Lurgi gasifier, for example, works at moderate pressures, uses large quantities of expensive oxygen, and may only be able to work on western noncaking coals. It also cannot handle small pieces of coal, called *fines*.

More advanced gasifiers are being developed to handle a wider range of coal, particularly the caking coals of the Midwest and East. To prevent the coal from sticking together, newer designs partially char the coal prior to gasification, or speed the coal's movement in the gasifier through the critical temperature range that causes caking, or use mechanical means such as rotating stirrers to break up caked particles of coal. The newer gasifiers also operate at increased efficiencies with the capability of processing more coal (including fines) in the same reactor volume with a lower product cost.

Several processes have emerged from laboratory research as promising "second-generation" technologies. In the late 1960s many of these techniques passed their first engineering tests in "process development units" (PDUs) (where coal is processed at rates up to a few tons per day). In the 1970s, those with the most potential were scaled up to "pilot plants" (which process coal at rates of about 100 tons per day).

Today, some of these technologies are moving into the last stage before they are deemed commercially ready: the demonstration plant (processing over a thousand tons of coal per day).

Synthetic gas from coal could be a large contributor to the nation's domestic energy mix both as a gaseous fuel and as a source of valuable hydrocarbon liquids. Facing the potential investor in synthetic gas, however, are the high production costs compared to today's conventional supplies, and the regulatory and institutional barriers. If these can be overcome, synthetic gas from coal can translate into direct reductions in oil imports.

Gasification Process Descriptions

This section provides descriptions of various coal gasification processes and a brief overview of process development.

HIGH-BTU GASIFICATION

In the high-Btu, or pipeline-quality, gas category, some of the major pilot plants include

- Hygas—Operated for DOE and the Gas Research Institute by the Illinois Gas Technology in Chicago.
- BiGas—Operated for DOE and the Gas Research Institute by Bituminous Coal Research working with Phillips Petroleum Co. at Homer City, Pennsylvania.
- CO_2 Acceptor—Operated for DOE and the Gas Research Institute by the Conoco Coal Development Company in Rapid City, South Dakota. Operations completed in 1977.
- Synthane—Operated for DOE by the CE-Lummus Company at Bruceton, Pennsylvania. Operations ceased in December 1978.

DOE-funded designs for two large demonstration plants which could produce pipeline-quality gas using the

- British Gas/Lurgi Slagging Gasifier—An advanced version of the conventional Lurgi specially designed to accept caking coals. Being developed by the Conoco Coal Development Company in cooperation with the British Gas Company, based on data from pilot testing in Westfield, Scotland.
- Cogas Gasifier—Being developed by the Illinois Coal Gasification Group, a consortium of five gas utilities. Process would be linked with the COED pyrolysis process which produces heavy coal liquids and a char, which is then gasified. Tested at pilot plant in Leatherhead, England.

DOE planned to select one of these demonstration projects for actual construction. The Conoco plant would be built in Noble County, Ohio, and the Illinois Coal Gasification Group plant would be built in Perry County, Illinois.

Further into the future is a new generation of coal gasification processes which may offer even greater improvements in both cost and the flexibility of using a wider variety of coals. Although often called "third-generation" techniques, these gasifiers are novel departures from the more conventional units in use or under large-scale development today. Instead of requiring that much of the methane be made in separate equipment, these gasifiers manufacture almost all their methane in the gasifier itself—which could reduce the capital expense of shift and methanation equipment.

Like the second-generation gasifiers, the newer techniques are also being designed for handling caking coals. Two of these processes are

- Hydrogasification—Uses many of the same principles as a rocket engine to mix pulverized coal directly with hot hydrogen to produce methane. DOE-sponsored work with Rockwell International and Cities Service Co.
- Catalytic Gasification—Developed by Exxon Research and Engineering Company. Uses a catalyst made from potassium to speed the chemical reaction of coal and steam at lower temperatures than conventional techniques. Product is continually recycled until it is of pipeline quality.

LOW-/MEDIUM-BTU GASIFICATION

As in the high-Btu category, several processes that produce low- or medium-gas are also being tested at the pilot or process development unit scale, including

- Pressurized Fluidized Bed—Testing of gasifier has been underway by Westinghouse since 1978 at small unit near Madison, Pennsylvania (Waltz Mill).

- Atmospheric Entrained Bed—Testing of gasifier has been underway since 1978 by Combustion Engineering at pilot plant near Windsor, Connecticut. Early in 1980, the Department of Energy announced plans to begin contract talks with the city of Memphis, Tennessee, to complete its design and build a demonstration plant using a third type of low-/medium-Btu gasification process.

This technology, which Memphis has been using since 1977 to prepare a preliminary design, is termed

- U-Gas—Developed by the Institute of Gas Technology in Chicago. Pending completion of environmental studies, construction could begin on the demonstration plant in fiscal 1981 and operations could begin in 1984. (Process is sometimes called "ash agglomerating gasifier.")

Also in 1977, W. R. Grace & Co. was selected to design a similar-size plant to produce a medium-Btu gas using the Texaco Coal Gasification process. Unlike the Memphis project, where the gas will be used as fuel for nearby industries, W. R. Grace planned to convert it into ammonia. In 1979, however, the firm proposed to revamp its project to a full-size plant that would use the gas to make methanol, which would then be converted to gasoline. DOE approved the concept in February 1980 and began negotiating a contract with W. R. Grace for a conceptual design of a facility planned for Baskett, Kentucky.

DOE began contract negotiations with Combustion Engineering to scale up their "atmospheric entrained bed" technology in a demonstration plant at West Lake, Louisiana.

GASIFIERS IN INDUSTRY

While no large commercial gasification plant exists today in the United States, several firms are turning to small-sized, on-site gasifiers as a reliable source of low-Btu fuel gas. In 1976, DOE selected several projects in this category to begin renewing industry's knowledge and confidence in these gas producers. Four projects are currently underway:

- Glen Gery Brick Co.—Low-Btu gas for brick kiln at York, Pennsylvania. Currently operating.
- University of Minnesota—Low-Btu gas for space heating boiler on Duluth campus.
- Pike County, Kentucky—Low-Btu gas for heating and cooling housing, a school, a nursing home, a fire station, and commercial facilities.
- Can-Do—Will use anthracite to produce low-Btu gas for industrial park at Hazelton, Pennsylvania.

DESCRIPTION OF PROCESS TECHNOLOGY

The following process descriptions include an overview, detailed process description, project development history, and progress report. The following coal gasification processes are described:

- Hygas process (steam−oxygen)
- Hygas process (steam−iron)
- BiGas process (pipeline gas)
- CO_2 acceptor process

- Lurgi (dual feed) process
- SRT hydrogasifier process
- Dilute-phase hydrogasification
- Atmospheric entrained-bed process
- Tri-Gas process
- Pressurized fluidized-bed process
- Molten salt gasification process

HYGAS PROCESS (STEAM–OXYGEN)

The Hygas process is being developed by the Institute of Gas Technology (IGT) as part of the $22,718,212 joint program of DOE and the Gas Research Institute (GRI). DOE is providing two-thirds of the funds and GRI one-third.

The objective of this contract is to operate the Hygas pilot plant facility with a mildly caking Illinois bituminous coal in order to develop an improved data base for the design of a conceptual commercial/demonstration plant based on the Hygas process.

Process Description

The Hygas process description is separated into two segments, including steam–oxygen using eastern coal and steam–oxygen using western coal.

Hygas (Steam – Oxygen) Eastern Coal. In the Hygas process, shown in Fig. 4-5, coal is reacted with hydrogen-rich gases to produce methane directly in a hydrogasifier. It is a high-pressure process for all types of coal. The hydrogasifier consists of a slurry drier and three high-pressure fluidized-bed reactors. Coal is dried, crushed, slurried with a process-

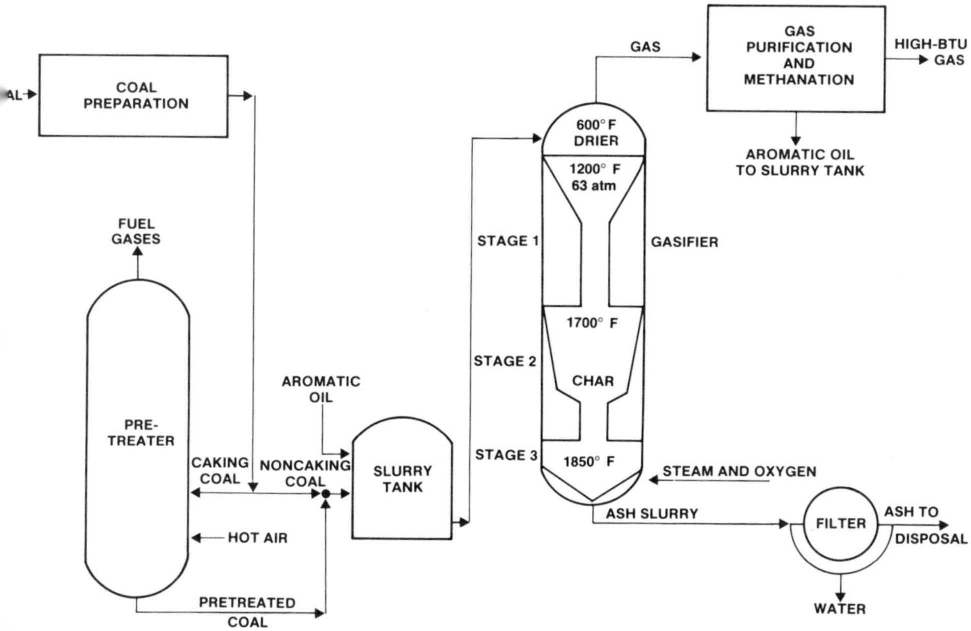

Fig. 4-5. Hygas (steam–oxygen) process diagram.

derived oil, and compressed to 1500 psi. The slurry is then moved into the slurry drier section of the hydrogasifier, where it is sprayed on a fluidized bed. The oil is vaporized and removed with the hot process gases rising through the bed from the lower stages of the gasifier. Vaporized oil is recovered by quenching and is reused. The dried coal is routed to the first stage of the gasifier, which is located below the drier and above the second stage. The coal enters the gasifier at mid-height and is heated to about 1200°F (649°C) by the hot reaction gas from the second stage. The first stage [1200°F (649°C), 1000 psi] favors the direct production of methane from the coal. Solids are channeled to the second stage, where rising gases contact the char for further gasification at about 1700°F (927°C). Unreacted char and ash drop into the steam–oxygen gasifier for a final gasification, producing a hydrogen-rich gas and steam mixture, at about 1850°F (1010°C).

The raw gas leaves at the top of the slurry drying section through a quench tower, which removes dusts, the slurry oil, and excess steam. Gas purification occurs in an acid removal system, which consists of an amine–water scrub followed by a caustic wash and a water wash. CO_2 and H_2S removal are essentially complete after the amine scrub. A Claus plant converts the sulfur products to elemental sulfur. After shift reaction, methanation takes place in two packed-bed downflow reactors. Heat-release control is by the IGT split-flow recycle method.

Hygas (Steam – Oxygen) Western Coal. When the Hygas plant is fed with western coal, the coal pretreater is eliminated from the process train. Western coal is noncaking and does not require pretreatment prior to entering the fluidized bed.

History of the Project

Development of the Hygas process by IGT under the sponsorship of the American Gas Association (AGA) began in 1946, and continued under joint sponsorship with the Office of Coal Research (OCR). Responsibility has since been transferred to DOE and the GRI. The pilot plant was designed and built to convert 75 tons of coal per day to 1.5 million cubic feet of clean high-Btu gas. During 1973, the plant produced pipeline-quality gas from the coal in sustained test operations. Since then, the duration of the test runs has increased, and longer, more stable operating periods have been achieved.

During 1975, 11 test runs were made in the Hygas pilot plant. The longest run (Test 37) was made in July, demonstrating the feasibility of using lignite. Fifteen days of self-sustained operation were achieved with over 1000 tons of lignite processed. The entire plant was on stream for 92% of the time, or 333 hours. Various individual steady-state periods were achieved which totaled 160 hours. The heat and material balances were within ±5% of closure. Subsequent to Test 37 with lignite, IGT was authorized to start testing with highly caking Illinois No. 6 bituminous coal.

Twenty-six runs were made with the ash-agglomerating gasifier during 1975, also. Process parameters were varied to determine the optimum operating conditions. Using coke breeze, agglomerates were successfully produced for a 200-hour steady-state period, and full dust recycle was maintained.

Concurrent laboratory research was directed toward the study of the penetration of slurry liquid into coal pores, effects of acidic condensing atmospheres on castable insulations, and methanation catalyst evaluation.

The liquid-phase methanation pilot unit was received at the Hygas plant in October, and all systems had been checked out by the end of December 1976.

Tests were conducted in 1976 in the Hygas pilot plant with pretreated Illinois No. 6 bituminous coal (highly caking) as well as with Rosebud subbituminous coal. Tests during 1977 used Peabody No. 10 mine coal.

Of major technical and economic importance to this process is the production of the hydrogen-rich gas required for hydrogasification. During the early part of 1974, shake-down runs and batch tests of the 2 MW electrothermal gasifier for producing hydrogen-rich gas were completed. Although this technique was demonstrated to be technically feasible, it was found to be too costly owing to the large quantity of electric power required. Therefore, experimentation using this technique was discontinued. A steam–oxygen unit, which gasifies char, was constructed and installed in the base of the Hygas pilot plant hydrogasification reactor during 1974 and is presently being used.

A fire occurred in June 1978 in Chem System's Liquid Phase Methanation Unit, with resultant damage to wiring and instrumentation. The fire did not affect the Hygas pilot plant. During the third quarter of 1978, however, the Hygas plant was shut down for maintenance and for annual vacations. Following repairs, Tests 74 through 77 were run in 1978, generating additional process data with different coals at different operating pressures.

HYGAS PROCESS (STEAM–IRON)

The steam–iron process is being developed by the Institute of Gas Technology as a possible alternative for producing hydrogen for use in the Hygas pilot plant and as a means of producing hydrogen for industrial or other coal conversion processes. The process is being developed under the auspices of DOE and the American Gas Association. DOE is providing two-thirds of the funds and AGA one-third.

The objectives of this contract are to design, construct, and operate a steam–iron pilot plant facility for the production of hydrogen. The system will provide process and design data for a commercial-size facility.

Process Description

The main units of this process, shown in Fig. 4-6, are a two-stage fluidized-bed hydrogasifier and a steam–iron process unit producing hydrogen for hydrogasification. Caking coal is first made nonagglomerating by pretreatment with hot air in a fluidized bed at 1 atm and 750°F (399°C) and is then mixed with light oil to form a slurry. This slurry is pumped into a fluidized drying bed, operating at 600°F (316°C) and 1000–1500 psi. The light oil evaporates, and coal from the drying bed passes successively through the first stage of the gasifier, where devolatilization and partial noncatalytic methanation occur at 1300–1500°F (704–816°C) in the presence of hydrogen-rich gas, and then as char into the second stage, where partial gasification at 1700–1800°F (927–982°C) occurs by reaction with the steam and hydrogen-rich gas from the steam–iron process. The product gases from the drying bed flow to the purification and catalytic methanation systems.

Char from the second stage of the hydrogasifier is fed into the producer vessel where it reacts with air and steam to generate a gas capable of reducing magnetite (Fe_3O_4) to iron. The reaction takes place at 1900–2000°F (1038–1094°C) and 1100 psi. The hot producer gas is fed to the reducer vessel in which it contacts a recirculating stream of magnetite to produce iron. The reaction is at 1400°F and 1100 psi. The iron is then fed to the oxidizer in which it contacts steam and is reoxidized to magnetite at the same temperature and pressure, producing a mixture of hydrogen and steam that is fed

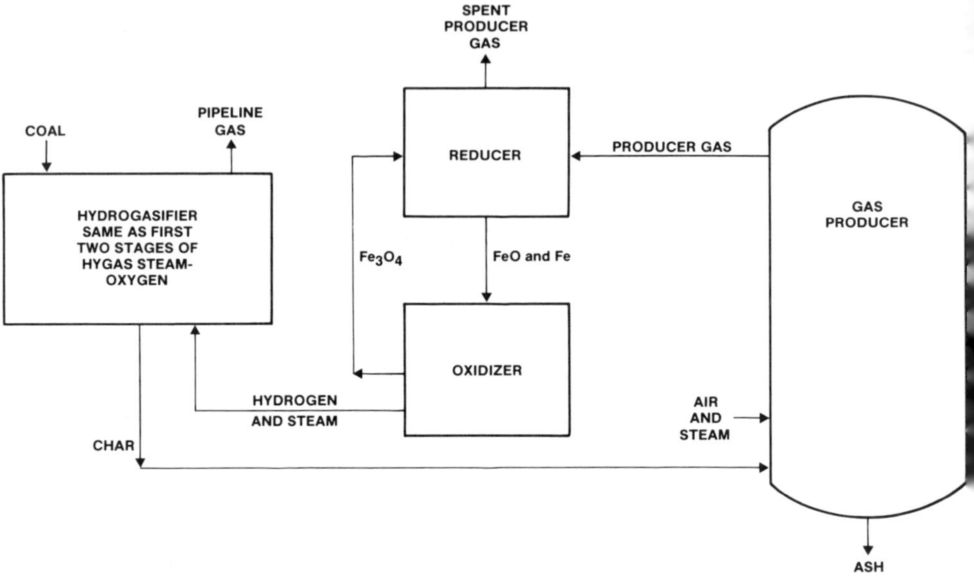

Fig. 4-6. Hygas (steam–iron) process diagram.

to the second stage of the hydrogasifier. The magnetite from the oxidizer is recycled to the reducer.

When the Hygas plant is fed with western coal, the coal pretreater is eliminated from the process train. Western coal is noncaking and does not require pretreatment prior to entering the fluidized bed.

History of the Project

Steam–iron process research has been conducted periodically for many years. The earliest use of the steam–iron process featured a fixed, packed bed of iron ore that was reduced to iron with gases containing carbon monoxide and hydrogen. The iron was then oxidized with steam to produce iron oxide plus a mixture of hydrogen and unreacted steam. Continuous hydrogen production was provided by using two beds that were cyclically reduced and oxidized. This process was neither economic nor compatible with high-pressure operations and was dropped in favor of steam reforming of natural gas as a method for producing hydrogen. The process was revised in the late 1950s by the U.S. Bureau of Mines for use in producing hydrogen for coal gasification. In 1961, IGT began working on the process and continued until 1971. This work was performed under the sponsorship of Fuel Gas Associates. In 1973, IGT resumed work on the project and has been developing the process under the joint sponsorship of the Office of Coal Research and AGA.

The pilot plant was completed in 1976, concluding the tasks under contract E(49–18)–1581. Subsequent operations were continued under a new contract. The first iron ore solids circulation tests were made in 1976. The first producer feed test was successfully completed in March 1977. Integrated operation of the producer and steam–iron reactor was achieved for a short period for the first time in October 1977.

The major objective throughout 1978 was to demonstrate the feasibility of the steam–iron process and to successfully produce hydrogen from this process during extended periods of operation. Eight runs were conducted in 1978, making a total of 20 runs over the 24-month project. The objective was only partially achieved, and the feasibility of the process was not demonstrated at design conditions.

During Test 20, which terminated at the end of the project period on September 30, 1978, the longest hydrogen production period was attained. Hydrogen was produced for 130 hours of continuous operation. On September 30, the operation of the steam–iron plant was discontinued. The final report for the project has been prepared.

During the 24 months of operating the steam–iron plant facility many accomplishments were achieved:

- An intensive development program was completed. A total of 355 days of operation was accumulated.
- The technical feasibility of continuously recirculating a stream of iron solids at high pressure and temperature was demonstrated. A total of 1393 hours of ore circulation were logged at rates up to 71 tons per hour. Circulation during the last few tests was smooth and easily controlled.
- A method of introducing dry coal to a high-pressure, fluidized-bed system was developed, using high-pressure mud pumps and a slurry vaporizer. Slurry concentrations up to 50 weight percent were processed through the system over a period of 944 hours.
- The technical feasibility of the combined steam–iron process for hydrogen production was indicated. A total of 342 hours of hydrogen production were obtained.
- The producer reactor was operated at design conditions. Char feed rates up to 2 tons per hour were maintained at reactor pressures of 500 psig and bed temperatures of 1800°F.
- Hydrogen production was increased to 5000 standard cubic feet per hour. Additional operating time would have been required to reach design temperatures.
- Hydrogen purity reached 90% and 95% on a nitrogen-free basis in Tests 17 and 19, respectively. (Nitrogen was introduced through instrument purges and aeration of seal legs.) The high purity was obtained without introducing any stripping gas into the reducer/oxidizer seal leg.
- Iron ore attrition rates were very low and consistent with the design data. No ore was added during each of the five continuous days of hydrogen production in Tests 17 and 20. During those periods no significant decrease in bed levels was noted.
- A reactor pressure control valve system was developed using oil-field choke valves.
- A number of solids control valve designs capable of operating in a fluidized bed of solids were developed and successfully tested.
- A high-pressure, high-temperature slip joint was developed and successfully tested in 16- and 20-inch-diameter service.

Important operating achievements during 24-month operating period are summarized in Table 4-1.

Table 4-1. Achievements Realized During 24-Month Operating Period of the Steam – Iron Process for Hydrogen Production

Total number of test days	355	
Hours of ore circulation	1,393	
Hours of char feed	944	
Hours of hydrogen production	342	
Char feed (tons)	864	
Average char feed rate (tons/hr)	0.92	(50% of design)
Maximum char feed rate (tons/hr)	2.00	(110% of design)
Char gasified in producer (%)	2 – 70	(Design conversion: 70%)
Reactor pressure (psig)	100 – 500	(Design operating pressure: 1000 psig)
Producer bed temperature (°F)	1200 – 1750	[Normal operating temperature: 1800°F (982°C)]
Steam – iron bed temperature (°F)	700 – 1300	[Normal operating temperature: 1500°F (816°C)]
Ore circulation rates (tons/hr)	6 – 71	(Normal circulation rate: 45 tons/hr)
Maximum hydrogen production (scf/hr)	5000	(Production rate at 500 psig: 22,000 scf/hr)

BIGAS PROCESS (PIPELINE GAS)

The BiGas process is being developed by Bituminous Coal Research, Inc. (BCR), in an effort to develop a method for producing high-Btu pipeline gas from coal. The pilot plant operational program is being managed by Phillips Petroleum Company with process engineering and operations personnel supplied by Stearns-Roger, Inc. Project funding is $66 million.

The principal objective of the program "Gas Generator Research and Development" is the development of the BiGas process for gasifying coal to produce high-Btu pipeline gas. Specifically, BCR is directed to

- Conduct laboratory-scale coal gasification experimentation, together with process equipment development, to verify the technical and economic feasibility of producing high-Btu gas using the BiGas process.
- Design, construct, and operate a multipurpose research pilot plant with the aid of engineering subcontractors.

BIGAS PRODUCTION PROCESS

The BiGas reactor, shown in Fig. 4-7, is a two-stage gasifier of the pressurized, entrained-flow, slagging type. The gasifier has three sections separated by formed restrictions: stage two (topmost) receives dense phase coal in gas-entrained flow through injectors located in a throat between stages two and one. Steam is also injected there. Process gas and char flow from the top of stage two of the gasifier to a cyclone. The char is recycled from the cyclone and fed to steam nozzle and then, with oxygen and steam, to the bottom of stage one. The products of char gasification in stage one are hot gas feeding into stage two and slag dropping into stage three, the slag quench section. Molten slag runs down the walls of stage one of the pressurized gasifier.

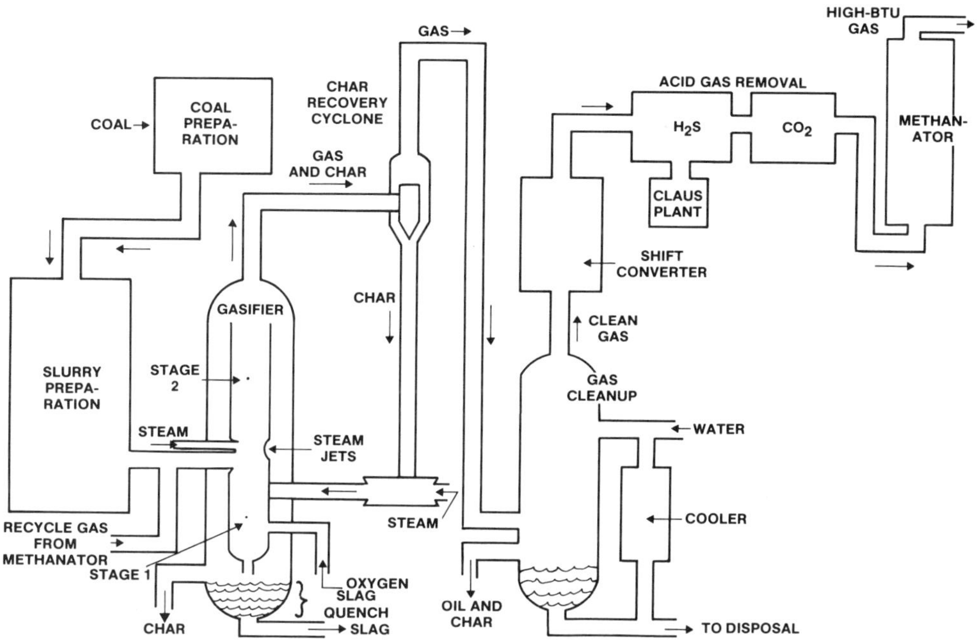

Fig. 4-7. BiGas production process diagram.

Raw gas leaves the char recovery cyclone, passes through a water wash column, and flows to the shift reactor where the H_2 to CO ratio is adjusted to the desired value. H_2S and part of the CO_2 are removed in the Selexol unit upstream of the methanator. H_2S gas goes to a Claus unit. The product gas from the methanator goes through final CO_2 removal in the Selexol unit. The methanator is a fluidized bed with embedded water-cooled tubes, and it offers great potential for commercial development.

The BiGas reactor can handle all types of coal without the need for a pretreatment step, and because of its high operating temperature of over 2500°F, there is no tar in the product gas, thus simplifying the gas cleanup step. High methane yields are achieved, and coals can be gasified without prior treatment, since the process uses an entrained rather than a fixed- or fluidized-bed system.

In summary, the BiGas process has several advantages:

- A high yield of methane is obtained directly from coal, minimizing subsequent processing of the product gas.
- All types of coal can be gasified without prior treatment, since the process uses an entrained rather than a fixed- or fluidized-bed system.
- The reaction conditions in the upper stage of the gasifier are such that no tar and oils are formed in the gasification process.
- All the coal charged into the process is consumed; the principal by-products are slag for disposal and sulfur for sale.

- Gas generated at high pressures is adequate for transport in existing gas distribution systems without further compression.

Potential operating problems are

- The production of naphthalene.
- The generation of excess coal fines during grinding of coal feed, especially with Rosebud Mine coal.
- Instrumentation relating to char flow control and stage 1 temperature measurement.

History of the Project

This project was started in December 1963. BCR's first experiments confirmed the basic assumption that a high yield of methane could be obtained directly from coal by reaction with steam at elevated temperatures and pressures. BCR next conducted continuous flow experiments in an externally heated reactor processing 5 pounds of coal per hour. The experimental data were extrapolated to design a process development unit that would simulate the conditions of stage two of the BiGas process. The unit was internally fired and could process 100 pounds of coal per hour. Experiments were designed to determine the optimum residence time, coal rank, and processing conditions (pressure, temperature, and hydrogen partial pressure) that would produce the greatest yield of methane. North Dakota lignite, Wyoming subbituminous C coal, and Pennsylvania high volatile A bituminous coal were used in the experiments. During these experiments, BCR found that the physical design of the stage two process development unit influenced methane yield. Therefore, a cold flow model of stage one and the bottom of stage two was developed to investigate methods for improving the flow patterns in stage two and to establish design criteria for the slagging section of the gasifier.

BCR developed the basic design criteria for a fully integrated pilot plant capable of processing 5 tons of coal per hour and producing 100,000 scf of clean pipeline-quality gas per hour. Stearns–Roger was awarded the contract for constructing the $34 million facility on July 11, 1972. The pilot plant management contract was subsequently awarded to Phillips Petroleum Company in November 1974. The gasification vessel to be used in the pilot plant was designed and built by Babcock and Wilcox, a Stearns–Roger subcontractor. The pilot plant was completed in 1976 and startup operations began. Formal acceptance of the plant from Stearns–Roger Construction Co. was made by Phillips Petroleum Company on September 15. By the end of the year, the flash drying system and gasifier system were operated successfully.

Testing of the gasifier was started in December 1976. In a base condition test, stage one was fired with natural gas, oxygen, and steam, and stage two was fed only purge gas and steam (no coal). The conditions were designed to match, as nearly as possible, the gasifier conditions that would exist when operating at 750 psig with a 3.5 tons per hour coal feed to stage two and char recycle to stage one. The objectives of the test were to determine gasifier heat loss and to establish reliability of equipment, instrumentation, and data acquisition system. The data indicated that the total heat loss from the system was about 7.3×10^6 Btu/hr under those conditions. With no slag coating on the cooling tubes of stage one, the heat loss represented about 47% of the total heat generated in stage one. Loss from stage one alone was about 5.1×10^6 Btu/hr, representing about 32% of the total heat generated. It was expected that, when char recycle was introduced to stage one and a layer of slag was formed on the tubes, the heat loss would be reduced.

After the base condition test was completed, Test G-1 was conducted. The objective was to feed coal to stage two while continuing to fire stage one only with natural gas. Coal was fed for a period of 51 hours, but was discontinued because of several plant problems, including high water level in the settling pond, a badly leaking rod mill in the coal grinding area, and a high concentration of hydrogen sulfide in the recycle purge gas. A spot material balance indicated that about 80% of the carbon in the coal feed was gasified in stage two. This high conversion would not leave enough carbon in the char for generation of heat and hydrogen in stage one.

The overall objectives of the next series of gasifier tests (G-2 through G-2L) were to operate stage one with char feed, while still using natural gas as a supplemental fuel, and to reduce stage two carbon conversion from 80% to 50–60%. The specific goal was 24 hours of lined-out conditions with 2000 lb/hr of coal feed and observable slagging. Although these objectives were not met, the major operating problems of the BiGas gasifier were identified and included

1. The inability to maintain continuous slag tapping.
2. Inadequate solids flow measurement and instrumentation.
3. Inadequate solids flow control.
4. The inability to maintain thermocouples in stage one over extended periods.

An effort was made to address the slag tapping problem in the next series of tests (G-2M and G-2P) by drawing hot stage one gases down through the tap hole while operating. Along with being ineffective in keeping the tap hole open, this procedure created a fog below the tap hole, which prevented visual verification of tapping.

The next series of tests (G-3 through G-36) addressed the slag tapping problem with (1) the addition of a fluxing agent (limestone), (2) the extension of the slag tap burner, and (3) the reduction of the cooling of the tap stone. The combination of these three effectively kept the tap hole clear and resulted in much longer runs. Several new problems surfaced during these tests, including

1. The formation of deposits in stage two.
2. Extensive damage to the char burners.
3. The inability to establish a level in the char vessel because of leaking control valves in the char line.

In the fifth series of tests (G-4 and G-4A), the main objectives were (1) to establish levels in the char vessel, (2) to maintain slagging without limestone flux, and (3) to test various solid flow indicators. In both tests, hot spots developed on the exterior of the char burners, probably caused by the backflow of hot (2700°F) stage one gases. On February 16, 1978, while attempting Test G-4B, the line between the "C" char burner and eductor ruptured. The gasifier depressured through the rupture, and a fire ensued. The resulting fire caused about $60,000 in damage and delayed further testing approximately 5 months.

After repairs were completed and instrumentation revised to prevent backflow of gases, the sixth series of tests (G-5 through G-5C) were conducted. Two major objectives were to (1) build a level in the char feed vessel and (2) control and measure the flow of this char into stage one. The ability to build a char level to feed steadily from a settled bed in the char vessel was demonstrated. Some of the solid flow indicators located in the char feed line responded satisfactorily to the flow of char, in particular the Argonne high-temperature microphone and the burner and eductor pressure differentials. A new problem surfaced during this test series—the formation of calcium-rich deposits in stage two. The buildup of material was thought to be related to the use of limestone as slag flux.

The next series of tests (G-6 through G-6F) were conducted without limestone flux, and the stage two deposits formed in the previous series were eliminated. The coal feed was also operated in an "as-designed" mode, i.e., with a level in the feed vessel. Progress was also made in improving the television picture of slag tapping and the factors which influence reception. An unsatisfactory condition which developed during this series was the inability to remove slag from the quench section, potentially limiting operation to 36–40 hours. Obtaining good material balance was also found to be more difficult than anticipated.

Progress

The project was managed by Phillips Petroleum Company, under subcontract to BCR until November 1979 when Stearns–Roger took over as prime contractor and program manager, as well as operator.

In January 1979, the BiGas pilot plant was revamped to correct problems with the slag tapping system and to modify the instrumentation. Run G-7 was initiated February 5, 1979. After problems with a plugged pressure instrument developed, the plant was shut down. Problems with cold weather effects delayed the start of Test G-7A until February 24. The test had to be concluded without any samples being taken for a material balance when the "A" char burner ruptured. The cause of the rupture was diagnosed as being due to a failure of the 321 stainless-steel expansion bellows in the fuel tube of the char burner. As a result, metallurgical experts representing Phillips Petroleum Co., Stearns–Roger, Inc., Pullman–Kellogg, and Universal Oil Products were called in to investigate the failure. A design was decided upon for the immediate future that would incorporate an 825 Incoloy bellows in the front section of the burner, which will give superior resistance to stress corrosion cracking. Engineering work was also escalated, leading to radical changes or redesign of the char burns.

Fire damage repairs were completed in June. Test G-8, the first gasifier test performed since the rupture of "A" char burner, which terminated test G-7A, was conducted during August. During Test G-8, coal was fed for 4.5 hours and char was fed for 2.1 hours. The test was terminated when the only remaining response of stage one temperature became erratic. In addition to the problems associated with the stage one temperature measuring equipment, difficulties were also encountered in lighting the slag tap and heating burners, pumping coal slurry to spray drying, keeping the coal and char legs from plugging, and obtaining a response from the coal and char vessels' level detection equipment. The use of a larger coal feed did appear to improve the operability of the gas washers; but because of the short duration of the test, insufficient information was obtained to reach any definite conclusions.

The addition of the new center tube in the char burners allowed smoother operation of the gasifier by avoiding char burner shutdowns when plugged gages occurred in the char burners. Fuel gas flow was not interrupted since it has a separate annulus, shared with the bypass steam.

The effects of adding the fuel gas heater could not be determined from Test G-9. "B" char burners did not feed well. The test was not long enough to vary the temperature of the fuel gas and observe the effects on "B" char burner.

For Test G-10, "A" char leg was removed and replaced with a 1-inch char leg, and the 4-inch vee-ball valve was replaced with a 1-inch Fisher vee-ball valve.

Test G-10 was conducted October 23, 24, and 25, 1979, with approximately 36 hours of coal and char feed. The test was terminated because of a leak in the packing of the valve controlling coal flow in the "A" coal leg.

The new 1-inch char system operated well. Flow was more stable than that through the 4-inch char leg and could be more sensitively controlled through the 1-inch vee-ball valve. Continued testing to the 1-inch char feed systems will be required to confirm this conclusion.

Because of the difficulty experienced in Test G-9 of transferring coal from the weigh belt to the bucket elevator, a screw conveyor was installed to feed the rod mill directly from the weigh belt. A new screw feeder in the coal-handling area improved the ability of coal feeding from the weigh belt to the rod mill.

Tests G-9 and G-10 were conducted in October 1979. The main difficulty in these tests was continual plugging in the char feed system. Plugging was minimized in Test G-10 when the 4-inch char feed system was replaced by a 1-inch system.

The major objective of Test G-9 was to solve the plugging problems in the char feed system. Two flowing mixtures, other than oxygen, were introduced into stage one via the char burner. Fuel gas (for flame stability and safety in the event of a char feed interruption) mixed with process steam and char mixed with motive steam were allowed to mix in the burner. A new center tube was added for Test G-9 so that these mixtures were separated within the burner. Additionally, a fuel gas heater was used to heat the fuel gas to 475°F before it entered the burner. Therefore, the first objective of Test G-9 was to determine the effect of these modifications on the problem of plugging in the char burner.

The tap stone at the bottom of stage one has had a very short lifetime, being completely gone after the four short tests prior to G-9. Therefore, another objective of Test G-9 was to reduce the firing rates of the slag tap and slag heating burners in order to lengthen the life of the tap stone.

Other objectives of Test G-9 were to achieve a gasifier material balance and collect additional calibration data for the char vessel level recorders.

Test G-9 was conducted on October 2 and 3, 1979, with approximately 22 hours of coal and char feed to the gasifier. The test was terminated because coal was causing plugging in the coal-handling area. Specifically, plugging was occurring in the transfer between the weigh belt and the bucket elevator. Also, a plantwide power loss was caused by a malfunction of electrical equipment associated with the recycle gas compressor.

Two material balance periods were chosen for Test G-9 of 2- and 3-hour duration. In the 2-hour period, the rate of carbon feed to the gasifier was estimated to be 3975 lb/hr, and the rate out was estimated to be 4472 lb/hr for a +13% error. The nitrogen was 1274 lb/hr in and 1345 lb/hr out for a +6% error; oxygen was 8854 lb/hr in and 7840 lb/hr out for a −11% error; and hydrogen was 1131 lb/hr in and 1053 lb/hr out for a −7% error. The material balance accuracy for the 3-hour period was similar to the above except for ash, which gave a −59% error in the 2-hour period but only −5% error for the 3-hour period. The reason for the difference was due to the slag agitator not cycling during the 2-hour period. This created an ash holdup in the quench section.

During Test G-8A, coal was fed for 25.7 hours and char was fed for 26.5 hours. Intermittent coal and char feeding problems were experienced throughout most of the test. Two sets of material balance samples were taken, but closure was not obtained. The repetitive occurrence of plugging problems finally led to the decision to terminate the test.

Tests G-9 and G-10 were conducted in October. Approximately 22 hours of coal and char fed to the gasifier were completed in Test G-9. An addition of a new center tube in the char burners before the test allowed smoother operation of the gasifier.

Before Test G-10, the "A" char leg was removed and replaced with a 1-inch char leg. In addition, the 4-inch vee-ball valve was replaced with a 1-inch Fisher vee-ball valve. Approximately 36 hours of coal and char feed were completed before the test was terminated owing to a leak in the packing of the valve controlling coal flow in the "A" coal leg.

In November 1979, Stearns–Roger, Inc., took over as prime contractor and program manager, as well as operator of the plant.

CO_2 ACCEPTOR PROCESS

In the CO_2 Acceptor process, shown in Fig. 4-8, lignite is devolatilized at 140 psia in the presence of steam, carbon monoxide, hydrogen, and dolomitic calcine ($MgO \cdot CaO$) in a fluidized-bed devolatilizer kept at 1500°F (816°C) by addition of calcine at 1870°F (1021°C); char from the latter is fed to a gasifier bed containing calcine and operating at 1520°F (827°C) and 150 psia in a fluidized-bed regenerator, which receives separate streams of partially carbonated calcine ($MgO \cdot CaCO_3$) from the devolatilizer and gasifier, returns regenerated calcine to the same units, and sends waste gas to an energy recovery system. The circulating solid material, introduced as dolomite ($MgCO_3 \cdot CaCO_3$), evolves

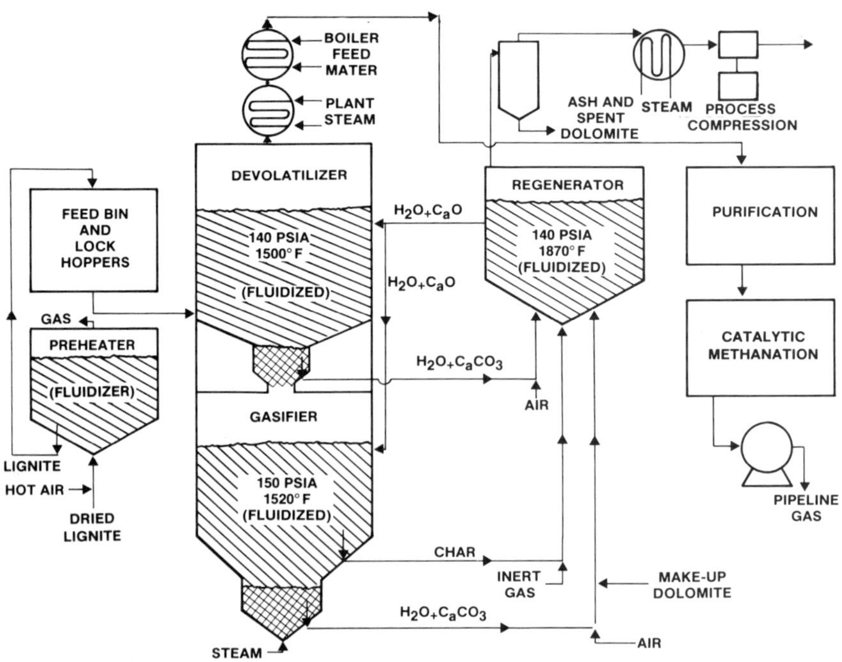

Fig. 4-8. CO_2 acceptor process diagram.

carbon dioxide with absorption of sensible and chemical energy in the regenerator, and accepts carbon dioxide and releases both sensible and chemical energy in the devolatilizer and gasifier. Gas from the gasifier, rich in hydrogen and carbon monoxide, is fed with steam to the devolatilizer, the gas from which is purified, catalytically methanated, and then compressed.

The process is also designed to operate at about 300 psia, in which case temperatures in the regenerator and gasifier change to 1940°F (1060°C) and 1575°F (857°C) and the gasifier operates with a recycle stream.

SYNTHANE PROCESS (COAL TO PIPELINE GAS)

Synthane (SYNthetic MeTHANE) is one of the promising processes for producing substitute natural gas from coal. The process consists of coal pretreatment, coal gasification, shift conversion, gas purification, and methanation systems. Synthane can operate with any coal—lignite, subbituminous, bituminous (including highly caking coals), and anthracite.

In the Synthane process, shown in Fig. 4-9, eastern coal is crushed to minus 20 mesh, is dried, is pressurized to approximately 70 atm, and is transferred into the fluidized-bed pretreater by means of high-pressure steam and oxygen. Pretreatment provides a mild oxidation of the coal particle surface so that caking coals will not agglomerate in the gasifier. The coal overflows from the pretreater into the top of the fluidized-bed gasifier about 10 feet above the bed level, falls through hot gases rising from the fluidized bed, and is devolatilized. This devolatilization contributes significantly to the methane yield. Steam and oxygen enter the gasifier just below the fluidizing gas distributor. The gasification reaction occurs within the fluidized bed. Unreacted char flows downward into a bed fluidized with steam and water sprays and is removed through lock hoppers. The product gas, containing methane, hydrogen, carbon monoxide, carbon

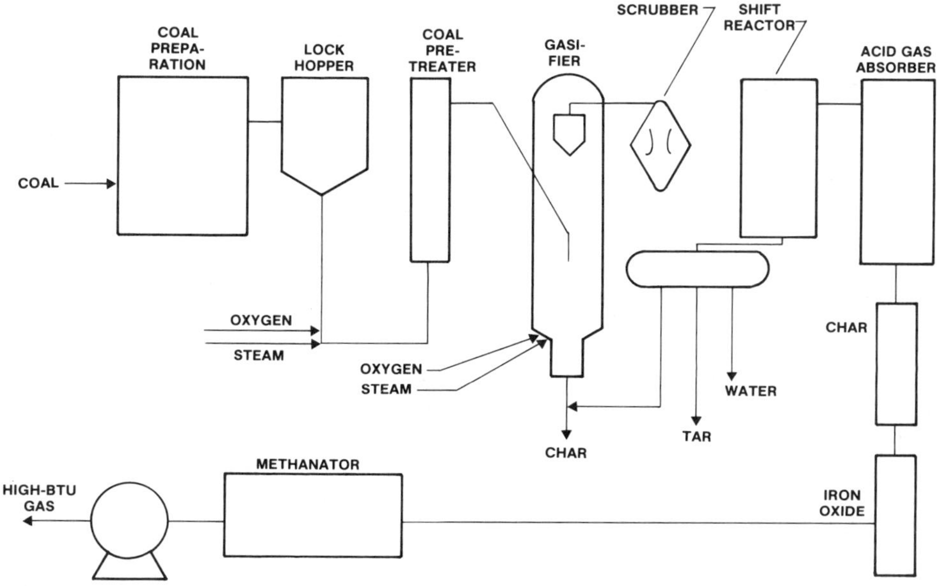

Fig. 4-9. Synthane process diagram.

dioxide, ethane, and impurities, is passed through a venturi scrubber and a water scrubber to remove carryover ash, char and tars. The concentration of hydrogen and carbon monoxide in the gas is then adjusted to a three-to-one ratio in a shift converter. The purified gas is then sent to a methanator for final upgrading.

Two methanation systems have been proposed: a tube-wall reactor, in which the catalyst is sprayed onto tube walls of an exchanger; and a hot gas recycle methanator, in which the catalyst is sprayed onto metal plates. Coal can be fed to the gasifier through a lock hopper system as shown, or as a coal–water slurry.

When the Synthane plant is fed with western coal, the coal pretreater is eliminated from the process train. Western coal is noncaking and does not require pretreatment prior to entering the fluidized bed.

Gas leaving the gasifier is scrubbed to remove fine particles, tars, and water. The cleaned gas passes through a shift converter and then goes to the hot-potassium-carbonate plant for removal of CO_2 and sulfur compounds. Hydrogen sulfide is used for commercial purposes or stored for later use.

Finally, the purified gas passes to the methanator for conversion of the hydrogen plus carbon monoxide to methane, increasing the heating value to that of natural gas. This process can be modified to make gasoline, methanol, or low-Btu gas.

Synthane Pilot Plant

The Synthane pilot plant at Bruceton, Tennessee, was designed and engineered for DOE by CE-Lummus Company. The plant was constructed by Rust Engineering Co. and CE-Lummus Co. and was operated under a government contract. The plant was designed to convert 72 tons of coal per day into 1.2 million cubic feet (34×10^3 m^3) of gas. The installed cost is about $16 million; annual operating cost is approximately $8.2 million; and about 200 people are employed in operating the plant.

The overall process includes coal sizing, pressurization, pretreatment, and gasification, and a series of cleanup and conversion operations. Coal is delivered by truck, loaded into a 240-ton raw-coal storage bin, then ground from the original minus ¾ in. (2 cm) to minus 20 mesh (0.8 mm) in a hammer mill and pneumatically transferred to a 100-ton pulverized-coal storage bin.

The pretreatment and gasification section is continuously fed from lock hoppers where the coal is pressurized to reaction pressure (600–1000 psig). In the fluidized-bed pretreatment, the coal particles are partially oxidized at 800°F (425°C) with steam and oxygen to prevent agglomeration; this step is not required for noncaking coals. The gasifier is a single-stage fluidized-bed reactor, operating between 1400 and 1800°F (760 and 980°C). The bottom section is a fluidized-bed char cooler from which there is a continuous char blowdown at reduced temperature [600°F (315°C)].

Raw gas from the gasifier, containing methane, hydrogen, carbon monoxide, carbon dioxide, steam, sulfur compounds (mainly H_2S), and several minor products, must be passed through a series of cleanup and conversion operations to produce clean high-Btu gas. These steps consist of water scrubbing, carbon monoxide shift conversion to adjust the hydrogen/carbon monoxide ratio to 3:1 for subsequent methanation, acid-gas removal using a Benfield unit (hot potassium carbonate), and finally methanation.

Methanation will be carried out using two methods. The first method utilizes a tube-wall reactor, which consists of a shell-and-tube heat exchanger with the internal surface of the tubes coated with Raney nickel catalyst. Dowtherm is used to remove the heat of reaction from this methanator. The other method involves hot-gas recycle through a bundle of stainless-steel plates coated with Raney nickel catalyst. Temperature in this methanator is controlled by diluting the fresh feed gas with recycle gas. Offgas from the Benfield unit will contain most of the sulfur originally in the coal and will be treated in a Stretford unit, which converts hydrogen sulfide to elemental sulfur.

Advantages of the Synthane process include its ability to handle any type of coal; about 60% of the methane is formed in the gasifier, thus reducing the quantity of oxygen needed; and its good operability and reliability characteristics. Scaleup is relatively simple because a single-stage fluidized-bed gasifier is used.

LURGI (DUAL FEED) PROCESS

The Lurgi process, shown in Fig. 4-10, takes noncaking coal and crushes it. It is then screened to remove fines, and fed to a bank of Lurgi fixed-bed gasifiers through a lock hopper system. The coal is distributed by rotating arms and travels downward in the gasifiers where it is dried, devolatilized, and gasified at 300–500 psi with steam and oxygen. Ash is removed by rotating grates through an ash lock at the bottoms of the gasifiers. Raw gases leaving the top of the gasifiers are quenched and passed through a shift convertor. Carbon dioxide and H_2S are removed in the purification system, and the product gas is methanated and dehydrated to produce pipeline gas.

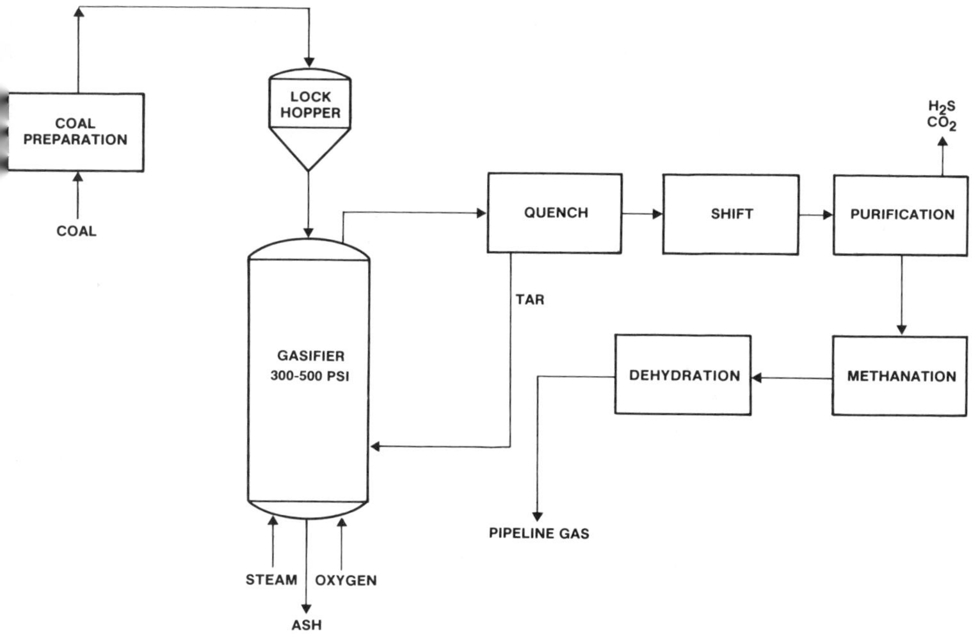

Fig. 4-10. Lurgi (dual feed) process diagram.

SRT HYDROGASIFIER PROCESS

Rockwell was awarded a $18,053,000 contract for a 3-year program in September 1978 to advance the development of the Rocketdyne single-stage short-residence-time (SRT) hydrogasifier to demonstration plant status. The program builds and expands on the foundation already established under predecessor Contract No. EX-77-C-01-2518.

A 4-ton-per-hour (coal feedrate) reactor facility will be designed, constructed, and operated at conditions applicable to a commercial SNG process with both a subbituminous coal and a highly caking bituminous coal as feedstocks. Scaling of reactant injection elements from the ¼-ton-per-hour element, investigated under the previous contract, to a commercial scale of a 4-ton-per-hour element will be investigated, and the technique of clustering elements of one size to feed a larger-sized reactor will be demonstrated.

Additional testing will be concerned with establishing economic reactor operating conditions and with building a viable process data base to support the next step toward commercialization with minimum risks. The 4-ton-per-hour hydrogasifier system will be capable of extended duration runs of up to 30 days.

In conjunction with an A&E firm subcontractor, a preliminary design will be generated of a full-scale commercial SNG plant utilizing hydrogasifiers of the type in the program.

Process Description

Raw coal (up to 8-in. size) is crushed at a rate of 15 tons per hour to less than 1½ in. size. The crushed coal is mechanically conveyed to a hopper from which it is fed to a mill where it is pulverized and dried to produce 7 tons of coal per hour sized for 90 weight percent passage through 100 mesh and containing less than 3 weight percent moisture. The dried, pulverized coal is pneumatically transported to a cyclone separator from which the fine coal is gravity fed to a 40-ton-capacity storage silo via a sizing screen that removes particles larger than 1⁄16 in. A 470-cubic-foot-capacity lock hopper is used to transfer the coal from the silo to the high-pressure reactor feeder in 6-ton batch quantities at a rate of 4 tons per hour.

The reactor is a continuously fed device that hydrogenates the pulverized coal at elevated pressures and temperatures. The reaction is terminated by quenching the product gas and char to 600°F (316°C) in two heat exchangers located downstream of the reactor. Gaseous hydrogen and deionized water are used, respectively, as the cooling media in the upper and lower quench heat exchangers. The deionized water is pressurized to 500 psig and operated at 420°F (216°C) minimum temperature to prevent moisture condensation on the product gas/char side of the cooler. (Moisture condensation could cause plugging of heat exchanger passages by buildup and agglomeration of char particles.)

The char removal system consists of a 200-cubic-foot-capacity char receiver, a 200-cubic-foot-capacity char lock hopper, a 400-cubic-foot-capacity char surge hopper, a char slurrolizer, a slurry storage tank, two product gas cyclones with solids collectors, and two vent gas cyclones with solids collectors.

The char receiver and product gas cyclones provide the primary means for separating the product gas from the char solids. The collected char is removed intermittently by dense-phase flow to the char lock hopper. The vent gases are directed to the product gas scrubber. Figure 4-11 is a flow diagram of the ¾-ton-per-hour coal hydrogasification facility.

History of the Project

Rockwell International Corporation (RI) initially translated the reactor concept into a practical high-throughout continuous-flow engineering-scale experimental reactor in a partial liquefaction R&D program. Experimental development in that program has been carried out in two hydroconversion test facilities: a ¼-ton-per-hour system capable of

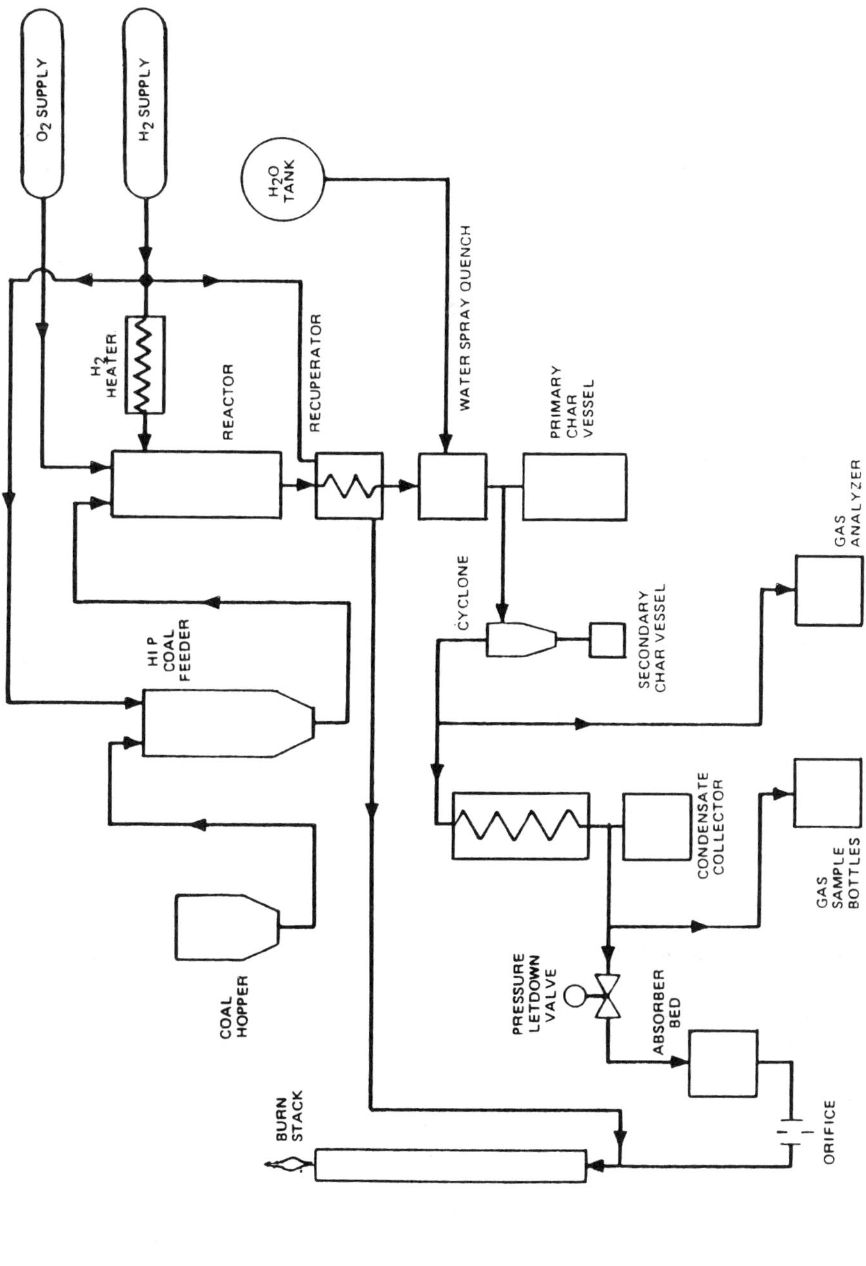

Fig. 4-11. Simplified flow diagram for the ¾-ton-per-hour coal hydrogasification facility.

operation for 10 minutes, and a 1-ton-per-hour system in which runs up to 1 hour duration have been made. Although maximization of liquid product yields was emphasized, a few tests at nearly zero-liquid conditions served as a prelude to development of the reactor as a hydrogasifier. This was undertaken in 1977 under Contract EX-77-C-01-2518 which extended from March 1977 through August 1978. The prime contractor was the Rocketdyne Division of Rockwell International Corporation, and Cities Service Research and Development Company (CSRD) was a major subcontractor.

The experimental studies comprised ¼-ton-per-hour short-duration (10 minute) reactor tests in the RI ¼-ton-per-hour hydroconversion facility, ¼-ton-per-hour long-duration (½–1-hour) reactor tests in the RI 1-ton-per-hour hydroconversion facility; and 2- to 4-pounds-per-hour reactor tests in the CSRD bench-scale laboratory facility.

Because the bench-scale equipment could be activated sooner and was less expensive to operate than the ¼-ton-per-hour unit, its function was to provide experimental guidance for the selection of test conditions for the relatively fewer ¼-ton-per-hour tests.

The work performed during the last quarter of 1978 consisted of six 8-minute tests on the ¼-ton-per-hour reactor using both Montana Rosebud and Kentucky No. 9 coal with a hydrogen-to-coal ratio of 0.24. After this, the unit was dismantled for moving to the Energy Systems Group (ESG) Energy Bowl Control Center.

A total of 42 short-duration ¼-ton-per-hour tests were made. Several conclusions may be drawn from these data:

1. Under approximately comparable reactor conditions, overall carbon conversions were lower for subbituminous than for bituminous coals.
2. Conversely, the lower-ranked peat exhibited appreciably higher overall conversion than did bituminous coals.
3. Reactor conditions yielding high values of overall carbon conversion also tended to give high carbon conversions to gases.
4. Conditions were established at which all carbon could be converted to conversion products for subbituminous and bituminous coal and for peat.

The main contributions of the tests in the 1-ton-per-hour facility were the experience gained concerning exposure of the reactor for longer periods, and achievement of more precise values for conversion and elemental balance.

Progress

Relocation of the ¼-ton-per-hour Rockwell hydrogasifier at the Canoga Park, California, Energy Bowl Control Center of Rockwell's Energy Systems Group was completed in April 1979. Modifications made to the unit increased its capacity to ¾ tons per hour. Full-scale testing at the ¾-tons per-hour rate commenced on April 25 and continued through May. During these initial tests, problems associated with injector and reactor hardware items were encountered. Because of these problems, modifications were made in hardware and in test procedures. Modifications included the installation of a larger (12-in.-diameter) pressure vessel, installation of preburners of improved design, revision of test procedures to avoid faulty sequencing, installation of additional reactor tubes, and installation of a three-tube, 11-ft-long recuperator.

Also during 1979, a 4-ton-per-hour hydrogasifier reactor development (PDU) was being designed for construction in the same location. However, in August DOE had directed that design work on the 4-ton-per-hour reactor system be terminated, and in its place efforts be devoted to the ¾-ton-per-hour IPDU. Most of the month of September was spent in replanning the ¾-ton-per-hour unit.

During the last quarter, all replanning effort associated with the design and construction of the ¾-ton-per-hour integrated process development unit (IPDU) was completed, and design of the facility installation was continued. This included an overall plan for implementation of the facility work. Most of the design features of the 4-ton-per-hour reactor train was carried over, except for the dimensions, which were scaled down.

Fifteen coal hydrogasification tests were conducted during the October–December 1979 reporting period using the 6.07-in.-ID, 15-ft-long reactor tube. During two consecutive tests, the injector was damaged at one of the oxygen orifices through the face because of a flaw in the fabrication of the orifice. A backup injector was mated to a reworked reactor tube, and eight more tests were successfully completed.

A ¾-ton-per-hour hydrogasifier IPDU is also being designed for construction in the ESG Energy Bowl. An existing large liquid-rocket-engine test stand, designated VTS-3, will be utilized as the site and structural base for the installation of the process system. The unit will be designed for continuous operation for periods up to 30 days, so it will have provisions for on-site coal storage; pulverization and drying; pneumatic lock hopper pressurization of the pulverized coal feed and letdown of the char produced; and walkaround, as opposed to remote, facility operation.

During this report period, all replanning efforts associated with the design and construction of the ¾-ton-per-hour IPDU was completed, and design of the facility installation was continued. This included an overall plan for implementation of the facility work.

The development of the design associated with the coal processing and coal feed subsystems has progressed as installation detail has been received from the suppliers of the coal processing equipment and is nearing completion.

A firm dimensional layout of the reactor train was completed in November. With this design base, an equipment layout of the product gas system and char system components was made.

DILUTE-PHASE HYDROGASIFICATION

The dilute-phase free-fall hydrogasification process (Fig. 4-12), based on earlier work on the Hydrane process, is a potential third-generation coal gasification process developed at PERC. Comparative cost studies show that it could be the most economical method of producing pipeline-quality gas from coal. It differs from steam–oxygen gasification processes in several important ways. One of its unique features is that it permits coal to be fed directly to the gasifier, circumventing the problems of caking and agglomeration, and thereby eliminating the expensive pretreatment steps common to other processes. In addition, a hydrogen-rich portion of the coal sometimes lost in pretreatment is retained. Almost all of the methane in the product gas is produced in the single-stage gasifier by direct reaction of coal with hydrogen. This process scheme results in a high-energy-conversion efficiency and requires only slight methanation to remove the small amount of carbon monoxide present in the gasifier product.

In the dilute-phase free-fall hydrogasification process, crushed raw coal and hydrogen are fed into the upper section of the single-stage hydrogenation reactor, which is maintained at process conditions of 1650°F (900°C) and 1000 psig.

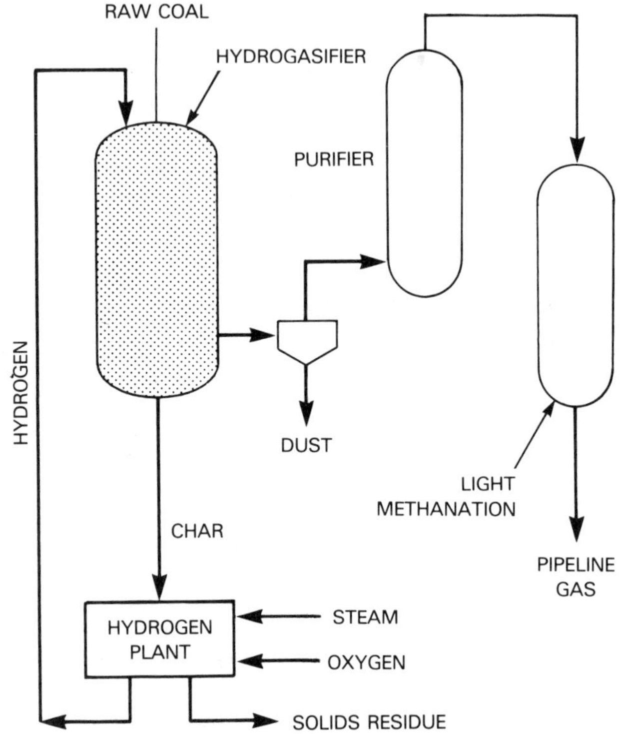

Fig. 4-12. Dilute-phase free-fall hydrogasification process flow diagram.

From the top of the reactor, the coal falls freely as a dilute cloud through the hydrogen. Volatile matter is released and the coal loses its agglomerating characteristics. About 40% of the carbon in the raw coal is converted directly to methane by reaction with hydrogen.

Product gas is drawn off near the bottom of the reactor and is cleaned of residual solids and unwanted gases. Finally, methanation of the small amount (2–5%) of residual carbon monoxide produces a high-Btu pipeline-quality substitute natural gas. Char formed during gasification is utilized to produce process hydrogen by reaction with steam.

Experimental studies are currently being conducted with a small-scale continuous-flow unit to optimize the reactor configuration and to acquire data applicable to the design of a larger-scale unit. Fundamental research related to the dilute-phase process is also being conducted. Fluid dynamic investigations of coal in free fall are carried out in a cold flow model to obtain information for improved reactor designs, and studies of the process kinetics are also in progress. Construction of a large-scale process development unit is being considered. This plant would be capable of processing 10–30 tons of raw coal per day and would employ coal handling and gas cleanup procedures similar to those used on other large-scale process development units.

As changes in process conditions can result in the formation of an appreciable amount of liquid product, dilute-phase liquefaction studies are also planned. A small-scale unit to demonstrate this novel liquefaction concept is the design stage.

ATMOSPHERIC ENTRAINED-BED PROCESS

Introduction

A method for the production of low-Btu gas from coal for electricity generation is being developed by Combustion Engineering, Inc., under the sponsorship of DOE and an industry team composed of Combustion Engineering and the Electric Power Research Institute. DOE is providing two-thirds of the $24,884,004 funds and the team, one-third. The overall objective of this program is to produce an economical low-Btu coal gas to generate electricity in an environmentally acceptable manner, with a minimum of process development. Combustion Engineering is conducting a three-phase program involving the design, construction, and operation of a coal gasification pilot plant. The objectives of this program are to

- Demonstrate the capability and suitability of the process and equipment to produce low-Btu gas of the predicted composition, heating value, and cleanliness from the design coal.
- Provide data to be used as the design basis for commercial-scale equipment that would process about 110 tons of coal per hour and result in an electrical generating capacity of 300 MW.

Combustion Engineering also has a continuing research program to refine the process and verify its technical and economic feasibility.

Process Description

The Combustion Engineering gasification process is based on an air-blown atmospheric-pressure entrained-bed gasifier. A schematic of the process is provided in Fig. 4-13. In the process, a portion of the pulverized coal and recycled char are fed to the combustion section of the gasifier and burned to supply the heat necessary for the endothermic gasification reaction. In the combustion section, nearly all of the ash in the system is converted to molten slag, which is then drawn off the bottom of the gasifier. The remainder of the pulverized coal is fed to the reduction portion of the gasifier where it is contacted with hot gases entering the reduction zone from the combustor. The gasification process takes place in the entrainment portion of the reactor where the coal is devolatilized and reacts with the hot gases to produce the desired product gas. This 1700°F (927°C) product gas is then cooled to 300°F (149°C). At this point, the gas contains solid particles and hydrogen sulfide that must be removed. Solids are removed and recycled by means of a spray dryer, cyclone separators, and venturi scrubbers. Hydrogen sulfide is removed and elemental sulfur is produced by the Stretford process. The clean low-Btu gas (about 120 Btu per standard cubic foot) can then be delivered to the burners of power boilers, gas turbines, or combinations of the two in a combined-cycle power generator.

Operating conditions will have a variety of effects on the cost and quality of the gas produced in this system. For example, oxygen could be substituted for air in the gasifier combustor, thereby increasing the heating value of the product gas from 120 to 285 Btu per standard cubic foot. Conversely, this change will also increase the cost of producing the gas, depending on the price of oxygen and the quantity used.

The value of the atmospheric gasification system is that development work is only necessary on the operation and control of the gasifier. All other components are commercially available items with predictable operating characteristics.

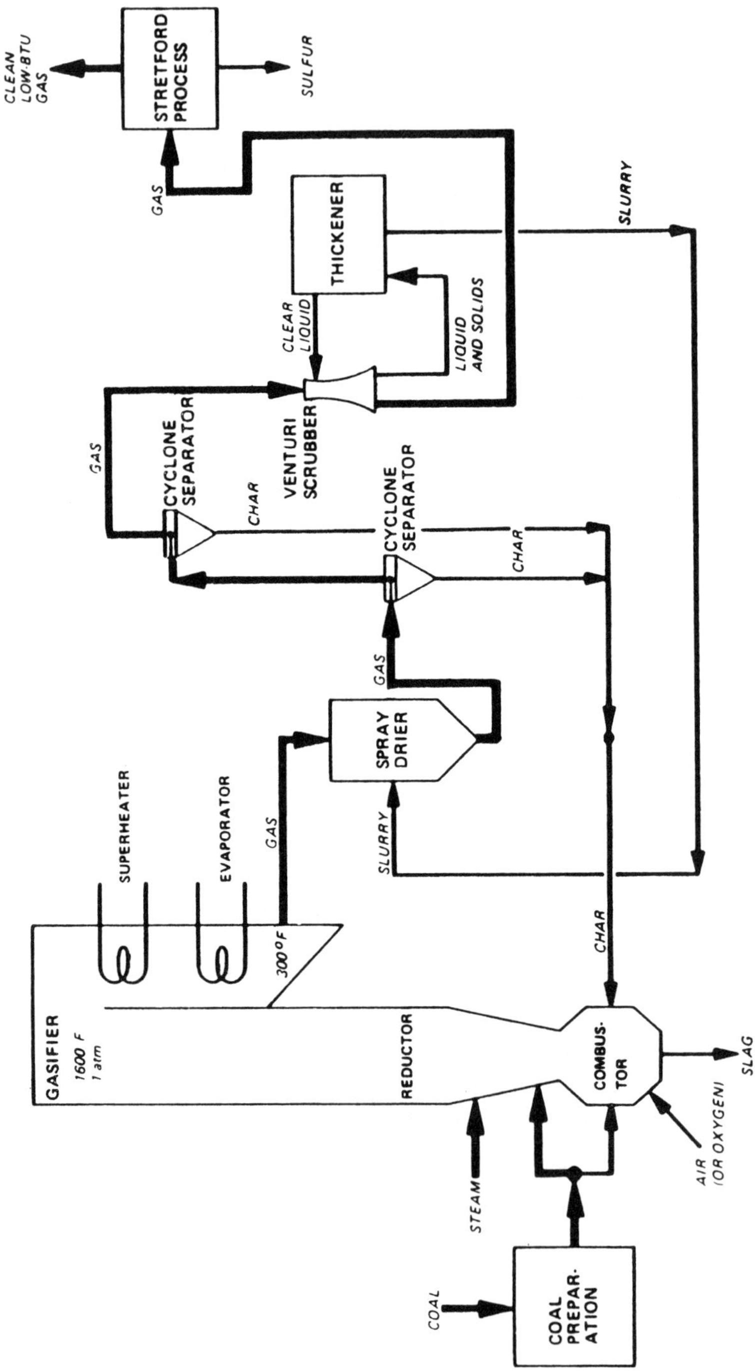

Fig. 4-13. Schematic of low-Btu gasification of coal for electricity generation.

History of the Project

Initial work on the development of a system for the low-Btu gasification of coal for electricity generation involved the economic evaluation of various coal gasification and energy conversion systems for the production of electricity. The primary criterion for the systems selected was the maximum use of proved processes. Of the systems selected for use in this process, the only unproved component was the gasification process. All other components were of established and proved design.

Two gasification processes, differing only in operating pressure, were considered: one at atmospheric pressure and the other at 10 atm. It was determined that if the gasification pressure were increased from 1 to 10 atm, the total plant cost would be reduced by 8%. This savings results from the tenfold reduction of gas volume and, hence, equipment size. Conversely, much of the process simplicity would be lost and additional development would be required for

- Lock hoppers for feeding solids into pressure vessels.
- A more complex gasifier to withstand more severe operating conditions.
- Heat exchangers to withstand greater erosion due to higher solids concentration in the gas stream.
- Techniques for char recovery at elevated pressures.
- Processes for removal of liquid slag from the pressurized gasifier.

Because of these complications, entrainment gasification of coal at atmospheric pressure was the process selected for development. The gasifier may be air blown or oxygen blown, depending on the particular requirements of the plant in which it would be installed. It was determined that the air-blown version would be developed initially because it appeared to be the most economical for application to the majority of electric power plants.

Design of the process development unit began in 1974. The unit, located at Windsor, Connecticut, has a capacity of 5 tons of coal per hour. The objectives of this phase include (1) development of components and process concepts and details necessary for the design and operation of gasification systems of this type, and (2) completion of the engineering required for construction of the unit. Construction was completed in the fall of 1977, and a dedication ceremony was held at the plant site on October 18, 1977.

Laboratory research included the investigation of various refractories with slag and gas and the development of a one-third scale plastic model for studying stream flow patterns. A mathematical analysis of the gasifier was carried out through the use of computer modeling of the unit. Plant metallurgical requirement studies and environmental studies were also conducted. Supporting studies consisted mainly of an analysis comparing the costs of producing electricity by various integrated gasification power plant systems. Two alternatives were selected for further study: a conventional steam plant with an atmospheric gasifier and a combined-cycle plant with a supplementary-fired boiler and atmospheric gasifier. In addition, specifications for an automated data-acquisition and reduction system were completed. The first PDU gas-making run was completed in June 1978. Four more runs were made during 1978. A sixth run was started but had to be terminated when a large amount of water was carried into the booster fan inlet duct. When this was cleared, the fan was found to be out of balance and the impeller had to be rebuilt.

Progress

Run 6 was started on January 13, 1979, and was completed in mid-March. Approximately 600 hours of gas production were logged and 460 million cubic feet of 40–75 Btu/scf gas were made. The PDU had a coal-firing availability of 63% and a gasification availability of 47% during the run. The production rate was limited to 2½–3 tons of coal per hour during Run 6 in order to minimize carryover of water through the scrubber's separator/demister. A new separator/demister was installed during the outage following this run, and it is expected to reduce the likelihood of carryover in future operation.

By March 1979, approximately three-quarters of a billion standard cubic feet of low-Btu gas has been produced in the PDU and over one thousand hours of gas-making time logged. It is believed that equipment-related problems in the PDU have been largely identified and that a significant number of them were fully corrected in response outages. If the particulate removal and other downstream systems are operable, and a continuous and uninterrupted supply of pulverized metered coal can be fed to the gasifier, the gasifier itself can be operated for extended periods.

Following Run 6, modifications were made to the equipment in the particulate removal system and other areas to further improve overall PDU reliability. Operations were conducted in May to demonstrate acceptable performance of the newly installed equipment.

Gas-making Run 7 began in late June to collect meaningful data to provide design information for scaleup of the gasifier and plant equipment to commercial size, to perform heat and energy balances, to demonstrate the operability of sulfur collection system and collect sulfur, and to demonstrate acceptable operation of the multiclone.

Four additional tests were made in July and 11 tests were made in August. Coal feed to the gasifier was progressively increased from about 5000 lb/hr to about 8000 lb/hr at the end of August. Material balance closures less than 5% were achieved on all 18 tests. Energy balance closures were not as good. Modifications are planned to achieve more accurate solids feed rate measurements, to further improve energy balance closure. The PDU was operated in the gas-making mode for 25 of the 31 days in August. Test data gathered during the month verified continued wastage of the small amount of refractory remaining in the high-temperature region of the gasifier, such that at month's end, the 3200°F design operating temperature could not be obtained.

Modifications made in October included change in feeder drop leg geometry, which will permit on-stream rezeroing and cleaning of the combustion coal feeders to provide better measurement accuracy.

After completion of the modifications, tests in November were made with bare combustor waterwalls, i.e., no combustor waterwall refractory remained. As such, heat absorption by the combustor waterwall was approximately four times design. Owing to the high heat loss, combustor temperatures were 200–400°F (93–204°C) below the 3200°F (1760°C) design temperature, unquestionably contributing to high char recirculation and low gas quality.

Coal firing was stopped on December 20 and the plant was shut down to prepare for installing new refractory in the gasifier. Significant improvement in material and energy balance closures was attained in the December testing.

Measurement and calculations were made during both Runs 7 and 8 to determine the performance of the PDU particulate removal system. Additional particulate loading tests will be made in Run 9 to establish dust loading when the gasifier is operating at design temperature conditions.

The post-Run 7 outage, begun in September 1979, was primarily directed toward modifying combustor coal and char fuel piping, as well as the combustor coal feeder drop leg geometry. The change in feeder drop leg geometry will permit on-stream rezeroing and cleaning of the combustor coal feeders to provide better measurement accuracy.

A new demister was designed during October and installed at the Stretford absorber outlet following Run 8. This resulted from the fact that during the 874 hours of gas-making associated with Run 7, it became obvious that sufficient carryover of liquid mist and particulates from the absorption tower was occurring, causing downstream problems.

Post-Run 7 outage items were completed in early November and plant start-up begun to verify acceptability of equipment modifications. During checkout operations in the first half of November, coal feed through the new combustor fuel firing geometry was demonstrated to be steady and consistent. In addition, the modification made on the combustor coal fuel piping to permit rezeroing coal feeders on line was demonstrated to work well. Run 8 was initiated in the second half of November.

The November tests were made with bare combustor waterwalls, i.e., no combustor waterwall refractory remained. As such, heat absorption by the combustor waterwall was approximately four times design. Owing to the high heat loss, combustor temperatures were 200−400°F (93−204°C) below the 3200°F (1760°C) design temperature, contributing to high char recirculation and low gas quality. This situation should be remedied by the installation of refractory brick in the combustor and diffuser.

Run 8, initiated in November, was continued into December. Ten tests, 8-10 through 8-19, were made in the first two weeks of December. Gasifier operation was continued after test 8-19 to gather temperature and gas composition data by traverses made in the combustor, diffuser inlet, and reductor outlet.

Significant improvement in material and energy balance closures was attained in December testing. Closure for most tests were less than +5%. The good closures indicate that improvements to the coal and char feed arrangements and data acquisition system have been successful. The gasifier was coal fired on all but two days between December 2 and shut down on December 20. Since inception on November 11 through shutdown, Run 8 logged 484 hours of coal firing and 315 hours of gas-making. Approximately 30,000 Mscf of product gas was produced.

Measurement and calculations were made during both Runs 7 and 8 to determine the performance of the PDU particulate removal system. Results of the evaluations have shown that the overall particulate removal efficiency varied over the range of 99.2−99.6%.

TRI-GAS PROCESS

Bituminous Coal Research (BCR) is developing a multiple fluidized-bed coal gasification process for producing low-Btu fuel gas from caking and noncaking coals. The project is being sponsored by DOE. The low-Btu fuel gas product will be suitable primarily as boiler fuel gas for electricity generation, although alteration of operating conditions could yield medium-Btu gas suitable for other uses. The research tasks include the design, construction, and testing of a process development unit, and laboratory investigations.

Process Description

The BCR low-Btu research effort is based on a three-stage fluidized-bed gasification process. Stage one of the three-stage system receives the raw coal and functions as the pretreatment step in which the coal is devolatilized. Stage-three flue gas is used as the fluidizing medium for stage one. The coal then flows by gravity to stage two, where devolatilized coal is gasified with air and steam to generate the desired product gas. Stage-one flue gas is fed to stage two where the entrained tars and oils are gasified. The remaining char from stage two is consumed in stage three by fluidized combustion. The ash discharge will contain only a minimum amount of carbon.

The advantage of this process is that both caking and noncaking coals can be used. In addition, low-Btu fuel gas is the only product; no wastes or by-products such as tars and oils are produced. The end use of the gasification product dictates the operating conditions as well as the gasifying medium. For example, gasifying with air and steam will yield a low-Btu fuel gas, and gasifying with oxygen and steam will yield a medium-Btu gas.

History of the Project

Initial development of the low-Btu fuel gas process involved laboratory investigations to obtain kinetic data for various chars, or devolatilized coal, that would be produced in the first step of a multiple fluidized-bed gasification process. A small fluidized-bed batch reactor system was then used to verify proposed rate equations, to determine the reaction rate between steam and char, and to provide physical data such as minimum fluidizing velocity, attrition rate, elutriation losses, etc. The successful laboratory studies led to the design of a process and equipment development unit with a capacity of 100 pounds of coal per hour. The design and definitive cost estimates were completed in December 1972. Responsibility for constructing the unit at BCR's Monroeville, Pennsylvania, laboratory was awarded to Blaw-Knox Chemical Plants Division of Dravo Corporation, and construction was completed in September 1975.

Mechanical checkout and pressure testing of the system was completed during the fourth quarter of 1975, and operation of the unit was started. A variety of problems were encountered during the pressure testing and startup operations; however, most of these have been resolved. Pretreatment testing of a western noncaking coal (Montana Rosebud) was completed in the stage-one reactor. Correlations were developed for the properties of the product char as a function of stage-one operating temperature. Subsequently, testing of a mildly caking coal (Illinois No. 6) was conducted in the stage-one reactor beginning in mid-1976. A noncaking product char was obtained at pretreatment temperature of less than 800°F (427°C) and in the presence of oxygen. Further testing at 900–1000°F (482–538°C) demonstrated the ability to devolatilize the coal while avoiding caking. The product char from this testing was free flowing and did not cake when subjected to laboratory testing at stage-two operating temperature. A data acquisition system was installed to sample selected tests points, convert the data to engineering units, and store the values on magnetic tape.

Concurrent with (PEDU) operations, BCR is conducting laboratory research in an effort to further refine the low-Btu fuel gas process. During 1975, laboratory research included using a cold flow model of the gasifier bed to study bed fluidization. Fluidization was improved when a 45° conical, rather than flat, distributor was installed in conjunction with special nozzles that direct the fluidizing gas downward toward the distributor base and horizontally toward adjacent nozzles. Among other fluidization studies were those to observe the behavior of char in a fluidized bed, determine whether the purge flows on the pressure taps were necessary, and study the effect of purge flows on pressure drop in the unit. A computer model to predict the conversion in the fluidized-bed gasifier was also developed. The results showed the effect of mass diffusion and bubble diameter on conversion and product gas concentration. Reactor testing has continued to the present time utilizing a 100 lb/hr PEDU.

Progress

During February 1979, two long tests were conducted in the PEDU using Montana Rosebud coal. The first of these (Test 3S-44) ended prematurely due to a rupture of the cyclone in the stage-two outlet line. After cleaning out the system and making necessary repairs, Test 3S-45 was successfully run.

Three tests were completed in the Tri-Gas PEDU during April and May of 1979, using Rosebud coal as feed. For the first time, the PEDU was operated totally in the design mode, i.e., with both solids and gas streams integrated. Operations were highly successful, except for minor equipment malfunctions. During June, no new tests were made in the PEDU, as the reactors were disassembled for inspection and for repairs to heaters.

During July, two tests were conducted in the fluidized-bed PEDU. Test 3S-50 was terminated after about 2 hours of feed owing to an inability to transfer char between vessels. The vessels were cooled, and inspection of the char present in them showed that a low-density (25 lb/cu ft) char had been used to change the vessels. As a result, the char filled the reactors well into the disengaging zones, covering the reactor feed points.

Test 3S-50A was then started. Very shortly thereafter it became necessary to shut down the system due to an inability to transfer solids between vessels. These transfer problems were later explained by the fact that Illinois No. 6 coal, which produces a much lighter char, was the feed to the reactor. Purge flows in the char transfer line were set to those values used in Rosebud coal operation.

During August, Test 3S-51 was conducted in the PEDU. The system was operated with both char and gas streams integrated in the design mode. Nitrogen flows to the system were reduced below previous levels, but, due to the burnout of the reactor heaters, desired reactor temperatures were not reached.

During September no integrated PEDU tests were conducted. Modifications, consisting of changes in four areas, were made to improve the seal between the reactor heat chambers.

During October 1979, Test 3S-52 was completed in the Tri-Gas PEDU with coal continuously fed for 39 hours with smooth char transfer between vessels for nearly the entire test. The stage-two heater failed as a result of char infiltration into the heater chamber. However, seals installed in the heater chambers of stages one and three effectively eliminated any infiltration. Inspection of the reactors after the test indicated an inadequate char transfer between stages two and three, which should be eliminated by an adjustment of timing and pressure.

Test 3S-53 was conducted during November in the PEDU. The integrated system worked smoothly for the entire 70-hour period. Char was transferred smoothly between vessels and the reactor heaters worked well for the first 24 hours of the test when the stage-one heater shut down due to char infiltration from a crack in the stage-one silicon carbide tube. Smooth char transfer was experienced during the test. Inspection of the reactors after the test resulted in a second probe being fabricated to eliminate dead spots on the stage-two grid plate.

PEDU Test 3S-54 was conducted during a 4-day period in December with the system operated in the integrated mode for a total of 45 hours. Heater failures resulted in lower bed temperatures than needed in stage two. Char was transferred smoothly during much of the test. However, a possible warpage of the PG-33 probe may have hindered the char from entering the transfer tube. Heater damage resulted from a warping of the support plates for the silicon carbide tubes in stages two and three which damaged the seals between the tubes and the heater chambers, allowing gas and char to infiltrate. Therefore, the design of the heaters, the silicon carbide tube, and heater chamber seal is being reevaluated. More evidence of dead spots at the stage-two grid has resulted in new probes, which are circular, being tested in the cold model.

PRESSURIZED FLUIDIZED-BED PROCESS

The advanced coal gasification system for the electric power generation is being developed by Westinghouse Electric Corporation. DOE is providing the major portion of the funds ($38,109,922) through the process development unit. Industrial contributions to date include expenditures of $4 million as well as various related test facilities. The industry team is composed of the following firms: Public Service Company of Indiana; Westinghouse Electric Corporation; Amax Coal Company; Bechtel, Incorporated; and Peabody Coal Company/Kennecott Copper Corporation.

The primary objective of this contract is to develop a fluidized-bed system for producing low- and medium-Btu gas from coal for application to combined-cycle electric generation and industrial fuel utilization. Westinghouse is to

- Conduct laboratory research to refine the advanced gasification process.
- Design and construct a process development unit with a capacity of 15 tons of coal per day.
- Operate the process development unit to establish the operating characteristics of the gasification system. This will provide data for developing scaleup factors for the design and construction of a future 250-ton-per-day pilot plant.

The successful operation of the pilot plant will demonstrate the technical feasibility of the process and provide data for determination of the economic feasibility of generating electric power from coal while meeting the anticipated environmental emission regulations. The overall program is directed toward the operation of a combined-cycle power plant integrating a gasifier with a state-of-the-art turbine.

Process Description

A schematic of the advanced coal gasification system for electric power generation is provided in Fig. 4-14. Production of low-Btu gas by this process requires completion of three operations: coal preparation, devolatilization, and gasification–combustion. In the coal preparation section, the coal is crushed to minus 6 mesh, dried, and transported to a reactor vessel for devolatilization and partial hydrogasification. A central draft tube is used primarily for recirculating solids. Recycled solids required to dilute the feed coal and temper the hot inlet gases flow downward in the fluidized bed surrounding the draft tube. The fluidizing agent is a portion of the gases entering the unit. Recirculating solids have flow rates up to 40 times the coal feed rate to prevent the agglomeration of the feed coal as it devolatilizes and passes through the plastic or sticky phase. Dense, dry char collects in the fluidized bed at the top of the draft tube and is withdrawn at this point. Heat for devolatilization is supplied primarily by the high-temperature fuel gas produced in the gasifier–combustor. After separation of fines and ash, product gas is cooled and scrubbed with water for final purification.

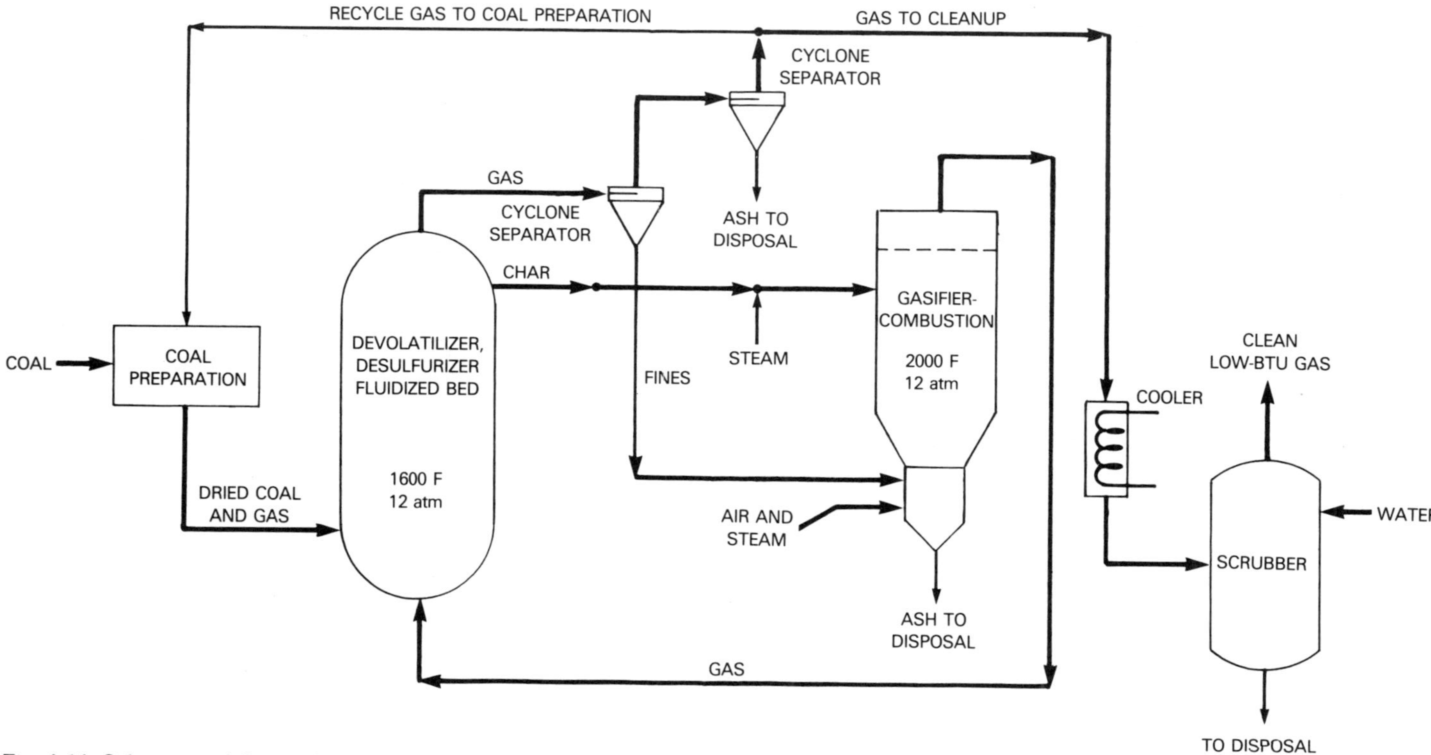

Fig. 4-14. Schematic of the gasification process development unit.

Final gasification occurs in a fluidized-bed gasifier–combustor. Char from the devolatilizer is partially oxidized with air in the lower leg of the gasifier at 1900–2000°F (1038–1094°C) to provide the heat for gasification. Heat is transported from the combustion zone to the gasification zone by gases flowing upward and by char circulating between the combustion and gasification zones. Ash from gasification/combustion of fines agglomerates on the ash from the char and segregates in the lower bed leg for removal.

There are several advantages to the advanced coal gasification system for electric power generation. For example, the system

- Uses a variety of coals, including caking and high-ash coals, without costly pretreatment.
- Utilizes a wide range of coal particle sizes including fines.
- Operates at high thermal efficiency.
- Produces a clean fuel that meets present and future environmental emission control standards.
- Reduces gaseous and particulate emissions from electric power plants.
- Eliminates the need for utility stack gas cleaning processes.

The potential economic advantages of the system included reduction of the rate of consumption of natural gas, reduced environmental costs normally associated with cheaper fuels, and reduction of the operating cost of an electric power generating plant by 20–30%.

History of the Project

Initial development of the advanced coal gasification system involved laboratory research, including the design and construction of a process development unit. The 15-ton-per-day PDU was mechanically completed in 1974, and a test program was initiated in 1975 to establish the operating characteristics of the system and provide data for developing scaleup factors for a pilot plant. Other laboratory research was involved:

- Fluidized-bed desulfurization with dolomite.
- Development of scaleup design correlations for fluidized beds.
- Combustion of coal-derived fuel in diffusion-flame gas-turbine combustors.
- Identification and testing of apparatus for high-temperature gas stream particulate control.
- Development of a specification for the tolerance of a modern utility gas turbine to particulates, vaporized alkali metals, and other gaseous contaminants.

In addition to laboratory research, work on developing conceptual designs of the combined-cycle generating plant was initiated in 1974. These designs will be used to estimate costs, system control requirements, mass balances, site requirements, etc., for a commercial-sized facility.

During 1975 the PDU was modified and repaired. The modifications included the installation of a computer control and data monitoring system. Corrective designs were completed and modifications were made to various process and utility subsystems including the cooling water and waste-handling subsystems. A number of design modifications were also made on the gas generator system.

Laboratory research in 1975 focused on providing design data for the PDU, and on providing data and predicting operating conditions for the conceptual design of a commercial-sized plant.

Studies of the gas-turbine combustor system were performed to determine the operational conditions necessary to ensure efficient, stable, durable, and environmentally acceptable performance. Tests were conducted on two dual-fuel coal gas/distillate oil combustors. Combustion stability was studied by measuring carbon monoxide emissions. It was determined that moisture content affected combustion stability and, consequently, required regulation. A new second-generation combustor was designed on the basis of these tests.

Efforts during the first two-thirds of 1976 were concentrated on tests in the PDU devolatilizer subsystem. Following the conclusion of these tests, much of the work effort shifted to completion of the gasifier PDU. Initial shakedown tests were completed late in 1976, and the first gasifier hot test was completed in December 1976.

During 1977, a series of shakedown tests, a reactor design test, and materials tests were performed to secure data and operational experience. The control of combustion in the unit was demonstrated; ash agglomeration and withdrawal were studied and successfully achieved; the reactor geometry was optimized to some extent; operating procedures were established; the ability to achieve high fines utilization and high carbon utilization was demonstrated; and operation was demonstrated with feedstocks with a wide range of characteristics in terms of composition, particle size, and reactivity. Design and procurement of hardware for integrated operation of the gasifier and devolatilizer reactors also continued throughout the year. Laboratory studies included support tasks for the PDU combustor–gasifier test program, preparation for integrated operation, coal behavior studies, ash behavior studies, and reactor analysis.

At the beginning of 1978, 16 PDU gasifier tests had been made, and the data from all 16 were correlated to evaluate process performance and scaleup criteria. Efforts were also initiated for a similar study with devolatilizer test data. Plant modifications for integrated operations continued. Final assembly of the gasifier reactor system and interconnecting piping was completed in March. Precommissioning of equipment and subsystems was completed in preparation for April startup, and instrumentation and control logic checked out.

Test TP-015 was conducted in April to validate the operability of the new expanded reactor configuration and system following redesign and modification. In the test, coke breeze was used, followed by FMC char and a short-duration Sub-C coal test for a total of 196 hours of operation.

Two air-blown gasifier tests were completed in May. The tests were conducted to gain additional operating data for the modified gasifier with direct feed of a highly caking Pittsburgh seam coal. Over 100 hours of operation were logged in each test, demonstrating operability of the gasifier. No material buildup occurred in the internal portion of the reactor and no measurable tar or oil was produced. In addition, a turndown capability of four to one was demonstrated.

The first oxygen-blown gasifier test with the expanded reactor configuration was initiated in June 1978. The series of tests utilizing oxygen continued, with four tests being completed during the third quarter and a fifth test being initiated. These tests were designed to measure the effects of steam flow and reactor temperature on the gasification of Pittsburgh seam coal using oxygen in the expanded gasifier configuration. The Pittsburgh seam coal was fed directly to the gasifier at 1800°F (982°C) (in test series TP-016) without the benefit of pretreatment or devolatilization.

Tests involving operating the gasifier as a single-stage reactor were concluded in December so that maintenance and modification could be started in January 1979, to prepare the PDU for integrated operation of both the devolatilizer and the gasifier.

MOLTEN SALT GASIFICATION PROCESS

The molten salt coal gasification process is being developed by the Atomics International Division of Rockwell International Corporation.

The objective of the project is to demonstrate the feasibility of the molten salt coal gasification process for use in environmentally acceptable generation of electric power. Atomics International is to design and construct a PDU that will convert 1 ton of coal per hour to a nonpolluting low-Btu fuel gas. The PDU will be located at the Santa Susana Test Site of Atomics International. In addition, Atomics International is to operate the plant to obtain data for evaluating the process and design a full-scale commercial plant.

Process Description for Gasification/Combined Cycle Power System

In the molten salt gasification process, coal is gasified to produce electricity through gas and steam turbines. Three major steps are required: (1) molten salt coal gasification, (2) electric power generation, and (3) sodium carbonate regeneration. Figure 4-15 provides a schematic of the process.

In the gasification section, coal and sodium carbonate are transported by compressed air (10–20 atm) into the bottom of the melt bed in the molten salt furnace. The molten pool is composed of sodium carbonate along with sodium sulfide, and sodium sulfate forms during the process. The process is designed so that gasification reactions (partial oxidation and pyrolysis) take place at 1800°F (982°C) and 20 atm. The fuel gas produced has a heating value of approximately 150 Btu per standard cubic foot and is predominantly carbon monoxide, hydrogen, and nitrogen. Because the melt retains the ash and sulfur from the coal, the metal must be continuously withdrawn from the furnace. The sodium carbonate is subsequently regenerated.

The hot fuel gas from the molten salt furnace will be combusted in a gas turbine which converts the energy of the fuel gas to electric energy. After passing through the turbine, the exhaust gas will be used to produce steam for operating a steam turbine, thereby producing additional electricity. Flue gas (primarily carbon dioxide) leaving the boiler heats incoming combustion air and is then used for sodium carbonate regeneration.

In the regeneration section, the melt is withdrawn from the molten salt furnace into a quench tank where the salt is dissolved in water. The resulting slurry is then passed through a filter for removal of the ash. The filtrate, containing dissolved sodium sulfide, is carried to the stripper where sulfur is stripped off as hydrogen sulfide and subsequently converted to elemental sulfur in a Claus plant. The remaining solution is carbonated, crystallized, centrifuged, and recycled. The crystals are dried in a rotary kiln and recycled to the molten salt furnace as regenerated sodium carbonate. The key advantages of the molten salt coal gasification process include the following:

- The process allows the use of a complete range of coal particle sizes, from ¼ in. down to fines. There are no requirements for close sizing of the coal, removal of

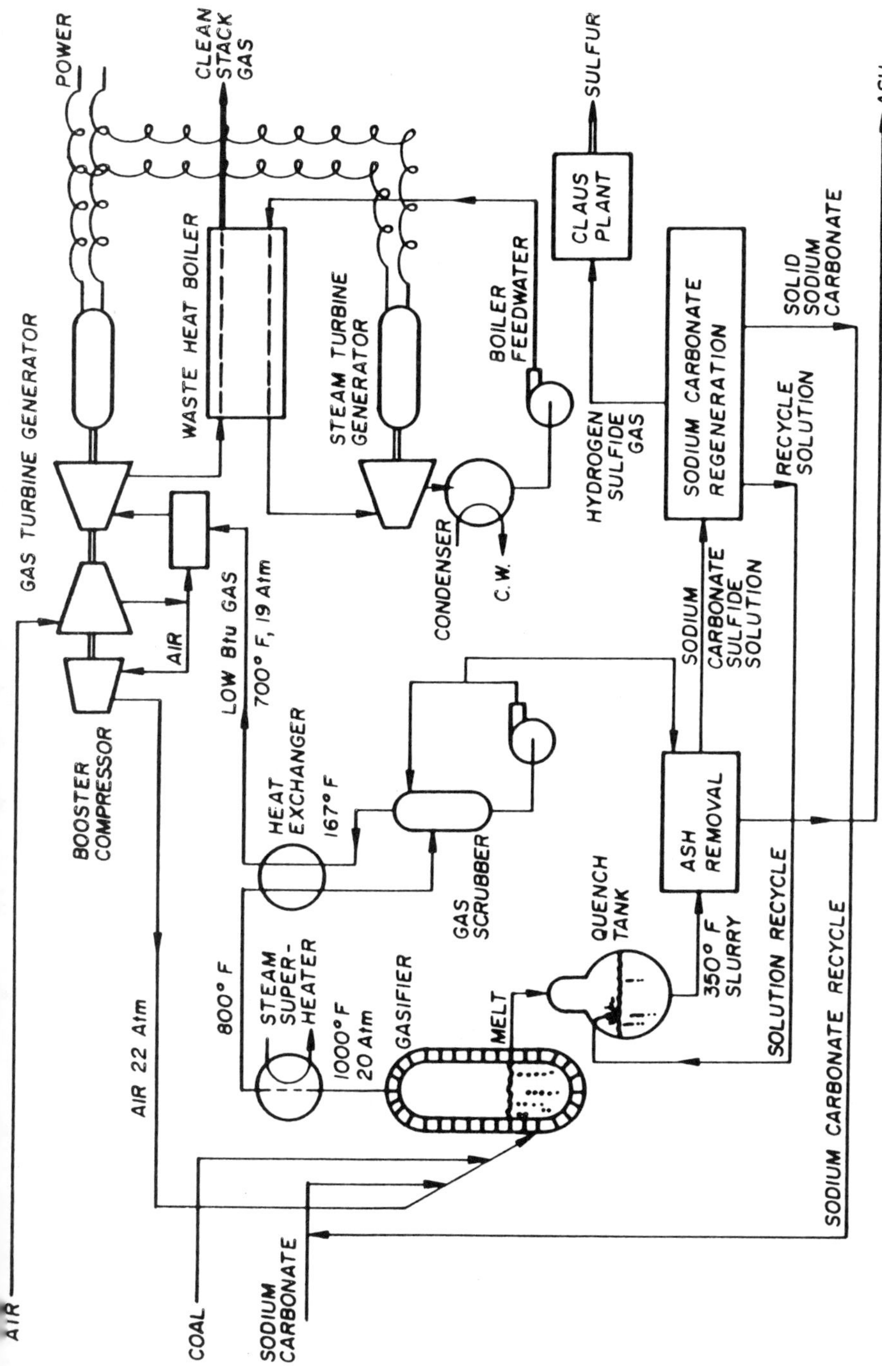

Fig. 4-15. Molten salt gasification/combined-cycle power system progress schematic.

fines, or coal pulverization. Only crushing is required to a size small enough to allow pneumatic transport of the coal (¼ in.).

- The process is well suited to the gasification of all types of coal, from anthracite through lignite. Pretreatment is not required for caking coals, and, therefore, there is no associated loss of heating value.
- The gasification is carried out in a thoroughly mixed, three-phase turbulent-bed environment, allowing excellent temperature control and precluding the occurrence of hot spots.
- Practically all the sulfur and ash of the coal are retained in the melt, minimizing the requirements for product gas cleanup.
- The sulfur of the coal is recovered in elemental form, minimizing storage, handling, and marketing problems.
- The molten salt gasifier operates in a stable manner over a wide range of air and coal feed rates, indicating the potential for good load following capability.
- The overall process has a high thermodynamic efficiency.
- The process shows a good potential for the economic generation of electric power.

History of the Project

Before December 1974 (when ERA, predecessor of DOE, and industry began cosponsoring the previous project), Atomics International did extensive experimental work to develop the molten salt combustion and gasification process. Tests were initially conducted in a laboratory-scale unit and then in a bench-scale combustor. The purpose of the experimentation was to determine combustion rates, conditions for obtaining different ratios of carbon monoxide to carbon dioxide, ash–melt interaction, ash capacity of melt, chemical reaction of sodium carbonate with ash, and suitable metal to withstand corrosion. The tests revealed that the combustion rates achieved for various feedstocks such as char, coal, and coke, with wide ranges of fixed carbon, volatile matter, sulfur, and ash contents were much greater than with conventional burning. In addition, the tests showed that the process was environmentally acceptable. Atomics International concluded that the molten salt process was sound and that a pilot plant should be constructed and operated to further explore and demonstrate its potential for commercialization.

To obtain the necessary design data for a pilot plant, Atomics International constructed a molten salt test reactor capable of processing 250 pounds of coal per hour. This reactor is 8 feet high and has an inside diameter of 3 feet. Experimental and analytical development work was also conducted to provide design data and confirm key assumptions. Some of the areas investigated are effects of pressure on gasifier operation and performance; air and solids feed system design and operating parameters; melt withdrawal and quench system design requirement; entrainment of particulates in the product gas and purification requirements; and optimum sodium carbonate regeneration process flow scheme.

In the previous project, the preliminary design of a pilot plant was completed during the third quarter of 1975 and the plant cost estimated. Because these new costs were considerably higher than the original estimates, alternative approaches were evaluated in order to select a program that would enable the objectives to be met at a lower cost. Based on these studies, it was decided to move the pilot plant from Norwalk Harbor, Connecticut, to the Atomics International field laboratory in Santa Susana, California. It was further decided that a 1-ton-per-hour plant would be constructed rather than the original 5-ton-per-hour facility. During the first quarter of 1976, the proposal to construct the 1-ton-per-hour plant was reviewed, and the development contract for the 5-ton-per-hour facility was terminated.

Phase I, preliminary engineering, was completed in August 1976. It was decided that Atomics International would be responsible for the entire detailed design, and this work began in November 1976. Ground breaking for the process development unit occurred on March 8, 1977. By the end of 1977, detailed design, most of the procurement, and a large part of the construction had been completed. By mid-1978, construction of the PDU was completed.

The first low-Btu gas production test was run in November 1978 at near-atmospheric conditions, using petroleum coke, but with no melt regeneration. The test was concluded on November 20 after 157 total hours of operation, 115 of which were under gasification conditions. Illinois No. 6 coal was gasified over a 4-hour period immediately prior to draining the system. The coal was gasified at rates up to 400 pounds per hour.

In preparation for the second planned series of PDU tests, modifications made to the PDU were as follows:

- Gas-fired burners for the melt withdrawal assembly to replace electric heaters were designed, fabricated, tested, and installed.
- Unsatisfactory capacitive-type level sensors in the solids feed hoppers were removed and replaced by instruments of a different type, which were on hand at the Rockwell International Molten Salt Test Facility.
- Impingement separator to eliminate excessive water carryover from the spray cooler was installed temporarily while high-efficiency cyclone separators were being ordered.

Progress

The product gas, regeneration, and sulfur removal systems were checked out and the plant startup activities were begun in February 1979. Initiation of gasification operations was delayed until March due to rains, which interfered with ongoing coal drying activities.

The gasifiers and the solids feed and combustion air systems were operated for 10 days during March. A total of 19 tons of coal, 16 tons of petroleum coke, and 15 tons of sodium carbonate were used. During periods of coal gasification, product gas heating values up to 125 Btu/scf were achieved.

Midway through the second test, the melt overflow system plugged up. The unit was shut down when the melt pool in the gasifier rose above normal levels. Full activation of the regeneration system during the test was prevented due to inadequate concentrations of sodium carbonate in the solution stream. Prior to initiating further tests, additional gas-fired burners were installed in the melt orifice area to keep the temperature hot enough to prevent the plugging observed in the previous runs.

The third gasification test was run during May. After 272 consecutive hours, the test was terminated when material failures resulted in a plugged system. A redesign was developed for the melt withdrawal assembly.

After removal of the solidified melt, which remained in the gasifier at the conclusion of gasification Run 3 completed in May, a detailed examination was made of the precast refractory block liner. The spalling damage observed during the preliminary inspection prior to cleaning out the gasifier was found to be more pervasive than predicted. Consequently, a decision was made to replace the entire Monofrax A liner.

Rebricking of the gasifier and bakeout of the liner were completed by December. Programmed heatup of the vessel for Run 4 was initiated by the end of the year.

Replacement of the Monofrax liner in the gasifier continued throughout October 1979. However, because of increased costs related to rebricking and the uncertainties associated with the receipt of supplemental program funding, other scheduled preparation activities for the next PDU gasification test run were curtailed.

By the end of October, all of the components for the modified melt withdrawal nozzle and shatter barrel system were assembled and ready for installation into the PDU.

As of November 1, preparations were completed for testing and calibrating the melt withdrawal system burners (and associated support systems) under PDU operating conditions.

During the month of November the rebricking of the gasifier with new Monofrax bricks and liner backout to remove water from the new mortar were completed. The four refurbished feed nozzle inserts were installed along with the melt withdrawal system, modified to incorporate thick-walled Monofrax components used during test Run 3. In addition, the new thermowells and preheat burner were installed and the liner backout program commenced. By the end of November, the vessel had not yet cooled down enough to remove the top and enter the gasifier.

In December, the backout of the newly installed Monofrax liner in the gasifier was completed and a detailed inspection of the interior was made. Some spalling was observed but further deterioration is not expected. By the end of December, all systems were operational, including the new metal withdrawal gas-fired system and metal sampler device, and on standby in anticipation of test Run 4 operations. However, because of the proximity of the year-end holidays, the test was postponed until December 30, 1979, at which time the preheat burner was ignited and programmed heatup of the vessel for Run 4 was initiated.

Following test Run 4, the entire PDU plant will be prepared for fully integrated low-Btu gasification run at 20 atm pressure while recycling sodium carbonate, collection ash, and harmlessly disposing of the sulfur content in the coal.

Current Project Status

By March 3, 1981, nine high-Btu projects had applied for assistance with the Synthetic Fuels Corporation (SFC). The project sponsors include American Natural Resources Co.; Panhandle Eastern Pipeline Co.; Memphis Light, Gas and Water; Tenneco Coal Gasification Co.; Mountain Fuel Resources Co.; Consolidated Natural Gas Co.; Northwest Corp.; the Crow Tribe of Montana; and Sante Fe International Corp. Fifty-two projects in other synfuel technologies had also applied to the SFC for assistance by March 31.

The SFC was established by Congress in 1980 through the Energy Security Act (P.L. 96-294) and was mandated to support synthetic fuels projects principally through loan guarantees, price guarantees, and/or price supports. Previously, the Alternative Fuels

Program (AFP), under the auspices of DOE, was involved in the funding of synthetic fuels projects. On July 9, 1980, DOE announced the first round of awards for feasibility studies and cooperative agreement funding under the AFP. Several high-Btu coal gas projects received awards. A second round of awards, announced on December 8, 1980, were later rescinded.

A new pricing formula for synthetic natural gas from the Great Plains project was issued through FERC on April 30, 1981, in Opinion 119 and agreed to by all involved parties. The agreement calls for the sale of gas to the pipelines at a price of $6.75/MMBtu, adjusted quarterly commencing April 1, 1981, in accordance with a factor which reflects an equal weighting of changes in the producer price index for all commodities and prices for No. 2 fuel oil. After the plant commences production of synthetic gas, the adjusted price would be limited during the first 5 years to the price of No. 2 fuel oil unless such price is regulated; during the fifth through tenth years of operation, to the higher of the prices paid by the pipeline affiliates for domestic natural gas or for Canadian and Mexican gas, but in neither case higher than the unregulated price for No. 2 fuel oil; and during the remaining life of the 25-year agreement, to such domestic natural gas prices unless such prices are then regulated, in which case the Canadian and Mexican price will set the limit.

The federal government has no further plans to financially support the synthetic fuel demonstration plants. As a result, the Noble County, Ohio, coal gasification plant sponsored by Conoco Coal Development Co. and the Perry County, Illinois, project sponsored by the Illinois Coal Gasification Group (presently completing the final stages of Phase I) are now considered inactive beyond present phases. Table 4-2 represents an update of the status of high-Btu coal gasification plants. The Reagan Administration has not actively pushed synfuel developments through the Energy Security Act financial incentive program.

DOE Coal Gasification Program

The Department of Energy (DOE) has continued the development of a broad-based research and development program for coal gasification. Their program is subdivided into two divisions: (1) surface coal gasification and (2) *in situ* coal gasification. The following material is presented to conform with the DOE program approach.

SURFACE COAL GASIFICATION

Successful development of the technology needed for surface gasification of coal will enable the United States to use its large deposits of coal to produce a wide variety of energy products and chemical feedstocks. The flexibility of this technology is that it can be used to convert coal into alternate products suitable for the demands of many markets, as shown in Fig. 4-16.

Accordingly, the surface coal gasification activity has as its objective the development and demonstration of that technology. The successful completion of the projects in this activity will provide the means to convert coal into a number of clean, synthetic gaseous and liquid fuels as well as a synthetic gas suitable for continued conversion into other fuels and chemicals. The synthetic gaseous fuels include high-Btu gas having a heating value of approximately 950 Btu per standard cubic foot and low- and medium-Btu gases with heating values that range from approximately 100 to 500 Btu per standard cubic foot.

Table 4-2. Status of Commercial and Demonstration High-Btu Coal Gasification Projects (As of July 1981)

Controlling Company	Site	Process	Coal Feed (tons/day)	Peak Output (MMcf/day)	Status
ACTIVE COMMERCIAL PROJECTS					
American Natural Resources Co., Peoples Energy Co., Tenneco, Inc. & Transcontinental Gas Pipe Line Corp.	Beulah-Hazen Area, Mercer County, N.D.	Lurgi gasification with methanation	12,328	137.5	The earliest completion date of this project is late 1984. ANR and Peoples have spent over $80 million and DOE has provided an additional $25 million loan for the project costs thus far. Construction has started on the project site. On August 6, 1981, DOE awarded the project a conditional commitment to guarantee up to $2.02 billion.
Panhandle Eastern Pipeline Co., Pacific Gas & Electric Co., Rhurgas A.G.	Douglas, Wyo.	Lurgi and Texaco gasification with methanation	16,400 16,400	150 (Stage I) 150 (State II)	Panhandle Eastern, through its subsidiary WyCoalGas Inc., plans to construct a facility to convert subbituminous coal to high-Btu gas. The plant is to be constructed in two stages. Construction on the first stage is to begin in the spring of 1983 with production from Stage I in early 1987. An application for a load guarantee from the SFC has been made.
Texas Eastern Corp., Texas Gas Transmission Corp.	Henderson County, Ky.	Lurgi gasification with Fischer–Tropsch and methanation	28,600	145 (and 16,000 bbl/day gasoline, 9,000 bbl/day fuel oil and LPG, 2,000 tons/day chemical products)	Project has been awarded cooperative agreement funds. Negotiations to obtain a site, coal and water rights are underway. Preliminary engineering work is in progress. Plant will produce a mixture of gas, liquids, and chemicals.
Memphis Light, Gas and Water Division	Memphis, Tenn.	U-Gas with methanation	3,158	4.3	DOE has entered into a joint venture with Memphis Light, Gas and Water to convert high-sulfur bituminous coal to 300 Btu/cf industrial gas at a rate of 167 MMcf/d. Approximately 17.5 MMcf/d will be methanated to yield 43 MMcf/d of pipeline quality gas. Phase II, which is the final design and construction of the plant is now underway. The Environmental Impact Statement was finalized and submitted for agency review. An application for a loan guarantee from the SFC has been made.
Tenneco Coal Gasification Co.	Wibaux, Mont.	Lurgi gasification with methanation and Texaco partial oxidation unit for liquids	37,000	280	Estimated total capital costs of the plant are $2.3 billion (1981 $). Production at half capacity is scheduled to commence 1989 with full production coming on line in 1990. An application for a loan guarantee from the SFC has been made.

FEASIBILITY STAGE OF COMMERCIAL PROJECTS

Texas Eastern Corp.	Northwest New Mexico	Lurgi gasification with methanation	29,000	142	Project has been awarded $3 million in feasibility study funds. Methanol will also be produced at this plant.
Texas Eastern Corp., Texaco	Wyoming	Lurgi gasification with methanation	38,000	127	Sources for private funding are currently being pursued. Methanol will also be produced at this plant.
The Columbia Gas System, Inc.	Illinois	—	—	300	Columbia Gas exchanged a 50% interest in some 43,000 acres of its 325,000 acres of West Virginia coal lands for a 50% interest in 35,000 acres of Illinois coal lands held by Exxon Coal Resources U.S.A., Inc. The Illinois coal is being held by Columbia for eventual coal gasification.
Mountain Fuel Resources Co., Mono Power Co., Conoco Coal Development Co.	Emery County, Utah	Lurgi gasification with methanation	11,000	68 & methanol	Planned completion in 1987. Economic and environmental feasibility studies are underway. State prequalification approval of the site has been received. An option for water rights has been signed. An application for a load guarantee from the SFC has been made.
Consolidated Natural Gas Co.	Mason County, W. Va.	Slagging Lurgi-Texaco	15,000	250	The company has purchased about 615 million tons of recoverable coal for gasification. The project will also produce methanol. A proposal has been submitted to the SFC for future support.
Northwest Corp.	Oregon	Texaco gasification	27,400	250	Project involves construction of a slurry pipeline from Wyoming to a plant located in Oregon. Forty percent of the coal would be converted to medium and high-Btu gas plus methanol, 20% would be burned for electricity and 40% would be exported to Japan. An application for a loan guarantee from the SFC has been made.
Texas Gas Transmission Corp.	Western Kentucky	Hygas	7,300	85	Coal reserves have been obtained. Location of the project may be modified.
Utah Resources International Inc.	Garfield County, Utah	Lurgi gasification with methanation	17,000	250	The company has applied to the SFC for a loan guarantee.
Washington Natural Gas Co.	McCone County, Mont.	Lurgi gasification with methanation	30,000	250	Private funding of the feasibility study is currently being sought.

Table 4-2. *(continued)*

Controlling Company	Site	Process	Coal Feed (tons/day)	Peak Output (MMcf/day)	Status
Crow Tribe of Montana and Pacific Coal Gas Co.	Crow Reservation, Mont.	Lurgi gasification with methanation	12,500 25,000	137.5 (Stage I) 275 (Stage II)	The Crow Tribe has been awarded DOE funds for a nine month feasibility study. Project will be built in two stages. An application to the SFC for a loan guarantee has been made.
Northern Natural Gas Co. (12 member consortium)	Oliver County, N.D.	Lurgi gasification with methanation	12,000	75 (18,000 bbl/d methanol)	Preliminary feasibility study underway. Plant would produce both high-Btu gas and methanol.
Northern Alabama Coal Gas Project	Murphy Hill, Ala.	Koppers–Totzek and/or Texaco	10,000	60	The project has received $125 million from the government for funding thus far. The initial module will cost about $1.5 billion (1980 $) and could start operating in 1986. Sante Fe International Corp. and Air Products & Chemicals, Inc. have proposed to sponsor the project. Kidder, Peabody & Co. will be the financial agent. An application to the SFC for a load guarantee has been made.
DEMONSTRATION PROJECTS					
Illinois Coal Gasification Group, DOE (Phase I only)	Perry County, Ill.	Cogas Process	2,200	23.2 plus 2,100 bbl/d fuel oil and naphtha	Phase I, which is the detailed engineering design of the demonstration project, is due to be completed by the end of July, on schedule and within the budget constraints.

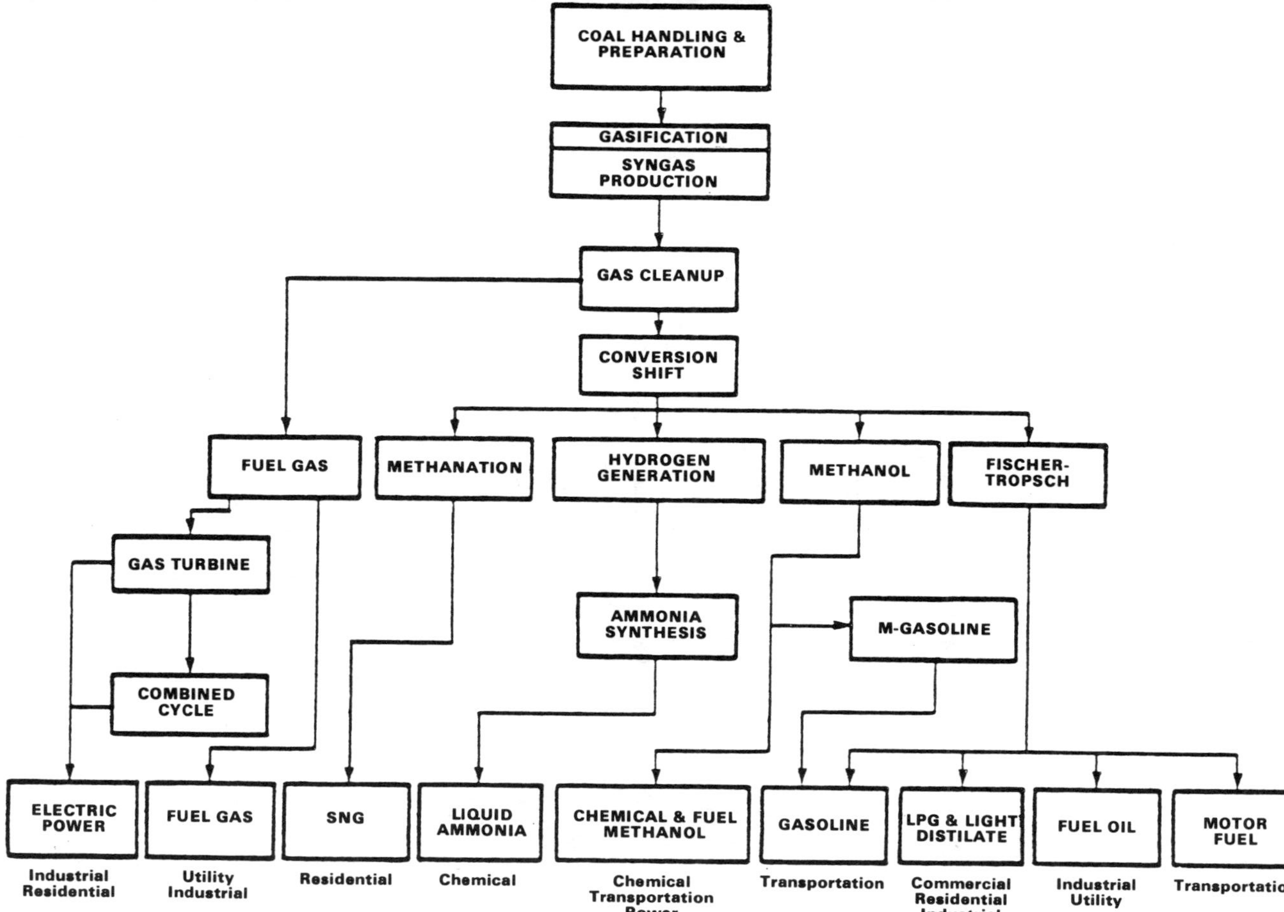

Fig. 4-16. Flexibility of coal gasification.

The high-Btu gas is intended for use principally as a fuel for residential and commercial heating as well as for industrial applications that require clean, high-energy fuel. The low- and medium-Btu gas is used for industrial process heat and electric power generation. As Fig. 4-16 shows, this same technology can be used as the first step in producing synthesis gas that can be subsequently processed into chemical feedstocks and liquid fuels.

Because the intended end uses for high-Btu gases differ from those for medium- and low-Btu gases, the objectives for these two efforts are different.

The objectives of the high-Btu coal gasification effort are

1. Evaluate the technological and economic status of existing first-generation processes, assess their suitability for meeting U.S. market needs, and provide needed technical support for improving these gasification systems.
2. Continue the development for new and improved second-generation technology for commercial-scale plants to convert domestic coals (caking as well as noncaking) to substitute natural gas for pipeline quality.
3. Continue the development for third-generation gasification processes.

The objectives of the medium-/low-Btu coal gasification effort are

1. Continue the development of new gasification technologies to convert coal to environmentally acceptable gaseous fuels for use in electricity generation, as an industrial fuel, and as a chemical feedstock.
2. Ensure that technological advances in medium-/low-Btu gasification are made available to users of existing gasification systems, and provide sufficient operating data to encourage other industries and utilities to use this technology.
3. Acquire operational experiences in specific applications or industries to better define such uncertainties as capital and operating costs, retrofit problems, and the effect of using gas from coal as the end product.
4. Establish the availability, through development or demonstration, of cleanup equipment and systems suitable for use with low-Btu gasifiers that satisfy current as well as projected environmental requirements.

As the above objectives for the high- and medium-/low-Btu gasification efforts are achieved, the technology needed for the production of a synthesis gas will also be developed.

Early identification and characterization of potential environmental contaminants is necessary to minimize adverse environmental effects. Standards of environmental performance serve as goals for development efforts to reduce environmental impacts, to minimize the severity for any impact that occurs, and to minimize the costs of technology and equipment required to accomplish these goals.

Environmental control is an integral part of this program and is accorded a high priority within each developmental project.

High-Btu Gasification

National patterns of growth in population and economic activity require increasing consumption of natural gas. At the same time, proven reserves have decreased despite intensified drilling activity. These forces have caused local and regional shortages in the short term. In the long term, the country ultimately will face resource depletion. Development of alternative domestic fuel resources of a quality and convenience equal to natural gas is a high national priority. The high-Btu gasification effort is meeting that priority by

- Identifying and accelerating the development of promising third-generation technology to improve processes that may be used on a commercial scale in the 1990–2000 period.
- Developing, in cooperation with industry, new and improved second-generation gasification technology that will permit the use of all coals (lignite, subbituminous, bituminous, anthracite, etc.) as feedstocks for conversion to alternate fuels and chemical feedstocks.

These efforts will produce the technology that can convert coal into a substitute natural gas with the following specifications:

- Heating value—950–1000 Btu per standard cubic foot.
- Purity level—Essentially free of sulfur and other potential pollutants and hazardous impurities.
- Process gas exit pressure—Approximately 1000 pounds per square inch (psi).
- Combustion characteristics—Similar to natural gas. Exhaust gases should meet Environmental Protection Agency regulations and other regulations.

Although many gasification processes under development have the same objectives, each is characterized by important differences that warrant concurrent development. These differences include reaction conditions, coal pretreatment, method of feed, reactor configuration, heat supply, gas purification, and methanation process. The Lurgi gasification process has been utilized as a benchmark for measuring economics of processes currently under development. Improvements in present technology will be measured primarily in terms of comparative capital and operating costs, as well as ability to operate successfully with U.S. caking-type coals. Development of each candidate process will continue through proof of concept or to the point that industry can be expected to construct, with an acceptable degree of risk, a commerical plant based on one or more of the concepts.

Gasification Pilot Plants

BiGas Pilot Plants. The BiGas pilot plant located in Homer City, Pennsylvania, is a facility designed to study a new process for converting coal into pipeline-quality gas. The principal feature of the plant is a two-stage, entrained-bed, high-pressure [1500 pounds per square inch gage (psig) maximum], oxygen-blown, and ash-slagging gasifier unit. The gasifier has three stages separated by formed restrictions (Fig. 4-17). Stage two (topmost) receives dense-phase coal in gas-entrained, flow-through injectors located in a throat between stages two and one. Steam is also injected there. Process gas and carryover char flow from the top of stage two of the gasifier to a cyclone. The char is recycled from the cyclone and fed to a steam injector nozzle. It is subsequently fed, along with oxygen and steam, to the bottom of stage one. The products of char gasification in stage one are hot gas feeding into stage two and slag dropping into stage three, the slag quench section. Molten slag runs down the walls of stage one for the pressurized gasifier.

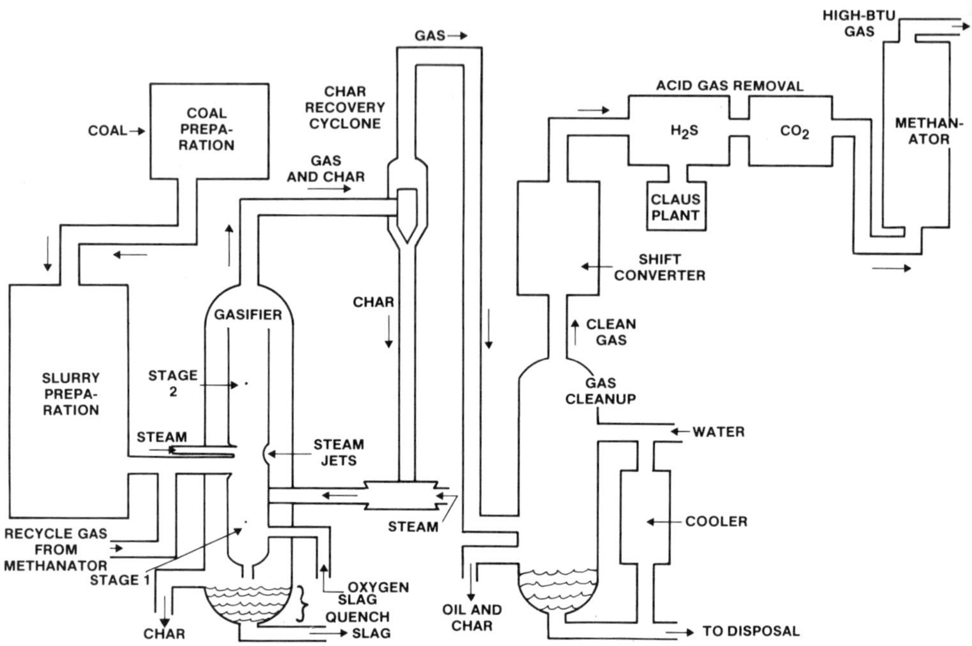

Fig. 4-17. BiGas process.

Raw gas leaves the char recovery cyclone, passes through a water wash column, and flows to the shift reactor, where the H_2/CO ratio is adjusted to the desired value. H_2S and part of the CO_2 are removed in the Selexol unit upstream of the methanator. H_2S gas goes to a Claus unit. The product gas from the methanator goes through final CO_2 removal in the Selexol unit. The methanator, a fluidized bed with embedded water-cooled tubes, offers great potential for commercial development.

The BiGas reactor can handle all types of coal. Because of its high operating temperature [more than 2500°F (1370°C)], there is no tar in the product gas, thus simplifying the gas cleanup step. High methane yields are achieved directly in the reactor, thus reducing downstream requirements for methanation. Since the process uses an entrained-bed rather than a fixed- or fluidized-bed feed system, caking coals can be gasified without pretreatment.

The BiGas project was initiated in December 1963 by the Department of the Interior's Office of Coal Research (OCR), and in 1973 became a part of the combined Program for OCR and the American Gas Association. Since December 1978, the Department of Energy alone has been funding the project. The development of this process by Bituminous Coal Research, Inc. (BCR) started with experiments in a rocking autoclave testing in 1965, followed by 5-pound-per-hour continuous (externally heated) flow reactor testing in 1967, 100-pound-per-hour internally fired process development unit testing in 1968, cold flow model testing in 1972, and, at present, operation of a 10,000-pound-per-hour, or 120-ton-per-day, pilot plant. The Stearns-Rogers Company completed the construction of the 120-ton-per-day pilot plant in July 1976 and startup operations commenced in August 1976.

Pilot plant support work is being conducted at the facilities of BCR in Monroeville, Pennsylvania. Stearns-Rogers has the overall technical and administrative management responsibility for development of the BiGas process, as well as the responsibility for operating and maintaining the plant.

Significant milestones achieved to date at the BiGas pilot plant include:

- Solving slag removal problems while using Montana Rosebud coal.
- Incorporating new char burners of different design into the gasifier that show promise for overcoming operational problems.
- Continuing evaluation of instruments and operational techniques to measure char flow and stage-one temperature.
- Solving operational problems in the areas of ammonia stripping, naphthalene deposits in the gasifier, Selexol carryover losses, and carryover of fines into the gasifier.
- Operating the gasifier for approximately five continuous days.
- Successful operation of the coal–slurry feed spray drying system at 750 psig.
- Successful demonstration of slag-tapping without the addition of flux.
- Attaining stage-two carbon conversions in excess for 40 weight percent (40 weight percent was anticipated from PDU studies).

During FY 1982, the development work on the BiGas concept for coal gasification was concluded, and the facility was placed in a standby status pending a decision as to its future role in the program.

Hygas Pilot Plant. In the Hygas process, hydrogasification (the carbon–hydrogen reaction) is promoted to produce methane exothermally. This heat is used to decompose steam on carbon to produce additional hydrogen. In the Hygas hydrogasifier, several stages are used to achieve the above reactions. The general processing scheme is shown in Fig.4-18.

Dry coal particles at about 600°F (316°C) flow by gravity from the slurry-drying section through a dipleg into a lift pipe. The lift pipe serves as the first stage for hydrogasification. Here, a dilute-phase contact occurs between the coal or pretreated char

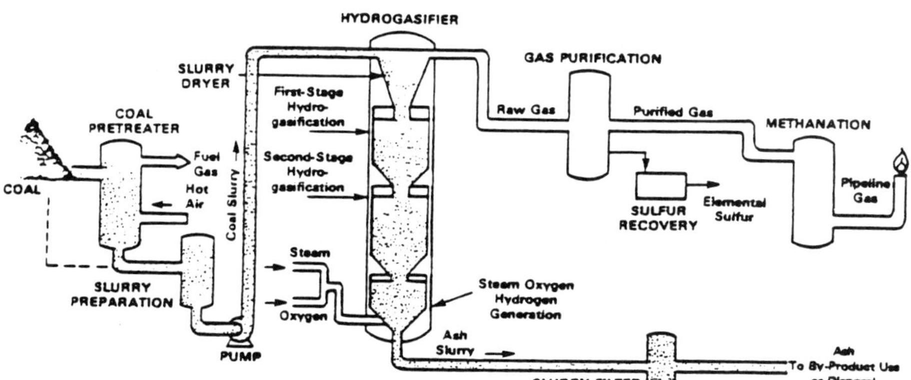

Fig. 4-18. Hygas pilot plant and configuration.

and the gases from the second stage. The temperature of the gases is approximately 1500°F (815°C). The gas lifts the solids to the gas–solid disengaging section. As the dried coal is lifted, it is flash-heated in the presence of hydrogen to a temperature of 1200–1300°F (649–704°C), and approximately 20% of the carbon in the coal is converted to methane.

The partially gasified char flows into the second-stage hydrogasification section and contacts the H_2-rich gas from the steam–oxygen gasifier. This stage uses a dense-phase, fluidized-bed reactor operating at 1600–1700°F (871–927°C). Here, methane is formed simultaneously with the H_2 and CO produced by the steam–carbon reaction, and approximately 25% more of the available carbon is converted. Thus, 45–50% of the carbon in the total feed coal is converted by hydrogasification reactions. In this stage, with steam present, any rise in temperature accelerates the endothermic steam–char reaction; a temperature drop slows the reaction. The hot gases rise to the first stage and to the dryer, where much of the heat is used to dry the feed coal.

The partially depleted coal char leaving the second-stage hydrogasification zone is used to produce hydrogen in the steam–oxygen gasification section. The steam and high-purity oxygen convert the char into hydrogen and carbon oxides. Ash is discharged from this stage without being slagged. The ash is discharged into a tank, where water is added to make a slurry that is then depressurized. After depressurization the ash is recovered by filtering, and the water is recycled.

The raw gas leaves from the top of the slurry-drying section through a quench tower, which removes dust, the slurry oil, and excess steam. Gas purification occurs in an acid removal system that consists of an amine and water scrub followed by a caustic wash. Removal of CO_2 and H_2S is essentially complete after the amine scrub. A Claus plant converts the sulfur products to elemental sulfur. Methanation occurs in two packed-bed, downflow reactors. Heat release control is carried out by the Institute for Gas Technology (IGT) split-flow, cold-gas recycle method.

Before 1964 research on the Hygas process was sponsored by the American Gas Association. Beginning in 1964, these studies were continued under the joint sponsorship of the Office for Coal Research and the American Gas Association. Since June 1979 the Hygas project has been funded solely by the Department of Energy. Early efforts were limited to bench-scale and process development unit studies of the basic process parameters and their effects on the reaction chemistry. In 1967 the design and construction of a large-scale integrated pilot plant was initiated and plant operation began in 1971. During 1974 a larger steam–oxygen gasification unit was designed, constructed, and integrated into the bottom of the Hygas reactor. The unit has been used successfully since that time.

The Hygas pilot plant has processed various coals—lignite, Illinois No. 6 bituminous (highly caking Sahara), Montana Rosebud subbituminous, Illinois No. 6 bituminous (washed, medium-caking), Illinois No. 6 bituminous (run-of-mine), and western Kentucky bituminous (washed)—into high-Btu gas.

In July 1977 the major program emphasis was directed toward generating data to support the design concepts of a conceptual commercial demonstration plant. The conceptual design was completed in April 1979.

Tests conducted during FY 1980 collected further data for optimizing the Hygas process design. The tests were performed on washed, caking, eastern Kentucky bituminous coal in order to expand the existing data base. These data were used to confirm and refine variables used in the conceptual commercial design.

Operation of the Hygas pilot plant was terminated on July 15, 1980, after Test 87. The data developed over the period of time the pilot plant was operated will be evaluated and used to determine the possible future role of this gasification technology in achieving the objectives of surface coal gasification. In FY 1981, this facility was considered for utilization in the peat gasification program.

Peat Gasification

The overall strategy of the peat gasification program is to develop a technology for converting peat to substitute natural gas (SNG) in an economic and environmentally acceptable manner. The following objectives address the technical problems and issues that must be resolved satisfactorily before committing the peat resources in the United States to any large-scale program for converting peat to SNG alternate fuels:

- Determine more precisely the locations and amounts of fuel-grade peat reserves in the United States.
- Determine the environmental and socioeconomic impacts of large-scale peat utilization.
- Develop an economic peat harvesting/dewatering technology.
- Develop an economically feasible method for reducing the moisture content for peat to 50% or less.

Resource estimation projects are underway to determine the amount and location of fuel-grade peat that may be harvested and utilized in an environmentally acceptable manner. DOE has been encouraging participation in the peat program by states that have significant peat deposits and interest in determining the energy potential for these resources. At the present time, 12 states (Louisiana, Florida, Georgia, New York, Rhode Island, Massachusetts, Maine, Minnesota, Michigan, North Carolina, South Carolina, and Alaska) are participating in joint DOE/state peat resource estimation projects.

The most significant technical obstacle to gasifying peat is the reduction of its moisture content to the acceptable level of approximately 50%. Peat dewatering is a significant technical obstacle that must be solved in order to use peat as a feedstock for gasification and for direct combustion, liquefaction, and other energy applications as well. The Department of Energy is supporting the development of alternative peat dewatering techniques, which include

- Solvent extraction—The major focus of this investigation is to test the performance of various organic solvents in removing water from peat. Preliminary economics will be estimated as part of the program. The Minnesota Gas Company (Minnegasco), Northern Natural Gas Co., and the Gas Research Institute (GRI) are providing joint support for these studies.
- Wet carbonization—Minnegasco, Northern Natural Gas Co., and the Gas Research Institute (performed by IGT) are studying the chemistry (kinetics, selectivity, etc.) and heat transfer of wet carbonization at a continuous PDU scale.
- Long-range research and development—DOE and GRI are sponsoring studies by the University of Pennsylvania for determining the nature of the affinity of peat for water (for example, colloidal, hydrogen bonding) and the development of novel dewatering techniques.

Peat Gasification Development. It has already been determined that peat is very reactive to gasification. In addition, a considerable amount of information derived from the operation for coal gasification pilot plants can be judiciously extrapolated and utilized for the special case for peat gasification. It is expected that the equipment required downstream of a peat gasifier to process the raw product gas will be virtually identical to the downstream processing requirements for coal-derived, raw product gas.

The following gasification projects directly addressed the technical concerns associated with gasifying peat and reflect the DOE position to develop promising long-range alternative technologies for converting peat to substitute natural gas.

- A study is being performed by IGT to determine the kinetics and fluid mechanics of peat gasification with peat feedstocks harvested in Alaska and Florida. Another study on wet carbonization is being performed by DOE/Minnegasco. Similar studies were performed previously with peat from Minnesota, Maine, and North Carolina.
- The DOE/Minnegasco low-severity hydropyrolysis study determined the chemistry of peat gasification at lower temperature and pressures. It is particularly important to understand the effect of gasification conditions upon the yield of liquids, that is, benzene and other gasoline blends.
- The DOE/Minnegasco study for peat biogasification, as a laboratory-scale project, is being performed to develop the concept of converting peat directly to methane using a partial oxidation reactor followed by a biological reactor (bacterial digestion). If this concept is developed successfully, no dewatering step will be required prior to treatment. The study is currently focusing on the effect of the specificity of the partial oxidation reactions on the performance of the biological reactor.
- The Rockwell peat gasification study was initiated to study the gasification of peat in the Rockwell short-residence-time, entrained-bed gasifier. The results will complement the IGT kinetic work, as the experiments will be performed at severe conditions of temperature and pressure. In addition, Rockwell has performed preliminary design of a peat gasification plant using the short-residence-time reactor.

DOE will terminate its large-scale activities related to peat gasification in FY 1982. This redirection principally concerns stopping efforts to utilize the Hygas pilot plant to gasify peat as these activities should be continued by industry if determined to be sufficiently promising for near-term application.

Environmental and Socioeconomic Impact Assessments. Assessment of environmental and socioeconomic impacts will be performed in three phases.

- Phase 1: Identification of Problems and Issues—Universal Oil Products (UOP), Systems Development Corporation (SDC), and Radian Corporation, respectively, have identified the environmental and socioeconomic problems and issues associated with large-scale peat gasification. A list of items to be considered has been prepared. UOP, SDC, and Radian Corporation also established a mechanism whereby the states participating in joint projects on resources estimation can address the identified issues.

- Phase 2: Gathering Data on Identified Issues—The joint projects of DOE and the states will be expanded to include tasks on gathering the environmental and socioeconomic data for issues identified in Phase 1. These data will serve as input for Phase 3.
- Phase 3: Overall Environmental and Socioeconomic Impact Assessments—Studies will be performed to assess the environmental and socioeconomic impacts of peat utilization in the United States. The approach is designed specifically to provide a reasonable cross section of environmental issues. This will enable localities to interpolate between case studies and to extract information applicable to their peat land as required for site-specific environmental and socioeconomic impact assessments. The intent is for the complementary environmental and socioeconomic impact assessments to provide information that would be required for a locality to decide whether or not to utilize its peat resources. The contracts to perform these assessments will be awarded in response to the results of competitive procurement.

A review of the program during FY 1981 resulted in a redirection of the efforts underway. An attempt will be made to establish a broad technology data base that is appropriate to a subsequent utilization of peat for many alternate energy users. As a result, no funds are requested specifically to support further development of technology to produce high-Btu gas from peat. In addition, certain elements of the FY 1981 program considered contrary to the new program direction have been proposed in a rescission.

Low-Btu Gasification

The potential market for clean-burning fuel gases as an alternate energy source for electric utilities and industrial users has increased as a result of the energy crisis, the decrease in natural gas reserves, and more stringent environmental regulations. A low- and medium-Btu gasification effort is being conducted to (1) develop and demonstrate new and improved gasification processes for the production of either a medium- or low-Btu gas as well as a synthesis gas, (2) establish the required techniques for utilizing these coal-derived gaseous fuels for utility power generation, chemical feedstocks, and industrial fuels, and (3) develop in cooperation with industry the technology needed for construction of commercial-scale plants to convert coal to environmentally acceptable gaseous fuels with a heating value of 100–500 Btu per standard cubic foot.

Medium-/low-Btu gasification is a promising method of using coal as both a utility and industrial fuel, that is, it is economically and environmentally acceptable. The generation of medium-/low-Btu gas from coal requires a less complex plant than that required for production of high-Btu gas. As a result, capital and operating costs are lower per equivalent Btu. In addition, high-Btu coal gasification technology can be used to produce a medium-Btu gas. Coal-derived fuel/synthesis gas, with a heating value of 240–375 Btu per standard cubic foot (i.e., medium-Btu gas), can be produced at lower cost than pipeline gas. It may be used with only minor modifications in most existing facilities designed for burning natural gas or fuel oil. These fuel gases can be used as a synthesis gas feedstock for several conversion processes that can convert the gases into a number of liquid fuels and chemicals.

As part of the medium-/low-Btu gasification effort, promising new concepts for improved gasification techniques are actively being developed at the PDU level. These include fixed-bed, entrained-bed, and fluidized-bed systems.

Fixed-Bed Gasification

Morgantown Fixed-Bed Gasification Process. At the Morgantown Energy Technology Center, a 42-in.-diameter, stirred fixed-bed gasifier has been in operation for several years. Continuing improvements in the gasifier and its gas cleanup system have resulted in a versatile system for the study of fixed-bed gasification phenomena.

In the fixed-bed gasifier, the coal is fed from a lock hopper into the top of the gasifier and moves slowly downward countercurrent to the gas streams moving upward. The gasifier has three distinct zones. They are, from bottom to top, carbon combustion, gasification, devolatilization, and drying. The steam and oxygen are introduced below an ash-discharge grate near the bottom of the gasifier. Gasifier temperatures range from 2400°F (1316°C) just above the grate to 800–1200°F (427–649°C) at the gas exit. In most cases, a water-cooled stirrer is used to control caking of the coal.

From the gasifier, the off-gas is passed through the gas cleanup system, which consists of the following components arranged in series: dust separator cyclone, humidifier, tar separator cyclone, venturi scrubber, electrostatic precipitator, direct cooler column, Stretford desulfurization unit, and a packed column to remove trace metals. The clean, pressurized fuel gas is suitable for a number of end uses or further processing.

In FY 1980, installation of the fixed-bed gasification system at Morgantown was completed. This included a complete new control system, expanded data acquisition capability, gasifier improvements, and completion of the gas cleanup system, including a pressurized Stretford desulfurization unit. Operability of the entire installation has been demonstrated during the acceptance tests and, although yet incomplete, preliminary data indicate a successful system design.

In FY 1981, Morgantown completed the acceptance tests for the entire full-flow cleanup installation, including the Stretford desulfurization system. A high level of data acquisition and evaluation activities will be continued during scheduled system characterization and tar utilization tests. Increasing emphasis will be placed on assessing and quantifying contaminants in the working environment and on supporting the environmentally acceptable management of gasification system effluent streams. Fixed-bed coal gasification efforts are aimed at expanding the technological data base required for the evaluation and scaleup of these systems.

Efforts in FY 1982 will include continuation for data acquisition to obtain engineering and operability information necessary for equipment evaluation and scaleup for fixed-bed gasifiers. In addition, Morgantown will continue to test the tar combustor and dirty-water evaporator.

Slagging Fixed-Bed Gasification Process. In a slagging fixed-bed gasifier, the coal from the lock hoppers moves by gravity flow into the reactor. The coal moves through the gasifier in the same manner as in a dry ash system. However, in the slagging gasifier, the coal is continuously gasified by an oxygen–steam mixture introduced through four water-cooled tuyeres located near the bottom of the gasifier. In this system, higher temperatures at the hearth form molten slag that flows out of the gasifier through a central tap hole into a water-quench bath.

The product gas leaving the gasifier at temperatures similar to those from a dry ash system [800–1200°F (427–649°C)] contains water vapor and tar products. To remove these condensable impurities, the gas is scrubbed in a spray cooler, and the gas liquor collected in the spray cooler is periodically discharged to a settling tank. The washed gas is cooled to 60°F (16°C) before being metered, sampled, and flared. The heating value of the gas is approximately 350 Btu per standard cubic foot. The advantages of slagging fixed-bed gasification as compared to a dry ash system include a fourfold increase in gas production for a given reactor volume, reduction in steam consumption by a factor of four, and reduction in the amount of liquid effluents produced.

The slagging fixed-bed gasification facility at the Grand Forks Energy Technology Center was constructed by and operated under the Bureau of Mines, Department of the Interior, from 1958 to 1966 to demonstrate the feasibility of the slagging operation and to assess operational parameters. Operations were resumed in 1976 to investigate the environmental concerns associated with commercial-scale slagging fixed-bed coal gasification. In September 1978 gasifier operation was suspended to update the existing equipment and to expand the operational capabilities. The 24-ton-per-day gasifier is expected to handle both caking and noncaking coals in test programs that can extend over periods of operations with a duration of 5–7 days.

The modification and installation of the slagging gasification system was completed in 1979 to include the dual lock hopper, a stirrer used to break the agglomerates, and new tar/water separator. Since then, numerous gasification tests with North Dakota lignite, each with 20 hours of slagging operation, have been completed. For example, in one recent test, 31,000 standard cubic feet of 330 Btu gas were produced under the following operating conditions: 300 psig, 1200 pounds per hour per square foot coal feed, 6000 standard cubic feet per hour oxygen injection, and a 1:1 mole ratio of steam to oxygen. A large quantity of waste water was collected for treatment studies. Currently, the function of the stirrer is being tested. Following that, gasification tests will be performed on coals for increasing caking tendency. Effluents will be characterized to support environmental assessment and industrial hygiene studies. Laboratory studies on biological and physical/chemical treatment of gas liquor will be continued.

During FY 1982, Grand Forks will continue testing a wide range of coals, including agglomerating bituminous coal, and will perform detailed characterization studies of liquid and gaseous effluents to establish effluent treatment methods. Emphasis will be placed on the environmental aspects of the fixed-bed gasification technique.

Fluidized-Bed Gasification

This task focuses on development of a low- or medium-Btu gasification process capable of using caking coals without pretreatment. The conceptual process consists of a single vessel, fluidized–bed gasifier–agglomerator.

The process was conceived as a two-stage process with separate fluid-bed devolatilizers and gasifier vessels. Testing has demonstrated, however, that these two functions could be accomplished in one vessel.

In this process, crushed coal is dried and introduced to the gasifier through an axial feed tube. The coal is carried upward into the gasifier by recycled product gas. Air or oxygen is fed into the gasifier by a tube concentrically located around the coal feed tube. Steam is injected into the same area for temperature control and to provide hydrogen for gasification. Fines removed from the product gas are returned to the lower area of the gasifier for further reaction. Ash particles from the feed coal and the fines agglomerate; these larger, denser particles are separated in the lower leg and are removed through the bottom of the leg. Gasification occurs in a fluidized bed in the upper portion of the gasifier–agglomerator. Caking of the feed coal is apparently prevented by the vigorous mixing caused by the jet of air or oxygen and by the partial oxidation of the coal by the oxidant in that jet.

The product gas from the gasifier is stripped of fines and ash in cyclones, cooled, and then scrubbed with water. The product gas higher heating value is 120–150 Btu per standard cubic foot in the air-blown mode of operation. In the oxygen-blown mode, gas higher heating values range from 240–300 Btu per standard cubic foot.

During calendar year 1980, the PDU operated in the single-stage gasifier mode, and seven tests were run. In March 1980, a new oxidant tube configuration was tested and the first test of the raw gas heat recovery unit was made. The first test with higher pressure (230 psi) oxygen was run in April. During May, June, and July, two other oxygen-blown tests at this pressure were accomplished. Previous oxygen-blown testing had been done at only 130 psi. Ohio No. 9 and Pittsburgh No. 8 seam coals were gasified in these tests. In August, the addition of a coal grinding and drying capability at the PDU was completed. The grinder–dryer was used to prepare the Wyoming subbituminous and western Kentucky coals that were gasified in the August test. Further gasification of Wyoming coal as well as Texas lignite was accomplished in the October test. The lignite was gasified at rates up to 2900 pounds per hour. The final test of the year was completed in December.

Accomplishments for calendar year 1980 include successful demonstration of high-pressure oxygen-blown operation, collection of significant process performance and optimization data for a half-dozen different coals, and demonstration of a capability for sustained production and withdrawal of high-ash agglomerates (high carbon utilization). The latter is important for its impact on process economics. Construction for the Large-Scale Cold Flow (LSCF) model was completed, and the test program with the model was initiated. Because there are no plans to build an intermediate-size pilot plant for development of this process, data from the LSCF model are essential for scaleup to a demonstration or commercial plant gasifier. Air-blown testing of the gasifier was deferred to 1981. Objectives of the calendar year 1981 test program include process performance data and data on behavior of the gasifier in transient and failure modes. Testing of raw gas heat recovery equipment will continue.

During FY 1982, the operational test program for the development of the pressurized fluidized-bed process will be completed. Design and scaleup data obtained on the gasifier system—including components, load following characteristics, transient and failure mode behavior, and control system characteristics—will be evaluated as well as used in an economic study.

Tri-Gas Process. This effort on a three-stage pressurized fluidized-bed concept of coal gasification was ended in October 1980, when the contract expired. No further developmental effort is anticipated at this time.

Entrained-Bed Gasification, Atmospheric Process

This effort includes the development of a gasification process and its associated cleanup system that is particularly suited for integration with an electric generation plant.

The Combustion Engineering, Inc. (CE) gasifier consists of two sections: (1) a combustion chamber that burns a portion of the pulverized coal and the recycled char with a substoichiometric quantity of air to provide heat for the endothermic gasification reactions of the coal in the reducing chamber, and (2) a reducing chamber where the remainder of the pulverized coal is reacted with hot gases from the combustion to produce low-Btu gas. Nearly all the ash in the system is withdrawn from the combustor bottom as molten slag. The gasifier is an entrained-bed reactor operating at atmospheric pressure.

The raw gas, with a higher heating value of approximately 100 Btu per standard cubic foot, leaves the gasifier at 1800°F (982°C) and enters a waste-heat-recovery unit where the temperature is reduced by tubular heat-transfer surfaces. These transfer surfaces recover heat in the form of saturated, superheated steam to be used in a steam turbine. The gas leaves the waste-heat-recovery unit at less than 300°F.

The gas cleanup section includes a spray dryer, cyclone, venturi scrubber, and a Stretford sulfur removal unit. Chars and fines from this operation are collected and recycled back to the gasifier's combustor. H_2S is removed and converted to elemental sulfur in the Stretford unit. The sulfur-free product gas is burned with preheated air in a boiler providing heat to produce additional high-pressure steam for the steam turbine.

The atmospheric entrained-bed gasifier offers the following advantages:

- All coals can be processed without pretreatment.
- There is no net char production.
- Fused ash is produced, minimizing disposal problems.
- Virtually all carbon in the coal can be consumed.
- All components except the gasifier are commercially available items with predictable operating characteristics.

CE has underway a three-phase project consisting of design, construction, and operation of a 5-ton-per-hour PDU at Windsor, Connecticut. In Phase 1, the PDU was designed and evaluated to determine technical and economic feasibility for the process. This phase included the detailed engineering design of all components, an evaluation for construction materials, a safety analysis, sampling requirements, an automatic data logging system, an environmental review, and mathematical modeling of the process. R&D support consists of studies on corrosion and erosion of refractory material, and a dynamic flow model to provide design information for the optimization of gasifier geometry and for nozzle design and placement. A study of expected operating characteristics and control requirements of a commercial-size system was conducted to develop control system designs. A test plant for operation of the PDU was defined.

Phase 2, PDU construction, was completed in FY 1977. This phase included site preparation, purchase, and fabrication of equipment, and installation. Phase 3, shakedown testing, started in December 1977, and the parametric test program was initiated in June 1978. During operation through 1979 and 1980, only one coal has been tested; however, the CE system has already demonstrated its capability for successful application to electric power plant use. No tar has been detected in the product gas at the PDU, and good ash-slagging performance of the gasifier's combustor has been demonstrated. Typically less than 0.5% carbon has been found in the ash from the gasifier. In addition, the gasifier has demonstrated the capability for quick shutdowns and restarts and thus is expected to easily accommodate changes in load. The Stretford H_2S removal system at the PDU has demonstrated the ability to reduce H_2S concentration in the product gas to the 20- to 50-parts-per-million level. Testing of COS removal equipment to further reduce sulfur emissions is now in progress. Finally, the atmospheric pressure construction and operation of the system at atmospheric pressure have proven that it is easy to build and maintain as well as safe to operate with minimum personnel.

Difficulties with the combustor chamber refractory were experienced during calendar year 1980. In July 1980 it was decided to install the type of refractory that would be used in a commercial-size gasifier. This decision was made even though it was anticipated that its installation would result in disproportionate heat losses in the PDU gasifier. After installation of the new refractory in August 1980, testing continued for the rest of the year without further refractory problems.

The air-blown test program with the design coal was completed during January 1981. Preparations were made for testing additional types of coal and for oxygen-enriched, air-blown gasification testing during 1981. This test program will be completed in FY 1982 and the engineering data will be reviewed and utilized in an assessment of the concept.

Gasifiers in Industry

This task promotes the use of low-Btu gas in one or more industrial processes, for steam boilers and chillers, for direct process heat, and in industrial dryers and kilns. Its objective is to demonstrate the integration of a low-Btu gasification system with industrial end uses, using present-day state-of-the-art components.

In these systems coal will be gasified at the industrial plant, and the resulting gas will be cleaned as required for the specific application. Since transportation costs dictate that low-Btu gas be utilized near its point of production, those factors controlling the connection of the gasifier to an end user of the gas must be evaluated as part of each installation. These factors will determine acceptability for near-term advances in gasification techniques. (See Fig. 4-19.)

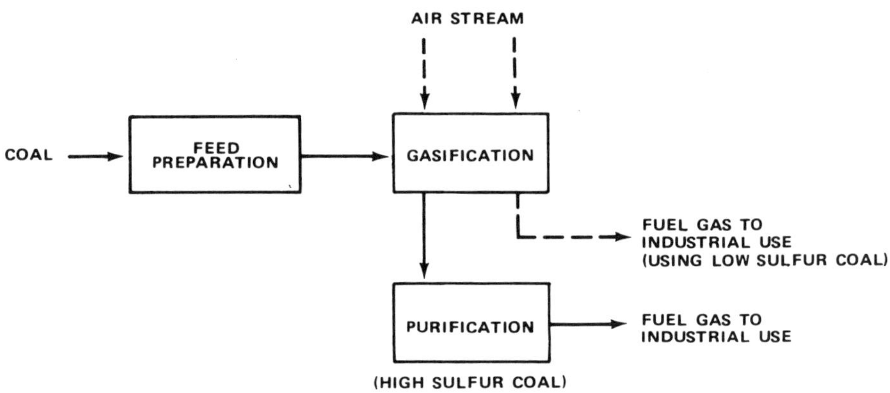

Fig. 4-19. Gasifiers in industry.

Small-sized individual gasifiers are commercially available. They are the forerunners of conversion from coal to synthetic fuel in industry. The federal government's program is of short duration and focuses on the use of caking coals, use of two-stage gasifiers not yet demonstrated on U.S. coals, modernization of single-stage gasifiers, and evaluation of the environmental acceptability of such systems. Therefore, no further funding has been requested for this project since FY 1979. This program was initiated by means of a Program Opportunity Notice calling for proposals for projects to achieve these objectives. Seven contractors were selected.

The Glen Gery Corporation, with Accurex-Aerotherm Inc., instrumented and operated an existing Wellman–Galusha gasifier. This gasifier is producing low-Btu gas from anthracite coal, and the gas is being used to fire a brick kiln at Glen Gery's plant in York, Pennsylvania. Valuable information has been obtained on selection and operation of instruments for measuring gasifier performance. Operational data, including material and energy balances, have been collected. Successful operation of this gasifier under the DOE agreement was completed in June 1979, and an operational report was available in January 1980. Total estimated cost was $ 1.47 million. DOE's share was $0.72 million, or 50% cost-sharing.

The University of Minnesota, with the aid of the Foster Wheeler Development Corporation, plans to use a low-sulfur Wyoming bituminous coal to produce low-Btu gas as fuel for an existing natural-gas-fired boiler. A two-stage STOIC gasifier is used to produce a gas low in particulate matter. The heavy oil contained in the top gas is removed by an electrostatic precipitator and collected. This oil has properties similar to No. 6 oil and is stored for use during the coldest months.

The design effort has shown that substantial modifications for existing foreign designs are required to provide reliable, environmentally acceptable operation in the United States. Startup of the gasifier was completed in mid-1979. The first year of continuous operation and detailed data gathering was completed in FY 1980. Operating and maintenance data will be collected for two additional years. The total estimated cost of this project is $5.5. million. The DOE share is $2.5 million, or 46% cost-sharing.

DOE has initiated an environmental monitoring program for the Minnesota project that will define and characterize emissions and their effects on the environment. It is being coordinated with EPA and the National Institute for Occupational Safety and Health.

Pike County, Kentucky, with the Stearns-Rogers Company as a subcontractor, planned to produce gas from a Kentucky bituminous coal, a caking variety that requires a gasifier with an agitator. Thus, a Wellman–Galusha gasifier equipped with particulate-removal devices was selected for use. The gas produced was to be used to fire a steam boiler to generate steam for heating and cooling several facilities, including housing, a school, a nursing home, a fire station, and commercial facilities.

Detailed design, procurement, and installation of the gasifiers, chillers, and boilers have been completed. However, owing to an expansion of the original job scope to include a required gas desulfurization unit, construction was halted until an integrated plant design could be completed and a new project baseline established. The results of the baseline indicated that significant increases in cost could be expected if the project was continued to completion. Discussions with other participants regarding future actions to be taken are now underway.

The Can-Do (EBECO) project uses anthracite to produce a low-Btu gas as a fuel for an industrial park in Hazelton, Pennsylvania. A single-stage Wellman–Galusha gasifier is being used to produce a cold and clean fuel gas. Design and construction of the facility was completed during FY 1980, and operation will continue to February 1983.

General Refractories, with its subcontractors, was to have used gas to fuel kilns and dryers at its Hitchins, Kentucky, plant, but the project has been terminated. A Woodall–Duckham two-stage gasifier was planned for this project. The cost estimates and process and preliminary designs were completed during FY 1978. Owing to the increasing cost of retrofitting the kiln burners and the need to install additional pollution control equipment, General Refractories decided to terminate the project. Estimated project cost was $4.2 million, and the DOE share was to have been $2.1 million, or 50% cost-sharing. DOE money expended prior to termination was $250,000.

Land O' Lakes, also a terminated project, was to use low-Btu gas to fire a boiler and a spray dryer at its Perham, Minnesota, plant. The use of low-Btu gas to spray-dry dairy products was unique and required careful design and extensive testing. A two-stage gasifier was proposed. The design was to include extensive particulate removal for the stream used in spray drying. Particulates and condensables were to be removed from the remainder of the gas. In addition, a low-sulfur nonagglomerating bituminous coal from Utah was to be used. The project was terminated when it became apparent that the projected return on investment would not justify the anticipated capital investment. Total estimated cost was \$6.5 million. DOE's share was to have been \$3.25 million or 50% cost-sharing. DOE expended \$50,000 prior to termination.

Irvin Industries, aided by Mason and Hanger Co. and the Kentucky Center for Energy Research, proposed to use gas as fuel for a 172-acre industrial park at Georgetown, Kentucky. DOE evaluated the project and decided against its continuation into FY 1980. This project has also been terminated.

The objectives of the gasifiers-in-industry task effort have been achieved in that the industries, universities, and industrial parks that have installed gasifiers are continuing to operate them. The technology transfer initiated under this task will continue. Data acquired through the continued observation of projects at the University of Minnesota at Duluth, Minnesota, and the Can-Do project at Hazelton, Pennsylvania, will be made available to other interested industrial groups.

Third-Generation Process

Several new advanced gasification concepts are now at the bench-scale or PDU stage of development. These concepts, if successfully developed, offer the opportunity to reduce the costs associated with generating a substitute natural gas (SNG) or synthesis gas from coal. The magnitude of the potential cost reduction serves as the incentive for the continued development of these concepts. The following technology options for pipeline gas applications are being developed: (1) hydrogasification and (2) catalytic coal gasification. In a similar manner, a molten salt gasification process is being developed for medium-Btu or synthesis gas applications.

Hydrogasification. Hydrogasification produces high-Btu gas by the direct reaction of hydrogen with coal. From a thermodynamic standpoint, this method has excellent potential for maximizing thermal efficiency and producing economic pipeline gas. The hydrogasification of coal to produce methane has an inherent thermochemical advantage when compared with the catalytic methanation of synthesis gas produced by gasifying coal with oxygen and steam. This advantage was exploited in the early hydropyrolysis experimentation at the Bureau of Mines, which led to the development of the two-stage Hydrane process. Two independent developments occurred as this process matured to pilot plant status. First, Cities Service demonstrated a laboratory-scale, entrained-bed, single-stage hydrogasifier with short-residence-time characteristics. Then, Rocketdyne showed potential for using rocket combustion technology in a hydrogen/coal reactor to achieve improved mixing with high feed throughout. (See Fig. 4-20.) It was determined that the two-stage Hydrane reactor was too complex for commercial application, and that a single-stage, short-residence-time hydrogasifier based on rocket principles was a more attractive option. A conceptual flowsheet for the overall process appears as Fig. 4-21.

Exploratory development work on the concept was completed in August 1978, with work performed at two levels of operation: 2–4 pounds per hour and ¼ ton per hour. Advanced development studies and tests were planned at the ¾- and 4-ton-per-hour levels. Engineering analyses serve as a basis for continuing the ¾-ton-per-hour work into the design, construction, and operation of an integrated PDU facility.

During FY 1981, the Rockwell short-residence-time engineering-scale hydrogasifier was operated at the ¾-ton-per-hour level to generate data needed for process design. The construction of a minimal, integrated PDU was continued, and operation of duplicate tests in this PDU were used to establish a reliable design point for a preliminary commercial design. The Lummus Company was selected as the architect and engineering firm for this study. Lummus has completed the work on the relative effects of hydrogasifier-operating parameters on the net cost of gas for a commercial-scale coal hydrogasification SNG plant.

During FY 1982, Lummus completed the preliminary commercial design with Rockwell support, and construction of the PDU was completed.

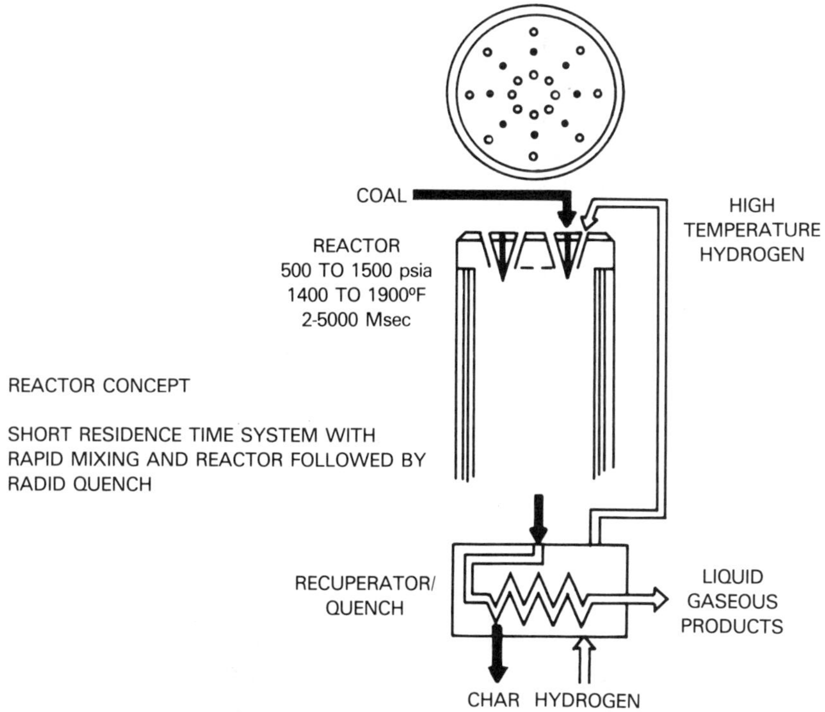

Fig. 4-20. Rockwell coal hydrogenation reactor.

Catalytic Gasification Process Development Unit

In the Exxon Catalytic Coal Gasification process (Fig. 4-22), coal entering the system is impregnated with a potassium hydroxide catalyst prior to entering the reaction vessel. In the gasifier the coal is gasified with steam at 1200–1300°F (649–704°C) in the presence of equilibrium concentration of CO and H_2. The gasifier product gas exchanges heat with incoming water, generating steam. The steam and recycled CO/H_2 are preheated to about 150°F (66°C) above gasification temperatures prior to injection to balance system heat loss.

The carbon monoxide and hydrogen are recycled to the reactor to keep the CO + H_2 content as high as possible, thus driving the net products of gasification reactions toward carbon dioxide and methane. The recycle rate is set so that there is no net yield of CO and H_2 in the gasifier. Methane is separated cryogenically from the CO and H_2 and is removed as the only product. The Selexol process is used to separate the acid gases (H_2S, CO_2) from the raw gas.

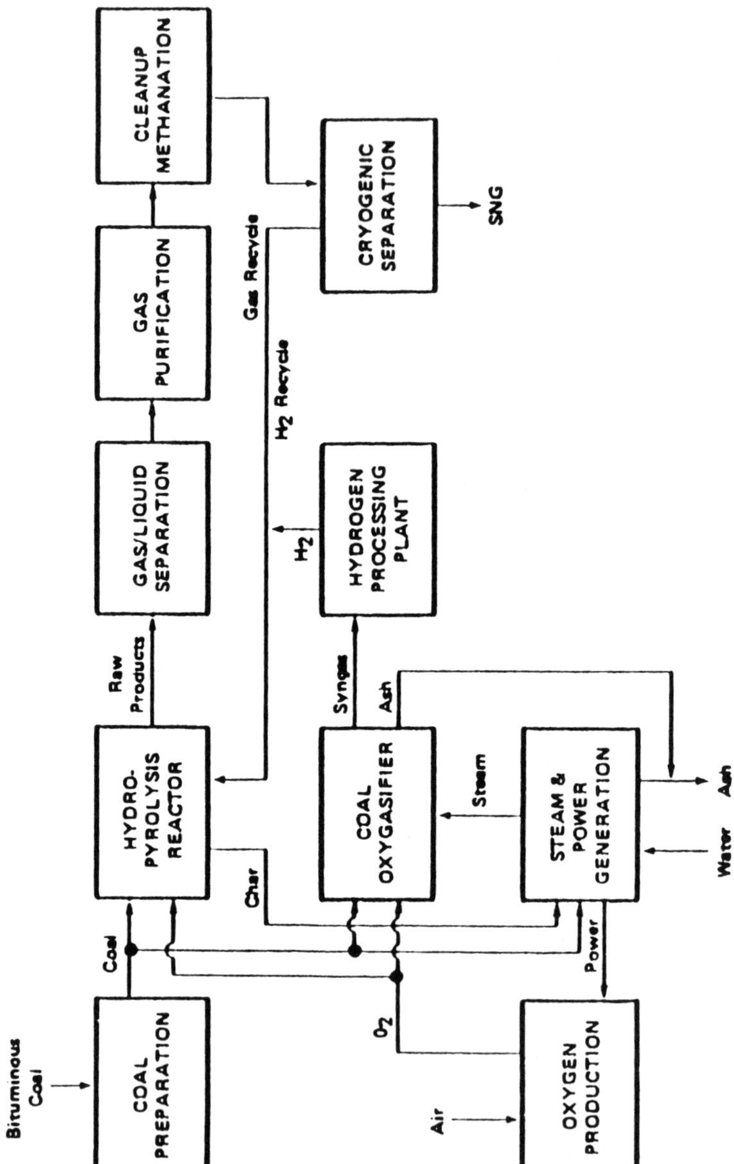

Fig. 4-21. Hydrogasification process.

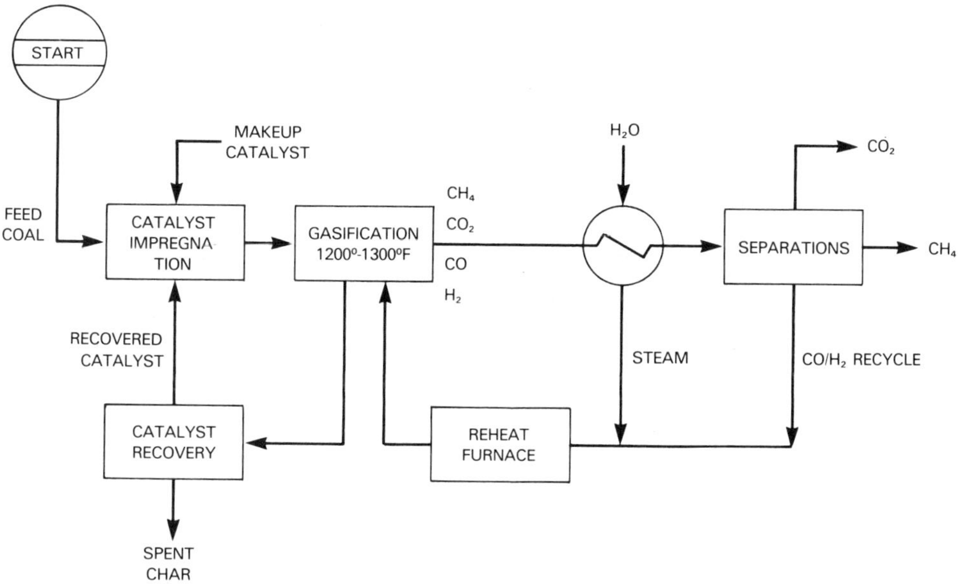

Fig. 4-22. Catalytic gasification process.

Char/ash-residue-containing catalyst is removed from the bottom of the gasifier. Catalyst is recovered by means of a countercurrent water leaching operation. It is estimated that up to 90% of the soluble catalyst may be reclaimed in this manner. Some catalysts react with coal–ash to form an insoluble potassium aluminosilicate. It is estimated that about 5 weight percent of the feed catalyst is lost in this insoluble form. Recovery of up to 90% of the total potassium by methods such as lime treatment and acid wash of char will undergo further investigation.

After cooling the product gas and generating the steam required for the process, the gas is sent to a series of separation steps to separate CO_2, produce methane, and recycle CO/H_2.

Advantages of the catalytic SNG process include the following:

- Caking coals can be fed in this process.
- The need for oxygen as a means of providing high-level heat directly in the gasifier is eliminated, so no air-separation plant is required.
- Gasifier temperatures are reduced to moderate levels, which results in less corrosive and less severe operating conditions for the gasifiers.
- Shift and methanation steps are not required.
- Potentially higher thermal efficiency than that of thermal coal gasification processes is possible because of a reduced need for high-level heat input.

The Catalytic Coal Gasification process is under development at Exxon's research facilities at Baytown, Texas. Exxon has constructed a 1-ton-per-day integrated PDU, and also has benchwork support capabilities at this location. DOE has taken part in the predevelopment activities of the project, and, with GRI, is currently cost sharing the operating costs of the PDU and supporting R&D as part of the combined high-Btu coal gasification program. PDU operations were initiated in 1978.

Recently, Exxon announced plans to design, construct, and operate a 100-ton-per-day integrated pilot plant at Rotterdam, The Netherlands. Although Exxon will provide the funds for the pilot plant effort—estimated to cost over $500 million—DOE will continue to take part in the PDU effort at Baytown, Texas. DOE's participation in this manner will ensure that data on this process are available for its commercial development in the United States. DOE and Exxon will work under a cooperative agreement where Exxon will assume the major share of the cost and DOE will cost share at a low level. In addition, GRI is expected to cost share the DOE portion of this effort. Data from the Exxon pilot plant will be available to DOE, and Exxon will provide third-party licensing and specialized specific support to the first commercial plant effort in the United States.

The PDU and supporting R&D effort will take place in three phases: Phase 1 will cover April 1981 through March 1983, Phase 2 will cover April 1983 through March 1986, and Phase 3 will run though April 1986 to March 1989.

Process Research/Technology Development

High-Rate Entrained-Flow Gasification. A pressurized high-rate entrained downflow gasification process is being developed. This work has progressed from extensive laboratory testing to the detailed design of a 30-ton-per-day PDU. The laboratory work began at the Energy Research Institute in 1974 under the sponsorship of the Office of Coal Research, Department of the Interior. An integrated system capable of continuous operation at feed rates up to 0.5 ton per day was constructed, and extensive testing was performed.

DOE is currently working with Mountain Fuel Resources, Inc. (with Ford, Bacon and Davis Utah, Inc., as a subcontractor) on an 80/20 cost-sharing basis, for the construction and operation of a 30-ton-per-day PDU located at the Interstate Brick Company near Salt Lake City, Utah.

The work completed to date indicates that the process offers the potential for two significant advantages over the current coal gasification technology. These advantages are (1) lower capital costs and (2) higher overall energy conversion efficiency. The principal technical questions that remain to be resolved during continued development are the design of feed nozzles and a high-temperature heat recovery unit that will provide reliable long-term service. The project seeks to resolve these problems as well as to provide scaleup and long-term operational data. Figure 4-23 is a simplified flow design of the overall PDU.

The detailed design of the 30-ton-per-day PDU is continuing. The cofunded project will take approximately 2½ years and will cover two phases: (1) construction and (2) operation. It is anticipated that contract negotiations will be completed during the second quarter of FY 1981 and construction of the pilot plant will start thereafter.

Molten Salt Pressurized Process. The molten salt coal gasification process offers a number of potential advantages:

- The process can handle caking coals without pretreatment.
- Finely pulverized coal is not required, yet fines can be handled effectively.
- As it comes from the gasifier, the gas is essentially free of sulfur compounds, ash, coal, air, and nitrogen oxides.
- The gasification reaction zone, and hence the gasifier vessel, is small because of the catalytic effect of the molten salt.

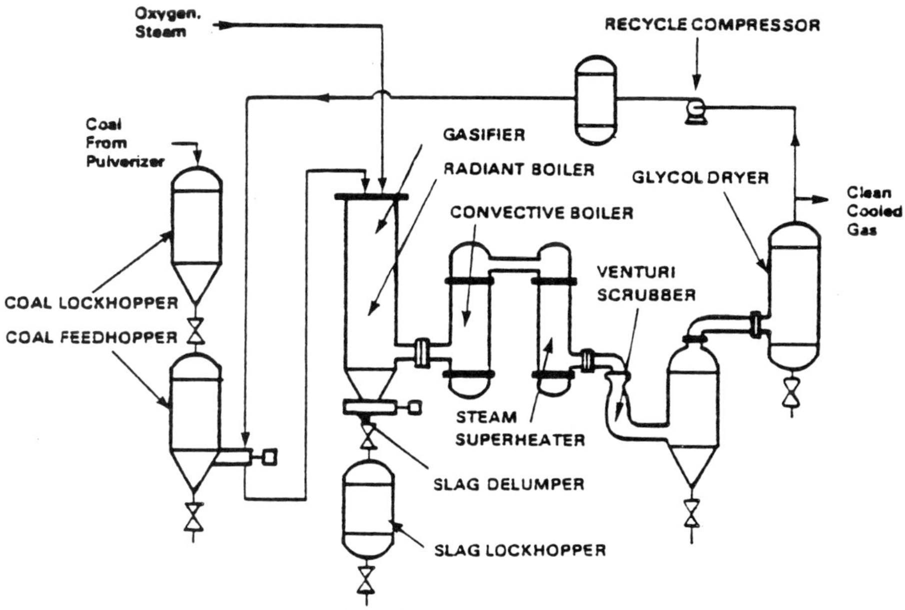

Fig. 4-23. Thirty-ton-per-day PDU.

- The process had good load-following capability because gasifier operation is not strongly affected by gas velocity.
- About 90% of the heating value of the coal is retained in the product gas.

In the molten salt process (Fig. 4-24), coal is crushed, dried, and sent to storage. From storage it goes to the feed hopper where it is conveyed by nitrogen and air to the molten salt reactor. The sodium carbonate is also conveyed by air from the carbonate feed hopper to the reactor.

In the molten salt reactor, coal is gasified in a molten pool of sodium carbonate, sulfide, and sulfate when air is blown into the system. Conducting gasification reactions in the molten salt medium permits very high oxidation rates (about five times higher than the corresponding gas–solid reaction rates) and results in trapping ash and sulfur in the melt. Process economics favors operation of the gasifier at elevated pressure, up to 30 atm, with recovery of thermal energy from the product gas.

Because ash and sulfur are retained in the molten salt pool, the melt must be continuously withdrawn from the gasifier and fresh sodium carbonate added. The melt stream is regenerated by an aqueous process where the melt is quenched and mixed with the water to dissolve the salt. Insoluble ash components are removed by settling or filtration, and the dissolved sodium sulfide is converted to hydrogen sulfide and sodium carbonate, or bicarbonate, by reaction with carbon dioxide. The resulting hydrogen sulfide is oxidized to elemental sulfur in a Claus sulfur plant, and the regenerated sodium carbonate is returned to the gasifier vessel.

The product gas has a heating value of about 150 Btu per standard cubic foot and is predominantly carbon monoxide, hydrogen, and nitrogen. After the product gas is cooled and cleaned of entrained-salt particles, it can be used as a low-Btu feed to a conventional boiler or a combined-cycle system.

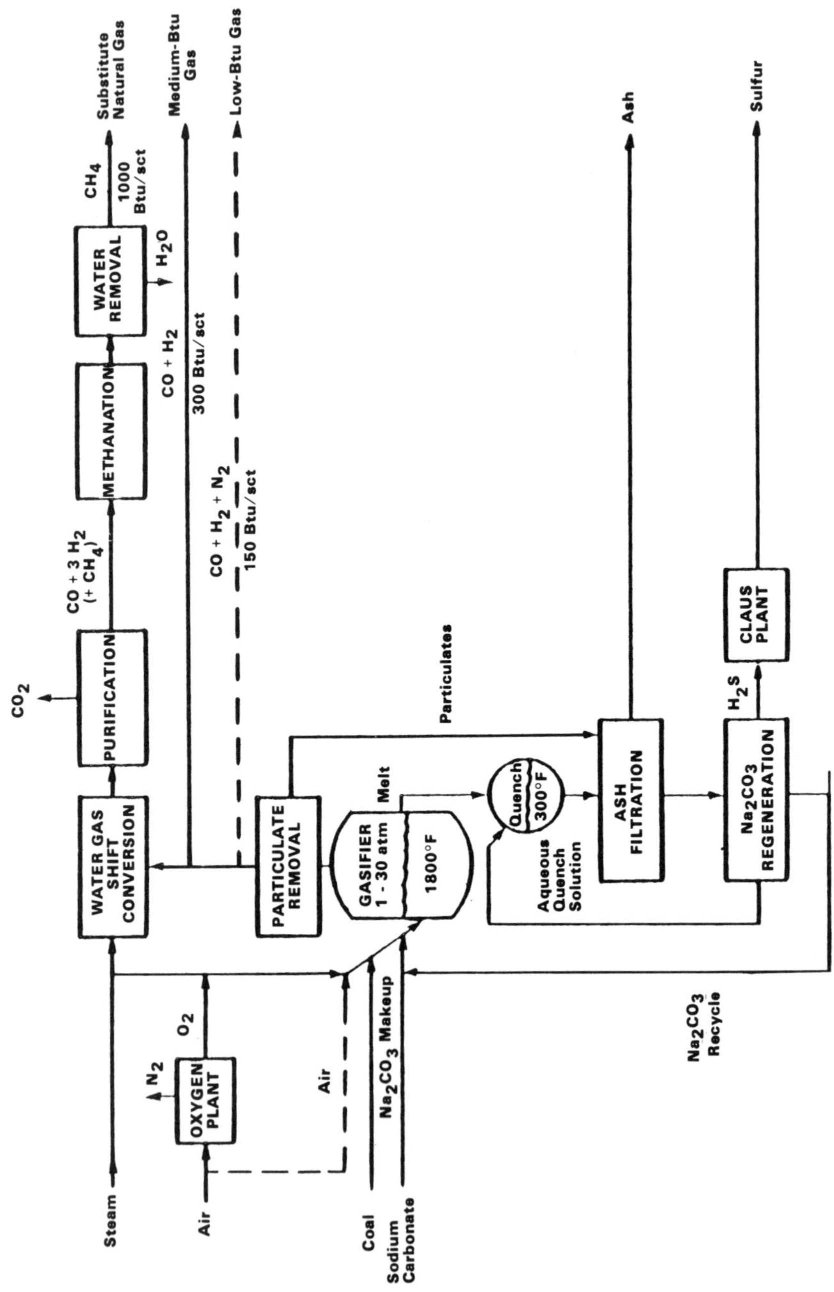

Fig. 4-24. Molten salt gasification PDU.

Prior to December 1974, Atomics International Division of Rockwell International did extensive experimental work to develop the molten salt gasification process. Tests were conducted first in a laboratory and later in a bench-scale unit. Based on these experiments, Atomics International concluded that the molten salt process was sound and a pilot plant should be constructed to further explore and demonstrate its potential for commercialization.

The development effort involved conducting a test program that included the operation of Rockwell International's 3-ft-inside-diameter test reactor. This was accomplished during FY 1976. The preliminary engineering of a 120-ton-per-day PDU was conducted concurrently with this activity. During one of the first engineering phases of the project, preliminary sizing of components, identification of long-lead items, and preliminary mass and energy balances were completed.

A new phase was initiated in March 1976 to design, construct, and operate a 1-ton-per-hour PDU at Santa Susana, California. This facility is designed for operation at 1800°F (982°C) and pressures up to 20 atm. Preliminary engineering was completed in August 1976. The detailed design effort began in June 1976 and was completed by June 1977. To avoid delay in construction of the PDU, key components were ordered in the first quarter of FY 1977. Construction of the facility began in March 1977 and was completed in FY 1978. Since that time several runs of the PDU have been made to produce a low-Btu gas from coal.

The air-blown development phase of the molten salt gasification will be completed during the operation of the facility in FY 1981. If successful, the design and procurement of the equipment needed to modify the facility for use with oxygen instead of air to produce a medium-Btu gas will also be started. During the first part of FY 1982, these modifications will be made and a test program utilizing oxygen will be initiated. As part of an overall evaluation of this technology, an associated material and component evaluation program will be continued.

Special Projects and Support Studies

Fixed-Bed Operational Studies. A commercial-size Wellman–Galusha gasifier has been tested at the Bureau of Mines Twin Cities Research Center near the Minneapolis/St. Paul, Minnesota, airport. The construction and operation of this facility has been cofunded by DOE, the Department of the Interior Bureau of Mines, and industrial participants. Thus far, the gas produced has been used in a kiln to harden powdered iron ore pellets for blast furnace feedstock as well as to test low-Btu gas burner designs in a 22-ft combustion chamber. Both the kiln and combustion chamber are connected to the gasification structure.

Practical applications of established gasification techniques are emphasized at the research center. The objectives of the project are to

1. Develop and demonstrate low- and medium-Btu gasification technology.
2. Establish operating and design parameters for a variety of coal types.
3. Develop and demonstrate gasification of fuels other than coal (for example, peat, combustible waste, etc.).
4. Define gasification operating conditions, and, as a result, gas output and characteristics for different end-use processes.
5. Gain hands-on operating experience.
6. Establish environmental baselines for actual coals of interest.

The Twin Cities Research Center gasifier is a well-instrumented, state-of-the-art, 6½-ft-diameter fixed-bed, atmospheric unit with a water-cooled agitator arm and a rotating grate. It has a nominal bituminous coal consumption rate of 3000 pounds per hour and is fed from above by a 10-ton storage hopper. Moist, warm air is generated for the gasification process by passing air over water heated in the gasifier cooling jacket. The air and steam react with heated coal to form the low-Btu gas, often called producer gas or coal gas.

The test projects completed to date have used a number of coals as feedstock: eastern Kentucky bituminous, Colowyo (Colorado) subbituminous, Absaloka (Montana) subbituminous, Indianhead (North Dakota) lignite, and coke briquettes made from Decker subbituminous coal.

The program will continue to test coals and other fuels in the gas producer and to test the producer gas in selected process applications. These tests include, but are not limited to (1) a 4-week lignite test to demonstrate gasifier stability, operability, and continuous capacity (the feasibility of using lignite low-Btu gas for iron ore pelletizing will be part of this test); (2) a 10-day test of gasifying briquettes of coal and municipal solid waste.

A long-range plan is currently being developed to address project objectives.

Evaluation of Steady-State and Dynamic Performance. Coal gasification/gas cleanup fuel plants for synthetic fuel applications must be developed to operate within the constraints imposed by dynamic load response to variations in end-use demand and in compliance with environmental regulations. The overall objective of this effort is to develop the technology base that is required to permit coal-conversion systems to operate within the constraints imposed by end-use applications. The gasification plant performance plus component and end-use equipment interactions will be characterized in both steady-state and dynamic operations. The information required to establish system control logic will be developed.

An integrated coal gasification system composed of proven process units—that is, a process evaluation facility (PEF)—existing at the General Electric Research and Development Center site will be used for generating an experimental information base to accomplish the following:

- Establish individual PEF performance parameters.
- Define dynamic interactions of PEFs operating in a variable load situation.
- Characterize effluent flow streams from the process.
- Develop a simulation model for integrated component operations.

The combined experimental and analytical approach will yield empirically verified performance projects for full-scale systems. The PEF that is available at the General Electric Research and Development Center contains the key elements—gasifier plus physical and chemical gas cleanup process units—common to many synthetic fuel systems. The facility has been designed to simulate the operational characteristics of full-scale fuel gas plant components.

To evaluate the fuel gas plant performance with imposed realistic conditions, this program will couple analytical models developed to simulate the dynamic and steady-state performance of a gasification plant and one developed to simulate a specific end-use application (that is, fuel gas for power generation) with the PEF results. This approach will yield experimentally verified techniques for prediction of steady-state and dynamic performance and assist in establishing control requirements for full-scale systems.

The contract for this project was established at the end of calendar year 1980. A 3-year program of testing, analysis, and system/component modeling is planned. Because the facility is operational and a theoretical gasification plant model has been developed, the technical activity will be initiated in the first year. The first year of operation will focus on evaluating unfired dynamic interaction of components and determining fired steady-state system and component performance. The second and third years will focus on characterizing the dynamic performance of the system and verifying specific component performance. No funds have been requested for this project since FY 1980.

Technical Support

Coal gasification technology is characterized by the development of a number of gasification techniques that offer the opportunity to convert coal into alternate liquid and gaseous fuels, as well as other products, that will meet the demands of a complex and varied market. Each project is directed specifically at achieving success in resolving key process development problems that, in most cases, are associated with the development of the gasifier or its associated key unit operations. However, a number of development issues common to the technology of coal gasification can be studied as individual projects. These studies can provide applicable data prior to the continued development of the individual gasification concepts.

Concurrent studies can acquire data that will greatly reduce the time and cost of the development of a gasification concept and its associated processing equipment and materials. The design and technology support program studies underway will provide data for:

- Process evaluation, system trade-off analysis, and estimates of economic viability.
- Understanding of observed kinetic and thermodynamic behavior patterns.
- Design, development, and reliability of key items of equipment and instrumentation.
- Testing the validity of computer models and thereby establishing their use as design tools and diagnostic devices for understanding and resolving process problems encountered in coal gasification.

In addition, this subactivity provides the support required to accumulate, correlate, interpret, and record the immense amount of data being generated in a number of coal gasification programs sponsored by industry and government. This support effort has resulted in a technical data book on coal conversion systems, one of the first attempts to provide a systematic method for recording and recalling gasification data. The data will be usable both by system designs and by process developers.

Projects supported as part of the technical support subactivity in FY 1980 have been continued in FY 1981. FY 1981 efforts include engineering evaluations, equipment development, experimental projects, design studies, and environmental support analyses of a generic nature rather than of a process-specific nature. FY 1982 efforts include continuation of the many studies already underway plus completion of Phase 1 studies on slagging gasification, completion of environmental crosscuts of pilot plant data, continuation of the data book development, and evaluation of completing process.

Gasification Demonstration Plants

As part of the President's program for economic recovery, DOE involvement in the gasification program demonstration subactivity will be discontinued during FY 1981. This change is consistent with the administration's policy to rely on private market forces to set the pace and choice of commercial introduction of synthetic fuel technologies. The administration's recent decontrol of oil and support to deregulate gas should stimulate industry's participation in the development and commercialization of near-term technologies to convert coal to synthetic fuels. As a consequence, DOE is redirecting its R&D thrust to primarily long-range and high-risk activity with high potential payoff.

Although DOE will end its program of major technical demonstration, this does not preclude the industrial partners from pursuing the development and commercialization of these technologies with funding solely from the private sector nor from applying for financial assistance from the Synthetic Fuels Corporation.

High-Btu Demonstration Plants

The prime objective of this task is to design high-Btu synthetic pipeline gas demonstration plants to convert coal into pipeline-quality gas. These designs could be used by the private sector to determine risk factors and projected costs for participation in future programs through plant construction and operation. Two competitive contracts were awarded—one to Conoco and the other to the Illinois Coal Gasification Group (ICGG). The Conoco design includes a fixed-bed British gas/Lurgi slagging gasifier; ICGG's design includes a series of fluidized-bed gasifiers (Cogas system). Both processes operate on eastern high-sulfur caking coals, and both offer an improved conversion efficiency over any commercially available gasification technology.

The administration's position concerning the high-Btu pipeline gas demonstration task is to utilize unobligated prior-year funds to complete the design work of both processes (Conoco and ICGG) in FY 1981 and not request additional funds in FY 1982. Therefore, the administration is asking Congress to rescind $2 million for project 76-1-b and $42 million for project 77-1-b and to deauthorized the respective projects. Congressional concurrence on these issues will effectively withdraw government involvement from the projects with the exception of continuing design efforts until the rescission action becomes final and termination cost payments are completed.

In conjunction with the administration's position, design efforts only will be continued until end of Phase 1. The projects would be terminated immediately on the completion of this design activity for the convenience of the government. Sufficient funds are available from prior years funding to terminate both projects.

Low-/Medium-Btu Demonstration Plants

Memphis Low-/Medium-Btu Industrial Fuel Gas Demonstration Plant. This project is a joint effort between DOE and Memphis Light, Gas and Water Division (MLGW). The objective of this project is to design a fuel gas demonstration plant in Memphis, Tennessee. Natural gas availability in the Memphis metropolitan area has diminished in recent years. Since 1970, industrial and residential allotment of natural gas in Memphis has been curtailed by the Texas Gas Transmission Corporation, from 93.4 billion cubic feet to only 45 billion cubic feet of natural gas per year in 1980. The industrial fuel gas demonstration plant will provide fuel primarily to sustain industries that will be the first to be curtailed under federal guidelines. Additionally, since fuel gas is expected to be competitive with fuel oil in the 1985 timeframe, new industries designed to use either natural gas or fuel oil, thereby reduce U.S. reliance on imported liquid fuels.

The administration's position concerning the MLGW Industrial Fuel Gas Demonstration Plant Project was to utilize unobligated carryover funds for the design work in FY

1981 and not request any additional funds in FY 1982. The administration asked Congress to rescind the $45 million for FY 1981 and will not request funds in FY 1982. This effectively withdraws government support from the project.

The project has reached the final design stage. Engineering to complete the drawings, specifications, and procurement packages necessary for the construction of the plant has been completed.

In conjunction with the administration's position, and assuming a favorable congressional decision the following plan will be pursued:

- Continue design efforts, through FY 1981.
- Continue technical support and pilot plant efforts in support of final design.
- Delay all long-lead procurements (except for vendor engineering information in support of final design); cancel after congressional rescission/deauthorization.
- Do not initiate any construction efforts, including clearing, grubbing, dredging, filling, or any other site work.
- Complete the National Environmental Policy Act, 1970 (NEPA) process by finishing the final EIS, which was made available to support a Record of Decision in April 1981.

W.R. Grace Medium-Btu Fuel Gas Gasoline Plant. This project was initiated to support the development and rapid commercialization of processes to produce fuel for transportation use from medium-Btu coal gasification technology. The current project is a rescoping of the earlier W.R. Grace and Company Low-Medium-Btu Industrial Plant B project. In the earlier project the Texaco Coal Gasification process was to convert 1900 tons of coal per day to synthesis gas from which 1200 tons of ammonia per day would be produced. The medium-Btu gasoline plant design was initiated in late FY 1980 after congressional approval of a $6 million reprogramming to provide a source of funding.

The prime objective of the rescoped project is to perform design work for a commercial-scale plant. This design effort will integrate the necessary technology to convert high-sulfur coal into methanol and gasoline. The end product will be 50,000 gallons per day of gasoline formed from 16,000 tons per day of methanol that will be produced from a plant feed of 29,000 tons per day of high-sulfur coal. W.R. Grace will manage this effort as prime contractor and subcontract with an architect and engineering (A&E) firm to develop the process utilizing commerically available process data, components, and equipment to the maximum possible extent. The project has reached the final approval stage of the major subcontractors for the conceptual design of the gasoline plant. In this regard, there will be participation by Texaco Development Corporation and the Mobil Research and Development Corporation.

The administration's position concerning the W.R. Grace gasoline plant is to limit government funding for this project to $3 million in FY 1981 and not to ask for additional funding in FY 1982. Therefore, the administration is asking Congress to rescind $3 million from the FY 1981 appropriation. This will effectively withdraw government funding support from the project with the exception of prior-year funding, which will carry the project until such time as the rescission action becomes final and will also cover termination costs.

Since the Congress concurred in the decision to rescind the FY 1981 funding, the project was terminated for the convenience of the government. The basic conceptual design/A&E effort was about 20% complete.

Low-Btu Demonstration Plants

Fuel Gas Demonstration Plant, Utility. This task was initiated to demonstrate the technical and economic feasibility of substituting coal for natural gas and/or oil in utility power plant applications in an environmentally acceptable manner. Coal is already the basic fuel for the electric utility industry in the United States. Low-Btu gasification of coal is a low-cost alternative for clean burning of coal to best meet environmental restrictions. It also is a method to convert existing oil- and gas-fired power plants to burn coal without prohibitive down-rating of plant capacity.

Based on a competitive procurement, Combustion Engineering, Inc. (CE) and Foster Wheeler Development Corporation were selected to compete in the design of a low-Btu utility demonstration plant. After selection, negotiations with Foster Wheeler were terminated in September 1980, and a contract was signed with CE on September 17, 1980, to commence Phase 1, conceptual design. Sufficient funding is available from prior years to complete the single conceptual design. No additional funding is requested.

The CE site is at the Gulf States Utility Westlake Power Station near Lake Charles, Louisiana. It utilizes the atmospheric entrained-bed gasification process developed by Combustion Engineering at Windsor, Connecticut. The Foster Wheeler proposal was sited at Springdale, Pennsylvania, and it would have employed the chemically active fluidized-bed process developed by ESSO at Abingdon, England. (A pilot-plant scale model of this process is located at San Benito, Texas.)

Fuel Gas Demonstration Plant, Small Industrial. After completion of the Phase 1, conceptual design, effort in the third quarter of FY 1979, the government and its industrial partner, Erie Mining Company, agreed not to proceed beyond Phase 1. This decision was prompted by the growth of construction and operating costs, in addition to problems concerning the ability to process the candidate coal.

In Situ Coal Gasification

Approximately 93% of our nation's vast coal resource is not technically nor economically recoverable by conventional underground or strip mining. Development of viable underground conversion processes could increase the recoverable portion to nearly 28% by allowing conversion of the coal to clean-burning gaseous and liquid fuels. (See Fig. 4-25.) Development of these processes could significantly expand the economic extraction of energy from coal in environmentally acceptable ways, and could provide a viable domestic source of competitively priced transported fuels.

Underground coal gasification processes now under study in the United States basically consist of the following steps. The coal is accessed by drilling wells through the coal seam from the surface. Gas flow is then established between injection and production wells, the coal is ignited, and a burn zone is caused to progress from injection to production wells, with gaseous products removed at the production wells. The burn zone combustion is maintained by injection of air or oxygen/steam.

Variations in specific processes involve the method of producing the required flow path. The linked vertical wells (LVW) process utilizes natural or induced fractures enlarged by severe combustion. A major variation utilizes directional drilling where drilling holes between wells provide the flow path.

Underground coal gasification (UCG) can produce low- and medium-Btu gas for electric power generation and industrial heating, as well as feedstock for production of various chemicals, substitute natural gas, and such transportation fuels as methanol,

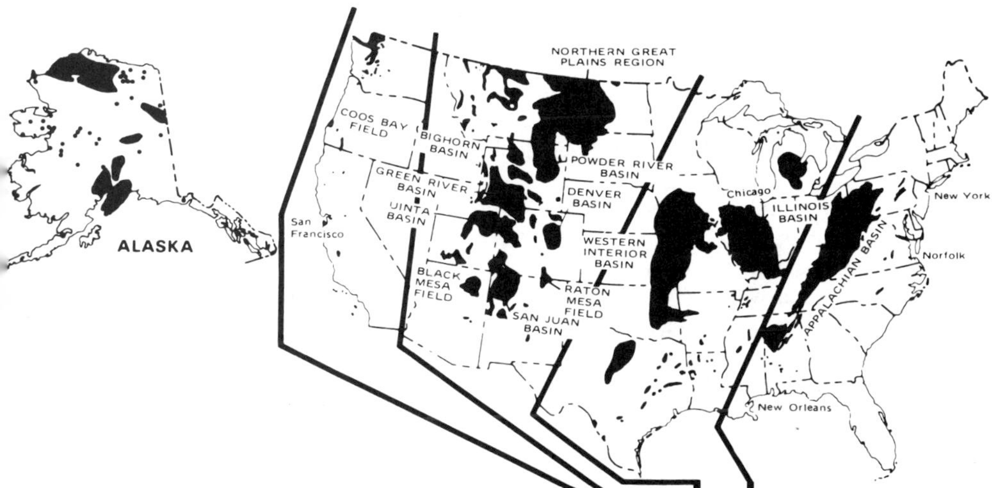

TOTAL COAL TO 6000 FEET (BILLIONS OF TONS): *	2030	100	3260	620	350 - 6360	TOTAL
ESTIMATED MINEABLE RESERVE BASE (BILLIONS OF TONS):*	10	2	200	110	110 - 432	TOTAL
ESTIMATED UCG RESOURCE (BILLIONS OF TONS): *	640	30	920	140	80 - 1800	TOTAL

* DIVIDE THESE NUMBERS BY 1.1 TO OBTAIN TONNES.

Fig. 4-25. Coal fields of the United States.

gasoline, or diesel. One recent study by the Stanford Research Institute[1] has shown that UCG-derived power can capture up to 1.0 quadrillion Btu per year of low-Btu gas and 0.5 quadrillion Btu per year of SNG in 19 states in the Rocky Mountain region. That transportation fuel option is also very attractive.

DOE's *in situ* coal gasification activity is designed to accelerate development of economically competitive, technically feasible, and environmentally acceptable underground processes to produce liquid and gaseous fuels. This is being met through joint efforts with industry and universities, as well as through an in-house program utilizing energy technology centers and national laboratories.

A program goal is development of UCG technology for production of medium-Btu gas with emphasis on utilization as liquid transportation fuels. Development of the technology includes studies of process performance and reliability, performance predictions, and application to coals with widely varying geologic and hydrogeologic conditions. The mid-term goal is the development of advanced concepts such as increased resource recovery and reduced dependence on geologic/hydrogeologic conditions. The end-use application for these fuels is in generating electric power, or in producing synthetic liquid fuels and high-Btu SNG.

Underground coal gasification offers a number of significant advantages. Some are yet unproven, but the DOE program is designed to show that, in addition to quadrupling our recoverable coal reserves, UCG minimizes health and safety problems associated

[1] Molland, A.J. and Olsen, D.L., "Preliminary Evaluation of Western Market for UCG-Derived Coal Gas," *Proceedings of the Fourth Annual Underground Coal Conversion Symposium*, Steamboat Springs, Colorado, July 17–20, 1978.

with conventional coal extraction and utilization techniques. It requires no mining, produces less surface disruption than strip mining, brings less solid waste to the surface, generates less atmospheric pollution, and reduces the socioeconomic impact. UCG also has the potential to reduce both capital investment and gas costs for coal gasification by at least 2% and may be deployed as small plants supplying local energy needs.

Potential problems posed by UCG are contamination of surface water and underground aquifers, gas leaks, and subsidence of the surface. To minimize the impact of these potential problems, extensive environmental monitoring before, during, and after each field test, along with the development of environmental control technology, laboratory simulations, and analytical modeling, are included as integral parts of each field project.

The technical feasibility of UCG has been proven by the British and, more notably, the Russians, who have had commercial UCG plants in operation for 20 years. Building on the available UCG technology, DOE's Laramie Energy Technology Center, Morgantown Energy Technology Center, and Lawrence Livermore National Laboratory have conducted highly successful subpilot-scale field tests at Hanna, Wyoming, Pricetown, West Virginia, and Hoe Creek (near Gillette), Wyoming, respectively. Sandia National Laboratory provided instrumentation support to the Laramie tests. The Gulf Research and Development Company, under contract to DOE, has run the first U.S. steeply dipping beds test near Rawlins, Wyoming. This test was also very successful. Low- and medium-Btu gases have been produced for sustained periods, and successful control of important process variables has been demonstrated.

The key to commercialization of UCG in this country is proof of economic feasibility. This will require demonstration of efficient use of the coal resource, control of the underground process (including environmental effects), reliable long-term operation at high production rates and using multiple process modules, and verification of the process design at several sites. These requirements are being fulfilled through an integral program of field tests, instrumentation and process model development, and laboratory supporting research.

The current DOE *in situ* coal gasification activity consists primarily of technology base development efforts addressing the remaining generic technical uncertainties. These include the relationship of site geology to process performance, formation of a flow path in the coal seam that links the process wells, early cavity growth, and determination of optimal operating conditions for a specific site. Also being addressed are such issues as environmental effects and impacts, process modeling, economic studies, and collection or development of all other relevant data needed for a valid assessment of the technology by industry.

Low-/Medium-Btu Gasification

The Laramie Energy Technology Center conducted low-Btu gas production research tests in a 30-ft-thick subbituminous coal seam in the Hanna Field. The objective was to develop a process for gasifying coal in place to produce a low-Btu gas for utility power generation of industrial use.

The linked vertical wells (LVW) process with air injection was one of the processes studied; it requires minimal seam preparation and has a good chance to succeed in coal seams with adequate natural permeability. The process is applied by drilling vertical wells into the coal seam to provide for injection of air and collection of product gases. Reverse combustion links the wells through the coal seam, followed by forward gasification of the coal between the linked wells.

Figure 4-26 shows the center and right wells are linked by reverse combustion through the following sequence. High-pressure air is injected in the right well and flows to the previously ignited coal at the center production well through naturally occurring cracks and paths in the coal. The combustion zone advances from center well to right, or injection, well against the gas flow, creating a hot char channel between the wells that can accommodate high-volume flow with low-pressure loss. Then, the link is ready for the second, or forward, gasification stage (shown between left and center wells in the figure). Forward gasification proceeds along the linked passage, which widens and caves in to expose more coal.

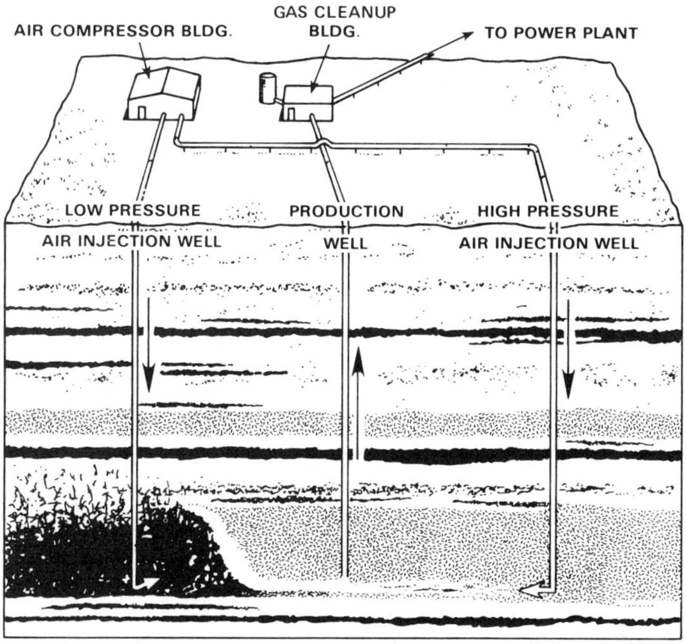

Fig. 4-26. Linked vertical wells process.

As illustrated, one pair of wells, making up one module, can be linked as the coal between a second pair is gasified. When the gasification zone passes the center well, that well is switched to injection and the right well is switched to production. A link is then started to a new well farther to the right. Thus, the gasification proceeds along a line, although it expands appreciably to either side of the line. Lines of wells are drilled closely enough so the gasification cavities overlap, using as much of the coal as possible. Success of the process depends on controlled movement of the reaction zone between wells and efficient use of the heat from the reacting coal to promote the chemical reactions that produce high-quality gas.

Development testing of the LVW process in the United States started in 1972 with the tests at Hanna. The product gas from this test had an average heating value of 126 Btu per standard cubic foot and was produced for about 6 months. Subsequent tests at Hanna resulted in production of gas averaging about 8.5 million standard cubic feet per day of 172 Btu per standard cubic foot of gas. This is equivalent to the electrical requirements of a town with a population of 6000, that is, from 6 to 8 MWe. Environmental tests completed by July 1977 at Hanna were primarily designed to measure the effects of the process in underground water flows and compositions, and to refine further the LVW operating and control procedures. Monitoring of water in and above the burn zone is continuing.

During the early phases of Hanna 2 testing, a major instrumentation effort was fielded by Sandia National Laboratories; it made the test series the most highly instrumented UCG test ever conducted. This was done to measure the character and progress of the underground process so that it could be understood and therefore controlled and improved. Field and laboratory data, plus theoretical considerations, are being used to develop a detailed mathematical model of the process. Refinement of this model is expected to provide a design capability for applying LVW to other sites.

Subsequent Hanna tests were designed to determine the relationship among other process parameters such as well spacing and sweep width over longer well spacings (effects of scaling). Difficulties were encountered during the Hanna 4 test owing to improper well completions and the natural fracture faulting of the coal seam. These problems resulted in a shift of the desired air-flow pathway at the bottom coal–overburden interface, which effectively bypassed a substantial amount of coal. The test was started again in FY 1979, at which time multiple links were successfully completed over a 112-ft well spacing. Subsequent gasification tests confirmed the existence of multiple air-flow pathways. After three weeks of acceptable gas quality production, the gasification cavity intersected an air-flow pathway near the top of the coal seam and the product gas quality deteriorated. The test was terminated at the end of FY 1979.

Because of Laramie's increased responsibilities as lead center for UCG implementation and the commitment of resources to technology base development work at the Washington site, no gasification tests at Hanna were planned for FY 1981–1982. Meanwhile, an extensive postburn analysis of the past Hanna test sites was initiated in late FY 1980, continuing through FY 1981. Coring operations were conducted at the Hanna sites to define more accurately the extent of the gasification cavity, and to relate the chemical and physical analyses of the cores to definition of the gasification cavity and site characterization.

The goals of the Sandia program for instrumentation and process control development are to obtain a thorough measurement and understanding of the chemical and physical mechanisms of the *in situ* reaction to improve and to predict process control. Preliminary instrumentation efforts on the Hanna 1 experiment, 1973–1974, demonstrated that process-related effects could be monitored by electrical, acoustic, and thermal techniques. These results led to a comprehensive instrumentation effort designed as an integral part of the later Hanna experiments conducted in 1975 and 1976. Sandia is now analyzing temperature data from past tests and is combining all results into a thermal data file for use by industry and other UCG experimenters. This effort will continue in FY 1981.

The Lawrence Livermore National Laboratory has conducted a medium-Btu gasification project in a 25-ft-thick subbituminous coal seam at Hoe Creek, Wyoming. The objective was to develop a commercial process using oxygen and steam injection to produce medium-Btu gas for surface upgrading to liquid fuels or SNG. Various well-linking methods are being developed for applying this technology to a wide variety of resource and process combinations, especially very deep, thick coal seams.

Livermore has been managing the medium-Btu gasification project since its inception in FY 1975. In FY 1977, the Hoe Creek 1 test of forward gasification with air was conducted after the seam had been explosively fractured to increase its permeability. Although the desired permeability distribution was not achieved, the reverse linking step was successfully omitted. Data reduction, groundwater monitoring, and postburn coring activities are now complete.

The Hoe Creek 2 test, using the LVW program process, was completed in FY 1978. This was a comparison of LVW performance in coal that was more uniform, contained more water, and was more reactive than the coal at Hanna. During this test, oxygen and steam were injected for 2 days. Medium-Btu gas was successfully produced with a heating value that averaged 260 Btu per standard cubic foot and ranged up to 300 Btu per standard cubic foot.

In FY 1979, the Hoe Creek 3 test was completed using oxygen/steam injection. The 57-day experiment consumed about 3400 tons of coal. A new technique was tested using a directionally drilled borehole to provide the link between the vertical wells in the bottom third of the coal seam. During the first 6 days of gasification, air was injected to create the initial gasification cavity. Then injection was switched to steam/oxygen, and gasification conditions were maintained for 48 days. Average gas quality was 210 Btu per standard cubic foot, lower than the expected value of about 265 Btu per standard cubic foot. This was probably the result of unfavorable Hoe Creek geology and hydrology, and emphasized the need for better site characterization.

As the coal was gasified during this experiment, the wet overburden collapsed into the coal seam. The influxing water cooled the gasification zone, absorbing heat energy that was needed for gas production. The development of improved diagnostic tools for characterizing sites will help avoid such problems in the future.

Other major tasks in FY 1980 included the cleanup and restoration of the Hoe Creek sites and analysis and documentation of the data collected. While there will be no further gasification tests at Hoe Creek, environmental monitoring will continue for a total of 5 years. Within 1 month of the completion of Hoe Creek 3, there were two incidents of subsidence between the process wells. DOE is working with the Wyoming Department of Environmental Quality to study and correlate the observations from these events. During FY 1981, the LLL team will assemble and disseminate the process data base tapes for use in analyzing the efforts of the entire UCG program to date. LLL will continue laboratory experiments with coal blocks. The results will help interpret the data from all UCG tests.

In FY 1981, the several in-house field teams were reorganized into a single team to manage technology base development tests at a new site in an unminable coal seam. In FY 1980, the Sandia and Livermore laboratories conducted a study to identify sites in the state of Washington that might be suitable for UCG. Choices were narrowed to a site near Centralia; they contained multiple seams of bituminous coal at depths of 500 feet and greater and 40% ash in the form of shale layers in the coal.

The coal is owned by Washington Irrigation Development Company, which has no plans to mine the coal. Negotiations are underway with the company to use the site for *in situ* testing. The site is attractive because it would give researchers a chance to examine what effect shale strata have on *in situ* gasification. Laramie will be responsible for managing these activities and ensuring that high-priority technical issues are addressed in a timely manner. Laramie will start large (20 × 20 × 20 ft^3) coal block tests in an outcrop on the site in FY 1981. The site is readily accessible so researchers will be able to mine or excavate the blocks after the tests are done to measure cavity growth directly. Site characterization will be completed and design of a single-module test will begin, using a directionally drilled link. The first field test would be conducted in late FY 1982 or FY 1983.

Much of this technology base development work received funding assistance under the terms of a cooperative agreement with the Gas Research Institute, which committed $1 million toward this work in early FY 1981.

Steeply Dipping Beds

The steeply dipping beds (SDB) process is designed for seams that dip greater than 35° relative to the earth's surface. This coal is generally not recoverable by existing mining methods. The process wells use slant drilled holes that require sophisticated directional drilling techniques.

Over 100 billion tons of SDB coal are estimated for the United States. A large proportion of the Pacific coast coal is steeply dipping, so commercialization of an SDB process could contribute to the future energy supply of the populous west coast. There are also substantial SDB deposits in the Rocky Mountain area and lesser amounts in the Appalachian and interior coal regions.

DOE cost shares an SDB project with the Gulf Research and Development Company to study the production of low- and medium-Btu gas. To develop the necessary cost data for process scaleup, the project incorporates several field tests. The first test, Rawlins 1 (Fig. 4-27), helped develop the necessary skills and procedures for demonstrating gasification using a single-module configuration at the most convenient location.

The Rawlins 1 test was designed, fielded, and completed in FY 1978–1980. The injection and production wells were completed using slant drilling techniques and were linked by reverse combustion. Forward gasification with air injection continued for 30 days, followed by 5 days of steam–oxygen injection for production of medium-Btu gas. Air injection produced 150 Btu per standard cubic foot gas, and the steam–oxygen injection produced gas averaging over 250 Btu per standard cubic foot.

The major FY 1980 tasks included completion of the Rawlins 1 test, analysis and documentation of the data, and the design of the Rawlins 2 test. The fielding of the Rawlins 2 test is currently underway and options will be in late FY 1981 or early FY 1982. This test is a 90-day test to produce medium-Btu gas from a larger single module extruding to a depth of 600 ft. After the second test, a decision will be made as to whether the SDB technology is sufficiently advanced for testing in other areas or with different coal types. No funds were requested in FY 1981 or in FY 1982.

FY 1982 funding for the last year of the modified contract with the Gulf Research and Development Company on SDB will complete this larger-scale longer-term test near Rawlins for production of medium-Btu gas. The 90-day gasification test should be completed during the first quarter of FY 1982 with postburn evaluation and reporting completed by the end of the fiscal year. This will be the second steeply dipping beds test in the United States and the longest production of medium-Btu gas from UCG in the world.

Supporting Research

The program's approach to determining the technical and economic feasibility of UCG in coal seams includes laboratory experimentation, process modeling, and investi-

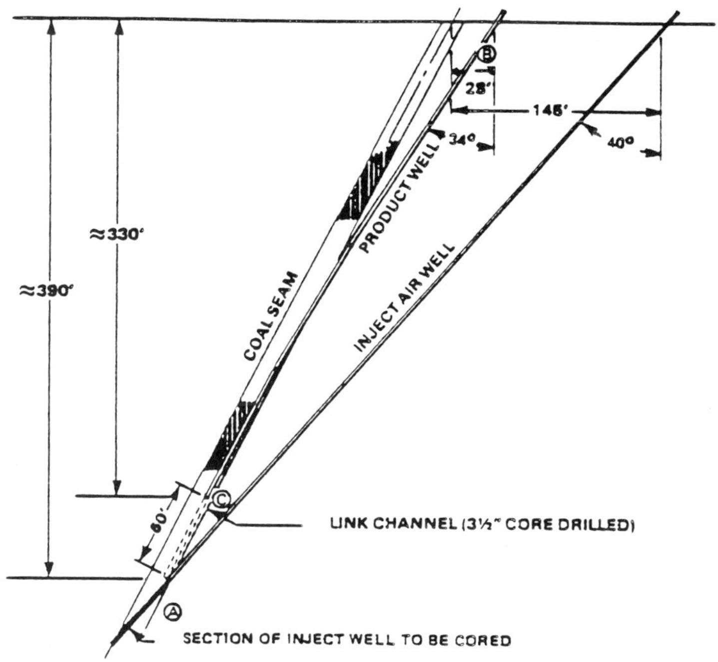

Fig. 4-27. Steeply dipping beds, Rawlins 1 experiment.

gation of alternative linking techniques and gasification concepts in well-instrumented field tests. The objectives of the laboratory research and modeling efforts is to support the field program by investigating process parameters and alternative linking methods and ignition techniques, and by developing instrumentation and software for field use. Some past work is described below.

Morgantown's laboratory and modeling program focused on determining the mechanical and physical properties, especially natural permeability and fracturing characteristics of coal under overburden stress.

Past projects at the Argonne, Oak Ridge, and Los Alamos national laboratories provided kinetic data on the important chemical reactions that occur during *in situ* gasification, experimental data on the thermal conductivity of coal blocks and coal pyrolysis, and feasibility of a two-stage carbon dioxide/oxygen underground coal utilization scheme with minimum water requirements.

Projects conducted by universities in Wyoming, Texas, West Virginia, and Washington have developed process models to describe various aspects of *in situ* operations.

In FY 1981, supporting research was redirected toward projects supporting the Centralia technology base work that address early cavity growth and roof subsidence.

Additional work will be conducted to investigate scaling effects that would arise for relatively small UCG plants. UCG should scale favorably to small sizes because the underground module quickly achieves maximum efficiency at small sizes. It may be feasible to trade off some of the cost advantages shown for large UCG plants against higher costs at small sizes, and possibilities may exist for unique financing and siting near small communities to meet their energy needs. The initial target would be a 4000-barrel-per-day methanol plant using medium-Btu gas from UCG.

Another supporting research effort has been the eastern coal technology project run by the Morgantown Energy Technology Center to develop a method of gasification for eastern coals that cannot be easily mined. Such reserves amount to approximately 220 billion tons. Thin, low-permeability, high-swelling, eastern coal seams are the most difficult to gasify by *in situ* methods. Rugged terrain and densely populated areas in the East make it very difficult to drill vertical wells spaced less than 110 ft apart. Directional drilling may solve this problem. Deviated holes provide the best means of developing the underground modules if they can be drilled economically.

Field experimentation has been concentrated on site, resource, and environmental assessments. A site near Pricetown, West Virginia, was selected for the initial test in the bituminous Pittsburgh seam. Using vertical process wells, Pricetown 2 was operated in FY 1979.

The Pricetown 1 test was a three-well pattern in a 6-ft-thick seam approximately 900 ft deep. The wells were linked by reverse combustion (widening cracks in the coal by partial burning), and gasification was accomplished using air injection. The average gas quality during linking was in excess of 200 Btu per standard cubic foot. The gas quality during the steady-state gasification period averaged 130 Btu per standard cubic foot. This feasibility test showed that (1) coal seams can be ignited at depth; (2) linked vertical wells and forward gasification are technically feasible in thin, deep bituminous coal seams; and (3) environmental effects are minimal.

Major tasks in FY 1980 were the analysis and documentation of the data from Pricetown 1, and the continued environmental monitoring of the site. A competitive contract for an assessment of U.S. bituminous coal suitable for UCG was awarded.

FY 1982 was highlighted by the completion of the Rawlins 2 SDB test; completion of large block tests at Centralia, Washington; and preparation for the first field test at that site.

Environmental Support

Environmental problems associated with *in situ* coal conversion are seen to be less serious than those with mining and surface gasification. The biggest problem could be disruption and contamination of local aquifers; this requires further research. Overall water use should be lower for *in situ* than for surface gasification. Air emissions should be no worse than those from surface gasification, and will be controlled by similar methods. Surface disruption due to subsidence should be minimal and easily repaired compared with strip mining, provided that the subsidence is not delayed over a long period.

Environmental support determines the true impact of environmental problems associated with *in situ* coal gasification, as well as methods for their control. Because environmental problems are common to all *in situ* concepts, environmental support crosscuts all major *in situ* projects.

The UCG process inherently has potential environmental advantages. It may be possible to further minimize surface disruption due to subsidence by adjusting the process design or by appropriate site selection. Most of the solid wastes remain underground. Because UCG uses less equipment than surface gasification, demand for cooling water is also less, an important factor in the arid west. In addition, the seam water will enter the underground reactor and be produced; part of this water will be recovered and can be used as process water. Because coal mining is avoided or reduced, mine safety problems will also be lessened.

On the other hand, uncontrolled subsidence could leak gases to the atmosphere through cracks to the surface. Subsidence can affect overlying aquifers and allows their water to enter the reaction zone during and after gasification. Pollutants can be picked up and carried away by the subsurface flows. Development of UCG resources in an area with limited potable water and population could lead to socioeconomic problems. Western deposits in regions with almost-pristine air quality protected by strict laws will require stringent control technology.

Baseline water quality and postburn changes have been measured in both the Hanna and Hoe Creek tests. Detailed hydrologic and geologic studies were also made in the Pricetown site area to provide baseline environmental data. Hydrology monitoring will continue. Livermore's laboratory tests indicate that coal surrounding the reaction core can cleanse surface water of harmful pollutants. Livermore has had a cooperative agreement with the Environmental Protection Agency to conduct a more extensive laboratory program to determine the amount of water cleanup required to remove organic and inorganic contaminants. The tar produced in the Hanna 2 test revealed potentially carcinogenic compounds, but very heavy components were absent. Evidently, these components were trapped and reacted underground. The primary objective of Hanna 3 was measuring underground water quality and determining how it is affected by *in situ* coal gasification. Longer and more-detailed air-emission measurements were planned on Hanna 4.

An environmental development plan has been issued. It provides for the implementation of a comprehensive environmental plan for *in situ* coal gasification projects to ensure that all environmental issues are fully considered in making decisions to use this technology.

Chapter 5

Coal Conversion System Comparisons

Scope of Comparison

For many years the federal government has supported the development of processes for converting coal to more desirable solid, liquid, and gaseous fuels under the auspices of the Bureau of Mines (BOM), Energy Research and Development Administration (ERDA), and the Department of Energy (DOE). This chapter is based on a report entitled *Coal Conversion Comparisons*, Report Number FE-2468-51 prepared by The Engineering Societies Commission on Energy for DOE.

A considerable amount of R&D support work for many different processes has been partially or fully funded by DOE. This support has included concept designs and cost estimating for industrial operation of many of the processes together with detailed economic studies. In addition to assistance to the private sector for energy planning, many of the studies also assist DOE in budgeting and program planning. Most of the detailed feasibility and economic studies have been done by different private firms and have been based on various economic and technical assumptions. The quality of the engineering input data and engineering designs and thus the quality of the conclusions have varied considerably.

This chapter starts with adoption of uniform assessment methods. These methods are then used with available data and published studies to prepare analytical comparisons for a variety of coal conversion processes.

It is important to note that this chapter does not recommend the selection of any single process for development or application. There are many different markets to be served and various coals available. Several of the processes under development may be chosen for commercial application at different sites within the United States before the end of the century.

Because of the U.S. reliance on foreign energy sources, the need is greater than ever to develop energy alternatives for the nation. A similar need was voiced at the beginning of this century. The need was again expressed during World War II when Congress passed the Synthetic Liquid Fuels Act of 1944. In each case, a cheaper energy source halted the development of alternatives and lulled the nation back into a complacent attitude; this can happen again as the current prices for petroleum are reduced or stabilized.

This chapter consists of uniform technical comparisons of 14 processes that convert coal to other energy forms. The major energy resources—fuel liquids, fuel gas, and electric power—were chosen intentionally to provide a broad comparison on a consistent basis. The techniques used may be easily applied to other processes not included here.

A study by C.F. Braun, Inc.,[1] completed in 1976, compared a number of gasification processes on a consistent basis. Since the Braun study was concerned only with pipeline-quality gas, reporting cost solely as dollars per million Btu ($\$/10^6$ Btu) was completely satisfactory. A cost per Btu basis has also been used for coal-derived liquid products in many preliminary studies. It is useful approximation for preliminary studies, and it also avoids the problem of allocating costs to the variety of liquid products made by coal liquefaction plants. However, simply calculating total product cost as $\$/10^6$ Btu is not satisfactory from a realistic market standpoint. Assigning product cost solely on a heating value basis gives relative fuel prices which never have, and probably never will, prevail in the marketplace. Even with federal price control, gasoline had a wholesale price higher than home heating oil, which in turn had a wholesale price higher than residual fuel oil. Even worse, a fixed price per Btu of heating value would invert the price ranking of these three fuels which are sold on a volumetric basis, since heating oils have a higher fuel value per gallon than does gasoline.

Because fuel gas, liquids, and electricity are traditional fuel competitors, and will continue to compete in the future, it was necessary to develop methods for objective economic comparison. Any method that does not consider inherent differences in values and market preferences for alternative fuels falls short of a realistic appraisal.

The processes that are included in this chapter and the abbreviations used are as follows:

Solid Solvent-Refined Coal (SRC-I)
Liquid Solvent-Refined Coal (SRC-II)
Exxon Donor Solvent (EDS)
H-Coal, Fuel Oil Mode (H-FO)
H-Coal, Syncrude Mode (H-Syn)
Fischer–Tropsch (FT)
Methanol
M-Gasoline (M)
CO_2 Acceptor, Substitute Natural Gas Mode (CO_2-SNG)
CO_2 Acceptor, Syngas Mode (CO_2-Syn)
Hygas (HyG)
Bigas (BiG)
Synthane (Synth)
Lurgi (Lur)
Westinghouse, Syngas Mode (West-Syn)
Westinghouse, Combined Cycle (West)
Combustion Engineering, Combined Cycle (CE)

For a technology process to be included in this comparison, it was decided that:

(a) Process must now be under development or considered a viable candidate

(b) Federal funds must be supporting basic R&D

(c) Process is not a combination of other established technologies

(d) Sufficient process data are available

There are several established coal processes that do not meet these criteria and, therefore, are not included in this comparison. However, even though the Lurgi process has never been supported by federal R&D funds, it is included as a reference standard because it is one of the few gasification processes that has been commercially available for some time.

[1] C.F. Braun & Co., *Factored Estimates for Western Coal Commercial Concepts*, also Technical Appendixes I and II, Interim Report. Prepared for the U.S. Energy Research and Development Administration, and American Gas Association, FE-2240-5. C.F. Braun & Co., Alhambra, CA, Oct. 1976.

The Cogas process was not included because it is a combination of other established technologies that have been developed and modified by a private group. The Exxon catalytic gas process is still in the early development stage, so data of sufficient reliability were not available. None of the omitted processes are expected to have viability significantly different from the range encompassed by the included technologies.

No separate gasifier units are included because the gasifiers are components within integrated processes. These reactors are important and serve as key operations within Fischer–Tropsch, methanol, and most of the other processes studied, either as primary reactors or as sources of hydrogen. Gasifiers warrant thorough study because of their significant contribution to process efficiency. Such studies are best limited to the gasifier only. Electric Power Research Institute has sponsored several good reports on this subject.[2]

Coal Resources

Significant consumption rates for coal are being projected through the end of this century. Projections range from 17 to 20 quadrillion Btu per year. Additional production beyond present rates is necessary to partially displace imported oil and to meet new energy demand. Reserves to meet these projections are proven. The location of these reserves is sufficiently well established that it is felt that economic production will be a reality.

Coal, however, is a heterogeneous raw material; this source of variation has existed ever since coal has been in use. Reducing the unwanted constituents of any particular coal can be accomplished in a number of fundamentally different ways. For coal from some sources, a portion of the ash and sulfur can be removed by relatively simple mechanical procedures, often including relatively simple water washing. Yet there are severe limits to the improvements of coal by these means.

Chemical Processes

Chemical methods hold the only promise for economic conversion of raw coal into more desirable end products. Coal is a relatively inert material and, while it burns readily, most other reactions proceed slowly. The chemistry of coal conversion is extremely complex and not thoroughly understood: many mechanisms and intermediate compounds occur; many reactions involving hydrogen and steam are involved; some of the reactions are endothermic and others release significant amounts of heat.

When coal is subjected to high temperature, it passes through stagewise changes and decomposition. Volatile compounds are released as vapors, and the remaining solid becomes plastic, swells, and then resolidifies. The plastic behavior is a function of heat, particle size, pressure, and type of coal. As heating continues a second variety of volatile matter is released leaving a semicoke residue. The volatile matter from both stages consists of gases and condensable liquids. Subbituminous and lower coals behave differently than described above.

Dissolving coal in a solvent was a laboratory technique used more than a hundred years ago to study its structure. The bitumen fraction is first to dissolve. The presence of free hydrogen or use of solvents able to donate hydrogen to the coal molecule was found

[2] Electric Power Research Institute, *Economics of Fuel Gas From Coal—An Update*, EPRI AF-782.

to increase the fraction of coal liquefied. Hydrogen has the effect of opening ring compounds and cracking long hydrocarbon chains. Liquid solvent may also serve as a transfer medium for hydrogen.

The liquids derived from direct breakdown and hydrogenation–fragmentation of the coal molecule are principally aromatic hydrocarbons with naphthenic and aliphatic compounds comprising the remainder. Many of the compounds contain oxygen, nitrogen, and/or sulfur atoms.

When gas is the desired product, the liquid may be hydrogasified, which is a pyrolysis aided by hydrogen. Hydrogasification of aromatics gives gas plus coke. Coke reacts to a lesser degree with hydrogen to give more gas plus a residue char.

Whether gas or liquid products are desired, the properties of the products are all important. The scientific tools available to effect and control conversion are pressure and temperature and the use of heat rate, hydrogen, steam, and catalyst.

There is a thorough review and discussion of coal conversion reactions in the dissertation by Li.[3]

Higher temperature and pressure are known to give a higher degree of conversion to liquids and gases and thereby leave less char residue. Hydrogen is an active reagent for breaking large molecules and forming more gas products. Steam has the effect of contributing hydrogen to react with the coal and supplying oxygen which reacts with carbon. It is the effective amount of hydrogen present in the reactor that is the main factor in determining the product slate.

Catalysts are materials that accelerate reaction rates, and, since they typically affect particular reactions, they can strongly influence product slate in a multireaction system. There are problems with maintaining catalyst activity since their activity is normally dependent on surface effects involving adsorption mechanisms. The catalyst surface is subject to two types of interference problems: The first is mechanical, e.g., masking by a coating such as coke or fusion by high temperature; the second is a chemical surface change due to combination with any of many *poisons*. Coals contain many potential catalyst poisons including sulfur and most of the heavy metal elements. All of the liquefaction processes involve catalysts in some way to accomplish and accelerate the conversion. Independent research continues to develop catalysts which will improve coal conversion.

Liquefaction technologies are often categorized as "direct" and "indirect." Indirect liquefaction technologies first convert all coal to a synthesis gas and then convert the gas to desired liquid products. Direct liquefaction technologies use processes that work directly with the solid and liquid phases, although they usually add hydrogen as a gas.

Desired results from a successful coal conversion process include

- High product efficiency—energy value of products close to input energy
- Ability to handle variety of coal types—particularly including caking coals
- Preferred product characteristics to replace either present petroleum-derived products or natural gas
- Low product cost

[3]Electric Power Research Institute. *Screening Evaluation—Synthetic Fuels Manufacture*, EPRI AF-523.

The fact that coal, as a raw material, has increased in price dramatically with increases in petroleum's price has confused the outlook for coal conversion processes. It is obvious that the end product must always cost more per energy unit than the raw material. The gains because of the conversion processes are decreased pollution and increased convenience due to better handling characteristics of the derived synthetic fuels.

Conversion Process Comparisons

The many different coal conversion processes that are currently being developed have significant differences in design concepts and product mixes. Some have limitations with respect to the type of coal that may be used. The general problem here is to develop and apply useful comparison criteria. Because the ultimate objective is commercial use, the criteria that will be most meaningful are those which recognize the market potential for the conversion products.

All of the processes studied are believed to be capable of meeting existing environmental regulations. All costs used in this chapter include systems within the process plant needed to achieve legal requirements.

Before comparing the various coal conversion processes, it is necessary to be sure that key standards for comparison are well understood and have the same meanings for all users. Aside from simple costs, the only common standard which applies is efficiency.

EFFICIENCY

There is a universally recognized definition for thermodynamic efficiency which can be applied to any of the conversion processes considered. The definition, in simplest terms, is total output energy divided by total input energy. This definition includes all utility inputs and credits for temperature and pressure of products above ambient conditions. This definition does not relate to the basic purpose of providing fuel products as well as a modified efficiency definition.

The modified definition, which is called *product efficiency* in this chapter, has been in common use by others for years. Others have also called it by other names such as *cold products* or *fuels* efficiency. This definition is the *higher heating value* (HHV) of all fuel products divided by total energy input of coal feed and electric power. As a practical consideration, where electric input is required for the energy balances, this chapter, as well as many other investigators, use 10,000 Btu/kWh rather than 3413 Btu/kWh to allow a realistic conversion loss for electric power generation.

PROCESS COMPLEXITY

When comparing processes, it is desirable to arrive at some measure of how complicated systems are relative to each other. If the same product can be made by several processes, it is axiomatic that, given the same feed materials and final products, the simpler the process, the less prone it is to process upsets, the less the plant should cost, and the less the final product should cost. However, it is difficult to provide a simple definition or criterion of process complexity. Process complexity involves many intangible elements that require a careful and realistic appraisal. These include:

Reaction complexity. A process which requires several consecutive reactions is less desirable than a process involving fewer reactions. The sensitivity of any one reaction to changes in any of the important variables, such as temperature, concentration, etc., may have strong effect on quality control and reliability. Coal is a heterogeneous material, and composition from a given mine often varies with time. This further aggravates the reaction problems.

Operational complexity. A process with many steps, which entails multiple handling of solids and fluid streams, will be prone to more equipment failures and, consequently, greater downtime. As detailed in the section on reactor complexity, the methods used for gas/solids contacting and catalytic conversion can also greatly increase process complexity. As a general rule, solids cause more problems than fluids, and liquids are more troublesome than gases.

Operating regime. The chemistry of coal conversion processes normally involves high operating temperatures and pressures. Very high pressures or temperatures involve more difficulties. Special materials and equipment such as high pressure solids feeders and nonstandard items must be built and maintained with much higher standards than required for simpler conditions.

Auxiliary facilities. The number of required support facilities such as catalyst reclaiming, by-product recovery plants, and special utility services will make the process more complex. Each auxiliary service brings with it its own complexity factor with an influence on cost and reliability.

COST

The final, and usually most important, criterion for selecting one coal conversion process from among a group of proven processes is based on cost projections. This is why cost analyses typically begin while a process is still bench scale and continue to be revised and refined throughout development.

Cost Per Unit of Energy

The simplest cost calculation is the determination of cost per energy unit, typically $\$/10^6$ Btu. This is sufficient for processes producing only SNG but has limited value for the other processes. To make such a calculation, all that is required are estimates of capital investment and some reasonable assumptions for operating cost, fuel cost, and product efficiency. This is one of the methods used in this chapter. Costs are shown for both private and utility financing. All data are presented so that calculations on different bases can be made if desired.

Energy Products Market

The problem becomes more difficult when product costs are to be determined for the purpose of comparing processes, each with a variety of products of different properties. For liquid products, assigning cost directly proportional to heating value is not realistic and is not supported by marketplace history. Price for any commercial fuel is dependent on availability, convenience, environmental suitability, production costs, and demand, i.e., value in a competitive market. Gasoline should and does cost more than boiler fuel because of its convenience in the very large transportation market. Natural gas is more valuable than coal because it is clean burning and now, after many years of extremely low regulated prices, is being priced at a premium value above coal. Fossil fuels have no more reason to be priced the same on an energy-content basis than should various foods be priced the same by such an oversimplified system.

Coal-derived fuels are going into an existing market dominated largely by natural gas and petroleum-derived fuels. Initial efforts are to tailor the new products to serve as replacements. So long as petroleum is available as the major energy source, it will tend to control the market.

Product Value Technique

The value method of determining estimated costs for individual products is based on the assumption that future energy prices for particular products will maintain a fixed ratio to each other. All prices are normalized relative to a reference product. Although we have used the relative values of products as existed in 1978 as the assumed relative values in the future, any other slate of relative values can be assumed.

Since all plant production costs must be recovered, the sum of all product sales in a particular time period must equal all costs (including profit) attributed to that time period. Thus

$$\Sigma P_i B_i = F + M + kC \tag{5.1}$$

where

P_i = individual product price in $\$/10^6$ Btu

B_i = annual production of product i in 10^6 Btu

F = annual coal cost in $

M = total annual cost for labor and all other operation expenses in $

k = capital factor which allows for all interest, taxes, return, and recovery of capital investment

C = total invested capital for project in $

For each individual product, the value factor f_i is defined as:

$$f_i = P_i/P_r \tag{5.2}$$

where P_r = the price of the reference product. Substituting P_i from Eq. (5.2) into Eq. (5.1) yields

$$P_r \Sigma f_i B_i = F + M + kC$$

or

$$P_r = \frac{F + M + kC}{\Sigma f_i B_i} \tag{5.3}$$

Thus, the necessary market price for the reference product if the subject technology is to be competitive is, from Eq. (5.3), a simple function of annual costs, annual production of the various products, and the relative market values of the various products.

Once the reference product price is calculated, the individual product prices are determined simply, from Eq. (5.2), by multiplying the reference product price by the respective value factors f_i. For this chapter, premium gasoline has been chosen as the reference product.

To calculate value factors f_i, it is first necessary to select a complete set of market values for all products. This price set is for a particular point in time, and the choice of the time is reflected in the value factors. For this chapter, f_i for liquid products were determined from mid-1978 published market prices. All liquid fuel products were subject to federal price controls to some degree in 1978, and this is reflected to the same degree in all value factors.

Natural gas prices are just now being released from crippling price controls over the last 25 years (the regulated well-head price for new natural gas in the interstate market has increased some eightfold in the last 6 years and will be deregulated in the near term); however, because of its convenience for most stationary uses and its high environmental advantages, we have assumed market value for natural gas equivalent to premium gasoline on a Btu basis. The slopes of recent price curves for oil and natural gas also support this as a reasonable assumption. Since there has been no significant production of lower-Btu fuel gas in recent years in the United States, there is no market and thus no 1978 market price. ESCOE assumed that all gas fuels have the same value per Btu regardless of heating value per volume.

There has been no market for SRC-I clean solids since there are no commercial production facilities.

The solid fuels are priced on their heating value relative to fuel oil. Clean SRC solid fuel is discounted 10% to allow for the added handling costs. Char is further discounted to allow for higher ash contents. The value shown for char would be too high for a material with a significant sulfur content. The amount of sulfur contained in a char depends strongly on the feed coal analysis and the process used.

Tar oil is a material peculiar to gasifiers. There is a potential for recovery of valuable chemical by-products.

Fuel oil in this chapter refers to industrial fuel commonly referred to as No. 6 Bunker C or residual fuel. There are variations in the physical properties for the oils sold in this category. It is not uncommon for residual fuels with a higher heating value to sell for less per barrel due to high viscosity and resultant extra handling cost. No. 2 oil is considered synonymous with diesel fuel in this chapter.

Naphtha is a generic term for a distillate product, which is not commonly sold on the open market. It is an intermediate refinery stream which ends up as gasoline, diesel fuel, and specialty products. Naphtha's physical properties and hence value varies considerably depending on crude oil source and pretreatment. The naphthas from coal conversion processes also vary both between processes and even from a given process depending on feed coal and operating conditions. For this chapter, all naphthas were assigned a single f_i value related to cost necessary to upgrade the product to gasoline.

Liquid petroleum gas (LPG) is a commercial product consisting mostly of propane and butane. It is shown as a lower energy price than the principal components, which reflects market conditions and value of the purer materials as feedstocks for higher value chemicals.

The gasoline shown in Table 5-1 is low octane such as obtained from a simple Fischer–Tropsch process; it is lower quality than the regular grade sold in the U.S. market. The premium gasoline is 93 Research Octane Number and the same as sold in the United States.

Methanol is sold commercially but not on the scale required for the transportation fuels market. Thus the mid-1978 commercial price has no relation to future cost as a major fuel. For this study pure methanol was assigned an energy value equal to premium gasoline since it can be used in the same type engines as an acceptable replacement. Methyl fuel is priced proportional to the heating value for pure methanol. It is discounted to allow for water.

Table 5-1. Value Factors for Energy Products (Reference to Premium Gasoline)

Product	1978 Price ($)	Heating Value (10^6 Btu)	Energy Price ($/$10^6$ Btu)	Value Factor f_i
SRC solid[a]	—	32.0	1.75	0.50
Char	26.32/ton	16.0	1.65	0.47
Tar oil	14.88/bbl	5.0	2.98	0.85
Fuel oil	12.35/bbl	6.3	1.95	0.56
No. 2 oil (diesel fuel)	14.92/bbl	5.2	2.87	0.82
Naphtha	14.92/bbl	5.2	2.87	0.82
LPG	12.04/bbl	4.0	3.01	0.86
Gasoline	15.75/bbl	5.0	3.15	0.90
Premium gasoline[b]	17.50/bbl	5.0	3.50	1.00
Methanol[c]	—	2.7	3.50	1.00
Methyl fuel[d]	—	2.6	3.36	0.96
Butane	13.11/bbl	3.5	3.75	1.07
Propane	12.10/bbl	3.2	3.78	1.08
All fuel gases[e]	—	—	3.50	1.00
Electricity	31.06/10^6 Wh	3.4	9.10	2.60

[a] No sales in 1978. Energy price assumed 10% less than fuel oil.
[b] Premium gasoline is reference fuel: $f_i = 1.00$.
[c] No significant market in 1978. Energy value assumed same as premium gasoline.
[d] No market in 1978. Energy price 4% below methanol for water content.
[e] Severe price regulation for natural gas in 1978. Energy value set at 1.00. (See text.)

The value factor for electricity was determined from a typical mid-1978 cost to medium size users.

Process Description Comparisons

The descriptions which follow are somewhat brief and intended to describe the reaction system for the actual conversion of the coal. Subsequent purification operations are generally similar among coal conversion processes and, in practically every case, the same kinds of processing are used in industrial chemistry operations. The various coal conversion processes are in different states of development.

SOLID PROCESSES

Two solid fuel products from coal are considered in this chapter. These come from completely different processes. The first is solvent-refined coal (SRC-I) from a federally sponsored development project. The intention for this product is to have a material which is very similar to coal but with much less pollutant content, particularly sulfur and ash. This material may eliminate the alternative for existing coal-fired equipment, which is to process all of the flue gas to remove the sulfur and solid materials to make the plants environmentally acceptable. SRC-I boiler fuel is a solid material, with low ash and sulfur, capable of burning clean. The product specifications for the solid process are shown in Table 5-2.

The other solid fuel material derived from some coal gasification processes is char. Char is principally carbon and the ash residue of the coal. Because much of the volatile matter has been removed and the coal molecule has been broken down into much smaller units, char can be a more-reactive form of carbon than raw coal. Char can be used as a fuel or as a feed for a gasifier system. Gasifier systems use the char feed with air or

Table 5-2. Synthetic Fuel Products Specifications

Characteristic	Process Specification	
	SRC-I	Synthane Char
HHV (Btu/lb)	16,000	8,000
Density (g/cc)	1.25	—
Melting point	over 355°F (180°C)	—
Boiling point	over 500°F(425°C)	—
Ultimate Analysis (wt%)		
C	87.14	52.1
H	5.66	1.0
O	3.91	5.1
S	0.88	0.2
N	2.31	0.3
Ash	0.10	41.3

oxygen to provide a mixture of carbon monoxide and hydrogen. Carbon monoxide and hydrogen mixtures have wide application and have been used for many years in various ratios to synthesize more valuable chemical compounds. This same gas mixture can also be used to synthesize methane, the chief ingredient of natural gas and the most desirable pipeline gas. (For all practical purposes methane may be considered equivalent to natural gas.) Most of the gasification processes produce some char. The sulfur content of the char is an important factor dependent on feed coal and the process. A high sulfur content reduces char value considerably.

Solvent-Refined Coal (SRC-I)

Development of the SRC-I process began in 1962. A pilot plant began operation in 1974. A second pilot plant operates completely independently on process development. The process was designed to dissolve the coal, remove the impurities, and produce a nonpolluting solid fuel. Hydrogen is added, and liquid and gas are produced as by-product fuels.

The coal is mixed with a process-derived solvent, pumped to reaction pressure, and combined with hydrogen, preheated, and fed to a dissolver vessel. Well over 90% of the carbon material in the coal dissolves in the reactor. Most of the sulfur in the coal combines with hydrogen to form hydrogen sulfide, which leaves the reactor as a gas. The effluent slurry from the reactor is depressurized and then goes to a solid separation step. In this step ash plus unreacted coal are separated from the solvent and liquids. The clean liquid is distilled. A portion of the distillate liquid, process-derived solvent, is recycled back to the dissolver. The SRC-I fuel is allowed to solidify. The melting point of the solid is about 355°F (180°C). The major use for this material is projected to be boiler fuel.

Development was underway to determine the best system for removing solids from the slurry. The best possibilities include filtration, antisolvent processing, and critical solvent deashing. Filtration has been used in the pilot plant.

There have been successful tests of SRC-I product in a commercial power plant. There were no problems handling or using SRC-I fuel, and performance was better than coal. Both the sulfur reduction and oxides of nitrogen limitation were achieved in the commercial test. It is likely that SRC-I solid could find wide application in utilities and possibly very large commercial boilers if cost incentive exists. This is a very large potential market. SRC-I properties depend on process conditions.

LIQUID PROCESSES

Fluids are more desirable from a handling standpoint than solids. Liquid fuels are desired to replace liquid petroleum products. Liquids have the general advantage over gases of lower storage cost. Gas fuels with characteristics as close to natural gas as possible are desirable for the purpose of replacing existing natural gas. There are a vast variety of liquids produced by several approaches to the conversion of coal. Pyrolysis is an historic method for producing liquids and gas from coal, but it is difficult to control the kinds of products which result. The COED process is a successful improvement which pyrolyzes in stages.

Aside from pyrolysis, the two different basic routes to coal liquefaction are:

Direct Liquefaction: Crushed coal is slurried in a process-derived oil and reacted directly with hydrogen at high temperature and pressure conditions. Some solid always remains in the reactor effluent. This solid consists of unreacted coal and ash. The separation and subsequent treatment of this material is an important area of process development for each of the liquefaction processes. The liquid produced has properties similar to certain petroleum hydrocarbons. The actual yields and physical properties depend directly on reactor conditions and degree of hydrogenation. The direct liquefaction processes include SRC-II, EDS, and H-Coal, all described in further detail below.

Indirect Liquefaction: Coal is first gasified to produce a hydrogen and carbon monoxide mixture. These materials are then recombined with the aid of catalysts to produce liquid compounds. A particular advantage of indirect liquefaction is that essentially all of the sulfur and nitrogen present in the coal can be separated in the gaseous phase and thus eliminated from the liquid products. These materials are difficult and expensive to remove to a very low concentration with direct hydrogenation processes. The practical processes for indirect liquefaction include Methanol and Fischer–Tropsch. Process descriptions follow.

Each of the liquefaction processes produces several types of product and some gas. In some cases the gas is used within the plant. Product properties are given in Table 5-3.

The liquefaction processes share many of the same production problems. Some are specific to the particular coal feed. An example of this is calcium carbonate deposition in the reactor. Alkali content tends to be higher in western coals. This can be an advantage for indirect liquefaction as calcium catalyzes the gasification reaction. Otherwise the deposition can be avoided by feed coal pretreatment with sulfuric or sulfurous acid. Other problems such as handling of residue slurries are common to many processes. The particular techniques found satisfactory by any one process developer can often be applied to another process.

Solvent-Refined Coal (SRC-II)

The SRC-II process was developed in the mid-1970s to eliminate two problems considered inherent with SRC-I: The first was the solids separation problem and the second was market acceptability for the product.

SRC-II is the same as SRC-I except proportionally more hydrogen is used and a portion of the slurry is recycled to increase the degree of conversion. Fuel oil is separated

Table 5-3. Synthetic Fuel Liquid Product Specifications

Naphtha[a]	Process				
Property	SRC-II	EDS	H-Coal	Hygas	Lurgi
Boiling range	C_5 – 390°F	C_5 – 390°F	C_5 – 390°F	Benzene to C_9	Footnote 2
HHV (MMBtu/bbl)	—	5.2	5.5	5.4	5.2
Density (g/cc)	0.84	0.82	0.80	0.88	—
Sulfur (wt%)	0.3	1.0	0.06	—	—
Nitrogen (wt%)	0.6	0.2	0.07	—	—
H/C ratio	0.14	0.14	—	—	—
Viscosity (centistokes @100°F)	0.9	0.9	—	—	—

Fuel Oil	Process			
Property	SCR-II	EDS	H-Coal Syncrude	H-Coal Fuel Oil
Boiling Range, °F (°C)	390 – 797 (200 – 425)	390 – 1000 (200 – 540)	390 – 940 (200 – 525)	390 – 1000 + (200 – 540 +)
Flash point, °F (°C)	167 (75)	198 (92)	—	—
Density (g/cc)	1.04	1.085	0.93	1.09
HHV (MMBtu/bbl)	6.27	6.31	6.13	6.70
Sulfur (wt%)	0.3	0.8	0.08	0.43
Nitrogen (wt%)	0.9	1.1	0.37	0.84
Viscosity (SSU @100°F)	50	102		
H/C ratio	—	0.08		

[a]Many properties will vary with different feed coals.

by vacuum distillation and residual materials are gasified. Only liquid and gaseous products result from the process.

A feed-coal-type limitation has been suggested for this process because certain trace minerals are required in the liquefaction reactor to act as catalysts for the breakdown of solid to liquid, and these elements are found in the ash of particular coals. Where the problem is concentration rather than kind of trace elements, the process developer has suggested that a recycle of residue would seem capable of broadening allowable coal feeds.

Exxon Donor Solvent Process

The Exxon Donor Solvent process development by Exxon began in 1966. The process is similar to SRC-II described above with the exception that a major portion of the hydrogen is supplied as a part of the solvent, chemically combined rather than as a free or dissolved gas. The process-derived solvent literally contributes hydrogen atoms to the coal molecule in the dissolver vessel. The solvent is rehydrogenated using a proprietary catalytic process. As a part of their process development, Exxon has been designing their proprietary Flexicoking process for the treatment of residual slurry produced in the liquefaction reactor.

H-Coal

Work on the H-Coal process began in 1964 as a natural extension of the commercially successful H-Oil process. In the H-Coal process, crushed and dried coal is slurried with recycle oils, mixed with hydrogen, and liquefied in direct contact with catalyst in an ebullated-bed reactor. This reactor utilizes an upward flow of liquid to expand the catalyst bed and maintain the suspended catalyst particles in random motion. The finely divided coal and ash particles flow through the ebullating bed and are removed with the liquid and vapor products.

The reactor effluent is separated into recycle and net product streams in conventional processing equipment under conditions more severe than existing applications. Conversion and yield structure are determined by reactor conditions, catalyst replacement rate, and recycle slurry oil composition.

Two different modes of operation for H-Coal are used in this chapter. The one emphasizes distillate production and the other maximizes total fuel oil. Product from the distillate mode is termed syncrude as the liquid material produced is similar to low-grade petroleum crude oil. Considerable upgrading is required to prepare this material for use as a refinery feedstock.

Fischer–Tropsch

The Fischer–Tropsch reactions were discovered in 1923. They include a range of combinations using iron/cobalt catalyst to react hydrogen and carbon monoxide to form paraffins, olefins, and alcohols. These reactions are exothermic and not specific to the formation of any single compound.

For a complete process to convert coal to hydrocarbons, a gasification section is required to precede the Fischer–Tropsch reactor section. There are a number of possibilities for the gasifier system. The choice is important because of the effect on final cost. The data used in this study are based on use of a British Gas Corporation (BGC) Slagging Lurgi gasifier. This choice is based on high conversion efficiency and ability to handle a broad range of coal types. This unit is still in the final development stage and so not yet in commercial use. Other gasifiers are under development which should be competitive with the BGC Slagging Lurgi.

The Fischer–Tropsch process has the advantage of being in commercial use with conventional Lurgi gasifier. It has the disadvantage of producing a wide variety of compounds that do not fit well into existing markets. Mobil Oil Corporation has developed catalysts that improve the quality and quantity of gasoline from the Fischer–Tropsch process.

Methanol and M-Gasoline

The production of methanol from a gas mixture of carbon monoxide and hydrogen is a specialized application of a Fischer–Tropsch reaction. The yield of methanol is maximized by a combination of optimum reactor conditions, catalyst, and recycle of the unreacted gases.

The conversion of methanol to M-Gasoline is a separate catalytic conversion, which can be done in fixed or fluid beds. This reaction is also exothermic. The thermal efficiency for the final M-Gasoline conversion step can be as high as 93%. Methanol has high potential as a fuel for peak-shaving turbine-driven generators and possibly as a boiler and vehicle fuel.

Gas Processes

While a variety of solids handling, gas contacting, and gas-quality-upgrading techniques may be employed in the various gasification processes, the primary goal is the same: to convert a sulfur-containing solid to an essentially sulfur-free gaseous product. The gaseous product may be a mixture of hydrogen and carbon monoxide with nitrogen. The definitions are not rigorous, but high-Btu gas is generally over 900 Btu/scf and low-Btu gas is generally below 350 Btu/scf. Medium- or intermediate-Btu gas falls between these values. High-Btu gas production is the more difficult task as it requires additional upgrading of a low- to intermediate-Btu gas stream to produce a commercially acceptable product. The carbon/hydrogen ratio is adjusted either by rejecting a portion of the carbon, by increasing the hydrogen, or by a combination of both. The chemical processing steps include production of a gas containing carbon oxides and hydrogen by the reaction of carbon and steam, gasification; reaction of hydrogen and coal to form methane, a hydrogasification process; reaction of hydrogen and carbon monoxide to produce methane, methanation; and the removal of impurities. Table 5-4 summarizes the synthetic fuel gas product specifications.

The gasification processes share a number of common techniques. The coal received at the plant site must be further cleaned and crushed or ground before it can enter the gasifier. Conventional cleaning methods are used to remove extraneous materials such as metal, and rocks. For some processes the coal must be finely ground so that it may be fluidized or entrained in a flowing gas stream. For other processes crushing and sizing is all that is required. Agglomerating coals may be pretreated by heating, mild oxidation with air, or other procedures to ensure smooth gasifier operation and avoid reactor plugging. Nonagglomerating coals such as lignite and subbituminous coal normally do not require pretreatment. Feeding a solid material such as coal to a higher pressure reactor is a difficult task. Dry feeders using a series of pressurized lock hoppers or wet systems in which a slurry is pumped into the reactor are used.

The gasification step is the heart of the process. Here the solid feed material is contacted with synthesis gas (carbon monoxide and hydrogen). The coal is devolatilized and a portion of the fixed carbon is converted to methane. The gas–solids contacting methods used include fixed beds, fluidized beds, moving beds, entrained beds, and

Table 5-4. Synthetic Fuel Gas Product Specifications

Gas Product	Process						
	Hygas	Synthane	CO_2 Acceptor SNG	CO_2 Acceptor Syngas	Lurgi	Bigas	EDS
HHV (Btu/scf)	990	940	950	378	980	908	1,300
Molecular Weight	16.2	17.0	15.8	10.2	15.8	15.0	
Analysis (vol %)							
H_2	0.8	1.8	4.6	65.5	3.0	10.0	
CO	0.03	0.1	0.01	15.7	0.05	0.1	
CH_4	97.9	92.4	92.6	11.4	96.0	86.7	
C_{2+}	0	0	0	0	0	0	
N_2	0.4	1.9	2.0	0.8	0.6	0.6	
CO_2	0.8	3.8	0.5	4.7	0.4	2.6	
H_2O	trace	trace	trace	1.9	trace	trace	

Methanol (per Badger Conceptual Design[a])

Methanol:	Purity	99.85	wt%
	Density	0.7954	g/cc
	HHV	9783	Btu/lb
Methyl Fuel:	Methanol	96.1	wt%
	Water	3.5	wt%
	Density	0.8019	g/cc
	HHV	9407	Btu/lb

M-Gasoline: 9 psia Reid Vapor Pressure
93 octane clear (RON)
57° API

[a]Badger Plants, Inc. *Conceptual Design of a Coal to Methanol Commercial Plant*, prepared for U.S. DOE, FE-2416-24, Cambridge, MA.

free-fall systems. The gasification process, overall, is endothermic, and the heat required to maintain reaction rates is usually supplied by burning a portion of the coal feed. External heat sources such as electric-resistance heating, molten salt or solid heat carriers, and introduction of a hot hydrogen stream have also been tested. Following the gasification step the particulate materials and heavy hydrocarbons contained in the reactor effluent are removed by a water-quench and solids-separation system. If low-Btu gas or synthesis gas is the desired product, the raw gas is scrubbed to remove carbon dioxide and sulfur compounds and then used directly. However, if high-Btu gas is desired, methanation is necessary.

In most gasification processes the hydrogen content of the gasifier effluent is too low for effective conversion to methane; therefore, the shift reaction is employed to react a portion of the carbon monoxide with steam to form more hydrogen. Fixed-bed reactors charged with catalysts are used in this process. The CO_2 formed is then removed. A number of commercially available processes are effective in removing both CO_2 and sulfur compounds. The off-gas from the purification processes is treated to recover the sulfur and minimize objectionable effluents.

Synthesis gas, which is mostly hydrogen and carbon monoxide, has a heating value much too low for direct injection into the natural gas network. The heating value of the gas is increased by catalytically reacting the hydrogen and carbon monoxide to form methane. This methanation process is highly exothermic and a number of techniques have been developed to control reactor bed temperatures. Fixed beds and fluidized beds with reactor system variations have been tested. The final step to prepare high-Btu gas for transmission is to remove water to specification levels and compress if necessary.

CO_2 Acceptor

The CO_2 Acceptor process was designed to accomplish two important functions within the same reactor: coal gasification and carbon dioxide (CO_2) removal. The commercial process uses three fluidized-bed reactors. The gasifier achieves coal gasification and the combination of the CO_2 formed with acceptor solid. The acceptor solid is regenerated in a second vessel. The acceptor is a lime material that reacts with CO_2 in the reactor and incidentally provides reaction heat which contributes to the heat input required for the coal gasification reaction. Heat for the acceptor regeneraton is generated by burning residual char withdrawn from the top of the gasifier bed. The combustion takes place within the acceptor regenerator.

Lignite and subbituminous coals are preferred feeds because of their high reactivity. With these, the gasification temperature is sufficiently low to avoid ash fusion and consequent solid deposition and particle agglomeration.

The hydrogen-rich gas mixture from the gasifier is purified, methanated, compressed, and dehydrated to produce pipeline-quality gas. The CO_2 Acceptor process is unique among the high-Btu gas processes as it does not require a separate acid gas removal system; it also has the lowest water requirement. However, it has the particular disadvantage of poor operability on feeds other than lignites or subbituminous coals.

A second mode of operation for the CO_2 Acceptor process produces syngas by not going through the methanation step. Syngas is a useful medium-Btu intermediate which may be used to synthesize methanol, other hydrocarbons, or alcohols.

Hygas

Development of the Hygas process began in 1954. Pilot plant operation began in 1972.

The Hygas reactor is a single vessel containing four separate, vertically stacked contacting zones. Coal is pumped into the reactor as a slurry containing up to 50 wt % solids in recycle oil. The slurry is dried in the top zone with hot gases rising from the lower reactor zones. The dried coal is then gravity fed down through the remaining reactor zones, countercurrent to the ascending reacting gas stream. There are two hydrogasification zones; the first, a short-residence-time (1–10 seconds) low-temperature [1200–1400°F (650–760°C)] dilute-phase transport reactor, followed by a longer-residence-time (20–40 minutes) higher-temperature [1600–1700°F (870–925°C)] dense-phase fluid-bed reactor. Sixty to seventy percent of the total product methane is produced in these two hydrogasification reactors. In the bottom zone the remaining residential char is converted to high-hydrogen-content synthesis gas by dense-phase, fluid-bed steam–oxygen gasification at temperatures of 1740–1850°F and 10–20 minutes residence time. This hot synthesis gas is the hydrogen source for the two upper hydrogasification zones. Ash is withdrawn from the bottom zone, water quenched, and removed from the process as a water slurry.

The raw gas mixture from the reactor is scrubbed, then passed through an acid gas removal system. A final methanation process converts most of the unreacted hydrogen and carbon monoxide present to methane.

A pretreatment step is required for caking coals, which would otherwise agglomerate in the reactor. The pretreatment consists of mild air oxidation which destroys the coals' agglomerating properties and reduced product yield.

Bigas

Process development began in late 1963. The process hypothesis was that a high methane yield would result from reactions of coal with steam at high temperature and pressure.

Coal is fed into the first gasifier stage as a water slurry. Steam is injected separately and combines with the coal, and the mixture is contacted by hot gas from the oxygen-fed lower stage causing conversion to methane, carbon monoxide, carbon dioxide, hydrogen, and char. This mixture rises through the gasifier and is discharged from the gasifier to a water quench. The gas is scrubbed further, then treated in a shift reactor which combines carbon monoxide and steam to form more hydrogen. The gas is then passed through acid scrubbers to remove hydrogen sulfide and carbon dioxide. This is followed by a methanation reactor which combines most of the carbon monoxide and remaining hydrogen and forms more methane raising the heating value of the product gas significantly.

This reactor system is able to handle all types of coals without pretreatment. There are material problems owing to the higher operating temperature. The slagging section is particularly prone to flow interruptions owing to solidification of the molten ash. Solids circulations throughout the system are a difficult problem. The process has never operated satisfactorily.

Synthane

Development of the Synthane process began in the early 1960s.

The sized coal is fed by a weigh-bin lock hopper system into the fluidized-bed pretreater where the coal is dried before flowing to the single-stage fluidized-bed reactor. Steam and oxygen are injected below the fluidized bed. Gasification occurs within the fluidized bed. Unreacted char flows downward into another bed fluidized and cooled by steam. Char is removed from the pressurized system as a water slurry.

Product gases are scrubbed prior to shift conversion, which gives the proper carbon monoxide-to-hydrogen ratio for methanation.

Large quantities of char are produced and exported from the plant as boiler fuel. This solid fuel is undesirable compared to fluid fuels and has been assigned an appropriate value factor in this chapter. In late 1978, DOE made a decision to cease operations of the Synthane pilot plant.

Lurgi Process, Dry Bottom

The Lurgi process employs a moving bed technique wherein the coal charge moves downward as a continuous bed and is contacted by an upward flowing steam–oxygen mixture. The feed coal is crushed and screened to size range of about ⅛–1¼ in. size. Fines are separated and used as fuel. The sized coal is continuously fed to the top of the reactor through a pressurized lock hopper system. As the coal moves downward through the reaction zone the steam–oxygen mixture reacts with the coal to form synthesis gas. The hot synthesis gas hydrogasifies and devolatilizes the raw coal in the upper zones and is withdrawn from the reactor for further cleaning and upgrading. The coal bed is supported on a revolving grate. Ash is discharged as a dry granular solid to the ash disposal system.

While the Lurgi reactor system has principally been used to gasify nonagglomerating or slightly agglomerating coals, the range of usable coals has been extended by the addition of stirrers to break up the agglomerates. In order to avoid sintering of the ash and plugging of the bed, the reactor normally operates with a steam/oxygen volume ratio of 6 to 10. This leads to high residual unreacted steam in the off-gas stream and increases the oxygen requirement. Slagging Lurgi process has reduced these requirements and improved overall process efficiency.

The off-gas is cleaned for use as low-Btu gas (heating value in the range of 300 Btu/scf) or cleaned and methanated to produce a high-Btu gas. The Lurgi process gas clean-up system includes quench vessels, steam generators, and coolers. A heavy tar oil is recovered from the quench water while other gaseous contaminants are removed in subsequent gas clean-up processes. If necessary, the hydrogen/carbon monoxide ratio is adjusted in a shift reactor for methanation. The shifted gas is scrubbed to remove the carbon dioxide. The hydrogen and carbon monoxide in the purified gas stream are then catalytically converted to methane to produce a final product gas with a heating value near 950 Btu/scf.

Slagging Lurgi Process

The Slagging Lurgi process utilizes a moving bed system in which downflowing coal is contacted with a rising gas stream in a manner quite similar to the conventional dry bottom Lurgi reactor. The principal difference between the two is the method of discharging ash. The Slagging Lurgi operates at a higher bed temperature and the ash is melted. The molten ash then flows from the bottom of the reactor through a slag discharge system. The primary advantage of the Slagging Lurgi is the reduction of steam input per unit of gas produced and the much higher rate of conversion obtainable at the higher bed temperature. Process efficiency is higher, and the cost of the reactor section of the gasification plant may be significantly reduced. While the need to maintain gasifier operability requires care in selecting coals with the proper ash fusion temperature, this process is expected to extend the range of coals that can be successfully gasified in the Lurgi reactor. Coals with lower ash fusion temperatures and lower reactivity can be gasified under slagging conditions, while the higher ash content coals with high ash fusion temperatures are gasified in a standard dry bottom Lurgi.

The gas produced from the Slagging Lurgi is roughly similar in composition to the dry bottom Lurgi product gas. Downstream treating and upgrading process steps are essentially the same. As stated above, process efficiency is better than dry Lurgi.

Other Processes

The final group of processes compared use both chemistry and thermal energy conversion to produce electricity as the final product. These processes convert the coal to the most convenient gaseous products and then burn the gas in simple efficient turbine-driven generator equipment to utilize pressure, thermal energy, and chemical energy to achieve conversion efficiencies potentially better than modern coal-burning electric power plants.

The following two power processes were included in this chapter for a broader comparison. There are many other systems for electric power generation with all of the same potential advantages that are being developed in the United States and abroad. These systems could have also been evaluated as low-Btu gas producers. They have potential as medium gas producers if oxygen instead of air is used in the reactors.

Combustion Engineering. Development of the Combustion Engineering system began in 1974. This process is a method for production of low-Btu gas for electricity production. It is based on an air-blown, atmospheric pressure, entrained gasifier.

A portion of the coal is burned in a combustion section of the gasifier to supply heat for the gasification reaction. The ash from the combustion section is removed as molten slag from the bottom. The remainder of the coal is fed to the reduction section where it contacts hot gas. The conversion process takes place in an entrained area where fresh coal contacts hot gas and devolatilization and gasification occur. The hot gas is cooled and solids are separated. After Stretford process sulfur removal the gas is ready for use as turbine or boiler fuel.

The ability to treat all coals without pretreatment and the complete carbon conversion are advantages. The low pressure operation necessitates relatively large vessels and piping for a given fuel rate. Low-Btu gas must be recompressed and utilized close to the source for the same reasons. Design and material-of-construction problems are still under development for this process.

Westinghouse. This process is called the "Advanced Coal Gasification System for Electric Power Generation." Development began in 1972 and the 15 ton/day system has been operating since 1975.

The initial concept for the Westinghouse coal gasification system was a two-reactor system. Both reactors, the devolatilizer and the gasifier, contain fluidized beds. The first reactor is used to devolatilize and decake coal as required. Devolatilizer char circulates around a draft tube and prevents coal particles from agglomerating as they pass through their sticky stage during heating. The second reactor is used to gasify char from the devolatilizer and agglomerate the residual ash at high temperature. The gasifier also provides heat for the process.

The initial concept also included adsorption of hydrogen sulfide with dolomite in the devolatilizer. Hydrogen sulfide removal with dolomite in the devolatilizer is currently not under consideration for initial commercialization; however, work in this area is still being explored at the Westinghouse Research laboratories.

Initial testing of the Westinghouse process at the 15 ton/day PDU located near Madison, Pennsylvania, began in 1975. Testing philosophy was aimed at demonstrating the technical feasibility of each reactor, the devolatilizer and then the gasifier, before running them together as an integrated system.

Tests on the devolatilizer ran from mid-1975 to mid-1976. Gasifier tests began late in 1976. The tests on each reactor demonstrated the feasibility of each reactor's conceptual design. Tests on the integrated system, devolatilizer and gasifier combined, began in 1979.

Extensive development work on the gasifier has been done. Tests on the gasifier have been successful with run times in excess of 200 hours. In addition to char, caking and noncaking coals without pretreatment have been successfully processed in the gasifier. Also, the gasifier has been run on air and on pure oxygen. From the development work on the gasifier, the concept of a single-stage system has evolved in addition to the original two-reactor system.

This coal gasification process is being developed for a variety of low-Btu and medium-Btu applications. These include (a) combined cycle power generation, (b) fuel cell power generation, (c) industrial fuel gas, and (d) chemical feedstock. The size used in chapter study would require multiple gasifiers.

The pressurized gasifier operation should give an efficiency advantage over gasifiers that operate at atmospheric pressure. This theoretical advantage is slight as evidenced by the higher product efficiency shown for the Combustion Engineering combined cycle process. The advantage of the latter process is due to a selection of a better steam cycle for the waste-heat-recovery part of the system.

Technology Comparison

One of the goals of this chapter is to make a technical appraisal of all processes noted in the scope section of this chapter. Since the processes are in various stages of development, this part of the chapter should lend further insight into the ultimate prospects for commercial success.

As a general comment many technologists feel that any fully integrated process that can be operated continuously on a small scale can be made to succeed on a large scale. There are no apparent impediments to commercialization due to lack of control or analytical equipment. At present, no needs exist to develop special devices to accomplish coal conversion. There will always be possibilities for improvement but no noticeable change in coal conversion viability or economics is expected from instrumentation developments.

The areas where significant development could affect commercial prospects include

1. Methods to allow more coal types to be used effectively
2. Improved practical systems to achieve maximum recovery of usable energy
3. Reactor and process improvements to achieve better yields and greater product selectivity
4. Improved chemistry to reduce environmental problems further and more cheaply
5. Improved high-temperature gas-turbine materials technology
6. Better hot gas cleaning processes

The first item would enhance application. All the subsequent improvements would result in lower cost for the product.

PROCESS COMPLEXITY

All of the processes compared in this chapter were analyzed to determine both the common and the singular features. Most had many common unit operations and comparable overall complexity. A list of significant considerations that have been factors in process development to date was prepared. To this were added items that do differentiate some of the processes to a degree that labor requirement or reliability for an actual plant could be affected. The list used and values for process complexity are shown in Table 5-5.

Table 5-5. Process Complexity[a]

Process Complexity	Process												
	SRC-I	SRC-II	EDS	H-Coal	FT	M-Gas	HyG	Synth	CO_2 Acc.	Lur	BiG	West	CE
Highest pressure of group				X							X		
Highest temperature of group							X	X	X			X	X
Ebullated or fluidized bed				X			X	X	X			X	
Multifluidized beds							X	X	X				
Ash fusion problems										X	X		
Cumbersome recycle pump or compressor				X		X	X						
Solid – fluid separation problem	X						X	X			X		
Significant solids handling	X							X	X		X		
Power turbine critical to cycle												X	X
Oxygen used in process							X	X		X	X		
Tar products					X			?		X			
Chemical reaction complexity			X		X	X							
Fired preheater used	X	X	X	X									
Complexity value	3	1	2	4	2	2	6	6	4	3	5	3	2

[a]Note that a lower value indicates less complexity. The items shown do not have equal importance and thus the Complexity Value scale is not linear and may not be monotonous.

It is apparent from the list of considerations used to develop the process complexity index that the items are not of equal weight. Thus the complexity index values show a number for identified complexities but should not be construed to be directly proportional to operational difficulty, reliability, or labor requirements. It is apparent that many unit operations common to practically all of the processes are not included on the list. This is particularly true for proven operations used in existing process industries. There are many examples of distillation columns, acid gas removal–regeneration systems, and Claus plants that operate with relatively little attention.

The reasoning behind the items used to determine complexity is as follows:

- High Pressure and Temperature: Relatively more difficult conditions and possible material-construction problems.
- Ebullated or Fluidized Bed: More sophisticated system with greater start-up difficulty and less forgiving to changes in operating conditions.
- Ash Fusion: Can lead to unreliable performance and reactor shutdown.
- Recycle Pump or Compressor: Recycle fluid streams can contain higher residual solids leading to seal leakage and erosion. In the case of H-Coal, the internal recycle pump can give severe maintenance problems. These items are high energy users if heads are high.

- Solid–Fuel Separation: Not always a problem—depends on actual values for density difference or fluid properties.
- Solids Handling: Solids handling always entails higher maintenance and less reliability than comparable for fluids.
- Power Turbines: These equipment items are still being developed and subject to damage by process upsets.
- Oxygen: A potentially dangerous material requiring extreme and continuing safety precautions.
- Tar Products: Viscous liquids that sometimes cause line plugging problems and general handling and disposal problems.
- Chemical Complexity: Highly sensitive reactions or fragile catalysts.
- Fired Preheater: Subject to coking and tube damage directly affecting equipment reliability.

Even though the table shown for complexity can have qualitative value, it is clear that this comparison alone cannot be used to choose or eliminate a process.

Reactor Comparison

Because the reactor system is the chief area where differences exist between competing processes, the subject requires special study. The reactors are where the highest temperatures and pressures occur and where most of the materials-of-construction problems are found.

Because of the higher mobility of gases than liquids within a reactor vessel, and the presence of three phases, from a transport-phenomena standpoint, all gas reactors tend to be more complex internally than liquid-phase reactor systems. This is in addition to the higher operating temperature associated with coal gasification reactors. Thus, generally all of the gasification reactors are more complex than any of the liquefaction reactors. Within the group of gasification reactors, those with all three phases normally present must be judged the most complex. This touches on the key to coal-handling problems (due to caking) and the belief that the Bigas process is probably not viable due to continuing slag-removal difficulties.

Fluid beds are considered more complex than other types of solids contacting systems. They are, furthermore, constrained with respect to process modulation compared with non–fluid-bed reactors. The other general reactor problem is bypassing within the vessel. This is a recognized area where good design is required; however, simple and economical methods such as baffle use or nozzle modification can reduce problems from this source.

Particular reactor problems encountered with some of the various processes will be discussed in the following sections. Table 5-6 shows conditions in the reactor. The actual values during operation may vary for different coals. Data shown are from reports by the process developers.

PROCESS COMPARISONS

The coal conversion processes must be compared and evaluated from both a technical and an economic standpoint. The two approaches will necessarily be interdependent and the ultimate goal is to develop the most inexpensive substitutes for raw coals, natural gas, and petroleum fuels. The purpose for process comparison is to determine which of

Table 5-6. Reactor Conditions

Process	Reactor Type	Temperature (°C) in/bulk/mix	Pressure (atm) normal/max	Recycle Phase	Coal Residence Time	H_2 Consumption
SRC-I	Empty dissolver	450	120	Liquid solvent and H_2 gas	0.5 hr	8 Mscf/ton coal
SRC-II	Empty dissolver	/460/	130/	Liquid with unreacted solids and H_2 gas	1.0 hr	17 Mscf/ton coal (oxygen = 5.3 Mscf/ton coal)
EDS	Empty dissolver	450/450/450	135/	Unspecified H_2 gas stream	36 min	11.1 Mscf/ton coal
H-Coal, Fuel Oil	Ebullated bed of solid catalyst	355/455/	205	Liquid and gas	N/A	13.8 Mscf/ton coal
H-Coal, Syncrude		355/455/	205	Liquid and gas	N/A	18.5 Mscf/ton coal
Fischer–Tropsch	Fluid bed or fixed bed of catalyst	180/335/	28/	Gas internal recycle	Depends on gasifier	H_2 used for catalyst reduction only
M-Gasoline	Fluid bed or fixed bed of catalyst	N/A	N/A	None	Depends on gasifier	No separate H_2 production
CO_2 Acceptor	Fluidized bed High Btu Low Btu	830/1010/	8/10	Solid acceptor	N/A	None
Hygas	Fluidized bed four stages in one vessel	300/725	69/	Oil slurry	75 min	O_2 = 141 lb/ton coal (maximum)
Bigas	Two Stage Entrained Bed	900	83/		N/A	O_2 = 249 lb/ton coal
Synthane	Fluid bed	980	68/	Steam	N/A	O_2 = 254 lb/ton coal
Lurgi	Fixed bed	620/850	24/32	None	1 hr	O_2 = 315 lb/ton coal
Combustion Engineering	Entrained bed	920/1760/	1	Solid ash and carbon	2½ sec	None
Westinghouse	Fluid bed	700/1100/1500	18	None	N/A	(Syngas O_2 = 1148 lb/ton

Table 5-7. Process Summary

Process	Developer	Primary Product	Secondary Product	Feed Coal Type/Size	Year Begun	Confidence Index	Percent Efficiency Process/Product	Remarks
SRC-I	Southern Co. Services + EPRI + DOE + Gulf	Solid boiler fuel	Naphtha	All types	1962	B-3	71/70	
SRC-II	Gulf + DOE	Liquid boiler fuel	Gas LPG Naphtha	All types with ash restrictions	1976	B-4	70/70	
EDS	Exxon	Liquid boiler fuel	LPG Naphtha Gas	All acceptable	1966	C-3	66/64	
H-Coal, Fuel Oil	HRI	Fuel oil	Naphtha Gas			C-2	/74	
H-Coal Syncrude		Syncrude				C-2	/69	
Fischer–Tropsch	Standard technology used for over 50 years	Range of hydrocarbons	LPG Alcohols No. 2 Oil Fuel Oil Gas	All coal is gasified	Before 1930	A-2	/48	Depends on gasifier efficiency
M-Gasoline	Mobil for methanol to gasoline conversion	Premium gasoline	LPG	Any gasifier to methanol process may be used		C-3	/52	
Methanol		Methanol	—	Depends on gasifier	—	A-2	/57	

Table 5-7 (*continued*)

Process	Developer	Primary Product	Secondary Product	Feed Coal Type/Size	Year Begun	Confidence Index	Percent Efficiency Process/Product	Remarks
CO_2 Acceptor	Conoco Coal Development Co.	High-Btu gas Low-Btu gas	None	Lignite or Subbituminous /8 × 100 mesh	1968	B-2	68/67	Claimed to be the only fluidized-bed process that can handle lignites
Hygas	IGT	High-Btu gas	Naphtha	All with pretreat	1954	C-3	78/66	
Bigas	Bituminous Coal Research	High-Btu gas	None	All	1963	C-4	64/60	
Synthane	BuMines PERC Lummus	High-Btu gas	Char	All/20 to max 20% – 200	1961	C-3	65/63	
CE	Combustion Engineering	Low-Btu gas Electric power	None	All	1974	C-4	/40	
Westinghouse	Westinghouse	Electric power Medium-Btu gas	None None	All	1972	C-4	/38.4 /83	Efficiency shown with induction
Lurgi	Lurgi	High-Btu gas	Tar, oils, char, naphtha, coal fines	Non to low caking	Before 1930	A-2	72/59	Based on 1975 FPC application by ANG—analyzed by Stearns–Roger

the options has the best prospect and should be pursued and optimized. The national needs are for processes that (1) provide affordable, abundant fuels; (2) minimize waste of existing resources; and (3) accomplish the conversion with acceptable environmental impact.

The required additional processing to meet existing environmental standards is included in all processes and costs shown. For a number of the cases, the primary reason for the processing of the coal is to produce a fuel which will be more environmentally acceptable and more economical than conventional coal burning with flue gas scrubbers.

It must be assumed at this stage that no improvements or cost savings will be achieved. It is expected that some improvements will be achieved, but these will be largely confined to internal systems and future costs of machinery due to experience and production economics. There are many who believe that increased demands for further environmental restriction will offset all other hopes for product cost decreases. It is hoped that some progress will be made in the assessment of values achieved and in the total effects on environment, economy, and living standards.

No single measure can pinpoint a best process for coal conversion. Some major users have flexibility in their operations and can utilize any of the energy forms listed in Table 5-9. In other cases the choice is limited to one type of fuel and a product cost from a site specific study is needed to choose the best process.

Efficiency

Process efficiency is shown in Table 5-7 if it was reported by the design firm for the particular source study used for this chapter.

The product efficiency was calculated from the product data used for this chapter. The reason for showing both values is that a significant difference between the two values allows the implication that there are potential process improvements, most likely waste heat recovery and possible conversion to electric power. It should be noted that if all light oil naphtha products were upgraded to gasoline, the product efficiency would decrease.

Process Development Status

The confidence index column in Table 5-7 lists a two-part alphanumeric identifier, which assigns a level for both process development and economic reliability. As shown in Table 5-8, there are four possible choices for each category.

The fact that few of the processes studied are rated higher than C-3 is a warning that numbers shown for product cost are not of high reliability. While this often automatically triggers a tendency to add a contingency or "safety factor," it can also hopefully indicate that at least some of the processes should have lower costs with more development work.

As suggested above, costs are very important criteria. The cost for product takes into account all significant money and time requirements including capital investment, financing charges, construction period, plant maintenance, and expected life.

Further description of the amounts and specifications for delivered products is given in Table 5-9.

Products

As previously discussed, all of the main products from coal conversion processes are replacements for existing commodities now in the marketplace. The immediate need is to produce products compatible with existing equipment to perform in conformance with all environmental constraints.

Table 5-8. Process Confidence Index

Process Development		Economic Reliability	
D	Exploratory stage—not beyond simple bench tests	4	Screening estimate, very approximate
C	Development stage—operated on small integrated scale only	3	Incomplete definition for estimates used
B	Precommercial—successful pilot plant operation	2	Firm basis for values developed
A	Complete—process demonstrated sufficiently to ensure commercial success	1	Values considered to be satisfactory for commercial venture

Table 5-9 gives a listing of all products produced for each of the processes studied. Any fuels consumed within the plant are not listed. Fuel amounts are shown as barrels (42 gallons/barrel) per operating day for liquids, tons per operating day for solids, and million standard cubic feet per operating day for gases. The last column gives value factors. The value factor f_i is the ratio of market value of the actual product produced to gasoline as the reference fuel.

Most of the processes included here are covered by several published reports showing much laboratory data and projected product qualities and quantities. Where the reported results were recent and appeared compatible with the purposes of this chapter, they were adjusted to the same basis and used. Source reports are identified. For those processes where either recent material balances were not published or if those available were not based on optimum design, the process developer was contacted. An "A" in the data source column of Table 5-9 indicates the information used for energy and material balances was received directly from the process developer. Even though a few of the processes may include an optimistic bias as to product yields, no basis for adjustment is available and thus the data were used as presented.

SOLID PRODUCTS

The general public will undoubtedly continue the aversion to the use of solid coal and solid derivatives for residential heating. This attitude was a factor in the 40-year shift to oil and gas. Labor savings and better system reliability are major reasons for the preference. It would require a significant cost incentive to encourage greater use of solvent-refined-coal solid or char for home heating. On the other hand, the utility and large industrial market is a realistic goal for solids. Far less cost incentive is required to win these markets.

LIQUID PRODUCTS

The major markets for liquid coal derivatives are boiler fuel and transportation fuels. The specifications for the products will necessarily make their use environmentally acceptable. The major national benefit will be reduced dependence on imported petroleum.

Liquids have advantages over other energy products in being both economical to transport and store. The liquids produced are close enough to petroleum products to be able to totally replace conventional crude oil by coal liquids. Some additional processing would be required to make coal-derived liquids completely interchangeable with all refined petroleum products.

Table 5-9. Products Summary
[Basis: Plant Feed Rate = 25,000 tons/day dry coal[a,b]]

Process	Data Source[c]	Product	Quantity B_i (Amount/day)	Value Factor (Gasoline f_i Reference)
SRC-I	B	SRC solid	10,880 tons	0.50
		Fuel oil[d]	7,500 bbl	0.56
SRC-II	A	LPG	5,500 bbl	0.86
		Naphtha	10,700 bbl	0.82
		Fuel oil	45,300 bbl	0.56
		Gas	23.1 MMscf[e]	1.00
EDS	B	Propane	3,270 bbl	1.08
		Butane	3,500 bbl	1.07
		Fuel Oil	19,900 bbl	0.82
		C_2-gas	28,700 bbl	0.56
			41.9 MMscf	1.00
H-Coal, Fuel Oil	A	Naphtha	18,200 bbl	0.82
		Fuel oil	42,200 bbl	0.56
		Gas	19.7 MMscf	1.00
H-Coal, Syncrude	A	Naphtha	31,900 bbl	0.82
		Fuel oil	24,300 bbl	0.56
		Gas	56.3 MMscf	1.00
Fischer – Tropsch	B	Gasoline	18,200 bbl	0.90
		LPG	18,800 bbl	0.86
		No. 2 oil	1,200 bbl	0.82
		Fuel oil	2,000 bbl	0.56
		Medium-Btu gas	90.7 MMscf	1.00
		C_2-gas	37.2 MMscf	1.00
M-Gasoline	B	Premium gasoline	52,700 bbl	1.00
		LPG	7,300 bbl	0.86
Methanol	B	Methyl fuel	113,400 bbl	0.96
		Methanol	8,400 bbl	1.00
Hygas	A	SNG	333.8 MMscf	1.00
		Naphtha	6,800 bbl	0.82
Synthane	B	SNG	350.0 MMscf	1.00
		Char	1,344 tons	0.47
CO_2 Acceptor	B	SNG	397.9 MMscf	1.00
Lurgi	B	SNG	326.4 MMscf	1.00
		Tar oil	4,500 bbl	0.85
		Naphtha	2,200 bbl	0.82
Bigas	B	SNG	388.2 MMscf	1.00
CO_2 Acceptor	B	Syngas	1,143 MMscf	1.00
Westinghouse	A	Syngas	1,272 MMscf	1.00
Westinghouse	A	Electric power	2,625 MW	2.60
Combustion Engineering	A	Electric power	2,838 MW	2.60

[a] Source data have been corrected to zero electric power requirement using 10,000 Btu/kWh for on-site generation.
[b] Products have been adjusted for source coal HHV to a common basis of 11,200 Btu/lb-dry.
[c] Data sources: A—direct data from process developer; B—obtained from published report cited.
[d] Fuel oil means low-sulfur boiler fuel comparable to No. 6 oil.
[e] Gases shown as MMscf = million standard cubic feet (20°C, 1 atm).

GASEOUS PRODUCTS

The final properties of the product from gasification processes can be varied depending on the use to be made of the gas. The heating value of gas destined to be used as industrial boiler or process heating fuel can be as low as 100 Btu/scf, while high-Btu gas for general utility service must have a heating value over 900 Btu/scf. Similarly, the specifications governing contaminants and diluents, including sulfur compounds and nitrogen, are more flexible for industrial fuel gases. Nevertheless, good practice requires that the gasification process produce a gas that will not excessively corrode the transportation and utilization equipment and not create unacceptable stack emissions when burned.

While no specific product quality standards exist for comparing low-Btu and intermediate-Btu processes, the gas produced must be coordinated with the proposed use. Thus low-Btu gas that would be used directly in combustion processes must either be of sufficiently high heating value to independently sustain combustion or it must be delivered to the combustor at an elevated temperature. A gasifier system used with a gas turbine must deliver a product free of particulate matter, which might damage turbine blades. Synthesis gas used as a chemical feedstock would normally contain a minimum of diluents such as methane or nitrogen in order to maximize the reactive gas (CO and H_2O) content. Chemical feedstock gas would also normally be scrubbed to remove all sulfur components to avoid catalyst poisoning.

The quality specifications for high-Btu gas have been much more rigidly defined. The primary standard is that the gas must be fully interchangeable with pipeline-quality natural gas. Heating values as low as 900 Btu/scf may be permissible if other quality standards are met and the gas is mixed with natural gas. (Some states set minimum quality standards.) In addition to heating value, other combustion characteristics must match those of natural gas. These properties—lifting, flashback and yellow tipping—are affected by the chemical constituents but, so long as methane constitutes most of the combustible gas content, these requirements are easily met. Sulfur compounds and carbon monoxide are the contaminants most often found in high-Btu gas. Carbon monoxide concentration is limited to 0.1 vol % while total sulfur content may not exceed 10 grains/100 scf. Hydrogen sulfide is objectionable even in low concentrations; the allowable level is 0.25 grains/100 scf. Water is objectionable in that it may cause corrosion in pipeline systems; normal moisture content acceptable for pipelines is 7.0 lb/10^6 scf.

BY-PRODUCTS

By-products are produced in significant quantities by conversion plants processing 25,000 tons/day of coal. Except for a few special cases, most of the processes export pure sulfur and ammonia as convenient forms to dispose of the sulfur and nitrogen from the raw coal. These are salable products with ammonia in particular having a place in the fertilizer market.

Some capital investment and expense will be involved in marketing and shipping these materials. At the least, the net disposal of by-products should not add to the cost of plant fuel products. At most, the amount of by-products is small, and even if high market prices are obtained by current standards, the effect will not reduce major product price from the plant by as much as 5%.

POWER

Two of the processes studied, CE and Westinghouse, are specifically designed to produce electrical power as the final product. Electricity is a premium form of energy.

For coal conversion process designs that require electric power import, adjustment of products and capital are made to allow for on-site generation of all required power via coal burning to generate steam for driving turbine–generators. Most of the processes were essentially power independent owing to effective waste heat recovery.

Market

The goal of federal efforts is commercialization of all appropriate coal conversion processes. Some of the impediments to market acceptance have already been touched upon. Additional particular considerations follow.

COMMERCIAL COMPATIBILITY

Ideally, for retrofit applications, the fuel products derived from coal should be completely interchangeable and miscible with the fuels they are replacing. In addition, for all fuels, sulfur and other regulated emissions should be within legal limits without use of flue gas scrubbing. At this time, these requirements can be met by the coal-derived products. A remaining question involves liquid miscibility because certain of the heavy coal liquids produce small amounts of sediment when mixed in certain proportions with petroleum products. This is not a major impediment, and several solutions are available, which include segregating fuels and avoiding critical proportions when blending.

Gas fuels from coal would find ready applications if their cost was truly competitive with fuel now being used. All processes described are considered to be for base load applications.

SOCIAL COMPATIBILITY

One of the primary, if not controlling, goals of the nation's technical and economic system is to provide each citizen with the highest possible standard of living that can be achieved in an environmentally acceptable manner. The development and commercialization of coal conversion technology can assist in meeting that goal. However, the production of significant quantities of products would require the creation of a totally new industry. Demographic shifts, triggered by the need to develop coal resources and build and operate conversion plants in hitherto sparsely populated regions, will create the need for new social, governmental, and economic infrastructures.

The social problems arising from commercial development of coal conversion technology are not greatly different among any of the proposed technologies. The problems of coal mine development may be more obvious in the western coal fields where little industrial development has occurred. But expansion of mining operations in the historic regions of Appalachia and the Midwest will also precipitate social displacements and will require careful planning. Construction of vast coal conversion plants will necessitate temporary employment of large numbers of construction workers, sometimes at remote sites with inadequate educational facilities and few social amenities. Local governments may find it difficult to provide social services for a transitory activity when the tax revenues may not be available until plant construction is completed and operations started.

After mine and plant operations have commenced, many of the social problems will be greatly reduced or eliminated. Assuming conventional taxation for coal mining and conversion facilities, the revenues will help fund the necessary programs and facilities needed to maintain a socially acceptable working and living environment. Careful attention to environmental control systems will ensure minimum environmental intrusion by the facilities.

There are positive social benefits as well. Payrolls, improved rail and highway transportation systems, upgraded educational facilities, and broadened cultural opportunities are some of the more evident and potentially realizable social benefits.

UTILITY APPLICATIONS

The gas and electric utilities in the United States have shown an interest in commercializing coal conversion technology. Dwindling gas resources have led gas companies to seek supplemental gas supplies from foreign suppliers and to develop marginal natural gas resources that previously were considered uneconomic. Electric utilities have had to find clean fuel replacements for fuel oil and natural gas. A keystone of national energy policy is increased use of coal. The larger utilities have the technological capability, the economic and reliability incentives together with the capacity to commercialize emerging coal conversion technology when conditions determine this to be their best choice.

The first commercial high-Btu gas plants are likely to be jointly funded by several gas companies. The gas would either be shared equitably among all of the participants or sold to an adjacent interstate pipeline. The national high-pressure pipeline network is sufficiently integrated to allow any pipeline company to receive gas either by direct delivery or by displacement. The high-Btu gas would be fully interchangeable with natural gas, so it could be injected into the pipeline system in any proportion. The mixture would be indistinguishable from conventional natural gas except for a slight, though tolerable, reduction in heating value in some instances. It would be transported and marketed through the same facilities and in the same manner as natural gas. A local distribution company could build and operate a high-Btu gas plant but only a few of the very largest utilities require the continuous output of an economically sized plant.

Low- and intermediate-Btu industrial fuel gas plants also offer gas utilities the opportunity to extend existing gas supplies. It is necessary to segregate those gases from normal systems. In the future, industrial customers, in a limited geographical region, may be able to purchase industrial fuel gas from their local utility to replace natural gas. The natural gas thus released would be available for other customers. Utilities can provide the capital, technical capability, and marketing knowledge needed to develop a single or multiple customer supply system. Geographical dispersion of customers is limited by transportation economics. A balance must be struck between the savings from large central plant operations and the cost of transporting gas. Some industrial customers may be able to justify an economically sized plant solely for their own needs, in which case the utility could provide technical and engineering services or could own and operate the plant for the industrial user.

Electric utilities see coal conversion technology as a possible means of utilizing high-sulfur coal. Solid, liquid, and gaseous fuels have been considered for electric utility applications. Gaseous fuels may be used in a gas-turbine/steam-turbine combined cycle unit with higher overall generating efficiency. Gaseous fuels have a limitation in electric utility applications—they are difficult to store. Therefore, the gasification system must have good load-following capability. Also good efficiencies are desirable at both part load and full load. Liquid fuels or solid fuels, on the other hand, can be stored so that the conversion plant can operate at full capacity for best economy, while the power generating unit follows the electric load demand.

Combined cycles offer the possibility of higher fuel efficiency, lower generating costs, and improved system performance. Because of the inherent high efficiency, combined cycle systems would be used for base load, insofar as possible. However, it is desirable that they too have good load-following and turn-down capability.

TIME REQUIREMENTS

For a pioneer plant of significant size, meaning a feed rate of at least 6000 tons/day of coal, the design and field construction time for the prototypes will require at least 3 years. This includes no allowances for unusual weather conditions or geographic or legal obstacles. The same process in a larger unit could easily require 5 years.

Time is an important dimension in project economics. Any real project financial schedule must include all time considerations for construction and start-up. Construction and start-up delays cause permanent increases in product cost.

CONSTRAINTS ON COMMERCIALIZATION

Commercialization constraints may be generally classified in five categories: technical, economic, financial, institutional, and social. Institutional constraints include limits imposed by federal, state, and local regulations. Social constraints include both demographic limits on availability of manpower and environmental requirements which limit available sites. There are strong interactions between many of the constraints; for example, the institutional constraints on product price may limit the availability of capital.

Technical constraints on commercialization of coal conversion relate primarily to the state of development. Only a very few of the conversion processes have been operated on a full commercial scale. Some may not perform well with all U.S. coals. Pilot plant and larger-scale tests of gasification and liquefaction systems have been integrated with auxiliary functions to demonstrate the technical feasibility and operability of several of the processes. But uncertainty still exists, and there are needs for performance assurance which can only be resolved in a true commercial-scale operation. All unresolved risks impede commercialization. Equipment builders are sometimes reluctant to provide performance guarantees. Potential customers are unwilling to enter into long-term contracts when both reliability of service and cost are uncertain. Numerous methods of risk sharing, such as loan guarantees, firm purchase prices, and tax incentives, have been delineated. Some method of reducing the risk and improving the return on unproven technology during the early stages of industry development is essential if the concepts now being developed are to be commercialized.

Economic impediments are related to both technical and market risk. The costs for coal-derived fuels will be significantly higher than conventional fuel today. Customers are uncertain about the future cost of energy. It seems clear that future costs will be higher and availability of conventional oil and gas less. The economic constraint on commercialization of coal-derived fuels is more complex than the simple belief that a coal conversion industry will suddenly become viable when the price for petroleum rises to a predetermined cost. The implication that such a situation exists, given in many cost studies of the past, has damaged the credibility of the technical community.

In *Barriers to Commercialization* presented at the Coal Dilemma II Symposium in February 1979, R.F. Hill made reference to the "receding breakeven point." The concept that the conceptual cost for coal-derived fuels is some undefined function of the cost of oil is well supported by a study of conventional fuel prices and estimated clean manufactured fuel prices over the past decade. The increases in estimated cost for coal-derived fuels are significantly greater than the corrections required for inflation. Any uncertainty that a project will "break even" serves as an absolute barrier even though such uncertainty may not be apparent to the public.

The most evident impediment to commercialization has been the difficulty in financing commercial projects. Utility financing has traditionally included a large proportion of debt funding which reduced the cost of capital and thereby reduced the cost of utility service. But in order to maintain reasonable debt/equity ratios, utilities must now raise billions of dollars of both debt and equity capital if they are to finance coal conversion facilities. A single facility could represent a majority of the capitalization for many companies. Thus they would need to form joint ventures or find other means for reducing their financial exposure.

Lending institutions require assurance, through assured cost recovery from ratepayers, loan guarantees, or other risk minimizing devices, that debt costs will be recovered. The alternative to some form of assurance to the bondholders is additional equity capital, or a higher interest rate for debt capital. Either alternative could significantly increase the project and product costs.

Institutional constraints are manifested primarily in the regulatory problems encountered. The multiplicity of government permits required has increased the problems of siting and planning coal conversion facilities.

Social and environmental implications must be carefully considered. The coal conversion industry will create massive new production centers which, if not carefully planned, could create social and environmental disturbances. There will be some areas that cannot tolerate even carefully planned developments. The screening process, which ensures problems will be avoided, is time consuming and, for the near term, has caused delays in commercialization. As planning techniques improve and guidelines for siting of energy facilities develop, many of the delay causes can be eliminated.

REGULATORY PROSPECTS

The coal conversion industry and the companies that participate in the construction and operation of coal conversion facilities will be faced with several types of government regulation at federal, state, and local levels. At the federal level, regulation will primarily be concerned with the siting, pricing, and safety of plants producing substitute gas which will be transported in the interstate pipeline network. State regulators will concern themselves with the siting and environmental impact of facilities which serve markets wholly within a state, while local jurisdictions will deal with siting and land use impacts of facilities. Although jurisdictional limits are not clearly drawn, electricity production and pricing regulation will fit into the historic pattern of FERC regulation of interstate sales and state regulation of intrastate operations.

SNG Regulation

Federal authority to regulate the transportation of SNG is based on the Natural Gas Act which requires the FERC to regulate facilities and rates for the transportation of natural gas and mixtures of natural gas and substitute gas (SNG) in interstate commerce. If the gas from a gasification process is sold wholly within the state where it is manufactured and is not mixed with interstate gas, the FERC would not be involved. This is likely to be the case for medium- and low-Btu gas. However, for high-Btu gas production, if it is mixed with natural gas for interstate sale, the FERC regulates the price of the gas and the rates charged for transportation of the gas to consumers. At present, in order to construct a high-Btu gas plant the builder must submit extensive technical, economic, environmental, and financial data to the FERC. Public hearings are held to allow all interested parties to submit information in support of, or in opposition to, the proposed plant. A major concern has been the length of time required for the hearing procedure. The principal issues which are evaluated by the FERC are

Technical feasibility of the process. Has it been commercially proven? What is the possibility of technological failure or obsolescence?

Economic feasibility. Will the plant design result in the minimum cost for safe, reliable, and environmentally acceptable operations?

Financial feasibility. Is the proposed financing plan sound? Will it result in the lowest reasonable cost to the purchasers of the gas?

Environmental feasibility. Will the plant have minimum impact on the environment? Are the proposed environmental safeguards adequate?

Impact on ratepayers. How will the construction and operating costs be shared by the ratepayers? Will the cost of the gas be "rolled in," that is, averaged with all other gas sources, or will the gas be sold under incremental rates to certain groups of customers?

Based on the voluminous documentation provided by both advocates and opponents of the plant, the FERC issues a decision which either denies the application or grants the company the right to sell SNG in interstate commerce under fixed contractual conditions. The initial sale price, escalation allowances, price adjustment for part load operation, and other price modifications are controlled by the initial authorization. The SNG may also be subject to proposed incremental pricing rules that would shift most of the higher costs to lower priority users.

Construction and pricing of high-Btu utility gas from wholly intrastate facilities are the responsibility of the state utility regulatory agency. Most states exercise only limited authority through the setting of rates for utility service. Preconstruction authorization is not required in many states, but the sale price for gas is normally scrutinized to ensure that the costs are reasonable.

Regulatory authority over intermediate- and low-Btu gas plants has not been defined clearly. If the product gas is sold to a number of customers solely by the producer or sold to a small number of customers in a limited area, utility-type regulation may not be applicable. But regardless of the utility regulatory stance, environmental regulations will still apply. A combined gas/electric plant poses a special situation. A portion of the product gas might be sold directly to users while the balance is used for power generation. Even though the direct sale of gas may not be regulated, the price of the gas to the electric utility would almost certainly be regulated by state authorities.

ENVIRONMENTAL IMPACT

Coal conversion facilities will create stresses on the environment. The industry will cause changes in the social and physical environment that must be recognized. Potential changes must be understood and actions must be taken to minimize the impact. The social effects have already been discussed in this chapter. The industrial and commercial infrastructure needed to support the new industry will itself spawn many environmental effects. The current developmental effort is structured to identify potential environmental problems and to develop effective means of making the processes environmentally acceptable. With this complete systems approach to solving the problems as they are created, ultimate acceptance of the industry should be accelerated.

The environmental problems of coal mining are not unique to the coal conversion industry. As the nation's use of coal for power generation and industrial applications grows, the concerns will increase. The rapid evolution of a gasification and/or liquefaction industry could increase the scope of the problems. Surface mining to utilize the large western coal deposits will require an irreversible commitment of the natural resources underlying extensive areas previously used only for limited agricultural purposes. Land reclamation is vital to avoid permanent loss of this natural resource. Mine drainage water and fugitive dust can create both nuisance and health hazard problems unless carefully controlled. Deep mining of eastern coals for conversion facilities may result in less visible effects, but it can result in land subsidence which would affect the value of surface land. Mine drainage and downstream contamination of water supplies is a problem unless careful water control practices are followed. Surface facilities such as cleaning plants and storage and loading areas could also create air and water pollution hazards unless carefully controlled. The technology for adequate water, air, and land protection exists, and strict enforcement laws will minimize the intrusion of mining operation for the coal conversion industry on the environment.

Within the conversion facilities, environmental problems are caused largely by the sulfur and ash content of the coal; the disposition of trace elements in the coal; and handling and disposition of liquid hydrocarbons and waste water streams. Coal storage and handling will create dust and drainage water disposition problems, but careful water management and drainage control should minimize external effects. Coal conversion processes usually operate at elevated pressures so gas streams are carefully contained to minimize product loss and ensure process integrity. The processing concentrates the sulfur compounds and makes recovery somewhat easier than from power plant stack gas streams. However, in conversion facilities where some or all of the coal is burned as fuel, stack effluents must be monitored and contamination reduced to within required limits.

Most of the processes that are being considered produce some liquid effluent streams. Liquid-handling processes require careful handling procedures to ensure that leaks or spills do not occur or are controlled without environmental damage. Gaseous fuel processes produce by-product streams that may contain phenols and other potentially harmful materials. Contaminated waste water streams must be treated before disposal. Coal contains a wide variety of trace elements including heavy metals. Studies have indicated these materials are discharged from the process either in the ash or foul water streams. Although indications are that the elements in ash are chemically inert and the possibility of leaching into the soil is limited, care must be taken in ash disposal. Wash water treatment can reduce trace element content of effluent streams to acceptable levels.

Coal conversion plants will create concerns about land use and aesthetics. Plants constructed in areas that previously had little or no industrial development will intrude upon scenic vistas, and the coal transportation network may detract from aesthetic values. These impacts can be minimized by careful selection of plant sites which utilize topographical screening. But it will not be possible to avoid all effects when these facilities are built. In more heavily industrialized areas the aesthetic impacts will be less but still evident.

Summary

The heterogeneous nature of coal and the lack of precedents for the sudden emergence of a new major coal conversion industry make any generalizations or predictions subjective and dependent on many qualifying assumptions. Since neither federal legislative nor administration policy has been firmly established yet, it is not surprising that implementation of solutions for the vaguely defined problems have not proceeded beyond a wide variety of uncoordinated discussions and fragmented proposals.

The selection of processes in this chapter is representative of the range of available technologies but is not exhaustive of all technologies. The exclusion of a particular technology is not necessarily indicative of its quality.

As a part of this chapter, data have been collected from many sources to allow computation of fuel costs on several different bases. These costs show some processes that have product price advantages. A variety of processes await commercial demonstration.

In using the information presented in this chapter, it must be kept in mind that data quality varies from the different sources. The confidence index reflects this to a limited degree with respect to the state of process development and quality of estimates. Added to these uncertainties are optimistic data from several of the developers of particular systems. This type of bias is not uncommon either within government or the private sector.

The technical review disclosed certain processes, such as Bigas, with more problems than others but none which are considered infeasible for solely technical reasons. All significant process development problems are ultimately reflected in plant costs. The combination of continuing operating difficulties and increased plant cost is sufficient to eliminate a process if better alternatives exist.

Among the gasification processes, the work done thus far may prove useful for specific projects to provide low- or medium-Btu gas to industrial systems. None of the high-Btu gas processes appears to have a clear advantage over the Lurgi plus methanation process, which is proven and undergoing continuing development.

All of the liquefaction processes included in this chapter have potential for some particular segments of the fuels market and could displace imported petroleum. No foreign process developer seems to have any attractive alternative to processes that were sponsored by DOE.

The solid SRC process may find application on a very limited scale supplying acceptable fuel to electric utilities. However, the threat of constantly increasing environmental regulation could affect prospects for this process.

The electric power systems described in this chapter could find application, particularly if they soon demonstrate high efficiencies and environmentally acceptable operation. Many combined cycle systems are under development and the two described in this chapter have no clear advantage over some others not included.

The need appears clear for a variety of coal-derived fuels to preclude near-future shortages. Economic incentives will continue to increase. Much of the quantitative information included in this chapter is preliminary and, as such, is reflected in the accuracy of the product cost estimates. As R&D efforts proceed, somewhat better results, meaning higher yields at lower costs, should be obtained.

Chapter 6

Oil Shale Processing

Introduction

Oil shale is neither oil nor shale; it is a fine-grained rock called marlstone that contains varying amounts of a brown to dark gray organic material called *kerogen*. When heated to about 900°F (480°C), the kerogen reacts to form synthetic oil and gas—two fossil energy sources which, in their conventional forms, are becoming increasingly scarce and expensive.

A vast deposit of oil shale exists in the Colorado–Wyoming–Utah area in a region known as the Green River Formation, as shown in Fig. 6-1. There, the estimate of recoverable oil is about 600–700 billion barrels, almost 40% of the 1.8 trillion barrels of oil equivalent in place.

These deposits contain 25 gallons or more of shale oil per ton of rock. Two tons of shale—about the volume of an office desk—would yield more than a barrel of oil.

Shale formations also exist in the East, in roughly a triangular region from Michigan to western Pennsylvania to Mississippi. This area, known as the Devonian and Mississippian deposits, is predominately of lower grade—less than 15 gallons per ton—and gasification is considered the most applicable technology. The potential yield, however, is significant. Some 3 trillion barrels of oil equivalent might be contained in these deposits, although uncertainties still exist as to the amount that is economically recoverable.

Next to coal, oil shale is the nation's largest fossil fuel resource.

BACKGROUND

Since the earliest commercial interest more than 100 years ago, the history of oil shale has been one of ups-and-downs. In almost cyclical fashion, the shale industry has appeared to be on the verge of expanding rapidly, economics have appeared potentially viable, and the problems have seemed minimal. But, then, a combination of factors, such as jumps in construction costs, an onset of drought conditions, the requirement for new environmental data, or the discovery of new conventional oil resources, have led to delays and, in some cases, to project cancellations by companies.

During World War II, a high demand for liquid fuels prompted Congress to pass the Synthetic Liquid Fuels Act of 1944. The Department of Interior's Bureau of Mines (BOM) then began development of a surface "retort," or heating vessel, in 1945 at the Anvil Points Facility, a research complex on the Naval Oil Shale Reserve near Rifle, Colorado.

Laboratory and pilot scale tests were performed at the facility until 1958, when it was shut down. A six-company consortium leased it from the BOM, modernized the retort, and conducted further experiments starting in 1964.

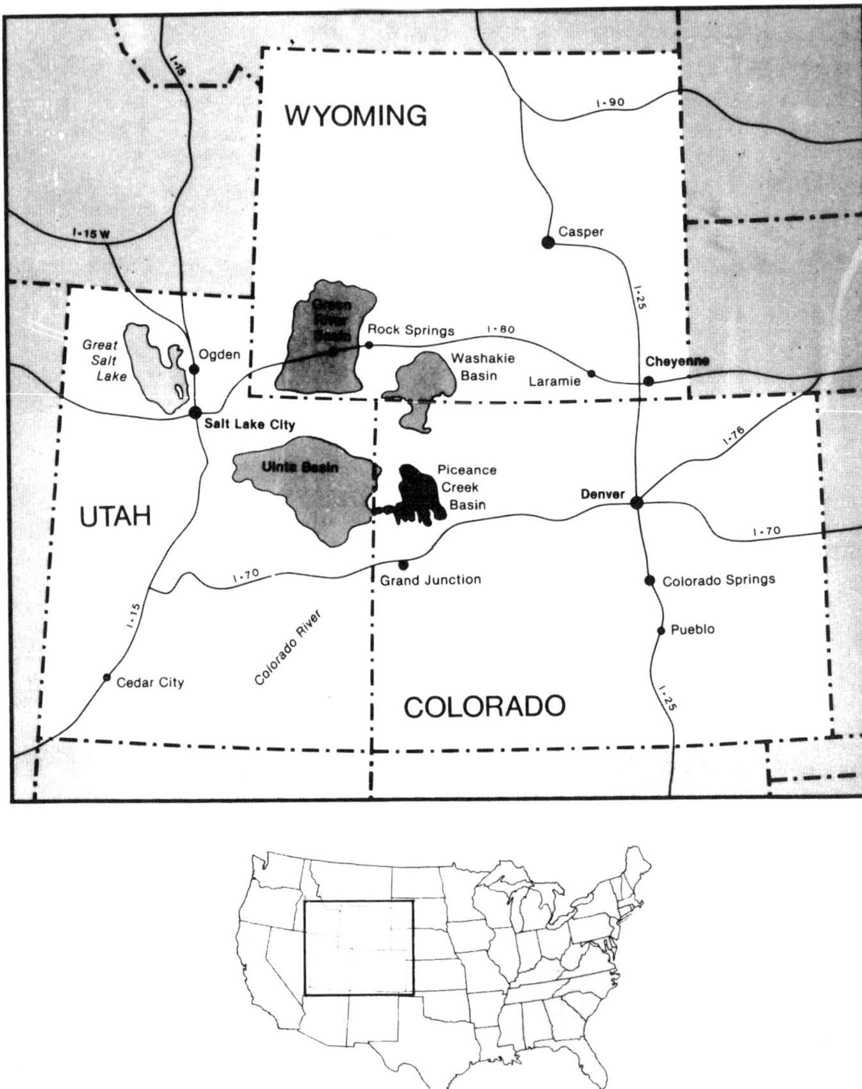

Fig. 6-1. Western oil shale region.

Anvil Points was transferred in 1975 from the BOM to the newly organized Energy Research and Development Administration (ERDA), which was subsequently merged into the new Department of Energy (DOE) on October 1, 1977.

DOE's surface retorting activity at Anvil Points consisted of observing or monitoring the operation and conducting instrumented measurements and environmental studies. The facility was leased by Development Engineering, Inc., a subsidiary of the Paraho Development Corporation, to develop a surface oil shale retorting process. Subsequently, the Department of Defense (DOD) contracted with the firm to produce a total of

100,000 barrels of shale oil, including that previously produced, to be used in refining research.

Much of the technology for surface retorting of shale can trace its heritage to the early experiments of researchers and engineers at Anvil Points. And once again, industry appears at the threshold of a commercial shale oil industry.

OIL SHALE MINING

To bring the raw oil shale to the surface for processing, either underground or surface mining can be used, depending on the extent and richness of the formation.

A room-and-pillar method is believed to be the most practical underground mining system. In this technique, interconnecting tunnels are mined, leaving a checkerboard pattern of pillars. As a second step, once the initial production is completed, some of the remaining pillars are mined. Initial underground mining of oil shale will probably be confined to the 30–90 ft thickness of 30 gal/ton and richer oil shale in the Mahogany zone of the Green River Formation.

Surface mining techniques such as open pit and strip mining may also be useful in areas containing significant oil shale deposits. Generally, in mining oil shale by surface techniques, both overburden and oil shale will require drilling and blasting and use of large "jumbo" rotary drills. Large shovels, draglines, and trucks will be required to move the blasted material to the disposed area or to crushers from which ores or wastes would be conveyed to either a dump or a processing area.

THE NEAR TERM

Although several companies have conducted pilot scale tests of different technologies for producing shale oil, there remains uncertainties concerning the costs of commercial production and the price of alternative petroleum supplies.

These uncertainties, combined with strict environmental standards, have discouraged companies from making the large-scale $1 billion-plus investments which will be necessary to achieve commercial production.

Production levels of oil shale could reach as much as 400,000 barrels per day by 1990. The majority of the initial production would likely come from surface retorting.

SURFACE RETORTING

Currently, the most practical method of recovering shale oil is to mine the shale, crush it, and then retort it on the surface. Two general methods of converting kerogen from its natural solid form to oil and gas are:

Direct Heating—Either air or oxygen is introduced into the retorting vessel containing the broken shale. Some of the potential oil and gas and some of the residue on the shale are burned to supply necessary heat.

Indirect Heating—A heat-carrying medium such as recycled gas, inert gas, or a reactant gas is passed through the bed, or hot solids are mixed with the shale.

Aboveground retorting, while potentially of significant benefit, particularly in the near term, does have limitations. Foremost is the expended shale which must be disposed of. A 100,000 bbl/day plant would require the daily disposal of 150,000 tons of shale rock. Since the rock has been crushed and sized for the retorting operation, it has a greater volume than the unmined raw shale. Therefore, only part of it can be returned to the mined-out areas. Not only must the material be moved and stored, but measures must be taken to ensure that minerals contained in the disposed materials cannot seep into groundwater or nearby streams.

In addition to space, the entire surface operation consumes large amounts of water—1–3 barrels per barrel of oil produced.

IN SITU PROCESSING

The drawbacks to surface retorting can be reduced by processing the shale underground, or so-called *in situ*. *In situ* processing can also be applied to lower grade shale, which is particularly important since two-thirds of the total estimated resource of 1.8 trillion barrels of shale oil in the West are in low-grade deposits that may never be recoverable by techniques that involve conventional mining.

Although *in situ* conversion offers several potential advantages, no single version is applicable to the full range of oil shale deposit types. Consequently, DOE's research and development program had been focused on several variations of *in situ* retorting. In general, the variations can be grouped under two broad categories: (1) modified *in situ*, where a void is created by mining or leaching; and (2) true *in situ*, where the zone is rubblized and retorted without creating a man-made void.

Vertical modified *in situ* techniques are believed to offer the most promise for early development and are especially attractive for deeper and very thick (more than 300 ft) deposits. In October 1977, DOE signed a contract for a multiphase program with Occidental Petroleum Corporation to develop a vertical modified *in situ* process at Occidental's Logan Wash site in Garfield County, Colorado. In Occidental's process, shown in Fig. 6-2, about 20% of the shale is mined to form an underground compartment. Then surrounding shale is broken up and expanded by explosives to create a kind of underground chimney full of rubblized shale. From the top of the chimney, a small amount of oil is ignited and the retorting begins. Combustion of the chimney moves slowly downward—typically 1–2 ft/day—and the heavy oil produced flows by gravity to the bottom where it is collected and pumped to the surface.

A horizontal true *in situ* technique being developed by DOE and the Geokinetics Oil Shale Group is currently applicable only to very shallow depth formations (less than 100 ft of overburden). Conventional mining is not used. Instead explosives expand and break up a bed of shale lifting the surface of the ground noticeably. Geokinetics, with DOE support, is conducting laboratory investigations and field development and demonstration tests through a series of progressively larger retorts in Uinta County, Utah.

A vertical true *in situ* process is also under development by DOE for processing the leached zone in deep, thick shale beds in Colorado. The process uses the void space created when the water-soluble states originally distributed throughout the 5000-ft-thick seam were dissolved by naturally occurring groundwater. (These leached zones can be enlarged by injecting a dissolving agent in a process called solution mining; if this is done, the process becomes a modified *in situ* process.) Because the zone is a saltwater aquifer, direct combustion methods are not applicable. The concept being tested under a contract with Equity Oil Co. involves passing superheated steam through the full thickness of the leached zone to achieve the required retorting temperatures. Tests of this process have been conducted in Rio Bianco County, Colorado.

Some true *in situ* techniques rely on an interconnecting system of cracks artificially produced by hydraulic and/or explosive fracturing to provide the required permeability. No mining is required. While true *in situ* techniques offer lower environmental impacts than most modified *in situ* methods, this will probably be the most difficult *in situ*

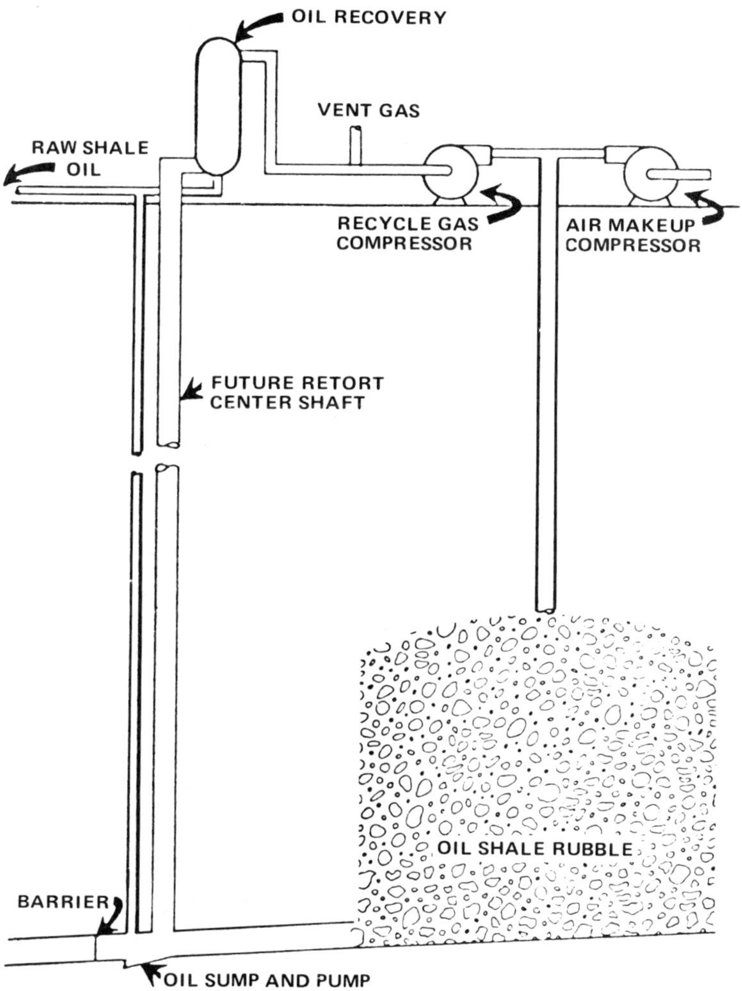

Fig. 6-2. Occidental oil shale process retort.

technology to develop successfully. In August 1977, DOE signed a contract with Talley Industries to develop a true *in situ* process near Rock Springs, Wyoming. Following initial fracturing, however, the shale bed was found not to have suitable characteristics for a good retorting test, and the Talley project was not continued into fiscal 1980.

DOE is also conducting ongoing research on advanced concepts including microwave and radiofrequency processing and vacuum retorting techniques which are today emerging from the laboratory.

EASTERN SHALES

DOE has also contracted with Dow Chemical Company, Midland, Michigan, to study underground gasification of the Antrium oil shale in Michigan. For the past 20 years, Dow has conducted an experimental program with this type of shale. Supported by a Michigan oil shale development consortium, the firm has subjected samples of Antrium shale to careful analysis and determined there was sufficient promise to field test various methods of extracting the shale's energy content.

Under the 4-year program with DOE which began in 1976, Dow is assessing four methods for the chemical explosive and hydraulic fracturing of Antrium shale in place. The retorting will take place underground with a minimum of disturbance to the environment.

DOE has also begun a program to test a process for adding hydrogen to lower quality shale during the retorting process. Using this technique, useful fuels may be obtained from shale formations previously considered unlikely candidates for development due to low oil yield under normal conditions. This technique could be particularly beneficial in the East.

At the end of this chapter, detailed descriptions are provided for each major oil shale project. Information for each project includes plant type, management and finance, major permit actions/decisions, product markets, environmental impacts, socioeconomic impacts, project history and status, project/process description, and project schedule.

Oil Shale Technology

Converting shale in the ground to finished fuels and other products for consumer markets involves a series of processing steps whose number and nature are determined by the desired mix of products and by-products and by the generic approach that is followed in developing the resource. The alternative approaches are:

- True *In Situ*: Processing in which the shale is left underground and is heated by injecting hot fluids.
- Modified *In Situ*: Processes in which a portion of the shale deposit is mined out, and the rest is broken with explosives to create a highly permeable zone through which hot fluids can be circulated.
- Aboveground Retort: Processes in which the shale is mined, crushed, and heated in vessels near the minesite.

Figure 6-3 is a flow sheet for the steps common to all three options. How the steps would be integrated in an aboveground retort is shown in Fig. 6-4. In the first step, the oil shale is mined and crushed for aboveground processing, or the deposit is fractured and rubblized for *in situ* processing. The main product is raw oil shale with a particle size appropriate for rapid heat transfer. Nahcolite ore can be one of the various by-products from this step; dust and contaminated water are among its wastes. In the retorting step, the raw oil shale is heated to pyrolysis temperatures about 1000°F (535°C) to obtain crude shale oil. Other products are the spent shale residue, pyrolysis gases, carbon dioxide, contaminated water, and, in some cases, additional nahcolite and dawsonite ore. The crude shale oil may be sent to an upgrading section in which it is physically and chemically modified to improve its transportation properties, to remove nitrogen and sulfur, and to increase its hydrogen content. (Crude shale oils from some *in situ* processes may not need upgrading before transportation.) Contaminated air and water, and in some units refinery coke, are the wastes produced along with gases that contain sulfur and nitrogen compounds. Depending on the extent of the treatment, the upgraded product—shale oil syncrude—can be a high-quality refinery feedstock, comparable with the best

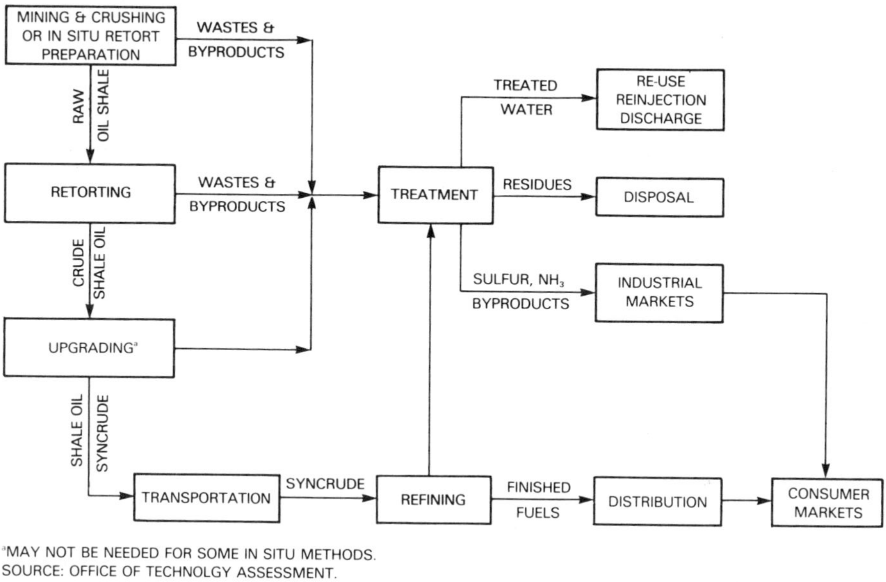

Source: Office of Technology Assessment.

Fig. 6-3. Oil shale utilization.

grades of conventional crude. In the refining step, which may be conducted either onsite or offsite, hydrogen is added to convert the syncrude to finished fuels such as gasoline, diesel fuel, and jet fuel. The syncrude, or the crude shale oil, could also be used directly as boiler fuel. After refining, the fuels are distributed to consumer markets. Refining also produces waste gases and various contaminated condensates.

To protect the environment, contaminated water must be treated for reuse in the oil shale facility, for reinjection into the groundwater aquifer source, or for discharge into surface streams. Contaminated air and process gases must be purified to meet federal and state air pollution standards before they can be discharged to the atmosphere. The waste gases from retorting, upgrading, and refining are potential sources of ammonia (for fertilizer and other uses) and sulfur (for sulfuric acid and many other materials). These can be recovered during the treatment steps and sold to industrial processors. In some portions of the Green River formation, the solid residues from mining and retorting will contain nahcolite (which can be used to produce soda ash and for stack gas scrubbing) and dawsonite (a source of aluminum metal). In any case, spent shale from aboveground processing must be moistened, compacted, and revegetated to prevent erosion and leaching. Retorted shale in *in situ* retorts, and spent shale from surface operations that is backfilled into underground openings, must be protected from leaching by groundwater.

This section deals with the processing technologies used in mining, retorting, upgrading, and refining oil shale.

OIL SHALE MINING

Both modified *in situ* and aboveground retort require mining. In the case of aboveground retort, the mined shale is conveyed to retorts where it is processed to recover shale

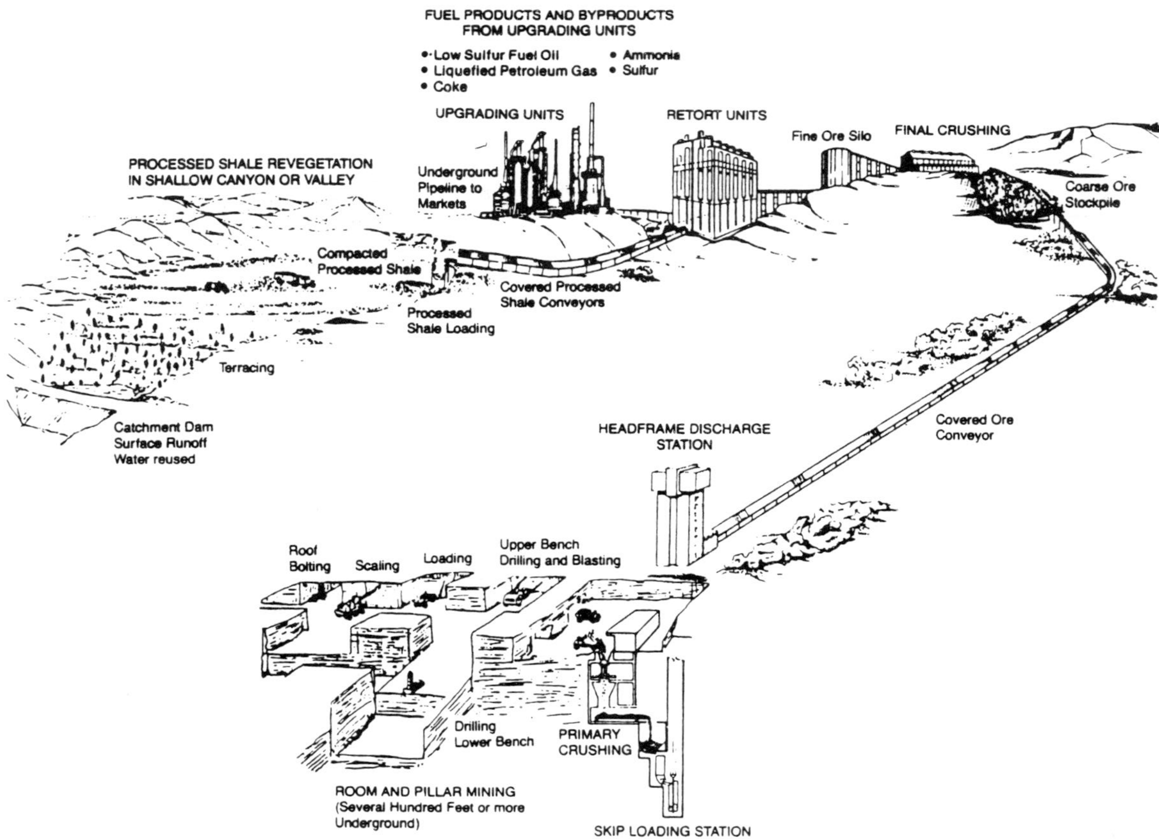

Source: C-b Shale Oil Project, *Detailed Development Plan and Related Materials*, Ashland Oil, Inc., and Shell Oil Co., February 1976, p. IV–11.

Fig. 6-4. The components of an underground mining and aboveground retorting oil shale complex.

oil. With modified *in situ*, the shale may also be retorted aboveground, or it may be discarded on the surface as a solid waste.

Green River oil shale deposits are characterized by their extreme thickness and by their extensiveness. The richer shale zones in the Piceance basin, for example, are more than 1000 ft thick, in places, and are continuous over an area of 1200 square miles. Deposits of this nature could be amenable to either surface mining (strip or open pit) or to underground mining methods (such as room and pillar), depending on topographical features, accessibility, overburden thickness, presence of groundwater in the mining zone, and many other factors. Surface mining may be feasible for very thick oil shale zones that are not deeply buried, especially if their average oil yield is not high. Because of the thickness of the overburden, only a limited area of the Piceance basin and somewhat more of the Uinta and Wyoming basins are amenable to surface mining. In other areas, streams have eroded gulleys and canyons through the shale beds, exposing some of the richer shale zones. Shale that outcrops in these areas, plus the shale in all deeply buried beds, will probably be extracted by underground mining.

Despite the high price of crude oil, shale is a lean ore compared with many ores that contain valuable metals. Economies of scale encourage massive mining installations, regardless of the mining method selected. A prospective developer once characterized commercial-scale oil shale mines as "prodigious," because it connotated a larger size than "giant." A sense of this can be conveyed by comparing their capacities with those of more conventional mines. At present, the largest surface mine in the United States is Kennecott Copper's Bingham Canyon pit in Utah, which produces about 110,000 ton/day of copper ore. The largest underground mine is the San Manuel copper mine in Arizona, which yields about 50,000 ton/day of ore. About 70,000 ton/day of 30 gal/ton oil shale would have to be mined to support a single 50,000-bbl/day plant that used aboveground retorts.[1] This mine would be substantially larger than the San Manuel mine. A 400,000-bbl/day industry of abovegound retorts would have to mine about 560,000 ton/day—the equivalent of 5 Bingham Canyon pits or 11 San Manuel mines. If the same industry used only modified *in situ*, about 230,000–460,000 ton/day would have to be mined—the equivalent of four to seven San Manuel mines.[2] Some of the mining techniques that could be used to achieve these levels of production are described below.

Surface Mining

The two principal types of surface mining—open pit and strip—both have been used widely to develop coal seams and deposits of many other minerals, as noted in Chapter 2. Their technical aspects are fairly well understood for these minerals. However, their feasibilities and effects vary with the nature of the ore body. Neither technique has yet been applied to the oil shales of the Green River formation.

Surface mining is economically attractive for large, low-grade ore deposits because it permits high recovery of the resource and allows sufficient space for very large and efficient mining equipment. An open pit mine could recover almost 90% of the oil shale in a very thick deposit. Strip mining could provide even higher recoveries. In contrast, underground mining would recover less than 60%. One of the reasons that industry's bids

[1] This assumes recovery of 100% of Fischer assay oil, a recovery efficiency that has been achieved in tests of the TOSCO II technology.

[2] This assumes mining of 20–40% of the shale in the retort volume, and an oil recovery of 60% of Fischer assay.

on a lease for federal tract C-a were so high during a recent auction was that the deposit could be mined by open pit methods. About five times as much shale could be recovered by open pit, which could have mined the entire deposit, than by underground room-and-pillar mining, which would have been limited to a relatively thin zone.

A mature open pit mine is shown in Fig. 6-5 and the steps in its operation in Fig. 6-6. In the first step, overburden is drilled and blasted loose over a large area above the oil shale zone. The burden is carried by trucks or conveyors to an offsite disposal area. When enough burden is removed to expose the shale beds, the shale itself is drilled and blasted, and is hauled from the pit for processing in aboveground retorts. As mining proceeds, a huge hole is formed, extending from the top of the overburden to deep into the oil shale deposits. The walls of the pit are under pressure from the overburden, and must be angled outwards to transmit the pressure without collapsing.

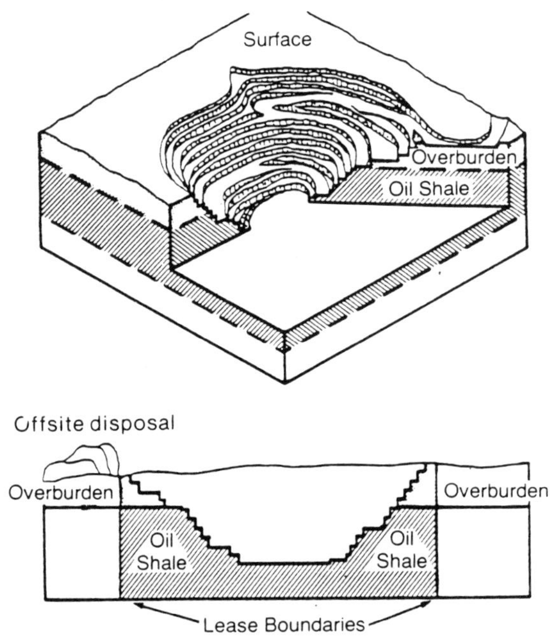

Source: *Hearings on Oil Shale Leasing*, Subcommittee on Minerals, Materials, and Fuels of the Senate Committee on Interior and Insular Affairs, 94th Cong. 2d sess., Mar. 17, 1976, p. 84.

Fig. 6-5. An open pit oil shale mine.

Open pit mining was originally proposed for tract C-a by the Rio Blanco Oil Shale Co., the lessee. The pit was to have started in the northwest corner of the tract, and was eventually to have covered its entire surface. After a few decades, freshly removed overburden and spent shale from the aboveground retorts would have been returned to the pit as backfill. In the interim, the solid wastes would have been disposed of on a

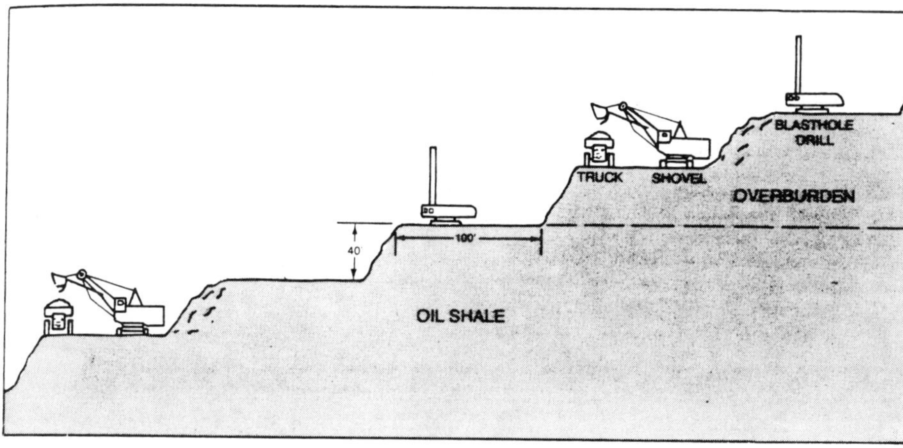

Source: *U.S. Energy Outlook—An Interim Report,* The National Petroleum Council, Washington, D.C., 1972.

Fig. 6-6. Open pit oil shale mining.

highland to the northeast of the tract boundaries. This concept was abandoned when the Department of the Interior (DOI) refused to allow offtract waste disposal. Rio Blanco later switched to modified *in situ* techniques because the alternative—underground mining—would have reduced resource recovery and threatened profitable operations. At present, there are no plans for any open pit mines, although Rio Blanco may reconsider its original plan if offsite disposal were allowed.

In strip mining, overburden is removed with a dragline—a massive type of scraper shovel. When the dragline has filled its scoop, it pivots and dumps the burden into an adjacent mined-out area. One difference between open pit and strip mines is that in strip mining, the burden is simply cast into a nearby area; in open pit, it must be moved far from the minesite to prevent interfering with the development of the pit. Strip mining has not been proposed for any of the Green River deposits.

Surface mining of most oil shade deposits is made difficult by the great thickness of the overburden that covers them. In the center of the Piceance basin, for example, the 2000-ft-thick oil shale zones are buried under about 1000 ft of inert rock and very lean oil shale. This does not necessarily preclude surface mining, because the deposits are generally characterized by a favorable stripping ratio—the ratio of overburden thickness to ore-body thickness. The thick beds in the center of the Piceance basin have 1 ft of overburden for every 2 ft of oil shale—a stripping ratio of 1:2. With coal, a stripping ratio of 10:1 is often economically acceptable. A study by the National Petroleum Council indicated that open pit mining would be favored for stripping ratios between 2:1 and 5:1, strip mining for smaller ratios, and underground mining for ratios larger than 5:1.

However, the economic principles of coal mining should be applied with caution to oil shale. Removing 1000 ft of overburden to recover 2000 ft of shale might be possible in theory, but the pit's boundaries would be so extensive that it would have to be located on a very large tract. Furthermore, the large "front-end" investment in removing overburden many years in advance of retorting would probably make open pit mining of deep deposits uneconomical. Also, strip mining would not be feasible in many parts of the oil shale region, even those with favorable stripping ratios, because the dragline would have to reach to the bottom of a 3000-ft-thick layer of overburden and oil shale. It is not clear that such a machine could be built.

Underground Mining

Many underground mining procedures have been proposed for oil shale deposits, but to date only room-and-pillar mining and mining in support of modified *in situ* retorting have been tested at any significant scale. In room-and-pillar mining, some shale is removed to form large rooms and some is left in place, as pillars, to support the mine roof. The relative sizes of rooms and pillars are determined by the physical properties of the shale, by the thickness of the overburden, and by the height of the mine roof. Most of the deposits of commercial interest are very thick and have relatively few natural faults and fissures. The ore itself resists compression and vertical shear stresses. These properties allow the use of large rooms, and relatively little shale needs to be left as unrecoverable pillars.

The BOM studied underground mining of oil shale at the Anvil Points Experimental Station in the late 1940s and early 1950s. The primary purpose of the mining program was to supply raw shale for the Bureau's retorting experiments, but the program was also designed to develop a safe, low-cost mining method. The room-and-pillar technique was adopted after extensive testing. From these studies, which included enlarging the rooms until the roof failed, the investigators concluded that for safe mining conditions the rooms should be 60 ft wide with pillars that are 60 ft on a side.

Many of the modern mine designs have been patterned after the BOM experimental mine. The design depicted in Fig. 6-7 is the mining plan proposed for Colony Development's 46,000-bbl/day facility in the Piceance basin. The rooms are 60 ft wide; the pillars are 6 ft square; and the mine roof is 60 ft high. Mining is conducted in two 30-ft-high benches. The upper bench is mined first by drilling blastholes into the walls of a production room, and breaking the shale loose with explosives. The broken shale is carried by trucks to the crushers where it is crushed to the size range required by the TOSCO II aboveground retorts. The walls and the roof of the new room are then "scaled" to remove shale that was loosened by the blasting but did not fall. Holes are drilled into the roof, and roof bolts are inserted to ensure its integrity, thus protecting the miners from roof falls. The BOM studies indicated that roof bolts would have to be installed in the access passageways but not in areas that were actively being mined. These production rooms would be vacated long before there was any serious danger of roof falls.

The lower bench is mined next. The sequence is similar except the blastholes are drilled into the floor of the upper bench, and additional roof bolting is not needed. The cycle of drilling, blasting, loading, scaling, and roof bolting was designed to produce about 66,000 ton/day of shale. About 60% of the shale in the mining zone was to be removed for processing in the aboveground retorts. The rest was to remain in the support pillars. Colony estimated that enough shale is present in the 60 ft seam to support the retorting facility for at least 20 years.

The same type of mine was proposed by the Colony partners for development of tract C-b, after considering longwall mining, long-hole blasting, continuous mining, block

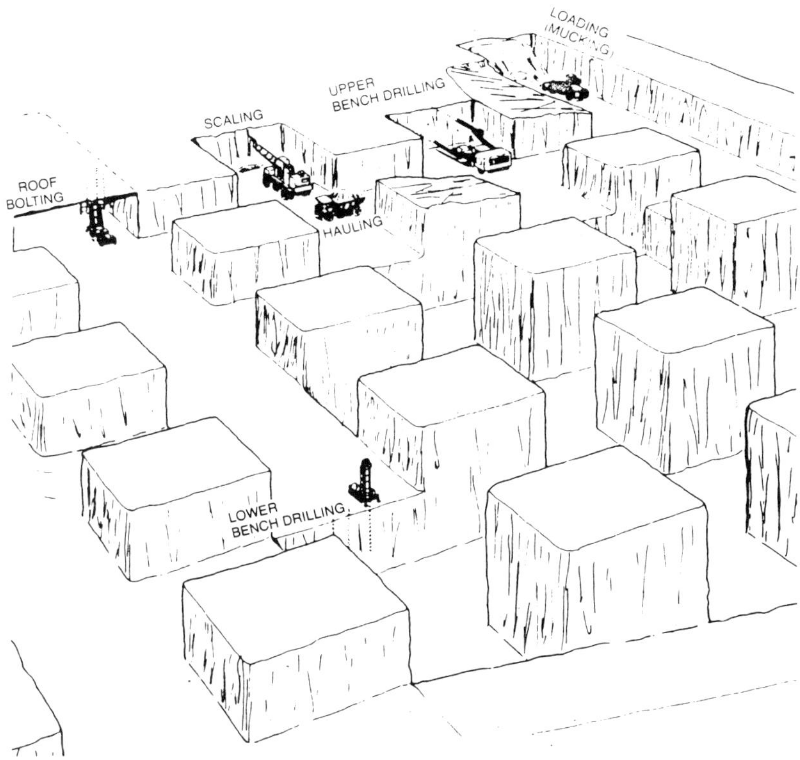

Source: R.W. Marshall, "Colony Development Operation Room-and-Pillar Oil Shale Mining." ***Quarterly of the Colorado School of Mines*****, 69, No. 2, April 1974, pp. 171–184.**

Fig. 6-7. The Colony room-and-pillar mining concept.

caving, open pit mining, *in situ* processing, and many other options. The major advantages of the room-and-pillar method were identified as the following:

- it is highly flexible and can easily be modified to accommodate new conditions, new equipment, or technological advances;
- it can be highly mechanized, and high overall production rates can be achieved because many areas can be mined simultaneously;
- the mine openings are relatively easy to ventilate;
- development of access passageways is also a production operation because oil shale would be removed;
- the mine could be design to minimize surface subsidence.

The disadvantages were the high cost of the roof support system and the relatively low recovery. In this regard, it was estimated that only 30–50% of the shale in a

74-ft-thick interval could be recovered. Higher recoveries would be possible if the pillars were subsequently mined out or if the mining were conducted on multiple levels. (See Fig. 6-8.)

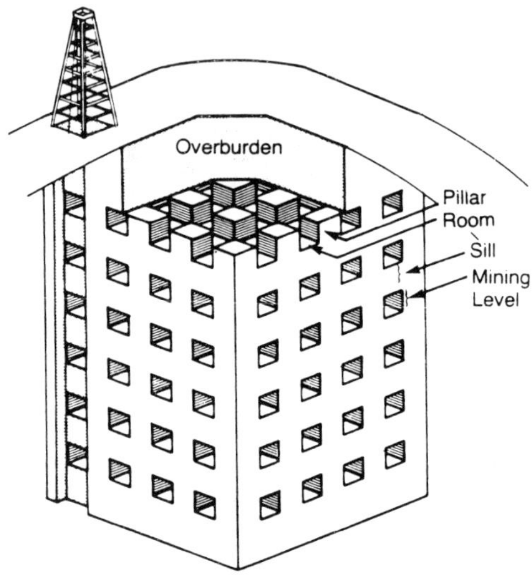

Source: *Hearing on Oil Shale Leasing*. Subcommittee on Minerals, Materials, and Fuels of the Senate Committee on Interior and Insular Affairs, 94th Cong. 2d sess., Mar. 17, 1976, p. 83.

Fig. 6-8. Room-and pillar mining on multiple levels.

The mine eventually would have extended under the entire surface of tract C-b, as shown in Fig. 6-9. However, the concept was abandoned when it was discovered that the shale in the mining zone was highly fractured—a condition that would have required larger support pillars, thus reducing resource recovery to uneconomic levels. All of the original partners subsequently withdrew from the tract, and it is now being developed by modified *in situ* methods by Occidental Oil Shale and Tenneco Oil Co.

Mining to prepare for modified *in situ* retorts is distinguished from room and pillar both by the volume of the deposit that is disturbed and by the configuration of the disturbed areas. As indicated, the underground mines both on the Colony property and on tract C-b would have been developed in a 60-ft-high mining zone; present plans for the operations on tracts C-a and C-b call for recovering oil from the shale in a 750-ft-high zone. However, the actual mining required to support this development will take place in relatively confined areas. The mine will consist primarily of fairly small underground openings in the region in which the modified *in situ* retorts will be created, plus shafts from the surface for ventilation, drainage, passageways for the miners, and transportation of combustion air and retorting products. Some of the mining plans are discussed later in the section on modified *in situ* retorting methods.

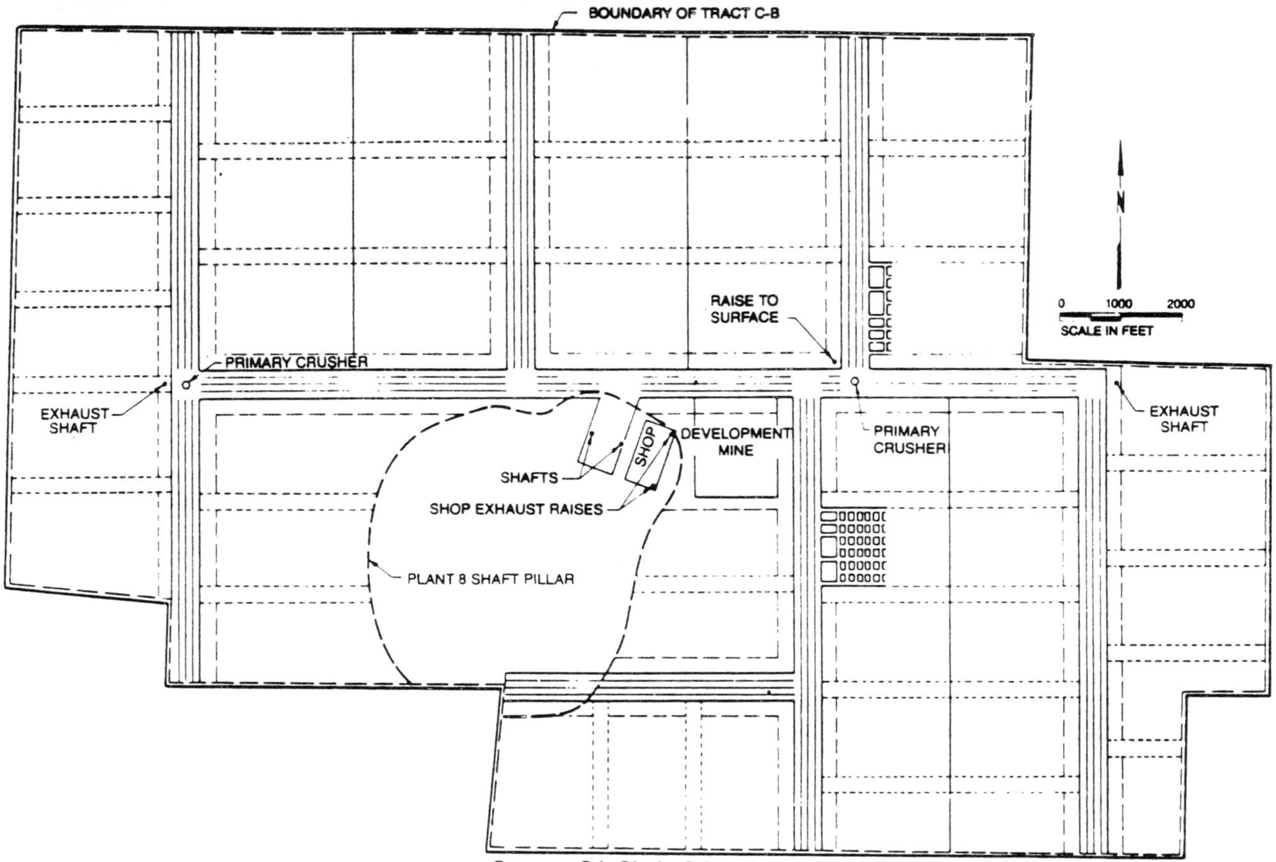

Source: C-b Shale Oil Project, *Detailed Development Plan and Related Materials*, Ashland Oil, Inc., and Shell Oil Co., February 1976, p. IV−11.

Fig. 6-9. The original mine development plan for tract C-b.

OIL SHALE RETORTING

The three generic methods for recovering shale oil from oil shale deposits are described below. Brief discussions of some of the more significant R&D programs that have been conducted as part of their development are included.

True *In Situ*

The sequential steps in true *in situ* processing are: (1) dewatering, if the deposit occurs in a groundwater area; (2) fracturing or rubbling if the deposit is not already permeable to fluid flow; (3) injection of a hot fluid or ignition of a portion of the bed to provide heat for pyrolysis; and (4) recovery of the oil and gases through wells.

The principles of true *in situ* processing are illustrated in Fig. 6-10. Several types have been proposed that differ from each other with respect to the methods for preparing and heating the deposit. All use a system of injection and production wells that are drilled according to a prescribed pattern. One that is commonly used is the "five-spot" pattern in which four production wells are drilled at the corners of a square and an injection well is drilled at its center. The deposit is heated through the injection well, and the products are recovered through the production wells.

For efficient true *in situ* processing, the deposit must be highly permeable to fluid flow, which is true of portions of the Green River oil shales. A good example is the Leached Zone of the Piceance basin where groundwater has dissolved the salt deposits to leave large rubble-filled zones. It is estimated to contain about 550 billion bbl of shale oil in place. There are interconnected fractures and voids in other areas of the formation, but these, in general, have only a very limited permeability. The permeability

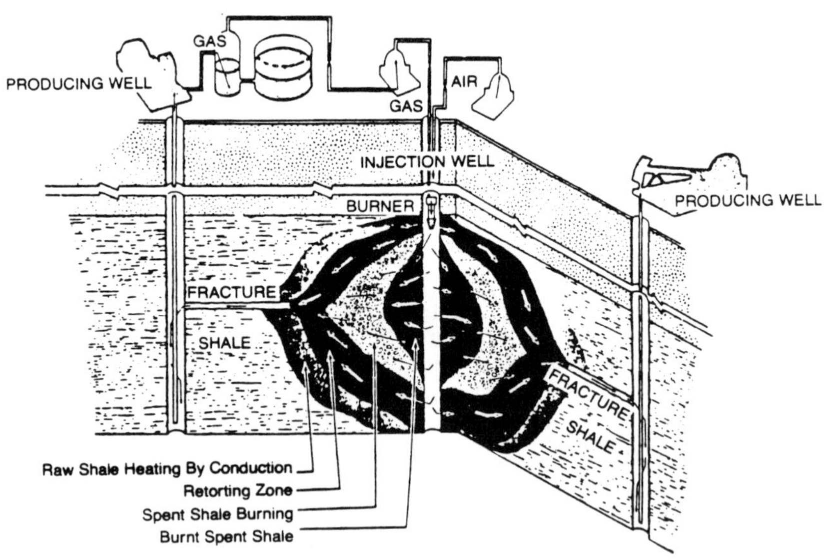

Source: B.F. Grant, "Retorting Oil Shale Underground—Problems and Possibilities," *Quarterly of the Colorado School of Mines*, 59, No. 3, July 1965, p. 40.

Fig. 6-10. True *in situ* oil shale retorting.

of most of the zones that appear to have commercial promise is essentially zero. Deposits that lie near the surface could be fractured by injecting water or explosive slurries, but mining would probably be needed to increase the permeability of deeply buried deposits. Modified *in situ* or aboveground retort processes are more appropriate for the deeper resources.

In 1961, Equity Oil Co. began developing a true *in situ* process for the Leached Zone, which was tested in the Piceance basin between 1966 and 1968. This process involved dewatering a portion of the zone followed by injecting hot natural gas. Pyrolysis gases and a small amount of oil were swept in the natural gas stream to production wells through which they were pumped to the surface. The gas was separated from the oil, reheated, and reinjected into the deposit. The oil had a much lower pour point, viscosity, and nitrogen content than oils from aboveground retorts.[3] These favorable characteristics may have been related to the solvent properties of the natural gas and to the absence of oxygen from the retorting zone. The quality of the retort gases was also good, partly because of their natural gas component and partly because during retorting combustion the decomposition of carbonate minerals was minimal.[4]

Process development was not pursued because too much of the natural gas was lost in the unconfined formation. In 1968, Atlantic Richfield Co. (ARCO) purchased an interest in the process and resumed its development. In 1971, a revised concept was announced in which superheated steam, rather than natural gas, was to be used to heat the deposits. In 1977, DOE contracted with Equity to test the concept in a 500-ft-thick seam in the Leached Zone of the Piceance basin. The seam underlies about 0.7 acre of surface, and contains about 700,000 bbl of oil in place. It was estimated that as much as 300,000 bbl could be recovered over a 2-year period, with about one-half of the oil produced in the first 7 months. No detailed estimates were released about production rates or retorting efficiency. If the process proves to be technically and economically feasible, it could be applied in portions of both the Piceance and the Uinta basins.

At present, no work is being done in slightly fractured formations, but much research is being performed to develop methods for increasing their permeability by enlarging natural fractures and creating new ones. Some of the fracturing techniques used have been chemical explosives, electricity, and high-pressure air and water injections. These methods have been used to enhance recovery of conventional petroleum; oil shale fracturing poses a similar problem. Nuclear explosives were also proposed in the 1960s, but were not tested because of their potential for harming the environment. A nuclear test—Project Rio Blanco—was conducted in the Piceance basin to fracture sand formations containing natural gas. The test failed.

The earliest true *in situ* work was by Sinclair Oil Co., between 1953 and 1966. A thin section of shale in the Mahogany Zone of the Piceance basin was fractured by injecting air. The bed was ignited, although with difficulty, and a small quantity of shale oil was collected before the hot shale swelled and sealed the fractures. After additional tests,

---

[3] These properties have economic significance. *Pour point* is the lowest temperature at which oil will flow. Oils with high pour points are hard to transport because they solidify at normal ambient temperatures. *Viscosity* is a measure of a fluid's resistance to flow. Oils with high viscosity are expensive to pump. Reducing the high nitrogen content of most crude shale oils consumes hydrogen, which is costly.

[4] Both of these processes produce carbon dioxide, which is a major constituent of gases produced by some aboveground retorts.

Sinclair concluded that the zone's limited permeability would not permit profitable operations.[5]

Research on true *in situ* processing began at BOM in the early 1960s. In 1974, the programs and personnel were transferred to ERDA, and in 1978, they moved to DOE. The R&D programs have included laboratory experiments, computer simulations, pilot plant studies, and field tests. Among the latter were tests of electrical, hydraulic, and explosive fracturing and combustion retorting near Rock Springs, Wyoming. These revealed some of the problems associated with the true *in situ* approach. In one early test, for example, after inert gas at about 1300 F (705 C) was pumped into a fractured formation for a period of 2 weeks, about 1 gallon of shale oil was recovered. In another experiment in a zone that contained 7800 bbl of shale oil in place, it was estimated that only 60 bbl of the close to 1000 bbl that were released were actually recovered.

Low oil recoveries are often associated with true *in situ* processing because of the large impermeable blocks of shale in the fractured formation. These cannot be fully retorted in a reasonable length of time. Irregular fracturing patterns that can cause the heat carrier to bypass large sections of the deposit are another problem. Oil shale that is located in the bypassed regions will not be retorted. And, even if all of the shale in a fractured area were retorted, much of the oils would not reach the production wells, but would remain trapped in the pores of the spent shale or would diffuse beyond the production wells, to be lost in surrounding areas.

To develop methods for improving recovery, an extensive R&D program has been conducted at the federal research center in Laramie, Wyoming. Aboveground batch retorts with capacities of 10 and 150 tons of shale have been used to simulate *in situ* retorting, but under more controlled conditions than exist underground. Oil recoveries of up to 67% have been achieved, and an experiment in a "controlled-state" retort achieved 95% oil recovery.[6]

In 1976 and 1977, the program was expanded to include field tests of different fracturing and retorting techniques under cost-sharing contracts with Equity Oil, Talley-Frac, Inc., and Geokinetics, Inc.[7] The Equity program has been described. The Talley-Frac program was terminated after the fracturing method failed. The Geokinetics process uses a fracturing technique called surface uplift in which an explosive is injected into several wells and detonated to fracture the shale and make it permeable to fluid flow. The shale is ignited by injecting air and burning fuel gas through one well and is pyrolyzed by the heat that sweeps through the bed in the gas stream. Oil and gases are pumped to the surface through outlying production wells. It is hoped that the technical feasibility of the process can be demonstrated for thin deposits that are covered by less than 100 ft of overburden. It has been estimated that there are at least 6 billion bbl of shale oil in place in this type of deposit. It is common in the Uinta basin, and at present, would be most economically developed by strip mining.

---

[5] The results of a mathematical simulation indicated that about 18 years of continuous steam injections would be required to heat the shale to pyrolysis temperatures within a 30-ft radius of a fracture.

[6] The large retorts resemble the classic Nevada–Texas–Utah (NTU) retort. Because of the higher percentage of void volume in the rubble, these retorts provided better simulations of modified *in situ* processing than of true *in situ*. The controlled-state retort was much smaller and was equipped for better control of retorting conditions.

[7] The procurement also includes a contract with Occidental Oil Shale, Inc., to develop its modified *in situ* process, which is discussed in the next section.

Geokinectics has been investigating the surface uplift approach since 1973, recently burning its 20th retort in Utah. The largest retort prior to 1979 measured 130 ft by 180 ft by 30 ft thick. A retort about 200 ft on a side and 30 ft thick will be required to demonstrate the technical feasibility of the process. By 1982, DOE and Geokinetics had hoped to have developed a commercial-scale operations with a production capacity of 2000 bbl/day.

Modified *In Situ*

In modified *in situ* processing, the permeability of oil shale deposits is increased by mining some shale from the deposit and then blasting the remainder into the void thus created. The two-step process is depicted in Fig. 6-11. In the first step, a tunnel is dug to the bottom of an oil shale bed, and enough shale is removed to create a room with the same cross-sectional area as the future retort. Holes are drilled through the roof of the room to the desired height of the retort. They are packed with explosives that are detonated in the second step. A chimney-shaped underground retort filled with broken shale results. The access tunnel is then sealed, an injection hole is drilled from the surface (or from a higher mining level) to the top of the rubble pile. The pile is ignited by injecting air and burning fuel gas, and heat from the combustion of the top layers is carried downward in the gas stream. The lower layers are pyrolyzed, and the oil vapors are swept down the retort to a sump at the bottom from which they are pumped to the surface. The burning zone moves slowly down the retort, fueled by the residual carbon in the retorted layers. When the zone reaches the bottom of the retort, the flow of air is stopped, causing combustion to cease.

Occidental Oil Shale, Inc., a subsidiary of Occidental Petroleum (Oxy), has demonstrated the modified *in situ* process. Oxy's work began in 1972 on the D.A. Shale property

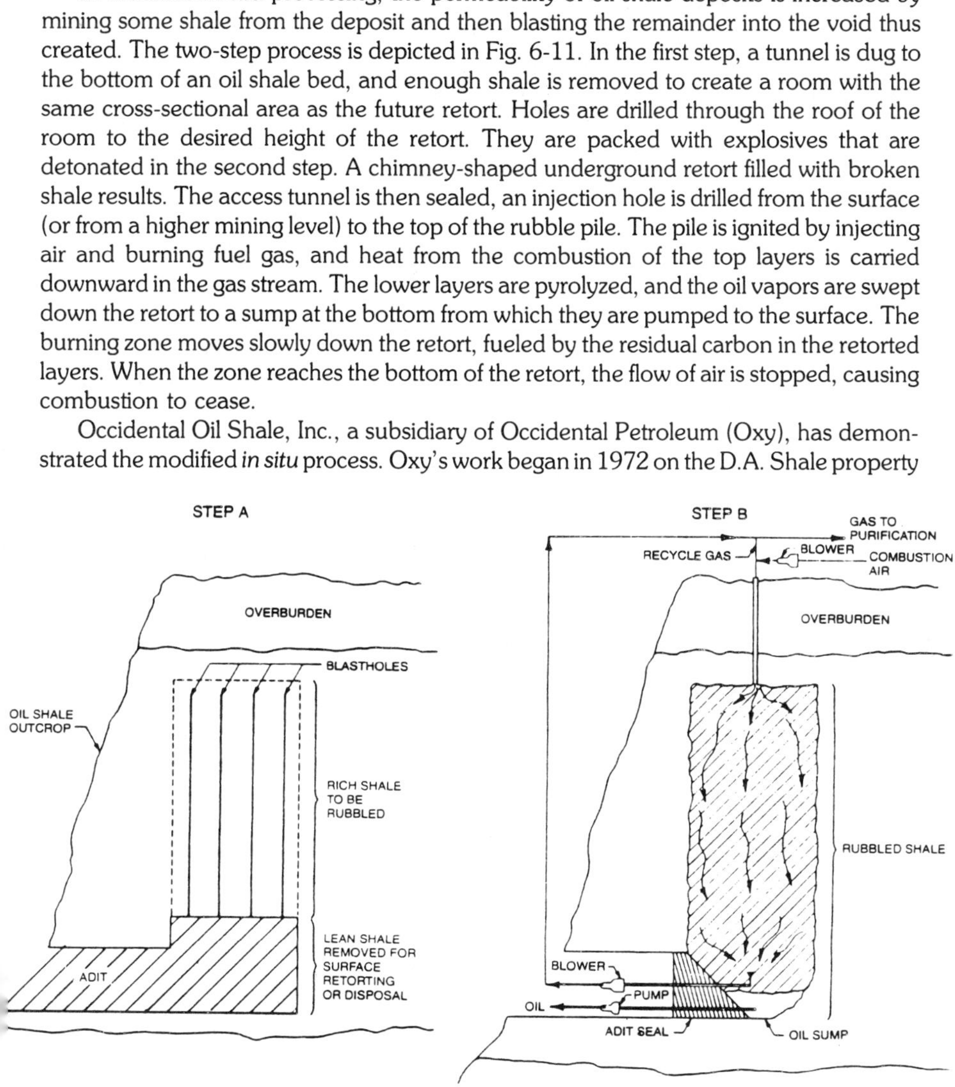

Source: T.A. Sladek, "Recent Trends in Oil Shale—Part 2: Mining and Shale Oil Extraction Processes," *Mineral Industries Bulletin*, 18, No. 1 January 1975, p. 18.

Fig. 6-11. Modified *in situ* retorting.

along Logan Wash—a private tract on the southern fringe of the Piceance basin about 50 miles northeast of Grand Junction, Colorado. To date, more than six retorts have been burned at the site, with varying degrees of success. The first retort was about 32 ft on each side and 70 ft high. About 60% (1200 bbl) of the shale oil in place was recovered, comparable to the best results from BOM's experiments at Laramie. The next two retorts were slightly larger and performed similarly. The fourth retort was of nearly commercial size—120 ft on each side and 270 ft high. It contained about 32 times as much shale as the first retort and yielded about 25 times as much oil. The fifth was the same size, but was designed for a much lower void volume. The rubbling step did not evenly distribute the void volume, and the performance was poor. Only 40% of the oil was recovered. The burning of retort 6 began in 1978, but the sill pillar—the layer of unbroken shale that caps a retort—collapsed into the rubble pile. Operations were disturbed while the top was resealed, but retorting was eventually completed, and about 40% of the oil was recovered. DOE shared the costs of retorts 7 and 8.

The oil shale at the Logan Wash site is not considered to be of commercial quality because of its low kerogen content. In 1976, Oxy acquired access to the higher-quality resources of tract C-b by exchanging its technical knowledge for a one-half interest in the lease. In 1978, Ashland Oil Co., Oxy's partner in the C-b shale oil venture, withdrew and left Oxy in full charge. In1979, Tenneco Oil Co. purchased a one-half interest in the lease for $110 million and is proceeding to develop the tract in cooperation with Oxy. If present plants are followed, Oxy's technology could be used to produce about 57,000 bbl/day by 1985.

Oxy's technique uses a vertical burn configuration—that is, the combustion zone progresses vertically through the shale bed. It is also theoretically possible to advance the burn front horizontally, in much the same way as it is done in true *in situ* processing. A crude version of this approach was implemented in Germany during World War II, when a few modified *in situ* retorts were created by digging tunnels into oil shale deposits and then collapsing their walls into the void volume. These operations were short-lived because oil recoveries were extremely low and they were very hard to control.

Horizontal modified *in situ* might be practical if a technique could be developed to remove large sections of oil shale strata. One possibility is to use solution mining; the injection of fluids into the formation to dissolve soluble salts from among the oil shale layers. The result would be a honeycomb pattern of voids that could then be distributed throughout the area to be retorted by injecting and detonating an explosive slurry. This method would be limited to areas like the Leached Zone or the Saline Zone that contain significant concentrations of soluble salts. Other methods, such as longwall mining or mechanical underreaming, could be used in other areas. It might be possible to operate mechanical underreaming machines by remote control from the surface, thus reducing or even eliminating the need for miners to work underground. None of these approaches has been tested in any oil shale deposit.

Other modified *in situ* techniques that use vertical-burn patterns have also been developed. For example, the firm of Fenix and Sciseson designed two systems for underground mining, rubbling, and retorting in the vertical mode. To date, they have been tested only with a computer model. DOE's Lawrence Livermore Laboratory has also developed a modified *in situ* technique called rubble *in situ* extraction (RISE) with the aid of computer simulation and laboratory experiments in pilot-scale aboveground retorts. (See Fig. 6-12.) Several levels are mined, and a portion of the deposit is removed at each. The remaining shale is broken with explosives. Sufficient broken shale is then

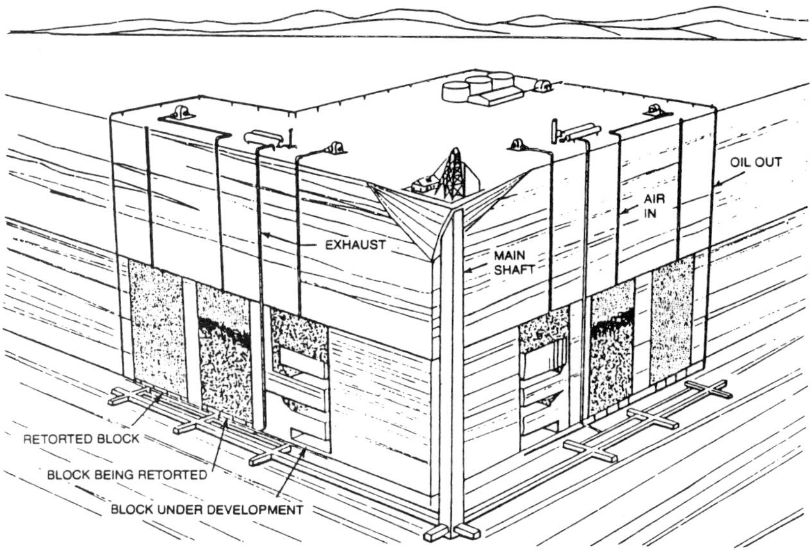

A. Isometric

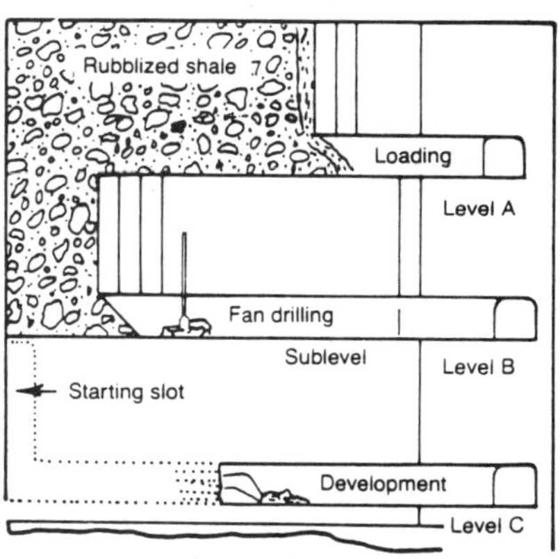

B. Elevation

Development proceeds simultaneously on several levels. At each level, horizontal drifts are driven the full width of the retort block, and a vertical slot is bored to provide void volume for blasting. About 20% of the broken shale is removed after each blasting operation. The rest is left in the retort volume.

Sources: ***Oil Shale Retorting Technology,*** **prepared for OTA by Cameron Engineers, Inc., 1978; and C.K. Jee** ***et al., Review and Analysis of Oil Shale Technology—Volume 3: Modified In Situ Technology,*** **report prepared for the U.S. Department of Energy by Booz-Allen & Hamilton, Inc., under contract No. EX-76-C-01-2343, August 1977, p. 6.**

Fig. 6-12. The RISE method of modified *in situ* oil shale retorting.

removed so that there is a total void volume of 20% in the retort area. The rubble is then ignited at the top, and retorting proceeds as in the Oxy system.

The RISE approach was originally proposed by Rio Blanco Oil Shale Co. for tract C-a, but the firm is now going ahead with its own process, which has benefited from technical information acquired under a licensing arrangement with Oxy. Livermore's modeling studies and laboratory experiments were continued.

The initial plans for developing tract C-a by modified *in situ* methods are shown in Fig. 6-13; they would have involved five precommercial retorts of increasing size. The present plant, which was adopted after purchase of Oxy's modified *in situ* technology, is shown in Fig. 6-14. In the new development plant, a small pilot retort (retort 0) will be followed by two demonstration retorts of increasing size. The largest (retort 2) will be close to commercial scale. Several options of different size are being considered for retort 2, with one option a retort with dimensions 60 ft by 150 ft by 400 ft high. The method by which the shale is rubbled and the fraction of the shale that is mined from each retort area are two of the major differences between the Rio Blanco approach and Oxy's technique. Oxy has tested several rubbling methods, including drilling blastholes up from a room at the bottom of the retort, drilling down from the top, and boring a central shaft the full length of the retort and then blasting the surrounding shale into the shaft. In Rio Blanco's approach, a room will be created at the bottom of the future retort, blastholes will be drilled from the surface into the roof of the room, and explosives will be detonated sequentially at different levels. The shale above the room will thereby be blasted loose in layers, with each layer of rubble falling to the bottom of the retort volume before the next higher layer is blasted. Through this technique, Rio Blanco hopes to obtain uniform size distribution in the rubble. This is believed to be a key technical requirement for efficient modified *in situ* retorting. Rio Blanco also proposed to mine twice the volume (40% vs 20%) as is contemplated by Oxy.

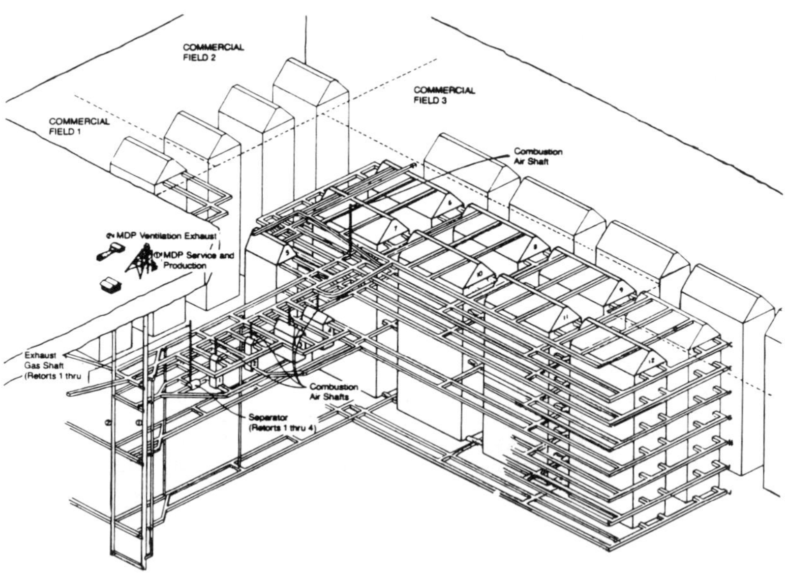

Source: Rio Blanco Oil Shale Project, *Revised Detailed Development Plan Tract C-a*, Gulf Oil Corp. and Standard Oil Co. (Indiana), May 1977, p. 1-2-5.

Fig. 6-13. Initial modular modified *in situ* retort development plan for tract C-a.

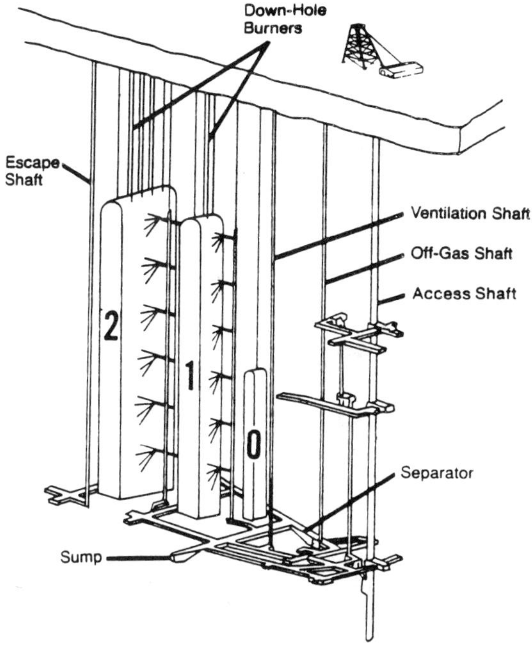

Source: The Pace Co. Consultants and Engineers, Inc., *Cameron Synthetic Fuels Report*, 16, No. 3, September 1979, p. 2–8.

Fig. 6-14. Present modified *in situ* retort development plan for tract C-a.

Another modified *in situ* method that is being developed is the integrated *in situ* technology proposed by Multi Mineral Corp. for recovering shale oil, nahcolite, alumina, and soda ash from oil shale deposits in the Saline Zone of the Piceance basin. This zone underlies the Leached Zone, is relatively free from groundwater, and contains extensive resources of nahcolite and dawsonite in addition to oil shale. Development is hindered because the zone is deeply buried (about 2000 ft) below the surface of the basin. To reduce the costs of its R&D program, Multi Mineral has proposed to use an 8-ft-diameter shaft that was drilled by BOM in 1978 through the Leached Zone and into the Saline Zone. In the first phase, mining and mine safety methods will be tested, and about 8000 tons of nahcolite and 11,000 bbl of shale oil will be produced. The nahcolite will be used for stack-gas scrubbing tests in a power plant. In the second phase, a retorting module will be used to produce up to 3000 ton/day nahcolite, 10,000 bbl/day of shale oil, 200 ton/day of alumina, and 3000 ton/day of soda ash. The modular retort and the access shafts and drifts are shown in Fig. 6-15. If technical and economic feasibility is indicated during the test program, Multi Mineral would proceed to a commercial-scale facility that would produce 50,000 bbl/day of shale oil, 10,000 ton/day of nahcolite, 1000 ton/day of alumina, and 20,000 ton/day of soda ash.

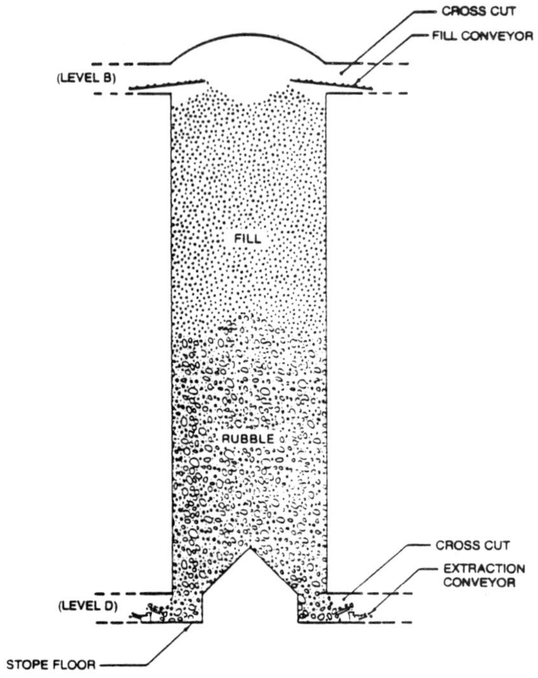

Source: B. Weichman, "*Saline Zone*" *Oil Shale Development by the Integrated In Situ Process*, Multi Mineral Corp. Houston, Tex., December 1979, p. 9.

Fig. 6-15. Retort development plan for the Multi Mineral modified *in situ* concept.

The Multi Mineral approach resembles the RISE technique in that mining and rubbling are conducted at several levels along the height of the retort. It departs from RISE in that, after rubbling, the broken shale in the retort is removed from the bottom, crushed and screened, and returned to the top in a continuous operation. This part of the retort development plan is shown in Fig. 6-16. The rubbled shale will contain free nahcolite and nahcolite that is still associated with the large blocks of shale. Crushing will liberate most of the associated nahcolite, and screening will separate it from the shale particles. The shale will be backfilled to the top of the retort; the nahcolite is transported to the surface for processing. The result would be a retort that is filled with uniformly sized oil shale particles. With this configuration, Multi Mineral hopes to avoid the channeling and bypassing problems that may occur in true *in situ* processing.

In a commercial-scale facility, the retorts would be operated in sets of three, as shown in Fig. 6-17. In retort 3, the oil shale is being pyrolyzed by injecting hot gases to produce shale oil (which is removed to the surface), residual carbon, and soda ash and aluminum oxide—the products of the thermal decomposition of dawsonite. The last three products remain in the retort rubble. In retort 2, the residual carbon is being gasified by injecting a mixture of steam and air. The heat from the gasification reaction is carried to retort 3. In

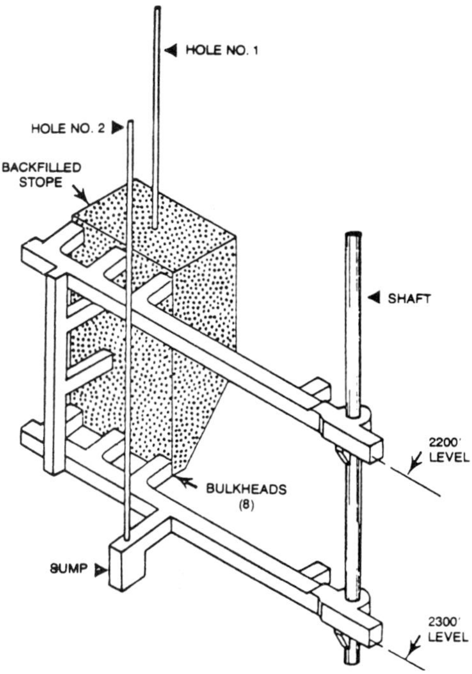

Source: B. Weichman, *"Saline Zone" Oil Shale Development by the Integrated In Situ Process*, Multi Mineral Corp. Houston, Tex., December 1979, p. 15.

Fig. 6-16. Preparation of a Multi Mineral modified *in situ* retort.

retort 1, the hot gasified shale is being cooled by passage of cold recycle gas. The heat thus recovered is conveyed to retort 2. After the shale in retort 1 is cooled, the soda ash and the aluminum oxide can be leached out with water. The leachate is then pumped to the surface and processed to recover its mineral values.

Temperature control is the key to the entire operation. Oil alone could be recovered with a combustion-type method, such as Oxy's, but because of the high temperatures, the decomposition of the dawsonite would produce compounds insoluble in water. The gas-recycle heating method would avoid this problem by maintaining lower retorting temperatures. The operation also depends on the ability of explosive rubbling techniques to produce broken shale that can be fed to conventional crushing equipment. Overall, the method is very interesting because of its potential for simultaneously recovering fuel and minerals from deposits that may not be accessible with any other approach. However, too little is known about the various steps to permit a thorough evaluation at this time.

Aboveground Retorting

Hundreds of retorts have been invented in the 600-year history of oil shale development. Most were never brought to the processing stage, but some were tested using

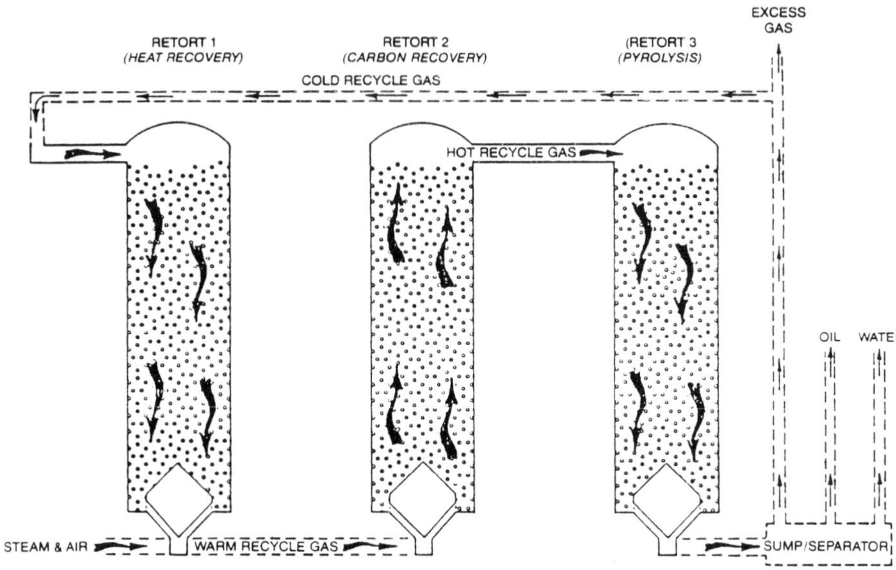

Source: B. Weichman, *"Saline Zone" Oil Shale Development by the Integrated In Situ Process*, Multi Mineral Corp., Houston, Tex., December 1979, p. 11.

Fig. 6-17. A set of Multi Mineral modified *in situ* retorts.

laboratory-scale equipment, and a few at pilot-plant and semiworks scales. None has been tested at a scale suitable for modern commercial operations. This section summarizes the technical aspects of seven retorting systems that offer the promise of being applicable in the near future. One obsolete technology that is the basis for several of the modern systems is also discussed.

Although aboveground retorts differ widely with respect to many technical details and operating characteristics, they can be grouped into four classes:

- *Class 1:* Heat is transferred by conduction through the retort wall. The Pumpherston retorts used in Scotland, Spain, and Australia were of this class, as is the Fischer assay retort that was developed in the 1920s; this is a laboratory device for estimating potential shale oil yields. Its oil yield is the standard to which the retorting efficiencies of all other retorts are compared. Because conduction heating is very slow, modern industrial retorts are in class 1.
- *Class 2:* Heat is transferred by flowing gases generated within the retort by combustion of carbonaceous retorted shale and pyrolysis gases. Retorts in this class are also called directly heated; this class includes the Nevada–Texas–Utah (NTU) and Paraho direct processes described below, and also BOM's gas combustion retort and the Union"A" retort. Class 2 retorts produce a spent shale low in residual carbon and low-Btu retort gas. Their thermal efficiencies are high because energy is recovered from the retorted shale. However, recovery efficiencies are relatively low—about 80–90% of Fischer assay.

- *Class 3:* Heat is transferred by gases that are heated outside of the retort vessel. Retorts in this class are also called indirectly heated. They include the Paraho indirect, Petrosix, Union "B," and Superior retorts discussed below, and also the Union SGR and SGR-3, the obsolete Royster design, the Soviet Kiviter, the Texaco catalytic hydrotort, and others. These retorts produce a carbonaceous spent shale and a high-Btu gas. Thermal efficiencies are relatively low because energy is not recovered from the residual carbon, but oil recovery efficiencies are high, from 80% to over 100% of Fischer assay.
- *Class 4:* Heat is transferred by mixing hot solid particles with the oil shale. This class includes the TOSCO II and Lurgi–Ruhrgas retorts described below, and also the Soviet Galoter retort. Class 4 retorts achieve high oil yields (about 100% of Fischer assay) and produce a high-Btu gas. The spent shale may or may not contain carbon, and thermal efficiencies vary, depending on whether the spent shale is used as the heat carrier. The retorts are sometimes referred to as indirectly heated, as in class 3, because they also lack internal combustion and produce a similar gas product.

Several other conversion methods have been developed that cannot easily be placed in these classes. These include microwave heating, bacterial degradation, gasification, and circulation of hot solid slurries.

The Nevada – Texas – Utah (NTU) Retort. The NTU retort is a modified downdraft gas producer similar to units used in the 19th century to produce low-Btu gas from coal. It is a vertical steel cylinder, lined with refractory brick, and equipped with an air supply pipe at the top and an exhaust pipe at the bottom. The top and bottom may be opened for charging a batch of shale and for discharging the spent shale after retorting, respectively. The unit was developed and tested for oil shale processing by the NTU Co. in California in 1926, and was also tested by BOM at Rifle, Colorado, from 1925 to 1929. Two units with nominal capacities of 40 tons of raw shale were built and operated by BOM at Rifle between 1946 and 1951. They produced more than 12,000 bbl of shale oil. Three units that were operated in Australia during World War II produced nearly 500,000 bbl. As noted previously, BOM used two units to simulate *in situ* retorting.

The operating sequence for the NTU retort is shown in Fig. 6-18. The unit is loaded with crushed oil shale and sealed. The gas burner is lighted, and air is blown in. Once the top of the shale bed is burning (step A), the fuel gas is shut off but the air supply continues. As the air flows through the burning layer, it is heated to approximately 1500°F (815°C). This hot gas heats the shale in the lower layers and induces the pyrolysis of the kerogen. The oil and gas produced are swept down through the cooler portions of the bed to the exhaust port (step B). The solid product from the conversion of the kerogen (residual carbon) remains on the spent shale and is consumed as the combustion zone moves down the retort, providing fuel for additional combustion and thereby heat for additional pyrolysis. When all the carbon is burned from the upper layers of the bed, the four zones shown in step C are formed. The top layer contains burned and decarbonized spent shale. The second layer is burning, releasing heat for pyrolysis in the third layer. In the bottom layer, the shale is being heated but is not yet at pyrolysis temperatures.

As time passes, the top layer expands downward and the lower three zones move uniformly down the retort. When the leading edge of the combustion zone reaches the

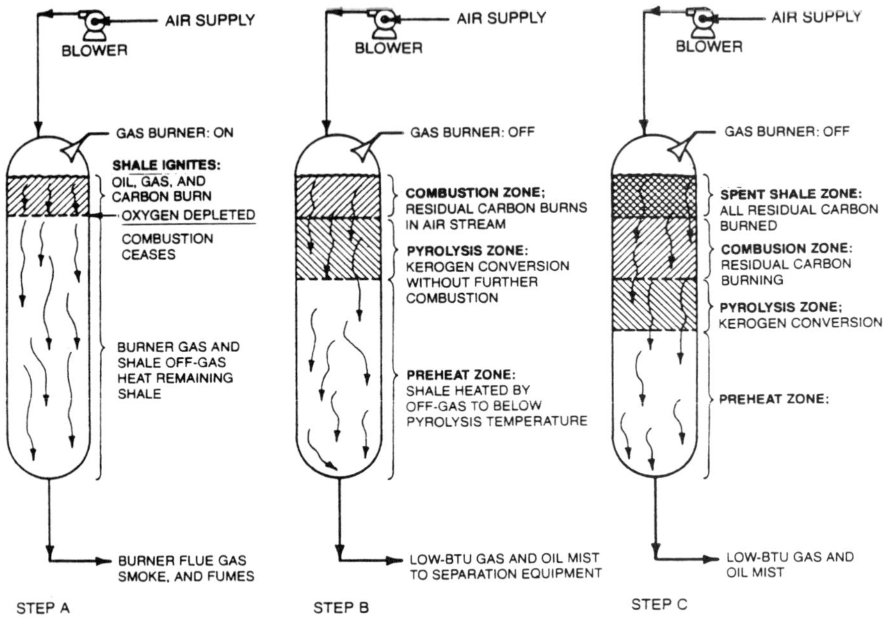

Source: T.A. Sladek, "Recent Trends in Oil Shale—Part 2: Mining and Shale Oil Extraction Processes," ***Mineral Industrial Bulletin,*** **18, No. 1, January 1975, p. 5.**

Fig. 6-18. The operation of an NTU retort.

exhaust port, oil production ceases, air injection stops, and the retort is emptied. The entire cycle, from ignition to dumping, takes about 40 hours.

NTU retorts are simple to operate, and require no external fuel except for small amounts of gas to start the retorting. They can process a wide variety of shales with oil recoveries ranging from 60% to 90% of Fischer assay. They are unsuitable for modern commercial applications because they are batch units with high labor costs and small capacities. Over 600 150-ton retorts would be needed to produce 50,000 bbl/day of crude shale oil. In contrast, plants using some of the continuous technologies described below would require only about six retorts for the same production capacity.

The Paraho Direct and Indirect Retorts. An NTU retort becomes a Paraho direct retort when it is turned upside down, made continuous, and mechanically modified. The Paraho retorts are based on BOM's gas combustion retort which, in turn, evolved from the NTU retorts tested at Anvil Points in the late 1940s. The gas combustion process was tested in 6-, 10-, and 25-ton/day units by BOM between 1949 and 1955, and by a consortium of six oil companies between 1964 and 1968. The Paraho direct retort is similar in design to the gas combustion technology but it is more likely to be commercialized. It was developed by Development Engineering, Inc. (DEI) in the late 1960s, and was tested for limestone calcining in three cement kilns in South Dakota and Texas. Each kiln had a capacity of 330 ton/day and was comparable to the largest gas combustion retort tested by the six oil companies.

After verifying the solids-flow characteristics of the Paraho technology in the limestone kilns, DEI leased the Anvil Points site from the federal government in May 1972, and began a 5-year program to develop the technology for oil shale processing. Funding was obtained from a consortium of 17 energy companies and engineering firms. In return for a contribution of $500,000, each company was guaranteed a favorable royalty arrangement in any subsequent commercial application of DEI's technology. The Paraho Development Corp. was formed to manage the project.[8] Two retorts, a pilot-scale unit 4.5 ft in diameter and 60 ft high and a semiworks unit 10.5 ft in diameter and 70 ft high, were installed and tested after August 1973. They were used to produce over 100,000 bbl of crude shale oil, some of which was used for refining and end-use experiments by DOE and DOD. Maximum throughput rates reached about 400 ton/s in the semiworks unit. Both direct and indirect heating modes were tested.

The Paraho retort is shown in Fig. 6-19. In its direct mode, the retort is very similar to the older gas combustion design. However, significant differences exist in the shale feeding mechanism, in the gas distributors, and in the discharge grate. During operation, raw shale is fed to the retort through a rotating distributor at the top. It descends as a moving bed along the vertical axis of the retort. As it moves, it is heated to pyrolysis temperatures by a rising stream of hot combustion gases. The oil and gas produced are swept up through the bed to collecting tubes and out of the retort to product separation equipment. The retorted shale retains the residual carbon. As the shale approaches the burner bars, the carbon is ignited and gives off the heat required for pyrolyzing additional raw shale. Passing beyond the burner bars, the shale is cooled in a stream of recycle gas and exits the bottom of the retort through the discharge grate. It is then moistened and sent to disposal.

The retort may also be operated as a class 3 retort. The configuration would resemble directly heated operations except that air would not be injected and the offgas stream would be split into four parts after oil separation: one part would be the net product gas; another would be sent through a reheating furnace and then reinjected into the middle of the retort; a third would not be reheated but would be reinjected through the bottom of the retort to cool the shale before discharge; and the fourth would be used for fuel in the reheating furnace. All heat for kerogen pyrolysis would be provided by the reinjected gases, and no combustion would occur in the retort vessel itself.

To date, the Paraho retorting technology has been tested at about one-twentieth of the size that would be used in commercial plants. Paraho would like to test a full-size module producing about 7300 bbl/day at Anvil Points. Both direct and indirect heating modes would be tested. Permission has been obtained from the Department of the Navy to use large quantities of shale from the Naval Oil Shale Reserves for the program. An environmental impact statement (EIS) was prepared for the project. Paraho has responded to a DOE Program Opportunity Notice for a $15 million engineering design study of modular oil shale retorting, and would base the design of the Anvil Points module on results of the study.

The Petrosix Indirectly Heated Retort. The Petrosix process was developed for the Brazilian Government by Petrobras, the national oil company, with the assistance of Cameron and Jones, Inc., a U.S. engineering firm. Russell Cameron, worked on the gas combustion program at Anvil Points, as did John Jones, later president of Paraho. The system is depicted in Fig. 6-20. The demonstration retort is 18 ft in diameter and has a capacity of 2200 ton/day of Irati oil shale. It was built in 1972, and was tested intermittently until quite recently. In 1975, a U.S. patent was issued for the process. A 50,000-bbl/day

[8] "Paraho" is from the Portuguese words *para homen*—for mankind.

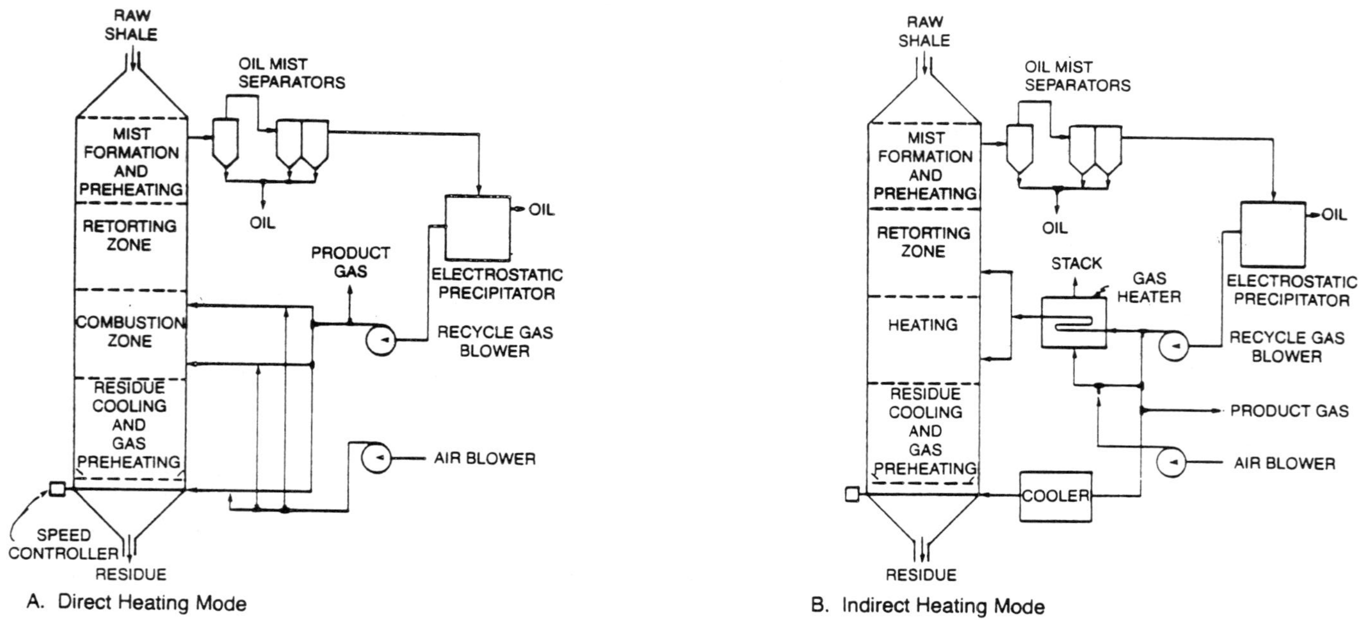

Source: C.C. Shih *et al.*, *Technological Overview Reports for Eight Shale Oil Recovery Processes*, report prepared for the U.S. Environmental Protection Agency by TRW under contract No. EPA-600/7-79-075, March 1979, pp. 20 and 23.

Fig. 6-19. The Paraho oil shale retorting system.

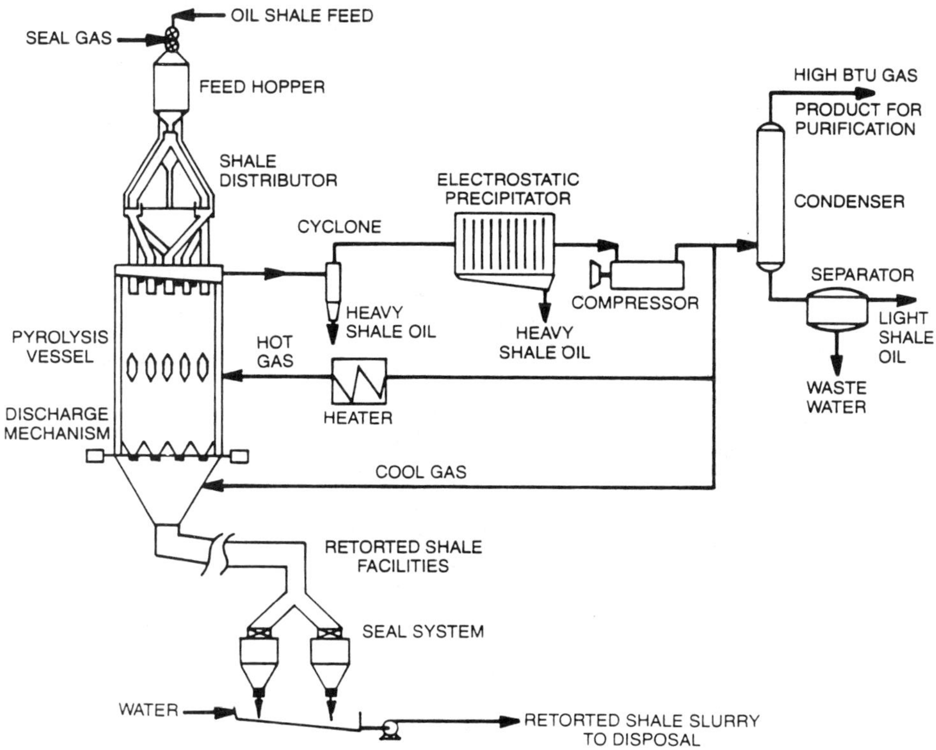

Source: *Oil Shale Retorting Technology*, prepared for OTA by Cameron Engineers, Inc., 1978.

Fig. 6-20. The Petrosix oil shale retorting system.

plant is planned by the Brazilian Government, to be developed in a two-stage project. The first stage would involve construction of a 25,000-bbl/day facility about 5 miles from the site of the demonstration plant near the city of Sao Mateus do Sul in the State of Parana. In the second stage, the full commercial plant would be built on the same site, to include 20 Petrosix units, each about 33 ft in diameter. Brazil is not committed to either stage, but if financing is obtained, the first stage could be completed by 1983 and the second by 1985. In addition to the shale oil, the plant would also produce sulfur and liquefied petroleum gases. Plans are also being prepared for two additional commercial-scale plants south of the present demonstration plant in the state of Rio Grande do Sul.

Except for mechanical differences, the Petrosix retort is very similar to the Paraho indirect retort described above. The similarity is not surprising in view of the shared engineering heritage of the two systems. One operational difference is that the Petrosix spent shale is discharged into a water bath and pumped in a slurry to a disposal pond. Paraho shale is discharged dry, with only a little water added prior to disposal.

Little information has been released about the demonstration program. Oil characteristics have been described, but these have little relevance to the processing of Green River shale. It can be predicted that the retort should have high oil recovery efficiencies and produce a retort gas with high heating value. The spent shale would be carbonaceous. In the demonstration plant, recovery of energy from the spent shale was not possible because of the slurry disposal method. In any commercial plant, it is possible that the shale would be burned in a separate unit to produce heat for pyrolysis.

The Union "B" Indirectly Heated Retort. The Union "B" retort is a class 3 retort that evolved from the Union "A," a class 2 retort, by the Union Oil Co. of California. Two other systems, the SGR and SGR-3, have also been proposed by Union. Union describes them all as continuous, underfeed, countercurrent retorts. The "B" has not been field tested, but the "A" was tested in Colorado in the 1950s at up to 1200 ton/day. Figure 6-21 is a sketch of the Union "B" design. It incorporates most of the design features of the "A."

During operation, shale is fed through the bottom of the inverted-cone vessel. The retorting process thereafter resembles that of a continuous NTU retort. Hot gases enter the top of the retort and pass down through the rising bed, causing kerogen pyrolysis. Shale oil and gas flow down through the bed. The oil accumulates in a pool at the bottom, which seals the retort and acts as a settling basin for entrained shale fines. Oil and gas are withdrawn from the top of the pool. The gases are split into three streams: one is reheated and reinjected to induce additional kerogen pyrolysis; one is used as fuel in the reheating furnace; and one is the net product. The shale is discharged from the top of the retort and falls into a water bath in the retorted shale cooler. From there it is conveyed to disposal.

The key to all of Union's retorting systems is the solids pump that is used to move the oil shale through the retort. In the "B" design, the solids pump is mounted on a movable carriage and is immersed in the shale oil pool. The pump consists of two hydraulic assemblies that act in sequence (see Fig. 6-21): while the cylinder of one assembly is filling with oil shale, the other is charging a batch of shale into the bottom of the retort. When this operation is completed, the carriage moves horizontally on rails until the full cylinder is under the retort entrance. A piston then moves the oil shale in this cylinder upward into the retort, while the other fills with fresh shale from the other feed chute. The carriage then returns to its original position and the process is repeated.

The "B" achieves high oil yields, and the retort gas has a high heating value, although much of it is consumed in the reheating furnace. The mechanical nature of the rock pump is troublesome because its moving parts would be subject to wear during operation. However, the pump is immersed in the shale oil pool, which should provide adequate lubrication. Union appears to be satisfied with its reliability.

In 1976, Union announced plans to build a demonstration plant on its private land in Colorado. The "B" retort was to be used to produce 7000 bbl/day of shale oil from 10,000 ton/day of oil shale. Later, the announced capacity was increased to 9000 bbl/day. The demonstration plant, called the Long Ridge project, would be a step toward a 75,000-bbl/day commercial-scale plant. Air quality permits have been obtained for the modular plant, which could be completed before 1985. Construction had not begun because Union was awaiting financial incentives from the federal government.

The Superior Oil Indirectly Heated Retort. The Superior retort is unique among the aboveground retorts discussed in this section because it is designed for recovery of

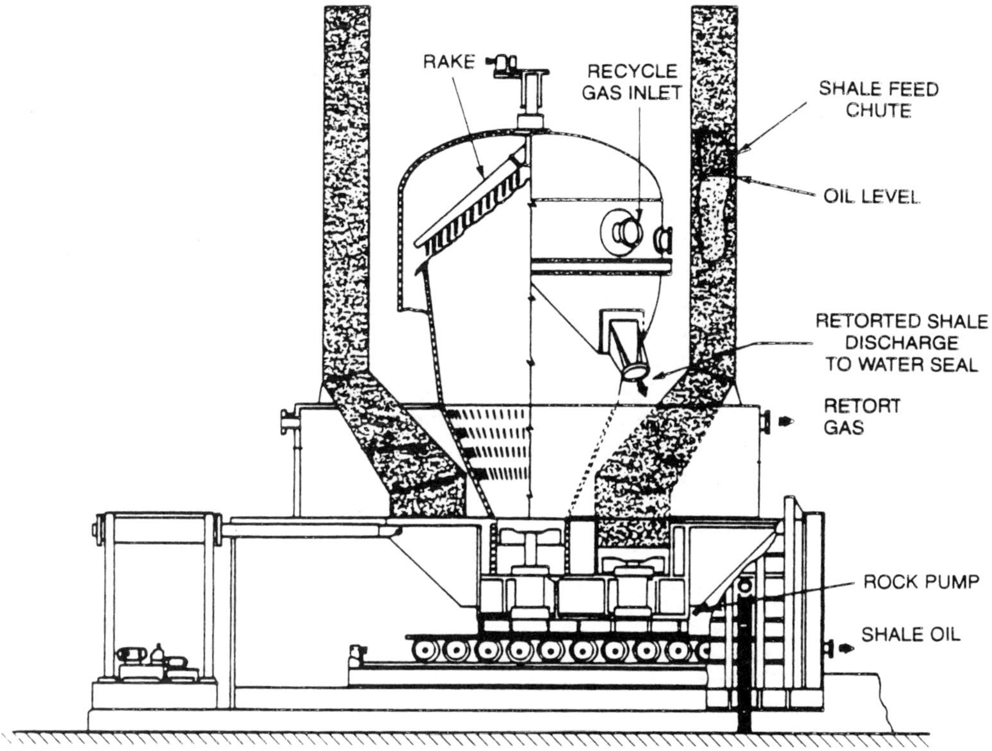

A. The Retorting System

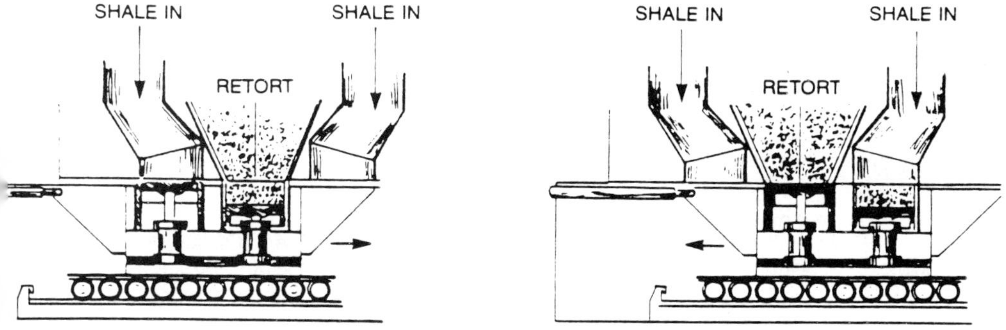

. Rock Pump Detail

Source: *Oil Shale Retorting Technology*, prepared for OTA by Cameron Engineers, Inc., 1978.

Fig. 6-21. The Union Oil "B" retorting process.

sodium-bearing minerals in addition to shale oil. The minerals nahcolite and dawsonite occur in substantial quantities in portions of the Piceance basin; they are potential sources of sodium bicarbonate, soda ash, and aluminum.

A block diagram of the Superior approach is shown in Fig. 6-22. In step 1, the mined oil shale is crushed to pieces smaller than 8 in. and screened. About 85–90% of the nahcolite in the shale is in the form of distinct, highly friable nodular inclusions that become a fine powder during the crushing operation. This is screened from the coarser shale in step 2. The shale is then further crushed to smaller than 3 in. and is fed in step 3 to a doughnut-shaped traveling-grate retort, which includes sequential stages for heating, retorting, burning, cooling, and discharging the oil shale feed. The retort is shown in Fig. 6-23. In the heating section, oil is recovered by passing hot gases through the moving bed. In the carbon recovery section, process heat is recovered by burning the residual carbon. In the cooling section, inert gases are blown through the bed and spent shale, cooling the shale and heating the gases for use in the heating section. After discharge, the cooled spent shale is sent to other units in which alumina is recovered from calcined dawsonite and soda ash is recovered from calcined nahcolite. The alumina is shipped to offsite aluminum refineries; the soda ash to glass plants; and the nahcolite to refineries and powerplants for use as a stack-gas scrubbing agent.

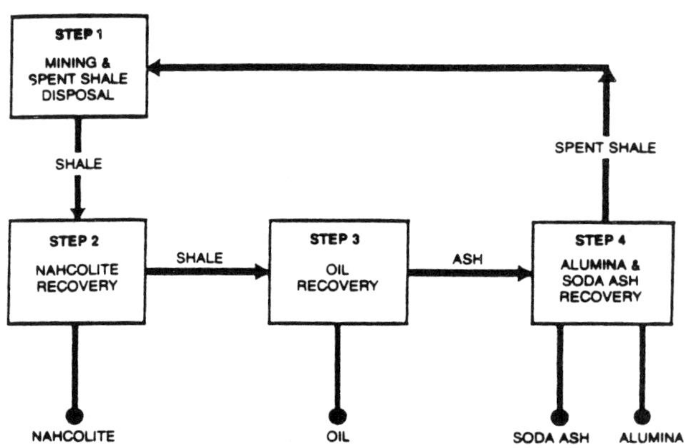

Source: *Oil Shale Retorting Technology*, prepared for OTA by Cameron Engineers, Inc., 1978.

Fig. 6-22. Block diagram for the Superior Multi Mineral concept.

Superior chose the traveling-grate retort because it allows close temperature control, which is important to maintaining dawsonite's solubility during the burning stage. Similar, simpler devices have been used to sinter iron ore for steel making and to roast lead and zinc sulfides. Superior's process has been tested for oil shale in pilot plants in Denver and Cleveland. A commercial-scale plant would consume 20,000 ton/day of oil shale to produce 10,000–15,000 bbl/day of shale oil, 3500 to 5000 ton/day of nahcolite, 500–800 ton/day of alumina, and 1200–1600 ton/day of soda ash.

In the early 1970s, Superior proposed building a commercial-size demonstration plant on its 7000-acre tract in the northern Piceance Basin. The deposit was to be

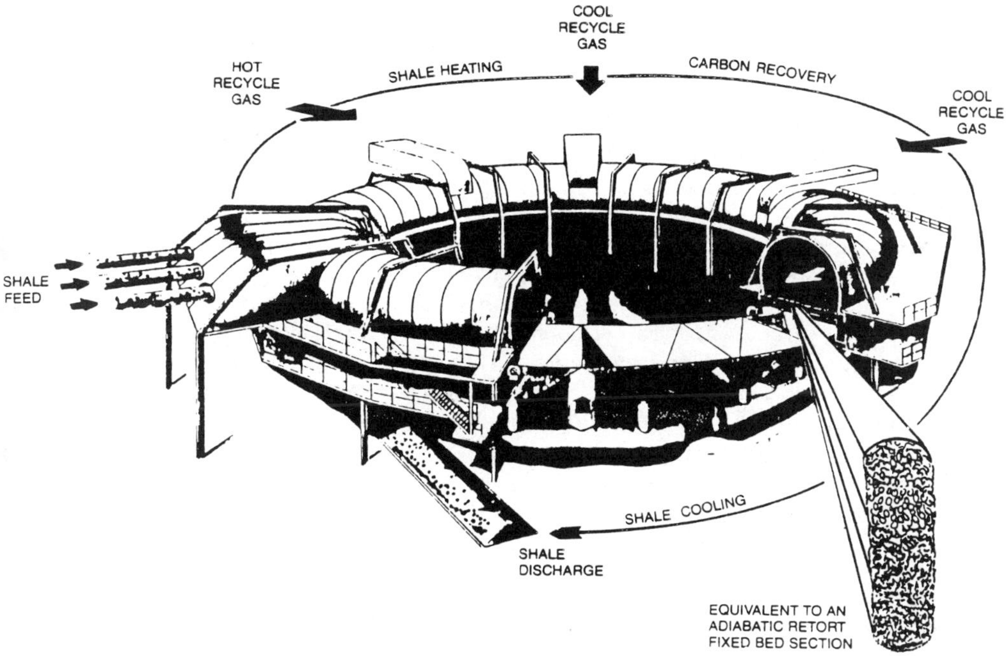

Source: *Oil Shale Retorting Technology*, prepared for OTA by Cameron Engineers, Inc., 1978 (after Arthur G. McKee & Co.).

Fig. 6-23. The Superior oil shale retort.

developed by deep room-and-pillar mining on several levels. The single retort was to produce 11,500 bbl/day of shale oil, plus the by-products described above. Although the tract's resources are extensive, its L-shaped configuration does not favor large-scale development. Superior therefore proposed to exchange portions of the tract for adjacent land controlled by the Bureau of Land Management (BLM). Approval was delayed by BLM review and by preparation of an EIS, a draft of which was recently released. In February 1980, BLM denied the exchange because the two tracts involved were not considered to have equivalent value.

The Tosco II Indirectly Heated Retort. The Tosco II is a class 4 retort in which hot ceramic balls carry heat to finely crushed oil shale. It is a refinement of the Aspeco process developed by a Swedish inventor. Patent rights were purchased by The Oil Shale Corp. (later Tosco) in 1952. Early development work was performed by the Denver Research Institute, including testing of a 24-ton/day pilot plant. In 1964, Tosco, Standard Oil Co. of Ohio (SOHIO), and Cleveland Cliffs formed a venture company to commercialize the Tosco II technology. Between 1965 and 1967, the group operated a 1000-ton/day semiworks plant on its land near Grand Valley, Colorado, next to the site of Union's semiworks operations. In 1968, Colony prepared a preliminary engineering design and cost estimate for a commercial-scale plant that would contain six Tosco II retorts, and convert 66,000 ton/day of oil shale into about 46,000 bbl/day of shale oil. In 1969, ARCO joined Colony, and a second semiworks program began to test scale-up procedures and to evaluate environmental controls. The program continued until April 1972. Between 1965 and 1972 the semiworks plant converted 220,000 tons of oil shale into 180,000 bbl of shale oil.

In 1974, the 1968 cost estimate, which had been updated in 1971, was further revised to incorporate operating data from the latter part of the semiworks program and to account for additional pollution controls. The resulting cost estimate was about three times as large as the previous estimate. This cost escalation raised doubts about commercial feasibility, and the project was postponed indefinitely. SOHIO and Cleveland Cliffs subsequently withdrew from the Colony group.

Early in 1974, Tosco, ARCO, Ashland, and Shell purchased a lease for tract C-b from the federal government. Initial development plans for the tract involved a plant similar to that proposed for Colony's private tract. These plans were also affected by the cost escalations, and in 1976 suspension of operations on tract C-b was granted by the government. In 1977 Tosco and ARCO withdrew from the tract. Shell withdrew in 1977. Ashland, the remaining partner, teamed with Oxy to develop the tract using modified *in situ* techniques.

The Tosco II retorting system is shown in Fig. 6-24. Raw shale is crushed smaller than ½ in. and enters the system through pneumatic lift pipes in which the shale is elevated by hot gas streams and preheated to about 500°F (260°C). The shale then enters the retort, a heated ball mill, and contacts a separate stream of hot ceramic balls. As the shale and balls mix, the shale is heated to about 950°F (510°C), causing retorting to take place. Oil vapors and gases are withdrawn. The oil is condensed in a fractionator. Some of the gases are burned in a heater to reheat the ceramic balls to about 1200°F (650°C). At the retort exit, the retorted shale and the cooled ceramic balls pass over a trommel, a perforated rotating separation drum. The shale, which has been thoroughly crushed during retorting, falls through holes in the trommel. It is cooled and sent to disposal. The larger balls pass over the trommel and are sent to the ball heater.

Oil yields exceeding 100% of Fischer assay have been reported for the Tosco II technology; however, overall thermal efficiencies are low because energy is not recovered from spent shale carbon, and much of the product gas is consumed in the ball heater. Tosco has patented processes to burn the retorted shale as fuel for the ball heater, thus increasing energy recovery, and freeing the valuable retort gases for other uses.

The retort's chief disadvantages are its mechanical complexity and the large number of moving parts. The ceramic balls are consumed over time. The need for a fine feed size results in crushing costs that are higher than those of systems that can handle coarser feeds. On the other hand, all crushing operations produce some fine shale that could be screened from the feed to coarse-shale retorts and converted in auxiliary Tosco II units. Disposal of Tosco II spent shale presents some problems because it is very finely divided and contains carbon.

Tosco was participating in two projects that were committed to its proprietary retorting technology. The Colony project has been mentioned previously. The other venture, the Sand Wash project, was being developed on 14,000 acres of land in the Uinta Basin leased from the state of Utah. The Sand Wash was proceeding toward commercialization in compliance with the due diligence requirements of the state leases.

The Lurgi–Ruhrgas Indirectly Heated Retort. The Lurgi–Ruhrgas retorting system uses a class 4 retort in which hot retorted shale carries pyrolysis heat to oil shale. The process was developed jointly by Ruhrgas A.G. and Lurgi Gesellschaft für Warmetechnik

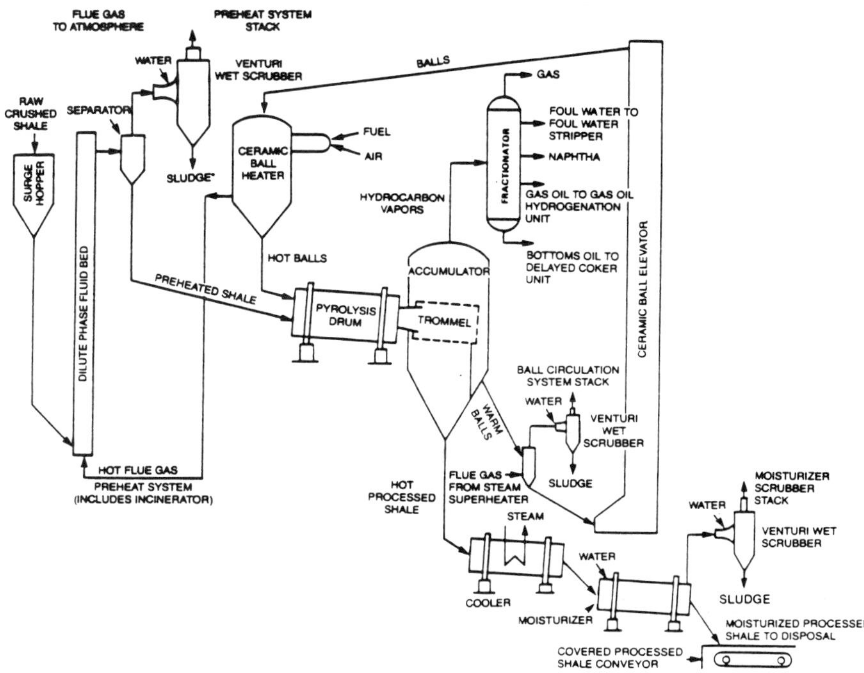

Source: ***Oil Shale Retorting Technology*, prepared for OTA by Cameron Engineers, Inc., 1978.**

Fig. 6-24. The Tosco II oil shale retorting system.

m b.H.—two West German firms that have been involved in synfuels production for decades. The process was developed in the 1950s for low-temperature coal carbonization. A 20-ton/day pilot plant was built in West Germany to test a variety of coals, oil shales, and petroleum oils. European shales were tested in the late 1960s, and Colorado oil shale in 1972. The plant has since been scrapped, but smaller retorts are available.

Coal processing plants using the Lurgi–Ruhrgas technique have been built in Japan, West Germany, England, and Argentina. There are also two Yugoslavian plants, each built in 1963, with a capacity of 850 ton/day of brown coal. The Japanese plant is also of commercial size, and uses the process to crack petroleum oils to olefins. No large-scale oil shale plants have yet been built.

The retorting system is shown in Fig. 6-25. Finely crushed oil shale is mixed with hot retorted oil shale in a mechanical mixer that resembles a conventional screw conveyor. Retorting takes place in the mixer, and gas and shale oil vapors are withdrawn. Dust is removed from these products in a cyclone separator, and oil is separated from the gas by condensation. Retorted shale leaves the mixer and is sent to a lift pipe where it is heated to about 1100°F (595°C) in a burning mixture of fuel gas and air. The hot retorted shale is then sent back to the mixer, and the process is repeated.

High oil yields have been reported for the retort, and the product gas is of high quality. Except for the mixer, the process is mechanically simple and has few moving parts. It should be capable of processing most oil shales, if they are crushed to a fine size. The two

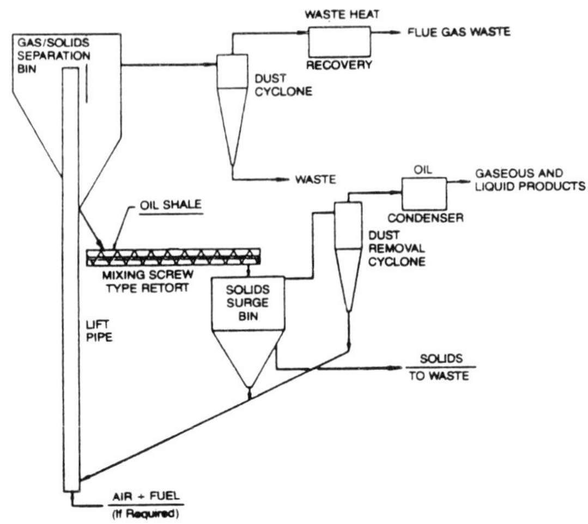

Source: *Oil Shale Retorting Technology*, prepared for OTA by Cameron Engineers, Inc., 1978.

Fig. 6-25. The Lurgi–Ruhrgas oil shale retorting system.

major problems with the system appear to be accumulation of dust in the transfer lines and dust entrainment in the oil. Dusty crude oil is not a severe problem because the dust is concentrated in the low-value residual product when the oil is subsequently refined. As with the TOSCO II process, requirements for a fine feed material will result in high crushing costs. These costs would be partially offset by the ability to process the fine fraction from any crushing process, including those used to prepare shale for coarse-shale retorts.

In 1972, Lurgi proposed to develop its retorting technologies with Colorado oil shale. Lurgi's interest in commercializing the technique has continued. In recent years Lurgi has been working with Dravo Corp. to interest U.S. firms in using the technology. At present, at least one company—Rio Blanco—has obtained a license to use the Lurgi–Ruhrgas retorts. Present plans call for constructing a modular Lurgi–Ruhrgas retort that will be close to commercial size (2200 ton/day). It will be used to retort the shale that will be mined during the preparation of modified *in situ* retorts on tract C-a.

Advantages and Disadvantages of the Processing Options

The greatest advantages of true *in situ* processing are that mining is not required and spent shale is not produced on the surface. The technical, economic, and environmental problems associated with aboveground waste disposal are thereby avoided. Modified *in situ* processing does involve mining and aboveground waste disposal, although to a lesser extent than with aboveground retort. However, the modified *in situ* waste is either overburden or raw oil shale. Both materials are found naturally exposed on the surfaces of deeper canyons in the oil shale basins. Although raw shale has low concentrations of the soluble salts, it does contain soluble organic materials that could be leached from the disposal piles. It should be noted that the presence of spent shale underground has the potential to cause environmental problems because soluble salts could be leached by groundwater. Therefore, environmental controls will also be needed for true *in situ* and modified *in situ* methods.

Surface Facilities. True *in situ* has another theoretical advantage in that the required surface facilities are minimal, consisting only of wells, pumps, gas cleaning and product recovery systems, oil storage, and a few other peripheral units. These facilities would probably resemble those for processing of crude oil and natural gas. Modified *in situ* requires more facilities to support the mine and the waste disposal program. Aboveground processing, which uses the largest number of facilities, causes the most surface disruption.

Oil Recovery

The advantage of aboveground retort is that the conditions within the retorts can be controlled to achieve very high oil recoveries—approaching or even in some cases exceeding the yields achieved with Fischer assay retorts. Retorting efficiencies are usually lower for modified *in situ* processing and much lower for true *in situ* because of the difficulty in obtaining a uniform distribution of broken shale and void volume, which, in turn, makes it difficult to maintain uniform burn fronts and leads to channeling and bypassing of the heat-carrier gases. The few estimates of the retorting efficiencies of true *in situ* operations that have been published have not been encouraging. (BOM achieved recoveries of 2–4% in its field tests.) Modified *in situ* retorting has not exceeded 60% recovery of the potential oil in the shale within the retorts. It is expected that yields from modified *in situ* retorts could be increased by injecting steam or hydrocarbon gases (as is done in Equity's true *in situ* process), but it is doubtful that recoveries can reach those of carefully controlled aboveground retorts. On the other hand, the present low efficiencies of modified *in situ* operations are partially compensated for by their ability to convert very large sections of an oil shale of a lower grade than would be practical for aboveground retort and by their lower cost of preparing the shale for retorting.

It is difficult to compare overall recoveries from modified *in situ* and aboveground retort without making numerous assumptions about the operating characteristics of both systems. To make a rough comparison, it could be assumed that aboveground retort (with room-and-pillar mining) and modified *in situ* were to be applied to two 300-ft-thick deposits with identical physical characteristics. The net recoveries from several development options are summarized in Table 6-1. The highest recovery (100%) is for full-subsidence mining in conjunction with aboveground retort, but it would result in extensive surface disturbance which could be reduced by backfilling the mined-out areas with spent shale from surface processing. The retorted shale in burned-out modified *in situ* retorts would also reduce the severity of subsidence.

The three generic approaches to oil shale processing also have various other advantages and disadvantages with respect to water needs, environmental effects, financial requirements, and social and economic impacts.

Crude shale oil (also called raw shale oil, retort oil, or simply shale oil) is the liquid oil product recovered directly from the offgas stream of an oil shale retort. Synthetic crude oil (syncrude) results when crude shale oil is hydrogenated. In general, crude shale oil resembles conventional petroleum in that it is composed primarily of long-chain hydrocarbon molecules with boiling points that span roughly the same range as those of typical petroleum crudes. The three principal differences between crude shale oil and conven-

Table 6-1. Overall Shale Oil Recoveries for Several Processing Options

Case	First-stage retorting technology	First-stage mining method	Second-stage processing	Overall shale oil recovery[a]
1	AGR	Conventional room-and-pillar mining on three levels. 60-ft rooms; 60-ft barrier barrier and sill pillars	None	36%
2	MIS	Conventional MIS mining. 40% of shale left behind in barrier pillars. 20% of the disturbed shale is removed to the surface to provide void volume in the retorts	None	29%
3	MIS	As in case 2	AGR processing of the mined shale.	41%
4	AGR	As in case 1	Barrier and sill pillars collapsed and retorted by MIS	74%
5	MIS	Full subsidence[b]	None	48%
6	MIS	Full subsidence[b]	AGR processing of the mined shale	68%
7	AGR	Full subsidence[b]	None	100%

[a] Assuming AGR recovers 100 percent of the potential oil in the shale retorted. MIS assumed to recover 60 percent of the potential oil.
[b] Entire deposit is mined.
Source: Office of Technology Assessment.

tional crude are a higher olefin content (because of the high temperatures used in oil shale pyrolysis), higher concentrations of oxygen and nitrogen (derived from oil shale kerogen), and, in many cases, higher pour point and viscosity.

The physical and chemical properties of crude shale oil are affected by the conditions under which the oil was produced. Some retorting processes subject it to relatively high temperatures, which may cause thermal cracking and thus produce an oil with a lower average molecular weight. In other processes (such as directly heated retorting) some of the lighter components of the oil are incinerated during retorting. The result is a heavier final product. Others may produce lighter products because of refluxing (cyclic vaporization and condensation) of the oil within the retort. One of the most important factors is the condensing temperature within the retorting system—the temperature at which the oil product is separated from the retort gases. The lower this temperature, the higher the concentration of lower-molecular-weight compounds in the product oil.

The properties of crude shale oil from several aboveground and modified *in situ* retorting processes are listed in Table 6-2. It is important to note that the oils that are characterized were produced in small-scale test runs under conditions that may not be representative of those that will be encountered in other areas and with larger processing systems. The oils from commercial-scale facilities in other parts of the oil shale region may have properties that are quite different.

The properties of the oil produced by different aboveground retort processes vary widely, but the difference between those oils and *in situ* oils are much more significant. *In situ* oils are generally much lighter, as indicated by their higher yields of material with relatively low boiling points, and would produce more low-boiling-point product (such as gasoline) and less high-boiling-point product (such as residual oil). In general, the low yields of residuum make shale oils attractive as refinery feedstocks in comparison with many of the heavy conventional crudes that are currently being processed in the United States. For example, *in situ* oil from Oxy's modified *in situ* process contains 1–4% of materials with a boiling range of over 1000°F (535°C), compared with 20% crude from Alaska's North Slope, and 18% for Arabian light crude. Other foreign crudes, such as Kuwait and Arabian heavy, contain even larger residuum which, although higher than the contents of *in situ* crudes, is substantially lower than that for many conventional crudes.

Coal-derived liquids are often regarded as alternatives to shale oil feedstocks. However, syncrudes from coal have a much higher yield of gasoline and low-boiling-point distillates than shale oil, with little or no material boiling at temperatures above 850–1000°F (455–535°C). The coal liquids would be well suited for gasoline production because their higher concentrations of lower-boiling-point constituents when refined would yield the desired light naphtha fractions. Shale oil, on the other hand, has a much higher concentration of high-boiling-point compounds, and would favor production of middle distillates (such as diesel fuel and jet fuels) rather than naphtha. Shale oil and coal-derived syncrudes should, therefore, be regarded not as competitive or substitutable feedstocks but rather as complementary feedstocks, with each yielding a different major fuel product from an equivalent amount of refining.

Among the negative characteristics of most crude shale oils are high pour point, high viscosity, and high concentrations of arensic and other heavy metals and of nitrogen. The pour point and viscosity are of economic importance because transporting viscous oil that has a high pour point is difficult and costly, thus suggesting the need for pretreatment prior to marketing.[9] As shown in Table 6-2 *in situ* oils with their relatively low pour points and viscosities could be marketed without pretreatment, but they would retain their high nitrogen contents. This would reduce their value as refinery feedstocks and boiler fuels.

High concentrations of arsenic and other metals are a disadvantage because they poison refining catalysts, especially in hydrogenation units. They must be removed prior to catalytic processing, and a variety of physical and chemical methods have been developed for this purpose. It should be noted that the concentrations of heavy metals in crude shale oil will vary with the location of the deposit from which the oil is recovered. Oils from some sites may be relatively free of such contaminants.

SHALE OIL REFINING

Shale oil has been successfully refined in oil shale operations in Sweden, Scotland, Australia, West Germany, the U.S.S.R., and other countries, although on a relatively small scale and under unusual economic conditions. In the United States, the initial refining research was conducted by BOM at the Petroleum and Oil Shale Experiment Station at Laramie, Wyoming. It was coordinated with the early development of the gas combustion retort at Anvil Points, Colorado. The results of this work, plus the findings of other investigators, allowed a preliminary assessment to be made of the economic aspects of shale oil utilization, and justified continued efforts aimed at its recovery. In recent years, refining R&D has been revived because refiners now consider the availability of shale oil to be a distinct possibility. There is also a need to perform more precise and up-to-date economic analyses.

[9] Some conventional crudes have comparably high pour points. For example, the Altamont crude from Utah has a pour point of about 100°F (35°C).

Table 6.2. Properties of Crude Shale Oil From Various Retorting Processes

Retort	Aboveground processes							
	Fischer assay	NTU			Gas combustion		Tosco II	
Retort, weight percent								
Carbon, weight %	84.59	—	84.61	84.58	83.92	—	85.1	—
Hydrogen, weight %	11.53	—	11.40	11.76	11.36	—	11.6	—
Oxygen, weight %	—	—	1.10	—	1.67	—	0.8	—
Nitrogen, weight %	1.96	2.20	2.10	1.77	2.14	2.12	1.9	1.85
Sulfur, weight %	0.61	0.92	0.79	0.76	0.70	0.68	0.9	0.67
Carbon/hydrogen ratio	7.34	—	7.42	7.19	7.39	—	7.34	—
Gravity, °API	—	19.4	20.3	25.2	19.8	21.1	21.2	—
Specific gravity	0.92	—	—	—	—	—	0.93	0.93
Pour point, °F	80	80	90	70	84	85	80	—
Arsenic, p/m[a]	—	—	—	—	—	—	—	—
Nickel, p/m	—	—	—	—	6.4	—	6	—
Iron, p/m	—	—	—	—	108	—	100	—
Vanadium, p/m	—	—	—	—	6	—	3	—
Distillation, volume %								
5 @ °F	—	—	—	—	378	445	200	—
10 @ °F	326	493	—	—	438	497	275	360
20 @ °F	430	560	—	—	529	589	410	480
30 @ °F	518	—	—	—	607	668	500	600
40 @ °F	—	—	—	—	678	742	620	690
50 @ °F	655	670	—	—	743	808	700	780
60 @ °F	685	692	—	—	805	865	775	860
70 @ °F	705	—	—	—	865	918	850	935
80 @ °F	—	—	—	—	935	984	920	1,020
90 @ °F	—	—	—	—	1,030	1,065	—	—
95 @ °F	—	—	—	—	1,099	—	—	—

[a]Parts per million.
[b]Vacuum distillation at 40 mm of mercury. All other results are from atmospheric distillation.
Source: Office of Technology Assessment.

Aboveground processes				In situ processes				
Union Oil		Paraho direct		Occidental oil shale			USBM	
"A"	"B"							
84.0	84.80	84.90	—	84.86	—	—	84.60	—
12.0	11.61	11.50	—	11.80	—	—	12.08	—
0.9	0.90	1.40	—	1.13	—	—	1.18	—
2.0	1.74	2.19	2.00	1.50	—	—	1.55	1.41
0.9	0.81	0.61	0.71	0.71	—	—	0.59	0.72
7.0	7.30	7.38	—	7.19	—	—	7.00	—
18.6	22.7	19.3	—	25	—	24.3	28.4	28.4
—	0.81	0.94	0.94	0.90	—	—	0.89	—
80	60	85	—	40	—	65	40	40
—	—	19.6	—	—	—	—	—	—
6	—	2.5	—	—	—	—	—	—
55	—	71.2	—	—	—	—	—	—
1.5	—	0.37	—	—	—	—	—	—
390	—	—	—	—	400	405	—	—
465	400	520	530	440	460	—	382	620
565	—	600	610	—	520	—	405	648
640	—	680	680	600	570	600	—	666
710	—	750	745	—	620	—	343[b]	681
775	731	810	805	700	680	—	375[b]	699
830	—	840	860	—	715	—	415[b]	732
980	—	—	910	770	765	—	464[b]	768
—	960	—	980	—	825	830	528[b]	809
—	—	—	1,040	920	910	—	—	867
—	—	—	—	—	1,050	1,000	—	914

To date, refining studies have been conducted on the upgrading of crude shale oil to a transportable product and on the total refining of shale oil into finished fuels. The difference between these operations lies in the nature of the desired final product. As discussed previously, some crude shale oils have pour points and viscosities that make transporting them difficult and expensive. In some situations, economic considerations may dictate that the crude shale oil be partially refined (upgraded) near the retorting site to improve its transportation characteristics. In other instances, a developer may desire to obtain a complete array of finished fuels from an integrated processing facility located near the minesite. In this case, a total-refining facility would be considered rather than a more simple upgrading plant.

To date, upgrading experiments have been carried out largely at the bench scale, in relatively small pilot plants. Theoretical studies and computer modeling have also been used to evaluate the expected performance of three types of upgrading processes: thermal, catalytic, and additive. Thermal processes include visbreaking (a relatively mild treatment) and coking (a severe treatment). Mild thermal treatment will reduce pour point and viscosity, but the oil will retain its initial amounts of nitrogen and sulfur. In contrast, severe thermal treatment reduces pour point, viscosity, and sulfur content in the heavier products. The properties of the lighter products will thus be considerably improved.

In catalytic processes, the shale oil is reacted with hydrogen in the presence of a catalyst. Viscosity is reduced, and the nitrogen and sulfur are converted to ammonia and hydrogen sulfide gases that can be recovered as by-products. In additive processes, blending agents are added that reduce the pour point and allow the crude to be transported by pipeline; such pour point depressants have been added in several instances with success, but the technique is not yet highly developed.

Total refining studies have focused either on the needs of existing refineries that would have to be modified for processing shale oil, or on those of newly built facilities that could be designed specifically for shale oil feedstocks. These studies differ in their approach to the analysis of refining requirements. Studies of existing refineries must consider the equipment that is in place and must allow for the limited flexibility of this equipment for processing a feedstock that is different from the one for which the refinery was designed. Studies of specially built refineries, in contrast, need not be biased in this manner, and can draw upon any processing technique that is available within the refining industry. However, both types of studies must make assumptions about feedstocks characteristics and desired product mixes. These will vary with the location of the refinery, the nature of the market it serves, and the type of retorting facility that supplies its feedstocks. The optimal refining conditions for one set of assumptions will probably not be applicable to another set. For example, a refining method to maximize gasoline production from TOSCO II shale oil would not maximize diesel fuel production from Oxy's *in situ* oil.

Numerous computer studies and bench-scale refining investigations have been conducted for a wide range of shale oil feedstocks and operating conditions. The results of these studies can be extrapolated, with some degree of caution, to predict the performance of commercial-scale refineries. However, refining tests both in pilot plants and in commercial-scale facilities, because of much higher costs, have focused on only a few feedstocks, and have been conducted for particular sets of operating constraints. In general, each large-scale study has dealt only with oil from aboveground retorts or with oil from *in situ* operations, but usually not with both types of feedstocks. The conclusions of all studies are highly dependent on the combination of feedstock and refining conditions assumed. Caution must be used when applying the results to different conditions.

Shale Oil Upgrading Processes

The treatment techniques that can be used to improve the transportation properties of crude shale oil are briefly described below.

Visbreaking. This technique involves heating the crude shale oil to approximately 900–980°F (480–525°C) and holding it in this temperature range for from several seconds to several minutes. The product is then cooled, and the gases evolved during the heating are removed. There is little reduction in the contents of nitrogen, sulfur, and oxygen; therefore, the principal improvements are reductions in pour point and viscosity. This technique is simple but energy intensive. It could reduce the poor point of crude shale oil from about 85°F (30°C) to about 40°F (4°C).

Coking. This process involves heating the oil to about 900–980°F (480–525°C) and then charging it into a vessel in which thermal decomposition occurs. If the vessel is a coke drum, the process is called delayed coking. The coke—the solid product from thermal decomposition—is allowed to accumulate until it fills about two-thirds of the drum's volume. The feed is then switched to another drum while the coke is cleaned out of the first one.

In the fluid coking process, hot oil is charged into a vessel that contains a fluidized bed of coke particles. The particles become coated with oil, which then decomposes to yield gases and another layer of coke. The gases are withdrawn from the vessel. The coke is also withdrawn continuously, at a rate sufficient to maintain an active stock of coke within the bed.

The Flexicoking process, developed by Exxon, combines conventional fluid coking with gasification of the product coke. The advantage is that energy is recovered from the coke.

Catalytic Hydrogenation. In these processes, the crude shale oil is reacted with hydrogen in the presence of a catalyst. The sulfur in the oil is converted to hydrogen sulfide, the nitrogen to ammonia, the oxygen to water, the olefin hydrocarbons to their paraffin equivalents,[10] and long-chain molecules to smaller molecules. The hydrogenation reactions can take place in a fixed-bed reactor through which a mixture of oil vapors and hydrogen is passed, in a fluidized-bed reactor, or in an ebulliating-bed reactor. The latter technique is being promoted by Hydrocarbon Research, Inc. as the H-Oil process: Plants in several foreign nations have successfully used it for heavy petroleum feedstocks and for residuum fractions from a variety of crudes. In this process, a mixture is injected into the bottom of a reactor at a high enough velocity to cause catalyst ebulliation (a boiling motion). This movement reduces the possibility that the bed will become plugged by coke and by the liquid tars that are formed during the coking process. It also allows spent catalyst and coke to be removed, and fresh catalyst to be added, so as to keep the bed actively stocked.

Catalytic hydrogenation produces upgraded products of the highest quality, but it is relatively expensive. The use of fixed-bed reactors would probably be confined to the treatment of streams from an initial fractionation step: fluid-bed or ebulliating-bed processes could be used for either fractionator products or for the whole shale oil.

[10] Olefins are unsaturated (lower ratio of hydrogen to carbon) hydrocarbon compounds having at least one double bond. They are the source of synthetic polymers such as polyethylene and polypropylene used in the manufacture of fibers and other materials. Paraffins are saturated hydrocarbons (higher ratio of hydrogen to carbon) having only single bonds. Methane, ethane, propane, and butane are some of the paraffin hydrocarbons.

Additives. Chemicals may also be added to crude shale oil to improve its transportation properties. Pour point depressants have been successful in some instances, but this does not mean that they would always work. A chemical that is suitable for one type of oil may not work at all with oil from another retorting process. Furthermore, pour point depressants are disadvantageous because they only change the physical characteristics of the oil and not its chemical properties. Thus, their cost can be offset only if they save transportation costs.

Conventional petroleum crudes are other potential blending agents. Since shale oil (at least in Colorado) will be produced in an area that also contains petroleum reserves, and even active crude oilfields, the possibility exists that the light petroleum crudes could be mixed with crude shale oil to form a transportable blend. The feasibility of this concept is unclear because, in general, the blend would not be as valuable as a refinery feedstock on a per-unit basis as would petroleum alone. However, the decrease in unit value would be offset by the increased volume. In the case of a refinery that does not have a reliable supply of crude, this could be a significant advantage.

Total Refining Processes

The three primary factors that affect the design of a refining system are (1) the characteristics of the crude shale oil feedstock; (2) the desired mix of finished products; and (3) the constraints imposed by the equipment and operating practices of the proposed refinery.

The first factor probably will have the lowest effect because, except for the higher nitrogen and arsenic contents, the characteristics of shale oil are not widely different from those of conventional petroleum. The second factor is much more significant. This is evidenced by the changes that have occurred in the proposed configurations of shale oil refineries since the 1950s: the earlier studies placed much more emphasis on gasoline production. For example, early design by BOM called for the extensive use of middle-distillate cracking and reforming to yield gasoline. The ratio of gasoline to distillate yields was nearly 3 to 1. Most of the refinery configurations that have been proposed more recently indicate a gasoline-to-distillate ratio of about 1 to 4.

The third factor has become increasingly important in recent years. The modifications to convert a conventional refinery to shale oil feedstocks might not be economically justifiable unless the refiner could be assured of an adequate supply of shale oil. The economic desirability of building a refinery specifically for shale oil would be thoroughly scrutinized. Modular retorts, or even a few pioneer commercial plants, would not produce enough shale oil in the mid-term to justify a new refinery unless the refiner was assured that the operations would continue until his investment cost could be recovered. For this reason, the most recent studies have stressed modifying existing facilities to make them suitable for processing shale oil, rather than building new ones. In some cases, this entails only minor changes to installed equipment, in others, the adaption of an existing facility by adding new units.

The basic unit operations in crude oil refining are coking, hydrotreating, distillation, hydrocracking, catalytic cracking, and reforming. The various refining schemes that have been proposed for shale oil cannot easily be generalized because, depending on both the desired product mix and the possible operating conditions, many configurations could be designed that would achieve the same results. The one selected will largely depend on the availability of equipment and the individual economics of the particular refinery.

One significant difference among the various configurations described in the literature is the relative arrangement of distillation and thermal or catalytic treatment. Two general approaches have been investigated: (1) distillation of the whole crude into its components, then catalytic or thermal treatment; or (2) catalytic or thermal treatment of the whole crude, then distillation. In the first approach the properties of the finished products are better controlled. In the second, the net load on successive processing units is reduced, and, in general, the overall yield of high-value hydrocarbon products is increased. Most of the refining research to date has been focused on the second approach. Three versions are shown schematically in Figs. 6-26–6-28. The scheme in Fig. 6-26, which was used by BOM at Anvil Points in the late 1940s, has a gasoline-to-distillate ratio of 3 to 1. Figure 6-27, a configuration that was investigated by Chevron U.S.A.

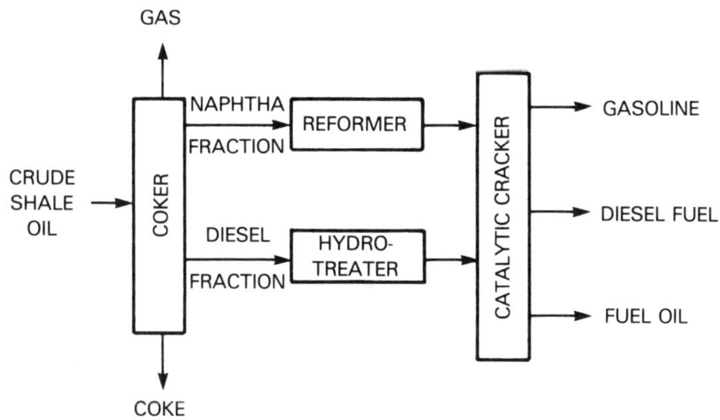

Source: G.L. Baughman, *The Refining and Marketing of Crude Shale Oil.*

Fig. 6-26. Refining scheme used by the U.S. Bureau of Mines to maximize gasoline production from shale oil.

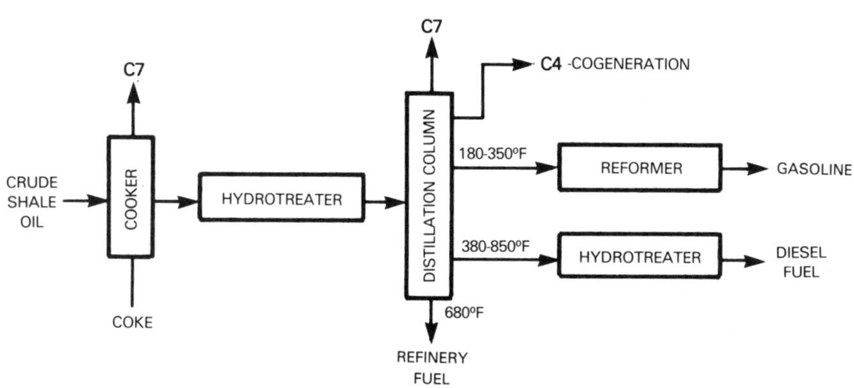

Source: G.L. Baughman, *The Refining and Marketing of Crude Shale Oil.*

Fig. 6-27. Refining scheme employing coking before an initial fractionation, used by Chevron U.S.A. in refining experiments.

during pilot plant runs on Paraho shale oil, had a gasoline-to-distillate ratio of 1 to 4. Both of these systems used an initial coking step to upgrade the feedstock and to supply a product stream more easily refinable into finished fuels. The disadvantage of coking is that there might not be a ready market for coke in the vicinity of the refinery, particularly if it were located in the oil shale region.

Another refining scheme that was investigated by Chevron is shown in Fig. 6-28. It uses a fixed-bed catalytic hydrotreater to upgrade the shale oil before distillation. No coke is produced because most of the heavier components of the crude shale oil are upgraded into lighter and more valuable liquid fuels during the hydrogenation process. This method may be more costly than the coking approach, but that can only be determined by operating experience and evaluating the economics.

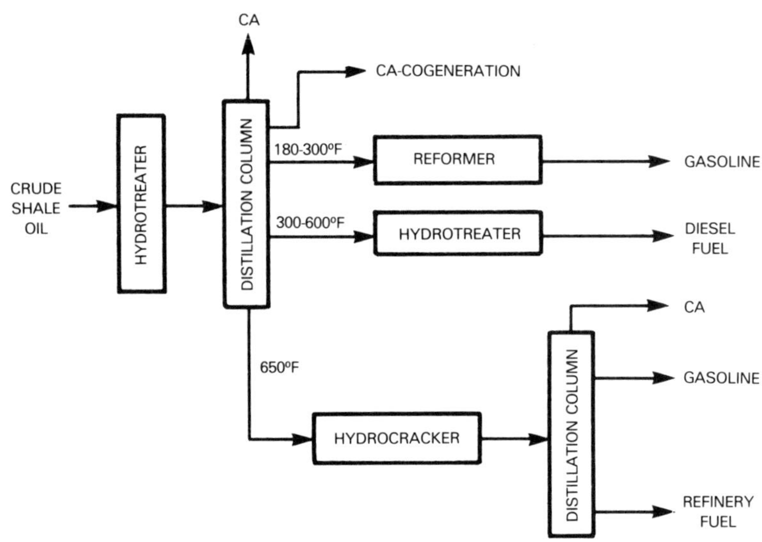

Source: G.L. Baughman, *The Refining and Marketing of Crude Shale Oil.*

Fig. 6-28. Refining scheme employing initial hydrotreating, used by Chevron U.S.A. in refining experiments.

Chevron also investigated the possibility of substituting a fluidized catalytic cracker for the hydrocracker in Fig. 6-28. A similar approach was used by SOHIO in processing 85,000 bbl of Paraho shale oil at its Toledo refinery. The chief difference between the SOHIO runs and the Chevron experiments is that SOHIO used an acid/clay treatment to upgrade the distillation products into jet fuel and marine diesel fuel for military applications. The residuum fraction from the column was used for fuel in the refinery.

The approach in which the crude oil is fractionated before hydrogenation or other treatment is shown in Fig. 6-29. This scheme was used by SOHIO during the prerefining studies carried out before 10,000 bbl of Paraho shale oil were refined at the Gary Western refinery in Fruita, Colorado. During the actual refinery run, a combination coker/fractionator was used rather than the separate units shown in the diagram.

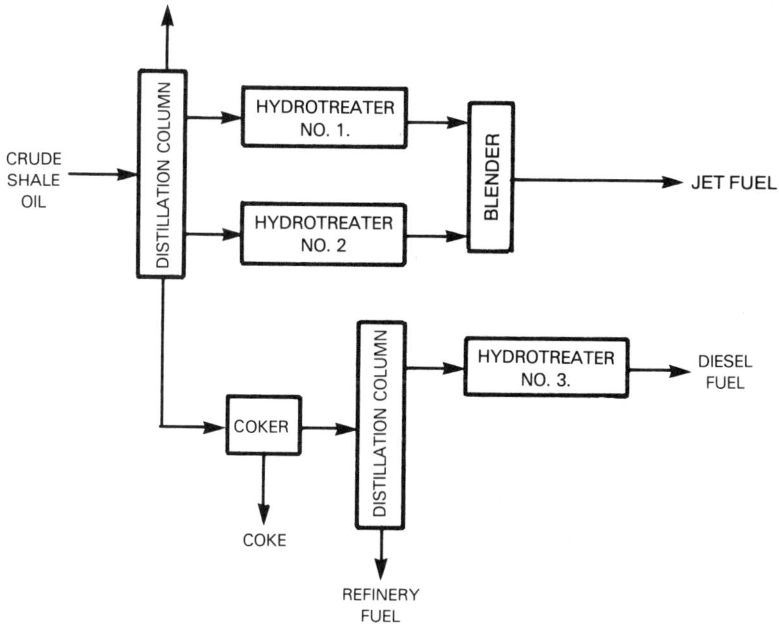

Source: G.L. Baughman, *The Refining and Marketing of Crude Shale Oil.*

Fig. 6-29. Refining scheme employing initial fractionation, used by SOHIO prerefining studies.

The shale oil must also be treated to remove excess amounts of water, ash, and heavy metals such as arsenic. Water must be removed because it can cause cavitation in pumps and explosions in processing units. Ash, or particulate matter, must be removed to prevent its deposition in pipes, heat exchangers, and catalyst beds. Recent studies have shown that heating the crude oil to about 165°F (75°C), then letting it stand for about 6 hours, allows the water and solid matter to separate from the oil. As noted, arsenic and other metals poison the hydrotreating catalysts. A variety of processes have been developed for their removal; consequently, their presence no longer presents a technical problem. ARCO has patented several catalytic techniques and methods for heat treating the oil in the presence of hydrogen. Recent studies by Chevron U.S.A. have shown that an alumina guard bed preceding the hydrotreater will effectively remove both arsenic and iron from shale oil.

The quantity and quality of the fuels produced will be determined by both the configuration of the equipment and the operating conditions used in the refining step. The fuels produced by BOM at Anvil Points in the 1940s were quite satisfactory. However, some of the fuels from the Gary Western refining run in 1975 failed to meet certain military specifications, principally those for stability. This has been attributed to the application of refining techniques unsuitable for shale oil feedstocks, specifically inadequate hydrogenation. Subsequent refining tests at SOHIO's Toledo refinery showed that, with appropriate refining, fuels can be produced that are of superior quality and that can meet all applicable specifications.

Cost of Upgrading and Refining

One of the most recent estimates of the cost of upgrading crude shale oil to a transportable refinery feedstock have been prepared by Chevron U.S.A. The retorting complex that was considered had a capacity of 100,000 bbl/day. Conventional hydrotreating was the upgrading technique evaluated. Chevron considered two possible locations for the upgrading facility: a newly built unit at the retorting site or a unit to be added to an existing refinery at some distance from the retorts. In both cases, the estimated cost for upgrading 1 bbl of crude shale oil was $6.50, in first-quarter 1978 dollars. The product would be a high-quality syncrude suitable as feedstock for most refineries in the United States.

It is more difficult to estimate the costs of the total-refining option for converting crude shale oil into finished fuels, because, in addition to the properties of the crude oil, consideration must be given to the location of potential refinery sites, the availability of refining equipment, the proximity and stability of potential markets, the ease of product distribution, and other factors. Chevron considered some of these in its analysis of refining costs, although in a relatively generalized manner. Three total-refining options and two refining capacities and refinery locations were considered. For a 100,000-bbl/day refinery located in an urban area in the Rocky Mountains (e.g., Denver), the refining cost was estimated to range from $8.00 to $10.00/bbl of crude shale oil. For a 50,000-bbl/day refinery located in a remote area of the Rocky Mountains (e.g., near the retorting facility), the refining cost would be approximately $10.00–$12.00/bbl.[11] These are somewhat higher than the costs for refining a high-quality conventional crude oil because of the additional amounts of hydrogen that would be needed to reduce the nitrogen content of the shale oil crude.

Another study compared the cost of shale oil refining with those of refining Wyoming sour crude oil and Alaskan crude. It was assumed that a refinery in the Rocky Mountain region was modified for these feedstocks. The increased costs to refine crude shale oil, rather than the other crudes,was in the range of $0.25–$2.00/bbl.

MARKETS FOR SHALE OIL

Crude shale oil has three major potential uses: as a boiler fuel, as a refinery feedstock, and as a feedstock for producing petrochemicals. The output from a mature oil shale industry will probably be used for all three purposes; however, the relative importance of the three markets will change with time as the industry develops. In the mid-1980s, when shale oil first becomes available in significant amounts, its most likely use will be as boiler fuel, with a smaller quantity directed to nearby refineries that could be modified to accommodate the feedstock without large capital expenditures. As more shale oil becomes available, its use as a refinery feedstock will increase as conventional petroleum becomes scarcer. The military will develop early efforts to utilize oil shale for aircraft and vehicle fuel. At a later date, when the market for boiler fuels declines, shale oil will begin to be used for petrochemical production.

Shale Oil as a Boiler Fuel

Shale oil will most likely first be used as a boiler fuel because of the relatively small capital investments and very short leadtimes that would be required. Because of government regulation, the current trend in the utility industry is to replace oil- and

[11] Costs are in first-quarter 1978 dollars.

gas-fired boilers with coal-fired units, thus freeing the natural gas for domestic consumers. In some areas, it will be a two-stage transition, with the gas first replaced by oil, and the oil later by coal. During the transition, there may be a market for about 50,000–80,000 bbl/day of crude shale oil near the oil shale region. In addition, the refining industry has a small but significant demand for boiler fuel because refineries are also changing from natural gas to oil. Therefore, refineries located near the oil shale region are likely to be near-term shale oil customers. In the long run, the largest market for shale oil boiler fuel is likely to be in the Great Lake states. Transportation distances will probably preclude its use in the other two major markets for boiler fuels—the east and west coast.

Shale Oil as a Refinery Feedstock

There has been relatively little research on the refining of crude shale oil because building an oil shale plant takes about 5 to 7 years, whereas a refinery operator can evaluate a new potential feedstock in a few days, develop a feasible refining strategy within a matter of weeks, conduct the necessary pilot plant refining studies in a few months, and modify the refinery to accommodate the new feedstock in less than 3 years. Thus, neither the developer nor the refiner has any inducement to study shale oil refining until they are sure that the oil will, in fact, be forthcoming.

Another reason is that, until quite recently, shale oil was not considered a highly desirable refinery feedstock. During the 1950s and 1960s, the refining industry tended to maximize gasoline yields at the expense of middle and heavy distillates. Because shale oil is a good source of heavier distillates, not of gasoline, it was not highly regarded. However, most projections indicate that gasoline demand will peak in the early 1980s and then decline slightly, even though total demand for refined products will continue to grow. This will be the result of the increasing efficiencies of gasoline engines in automobiles and of a greater use of diesel engines in automobiles and light trucks. Also, because the current supplies of conventional petroleum are becoming more like shale oil with respect to their distillate yields, the refining industry is being forced to adopt techniques that would be equally suitable for shale oil.

For these reasons, shale oil's desirability is increasing, and its potential availability as a premium feedstock has encouraged the refining studies that have been conducted by Chevron and other organizations. These studies have dealt with four general types of refining facilities: (1) a new refinery just for shale oil; (2) a new refinery for a mix of shale oil and conventional crude; (3) an existing refinery modified for, and dedicated to, shale oil; and (4) an existing refinery processing a mix of shale oil and conventional crude. The first two approaches are precluded for the near future because, as long as refined products continue to be imported, the United States will have excess refining capacity. At least through the 1980s, shale oil will most likely be refined in existing refineries, either by itself or as a blend with conventional petroleum.

The shale oil produced by demonstration facilities will probably be processed in local refineries[12] or in more distant refineries owned and operated by the energy companies that participate in the oil shale programs. The much larger output from a commercial-size industry will be more widely distributed; this it will have to compete with other feedstocks, at least regionally. Recent studies have indicated that the Midwest is the most likely market area for large quantities of shale oil. This includes the states in the Petroleum Administra-

[12] These could include the Gary Western refinery in Fruita, Colorado, the Little America refinery in Rawlins, Wyoming, the Chevron refinery in Salt Lake City, Utah, and others.

tion for Defence District 2 (PADD 2), as shown in Fig. 6-30. There will be secondary effects in other districts, as a consequence of the supplies of shale-derived fuels in PADD 2, because the conventional petroleum that is displaced will become available for use elsewhere.

The quantitative impact on the supply situation in PADD 2 can be determined by referring to Table 6-3, which indicates how the district's supplies of finished fuels were divided among domestic and foreign sources in 1978. As shown, the district consumed about 518,000 bbl/day of medium and heavy distillates (jet fuels, diesel fuel, and distillate fuel oil) from foreign sources. According to Chevron's studies, refining will convert about 74% of a crude shale oil feedstock to similar distillates. A 1-million-bbl/day industry would yield about 740,000 bbl/day of medium and heavy distillates. If it were marketed in PADD 2, this production would completely displace the foreign supplies and free an additional 222,000 bbl/day of the fuels for use in other districts. The same size industry would produce about 170,000 bbl/day of gasoline, which would be equivalent to about 17% of the district's gasoline obtained from foreign sources.

An alternative marketing strategy would be to sell the output from a major shale oil industry in PADD 4—the Rocky Mountain region—and to supply any surplus fuels to adjacent districts such as PADD 2 or PADD 3 (Texas and New Mexico). As shown in Table 6-3, the Rocky Mountain district consumes relatively little distillate fuel (about 20,000 bbl/day) from foreign sources. A 1-million-bbl/day shale oil industry could easily displace this entire supply. The surplus production (about 720,000 bbl/day) could displace about 95% of the supply of foreign-derived distillates in both PADD 2 and PADD 3. The gasoline derived from the shale oil could supply about 67% of the total gasoline demand in the Rocky Mountain states.

As indicated previously, the capabilities of the refineries in the Midwest and the Rocky Mountain states will strongly affect the willingness of the refiners to accept shale oil feedstocks. In some cases, shale oil could be accommodated without significant investments of capital. However, the receptivity of refiners to shale oil will also be influenced by the reliability of other feedstocks such as foreign petroleum. The area in which the supplies of crude are most uncertain is the northern tier of states, which includes Montana, North and South Dakota, Minnesota, and Wisconsin. These states have historically depended on refinery feedstocks from Canada, but, in recent years, a significant reduction in these supplies has led the refiners in this area to look elsewhere. The result has been the present interest in building a pipeline to transport crude from Alaska and from foreign nations into the area. An alternative—a pipeline from the oil shale region—could also be built as the oil shale industry developed.

The area that covers Iowa, Missouri, Illinois,Indiana, Michigan, and Ohio also does not have an adequate indigenous supply of crude. However, unlike the northern tier, these states have good pipeline systems with adequate access to both foreign and domestic crude supplies. The feasibility of marketing shale oil in this area will largely be determined by the cost differential between it and other crude supplies and by the differences in the reliability of its supply versus that of foreign crude. Recent marketing studies have identified several large refineries in this area that with only minor modifications would be able to handle crude shale oil. Furthermore, some of the refineries only have access to the heavier petroleum crudes at present, and their production is being limited by their capacity to process the large quantities of residuum from the distillation of these fuels. Shale oil, with its relatively small yield of bottoms fractions, would help alleviate this problem.

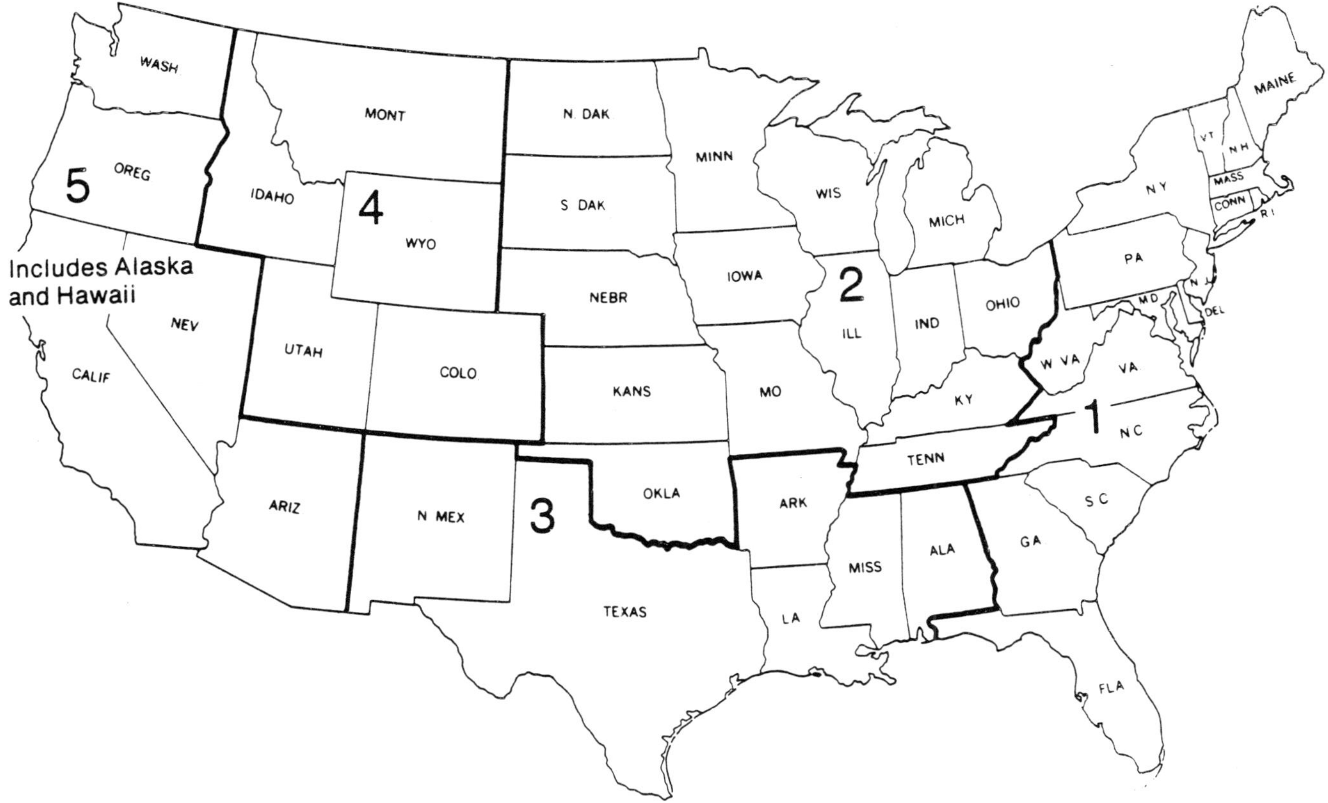

Source: *National Atlas*, Department of the Interior.

Fig. 6-30. The Petroleum Administration for Defense districts.

Table 6-3. Supply of Finished Petroleum Products in PADDs 2, 3, and 4 in 1978 (thousands bbl/day)[a]

	PADD 2: Midwest			PADD 3: Texas and New Mexico			PADD 4: Rocky Mountains		
	Foreign sources	Domestic sources	Total	Foreign sources	Domestic sources	Total	Foreign sources	Domestic sources	Total
Gasoline	976	1,540	2,516	422	604	1,026	33	219	252
Light ends	224	236	460	276	388	664	12	25	37
Jet fuel	81	131	212	50	71	121	7	29	36
Kerosine	17	27	44	19	27	46	0	2	2
Distillate fuel oil	420	672	1,092	194	279	473	13	113	126
Residual fuel oil	150	173	323	237	340	577	4	38	42
Petrochemical feedstocks	18	26	44	211	298	509	0	1	1
Special naphthas and still gas	60	98	158	114	165	279	2	14	16
Others	138	232	370	102	146	248	4	31	35
Total	2,084	3,135	5,219	1,625	2,318	3,943	75	472	547

[a]PADD = Petroleum Administration for Defense District.

Source: G.L. Baughman, *The Refining and Marketing of Crude Shale Oil.*

Shale Oil as a Petrochemical Feedstock

Three principal factors must be considered in evaluating the suitability of shale oil supplies for producing petrochemicals: (1) the yields of petrochemicals from shale oil feedstocks, (2) the ability of existing and future petrochemical plants to process the shale oil, and (3) the logistics of supplying the shale oil to the plants. Because shale oil is produced by pyrolysis, its olefin content is approximately 12%, which is appreciably higher than conventional crudes. Together with its fairly high hydrogen content, these characteristics make shale oil, and its hydrogenated derivatives, appropriate feedstocks for petrochemical production. Steam pyrolysis has been used to process crude shale oil, and the yields of olefin products have been comparable with those from many conventional crudes. Shale oil syncrudes, with even higher olefin yields, are considered to be premium petrochemical feedstocks. These conclusions are based on laboratory studies under carefully controlled conditions. The feasibility of marketing shale oil to the petrochemical industry depends on the ability to replicate these conditions in commercial chemical plants.

Historically, the primary feedstock for petrochemical plants has been natural gas liquids from the Gulf coast. The production of domestic natural gas and its associated liquids is declining, and the petrochemical industry is shifting to heavier feedstocks such as naphtha and gas oils. The supplies of these feedstocks are uncertain and irregular. The availability of naphtha has been affected by a growing demand for its use as a gasoline blending agent in response to the phasing out of tetraethyl lead. Gas oils are also being used more frequently for home heating fuels, which causes seasonal variations in the availability. Because of these supply uncertainties, new petrochemical plants are being designed to be highly flexible with respect to feedstocks. As using heavier feedstocks becomes more common in the industry, shale oil may become a highly regarded raw material.

The use of shale oil for petrochemical production is hampered by the distance between the oil shale region and the petrochemical plants. While refineries and oil-fired boilers are distributed fairly uniformly across the United States, the petrochemical industry is concentrated on the Gulf coast. This concentration will continue into the foreseeable future. Therefore, it will be necessary to either move the shale oil to the coast or build a new petrochemical complex in the Rocky Mountain region. In the latter case, the half-finished products from the new plant would still have to be transported to the coast for final conversion to commercial chemicals. The former approach is more likely but is impeded by the lack of a product pipeline system between the oil shale region and the Gulf coast, and by the high cost of alternative modes of transportation.

In summary, tests have shown that crude shale oil and its derivatives could be used to produce petrochemicals. However, these materials cannot be considered to be viable feedstocks in the near term because existing chemical plants are generally unable to process them and there is no economical transportation link between the oil shale region and the existing petrochemical plants.

SUMMARY

The technological readiness of the major mining and processing alternatives is summarized in Table 6-4. Estimated degrees of readiness are shown as judged by DOI in 1968, and as they appear under present conditions. There are significant differences between the two evaluations because much R&D work has been conducted in the interim and because two new processing methods—modified *in situ* retorting and concurrent

Table 6-4. Technological Readiness of Oil Shale Mining, Retorting, and Refining Technologies[a]

Unit Operation	Technological readiness in 1968	Technological readiness in 1979	Developments in the interim	Remaining areas of uncertainty
Mining				
Room and pillar	Medium	Medium	Mine design studies. Experience of Colony, Mobil, and the six-company program at Anvil Points.	Rock mechanics. Groundwater. Deep shales. Logistics.
MIS	Unknown	Medium	Oxy experience at Logan Wash. Mine design studies.	Rock mechanics. Groundwater. Deep shales. Logistics.
Open pit	Low	Medium	Established procedure for other minerals. Large-scale mines in West Germany. Mine design studies for tract C-a.	Rock mechanics. Logistics. Reclamation.
Other	Low	Low	Mine design studies. No field experience.	All areas.
Retorting				
Aboveground	Medium	Medium	Pilot studies for Lurgi-Ruhrgas. Semiworks programs by Colony and Paraho. Detailed designs by Colony and Union.	Effects of scaleup on stream factor and recovery. Characteristics of emission streams. Reliability of peripheral equipment. Materials handling.
TIS	Low	Low	Field work and lab tests by USBM. Field tests by Equity, Geokinetics, and Talley-Frac. National lab support.	Stream factor and recovery. Characteristics of emission streams. Rock mechanics. Deep shales.
MIS	Unknown	Medium	Oxy experience. Design studies for tracts C-a and C-b. DOE and USBM design studies. Modeling and lab support by national laboratories.	Effects of scaleup on stream factor and recovery. Characteristics of emission streams. Reliability of peripheral equipment. Use of low-Btu gas. Rich shales. Mixed shales. Fracturing. Rock mechanics. Deep shales.
Multimineral	Unknown	Medium	Pilot plant studies by Superior. USBM lab tests of product recovery methods. Nahcolite scrubbing tests.	Effects of scaleup on stream factor and recovery. Characteristics of emission streams. Reliability of peripheral equipment. Materials handling. Integration of recovery steps. Underground waste disposal. Marketability of byproduct minerals.
Upgrading and refining	High	High	Studies and refinery runs by SOHIO Paraho, Chevron, and others. Marketing analyses by Oxy and DOE	Effects of retorting conditions on crude characteristics. Cost effectiveness of alternate processes. Use of pour point depressants. Effects of metals.

[a] *Note:* MIS – modified *in situ*; TIS – true *in situ*.

recovery of associated minerals—have since entered the picture. As shown, room-and-pillar, open pit, and modified *in situ* mining methods are regarded as reasonably well understood. Open pit mining has not been tested with Green River shales, but it is highly developed for other minerals such as copper and iron ores. It was evaluated on paper for application to the shales on tract C-a. Some highly relevant experience has been obtained from the operation of large-scale lignite (a form of coal) mines in West Germany. In these operations, the lignite is covered by 900 ft of overburden, and stripping operations will soon extend to 1600 ft. This is comparable to the oil shale deposits, which are covered by a maximum of about 1800 ft of overburden.

Nevertheless, uncertainties remain with respect to the effects of shale stability and strength on mine design, mine safety, and resource recovery. The effects of large inflows of groundwater, such as have been encountered on tract C-a, could pose severe operational difficulties, especially with underground mines. In all mines, the logistics associated with moving many thousands of tons of raw material and solid wastes could present some formidable problems. Materials-handling systems exist that could be applied, but they have yet to be tested in commercial oil shale operations.

Aboveground retort is regarded as having a medium level of readiness, as it was in 1968. The understanding of the technical aspects of the aboveground retort has been improved because of extensive field tests conducted at the semiworks scale, which is about one-tenth of a commercial-size plant. Their results do not permit accurate cost projections for commercial-scale plants. Particular problems are noted with respect to the effect of scaling up the semiworks design to commercial size. The on-stream factor—the fraction of the time that the retorts could be expected to operate at design capacity—is unknown. The reliability of some associated systems (emissions controls, product recovery devices, materials-handling equipment) is also questionable.

Although understanding has increased since 1968, true *in situ* must still be regarded as being in the conceptual stage. Many uncertainties remain, especially with respect to economics and environmental effects.

Modified *in situ* retorting, a new concept since 1968, has advanced to a medium level of technological readiness, approaching that of aboveground retorts. This progress is largely a result of Oxy's development efforts in Colorado, but additional understanding has been obtained through simulations by BOM, DOE, and the national laboratories, most notably the Lawrence Livermore and Los Alamos laboratories. The remaining uncertainties are similar to those for aboveground retorts, except the materials-handling problems may be less substantial with modified *in situ* methods. The major uncertainties relate to the use of the large quantities of low-Btu retort gas for power generation and to the application of the technique to very rich and very deep shales.

The Multi Mineral concept was also largely unknown in 1968, although the response of sodium minerals was recognized. At present, the aboveground system of Superior Oil is regarded as having medium technological readiness. Its uncertainties are much the same as for other aboveground retort methods, with the addition of the potential difficulties with integrating the systems for recovering mineral by-products with those for recovering oil and gas. The marketability of the nahcolite, soda ash, and alumina has also not been well established. The modified *in situ* concept proposed by Multi Mineral Corp. is not specifically evaluated in the table. It would share the same uncertainties as conventional modified *in situ*, with additional potential problems introduced by the need to evacuate a large underground retort, and because the oil shale resource is deeply buried.

Upgrading the refining systems were given a high rating in 1968, which has been improved by additional study. No major problems are anticipated, although more needs to be known about the feasibility of using pour point depressants, the effects of retorting conditions on the characteristics of the crude oil, and the effects of metals (such as arsenic) on refining catalysts. The major uncertainty in the distribution area—the existence of an adequate pipeline system to the most likely markets—is not directly related to the nature of the refining technology and is therefore not indicated.

It should be noted that mining and retorting (the major subprocesses having only medium levels of technological advancement) require only about 35% of the capital investment that is needed to establish an aboveground retort complex. The remaining 65% is distributed among upgrading units, by-product recovery systems, utilities, sidewalks, and other well-established items. Yet developers hesitate to commit themselves to oil shale plants. The reasons given are an uncertain regulatory climate, an uncertain future for the price of conventional petroleum, and the fact that all of the components of a facility must work as designed—not just the well-established ones. With the present state of technological knowledge, it is not clear that an oil shale plant would perform as desired, nor that the oil would be sufficiently cheap to compete with its currently designated competitor—imported petroleum. Even though the future of oil shale looks bright, few companies are willing to build large-scale plants immediately; they prefer instead to follow conventional engineering practice by proceeding to an intermediate step—the so-called modular demonstration retort.

The modular retort is the smallest unit that would be used in commercial practice. A commercial-size oil shale facility would use several of these retorts in parallel to obtain the desired production rate. The capacity of a module varies with the developer. For Oxy, a modular modified *in situ* retort might have a capacity of only a few hundred barrels per day of shale oil; a commercial facility would have several dozen of these retorts operating simultaneously in modular clusters, each producing several thousand barrels per day. A modular Lurgi–Ruhrgas retort would have a capacity of about 2200 bbl/day. One or two such units might suffice for the mined shale from a modified *in situ* operation; a facility that used only aboveground retort might need a dozen comparable units. A commercial-sized plant that used Paraho retorting might have seven or eight individual retorts, each producing 7300 bbl/day. A comparable plant might use seven Union "B" retorts, each producing about 9000 bbl/day. Colony prefered to bypass the modular demonstration phase and to proceed directly to a commercial-size facility, claiming that the Tosco II technology was ready to be scaled up to such a capacity for demonstration purposes.

Regardless of these differences, the several "next steps" that are proposed by the developers have two points in common: the production units must be large enough to simulate actual commercial practice, and the equipment must be operated long enough to obtain reliable data on its performance under a complete spectrum of operating conditions.

Oil Shale Project Descriptions

The following oil shale projects are described in detail in this section:

- Buffalo Trace, Kentucky
- Cathedral Bluffs, (C-b), Colorado
- Chevron, Colorado
- Colony, Colorado

- Equity, Colorado
- Exxon, Colorado
- Gary Energy, Colorado
- Geokinetics, Utah
- Magic Circle, Utah
- Mobil, Colorado
- Multi Mineral, Colorado
- Occidental, Colorado
- Paraho-Anvil Points, Colorado
- Paraho-Ute, Utah
- Plateau, Utah
- Rio Blanco (C-a), Colorado
- Southern Indiana, Indiana
- Superior Multimineral, Colorado
- Superior Pacific, Colorado
- Syntana-Utah, Utah
- Thermo-Mist, Colorado
- Tosco Sand Wash, Utah
- Union, Colorado
- White River, (U-a, U-b), Utah

BUFFALO TRACE PROJECT

Plant Type

Product: Shale oil

Production: 10,000 bpd (minimum)

Process: Paraho IGT Hytort Processes are being considered

Location: Flemming and Lewis Counties, Kentucky

Management and Finance

Sponsor: Buffalo Trace Area Development District

Contractor: Davy McKee Corporation, Cleveland Cliffs, University of Kentucky Institute of Mines and Minerals

Manager: Stephen H. Miller
Phone: (606) 564-6894
Location: 327 West Second Street, Maysville, Kentucky 41056

Key Contact: Stephen H. Miller

Financing Plan: $50,000 to Buffalo Trace Area Development District for project administration and review of study findings; $300,000 to primary contractor, $350,000—TOTAL

Major Permit Actions/Decisions

Kentucky regulations are under preparation.

Product Markets

Shale oil, high-Btu gas, and electricity.

Environmental Impacts

Less than or equivalent to coal operations.

Socioeconomic Impacts

Expected to be positive for area selected.

Project History and Status

The purpose of the feasibility study was to provide information and more accurate cost estimates on commercial production of synthetic crude oil from eastern Devonian oil shale deposits by evaluating the resource quality of sample oil shales processed by alternative technologies. This study was an objective assessment of the procedures and costs required to obtain synthetic fuel from oil shale as efficiently as possible, and recommend commercial means to produce 10,000 barrels of synthetic fuel per day by 1985. The program sponsor is trying to coordinate local response to federal interest in the rapid development of fuel production from domestic natural resources— especially oil shale. The Buffalo Trace Area Development District sponsored, by the way of grant funds from DOE, a study that evaluated the practicality of mining and processing oil shale in eastern Kentucky to take advantage of the area's natural resources and labor availability. Mining and processing of oil shale and the abundant deposits of eastern Kentucky were assessed by a consultant without vested interest in order to demonstrate to both government and private industry that such a venture is feasible. The feasibility study report is available as guidance to industry, government, and the local participants working on similar projects.

Project/Process Description

The primary benefit from the study was to identify the potential oil shale processing technologies available and applicable to eastern Devonian oil shale, as evidenced by shales extracted and sampled from Lewis and Flemming Counties. The potential processes were identified and evaluated with standard criteria for evaluation. The two-volume study report is available from the Buffalo Trace Area Development District office in Maysville, Kentucky.

CATHEDRAL BLUFFS (C-b) PROJECT

Plant Type

Product: Shale oil

Production: 55,000 bpd—1986; 94,000 bpd—1990

Input Material: 166,000 TPD oil shale and 10 million GPD water required

Process: Oxy modified *in situ* and surface retorting

Location: Piceance Basin, Rio Blanco County, Colorado.

Current Status: Site preparation began: September 1977
Actual construction began: January 1978
Initial retorts: 1985
Full production: 1990

Management and Finance

Sponsor: Occidental Oil Shale, Inc. and Tenneco Shale Oil, Inc.

Contractor: Fluor Corporation

Manager: R.A. Loucks (Vice President & Project Manager)
Phone: (303) 242-8463
Location: P.O. Box 2687, Grand Junction, Colorado 81502

Key Contact: W.H. Love, Chairman of the Board and Chief Executive Officer, Occidental Oil Shale, Inc.
C.W. Rackley, President, Tenneco Shale Oil Company

Funds Expended:	$117.8 million (Tract C-b Lease Bid)
Total Estimated Capital Required:	$3 billion (1980 dollars)
Financing Plan:	SFC load guarantee requested for approximately 75 percent of total project cost

Major Permit Actions/Decisions

As required by the lease agreement, a Detailed Development Plan (DDP) for aboveground retorting was submitted to the Area Oil Shale Supervisor (AOSS) in March 1977 and approved on August 30,1977. An update allowing the combined aboveground and modified *in situ* retorting is currently in preparation. Major government permits needed for the commercial phase of this project are:

- Bureau of Land Management (BLM)—Rights-of-way for product pipelines and power lines
- U.S. Army Corps of Engineers—Pipeline river crossings
- Colorado Department of Natural Resources—Wells and water use; land reclamation; air and water quality to be obtained
- Colorado State Department of Health—National Pollutant Discharge Elimination Permit and air emissions permits
- Environmental Protection Agency (EPA)—Prevention of Significant Deterioration Permit

Product Markets

Current plans are to transport, by pipeline, raw shale oil to refiners and/or utilities.

Environmental Impacts

The "Final EIS for Prototype Oil Shale Leasing Program" (DOI August 1973) covers projects on all prototype lease tracts, including Tract C-b. An extensive 2-year baseline study has been completed. The present plan includes raw shale disposal on site. Revegetation of the raw oil shale disposal area is planned, and much of the original wildlife habitat will be protected. The land in the Tract C-b region was changed to improve natural vegetation for wildlife; thus, oil shale operations on C-b will have no impact on regional agriculture.

Socioeconomic Impacts

The Colorado communities of Meeker and Rifle are expected to experience socioeconomic impacts upon development of this project, as they are in closest proximity to the Tract C-b project site. Their current populations are about 3450 and 2350, respectively. At 94,000 bpd production, about 1600 permanent employees are expected (2200 at construction peak). Most of the new residents are expected to live in Rifle and Meeker, but some will live in other parts of Garfield and Rio Blanco Counties. Meeker and Rifle also could be impacted by other projects in the same area. Rifle is close to the Union, Colony, and Rio Blanco Shale Oil projects, and is located on the major highways and also is the railhead for the area. Meeker also will receive from the Superior and Rio Blanco projects.

Project History and Status

On February 21, 1974, the original project sponsors (Tosco, Shell, ARCO, and Ashland) purchased Prototype Oil Shale Lease C-b by closed bid sale from the U.S. Department of the

Interior in Denver, Colorado. The price of $177.8 million was to be paid in five equal installments (5094 acres at $23,110 per acre). Under terms of the lease, the Preliminary Development Plan (PDP) was submitted to the Area Oil Shale Supervisor (AOSS) in March 1974. A Detailed Development Plan (DDP) was submitted to the AOSS in March 1977. Commencement of site development activities in 1977 followed approval of the DDP. On December 19, 1975, ARCO and Tosco announced their withdrawal from the Tract C-b project, and in late 1976 Shell Oil also withdrew. On December 1, 1976, Ashland Oil of Colorado, the last remaining original sponsor, entered into a partnership agreement with Occidental Oil Shale, Inc., for joint development of the tract. In February 1979 Ashland withdrew, and in September 1979 Occidental entered into a 50/50 partnership with Tenneco for development of the tract with Occidental as the operator.

Project/Process Description

Federal Prototype Oil Shale Lease Tract C-b is located in the center of the Piceance Creek Basin of northwestern Colorado. The 5094-acre tract has a present potential recoverable shale oil reserve of approximately 1.2 billion barrels (27 GPT). Overburden thickness is about 1250 ft on the tract, requiring an underground mining plant. Site preparation work commenced on September 1, 1977, and shaft sinking operations in February 1978. Three large diameter shafts are in process: a 15-ft-diameter ventilation/escape shaft, a 29-ft-diameter production shaft, and a 34-ft-diameter service shaft. A fourth shaft (34-ft-diameter) will be completed in December 1986 to handle offgases from the commercial retorts. Aboveground retorting of the shale mined from the *in situ* retorts will result in a total estimated production rate of about 94,000 BPD.

CHEVRON PROJECT

Plant Type

Product: Upgraded shale oil

Production: 50,000 bpd—1989
100,000 bpd—1992

Input Material: Not available

Process: Chevron is considering the follow processes: Chevron (proprietary); Paraho; and Lurgi

Location: Garfield County, Colorado

Management and Finance

Sponsor: Chevron Oil Shale Company, CONOCO

Contractor: Morrison-Knudsen; Foster-Wheeler; Environmental Research and Technology

Manager: D. Roger Loper, President
Phone: (415) 894-5983
Location: 595 Market Street, San Francisco, California 94105

Key Contact: R.F. Schlecht

Funds Expended: $5 million in 1979; $20 million in 1980

Major Permit Actions/Decisions

Discussions have been initiated with state and local government agencies.

Product Markets

Chevron is pursuing the project for strategic reasons to ensure refinery feedstock for their refinery and distribution system.

Environmental Impacts

Background studies have been initiated.

Socioeconomic Impacts

Schedules and alternatives for development are being considered.

Project History and Status

1930s	Standard Oil Company of California began acquiring oil shale lands in western Colorado.
1948–1950	Socal worked with Bureau of Mines in developing the Gas Combustion Retorting Process.
Early 1950s	Socal started acquiring water rights.
Late 1950s	Socal conducted core drilling and isopach mapping and consolidated holdings through land exchanges.
1967	Socal (Pacific Oil) and Union Oil Company applied jointly for a 34,000 AF reservoir on Parachute Creek for storage of water from Parachute Creek and the Colorado River.
Mid-1970s	Chevron Research Company participated in the Paraho's retorting demonstration at Anvil Points near Rifle, Colorado.
Late 1970s	Chevron Oil Company was formed and conducted resource evaluation work. Chevron Research Company began pilot plant phase of retorting process development and upgraded raw shale oil at the pilot plant level. Chevron's Salt Lake Refinery processed, over a 2-day period, 3000 bbl of raw shale oil obtained from Occidental Petroleum.
1980	Chevron announced that a 350 TPD semiworks plant would be constructed in Salt Lake City.
1981	Chevron exchanged a 30% interest in their Clear Creek property for coal resources with Conoco.

Project/Process Description

Chevron Oil Shale Company currently owns 43,000 acres of oil shale property in the Roan Plateau area of northwestern Colorado. In February 1980, Chevron Oil Shale Company began conducting feasibility studies aimed at commercial development of its Colorado property. These feasibility studies were contracted to Morrison-Knudsen with Foster-Wheeler Energy Corporation as its prime subcontractor. These studies, which will extend over an 18-month period, consist primarily of retorting process analysis, mining studies, site selection, and socioeconomic impact evaluation. In addition to these studies, the baseline monitoring work required for permitting activities will be performed by Environmental Research and Technology, Inc. In the summer of 1980, a field core drilling program will be conducted to more accurately evaluate resources and to provide design information for mining activities. The site is expected to be producing 50,000 bpd of shale oil by late 1989 with a total 100,000 bpd by 1992. These dates are based on a successful demonstration of retorting process technology by 1985 and appropriate financial incentives being available. In order for Chevron to handle a project of this size, a concerted effort to expand its staff has begun. In April 1980, Mr. D. Roger Loper was named President of Chevron Oil Shale Company.

COLONY PROJECT

Plant Type

Product:	Hydrotreated shale oil
Production:	48,300 bpd
Input Material:	66,000 TPD of 35 GPT oil shale; 10,000 AF/Y of water
Process:	TOSCO II
Location:	Fee land at head of Parachute Creek, Garfield County, Colorado
Current Status:	Planning, detailed engineering design, and costs complete: construction stopped.

Management and Finance

Sponsor:	Colony Development Operation (Exxon 60%, Tosco 40%)
Contractor:	Colony Development Operation
Manager:	Lawrence Hayes
Phone:	(303) 245-6907
Location:	759 Horizon Dr., P.O. Box 308, Grand Junction, Colorado 81502
Key Contact:	Frank Barrow, (713) 656-1553
Funds Expended:	$75 million (January 1980)
Total Estimated Capital Required:	$1.7 billion (January 1980)

Major Permit Actions/Decisions

Final EIS approved and issued mid-1977. Permits status is as follows:

- Bureau of Land Management (BLM)—Right-of-way for products pipeline and power lines submitted; action pending announcement of project reactivation.
- U.S. Corps of Engineers—Colorado River Crossing by the products pipeline and plant water intake facility near Grand Valley permit issued.
- Bureau of Reclamation (BUREC)—Contract for delivery of water from Green Mountain and Ruedi Reservoirs; request to reopen negotiations submitted.
- State emissions permit issued.
- Environmental Protection Agency (EPA)—Prevention of Significant Deterioration Permit issued.
- Environmental Protection Agency (EPA)—Spill prevention countermeasure plan (must be filed within 6 months after completion of storage facilities).
- Environmental Protection Agency (EPA)—National Pollutant Discharge Elimination System for quality of water effluents to be submitted if needed.
- Colorado State Mined Land Reclamation Permit—Being prepared.
- Colorado State Engineer—Construction water well, construction and water uses—approved.

Product Markets

Product markets will be transportation and boiler fuel users primarily, with some heating oil and petrochemical usage. By-products of sulfur and ammonia will be used by traditional markets.

Environmental Impacts

A final EIS for the Colony Plant has been prepared by BLM and issued mid-1977. Bureau of Land Management (BLM) Environmental Impact Statement (EIS) includes consideration of 194-mile, 50,000 bpd pipeline to Lisbon, Utah, The Colony Plant has no discharge of liquid effluents. Air effluent discharge complies with all federal, state, and local regulations. Sulfur compounds generated with the process are routed by Claus and Wellman-Lord units for processing; elemental sulfur is produced for sale as a by-product. By-product ammonia is also produced by the Wastewater Treating Unit. Spent shale will be disposed of continuously at various locations in Davis Gulch, a high shallow valley in the northwestern portion of Colony's Dow West property. Topsoil may be removed ahead of filling and stockpiled for reuse in reclamation of the 800-acre disposal area. Revegetation (irrigation, fertilization, seeding) of the entire area is planned and would reestablish much of the original wildlife habitat disturbed during plant construction and operation.

Socioeconomic Impacts

The three counties expected to receive the major direct impacts of the proposed development had a population of 75,000 in 1975. Mesa County had about 55,000 residents, Garfield County had 15,000, and Rio Blanco County had 5000. Current population levels are substantially higher. A planned community for plant employees and supportive infrastructure has been designed for land purchased by the project near Grand Valley, Colorado. Peak plant/mine construction employment of 2400 is anticipated. Operation work force will be 1200. Population increases in Garfield and Mesa Counties resulting from this project range between 3000 and 4500 (6.0% net increase) with total population increase in peak construction ranging to 12,000.

Project History and Status

The Colony Group has demonstrated retorting technology at Parachute Creek in a 1000 TPD semiworks plant. Approximately $75 million in research and engineering has been invested to bring the project to commercial readiness. Detailed engineering design and cost estimates for a 66,000 TPD commercial plant have been completed. The Colony Development Company was formed in 1964 by Tosco (30%), Cleveland Cliffs (30%), and SOHIO (40%) to develop a commercial oil shale facility on 4400 acres of fee land known as the Dow West Property. The property, purchased in 1964, is located at the head of Parachute Creek, Garfield County, Colorado (R95W, T5S). Total reserves of property are not available. Other properties are owned by Colony or its venture participants. A prototype Tosco retort was operated on Parachute Creek between 1965 and 1967 as well as a development mine. In 1969 ARCO joined the venture and the semiworks plant (TOSCO II) was operated between 1970 and 1972 (Cliffs and SOHIO withdrew in 1971 and relinquished all interest in the property in 1974). Extensive environmental and engineering studies were conducted. The planned construction start was suspended in October 1974 and resumed in 1980, after Exxon purchased ARCO's share of the project in August. Construction of the site access road and mine bench commenced in June 1980, and is scheduled for completion early in 1981. Detailed design and cost estimates for the commercial-scale Colony plant were prepared by C.F. Braun & Co. in 1974 at a cost of more than $12 million and were updated by C.F. Braun in 1979.

Project/Process Description

The Colony project will utilize underground room-and-pillar mining, TOSCO II retorting, and conventional refining methods to produce a premium-quality synthetic oil. The TOSCO II process recovers 100 of the Fischer Assay oil contained in the mined shale. Colony Development Operation proposed to construct, on the Dow West property, the following: (1) An underground oil shale mine of over 4000 acres capable of producing 66,000 TPD of raw oil shale for a 20-year period, and a mine bench with primary crushing facilities. (2) An oil shale plant (TOSCO II) consisting of retorting and upgrading facilities which are capable of producing 47,000 bpd of low-sulfur fuel oil, 4330 bpd of liquified petroleum gas (LPG), 135 TPD of ammonia, 173 TPD of sulfur, and 800 short TPD of coke, for a 20-year period. (3) A disposal area of about 800 acres in Davis Gulch to hold processed shale. (4) Two small reservoirs, one each in Davis Gulch and Middle Fork, for drainages. Additional actions proposed by the applicant, but which may not be entirely on the Dow West Property, include: (5) a 194-mile, 16-in. pipeline for transportation of low-sulfur fuel from the plant site to a 120,000 barrel floating roof storage tank to be constructed from the Lisbon Valley Station in Utah. (6) Development of a 15-mile-long service corridor in Parachute Creek Valley from the plant site to associated facilities located north of the town of Grand Valley, Colorado. The corridor will include three pipelines (water, ammonia, LPG), and a paved two-lane highway. (7) Terminal/storage facilities to be located at the Grand Valley end of the service corridor, and a railroad spur line to these facilities. (8) Two 230-kV power lines which may be built by Public Service Company to service the plant and mine. It is necessary to provide 100% of the plant's ultimate electrical requirements. (9) An exchange with the federal government of approximately 337 acres of federal lands, consisting of isolated parcels located within or projected into the Dow West property, for private lands adjacent to the Naval Oil Shale Reserves. (10) Diversion of 12.5 cubic feet per second of water from the Colorado River at Grand Valley which will be piped to the plant site and planned community. This diversion is contingent upon a proposed contract with the U.S. Bureau of Reclamation for water from Green Mountain or Ruedi Reservoirs. (11) Plans for reclamation at the end of the projected 20-year plant life will be subject to state reclamation standards. Plans will be submitted to the Mined Land Reclamation Board in 1979. Some possibilities are complete removal of plant facilities and rehabilitation of the area, or "moth balling" of the facilities. Implementation of the preceding proposals would require a number of actions by government agencies.

EQUITY PROJECT

Plant Type

Product:	Shale oil
Production:	Not applicable at this time
Process	Superheated steam to retort *in situ*
Location:	Piceance Creek Basin of northwestern Colorado
Current Status:	Technical feasibility demonstration project commenced March 1977 and is scheduled to conclude in early 1982.

Management and Finance

Sponsor:	Equity Oil Company and the Department of Energy
Manager:	Paul M. Dougan, Project Manager
Phone:	(801) 521-3515
Location:	Suite 806, 10 West 3rd South, Salt Lake City, Utah 84101
Key Contact:	Dallas Goodrich, Project Field Superintendent
Financing Plan:	Under a cooperative agreement, Equity pays 14% of the cost of the project, with the U.S. government defraying the remaining 86% of the $11 million plus expense budget.

Project History and Status

Equity Oil Company's knowledge of the existence and characteristics of the leached zone was gained initially in the course of drilling oil and gas exploration wells in the central Piceance Creek

Basin during the early 1950s. The late J.L. Dougan, former president of Equity Oil Company, in 1961 conceived the idea that shale oil contained in the leached zone could be recovered by heating the shale in place using hot natural gas. To validate Mr. Dougan's theory, Equity Oil Company inaugurated an oil shale research program that was to continue for 10 years. This program included 5 years of laboratory research conducted at the Fuels Engineering Department of the University of Utah and 7 years of field work performed at the BX In Situ Oil Shale project site in Rio Blanco County, Colorado. [The initials "BX" derived from the project's location on Equity-interest Boies fee acreage and from its experimental character. The fee land, consisting of 1000 acres, is owned by Equity (50%) and Atlantic Richfield (50%). The interests are undivided and each company has right of access to and use of the property.] The BX project and attendant laboratory work were discontinued in 1971 because of a lack of funding.

A cooperative agreement resulted from Equity's submittal on March 23,1976, of a proposal in response to DOE's Program Opportunity Notice No. 2. That proposal described a process proprietary to Equity Oil Company for *in situ* treatment of a water-impregnated, naturally leached, and permeable zone of oil shale. The project was officially designated as Cooperative Agreement No. DE-FC20-78LC10747 between the U.S. Department of Energy (DOE) and Equity Oil Company for the BX In Situ Oil Shale Project. It became effective March 1, 1977, and its duration is estimated to be 55 months. Field tests of the BX In Situ Oil Shale Project commenced in September 1979 and will be conducted over a period of approximately 2 years. During that time, some 1000 billion Btu of heat will be injected into the leached zone as superheated steam. If the process can be perfected to the point of technical, economic, and environmental viability, the key will have been found to the recovery of millions of barrels of shale oil from a section of the Green River formation heretofore unavailable as an energy resource by any other recovery technology. Moreover, this recovery will have been made possible using equipment which for the most part is readily available from conventional oil field stocks. As of December 31, 1980, Tasks No. 1 and No. 2 had been completed and Tasks No. 3 and No. 4 are in progress. During the period September 18, 1978, to December 31, 1980, a total of 549,993 barrels of water as steam were injected into project wells at an average wellhead temperature of 552°F (289°C) and an average wellhead pressure of 1360 psig. Mechanical problems with steam generation equipment have resulted in the injection procedure at rates and temperatures below design; however, they were substantially resolved in October 1979. Evidence of leached zone heating has been observed in all production wells and in project temperature observation wells. In October 1980, the first oil was produced from the project wells.

Project/Process Description

The project is based on one basic premise: in the central portion of the Piceance Creek Basin of northwestern Colorado, a section of oil-shale-bearing rocks in the Parachute Creek member of the Green River Formation, commonly referred to as the leached zone, contains very large reserves of oil in place as oil shale. This section, or leached zone, has enough permeability and porosity to permit *in situ* retorting of the oil shale contained therein without resorting to mining and/or other fracturing techniques to create permeability and porosity before retorting. The project's purpose is to demonstrate the technical feasibility of using superheated steam at 1000°F (538°C) and 1500 psi as a heat-carrying medium to retort *in situ* oil shale of the leached zone, thus providing a mechanism for oil recovery with minimum environmental impacts. Specifically, the project involves retorting the oil shale by injecting superheated steam into the leached zone through an array of injection wells, and recovering the steam, water, oil, and gas produced from the leached zone through an array of production wells. During a 2-year course of field tests, it is intended that approximately 1000 billion Btu of heat will be injected as superheated steam into a leached zone about 540 ft thick, covering seven-tenths of one acre, and containing an estimated 636,000 barrels of oil in place as oil shale. Project field tests commenced June 14, 1979. Project success will be measured by the degree to which (1) the injection goals can be met, (2) the target amount of oil shale can be raised to the retorting temperature, and (3) a reasonable percentage of the retorted oil is recovered. The project's field test program borders Black Sulphur Creek at a point 35 miles southwest of Meeker, in the east half, southeast quarter, Section 6, Township 3 south, Range 98 west, Rio Blanco County, Colorado. Under terms of its cooperative agreement with the government, Equity, as operator of the project, agreed to perform four basic tasks: leached zone site evaluation, laboratory experimentation, field tests, and environmental research.

EXXON PROJECT

Plant Type

Product:	Shale oil
Production:	60,000 bpd
Input Material:	80,000 TPD of 30 GPT shale; 8500 AF/yr of water
Process:	Room-and-pillar mine, surface retort
Location:	Piceance Basin, Rio Blanco County, Colorado
Current Status:	Planning studies underway. Land exchange requested.

Management and Finance

Sponsor:	Exxon Co., U.S.A.
Manager:	J.P. Racz
Phone:	(713) 656-6341
Location:	P.O. Box 2180, Houston, Texas 77001
Key Contact:	J.P. Racz
Funds Expended:	$4 million
Total Estimated Capital Required:	$2 billion (1980 dollars)
Financing Plan:	100% equity

Major Permit Actions/Decisions

In December 1979, Exxon applied for an exchange of 11,814 fee acres for 10,840 acres of BLM land. BLM denied the land exchange request in 1980. Approval of this land exchange will result in consolidation of Exxon's scattered fee lands for an economic block of minable resources. It also will contribute significantly to increased federal control of the central Piceance Basin and allow BLM to consolidate additional minable blocks. Exxon's own senior water rights on the White River and an application for a right-of-way permit to inundate 40 acres of federal land in conjunction with development of the required water system has also been made.

Product Markets

Product will be shale oil suitable for further processing in a refinery.

Environmental Impacts

Data gathering for an environmental baseline program began in the spring of 1980, in cooperation with federal and state authorities.

Socioeconomic Impacts

About 2000 permanent employees are anticipated at the proposed level of production, with a peak construction employment of 4000. A study of community requirements for plant employees and support infrastructure is being planned.

Project History and Status

Starting in 1961, Exxon acquired mineral rights ownership of about 11,800 acres of land in the Piceance Basin. During this period 1963–1967, Exxon participated in the Bureau of Mines GC retort work at Anvil Points and in the Paraho Demonstration Program from 1974 to 1976. Shale oil upgrading was investigated from 1963 to 1967. Exxon also has investigated catalytic cracking of the upgraded shale oil and jet fuel producibility from shale oil in 1974 to 1977 for the Department of Defense.

Project/Process Description

The Exxon project will utilize underground room-and-pillar mining with surface retorting to produce shale oil suitable for processing in existing refineries. The underground mine will be capable of producing 80,000 TPD of raw oil shale for a 30-year period. Surface retorting facilities will be capable of producing 60,000 bpd of shale oil. Processed shale will be disposed of in gulches on the site. A water system consisting of diversion from the White River, a delivery pipeline, a reservoir at the confluence of 13-Mile and 14-Mile Creeks, to the plant will be developed. Other facilities include a product pipeline and a supply pipeline for the shale oil product, ammonia and sulfur storage, and loading facilities near Rifle, two 230-kV power transmission lines, and an access road from the Piceance Creek Road to the plant site.

GARY ENERGY PROJECT

Plant Type

Product:	Transportation fuels
Production:	10,000+ bpd liquid products
Input Material:	10,000 bpd raw shale oil
Process:	Hydrotreating/hydrocracking
Location:	Gary Refining Company, Fruita, Colorado
Current Status:	Feasibility study grant awarded.

Management and Finance

Sponsor:	Gary Energy Company (A Gary Refining Company Subsidiary)
Contractor:	UOP Process Division, UOP Incorporated, The Ralph M. Parsons Company, Environmental Research & Technology, Incorporated
Manager:	Stanley B. Weil, Vice President, Refining
Phone:	(303) 858-9811
Location:	Gary Community Rural Station, Fruita, Colorado 81521
Key Contact:	George Gerson, Manager, Economics & Process Services

Major Permit Actions/Decisions

Permit applications have not been filed. Information required to prepare pertinent permit applications will be developed during the study.

Product Markets

A shortage of crude oil in the refinery area has prevented expansion of the refinery to meet the existing market for refined petroleum products in Mesa and nearby counties. The products derived from the proposed shale oil refining facility can therefore be easily absorbed within the boundaries of Gary's existing marketing network, and new markets are being explored for military and commercial jet fuels. Potential markets have been identified for sulfur and ammonia, which will be produced in the new facilities.

Environmental Impacts

With the project being located on and adjacent to the existing Gary refinery site, many factors associated with site evaluation and impacts are inherently less critical to this project than for a grass roots refinery on a previously undisturbed site. In addition:

- The current refining operation is in compliance with environmental laws and regulations.
- The project involves commercially available technology having well-defined emission characteristics. No novel or unproven effluent control measures are required.

Socioeconomic Impacts

The existing refinery site is approximately 15 miles from Grand Junction, Colorado, the center of a population area in excess of 60,000 people. With a projected 300–400 labor force during construction and an additional 60–70 during plant operations, an impact on the surrounding communities is not expected.

Project History and Status

The Gary refinery was constructed in 1957 by American Gilsonite Company to process gilsonite, a solid hydrocarbon ore from northeast Utah. Gary Energy acquired the installation in 1973 after American Gilsonite discontinued refining gilsonite ore. Processing gilsonite into refined products was a forerunner of the synthetic fuel industry. Gilsonite has a very high nitrogen content. The problems associated with processing gilsonite ore are similar to those associated with shale oil.

Two substantial shale oil refining programs have been conducted at this site. In 1961, the facility was engaged by Union Oil Company to process over 13,000 barrels of crude shale oil product from the Union "A" retort semiworks project in the Piceance Basin. In 1975, 9956 barrels of crude shale oil were refined from the Paraho oil shale retort at Anvil Points, Colorado, into military fuels for the Navy. The refining experiments on Union and Paraho synthetic crudes provided experience in shale oil refining. They also established the Gary Refining Company as the first commercial refinery to process significant quantities of Green River shale oil. An existing common carrier pipeline, dedicated for crude oil delivery to Gary's refinery, runs adjacent to proposed shale oil producing areas. This pipeline could carry 10,000 bpd of shale oil to the refinery.

In 1980 Gary Energy Company submitted a proposal to the Department of Energy to conduct a feasibility study for upgrading 10,000 bpd of raw shale oil into transportation fuels. The upgrading facilities would be installed at the Gary Refining site in Fruita, Colorado, near the Colorado–Utah shale oil reserves. Gary was award the requested grant to conduct the study. Final grant negotiations between Gary,its contractors, and DOE have been completed. The study is expected to begin in 1981. The feasibility study is projected for completion in 15 months. An additional 42 months will be required for engineering, construction, and start-up of the new facilities.

Project/Process Description

The proposed facilities will refine shale oil into high-quality transportation fuels using proven technologies. Since the quality of the feedstock varies with the source and retorting method, the scheme must allow for processing a wide range of shale oils. The chosen method is hydrotreating followed by hydrocracking. The hydroprocessing/hydrocracking technology has been demonstrated on petroleum-derived feeds in many commercial units operating throughout the world. Pilot plant testing using shale oil has been performed to develop modifications to petroleum-derived predictive correlations.

GEOKINETIC PROJECT

Plant Type

Product: Shale oil and low-Btu gas

Production:	2000 bpd for each production unit
Input Material:	20 GPT shale
Process:	Horizontal modified *in situ*
Location:	70 miles south of Vernal, Utah.
Current Status:	Pilot operations; 15,000 barrels of shale oil produced through December 31, 1980.

Management and Finance

Sponsor:	Geokinetics, Inc., with DOE support
Contractor:	Geokinetics, Inc.
Manager:	J.M. Lekas, Project Manager
Phone:	(801) 353-4747
Location:	Box 889, Vernal,Utah 84078
Key Contacts:	M.A. Lekas, President, J.D. Downen, Vice-President (415) 676-4462, 280 Buchânan Field Road, Concord, California 94520
Funds Expended:	$9.0 million

Major Permit Actions/Decisions

Collection of data for permits began in 1977. Work is underway to secure necessary permits for full-scale operations.

Product Markets

Shale oil has been sold to Plateau Refining Company and Western Refining Co. for blending tests.

Environmental Impacts

Environmental assessment reports have been prepared in connection with a DOE Cooperative Agreement. The primary environmental concern is expected to involve combustion gases from the retorts. Additional concerns will include the need to dispose of waste water produced with the oil,and also revegetation of the surface.

Socioeconomic Impacts

The operation is expected to have a minimal impact on local communities. The work force for each production unit will be about 60 employees, most of whom will live on site in a mobile home village.

Project History and Status

Geokinetics, Inc., was organized in April 1969 as a minerals development company. In July 1972 Geokinetics organized a joint venture with a group of independent oil companies to develop *in situ* methods of shale oil extraction, and to acquire and develop oil shale leases. Work on the horizontal modified *in situ* process began in 1972. Design and cost estimates were made for a horizontal modified *in situ* operation in Tracts C-b, U-a, and U-b in preparation for bidding on the Prototype Federal Oil Shale lease sale. Small-scale pilot tests in steel retorts, to simulate a horizontal basis, were carried out in 1974 and early 1975. In April 1975 *in situ* field tests began in Kamp Kerogen and have continued without interruption to date.

During 1975 and 1976 the basic parameters of the process were established. In late 1976, a cooperative agreement was signed with DOE, with whose assistance progress was greatly accelerated. In 1977 and 1978 the process was scaled up substantially, and rock breaking designs were improved and tested. In1979 larger retorts were tested, up to one-quarter of full scale, and tests began to optimize recovery. The first full-sized retort was blasted in 1979. During 1980 a second full-sized retort was blasted, and equipment was installed for ignition of the first full-sized retort. Twenty-four experimental retorts have been blasted, 14 retorts have been burned, and 15,000 bbl of oil have been produced. Geokinetics expects to complete its R&D program in 1982, and begin immediately to design and construct a 2000 bpd commercial production unit. Thirty-thousand acres of oil shale leases were acquired between 1975 and 1980, representing in-place reserves of 1.7 billion barrels of shale oil.

Project/Process Description

Geokinetics is developing the horizontal *in situ* retort, using explosive fracturing of the oil shale. There are three basic applications of the process:

1. The LOFRECO process, where blast holes are drilled from the surface to fracture the oil shale bed.
2. Horizontal modified *in situ*, where part of the bed is mined out to provide expansion space for the broken rock.
3. Secondary recovery after room-and-pillar mining. After mining is completed, the pillars, roof, and floor are blasted to create a large volume of rubblized rock that is retorted using the horizontal *in situ* process

In the LOFRECO process, a pattern of blast holes is drilled from the surface through the overburden and into the oil shale bed. The explosion produces an upward movement of the overburden and fragments the oil shale. The bottom of the retort is sloped to provide for drainage of the oil to the production wells. Air injection holes are drilled at one end of the retort, and gas exhaust holes are drilled at the other end. The oil shale is ignited at the air injection holes, and air is injected to establish and maintain a burning front. The front is moved in a horizontal direction through the fragmented shale toward the gas exhaust holes at the far end of the retort. The burning front heats the oil shale ahead of the front, driving out the oil, which drains to the bottom of the retort, and flows along the sloping bottom to a sump, where it is lifted to the surface by conventional oil field pumps. As the burning front moves from the air injection to the gas exhaust holes, it burns residual coke in the retorted shale as fuel, and produces a large volume of low-Btu combustible gas.

MAGIC CIRCLE PROJECT

Plant Type

Product:	Shale oil
Production:	30,000 bpd
Input Material:	70,000 TPD, 27 GPT from room-and-pillar mine, 1500 ft below surface (surface retort).
Process:	T^3 retorting process (Science Applications)
Location:	Cottonwood Wash located in T10 South, Ranges 21 and 22 East Uinta County, Utah (approximately 11 miles southeast of Ouray).
Current Status:	Feasibility study has been completed. Preliminary engineering, full-scale permitting, environmental and socioeconomic programs, product marketing, project financing, and risk assessments are underway.

Management and Finance

Sponsor:	Magic Circle Energy Corporation
Contractor:	Science Applications, Incorporated

Manager:	Harry E. McCarthy
Phone:	(303) 279-0701
Location:	1726 Cole Boulevard, Suite 350 Golden, Colorado 80401
Key Contact:	Robert G. Hall, Project President, (405) 843-6654
Funds Expended:	$1,121,000 for feasibility study, initial field work, and permitting
Total Estimated Capital Required:	$900 million (commercial development—1980 dollars)
Financing Plan:	Magic Circle has applied for a loan guarantee for $1.45 billion, not to exceed 75% of total project cost, and a limited time-phased purchase agreement.

Major Permit Actions/Decisions

- Major permits obtained
- Project viable from feasibility study
- Availability of loan guarantee from U.S. Synthetic Fuels Corporation
- Completion of feasibility and engineering design
- Raising of equity capital

Product Markets

Pipeline shale oil to refineries in Fruita, Colorado, or Roosevelt, Utah. There are several pipelines in the area, and a market for the product seems assured.

Environmental Impacts

The environmental impact statement draft is being prepared to be submitted no later than January 1982 to the Bureau of Land Management. Preliminary indications are that the project will have negligible to moderate impacts on hydrology, water quality, vegetation and fauna. Special attention will be given to minimizing short-term impacts on local air quality of the processing complex and processed shale embankment. Effects on local geology will include surface alterations to accommodate the processing complex, surface alteration by the addition of a processed shale disposal embankment and catchment dam, and cutting and filling to create or improve access roads. The project will create visual intrusions in an isolated area, but careful design processes can minimize adverse visual impacts. Effects of the project on local climate and aquatic organisms are expected to range from negligible to minor.

Socioeconomic Impacts

The total on-site work force during the construction period is expected to peak at slightly over 2600 during 1984. The breakout of on-site personnel during this period is not yet available. However, during full commercial production operations, the classification of on-site personnel is approximately as follows: administration—220; maintenance—480; technical support—60; production—700; and mining—440. The plans for housing and transportation are being formulated in cooperation with local authorities. The total population growth is expected to be approximately 7200.

Project History and Status

The 76,000 acres of state of Utah shale leases currently owned by Magic Circle were acquired from Western Oil Shale Corporation through a stock exchange in July 1980. A segment of continuous lands equal in size to the federal lease site is available on the 5090-acre Cottonwood Wash Properties which will be developed here; the remaining leases are noncontinuous sections currently being evaluated for exchange purposes to provide for additional commercial properties. In December 1980, Magic Circle management decided to initiate a project to explore development of the Cottonwood Wash Properties. At that time, Magic Circle contracted with Science Applications to perform feasibility studies, initiate permit applications, and perform initial coring for better resource definition, mine design, and environmental evaluations. E.F. Hutton also has been selected to initiate development of project financing for a commercial project at Cottonwood Wash. With this team, Magic Circle is committed to going ahead with development, subject to obtain the necessary financing, and to a continued favorable economic outlook for the project.

Project/Process Description

Cottonwood Wash is a 5090-acre tract (of the total 76,000 acres leased by Magic Circle) in the heart of the Uinta Basin in Utah, located in TWP 10 South, Range 21 East (SLM). The estimated oil in place, averaged over the 15 GPT and 25 GPT shales, is 2.1 billion barrels at depths between 1500 and 2000 ft. Estimates of recoverable oil using surface processing is at least 200 million barrels. If modified *in situ* (MIS) technology were commercially available, recovery is estimated to be as high as 500 million barrels. At this time, commercial production is planned at 30,000 bpd using surface processing. System design, however, will be carried out to provide maximum potential for the future use of MIS, should it become a commercial technology. The surface processing technology initially selected for the project is based on the Moroccan T^3 process, selected because of its minimal risk in scaling to a commercial operation. Innovative T^3 design features, compared to previous technology, are lowest water consumption, improved safety, sensible heat recovery, increased process yield, efficiency, flexibility, and continuous process operations. Waste disposal is currently planned to be accomplished on 1200 acres of the tract. However, an effort is underway to obtain additional properties through exchange or leasing mechanisms to provide for potential future project expansion. Water rights applications have been made for water from the Green River. There are also two aquifers on the site—one at approximately 100 ft depth, and the Bird's Nest aquifer, at a depth of 850 ft. If necessary, these can provide additional water to the project.

Activity Schedule/Production Estimates

I. Feasibility Study (completed)

II. Construction
Startup: January 1983
Completion: October 1986 (Process Module No. 1)

III. Initial Production
Startup: October 1986 (Process Module No. 1)

IV. Full Scale Production
Startup: February 1986 (Process Module No. 4)

MOBIL PROJECT

Plant Type

Product:	Shale oil
Production:	50,000 bpd—1990; 1,000,000 bpd—1996
Input Material:	30–35 GPT oil shale (150,000 TPD would produce 100,000 bpd shale oil)
Process:	To be determined
Location:	Parachute Block, Garfield County, Colorado
Current Status:	Concept development and feasibility study.

Management and Finance

Sponsor:	Mobil Oil Corporation
Manager:	J.F. Trautschold, Jr., General Manager, Synfuels Division and W.H. Marshall, General Manager, Mining and Coal Division
Key Contact:	P.C. Fuselier, P.O. Box 1772, Denver, Colorado 80217
Total Estimated Capital Required:	$4 billion
Financing Plan:	Both loan guarantees and price guarantees are being requested from the U.S. Synthetic Fuels Corporation.

Major Permit Actions/Decisions

The permitting process for the project has begun.

Product Markets

Product will be pipelineable shale oil suitable for refining.

Environmental Impacts

The water supply would be a combination of a reservoir on Main Elk Creek, a diversion intake on the Colorado River near Parachute, and a delivery system involving several miles of above- or below-ground pipeline. It is anticipated that spent shale disposal would take place in Wheeler Gulch. Air quality monitoring has been conducted at the proposed site since the fall of 1980 and additional baseline studies will commence this spring.

Socioeconomic Impacts

Oil Shale Mine & Retort 12,500-bpd Test Module—50,000 bpd Commercial Scale: Tentative Direct Employment

Year	Facility Construction	Facility Operation	Main Elk Construction Operation
1	600	50	0
2	600	417	75
3	0	417	75
4	0	417	75
5	1800	417	4
6	200	915	4
7	0	915	4

Project History and Status

Mobil is considering oil shale development on the Parachute Block in Garfield County prior to the year 2000. Most development will be on lands held in fee by Mobil. Work currently is in a stage of concept development and feasibility study. Movement of personnel and equipment to the mine would be via an access road from the Parachute Creek Valley to the mine at the head of Wheeler Gulch.

Project/Process Description

Mobil will utilize underground room-and-pillar mining and surface retorting. The precise location of surface retorting facilities is not yet certain; however, they probably would be located on either the Parachute Block or the Mahaffey properties, the smaller parcel south of the Parachute

Block. The type of retorting technology is still under review. The ultimate capacity of the facilities may be 100,000 bpd; however, development to this level probably would be phased. A 12,500-bpd module would be built first, then a 50,000-bpd facility; and ultimately a 100,000-bpd facility may be developed if economics remain favorable. It may be possible to have operational a 50,000-bpd facility by 1990, and a 100,000-bpd facility by 1996. Primary crushing will take place underground with the raw shale being moved to the surface retorting facilities by conveyor. Secondary oil recovery through modified *in situ* methods also may be possible at a later date. Secondary crushing of raw shale would take place on the surface. Spent shale will be moved by conveyor. Raw fines will be compacted at the head of the surface drainage for potential future retorting. Spent shale will be compacted into terraces in Wheeler Gulch. Catchment basins would be installed at each disposal pile to capture surface runoff and seepage.

Activity Schedule/Production Estimates

Construction start: 1984–1985

Production from the first retort: 1986–1987

MULTI MINERAL PROJECT

Plant Type

Product:	Shale oil and associated minerals
Production:	50,000 bpd shale oil (1989)
Input Material:	30–35 GPT shale
Process:	Multi Mineral Integrated *In Situ*
Location:	Piceance Basin, Rio Blanco County, Colorado
Current Status:	Mining Research Program underway at USBM Oil Shale Mining Environmental Research Facility on Horse Draw located in the Piceance Basin, Colorado. Preliminary engineering plans are complete for a subsequent Processing Research Program. Preliminary engineering plans are underway for commercial plant.

Management and Finance

Sponsor:	Multi Mineral Corporation
Contractor:	Multi Mineral Corporation
Manager:	Hal Aronson
Phone:	(303) 243-9406
Location:	715 Horizon Drive, Suite 380, Grand Junction, Colorado 81501
Key Contact:	Hal Aronson
Total Estimated Capital Required:	Project cost is expected to exceed $500 million.
Financing Plan:	Source of financing is proposed to be from Multi Mineral Corporation.

Major Permit Actions/Decisions

Multi Mineral Corporation is seeking to acquire rights to mine oil shale from lands on which they currently hold sodium mineral rights. It is anticipated that such an action may require an Environmental Impact Statement (EIS) that would be followed by the major permit actions listed below:

- Bureau of Land Management (BLM)—A right-of-way for products pipeline and power lines
- U.S. Army Corps of Engineers—Pipeline river crossing
- Environmental Protection Agency (EPA)—Spill prevention countermeasure plan; National Pollution Discharge Elimination Plan
- Colorado Department of Natural Resources—Spent shale disposal; mine approval; wells and water use; land reclamation; air and water quality

Product Markets

This process generates four products: shale oil, nahcolite, soda ash, and alumina. Market research is currently underway to identify end users of these products.

Environmental Impacts

Multi Mineral Corporation has not prepared an assessment of the environmental impacts related to a commercial complex. An EIS is expected to be required on this project because of the anticipated nature of the land acquisition. Approval of such a statement should not pose the major constraint facing other *in situ* projects: degradation of subsurface water by leaching of spent shale left underground after retort abandonment. The reason is that mining will take place in the Saline Zone where the oil shale beds are relatively dry and unfractured. If water is encountered, the stopes can be sealed using the mining access ways constructed during stope development. No outside source of water will be required in this process. Makeup water for processing will be supplied from the Saline Zone.

Socioeconomic Impacts

It is estimated that a Multi Mineral Integrated *In Situ* Process operation will require approximately one-half as many employees per ton of rock retorted as an aboveground retorting process. Applied to a 50,000-bpd facility, this assumption projects a peak construction force between 300 and 400. Total new direct and indirect population influx to the Rio Blanco/Garfield counties area may approach 400 during peak construction and 800 on a permanent basis. Since the towns of Meeker and Rifle are about equidistant (approximately 40 miles) from the site, they probably will share the project's major socioeconomic impacts.

Project History and Status

In November 1978 the Multi Mineral Corporation submitted an unsolicited proposal to the Department of Energy and the U.S. Bureau of Mines (Department of the Interior) to jointly enter into a program which would involve constructing and testing a full-scale, minimum-size oil shale plant using the Multi Mineral Integrated *In Situ* process. A possible site is the 2400-acre U.S. Bureau of Mines Oil Shale Tract on Horse Draw in the Piceance Creek Basin.

The proposed program has two phases—the first involves conducting a pilot mining program in the Saline Zone; the second, sinking a larger diameter shaft and testing the Integrated *In Situ* process at full module size. As proposed, the program would take approximately 5 years to complete, after which the Multi Mineral Corporation would bring the tract to commercial-size development under normal lease terms. The federal government would acquire the use of all technical data on which the process is based. All new technology developed during the test program is proposed to be owned jointly by the federal government for use in planning the total resource development. Multi Mineral Corporation currently is conducting a 24-month mining research program at this site utilizing the existing 8-ft-diameter shaft. This program will provide geotechnical data to determine mining feasibility and bulk samples of the nahcolite/dawsonite oil shale horizons to provide material for process plant engineering. Since the proposed joint program with DOE and DOI was never approved, Multi Mineral Corporation has been seeking alternative sources of action that would allow this project to proceed.

Project/Process Description

The Multi Mineral Integrated *In Situ* process for the recovery of shale oil and associated minerals consists of three steps: (1) underground mining and nahcolite recovery; (2) retorting and hydrocarbon recovery; and (3) leaching and soda ash/alumina recovery.

Mining—The process involves creating large underground chambers (stopes) by sequential blasting and removal of oil shale from the stope. Gravity flow extraction is through funnellike openings at the base. The stopes are separated by walls of unbroken oil shale. Their dimensions are variable, depending on the rock character and zone thickness. Extracted oil shale is crushed and screened into two sizes with the larger size being backfilled into the stope. The smaller size is then crushed and screened to separate nahcolite from the oil shale. Rejects from this operation are stored on the surface for future processing.

Retorting—After backfilling with the top size, the stopes, full of sized oil shale, become underground retorts. The oil shale is retorted by hot gas introduced into the top of the retort and extracted at the base. The shale oil, formed along with the pyrolysis gas and water, is directed to a collection sump where the oil, water, and gas are separated. The oil is pumped to the surface for marketing. After the oil shale has been retorted, residual carbon is left on the spent shale. The carbon is converted into a low- to medium-Btu gas which will be used for on-site electrical power generation.

Leaching—After shale in the retort is cooled, soda ash and alumina are leached from the stope by spraying water over the top of the retorted shale and collecting this water as pregnant liquid at the base. This can be done in more than one stope simultaneously in a countercurrent process. The pregnant liquor is then pumped to the surface for recovery of the alumina and soda ash.

OCCIDENTAL PROJECT

Plant Type

Product: Shale oil

Production: Not applicable as field test site

Input Material: 15 GPT shale

Process: Modified *in situ*

Location: Between Parachute and Roan Creeks, Garfield County, Colorado

Current Status: Demonstration of *in situ* retort began December 20, 1975. The seventh commercial size retort was recently rubbled and is expected to be retorted with the eighth retort later this year.

Management and Finance

Sponsor: Occidental Oil Shale, Inc.

Contractor: Occidental Oil Shale, Inc.

Manager: R. Nelson, Vice President and General Manager
Phone: (303) 242-8463
Location: P.O. Box 2687, Grand Junction, Colorado 81501

Key Contacts: W.H. Love, Chairman
W.F. McDermott, Executive Vice President

Funds Expended: $63 million (through January 1, 1979)

Major Permit Actions/Decisions

A permit has been issued to Occidental to increase the size of its waste rock disposal area from 500,000 to 8.8 million cubic yards. Continued development at the present site was contingent on the issuance of this permit. To date, the primary federal approval needed for this project has come from the Mine Environmental Safety Administration for maintaining a subsurface combustion zone. Permits have been obtained for Retorts 7 and 8 from the following agencies:

- Bureau of Land Mangement (BLM)—Rights-of-way and other matters concerning use of public lands
- Mined Land Reclamation Board—Surface disposition of run-of-mine shale
- Colorado State Department of Health, Air Pollution Control Division—All air emission permits are approved
- Colorado State Department of Health – Water Pollution Control Division—Subsurface disposal permit allowing underground storage of spent shale and a National Pollutant Discharge Elimination System permit

Product Markets

Occidental has sold shale oil to refiners for refining tests and to a utility for a boiler fuel test.

Environmental Impacts

Environmental impact assessment reports have been prepared in connection with the DOE cooperative agreement. An environmental statement may be required in the event of a commercial-scale project. Such a statement has not been prepared. The abandonment plan for spent retorts requires further study at this time.

Socioeconomic Impacts

It is estimated that this modified *in situ* process operation will have an impact on the area. The three counties that may receive socioeconomic impacts from this project are Mesa, Garfield, and Rio Blanco Counties, Colorado, with present populations of 80,000, 20,000,and 5000, respectively (total 105,000). The impacts, if any, probably will be felt most significantly in Garfield County. In light of other state developments concurrent with this project in Garfield and Rio Blanco Counties, housing and community services will be in short supply during construction phases, but with presently ongoing community planning studies, social needs for the new population will be adequate.

Project History and Status

Occidental Petroleum Corporation entered into an agreement with D.A. Shale, Inc., in mid-1972 and acquired about 4000 acres of land between Roan and Parachute Creeks, Garfield County, Colorado. The 2400 shale-bearing acres contain an estimated shale oil reserve of 0.3 billion barrels, averaging about 17 gallons of oil per ton of shale.

Site preparation began in 1972 and construction of the first modified *in situ* retort, 1E, was completed by December 1972. Retort 1E contained approximately 4000 tons of broken shale at 25% average void volume and was ignited in June 1973. Since then, operations have been underway almost continuously, through a series of progressively larger retorts. Retort 6 has 100 times the volume of the first experimental unit and does not require further scale-up for commercial operations.

The initial three retorts were located off a single horizontal mine opening. Retort 1E was mined in the form of a small room with a vertical cylindrical rise providing the initial void volume. The retort operated successfully, producing over 1200 barrels of oil. In Retort 2E, the void volume was reduced, the blast pattern was modified, and the retort depth was increased 22 ft; the retort was fired in March 1974. Retort 3E tested an entirely different retort design which ultimately provided the basis for scale-up to commercial-size units. Retort 3E was ignited in February 1975 and produced 1600 barrels of oil. Following completion of the first three retorts, operations were transferred to a new large-scale development mine. Retort 4 was the first commercially sized unit, being 50 times larger than the first retorts. Ignited in 1975, Retort 4 produced some 27,500 barrels of oil, somewhat less than the full potential, with difficulties traced to geologic conditions resulting in inadequate ore rubblization. Geologic conditions were overcome in Retort 5 by design changes. However, the method of rubblization using a vertical "tapered-slot" void produced an uneven horizontal distribution of porosity, resulting in channeling of gas flow. Burned in 1977, Retort 5 produced 10,100 barrels of oil. Retort 6 was a scale up of the successful Retort 3E design, was half an acre in cross-sectional area, and was as high as a 30-story builing. The 24% void volume in the retort was created by mining horizontal rooms to provide more uniform permeability in the rubble zone. Retort 6 was ignited in August 1978, and operating conditions were upset soon after the startup by a partial collapse of the retort roof (sill pillar slumping into the retort). However, corrective actions were taken and 55,700 barrels of oil were produced from this retort, representing 46% of the oil in place.

Retorts 7 and 8, currently under construction, will utilize the Retort 6, three-level design, except that they will be operated from the ground surface rather than from a mine level separated from the retort by a sill pillar. Retorts 7 and 8 will be operated simultaneously to study conditions resulting from multiretort operation.

Occidental Modified *In Situ* Retorting Experiments

Retort	Year Production	Size (ft)	Oil Produced (barrel)
1E	1973	31 × 31 × 72	1,200
2E	1974	32 × 32 × 94	1,400
3E	1975	32 × 32 × 113	1,600
4	1975	120 × 120 × 271	27,500
5	1977	118 × 118 × 158	10,100
6	1978	162 × 162 × 254	55,700
7	1982	162 × 162 × 243	97,000 (est.)
8	1982	162 × 162 × 242	97,000 (est.)

The D.A. Shale property is at best only marginal for sustained commercial production. Occidental bid unsuccessfully for Prototype Oil Shale Leases C-a and U-a. Occidental has since acquired a major interest in the federal C-b Oil Shale Tract from the original lessees, and has limited commercial plans to that tract.

Project/Process Description

The Occidental modified *in situ* process for the recovery of shale oil consists of creating underground chambers of tightly packed but broken oil shale, in two steps: (1) underground mining and removal of a certain quantity of shale; (2) expanding the shale between the mined-out rooms into the rooms with conventional explosives. The mined-out area has the horizontal dimensions of the retort—the vertical dimension is proportional to the desired void space in the completed chamber. The chamber is sealed from the atmosphere before firing. The retorting of the broken shale is accomplished by drawing air with steam down through the broken shale. A combustion zone, in which residual carbon on the spent shale is burned, moves downward through the broken shale. Shale oil drains from the bed to the floor of the retort where it is collected in sumps and pumped to the surface. The gas produced in the process can be used to generate process steam or burned. This gas has a potential for power generation significantly in excess of the plant needs. The shale mined during initial retort construction is available for surface retorting. All waste rock from mining is to be revegetated.

PARAHO–ANVIL POINTS PROJECT

Plant Type

Product:	Shale oil
Production:	200 bpd (semiworks retort)
Input Material:	Approximately 300 TPD of 28 GPT shale
Process:	Paraho aboveground patented technology
Location:	Anvil Points Oil Shale Mine and Retorting Facility, Rifle, Colorado
Current Status:	Operational with research and development programs on foreign and domestic shales

Management and Finance

Sponsor:	Paraho Development Corporation
Contractor:	Development Engineering, Inc. (Paraho subsidiary)
Manager:	Harry Pforzheimer, Jr., President
Phone:	(303) 243-9550
Location:	300 Enterprise Building, Grand Junction, Colorado 81501
Key Contact:	Harry Pforzheimer, Jr., President
Funds Expended:	Approximately $25 million expended in research and development work, and construction and operation of a pilot and semiworks plant.

Major Actions/Decisions

Paraho has obtained all necessary permits for present research and development operations. Possible future scale-up to a full-size retort or module facility at Anvil Points would require various new permits including an Environmental Impact Statement (EIS) and congressional approval to use shale from the Naval Oil Shale Reserves. DOE began work on an EIS for Paraho's proposed module retort in 1975. The draft EIS was issued in the fall of 1980. The operator has applied to DOE for an extension of the lease on the Anvil Points facility beyond the present leases which expire in 1982.

Product Markets

Paraho has produced over 110,000 barrels of crude shale oil. The majority of the crude was commercially refined into military products under contracts from the Department of Defense. These fuels included jet fuels (JP-4, JP-5, and JP-8), gasoline, diesel, and marine diesel fuel. All products have undergone extensive testing. Paraho crude shale oil has been used by a utility in boiler fuel tests.

Environmental Impacts

Paraho's technology has provided factual data on many of the environmental impacts associated with aboveground retorting of oil shale. Various DOE, EPA, and industry environmental monitoring programs have been conducted on disposal of retorted shale. Fugitive dust problems have been minimized because shale is lump size (+½–3½ in.) when entering the Paraho retort and the same size leaving the retort. Retorted shale reclamation programs have demonstrated successful compaction and vegetation techniques. Paraho's retorting process consumes no water.

Water requirements for a Paraho commercial facility are projected to be one-half barrel of water for every barrel of shale oil produced. Water is additionally required for mining, reclamation, and plant infrastructure.

Socioeconomic Impacts

The counties which have received impacts from the Anvil Points Facility have a total of approximately 80,000 (Garfield and Mesa). Total population increase is presently limited because the facility contains only research and development operations. Employment has varied from 30 to 128 people. The work force has been drawn from the towns of Grand Junction, Rifle, DeBeque, and Parachute.

Project History and Status

Paraho Development Corporation is an oil shale technology company organized in 1971 and based in Grand Junction, Colorado. Paraho has conducted several research and production programs. The Paraho Oil Shale Demonstration was privately funded by 17 participants from 1973 to 1976, at a cost of $10 million, each participant receiving the right to enter into a nonexclusive, preferential license agreement to use the Paraho retorting technology. Oil yields averaged 90%. As a result of the successful demonstration program, the U.S. Department of Defense and Department of Energy awarded Paraho a number of production contracts to produce and ship up to 100,000 barrels of crude shale oil. Production was carried out at the Anvil Points facility from 1976 to 1978. A 105-day continuous on-stream operation and oil yields above 96% were obtained. The crude oil produced has been commercially refined into military products.

Paraho has processed foreign shales at Anvil Points and is planning additional foreign shale as well as domestic shale operations.

Project/Process Description

The Paraho retort is an aboveground vertical shaft vessel with a carbon steel shell and a refractory brick lining. Oil shale rock from the crushing and storage plant, sized from +¼ to 3½ in., is fed into the top of the retort and flows downward by gravity as a continuously moving bed. As the shale moves downward in the retort, it passes through four zones:

- *Mist Formation Zone*—As the ascending gas stream is cooled by descending raw shale entering the top of the retort, oil vapors condense into a stable mist which is collected and drawn off the top of the vessel.
- *Retorting Zone*—Descending preheated shale enters the retorting zone where it is heated by the hot gases rising from the combustion zone. When the shale is heated, the organic matter (kerogen) decomposes and forms oil vapors and fixed gases, and leaves a carbon residue on the retorted shale.
- *Combustion Zone*—The zone extends below the retorting zone. Air along with recycled shale gas is injected through the top and middle distributors and distributed across the shale bed where ignition occurs. The burning of some of the gas and carbon residue provides the required process heat.
- *Retorted Shale Cooling Zone*—The principal function of this zone is to recover heat from the retorted shale before it leaves the vessel. The recovery of the heat enhances the combustion process by preheating a portion of the ascending air/gas mixture which is entering the base of the vessel. The retorted shale leaves the bottom of the vessel at a low temperature. As a result, no water is required for cooling the retorted shale. Retorted shale is reclaimed and revegetated.

PARAHO–UTE PROJECT

Plant Type

Product:	Shale oil, product gas, ammonia, sulfur
Production:	30,000 bpd of hydrotreated shale oil
Input Material:	Oil shale, average of 28 GPT
Process:	Paraho direct heated mode of retorting
Location:	50 miles southeast of Vernal, Utah, in Uinta County
Current Status:	Construction of the Paraho–Ute commercial shale oil plant will begin as soon as the necessary permits are obtained, which is expected to be mid-1982.

Management and Finance

Sponsor:	Paraho Development Corporation
Contractor:	Davy McKee Corporation, The Cleveland-Cliffs Iron Company, VTN Consolidated, AeroVironment, Inc., Woodward-Cycle Consultants
Manager:	Harry Pforzheimer, Jr., President
Phone:	(303) 243-9550
Location:	300 Enterprise Building, Grand Junction, Colorado 81501
Key Contact:	Harry Pforzheimer, Jr., President
Funds Expended:	Over $25 million has been expended in the research and development phases leading to commercialization.
Total Estimated Capital Required:	Present DOE and/or industry sponsored programs leading specifically to the construction of the Paraho commercial plant would exceed $47 million.
Financing Plan:	Construction and operation phases anticipate government incentives (loan guarantees) and a venture of industry sponsors. Paraho–Ute submitted a proposal to the U.S Synthetic Fuels Corporation for price guarantees and loan guarantees.

Major Permit Actions/Decisions

Final environmental and construction permits for the full facility will have been obtained by late 1984.

Environmental Impacts

Paraho's experiments have produced important information on the environmental impacts previously associated with aboveground retorting of oil shale. DOE, EPA, and industry environmental monitoring programs have been conducted on aboveground retorting and disposal of retorted shale. Fugitive dust problems have been minimized because shale is lump size (+½–3½ in.) when entering the Paraho retort and the same size leaving the retort. Retorted shale reclamation programs have demonstrated successful compaction and vegetation techniques. Paraho's retorting process consumes no water. Water requirements for a Paraho commercial facility are projected to be one-half barrel of water for every barrel of shale oil produced. Water is required for mining, reclamation, and plant infrastructure.

Socioeconomic Impacts

Because the shale tract is 50 miles southeast of Vernal, Utah, and 27 miles west of Rangely, Colorado, several areas are expected to receive impacts from this proposed development Uinta and Duchesene Counties in Utah and Rio Blanco County in Colorado will be the areas from which a portion of the employment force is drawn. The highest employment of 1800 persons will occur during 1984–1986, and will level off to 1200 by 1987. Paraho is working closely with the Utah State Governor's Office, the State of Utah Planning and Energy Boards, Uinta County, and other organizations to identify and plan for associated impacts. Various plans are under discussion, including a new community near the White Dam Reservoir to house Paraho project employees and other persons. Paraho has received strong support from the state of Utah for its proposed project.

Project History and Status

Paraho Development Corporation is an oil shale technology company organized in 1971 and based in Grand Junction, Colorado. Paraho has conducted several research and production programs. The Paraho Oil Shale Demonstration was privately funded by 17 participants from 1973 to 1976, at a cost of $10 million, each participant receiving the right to enter into a nonexclusive, preferential license agreement to use the Paraho retorting technology. Oil yields averaged 90%. As a result of the successful demonstration program, the U.S. Department of Defense and Department of Energy awarded Paraho a number of production contracts to produce and ship up to 100,000 barrels of crude shale oil. Production was carried out at the Anvil Points facility from 1976 to 1978. A 105-day continuous on-stream operation and oil yields above 96% were obtained. The crude shale oil produced has been commercially refined into military products.

Phase I Design and Engineering (DE-FC03-80ET14103) and Feasibility Study (DE-FG01-80RA50385) are optimistically scheduled for completion by fall of 1981. Paraho has also received notification of acceptance by DOE of its Commercial Development Program proposal (DE-PS01-80RA-50413). This Commercial Development Program (a 435 million cooperative agreement) will complete all necessary steps leading to the point of beginning actual construction. The program will complete detailed design, identify and order long-lead times items of equipment, and apply for and obtain all necessary permits. Scheduled program start is during the first quarter of 1981; however, federal funding for the final engineering cooperative agreement is uncertain pending current budget discussions.

If current plans can be realized on schedule, construction of the Paraho facility could begin by mid-1982 with production of 10,000 bpd by late-1984, 20,000 bpd by 1985, and 30,000 bpd in mid-1986. Paraho expects to fund construction and operation of the facility through such government incentives as a loan guarantee, along with a venture of industry sponsors.

Project/Process Description

The Paraho retort is an aboveground vertical shaft vessel with a carbon steel shell and a refractory brick lining. Oil shale rock from the crushing and storage plant, sized from +½ in. to 3½ in. is fed into the top of the retort and flows downward by gravity as a continuously moving bed. As the shale moves downward in the retort, it passes through four zones, as described in the Paraho–Anvil Points Project section.

The Utah facility will contain three full-size commercial retorts and associated equipment including hydrotreating facilities. Each retort will be approximately 130 ft long, 25 ft wide, and 100 ft high, with capabilities of producing 10,000 bpd of shale oil.

PLATEAU PROJECT

Plant Type

Product: Refined products

Production: 20,000 bpd (7000 bpd of JP-4, 1000 bpd of gasoline, and 1000 bpd of military grade diesel fuel for the Department of Defense; the remaining shale oil products will be 7000 bpd gasoline, 2000 bpd diesel fuel, and 1000 bpd of jet-A fuel)

Input Material: Upgraded shale oil

Process: Refinery

Location: One mile west of Roosevelt, Utah

Current Status: Operating refinery

Management and Finance

Sponsor: Plateau, Inc.

Contractor: Ralph M. Parsons Company

Manager: A.C. Leonard, President
Phone: (505) 262-2221, Ext. 623
Location: P.O. Box 26251, Albuquerque, New Mexico 87110

Key Contact: Paul W. Liscom, Project Director

Total Estimated Capital Required: $5.016 million

Financial Plan: Feasibility study grant from DOE for $4.056 million. Submitted a proposal to the U.S. Synthetic Fuels Corporation for a loan guarantee, product purchase agreements, and price guarantees.

Major Permit Actions/Decisions

Plateau plans to complete a feasibility study for the project in early 1982. Field construction of the complex will start in early 1983, and the shale oil processing facilities will be producing products by late 1984.

Product Markets

The complex will include a raw shale oil pipeline from the supply sources to the refinery and a products pipeline from the refinery to Salt Lake City where it will connect with the Chevron products pipeline.

Environmental Impacts

The environmental, health, and safety aspects of this proposed project are similar to a crude oil refinery project. The use of pipelines to move raw shale oil to the refinery and most products to market will minimize any additional emissions from transporting the additional material.

Socioeconomic Impacts

It is anticipated that many of the construction workers who move into the area for this project will relocate to shale oil mining and retorting projects afterwards and hence represent permanent additions to the area population.

Project History and Status

On December 11, 1980, the U.S. Department of Energy selected Plateau, Inc., for a feasibility study grant of $4.056 million. This study will determine the commercial viability of modifying Plateau's existing refinery at Roosevelt, Utah, to process raw shale oil. Total project cost is $5.016 million.

Project/Process Description

The process which is proposed to upgrade raw shale oil is an existing refining technology, and the upgraded shale oil will then be processed into finished products using conventional petroleum units.

RIO BLANCO (C-a) PROJECT

Plant Type

Product:	Shale oil
Production:	76,000 bpd
Input Material:	98,000 TSPD—MIS 72,000 TSPD—surface retorting
Process:	Modified *in situ*—38,000 bpd; surface retort—38,000 bpd
Location:	Tract C-a in Piceance Basin, Colorado
Current Status:	Shaft sinking, surface construction activities

Management and Finance

Sponsor:	Rio Blanco Oil Shale Company, a general partnership of Gulf Oil Corporation and Standard Oil Company (Indiana)
Contractor:	Morrison-Knudsen Company, Inc., C.E. Lummus
Manager:	J. Blaine Miller, President
Phone:	(303) 695-2400
Location:	2851 S. Parker Road, Suite 500, Aurora, Colorado 80014
Key Contact:	Tom Ten Eyck, Vice President, Community, Government & Public Affairs
Funds Expended:	$260 million (Modular Development Phase through January 1980)
Total Estimated Capital Required:	$1 billion + (commercial phase)
Financing Plan:	Submitted a proposal to the U.S. Synthetic Fuels Corporation for loan guarantees and price guarantees

Major Permit Actions/Decisions

A Detailed Development Plan for modified *in situ* development for Tract C-a was approved by the Area Oil Shale Supervisor (AOSS), now the Oil Shale Office (OSO), in September 1977. Rio Blanco has acquired the following major permits for the MIS Modular Development Phase only:

- Environmental Protection Agency (EPA)—Prevention of Significant Deterioration Permit (PSD)
- Colorado Department of Natural Resources—Mined Land Reclamation Permit (MLR)
- Enviromental Protection Agency (EPA) and State of Colorado Department of Health—National Pollution Discharge Elimination System Permit (NPDES)
- Colorado Department of Health—Subsurface Disposal Permit (Water)
- Colorado Department of Health—Emission Permit (air)
- Rio Blanco County—Special Use of the Land. Similar permits will be required for a surface retort demonstration and commercial operations.

Product Markets

In the commercial phase, shale oil would be transported to existing Gulf or Standard refineries.

Environmental Impacts

Completion of 2 years of environmental baseline studies plus continuous environmental monitoring will provide considerable scientific background for surface effects resulting from C-a development. Although modified *in situ* was covered in the 1973 Department of the Interior EIS, Rio Blanco is researching potential pollution problems of subsidence, underground retort leaching following burning, and other environmental questions. Revegetation studies have been underway for several years on Tract C-a. Negligible impacts on wildlife, agriculture, or cattle or sheep grazing in Rio Blanco County are anticipated due to the MIS and surface retort demonstrations.

Two aquifers exist on Tract C-a. The upper aquifer runs through the retort area planned for the modular development phase and shaft sinking operations have required dewatering. The water primarily has been reinjected back into the aquifer away from the mine site during the demonstration phase. The lower aquifer would be encountered in the commercial phase and, with both aquifers present, a great deal of water will be available for development needs. If additional water is required, Rio Blanco holds an option which could be exercised on an industrial water right on the White River.

Socioeconomic Impacts

The Colorado communities of Rifle, Meeker, and Rangely are expected to experience growth from the development of Tract C-a. By existing roads, the tract is 58 miles from Rifle, 51 miles from Meeker, and 68 miles from Rangely. Engineering studies made in 1974 and 1975 demonstrated that a new road, which could be built southeast from Rangely to C-a, would reduce commuting distance to 25 miles. To date, the Colorado General Assembly has not approved funding the road from the state's Oil Shale Trust Fund, which was created by the placement of some $70 million from the nearly $200 million paid to the federal government by the lessees of the two Colorado oil shale tracts. The trust fund money is primarily for socioeconomic impacts caused by oil shale development, and more than $47 million of the total placed in the state fund came from payments made by the lessees of Tract C-a. During the initial modular development phase of construction through part of 1982, employment is not anticipated to exceed 380 workers, with fewer than that number for most of the period. A commercial-scale production level of 76,000 bpd could result in permanent employment for about 2500 workers. Rio Blanco is developing housing and transportation plans for its workers; the company now underwrites a mobile home park in Rangely and is planning another for Meeker. More than 90% of Tract C-a employees are transported to and from work by buses and vans. Rio Blanco Oil Shale Company works with state and local officials in identifying socioeconomic impacts and in planning mitigative measures.

Project History and Status

Tract C-a was offered for lease January 8, 1974, by closed bids by the U.S. Department of the Interior. Gulf Corporation and Standard Oil Company (Indiana) acquired the lease officially on March 1, 1974; the total amount bid was $210.3 million. The two companies later formed the Rio Blanco Oil Shale Company, a general partnership, to develop the tract. In accordance with terms of the lease, three annual payments of about $42 million each have been made. The final two payments were offset by the Area Oil Shale Supervisor (now Deputy Conservation Manager/Oil Shale) by approved tract development expenses. Originally, Tract C-a was to be developed by open-pit methods as projected in the Interior Department environmental impact statement. However, the department was not legally able to make additional federal land available for disposal of processed shale and overburden. Because of this and other constraints, the lessees decided to develop the tract by underground modified *in situ* methods. Following submittal of an open-pit Detailed Development Plan in 1976 and a 1-year suspension of operations, the lessees submitted a revised DDP, which was approved in September 1977. Shaft sinking began later that year and was completed in the fall of 1979. In February 1979, the Rio Blanco Oil Shale Company purchased the modified *in situ* technology of Occidental Oil Corporation. A technical modification involving the method of constructing the modified *in situ* retort and retort configuration was submitted to the Department of the Interior on June 28, 1979. This resulted from technical information developed by Rio Blanco Oil Shale Company and acquired since the Detailed Development Plan was approved on September 22, 1977. This first MIS retort was burned starting in October 1980 and was completed at the end of December 1980. Results presently are being evaluated. Construction is anticipated to begin on a surface retort utilizing Lurgi technology in 1981. Operation of the retort is scheduled to begin approximately mid-1983. Ore would be obtained from a small open pit located in the northeast corner of Tract C-a. The process plant and support facilities will be located on Rio Blanco's private property adjacent to and north of Tract C-a. The spent shale will be disposed of on this private property. A modification to the DDP to include the Lurgi Demonstration Project was submitted to the Oil Shale Office in early February 1981. Permit applications presently are being submitted to various government agencies. A decision to construct will follow Lurgi surface retort operation, approximately late 1983 through 1984. Operation of the first commercial increment of capacity could begin in the late 1987 through 1988 timeframe.

Project/Process Description

The C-a Federal Prototype Oil Shale Lease is located at the head of Yellow Creek on the western edge of the Piceance Creek Basin of northwestern Colorado (R99W between T1S and T2S). The 5090-acre tract has a potential recoverable shale oil reserve between 1.3 and 5.2 billion barrels, depending on extraction method (open pit, underground room-and-pillar, or modified *in situ*). Maximum daily production by open pit was estimated at 300,000 bpd. Maximum for modified *in situ* (MIS) is projected at 200,000 bpd. The Rio Blanco Oil Shale Company is engaged in a modular development phase to prove the commercial viability of the MIS process. Some $93 million initially was authorized by Gulf and Standard to burn a series of underground retorts. The program is expected to be complete by 1982. Commercial development efforts could begin shortly after completion of the MDP which would lead to a commercial production level of 76,000 bpd by 1987 at the earliest. Production would consist of 38,000 bpd by underground modified *in situ* mining and processing and 38,000 by surface retorting. The MIS method involves removing some 40% of the desired underground retort void space and bringing that ore material to the surface to be surface retorted. After removal of the material from the underground, the retort is rubblized and burned and the shale oil is pumped to the surface. Rio Blanco Oil Shale Company presently is studying the commercial viability of an open pit/surface retort operation. A commercial decision for developing Tract C-a is expected to be made in 1982–1983.

SOUTHERN INDIANA PROJECT

Plant Type

Product:	Shale oil
Production:	10,000 bpd, eventually expanding to 50,000 bpd
Input Material:	Eastern oil shale
Location:	Clark County, Indiana
Current Status:	A $6.5 million study grant was awarded by DOE

Management and Finance

Sponsor:	Southern Indiana Shale Oil Co., Inc.
Manager:	James M. Robison, President
Phone:	(317) 392-4886

Location:	45 West Washington St., P.O. Box 748 Shelbyville, Indiana 46176
Key Contact:	James M. Robison

Project History and Status

On December 11, 1980, the U.S. Department of Energy announced the selection of the Southern Indiana Shale Oil Co. for $6.48 million to assess the feasibility of construction of a plant to produce shale oil from eastern oil shale. Time requirement for the feasibility study is 18 months. Proposed subcontractors are Mittelhauser Corp., Ralph M. Parsons Co., A.J. Kearney, Inc., S.K. Meuchner Associates, Inc., Furgo, Inc., Bowser-Morner Testing Labs, Inc., and UOP, Inc. Initiation of the study has been delayed pending current federal budget discussions.

Project/Process Description

The shale oil plant under consideration would be located in Clark County, Indiana, and have a design capacity of 10,000 bpd, with later expansion to 50,000 bpd planned.

SUPERIOR MULTIMINERAL PROJECT

Plant Type

Product:	Shale oil, minerals
Production:	13,000 bpd shale oil, 4000 TPD nahcolite, 700 TPD Al_2O_3, 15,000 TPD Na_2CO_3
Input Material:	26,000 TPD of 25 GPT shale, 20% nahcolite, 7% dawsonite
Process:	Proprietary retorting process (traveling grate) and retorted shale leaching
Location:	Piceance Basin, Rio Blanco County, Colorado
Current Status:	Pilot studies completed; U.S. Department of the Interior evaluation and environmental statement are complete. DOI's dollar evaluation and exchange decision are being appealed.

Management and Finance

Sponsor:	The Superior Oil Company—60%, Sohio—30%, Cleveland-Cliffs—10%
Manager:	J.H. Knight
Phone:	(303) 761-5853
Location:	2750 S. Shoshone, Englewood, Colorado 80110
Key Contact:	J.H. Knight
Total Estimated Capital Required:	Project costs expected to exceed $300 million
Financing Plan:	Source of financing not determined. Superior contends that the multiproduct (shale oil, nahcolite and dawsonite) nature of the project may allow the venture to be financially independent of the federal government.

Major Permit Actions/Decisions

In November 1973, an application for an exchange of land of 2571 fee acres for 1769 acres of BLM land was made to the BLM by Superior. The land exchange was denied in February 1980. The approval of this land exchange will result in a more economic block of extractable resources.

Further permit action is suspended pending approval of the exchange by BLM. Other major government permits needed for the success of this project are:

- U.S. Army Corps of Engineers—Creek conduit
- Environmental Protection Agency (EPA)—Spill prevention countermeasure plan; National Pollution Discharge Elimination Plan
- Colorado Department of Natural Resources—Spent shale disposal; wells and water use; land reclamation; air and water quality
- Superior does not plan to utilize surface water, as sufficient water will be produced from mine drainage and salt aquifers

Product Markets

The four products of this process are crude shale oil, nahcolite, alumina, and soda ash. The minerals are roughly equal to three times the value of the oil in terms of financial return on capital and operating costs expended. Customers for nahcolite have been identified.

Environmental Impacts

The BLM has completed the final Environmental Statement (ES) as a portion of the information required for the land exchange. The appeal of the Department of the Interior's dollar evaluation and decision regarding Superior's land exchange timing is undetermined. Spent shale disposal will not be a problem at the Superior plant. The extraction of marketable nahcolite and dawsonite from the shale (as well as oil) will reduce the quantity of spent shale such that it can be backfilled in the underground mine. Superior would use nahcolite to eliminate sulfurous and nitrous emissions from its process, as well as marketing nahcolite to power companies and other customers for stack gas emission abatement.

Socioeconomic Impacts

Potential socioeconomic benefits and impacts will be discussed in the above-mentioned ES. Affected communities will be Rifle, Meeker, and Rangely, Colorado. Meeker and Rangely both have populations estimated to be 1800. New area residents are estimated to be employed at 900 new shale jobs and 1400 support jobs for continued operation of the plant (up to 1350 on shale jobs at construction peak). Some impact also will occur at Rifle, Colorado, which serves as a major railroad center for the area.

Project History and Status

Superior has acquired mineral right ownership of about 6750 acres of private land bordering the White River at the northern edge of the Piceance Basin, Colorado (6000 acres of tract are shale-bearing). In November 1973, Superior applied to the BLM for exchange of private lands within the leased tract for certain public lands being managed by BLM. BLM denied the land exchange on February 5, 1980. Superior has made no significant actions other than the above-mentioned application that moves the project closer to construction. Superior has been conducting laboratory and pilot testing of facility components and processes; it also has conducted feasibility studies in the area of mining, retorting, mineral processing, and product markets. In 1980, Superior consummated a shale resource agreement with Sohio and Cleveland-Cliffs in which Superior gained a 20% working interest in the Pacific oil shale project and provided a 30% Sohio and 10% Cleveland-Cliffs interest in Superior's mineral/oil shale resource.

Project/Process Description

Shale of ¼ × 4 in. (6 × 100 mm) size is fed to the continuously rotating grate through inert-gas-purged mechanical seals. The resulting bed of shale rotates into the heating zone where oil is evolved and carried away with the circulating gas heating media. The circulating gas with oil mist passes through oil removal equipment where oil and water are removed. The retorted shale is passed through the carbon recovery zone where residual carbon is partially oxidized and oil evolution is completed. It then continues into the shale cooling zone. Recycle gas is heated in the shale cooling zone by recovering sensible heat from the hot retorted shale. The shale is further cooled by ambient air, which is heated for use in the direct-fired burners or alternately in the indirect heater. This energy recovery improves thermal efficiency. Spent shale is discharged either by mechanical seals or a water seal which wets the shale for dust control. The rotating grate is sealed between stationary hoods on top of the bed and windboxes underneath by water troughs on both sides of the shale bed. Preheated recycled gas from the shale cooling zone is further heated by mixing with hot combustion gases of 2500–3000°F (1370–1650°C). The resulting gas mixture, at 1250–1500°F (680–815°C), is introduced into the shale heating zone completing its cycle. Product gas is taken as a slip stream from the recycling heating media gas. A medium-Btu gas of 400–600 Btu/scf (depending on grade of feed shale) is produced by using the indirect heating mode. The gas produced from direct heating is diluted by combustion gases (CO_2 and N_2) and has a heating value of 80–130 Btu/scf depending on shale grade. This gas product stream is compressed to 50 psig, and additional water–oil is condensed and removed prior to leaving the retorting process.

SUPERIOR PACIFIC PROJECT

Plant Type

Product: Shale oil

Production: 15,000-bpd demonstration plant; increasing to three, 15,000-bpd modules, 2.6 million GPD (per module)

Input Material: 22,000 TPD of 32 GPT shale (per module)

Process: Superior Oil/Davy McKee (circular grate)

Location: West of Rifle, Garfield County, Colorado

Current Status: Phase I engineering design and permit acquisition started

Management and Finance

Sponsor: The Superior Oil Company 20%; Sohio 60%

Manager: J.H. Knight
Phone: (303) 761-5853
Location: 2750 S. Shoshone, Englewood, Colorado 80110

Key Contact: J.H Knight

Total Estimated Capital Required: Demonstration project cost expected to exceed $300 million. 50,000 bpd plant cost estimated at $1.3 billion.

Financing Plan: Source of financing on engineering Phases I and II demonstration unit is DOE-Superior cost sharing. Construction (Phase III) financing is undetermined at this time. Cleveland-Cliffs has submitted a proposal to the U.S. Synthetic Fuels Corporation for a loan guarantee of up to 75% of its share of total project costs.

Major Permit Actions/Decisions

- Environmental Protection Agency (EPA)—Spill Prevention Countermeasure Plan; National Pollution Discharge Elimination Plan.
- Colorado Department of Natural Resources—Spent shale disposal; wells and water use; land reclamation; air and water quality.

Product Markets

The products of this process are crude shale oil of approximately 15,000 bpd from the demonstration module; 50,000 bpd or more could be produced upon expansion of the demonstration module to full plant size (Phase IV).

Socioeconomic Impacts

Superior is providing DOE additional socioeconomic data at this time. Potential socioeconomic benefits and impacts will be discussed in the above-mentioned EIS. Affected communities will be Grand Junction, DeBeque, and to some extent, Parachute and Rifle, Colorado. New area residents are estimated for employment at 400 new shale jobs and 700 support jobs in operating the demonstration plant. The construction employment for all three modules peaks at 2800 in 1988; for the first module, construction employment peaks at 2200 in 1984 or 1985. Operating employment is projected to be approximately 740 for the first module and 1400 for all three modules. Additional housing and community services will be needed. The exact extent of these needs will be further determined at a future date.

Project History and Status

Superior acquired 20% ownership of the Pacific property in 1980. The Pacific property is on the southwestern corner of the Piceance Creek Basin, Colorado (8000-acre tract is high grade shale-bearing). The 1980 resource agreement results in the Pacific oil shale property ownership as 60% Sohio, 20% Superior, and 20% Cleveland-Cliffs.

SYNTANA–UTAH PROJECT

Plant Type

Product: Shale oil and a low-heating-value combustible gas

Production: 16,500–50,000 bpd by the early 1990s

Input Material: 24,500 TPD and up, of 25 GPT shale—approximately 3000 AF/Y of water

Process: Deep mining in combination with surface retorting

Location: R25E, T9S, Section 16 in Uinta County, Utah

Current: Mine/surface facility engineering studies underway in combination with additional field drilling to further quality resource, assay and location

Management and Finance

Sponsor: Syntana–Utah, a joint venture organized by the Synthetic Oil Corp., Oklahoma City, Oklahoma, and Wuintant Minerals Corp., Houston, Texas

Contractor: Morrison-Knudsen Co., Davy McKee Corporation, and VTN Consolidated, Inc.

Manager: R.E. Lee, Project Director
Location: Syntana–Utah, 601 Jefferson Street
Houston, Texas 77002

Key Contact: Wayne E. Shannon, Assistant Project Director
T.C. Driscoll, Manager–Finance

Funds Expended: In excess of $1 million—combined prior and current year expenditures

Total Estimated Capital Required: $300 million (1980 dollars) for a 10,000-bpd facility

Financing Plan: Private sector financing in combination with the loan guarantee requested from the U.S. Synthetic Fuels Corporation.

Major Permit Actions/Decisions

- Syntana–Utah Unit #1 process shale oil mined from mineral leases held on state of Utah owned lands.
- A preliminary description of the Unit #1 development was approved by the Utah Department of Natural Resources. Utah Division of Oil, Gas and Minerals approved the current field drilling program to better define resource locations and assay.
- Syntana–Utah properties of current interest are in Uinta County, T9S, R25E, and R24E, close to White River Shale Project. Project sponsors find that the Department of Energy final EIS for the prototype oil shale program, August 1973, and the Environmental Baseline Report applicable to Tract U-a and U-b, and prepared in October 1977, are partially applicable to the Syntana–Utah activities. Thus prolonged on-site environmental monitoring and development of an environmental database will not have to be developed by this project. Permitting requirements associated with 15 separate categories of required project development activities have been identified, and these permit issuances will be pursued with federal, state, and local government units as appropriate.
- October 14, 1980—Presentation of Preliminary Detailed Development Plant by Syntana–Utah to the Utah Division of State Lands and Forestry.
- January 19, 1981—Environmental Protection Agency, Region VII, Denver, Colorado, preapplication meeting.

Product Markets

The product will be synthetic crude, suitable for tanker-truck, pipeline, and/or rail transport to refineries in the west and southwest, and a low- to medium-Btu gas.

Environmental Impacts

Site is surrounded by several proposed major projects including the White River Shale Project, the White River Dam, shale facility developments by Paraho and Tosco, and the construction of a new electric power plant, the DESERET Unit #1 of the Moon Lake Electric Cooperative. The Syntana–Utah facility development adds little to the overall regional environmental impact when considered in the context of the foregoing anticipated projects. With extensive environmental baseline efforts already performed as part of the Federal Prototype Lease Program, only a limited amount of site-specific environmental data will have to be developed by this project. The mine and surface retorting facility plant design will comply with best available control technology, and all requirements—federal, state, and local—as they relate to air and water quality standards.

Socioeconomic Impacts

Proposed location of Syntana–Utah Unit #1 is approximately 45 miles southeast of Vernal, Utah, and 25 miles southwest of Rangely, Colorado. Rangely has approximately 2000 residents and Vernal approximately 20,000. In view of the anticipated development programs soon to begin in the eastern Uinta County region, the Uinta Basin Energy Planning Council in cooperative with state and federal officials has proceeded with the examination of alternatives for creating the appropriate infrastructure to support the projected individual development. Syntana–Utah is aware of these activities and its development plant includes appropriate consideration of infrastructure development necessary to create a new grassroots facility in the design region.

Workforce and Population Generated by Syntana–Utah Project

Mining	500
Processing	325

Primary	1,500
Secondary	650
Population Expansion	4,000

Project History and Status

In 1979, the Synthetic Oil Corporation of Oklahoma City was formed to pursue the development of facilities leading to production of a refinable crude derived from Utah shale resources. The founders of Synthetic Oil Corporation, in turn, negotiated access to state-owned mineral leases in Utah as the basis for the available resource. In the third quarter of 1980, the Synthetic Oil Corporation was joined by Quintana Minerals Corporation of Houston, Texas, the two corporations forming a new joint venture known as Syntana–Utah. Syntana–Utah in turn was charted to pursue the same objective as initially defined by the Synthetic Oil Corporation; however, in pursuit of this objective, the joint venture now draws on a broader base of management, technical, and financial resources to develop a new private sector venture. An initial business plan of the joint venture was prepared in August 1980, and in October 1980 the Utah Department of Natural Resources, Division of Land and Forestry, was advised as to the long-term oil shale commercialization objectives of the joint venture. Subsequently, Syntana–Utah gave its approval for initial activities supporting the facility development, with fourth quarter 1980 activities concentrating on further quantification of the in-place resource location and assay as well as first efforts to develop an engineering definition of both the deep mine and surface retorting facilities. The project management for Syntana–Utah Unit #1 is currently seeking support from DOE under the provisions of P.L. 96-304 for support via grant award of continuing feasibility studies for the facility during CY 1981. The project management has submitted a proposal to the U.S. Synthetic Fuels Corporation with a request for financial incentive support to continue project development with total capitalization, to represent a combination of private financial sector funding with that of the U.S. Synthetic Fuels Corporation.

Project/Process Description

Syntana–Utah currently is evaluating several mature surface retorting technologies which might be used as the basis for design and construction of Unit #1, including those made available for license by Paraho, Superior Oil, Tosco, and Lurgi. The Unit #1 facility will consist of a deep mine and surface retorting plant. The surface retorting facility currently is anticipated to be based on the utilization of individual retorting modules having input capacities in the range of 4000–10,000 TPD of shale ore. An engineering/economic decision yet to be made is the selection of the specific numbers of modules to be associated with the Unit #1 facility. The business plan was structured around early evaluation and selection of a preferred surface retorting technology. This was to be followed by the detailed design of mine and surface facilities, with required permitting to be initiated in CY 1981, and initial site work, long-lead procurement, and surface facility and mine construction to be initiated in CY 1982.

THERMO-MIST PROJECT

Plant Type

Product:	Shale oil
Production:	Phase I—2000 BPD Phase II—Additional 2000–4000 bpd from patented land Phase III—2000–6000 bpd from unpatented land when patent is received
Input Material:	3000 TPD oil shale
Process:	Surface retort
Location:	T 5&6 S, R 99&100 W, Colorado
Current Status:	Awaiting financing

Management and Finance

Sponsor:	The Thermo-Mist Company, a California limited partnership
Manager:	Margaret W. Savage
Phone:	(213) 862-3223
Location:	9528 Lemoran Avenue, Downey, California 90240
Key Contact:	Margaret W. Savage
Total Estimated Capital Required:	$23 to $30 million

Major Permit Actions/Decisions

Necessary permits have been identified and environmental and socioeconomic studies planned. No permits have been applied for.

Product Markets

Possible on-site refining; otherwise sale to nearest refinery.

Environmental Impacts

Because of its small size and remote location, the project is not expected to have any significant environmental impacts.

Socioeconomic Impacts

As a small project which will not require a large work force either for construction or operation, it will not have broad socioeconomic effects. The short construction period also will minimize the duration of impact. Socioeconomic impacts will be mainly in DeBeque, Colorado, the nearest town, located 28 miles from the site. Improvements to the county road along Roan Creek that presently serves the Occidental Petroleum Company Logan Wash facility may be required between DeBeque and the Thermo-Mist site.

Project History and Status

The Thermo-Mist Company is a California limited partnership formed in1971 by six members of the Savage family, who own a 75% interest, and owners of the Ben Holt Company, an independent consulting, engineering, and construction firm located in Pasadena, California. The project involves Phase I development of the patented portion of a tract of oil shale land owned for many years by John and Joan Savage (Thermo-Mist partners) and co-owners. Continental Oil Company, Standard Oil Company of California, Texaco, and The Tosco Corporation also own land in this area of the Piceance Creek Basin. The project has a 60-year background of Savage family involvement in Colorado oil shale claims patenting, land acquisition, and retorting process development. The Thermo-Mist retort was invented and patented by Harry K. Savage, father of four Thermo-Mist partners. John Savage has been engaged in oil shale retorting research for over 30 years. The patented and unpatented Thermo-Mist oil shale property, an adjacent ranch, various water rights, and the Thermo-Mist oil shale retorting process constitute the project assets. The patented portion of the property is ready for development; development of the unpatented part, which is legally patentable, must await resumption of the issuance of patents by the Department of the Interior. Whether the DOI is required to do so has been at issue for several years. A feasibility study was made in 1968 comparing a large Thermo-Mist project with a similarly sized U.S. Bureau of Mines project. In September 1980 Thermo-Mist proposed to DOE the funding of a study to demonstrate the project's economic feasibility and to advance it to the point where money for final design, permit acquisition, and construction could be sought. Thermo-Mist did not receive this award and is now seeking alternative sources of project financing.

Project/Process Description

The Thermo-Mist property lies between Carr and Roan creeks near their confluence in Garfield County, Colorado (T &6 S, R 99& 100 W). The oil shale deposits are on parts of 4-A Mountain and Henderson Ridge. The property contains 3510 acres, of which 2510 are patented and 1000—contiguous to the south on 4-A Mountain—are unpatented. The estimated recoverable shale oil is 50 million barrels from the patented portion and 54 million from the unpatented. Phase I, for which funding is currently sought, includes an underground mine, a one-module, commercial-scale, 2000-bpd Thermo-Mist retort, a spent shale disposal site, and other needed facilities. The mine entry will be on the face of the cliff at the Mahogany Ledge on the north side of 4-A Mountain. The retort also will be on the north side. Spent shale will be placed in a carefully prepared side gulch. The Thermo-Mist retort is gravity-flow, vertical kiln type that more closely resembles the Paraho retort, which it antedates, than other types of surface retort. The owners feel it has features which should ensure superior heating of the shale, ease of operation, comparatively inexpensive construction and operation, and a relatively brief construction time. The retort process does not consume water, and Thermo-Mist's water rights should be adequate for the three development phases. Phase II will provide additional production from the patented land. Phase III will develop the presently unpatented land when patent has been obtained.

The owners' estimated cost of the 2000-bpd Phase I project is $25–30 million, depending upon the extent of on-site oil upgrading. The amount recently proposed to DOE for a grant to (1) prepare preliminary designs, (2) estimate costs, (3) study marketing, (4) determine economic feasibility, and (5) conduct all environmental and socioeconomic studies preparatory to applying for permits and determining the environmental controls needed was $1.84 million (in late 1981 dollars). Production of shale oil from the center of the Piceance Basin will require sinking deep shafts or excavating giant open pits, demanding many years of lead time and huge investments before operation can begin. The owners feel there are no technical reasons why the peripheral areas cannot be brought into production quickly with mine access from outcrops. Analysis shows that when rapid recovery (between now and the year 2000) rather than total reserves is considered, 80% of the near-term recoverable reserves lie in the peripheral area. These have greater potential for development as small operations.

TOSCO SAND WASH PROJECT

Plant Type

Product:	Upgraded shale oil
Production:	47,000-bpd project on 14,688 acres
Input Material:	66,000 TPD of 35 GPT shale, 12,000 AF/Y of water
Process:	Tosco II Surface Retort
Location:	Uinta Basin, Utah (35 miles south of Vernal)
Current Status:	Permitting, basic design

Management and Finance

Sponsor:	Tosco Development Corporation, Sand Wash Project
Contractor:	TRW, Harrison Western, Ralph M. Parsons
Manager:	George C. Kane, Project Director
Phone:	(303) 831-4567
Location:	1600 Broadway, Suite 1400, Denver, Colorado 80202
Key Contact:	George C. Kane

Funds Expended:	$4.3 million through 1980
Total Estimated Capital Required:	$1.5 billion (1980 dollars)
Status of Financing:	No lenders have been identified owing to early stages of project

Major Permit Actions/Decisions

Utah Board of State Lands was asked to Unitize 29 state oil shale leases totaling 14,688 acres in the Sand Wash area. This unit plan was approved by the State Land Board on January 7, 1976. Project planning, design, and permit acquisition for a commercial-scale facility is proceeding. Major permits to be acquired for the Sand Wash Project include:

- Utah State Board of Oil, Gas & Mining—Surface rehabilitation plan and mine plan
- BLM—Rights-of-way for products pipeline, power lines, conveyors, roads, and water supply
- EPA—PSD Permit
- EPA and state-approvals for any water discharge or reinjection
- State air permits

Potentially, permits may be required from the Bureau of Indian Affairs and the Ute Tribe owing to rights-of-way and close proximity to Uinta Indian Reservation.

Product Markets

Most likely a low-sulfur hydrotreated shale oil product will be produced to be pipelined to either Salt Lake City or a midwest market via Rangely. Local markets are preferred. Marketing outside of region is heavily dependent upon availability of exchanges.

Environmental Impacts

Actions leading to a site-specific EIS have been initiated. Detailed environmental impact analysis is proceeding. The plant should have no discharge of liquid effluents (similar in design to the Colony plant in Colorado); however, there may be temporary disposal of excess mine water. Air effluent discharge will comply with all current federal and state regulations. Sulfur compounds and ammonia generated are routed to processing units for recovery of elemental sulfur and ammonia for sale as by-products. The disposal of spent shale will cause land disturbance at the project site. Topsoil will be removed ahead of filling and stockpiled for reuse in land reclamation. revegetation of all disturbed areas is planned and would reestablish wildlife habitat disturbed during plant construction. No impact on the agriculture of the largely arid region is anticipated. Construction of a water impoundment facility on the White River to serve oil shale development in the Uinta Basin is the subject of an environmental impact statement now being processed.

Socioeconomic Impacts

The current estimated population of Uinta County is 20,200 based on the 1980 U.S. census household count. Vernal, the largest city in Uinta County, has a population of about 6500. About 10,000 new permanent residents will move into the region as a result of this project.

Project History and Status

The project is in a planning and environmental analysis phase. Planning work and the design for a shaft and experimental mine are in progress. A pilot hole was completed at the proposed site of the shaft. An estimate for a single module plant was also prepared. Tosco has demonstrated the retort process technology at Parachute Creek, Colorado, in a 1000-TPD semiworks plant. A detailed commercial design for a 66,000-TPD plant performed for Colony has been applied to the Sand Wash Site.

Project/Process Description

The Sand Wash project plans to utilize a form of underground room-and-pillar mining, TOSCO II retorting, and conventional refining methods to produce a premium quality synthetic oil. The TOSCO II process recovers 100% of the hydrocarbons contained in the mined oil shale as reasoned by Fischer Assay. The mine will draw 35 GPT shales from 30–40 ft intervals 2000 ft underground.

UNION PROJECT

Plant Type

Product:	Upgraded shale oil yielding high percentage of jet fuel (JP-4), military grade diesel fuel, and syncrude
Production:	10,000 bpd—1983; increasing to 50,000 bpd—1987
Input Material:	Phase I—2500 TPD, 34 GPT shale; Phase II—68,000 TPD, 34 GPT shale
Process:	Union Up-Flow Retort Process
Location:	Parachute Creek, Garfield County, Colorado
Current Status:	All permits for Phase I 10,000-bpd mine and retort have been issued. EPA PSD Permit, Colorado State Air Permit, and Garfield County Special Use Permit for the upgrading facility have been applied for and are expected by early 1981. Road construction, site preparation, and preliminary mine work are underway.

Management and Finance

Sponsor:	Union Oil Company
Contractor:	Union Oil Company
Manager:	John M. Hopkins, President, Energy Mining Division
Phone:	(213) 977-6437
Location:	P.O. Box 7600, Los Angeles, California 90051
Key Contact:	John M. Hopkins
Total Estimated Capital Required:	Phases I & II, $1.8 billion (1980 dollars)
Financing Plan:	Union proposed that the U.S. Synthetic Fuels Corporation assist in financing through advance payments and purchase commitments for Department of Defense fuel purchases.

Major Permit Actions/Decisions

All licenses and major permits necessary to start construction of the 10,000-bpd Phase I mine and retort have been obtained. Applications for a permit for the 10,000-bpd upgrading plant were made in September 1980. Permits critical to the Phase II 40,000-bpd plant are:

- Environmental Protection Agency (EPA)—PSD, RCRA, RSCA Permits

- U.S. Corps of Engineers (COE) Bureau of Land Management (BLM)—Environmental Impact Statement
- State of Colorado—Mine Land Reclamation and APC
- COE Section 404
- BLM Right-of-way

Product Markets

Phase I: Fractionation facilities will produce JP-4 at Parachute Creek, with additional volume available for further processing into military specification fuels at other refineries.

Phase II: In addition to military grade products, syncrude suitable for refinery feed will be produced.

Products will initially be shipped by rail and tank truck.

Environmental Impacts

Environmental baseline data gathering programs and engineering studies for Phase I have been completed. Applications for permits to construct and operate were initially submitted in April 1978. Environmental emissions will be in compliance with existing state and federal regulations. Disposal of spent shale will cause land disturbance at the project site. Revegetation of the entire area is planned and would reestablish much of the original wildlife habitat disturbed during plant construction and operation. Revegetation also will reestablish most of the grazing land for agricultural purposes. Union Oil Company and its consultants have performed environmental studies on the project site for most of the past decade. Results of these studies indicate there are no apparent environmental considerations that would preclude industrial development in the area.

Socioeconomic Impacts

The two counties expected to receive the major impacts of the proposed developments have a combined population of 70,000: Mesa County about 55,000 residents and Garfield County about 15,000. The cities of Rifle, Parachute, DeBeque, and Grand Junction also will be affected. These cities and counties also will be impacted by the Colony Development project which may closely follow the Union project. Severity of the impacts resulting from the proposed project will depend upon the interplay of Union's proposed activities (in particular, the size, earnings, and other characteristics of its work force), and the changing and complex nature of locally affected communities in the project region. New basic employment associated with the project consists of both temporary (construction-related) and permanent (operation-related) employment. Construction employment has begun and will continue for approximately 6½ years. Average construction employment will be about 2000 persons. Peak construction employment will reach about 4700 during the fourth quarter of 1986. The mix of required skills will change from quarter to quarter, but about 75% of the total work force will be skilled workers. Operational employment will begin in the first quarter of 1983 with 350 workers and increase to about 1100 starting in mid-1987. Operational employment will remain at this level until project termination.

Project History and Status

- 1920s—Union began acquiring oil shale lands in western Colorado
- 1955—A 3½-year research and development program began at Union's Parachute Creek, Colorado site to demonstrate oil shale mining and retorting and shale oil upgrading technology. Detailed engineering studies related to commercial operation commenced.
- 1967–1968—Union's "Upflow Retort" was successfully demonstrated at the 1200-TPD level.

- 1961—15,000 bbl shale oil (obtained from Parachute Creek runs) upgrading and refining demonstration run was made at the American Golsonite Refinery at Fruita, Colorado.
- 1974–1977—Prototype Shale Oil Plant. Process comparisons, economic evaluations, and capital cost estimates for various shale oil facilities were performed, ranging from a 1500-TPD demonstration plant up through a 50,000-bpd syncrude plant. Major improvements were achieved in the Union upflow process to increase capacity and reduce costs. A complete 10,000-TPD prototype oil shale plant, including facilities from the mine through unit train marshaling yard, was developed, together with capital and operating costs. In 1977, engineering studies were performed to upgrade the test unit.
- 1977–1979—Experimental Oil Shale Plant. Engineering Design Specifications and Cost Estimate for a 10,000-TPD experimental oil shale mine and retorting plant were developed. The plant was to be located in the Parachute Creek area near Grand Valley (now Parachute), Colorado. This plant was to test Union Oil's upflow rock feeder and retort, producing about 9000 bpd shale oil. Included with this development work was the preparation of the necessary documents to secure the major environmental permits.
- 1979—Semiworks Rock Feeder Test Facility. A semiworks rock feeder was designed, fabricated, and constructed on Union Oil Company's property in the Parachute Creek area near Grand Valley, Colorado. This semiworks rock feeder was operated in order to evaluate the performance resulting in the scale-up design capability to a full-size rock feeder.
- 1980—10,000-bpd Retort Plant. Based on information developed in the last 5 years, detailed engineering and design were started on the development of a 10,000-bpd retort plant. Included in this work are the development of a mine and upgrading plant to support the retort operation.
- 1980—50,000-bpd Oil Shale Plant. Based on the Experimental Retort Plant, work has started on the development of a 50,000-bpd oil shale retort facility complete with mine and upgrading plant auxiliary facilities.

Project/Process Description

Union Oil believes that commercial development of its property could produce shale oil up to the 150,000-bpd level for more than 25 years. Located on Parachute Creek, Garfield County, Colorado (T5, 6S, R8W), the property is estimated to contain 2 billion barrels of shale oil recoverable by today's technology. Union presently is planning a 50,0000-bpd facility to be started with one commercial module at the 10,000-bpd level, possibly by 1983.

Union's water supply will come from the Colorado River and will be impounded in a reservoir on Parachute Creek. The mine will be underground room-and-pillar and produce 68,000 TPD at the 50,000-bpd capacity.

Union's upflow retort is a vertical cone. Crushed shale is fed into the bottom and forced up through the vessel by the rock pump. Recycle gas, heated in an external fire heater to 950°F (510°C) is forced into the top of the retort by a compressor and flows downward through the shale. The hot gas decomposes the kerogen in the shale, producing oil and gas which flow downward over the incoming cool shale, preheating the shale and condensing and cooling the oil. The oil and gas are drawn off the lower part of the cone. The retorted shale flows off the top of the cone, is cooled, and then moved by conveyor to an ore pass for disposal.

WHITE RIVER (U-a, U-b) PROJECT

Plant Type

Product: Shale oil

Production:	50,000–100,000 bpd contingent upon commercial economics
Input Material:	179,000 TPD of 28-GPT shale rock; 13,000–27,000 AF/Y of water
Process:	Surface retorting
Location:	10,240 acres; contiguous Federal Prototype Oil Shale Lease Tracts U-a and U-b, Uinta County, Utah
Current Status:	Activities suspended pending clearance of property title. DDP update in progress. Environmental monitoring continuing

Management and Finance

Sponsor:	White River Shale Project (Phillips, Sunoco, and Sohio)
Contractor:	Bechtel
Manager:	C.E. Doney, Project Manager
Phone:	(801) 328-3700
Location:	1315 West Highway 40, Vernal, Utah 84078
Key Contact:	Rees C. Madsen, Administrative Coordinator James W. Godlove, Environmental Affairs Coordinator (801) 789-0571
Funds Expended:	Combined Bonus Payments, $72.4 million; Environmental Programs, $9 million; Engineering, $8 million
Total Estimated Capital Required:	$3.0 billion (1980 dollars) for 100,000-bpd facility
Financing Plan:	Phillips Petroleum Company and Sunoco Energy Development Co. are requesting loan guarantees in the amount of 60% of those project costs representing the commitment of the applicants, or 40% of total project costs.

Major Permit Actions/Decisions

As required by lease agreements, detailed environmental baseline monitoring studies have been conducted on tract. The Detailed Development Plan was submitted in 1976 and is being updated at this time and will be submitted in June 1981. Progress on the project is contingent upon approval of the DDP and receipt of other government approvals, examples of which follow:

- Environmental Protection Agency (EPA)—PSD Air Permit, National Pollutant Discharge Elimination System Permit (will be submitted in June 1981)
- Bureau of Land Management (BLM)—Right of Way for Support Facilities Approval of White River Dam Project
- Utah—Mined Land Reclamation Plan, Air and Water Discharge Permits

Product Markets

During modular operations shale oil product will be trucked to nearby markets. During commercial operations shale oil will be transported by pipeline to refineries principally in Texas. On-site product upgrading is being investigated.

Environmental Impacts

The Final EIS for Prototype Oil Shale Leasing Program issued by the DOI (August 1973) assesses impacts of all prototype lease tracts, including White River Shale Project (WRSP). However, through the completion of an extensive environmental baseline monitoring program (elements of which are continuing) and preparation of a Detailed Development Plan and various preconstruction permit applications, significant impacts associated with WRSP activities are being fully assessed. A detailed environmental monitoring program will be implemented to evaluate project impacts and guide efforts to reduce and mitigate any adverse impacts.

Socioeconomic Impacts

WRSP has published two reports (1975) dealing with prevailing socioeconomic conditions and impacts of oil shale development in the Uinta Basin. These reports are being updated at this time. WRSP also is supporting a current socioeconomic study of growth impacts in the basin, being conducted by consultants to the Department of Energy and the Utah Department of Community and Economic Development. These reports will provide a basis on which impact assistance decisions can be made by WRSP. It is currently planned that construction personnel will be housed in bachelor camp facilities located near the tracts. Operating personnel will reside in existing nearby communities, primarily Vernal, Utah (population 7000), and Rangely, Colorado (population 2500). The work force to construct the Phase I facility will peak at 1500 persons during 1983, dropping rapidly thereafter. The operating work force will be comprised of approximately 850 persons. Construction of later phases of the project will require a peak of 4000 persons by 1988 and an operating workforce of about 3200 persons after 1993. Construction of the facility will require the influx of a relatively large number of skilled construction workers into northeastern Utah. Project operations will produce steady, long-term growth in the area. To help minimize adverse socioeconomic effects in the area, a temporary campsite will be established to provide housing, services, and recreation for construction workers, and for those indirectly employed who will provide goods and services. The campsite will consist of modular, transportable bachelor units and eating, shopping, and recreational facilities. It will be naturally landscaped to fit the terrain and will provide modern utilities. Since the campsite will be nearly self-sufficient, there should be minimal increases in demand for other local services. Plant and mine operations begin to make a major contribution to employment in 1985, when the operations work force reaches 850. Since the operations employment at full commercial production will be stable over a long period of time, this population is expected to be assimilated into the local communities. This will result in an expansion of housing and population in the area. Every effort will be made to work with local, state, and federal officials to assure that the transition is smooth.

Project History and Status

On March 12, 1974, Sun and Phillips were awarded Oil Shale Tract U-a (5120 acres) by closed bid sale from the U.S. Department of the Interior for $75.6 million to be paid in five equal installments. Shortly thereafter Sohio joined Sun and Phillips to form the White River Shale Oil Corporation (WRSOC). On April 9, 1975, WRSOC was awarded Tract U-b (5120 acres) for $45.1 million, which is contiguous to Tract U-a. Tract U-a has been fully assigned to Phillips and Sunoco Energy Development Company. Tract U-b has been fully assigned to Sohio Shale Oil Company. The White River Shale Project was formed to represent the three companies and to plan for joint development of the two Utah tracts. Under the leaser terms, a Preliminary Development Plan was submitted to the Area Oil Shale Supervisor's (AOSS) Office in May 1974. The Detailed Development Plan was submitted to AOSS in June 1976. The lease terms were administratively suspended for 1 year in November 1976 because high levels of naturally occurring air pollutants were measured on the tracts. Prior to expiration of the administrative suspension, the terms of the lease were suspended by a court-ordered injunction issued in May 1977. This action was taken by WRSP in light of two severe title questions regarding legitimacy of federal ownership of the two tracts. This injunction is still in full force although progress is being made toward resolution of the issues which continue to cloud the title to Tracts U-a and U-b. Environmental monitoring activities have continued on the tracts during the suspension period. In April 1980 WRSP commenced engineering studies to update project development plans and prepare an updated version of the Detailed Development Plan (DDP). The DDP will be submitted to AOSS in mid-1981 for approval.

Project/Process Description

Federal Prototype Oil Shale Tracts U-a and U-b are located at the eastern edge of the Uinta Basin in northeastern Utah. The combined tracts have a total surface area of 10,240 acres and a potential recoverable shale oil reserve of 700 million barrels. The DDP being updated at this time summarizes WRSP's proposed development scenario for Tracts U-a and U-b as follows:

1. Joint development of both tracts.
2. Phase development of the resource to thoroughly investigate available mining, retorting, and processed shale disposal technologies:

   Phase I—Modular program mining up to 30,000 TPD of shale rock and producing up to 15,000 bpd of shale oil.

   Phase II & III—Progressive commercial development program culminating in the mining of up to 179,000 TPD of shale rock and producing up to 100,000 bpd (nominal) of shale oil.
3. Room-and-pillar mining technology to be used.
4. Surface retorting technology used to process both coarse and fine shale rock.
5. Surface disposal of processed shale on tract planned with appropriate land rehabilitation program implemented.
6. Product upgrading facilities are being investigated for the commercial phases of the project.
7. Primary water source would be the state of Utah's proposed White River Dam and Reservoir.

DOE Oil Shale Program

The Department of Energy has continued a broad research and development program for oil shale.

The United States is blessed with an enormous hydrocarbon resource in the form of oil shale. Oil similar to petroleum crude can be produced by heating (retorting) oil shale at about 900°F (480°C) to produce a condensible vapor (shale oil), a fuel gas, and a cokelike residue which remains on the spent shale or may be burned as a low-grade fuel. The amount of oil and gas produced depends on the retorting process and the richness or assay of the shale. The U.S. oil shale resource is divided into the western resource generally characterized by relatively rich deposits of shale of at least 25 gallons per ton and greater, and the eastern resource, generally characterized by relatively lean deposits of 10 to 25 gallons per ton.

The western resource is estimated in the trillions of barrels of shale oil, with the amount of recoverable resource estimated at between 80 and 600 billion barrels of oil. Actual estimates depend on assumptions made concerning the technology and economics of shale oil recovery. Even the lower estimate, however, is almost triple the conventional United States petroleum reserves.

The western resource varies greatly in quality, in depth, and in thickness of the seams and is found in three basins: the Piceance Basin in Colorado, the Uinta Basin in Utah, and the Washakie Basin in Wyoming. By far the richest reserves are found in Colorado, with lesser-quality shale in Utah and Wyoming. Also 80% of all of the oil shale resources (72% of the land) is owned by the federal government with the remaining 20% of the resources in the hands of private developers.

The eastern oil shale resource is somewhat equivalent to the western resource in terms of the overall magnitude of the deposit. It also numbers in the trillions of barrels of oil, although generally the shale is much leaner with a significant amount of the resource assaying as low as 5 gallons per ton. The eastern resource is found in many states, but the best deposits are thought to be in Tennessee, Kentucky, Ohio, and Indiana. Additional deposits are found in Michigan, Pennsylvania, West Virginia, New York, and Alabama. There are no generally recognized estimates for the recoverable reserves of eastern shale, although some have estimated a recoverable resource of approximately 400 billion barrels of oil.

Shale oil is of significant interest to the United States because, besides being an enormous hydrocarbon resource which is as yet untapped, shale oil can be converted with a high yield of transportation fuels. Specifically, shale oil is an excellent source of gasoline, diesel, and jet fuel.

Although the potential production of shale oil is very large because of the enormity of the resource, in actuality production is limited to a much lower value than the resource can support because of environmental considerations in the western regions. The oil shale areas of Colorado, Utah, and Wyoming are characterized by excellent air quality, with low population density and relatively little industry. Because of the concentration of this resource in limited geographic areas and existing environmental standards, shale oil production may be limited to only a few hundred thousand barrels per day even though the resource could support production in the range of millions of barrels per day.

The eastern oil shale resource has generally been neglected in terms of research and development activities over the years. This has been the case because the western resource has been viewed as much more commercially attractive owing to the higher quality of the shale. However, the eastern resource merits research and development support. The socioeconomic problems of very rapid population growth should not be as severe as those in the west. In addition, shale oil plants in the east will be closer to large manufacturing facilities and large population centers where obtaining skilled labor and process equipment availability may be a lesser problem than in the west. Finally, shale oil produced in the east will be closer to markets and to existing pipelines and other transportation systems.

The oil shale strategy of the Office of Fossil Energy takes into consideration the above factors and is aimed at developing the technology base needed to ensure that the very large oil shale resources in both the east and west can contribute liquid fuels proportionately to the resource base.

The goal for the oil shale program is to acquire a sound database to assist industry in the development of technology necessary to permit the total U.S. base of oil shale, including both western and eastern resources, to achieve its maximum potential contribution to domestic energy production within the unique environmental constraints of the oil shale regions. This involves the longer-range, high-risk, and generic R&D not normally supported by industry.

This goal will be achieved with the following objectives:

- Develop the technology base needed to assist in overcoming technological barriers to shale oil development.
- Foster development of innovative processes for all shales to reduce environmental impacts and to make the leaner shales in the east and the west exploitable. New processes characterized by reduced environmental effects are essential to raising the environmental "cap" in shale oil production.
- Obtain environmental data on impacts of oil shale development and demonstrate or develop environmental impact mitigation measures.

The first objective is to develop the fundamental data on shale chemistry and engineering to assist in developing shale oil. The second objective relates to the longer-term goal of effectively raising the environmental cap on oil shale production through the development of innovative process concepts, which are characterized by increased thermal efficiency and reduced environmental impact. The third objective relates to the critical concept that only through adequate environmental data can one accurately determine the capacity of the oil shale regions for oil shale production. In addition, it is necessary to develop appropriate environmental impact mitigation measures to ensure the development of technologies that can operate in an environmentally satisfactory manner.

There are three basic alternative technologies for shale oil production: surface production, true *in situ* production, and modified *in situ* production. In each method, the oil shale is heated to about 900°F (480°C), producing both oil and gas. The heat breaks down the kerogen, a solid hydrocarbon material found in the shale, by a process called "pyrolysis," releasing a liquid hydrocarbon (shale oil) with some combustible gases and leaving a cokelike residue of spent shale. The oil contains nitrogen, water, and other impurities, and must be upgraded before going to conventional oil products. The gas may be used for generating electricity for process and onsite power.

Surface processes involve mining and crushing the shale and retorting in an aboveground processing vessel (retort), whereas *in situ* retorting is carried out underground. Advantages of *in situ* methods are potentially lower water, manpower, and shale disposal requirements; and possible application to shale deposits which are not amenable to recovery by conventional mining and surface retorting, either because the deposits are too deep or are of lower grade. The potential application to lower-grade shales is important because two-thirds of the total estimated resource in the western Green River formation is in low-grade deposits that may never be recovered by conventional mining and surface retorting techniques.

Although *in situ* conversion of oil shale to liquid products offers several potential advantages over conventional mining and surface processing, no single version of *in situ* technology is applicable to the full range of oil shale deposit types.

True *in situ* techniques require no mining, but rely instead on either natural permeability (found only in the leached zone of Colorado) or artificial permeability created by hydraulic and/or explosive fracturing from wellbores. Success with the explosive fracturing technique has thus far been demonstrated only in very shallow Utah shale, in a project by Geokinetics, Inc., which continues to show promise. Deeper tests in western shale have been unsatisfactory. Final assessment of hydraulic/explosive fracturing for gas production in eastern shales was completed in FY 1980. It was found to have little potential for success and was discontinued.

Modified *in situ* techniques use partial mining prior to explosive fracturing to achieve enhanced void volume as well as improved shale particle size and void distribution. Present void targets are in the range of 15–40% of the total rubble mass.

Of the *in situ* methods, the vertical modified *in situ* (VMIS) technique is especially promising for deeper and very thick (more than 300 ft) deposits. Four major variations of modified *in situ* techniques are of interest, the choice of a particular variation being partially governed by the actual thickness of oil shale to be processed. Two of these variations are similar in that an initially mined void volume is subsequently redistributed throughout a much larger mass of rubblized shale. Physical dimensions of the completed retort modules have led to the general description of these processes as "vertical" or "horizontal." Two versions of the vertical process are under active development, with one having reached the stage of testing modules which are approximately the size required for commercial operations. Both utilize a downward-moving combustion zone to process the shale. Two processing concepts have been proposed for "horizontal" retorts, one using a downward-moving combustion zone and the other using a laterally moving combustion zone.

In situ retorting produces feedstock for refining into products for the commercial fuels market similar to those of surface processes:

- Shale oil syncrude that can be refined into products (gasoline diesel fuel, jet fuels, liquified petroleum gas, and heavy fuel oil).
- Low- to medium-Btu gas that can be used onsite for steam generation and in a gas turbine generator providing electricity for process operation.
- Ammonia and sulfur products.

IN SITU CONVERSION

During 1982 the oil shale budget was reduced $16,048,000, which resulted in the termination of three major *in situ* field projects and related supporting and project management activities. The major projects are the Occidental Oil Shale Vertical Modified *In Situ* Project in Colorado; the Geokinetics Horizontal Modified *In Situ* Project in Utah; and the Equity True *In Situ* Project in Rio Blanco County, Colorado. The decision to terminate these projects was made because the technologies being demonstrated had progressed to the stage where decisions to proceed toward commercialization could be made by industry without further government involvement. The orderly phaseout and termination of these projects began in FY 1981.

The FY 1982 budget request of $15,250,000 could provide support to the ongoing activities involved in the DOE *in situ* technology base program. The intent of this effort is to develop a sound technology base for *in situ* technology for the large-scale production of shale oil. *In situ* technology is considered to be advantageous principally because of the decreased volume of solids handling and disposal required, reduced water requirements, and decreased mining and capital equipment costs. This translates into fewer environmental impacts and decreased costs. In addition, overall resource recovery can be improved since *in situ* retorting could be used to process oil shale which cannot be economically processed any other way, such as shale which is too deep to mine or too lean (i.e., low gallons per ton of rock) to economically process.

The FY 1982 budget request covers all of the work at the Laramie Energy Technology Center and national laboratories in support of the oil shale program, including support of university R&D projects.

Results to date indicate that *in situ* technology has potential for significant production if improvements in resource recovery, production (retorting) rates, and retort preparation logistics can be achieved. Basic and applied research on the complex chemistry and chemical engineering of oil shale retorting are supported as part of the technology base activities. This support includes kinetics and mechanisms of shale retorting; mathematical process modeling; R&D on rock fracturing and rock mechanics; environmental characterization development of control technology; and oil shale geochemical characterization. Also to be developed and proven are retort abandonment strategies and other environmental controls and worker health and safety measures necessary for commercial operations.

Chemical and physics of oil shale conversion. Oil shale conversion is a complex chemical and physical process which is still only crudely understood. This activity will develop basic data on chemical kinetics and mechanisms needed to model complex, high-temperature reactions taking place during retorting of oil shales of varying chemical composition and physical properties. This involves analysis of oil shale mineral composition and organic content; determination of high-temperature reactions of mineral decomposition (especially carbonates) and subsequent interactions with char to produce valuable CO and hydrogen; degradation reactions that reduce oil loss; mechanisms of gaseous sulfur compound evolution; water–gas shift reactions; and stoichiometry of hydrocarbon combustion. Various sized laboratory and engineering-scale retorts will be operated with shale mixtures and gaseous atmospheres; samples will be withdrawn and analyzed from various points in the conversion cycle. Data will be integrated with existing computer models of retorting that simulate effects of particle size, porosity distribution, and other properties of *in situ* retorts to provide improved overall understanding of retorting mechanisms from which advanced process control techniques and new process concepts can be developed. Particular attention will be given to the development of new process concepts characterized by increased thermal efficiency and those which produce more oil with reduced emissions of pollutants.

Instrumentation and diagnostics. Retort diagnostics and instrumentation will be supported to develop successful techniques for monitoring and controlling *in situ* retorting. *In situ* processes require "remote" instrumentation techniques that allow determination of process variables by monitoring only a small number of parameters from a limited number of sampling points. Reliable and accurate instrumentation and analytical techniques for this purpose do not yet exist. Instrumentation needs include thermal sensors, as sampling devices, pressure probes, remote sensing devices, health and safety monitoring equipment, and instrumentation designed for environmental data gathering.

Oil shale fracturing. Correct fracturing of shale deposits is essential to the development of successful *in situ* retorting technologies. This activity supports continuation of research and development in oil shale fracturing for *in situ* retort preparation. Because oil shale deposits are generally impervious, fragmentation of the rock is necessary to provide void space through which fluids and gases can flow if an *in situ* retort is to function at even a minimum level. Optimum operation of the retort requires fragmentation to produce uniform porosity and particle size for the bed of broken shale. Under this project, parameters such as rubble size, permeability, and void volume distribution will be defined through laboratory and field experiments. It is necessary to develop the technology to achieve these defined values in a predictable manner. This requires a basic understanding of the characteristics of explosive fracture, such as rubble size and distribution as a function of explosive type, quantity, placement, and sequencing; bore size; and rock properties and discontinuities.

Efforts in FY 1982 will involve laboratory work to define further detonation wave behavior and to validate laboratory data on response of shale to static and dynamic loading through field experiments involving single and multiple boreholes. These parametric studies will provide the data inputs to develop a fragmentation prediction model.

Environmental impact mitigation. Environmental problems are the most severe constraint on the capacity of the oil shale region to produce oil. The objective of this activity is to develop solutions to environmental problems associated with the operation of *in situ* shale retorting processes. Additional R&D is needed in determining potential environmental impacts and developing appropriate means to mitigate the impacts. The R&D strategy is to develop the data necessary to analyze the magnitude of potential problems for each of five environmental areas: retort abandonment strategies, worker health and safety measures, air emission control procedures, water management systems, and solid waste management strategies. Utilizing these data, the applicability of existing environmental control technology will be evaluated. This will provide performance characteristics for existing devices and will determine the need for new or improved controls.

Key areas to receive priority in FY 1982 are procedures for abandoned "spent" *in situ* retorts that will minimize potential for contamination of aquifers at an acceptable control cost; detailed analysis of air emission streams to quantify and identify chemical species, with particular attention given to determining acceptable sulfur compound levels; and developing chemical analysis data for critical water quality parameters with regard to the intended use, for example, industrial or agricultural revegetation.

FIELD ACTIVITIES

The oil shale program management responsibilities have been delegated to the Laramie Energy Technology Center (ETC). The majority of the *in situ* research programs will be carried out at Laramie ETC and the national laboratories (that is, Sandia, Lawrence Livermore, Los Alamos Scientific Laboratory, and Lawrence Berkeley).

Laramie Energy Technology Center. Laramie conducts a broad-based research program in support of alternative methods of oil shale processing applicable to the varied natural characteristics of the total oil shale resource. The center also provides management for experimental and prototype demonstration contracts with industry. Laramie's in-house technology-base research includes (1) characterization of oil shale deposits and rock fracturization, (2) operation of bench-scale and pilot-scale retorts to simulate *in situ* processes, (3) interpretation and evaluation to simulate *in situ* processes, (4) interpretation and evaluation of processing data to develop a better understanding of processes and their control, (5) development of mathematical models to describe and predict retort processes, (6) development of analytical methods for determining product properties and process events, (7) evaluation and management of environmental monitoring and compliance activities, and (8) determination of industry requirements for storage, transportation, and refining of shale oil.

Lawrence Livermore Laboratory. The Oil Shale Project at Livermore is a multidiscipline group working on the technical problems of shale oil recovery. Currently under investigation are (1) pyrolysis, gasification, and mineral chemistry; (2) operation plans and control schemes for commercial retorts; (3) computer modeling of chemical and physical processes; and (4) chemical, toxicological, and mutagenic characterizaton of retort products. Expertise at Livermore includes chemistry, chemical kinetics, computer modeling of retort processes, retort operation, data acquisition and analysis, and biomedical studies. Unique tools available at the Livermore site are high speed computers, three retorts capable of simulating both modified *in situ* and fixed-bed retorts, a data acquisition and analysis system, a working model to simulate burn conditions in pilot and field retorts, a multifacility for environmental research, and a fully equipped fossil fuel research laboratory.

Lawrence Berkeley Laboratory. Berkeley's research objectives are to (1) determine the distribution of major and minor elements between shale oil, process water, off-gas, and spent shale as a function of retort operating conditions; (2) identify and quantify the organometallic species that are released to retort waters and off-gases during *in situ* retorting; and (3) develop and apply in place monitoring systems to measure gaseous emissions from an *in situ* retort. Data from these investigations will be used to assess the impact of field *in situ* operations on the environment.

Los Alamos Scientific Laboratory. Involved in a broad spectrum of oil-shale-related activities, the laboratory has lead role assignments for *in situ* retort bed preparation (including experimental rock mechanics, theoretical fracture mechanics, computer modeling, and intermediate-scale field verification experiments carried out in conjunction with industry), and environmental and health and safety studies (including the *in vivo/in vitro* determination of the effects of oil shale liquids and particulates on workers).

Sandia Laboratory. Sandia performs a dual role in the oil shale R&D project: (1) it evaluates DOE-supported oil shale field projects through the fielding of instrumentation and data acquisition systems, and the analysis of process data and results (evaluations were in progress at both Occidental Oil Shale Logan Wash and Geokinetics project sites), and (2) it develops instrumentation for measuring rock mechanics in processing areas. This includes measuring oil shales to determine shale grade, bedding plane orientation (anisotropic characteristics), temperature and strain rates (extending from the high rates associated with explosive rubblizing to the low rates associated with the creep and stability of pillars inside *in situ* retorts); developing strength and failure models for use in the design of systems and the analysis of field results; developing advanced instrumentation sensors and measurement techniques and their application to field testing; conducting processing and thermochemical studies to determine the minimum retort bed conditions (for example, size of rubbled shale) necessary for *in situ* processing, including void, permeability, and particle size distributions; and applying multidimensional process models to analyze and evaluate field experiment data.

Other Activities

Resources development planning research by Laramie in FY 1980, FY 1981, and FY 1982 will continue the development of a computer system to analyze oil shale data and produce geologic quality maps of Colorado, Utah, and Wyoming. Paralleling this effort will be development of an eastern shale resource database. Technical and economic planning for oil shale resource development will be integrated with commercial *in situ* designs and an environmental database. Laramie has recently published a new quantified assessment of the Colorado oil shales that are candidates for development.

An updated environmental development plan (EDP) has been formulated for the oil shale activity. This plan provides for a comprehensive oil shale environmental strategy and an implementation plan to ensure that all environmental concerns are addressed. Environmental impact assessments are prepared for all major DOE oil shale field projects.

The development of comprehensive retort abandonment plans is important for compliance with the National Environmental Policy Act. The R&D effort will collect data on retort abandonment and evaluate usefulness. Existing hydrological models will be used to estimate pollutant dispersion. Potential subsidence in the vicinity of an *in situ* retort will be assessed, including the potential for natural flooding in abandoned retorts. Based on information gained from the risk analyses, new controls will be developed and field tested.

The expected result of this effort will be an environmentally acceptable retort abandonment plan. This work in FY 1981 and FY 1982 included laboratory studies by Lawrence Berkeley Laboratory to assess abandonment strategies, such as grouting or preleaching, and comparative cost analyses of such strategies. Development of design criteria for atmospheric emission controls will be continued in FY 1982. Solid waste design criteria research will also be continued.

Los Alamos will continue a series of small-to-intermediate field tests on oil shale response to explosive shock in FY 1982 and will refine empirical models. The experiments will be performed in the Anvil Points Mine. These tests and data from major field projects will be used to complete the VMIS rubbling design.

Continued development of a commercially viable technology for controlling the *in situ* retort and burn operation will be supported by Lawrence Livermore Laboratory and Laramie, using data from contractor field operations and large laboratory retorts to provide kinetic, reaction chemistry, and transport phenomena data to a mathematical process model. This model has reached a stage of preliminary usefulness for predicting demonstration-size retort design performance. Final development and verification will require input from actual commercial-scale VMIS retorts. This started in FY 1981 and continued into FY 1982.

In addition to the *in situ* base technology program supporting oil shale development, the *in situ* area of the oil shale activity has conducted studies of advanced process concepts such as microwave or radio-frequency retorting of oil shale and the retorting of both eastern and western shales in a hydrogen atmosphere. Advanced retorting processes are those that show promise of substantial economic improvements and/or reduced environmental impacts compared to existing programs. Such advanced processes have typically undergone bench-scale development or are presently in an active laboratory or small process development unit (PDU) development phase that has provided promising data. Certain concepts require laboratory research to provide the first level of confirmation of potential benefits.

Work on the radio-frequency and hydrogen-atmosphere retorting process in FY 1981 resulted from contracts awarded to the Illinois Institute of Technology Research Institute (IITRI) and the Institute of Gas Technology (IGT) under DOE's FY 1979 Program Research and Development Announcement (PRDA) for innovative processes for retorting oil shale. This work was completed in FY 1981. The IGT project is discussed further in the surface conversion section.

In the production of oil from oil shale, major accomplishments include the following:

Occidental Oil Shale, Inc., field test. The concept behind this demonstration involves partial conventional mining in order to create an enhanced void space for redistribution. The process is applicable to impermeable oil shale at least 300 ft thick. This is done by explosive rubbling followed by direct combustion retorting.

This process was under active development by the contractor prior to DOE's Energy Research and Development Administration (ERDA) involvement in 1977. Final analysis of the results of phase 1 of the Occidental-DOE cooperative agreement were published in FY 1980. Phase 1 included the retorting of two demonstration-size retorts, Nos. 5 and 6. Final calculations indicate that Retort No. 6 produced about 54,900 barrels of oil during its ten and one-half month retorting period.

Phase 2 of the cooperative agreement with Occidental Oil Shale, Inc., continued development of the verical modified *in situ* VMIS technology. Phase 2 provides a significant increase in the amount of data available through this contract and reduced the DOE portion of the cost. Mining efforts were initiated in FY 1980 and continued into FY 1981 as preparation for the rubbling and ignition of a two-retort "cluster." The cluster demonstrated the VMIS technology at scales approximately that of Occidental's planned commercial venture at their federal lease C-b. Rubblization of the retorts is scheduled for mid-FY 1981 with ignition in late FY 1981.

Work in FY 1981 includes two miniretort start-up tests, rubbling of the full-scale, two-retort cluster, completion of diagnostic tests on the rubblized retorts, and ignition of the two demonstration-size retorts (Retorts 7 and 8) late in FY 1981. As discussed earlier, orderly termination of this project took place in FY 1981.

Geokinetics, Inc., Uinta County, Utah, field test. This project involves demonstration of the surface uplift technique for creating enhanced void volume and a rubblized bed for horizontal retorting. The technology is known as horizontal modified *in situ* (HMIS) and is applicable for thin- to medium-thickness shales with up to 200 ft of overburden. Geokinetics, during work prior to November 1976 and during the subsequent cost-shared cooperative agreement phases, prepared a series of 24 retorts of progressively increasing size and increasing depth and thickness, using three different blasting techniques.

In the horizontal *in situ* retorting tests in shallow Utah shale, No. 18, the largest retorting experiment to date in this project, was completed in May 1980. This horizontal retort is 108 ft wide, 156 ft long, and 17 ft thick at a depth of 27–44 ft, roughly one-fifth the volume of a full-scale commercial retort; it produced 5547 barrels of shale oil.

Retort No. 24 was ignited November 28, 1980; its dimensions are 217 ft wide, 230 ft long, and 30 ft thick (demonstration scale) under an overburden of 45 ft. Preliminary tests indicate good fracturing and permeability throughout the retort. Anticipated oil recovery is 25,000 barrels at a 50% recovery factor; the production rate should be about 100 barrels per day. As discussed earlier, orderly termination of this project took place in FY 1981.

Equity Oil Company, Rio Blanco County, Colorado, field test. This project is an outgrowth of private industry research in the 1960s that tested horizontal retorting with hot gases in the leached zone. This zone contains water that prohibits direct combustion retorting but can be used for steam generation and other process needs, thus eliminating external water requirements. The process also results in a high-Btu off-gas and has reduced overall off-gas handling requirements.

Project development since 1977 has identified a low-permeability zone near the center of the leached zone. Design modifications now allow simultaneous processing of the full thickness by separate upper and lower circulation patterns. Project work has also revealed that the zone contains a static (nonmigrating) aquifer of saline water; this conclusion is based on heat retention from experiments conducted 7 years previously.

The leached zone oil shale contains a natural honeycomb-porous structure created by the dissolution of salts. The zone contains kerogen equivalent to more than 1 million barrels of shale oil per acre at the project site. If successful, the 0.7-acre-square test is expected to produce at least 300,000 barrels over a 2-year period. The zone contains an estimated total of 550 billion barrels that can only be retorted by an indirect-heat process.

Injection of superheated steam began in mid-FY 1979; however, the superheater had several weld failures which caused significant loss of injection time. Also, electrical power to the site has been very intermittent because of other loading on the main line.

This first oil was produced during September 1980 and is of tarlike consistency with high water content. Oil production has been low to date but is expected to increase as the formation is heated to retorting temperature. As discussed earlier, orderly termination of this project took place in FY 1981.

Illinois Institute of Technology Research Institute, Chicago, Illinois, field test. This project resulted from a PRDA competitive solicitation for advanced retorting processes and supports research on an *in situ* process using radio-frequency heating. Laboratory and field testing of this unique process was carried out. In the field tests, a cubic meter of oil shale at a field site in Utah was retorted (about 4410 pounds of oil shale) *in situ* successfully producing oil. Equipment design was demonstrated and radio-frequency leakage, a key concern for this technology, was found to be low. This effort is part of a combined oil shale–tar sands contract which has substantial cost-sharing by Halliburton Services, a major petroleum service company. The current project was completed in FY 1981.

SURFACE CONVERSION

A PRDA for advanced processing concepts resulted in contract awards to IGT for further testing of a hydrogen retorting process (HYTORT) that may have application to shale oil production from eastern Devonian shales as well as increased yields from western oil shale.

The IGT effort has expanded our understanding of the geochemical makeup of the eastern shales, reaction kinetics and mechanisms in a pressurized hydrogen environment, and expected yields. Results to date indicated that the HYTORT process has promise for opening up the vast eastern shale resource as a major contributor to synthetic fuels production.

Phase 1 of this contract with IGT is aimed at establishing the technical and economic feasibility of the process as applied to the lean, hydrogen-deficient eastern shales and the richer western shales. The process involves retorting in a hydrogen atmosphere under pressure. It is expected to significantly increase liquid product recovery from the eastern shales and improve overall process efficiency for western shales. The current phase was completed in FY 1981.

On June 15, 1980, cooperative agreements were awarded to Paraho Development Corporation and Superior Oil Company for the design of commercial-scale retort modules base on their R&D efforts through semiworks operation; preparation of environmental health and safety documentation; and construction and operating plans requisite to proceeding with hardware for these designs.

These cost-shared cooperative agreements were initiated under a Program Opportunity Notice (PON) issued in FY 1979. The Paraho Development Corporation contract has a total value of $8.1 million, with the DOE share being $4.4 million. The Superior Oil Company contract has a total value of $7.5 million, with the DOE share being $5.6 million.

The objective of these projects is to stimulate the development of an oil shale industry by demonstrating the engineering, economic, and environmental feasibility of surface retorting processes at the unit scale necessary for commercial use.

These design efforts are scheduled for completion by January 1982. The discrete areas involved include development of a management plan, with schedules and cost estimates for developing site access for the demonstration program; identification of environmental, health, and safety permits required by federal, state, county, and local agencies; production of an environmental compliance plan depicting environmental work necessary for compliance with DOE regulations and permit requirements; compilation of data necessary to complete an Environmental Impact Study (EIS) to comply with the National Environmental Policy Act, 1970 (NEPA), and acquisition of necessary permits; creation of a detailed design for construction of a surface retort module and preliminary design of auxiliary facilities needed for the demonstration; and development of a management plan for scheduling and estimating costs associated with construction, operation, and maintenance of the demonstration facility.

In late FY 1981, a PRDA soliciting advanced process concepts for conversion of eastern shale to syncrude feedstock will be initiated. FY 1982 funding will support the continuation of this effort which will start in FY 1981 with a low level of funding. Process development is the key to utilizing this large, untapped resource. Eastern shales are far different than the western shales that have been targets for process development over the past 20–30 years. Although little research has been conducted, the resource is very large and its advantages mandate a significant effort in this area.

A waste management experiment of sufficient scale to extrapolate to commercial operations was initiated in FY 1982 to address spent shale disposal of western shales. No previous experiments have been of large enough scale to adequately address such factors as compaction, pile configuration, and revegetation, which are essential to stabilization assessment and run-off water control measure.

Chapter 7

Utilization of Synfuels

Introduction

This chapter broadly sketches out the essential facts of interest about the utilization of synfuels. This chapter deals more with the fate of synthetic fuel products after they leave the plant gate than their production.

Current trends in the international energy situation are rapidly increasing the probability that a domestic synthetic fuels industry will emerge in the 1980s. Because government incentives and private ventures in the synfuel arena are increasing in response to increasing world oil prices and decreasing reliability of oil imports, forecasters are now projecting earlier start dates, faster growth rates, and larger ultimate sizes for such an industry.

Several synfuel technologies are under consideration for commercial production. A wide range of synfuel products are expected to be produced, and they will be utilized in a broad category of end uses as shown in Table 7-1. Synfuel products will most likely be used largely as transportation fuel, including gasoline and diesel fuel from refined shale oil and coal conversion processes and jet fuel from refined shale oil. Utility and industrial boilers will utilize the fuel oils produced from coal liquids. High-, medium-, and low-Btu gases from coal will find use in commercial, residential, and industrial heating applications. The products from most synfuel processes will be used as chemical feedstocks in a large variety of industries.

The national environmental impacts of a large-scale synfuels industry could be significant. The environmental concerns of end use, including handling and transport, will have to be investigated in detail. Since there is limited information concerning the end-use exposure effects of synthetic fuel products and by-products, the nature of these future impacts is largely speculative. In fact, since synthetic fuel technology is highly evolutionary, even the composition and amounts of future industrial synthetic fuel products and by-products are not well known.

In this chapter the term synfuel product refers to primary products of the synfuel industry such as gasoline and high-, medium-, and low-Btu gas, whereas the term by-product has been used to identify secondary useful products that are likely to be produced from synfuels such as plastics, solvents, varnishes, and fertilizers.

A Synfuel Industry Is Emerging

The primary incentive for synfuels development is the imbalance between domestic supply and demand for petroleum liquids and natural gas. The long-term decline in domestic oil production coupled with increased demand has resulted in a level of oil

Table 7-1. Synfuels Market Overview

What Technologies Produce Synfuels?		What Major Products/By-Products Will They Make?	Where Will the Products/By-Products Be Used?	What Are the Relative Potential Exposure Levels to the Products?
Oil Shale	Numerous retorting processes, including Tosco, Paraho, Union, Occidental	Syncrude upgraded and refined to yield: LPG Gasoline Jet Fuel Diesel Fuel Residuals Lubricants[a] Waxes	• Commercial and military transportation, including highway vehicles, aircraft, ships • Utility and industrial boilers • Commercial and residential heating • Industrial lubricants	Low for transport of crude shale to refinery; moderate during refining and end use as boiler fuel; increased exposure level when used in transportation sources and space heating.
Direct Coal Liquefaction	SRC-II	LPG Naphtha Fuel Oil SNG[b] Tar Oils[a]	• Utility and industrial boilers • Commercial and residential heating • Chemical feedstocks	Low for LPG, SNG, naphtha, butane, moderate exposure for fuel oils at industrial sites with exposure increasing when used in space heating.
	Exxon Donor Solvent	Propane Butane Naphtha Fuel Oil Solvent[a]	• Utility and industrial boilers • Commercial and residential heating • Paint thinners	
	H-Coal	Naphtha Fuel Oil	• Utility and industrial boilers • Commercial and residential heating	

Table 7-1. *(continued)*

Indirect Coal Liquefaction	Fischer – Tropsch	Gasoline LPG Diesel Fuel Heavy Fuel Oil Medium-Btu Gas SNG Tar Oils[a] Phenols[a] Chemical Feedstocks[a] Pesticides[a] Fertilizers[a]	• Commercial and military transportation • Utility and industrial boilers • Commercial and residential heating • Chemical feedstocks • Agricultural uses	Low for LPG, SNG, and medium-Btu gas. Moderate exposure when fuels used in transportation sources and boilers. Low to moderate exposure is also estimated when products used as chemical feedstocks.
	M-Gasoline	Gasoline LPG	• Commercial and military transportation	
	Methanol	Methyl Fuel Methanol	• Commercial and military transportation • Chemical feedstocks	
High-Btu Coal Gasification	Numerous processes, including Lurgi, COED-COGAS, Texaco, Shell – Koppers	SNG[b]	• Commercial and residential heating	Very low—similar to current distribution of natural gas.
Medium/Low-Btu Coal Gasification	Numerous Processes	Medium-Btu Gas Low-Btu Gas	• Captive fuel use for industrial heating and chemical feedstocks	Very low since it is primarily captive use.

[a]Only representative by-products are indicated.
[b]Substitute natural gas.

imports of 9 million barrels per day (MMBPD) of oil or about 50% of U.S. consumption. The proven domestic reserves of natural gas are also declining, and demand is now being met with increasingly higher priced supplies.

A substantial market for liquid synthetic fuel products and chemical feedstocks is expected by 1990. A recent analysis concludes that about 2.9 quads of energy or about 1.5 MMBPD will have to be supplied from synthetic liquid fuels as shown in Table 7-2. As indicated in this analysis, use of synfuels is expected to be heavily directed toward transportation. Industry concern over potential interruptions in gas supplies has provided the incentive to develop coal gasification processes to supplement current gas supplies and for use as chemical feedstocks.

Table 7-2. Anticipated Liquid Fuel Products Demand in 1990[a]

	Supplied to Consumers (quads)	Petroleum Supplies (quads)	Synfuels[b] (quads)
Gasoline	14.2	12.7	1.5
Jet fuel	2.1	1.8	0.3
Kerosene	0.3	0.3	—
Heating oils	6.4	5.7	0.7
Residual	3.2	3.2	—
Asphalt	1.0	1.0	—
Miscellaneous Product	2.4	2.0	0.4
LPG	0.7	0.7	—
	30.3	27.4	2.9

[a] *Coal Technology Market Analysis,* ESCOE, January 1980. Assumes U.S. refineries will operate with the same mix as 1978.
[b] 1 quad/yr ≃ 0.5 MMBPD.

SYNFUELS UNDER CONSIDERATION

The term "synfuels" has become synonymous with any combustible nonpetroleum fuel source which may include coal- and shale-derived fuels and feedstocks as well as those derived from agricultural products such as grain, wood, and cellulose. However, industry has become increasingly more interested in synfuel technologies with products that are easily substituted, in a marketing and utilization sense, for petroleum and natural gas. These synfuel technologies are those relating to coal- and shale-derived products. The following discussion is limited only to these products.

Current Oil Shale Retorting Technology

As a direct substitute for large volumes of liquid fuels, oil shale technology is perhaps closest to commercialization in the United States. Several consortia and companies, listed in Table 7-3, with established shale oil projects have been engaged in the development of shale oil technology for some time. These projects are all located in prime shale areas of Colorado and Utah.

Many technologies are being developed and tested that are aimed at extracting kerogen, a waxy organic material, from shale. Most involve heating shale to about 896°F (480°C) and pyrolyzing the kerogen into a viscous liquid called shale oil. They differ in the manner in which this heating process is accomplished: surface retorting, *in situ* retorting, or modified *in situ* retorting.

Table 7-3. U.S. Oil Shale Projects

Project	Location[a]	Technology	Production Capacity Goal (bbl/day)	Status
Chevron	Piceance Basin	Undecided	50,000	Technical assessment phase.
Colony (Tosco, EXXON C)	Parachute Creek	Surface retorting	47,000	Construction delayed.
Equity Oil	Piceance Creek	Solution injection, modified *in situ*	—	Steam-injection feasibility.
Geokinetics, Inc.	Uinta County	Horizontal modified *in situ*	7 – 13 2,000 – 5,000	Several small retorts successfully burned; work on larger retorts in progress.
Getty Oil	Piceance Basin	Surface thermal extraction	—	Getty R&D proposal submitted.
Mobil	Piceance Basin	Undecided	50,000	May start module in 1987.
Occidental Oil	Logan Wash	Vertical modified *in situ*	70,000	Six retorts burned; 48,000 bbl produced. Retorts 7 and 8 scheduled for cluster burn.
Occidental Oil—Tenneco	Tract C-b, Piceance Basin	Vertical modified *in situ*	57,000	Shaft sinking in progress; construction of initial retorts scheduled.
Paraho (Development Engineering, Inc.)	Anvil Points	Surface retorting	150 – 200	Shut down due to lack of funding; 88,225 bbl produced over about 1-year period.
Rio Blanco (Gulf, Amoco)	Tract C-a, Piceance Basin	Vertical modified *in situ*, surface retorting	50,000	Modular program consisting of five retorts.
Superior Oil	Piceance Creek	Multimineral recovery, surface retorting	13,000	Company seeking land exchange with federal government which was denied in February 1980.
Tosco – Sand Wash	Uinta Basin	Modified *in situ*, surface retorting	50,000	Feasibility studies in progress.
Union Oil	Parachute Creek	Surface retorting	50,000	Construction delayed.
White River (Sohio, Sunoco, Phillips)	Tracts U-a and U-b, Uinta Basin	Modified *in situ*, surface retorting	100,000	Operations suspended due to legal proceedings on ownership of lands.

[a] State Locations: Piceance Basin, Colorado; Parachute Creek, Colorado; Uinta County, Utah; Logan Wash, Colorado; Anvil Points, Colorado.

In surface retorting, oil shale is mined, crushed to the proper size, and then fed to a large kiln for heating. Several surface retorting processes are under development, and they differ primarily in the heating method employed. Internal-combustion retorting heats shale by the circulation of hot gases that are produced inside the retort by the combustion of residual carbons in the shale. Gas-cycle retorts used by Union Oil heat the shale by circulating externally heated fluids. No combustion occurs inside the retort. In solid-heat-carrier retorting, shale is mixed with hot solids that are heated outside the retort and cycled through the shale. Tosco II is an example of this method, using ceramic balls as the heat carrier.

In situ retorting pyrolyzes oil shale while it is still in the ground. The shale bed is ignited and sustained by injection wells, the shale is pyrolyzed, and the oil produced is pumped out of the retort volume through a production well. The spent shale remains in place. For successful *in situ* retorting, the shale bed must be made permeable to the flow of heat and product oil; various techniques of bed leaching or fracturing are employed. The difficulty of creating a permeable shale bed has led to the development of modified *in situ* processes. Vertical modified *in situ* (VMIS) retorting removes a portion of the shale from the bottom of the deposit and fractures the remaining shale to create a chimney of shale rubble. The shale is retorted in this chimney from top to bottom. Occidental Oil Company has been testing VMIS retorting on shales at Logan Wash and Piceance Creek Basin in Colorado. Horizontal modified *in situ* retorting lifts the overburden in some cases, and fractures the shale seam to retort the shale from side to side. Geokinetics, Inc., is developing this technique in Utah.

The technology for surface retorting is more advanced than *in situ* retorting. Process variables are easier to monitor and control in aboveground retorts than in underground retorts. However, large-scale commercial surface retorting requires large-scale oil shale mining, hauling, and crushing, and large-scale disposal of spent shale. It is also limited to that portion of the shale resources that is mineable. *In situ* retorting without mining is applicable to a greater variety of shale beds, and eliminates the requirements for handling, crushing, and spent shale disposal. Attempts to demonstrate this technology have identified many development problems. Modified *in situ* processes present a compromise, requiring some mining and handling, but offer more process control and easier development.

The crude shale oil produced by retorting will be upgraded by further processing. This upgraded shale oil, or syncrude, will be used as a refinery feedstock or boiler fuel. It is well suited for refining into middle distillate fuels. If hydrocracking is chosen for the refining process, the yield and range of products is particularly desirable: motor gasoline, 17%; jet fuel, 20%; diesel fuel, 54%; and residuals, 9%.

Several oil shale projects, with identified participants, plan to begin operation during the 1980s. The technologies, which are proprietary in many cases, appear to be sufficiently mature to move ahead to commercialization. Several retorts have been operated successfully by Geokinetics, Inc., Occidental Oil, Paraho, Union, and Tosco. Colony, Union Oil, and Occidental Oil had announced plans to begin commercial development in the early 1980s. Current oil prices and high interest rates have deferred some of these projects. All technologies have been demonstrated at pilot scale or larger.

Both Direct and Indirect Routes to Coal Liquids Are Available

Coal, hydrogen, and a coal-derived oil are mixed at high temperature and pressure to accomplish direct liquefaction. Under these conditions, the coal decomposes, and the fragments react with hydrogen to form additional derived oil, which is separated from the unreacted solid and further refined to produce usable liquid fuels. Indirect liquefaction processes react the coal with oxygen and steam in a gasifier to produce a synthesis gas composed mainly of carbon monoxide, carbon dioxide, and hydrogen. After the carbon dioxide and other impurities are removed from the gas, the carbon monoxide and hydrogen are chemically combined in a catalytic reactor to produce liquid products for use as chemical feedstocks or liquid fuels.

There are three major direct coal liquefaction processes currently undergoing development: SRC-II, Exxon Donor Solvent (EDS), and H-Coal as shown in Table 7-4. These processes differ mainly in the manner in which the hydrogen is made to react with coal fragments to produce the unrefined coal liquids. In the SRC-II process, the coal feed and hydrogen are mixed with a process recycle stream that contains unreacted coal ash as well as coal-derived oil. The iron pyrite in the unreacted ash catalyzes the reaction between the coal fragments and hydrogen. In the EDS process, the coal feed and hydrogen are mixed with a specially hydrogenated coal oil called the donor solvent. The hydrogen added to the coal fragments is provided by the solvent and the hydrogen gas mixed in the reactor. The donor solvent is made by catalytically hydrogenating coal-derived oil using conventional petroleum refinery hydrotreating technology. In the H-Coal process, the unreacted coal and hydrogen are mixed with coal-derived oil and an added solid catalyst in a special reactor referred to as an ebullated bed.

Once the gases and distillate liquid products have been separated from the reactor effluent, the remaining "bottoms" material is processed. This material contains significant quantities of heavy hydrocarbons, which must be efficiently utilized to enhance process economics. The principal bottoms processing step under consideration for the EDS process is Flexicoking, which consists of thermal cracking of the bottoms to produce additional liquids and coke. The coke is subsequently gasified to produce plant fuel gas or hydrogen for the liquefaction step. Bottoms processing for the SRC-II and H-Coal processes probably will be partial oxidation (i.e., gasification) to produce hydrogen for the liquefaction step.

There are two major indirect coal liquefaction processes: Fischer–Tropsch, which is commercial now in South Africa, and Mobil-M, which is expected to be commercial in 1983–1984. In the Fischer–Tropsch process, the purified synthesis gas from the gasifier is reacted over an iron catalyst to produce a broad range of products extending from lightweight gases to heavy fuel oil. The broad product distribution from this process is generally considered as a disadvantage where large yields of gasoline are desired. Improved catalysts are currently being developed at the bench scale to maximize the yield of gasoline-range hydrocarbons. In the Mobil-M process, the synthesis gas is first converted to methanol using commercially available technology. The methanol is then catalytically converted to high-octane gasoline over a molecular-size-specific zeolite catalyst.

Indirect coal liquefaction is successfully operating on a commercial scale at the SASOL I plant in South Africa using the Fischer–Tropsch technology. The SASOL I plant produces gasoline, jet fuel, diesel oil, middle distillates, and heavy oil. SASOL II, producing 50,000 barrels per day of coal-derived liquids, has been completed and began operation in 1980. Active interest in this technology has developed and plans to license and construct similar plants in the United States are progressing. There is strong interest in the Mobil-M gasoline indirect process because of its attractive high-octane gasoline yield.

Table 7-4. Major Coal Liquefaction Processes

Process	Process Type	Products	Status
Solvent Refined Coal, SRC-II (Gulf Oil)	Direct liquefaction by solvent extraction: coal dissolved in solvent, slurry recycled, catalytic hydrogenation	LPG Naphtha Fuel oil SNG	Pilot plant under operation. 6700 ton/day of coal (20,000 barrels/day of oil equivalent) demonstration module under design
H-Coal (Hydrocarbon Research, Inc.)	Direct liquefaction by catalytic hydrogenation, ebullated catalyst bed	Naphtha Fuel oil	600 ton/day (1400 barrels/day of oil equivalent) pilot plant under construction. Plant is located at Catlettsburg, Kentucky
Exxon Donor Solvent, EDS (Exxon Research and Engineering Company)	Direct liquefaction by extraction and catalytic hydrogenation of recycled donor solvent	Propane Butane Naphtha Fuel oil	250 ton/day (500 barrels/day of oil equivalent) pilot plant under construction. Plant is located at Baytown, Texas
Fischer – Tropsch (M.W. Kellogg/Lurgi)	Indirect liquefaction, liquefaction of synthesis gas in a fluid-bed catalytic converter	Gasoline LPG Diesel fuel Heavy fuel oil Medium-Btu gas SNG	SASOL I, 800 tons/day, producing over 10,000 bbl day of liquids in commercial production since 1956. SASOL II, 40,000 ton/day, producing over 50,000 bbl day of liquids has been completed and will begin start-up in 1982. SASOL III with approximately the same capacity as SASOL II is currently being planned.
Mobil M	Indirect liquefaction, liquefaction of synthesis gas in fixed bed using molecular-size-specific zeolite catalyst	Gasoline LPG	Commercial-scale plant to produce 12,500 barrels of gasoline using reformed natural gas is planned for New Zealand in 1984 – 1985

A commercial scale plant producing 12,500 barrels per day of gasoline is planned for operation in New Zealand by 1985.

Direct coal liquefaction technologies are in various stages of development. SRC-I and II processes have been tested at the pilot plant level and are entering into the demonstration plant stage.

Large pilot plants are currently under construction for testing of the H-Coal and EDS processes. These plants are located at Catlettsburg, Kentucky, and Baytown, Texas, respectively.

SRC-I process produces primarily a solid product with a small amount of useful liquid product; however, SRC-II process produces primarily liquid products.

In addition to these major coal liquefaction technologies, several other processes have received attention, including the Dow process, Riser Cracking, Synthoil, and the Zinc Halide process. All have been tested in small-scale units.

Gaseous Fuels and Chemical Feedstocks from Coal

Most coal gasifiers react coal, steam, and oxygen to produce a gas containing carbon monoxide, carbon dioxide, and hydrogen. When air is used as the oxygen source, the product gas contains up to 50% nitrogen and is referred to as low-Btu gas since its heat of combustion is only 80–150 Btu/standard cubic foot (scf). Synthesis gas or medium-Btu gas ranges from 300 to 500 Btu/scf.

Low-Btu gas is used as a fuel gas near its point of generation since its low heating value makes it uneconomical to distribute over long distances. Medium-Btu gas can be used as a fuel gas and transported economically over distances of up to 200 miles; it can also be used as a chemical feedstock for the production of methanol or gasoline. Finally, it can be converted catalytically to substitute natural gas (SNG), having a heating value of about 1000 Btu/scf. Additionally, medium-Btu gasification is an integral part of all indirect liquefaction technologies.

There are many coal gasification technologies that differ in design and operation, depending on the type of coal used and the product desired. High- and medium-Btu gasification technologies using noncaking coals characteristic of U.S. western coals are relatively well developed. Severe operational problems are encountered with commercially available gasifiers in processing caking coal such as those found in the eastern United States. Several gasification technologies, shown in Table 7-5, for high- and medium-Btu gases are under active development. Many additional processes are being tested, but are at less advanced stages of development as noted in Table 7-6.

A fixed-bed gasifier, such as the Lurgi, feeds coal to the top of the gasifier. The descending coal is successively dried, devolatilized, and gasified in contact with gases rising from the bottom. Steam and oxygen are introduced at the bottom of the gasifier, and solid ash is removed through an ash lock. In some gasifiers, such as British Gas Company (BGC) Lurgi, the temperature at the bottom of the bed is sufficient to melt the ash, allowing its removal as molten slag. The slagging feature provides a distinct advantage in contending with the caking characteristics of eastern U.S. coals.

Lurgi high-pressure operation, in conjunction with relatively low gasification temperatures, favors the formation of significant quantities of methane in the gasifier, enhancing the heating value of the product. These conditions also favor production of by-products such as tars and impurities like phenols, organic nitrogen compounds, and sulfur compounds.

Table 7-5. Coal Gasifiers for High-, Medium-, and Low-Btu Gas

Process	Process Type	Potential Products	Most Suitable Products	Status
Lurgi Dry Ash	Pressurized fixed bed, dry bottom	Substitute natural gas (SNG, also known as high-Btu gas), medium-Btu fuel[a] gas, low-Btu fuel gas	SNG, medium-Btu fuel gas, low-Btu fuel gas	40 years of commercial development and 14 commercial plants located in Australia, Germany, England, India, Pakistan, South Africa, Korea. Average module size 800 tons/day (2000 BOE)[c]
British Gas Company (BGC) Lurgi	Pressurized fixed-bed slagging bottom	SNG, medium-Btu fuel gas, low-Btu fuel gas	SNG, medium-Btu fuel gas, low-Btu fuel gas	790 ton/day (of coal) (2000 BOE) pilot plant tested in Westfield, Scotland
Texaco	Pressurized single-stage entrained, slurry feed	SNG, medium-Btu synthesis gas, low-Btu fuel gas	Medium-Btu synthesis[b] gas	160 ton/day (400 BOE) plant operating in West Germany
U-Gas Institute of Gas Technology (IGT)	Pressurized fluid bed, ash agglomerating	SNG, medium-Btu fuel gas, low-Btu fuel gas	Medium-Btu fuel gas	14,000 ton/day of coal plant (35,000 BOE) producing medium-Btu fuel gas, under design for construction in Tennessee
Westinghouse	Pressurized single-stage fluid bed, ash agglomerating	SNG, medium-Btu fuel gas, low-Btu fuel gas	SNG medium-Btu fuel gas	15 ton/day (40 BOE) process development unit, under testing at Waltz Mill, Pa.
Shell Koppers	Pressurized entrained, dry feed	Medium-Btu synthesis gas, low-Btu fuel gas	Medium-Btu synthesis gas	150 ton/day (400 BOE) pilot plant in operation in W. Germany; 1000 ton/day scheduled in 1983/1984.
Koppers – Totzek	Atmospheric entrained, dry feed	Medium-Btu synthesis gas, low-Btu fuel gas	Medium-Btu synthesis gas	1000 ton/day (2500 BOE) plant in operation in South Africa for the production of ammonia

[a] Medium-Btu gas with significant concentration of methane is more suitable for use as fuel, and therefore identified as medium-Btu fuel gas.
[b] Medium-Btu gas with low concentration of methane is more suitable for chemical synthesis, and therefore identified as medium-Btu synthesis gas.
[c] BOE = barrels per day of oil equivalent.

Table 7-6. Status of Other Coal Gasification Processes

	Scale (tons/day coal feed)	Status
DEMONSTRATION PLANTS		
Hygas	7,340	Conceptual design[a]
COED-COGAS	2,210	Detailed design[b]
U-Gas	3,160	Detailed design
PILOT PLANTS/PDUs		
Bell High Mass Flux	6	Operational[c]
Bigas	120	Operational
Combustion Engineering	5	Operational
Dow	24	Under Construction
Exxon Catalytic	100	Proposed
Gegas	24	Operational
Hydrane	4	Proposed
Molten Salt	24	Operational
Mountain Fuel	12	Proposed
Synthane	72	Mothballed
Tri-Gas	1	Operational

[a] Conceptual design incorporates all important details of major unit areas in the plant. Material balances are provided around all major unit areas. (Unit area is a section of the plant consisting of several components integrated to perform a single transformation on the product stream. Examples are gasification, raw gas cooling, gas cleanup, or methanation.)
[b] All equipment and detailed pipeline diagrams are prepared as part of detailed design. In addition, detailed material balances are prepared for each piece of equipment.
[c] The plant is either operating or has operated successfully in the past.

In fluid-bed gasifiers currently under development, high-velocity gases pass up through the bed to fluidize the coal, providing excellent mixing and temperature uniformity throughout the reactor. Operability with caking coals (eastern United States), as well as low tar production and tolerance to upsets in fuel rates, has been demonstrated at the pilot scale for both the Westinghouse and U-Gas gasifiers.

The Texaco and Koppers–Totzek gasifiers are representative of entrained-bed technology in which the solid particles are concurrently entrained in the gaseous flow. Flame temperatures at the burner discharges are in the range of 2500–3500°F (1370–1925°C), resulting in melting of the coal ash with minimum production of impurities. Entrained-flow gasifiers may be favored for the production of synthesis gas for indirect liquefaction: they can operate with caking coals. However, compared to fluid-bed gasifiers, they have very low carbon holdup capability in the reactor and, therefore, have limited safeguards against possible formation of explosive mixtures in the reactor in case of coal feed interruption.

There has been extensive commercial experience in the United States with low-Btu coal gasification technologies operating near atmospheric pressure. However, these applications have been limited to small-scale captive applications for providing industrial process heat and space heating. For example, the Wellman–Galusha gasifier designed for atmospheric pressure operation was used extensively by industry years before pipeline-supplied natural gas was readily available at comparatively lower cost. Pressurized gasification processes capable of yielding high-Btu gas for pipeline use and medium-Btu gas for chemical feedstocks are less developed, with the exception of the Lurgi fixed-bed process. The Lurgi process is based on 40 years of commercial development at 14 commercial plants that are located in Australia, Germany, Britain, Korea, India, Pakistan, and South Africa. A great deal of interest in the Lurgi technology is emerging in the United States with several announced plans for SNG production by pipeline and gas utility companies. Several projects utilizing the Texaco process for captive applications (chemical feedstocks and on-site power generation) are in the planning and design stage with at least one project (Tennessee–Eastman) scheduled for construction.

Development of the Synfuels Industry over the Next 20 Years

Three scenarios or projections of synfuel industry buildup rates to the year 2000 have been developed to illustrate the potential range of synfuel product utilization:

- A "national goal" scenario driven by federal incentives
- A "nominal production" or most likely scenario
- An "accelerated production" scenario representing an upper bound for industry buildup

ACHIEVING THE NATIONAL GOAL—SCENARIO I

In July 1979, President Carter announced new energy initiatives for the United States aimed at reducing our dependence on imported oil. One of the key elements of this policy is the provision of federal funds to stimulate production of synthetic fuels at the rate of 2.2 million barrels per day (MMBPD) by 1992. Specifically, the national synfuel goals are the following:

Coal Liquids. To stimulate and accelerate the construction and operation of the first few plants to provide sufficient data on the competing commercial coal liquefaction processes so that industry, with its own investment, stimulated by government incentives if required, will build plants with sufficient capacity to provide upward to 1 MMBPD liquid fuels by the year 1992.

Shale oil. To stimulate shale oil production at the rate of 0.4 MMBPD by 1990.

High-Btu Gas. To develop and implement a program that enables the United States, by 1992, to produce significant quantities of pipeline-quality gas (0.5 MMBPD–oil equivalent[1]) from commercial high-Btu gas (HBG) plants in an environmentally acceptable manner. This is facilitated by the short-range goal of having two or three commercial HBG plants in operation by the mid-1980s.

Low-/Medium-Btu Gas. To stimulate an initial near-term commercial capability for several medium-Btu commercial plants in key industries as well as utilities, for energy and feedstock applications for both single and multiplant use, and for multiple applications of low-Btu gas in each of the prime industry markets. Commercial-scale development will depend on the long-term economics of this technology, vis-à-vis the price of domestic oil

[1]For easy comparison with petroleum supply/demand figures, synfuel production rates are expressed in barrels of oil equivalent. This does not imply that high-, medium-, and low-Btu gases from coal that are substituted for domestic natural gas will have any direct effect on the reduction of imported oil.

and natural gas. Once a capability has been established, capacity will be accelerated to achieve at least 0.29 MMBPD oil equivalent by 1992. Of this total, up to 0.04 MMBPD oil equivalent will be provided from 40 to 50 low-Btu facilities and up to 0.25 MMBPD oil equivalent from 25 to 30 medium-Btu plants. Again, it must be mentioned, that if this low- and medium-Btu coal gas is substituted for natural gas, there will not be a direct effect on the reduction of imported oil.

The key assumptions allowing achievement of these goals are (1) federal funds provided are sufficient to reduce investment risk by the synfuel industry through 1992, and (2) other requirements for industry development are satisfied, i.e., environmental permits, material, equipment, and labor. A likely buildup rate profile for the synfuel industries under this scenario is shown in Fig. 7-1.

For shale oil, several of the most advanced projects were selected as a basis. The planned operation start-up schedules and capacity buildup rates for these projects were used to generate the industry production buildup profile to about 0.4 MMBPD by 1992. The period beyond 1992 is viewed as one of technology consolidation: gaining a firm footing with regard to environmental and economic performance and technology improvements. This type of industry production profile is not without precedent, for example, the federal support of the synthetic rubber industry during World War II.

The goal of 1 MMPBD of coal liquids could be met predominantly by indirect coal liquefaction. At present, the only commercially demonstrated coal liquefaction process is the Fischer–Tropsch embodied in the SASOL plants in South Africa. The Mobil-M process should be commercially demonstrated within the next 5 years. Considering construction and permitting lead times, plants of this type could begin operation around 1985. To meet the production goal, 10–15 plants of a nominal 0.05 MMBPD capacity must be in operation by 1992. A potential drawback to the commercialization of SASOL technology in the United States is the broad product distribution, ranging from light hydrocarbon gases to heavy fuel oil. The Mobil-M technology, on the other hand, produces an all-gasoline product which would be particularly well suited to the U.S. market demands. Given this apparent advantage of Mobil-M technology over SASOL, it is believed that industry should favor commercialization of both Mobil-M and SASOL technology during the next few years, with the breakdown being roughly 50/50. Ap-

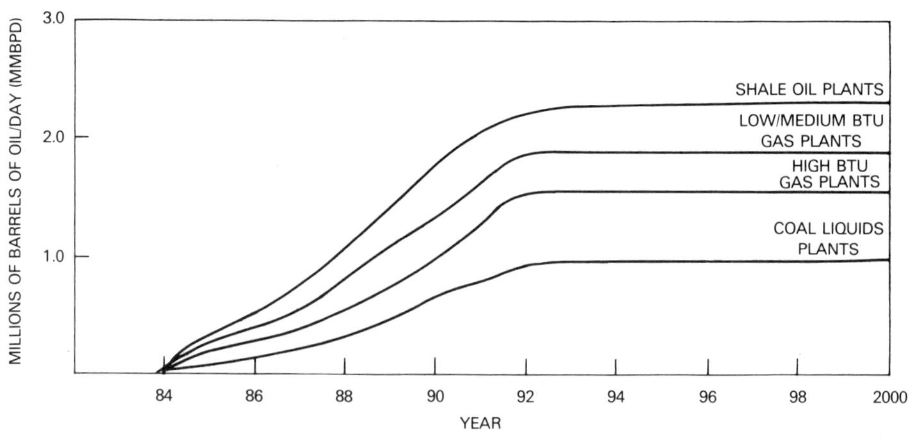

Fig. 7-1. Synfuel industry buildup for the national goal scenario.

proximately 75% of the coal liquids production will be due to these indirect liquefaction processes.

For the direct liquefaction processes, there will not be sufficient experience and information to attract any more than developmental interest over the next few years, under this scenario. By 1985 there should be sufficient information available from the operation of the EDS, H-Coal, and SRC-II plants to support a commercialization decision concerning these processes. Federal incentives will likely be distributed such that by 1982, three or four pioneer commercial-scale plants employing direct liquefaction will begin to appear. Of the total production goal of 1 MMBPD of coal liquids it is estimated that 25% will be produced by these first commercial direct liquefaction plants embodying the basic SRC-II, EDS, and H-Coal technologies, or improvements and modifications to these. It is projected that for the next few years after 1992, production will remain at 1 MMBPD while technological evaluations are performed. These direct liquefaction plants will be located near the major eastern U.S. coal areas.

The Lurgi fixed-bed process is the high-Btu coal gasification technology and has been demonstrated commercially outside the United States. It is expected to be utilized in all commercial plants constructed over the next 10 years. As the process requires noncaking coals, these plants will most likely be located in the western United States. Interest will continue in other high-Btu gasification technologies such as the Slagging Lurgi, which is capable of using eastern caking coals. At least one of these alternate or advanced processes probably will be supported under federal incentives but it is unlikely that a commercial plant will appear until the early 1990s, and this would probably be located near a midwestern coal resource.

The Lurgi fixed-bed medium-Btu process is the lead technology for medium-Btu gas. Texaco partial oxidation gasification or similar pressurized entrained-bed gasifiers, such as pressurized Koppers–Totzek, will be under development and demonstration during the early 1980s and will likely serve as the prime medium-Btu gasification process for eastern coals. To 1992, however, the major buildup in medium-Btu gasification will come from Lurgi plants located in the western United States.

For low-Btu gasification, the several technologies that are currently available and providing commercial service are assumed to be easily applied, under the incentives existing to 1992, to generate the 0.04 MMBPD production rate goal. Low-Btu gas will generally be captively employed as fuel gas or used on-site for combined-cycle power generation.

The production buildup profile for major synfuel products resulting from the synfuels industry buildup in Scenario I is shown in Fig. 7-2. These product quantities are projected to enter commercial use and are to be considered in assessments of potential environmental impacts from synfuels. Naturally, these major products are presented for the sake of clarity, but there are many other products and by-products that will be produced and distributed into the market place. These products and by-products will also vary in greater or lesser quantities in Scenarios II and III which follow.

It is very unlikely that the Scenario I goals could be accomplished.

PRODUCTION AT A NOMINAL RATE—SCENARIO II

Recent studies of the technical capability of the United States to meet the synfuel national goal point out that there are significant concerns regarding achieving this goal. They include

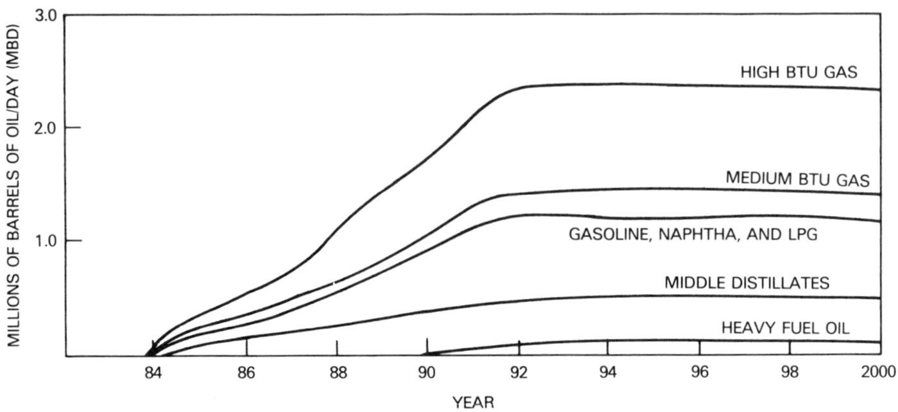

Fig. 7-2. Major synfuel product buildup for the national goal scenario.

- Availability of skilled manpower: it is expected that the supply of engineers and construction labor will be severely taxed to meet the synfuel production goal set forth in Scenario I.
- Availability of critical equipment: certain critical equipment for the synfuel industry such as compressors, heat exchangers, and pressure vessels are expected to be in short supply unless corrective measures are taken now, thus slowing the synfuel industry buildup rate indicated in Scenario I.
- Diversion of investment to competing technologies: demand on the limited capital available in the economy by competing energy supply technologies, such as coal liquefaction, coal gasification, oil shale, geothermal, and solar technologies, could result in the slowing of buildup rates for some technologies.
- Environmental data: lack of environmental data needed for regulatory approvals could slow down the buildup rate.
- Licensing: time and construction schedule constraints imposed by state and federal licensing and permitting requirements could hinder synfuel industry buildup rate.

Taking these concerns into consideration, a nominal synfuel production buildup—Scenario II—has been developed, as indicated in Fig. 7-3. A production rate of about 2.1 MMBPD is estimated by the year 2000, instead of 1990 as indicated in Scenario I. The technologies expected to contribute to both Scenarios I and II are the same; the major difference is in the rate of buildup: it is slower and delayed in time.

For shale oil, a nominal production rate of 0.4 MMBPD should be achieved by the year 2000. The buildup rate is estimated to lag about 4 years behind that of Scenario I and is based on the following observations: some technologies are still considered developmental, such as the modified *in situ* process; land problems, including availability of off-tract disposal sites, may take longer to resolve.

Under this scenario, no large-scale commercial coal liquefaction plants are projected to be on line until 1992 with a growth rate beyond yielding 1 MMBPD by the year 2000. It was believed that federal incentives will be applied to support construction of one each of the indirect liquefaction plants and a direct liquefaction plant only after sufficient assess-

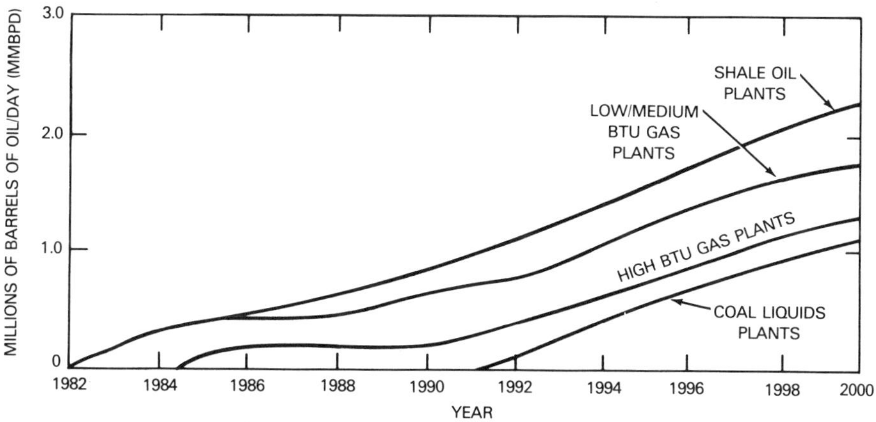

Fig. 7-3. Synfuels industry buildup for the nominal production scenario.

ment has been made of the operations of the EDS and H-Coal pilot plants and the SRC-II demonstration plant. Rather than commit sizable resources to the commercialization of indirect liquefaction, a decision probably will be delayed resulting in no operating commercial liquefaction plants before 1992 under this scenario. During the 1980s it is believed that improvements will be made in both the operating indirect liquefaction plants and the designs of the direct liquefaction processes. These "advanced" technologies with product slates yielding primarily transportation fuels will be sufficiently attractive to encourage development of 1 MMBPD of coal-derived liquid production by the year 2000.

Currently there is a great deal of interest in SNG technology. Several gas utility and pipeline companies have announced plans to construct high-Btu plants. With incentives, several of these plants will be constructed and in operation by 1985. However, as a result of the projected improved outlook for gas supplies, including potential from unconventional sources, the availability of "imported" conventional natural gas (Alaskan, Canadian, and Mexican), and the current unfavorable rate-structure pricing policy, the complete commercialization of HBG will be hampered. Its production rate is not likely to expand beyond the 0.25 MMBPD level attained around 1992 under this scenario.

The buildup of medium-Btu gas plants will also be impeded by the availability of natural gas; however, for certain industrial applications requiring large volumes of uninterrupted supplies (e.g., chemical feedstocks, cogeneration) low/medium-Btu plants will remain attractive. It is estimated that production of low/medium-Btu gas will reach a level of 0.45 MMBPD by 1992.

ACCELERATED PRODUCTION—SCENARIO III

The accelerated production scenario is based on the assumption that federal incentives are sufficient to synfuels production to meet the national goals in 1992, that operation of synfuel plants up to 1992 is successful to the extent confidence in processes is gained, and all resource requirements are satisfied. Licensing and permitting procedures must also be streamlined. It is assumed that demand for coal-derived synfuels remains at a level such that new plant capacity continues to be added to 2000 at about the same rate as the buildup to 1992. For shale oil, the production of 0.9 MMBPD by the year

2000 is based on a survey and analysis of the desired goals of each industrial developer. As indicated in Fig. 7-4, a total synfuel production rate of 5 MMBPD may be reached by the year 2000. This includes 2.6 MMBPD of coal liquids, 1.5 MMBPD of SNG, and 0.9 MMBPD of shale oil.

However, in view of the limitations facing the synfuel industry, some of which were discussed earlier, the accelerated production scenario is highly unlikely. The synfuel industry buildup rate shown in Fig. 7-4 for this scenario can be considered an upper bound to synfuels utilization over the next 20 years.

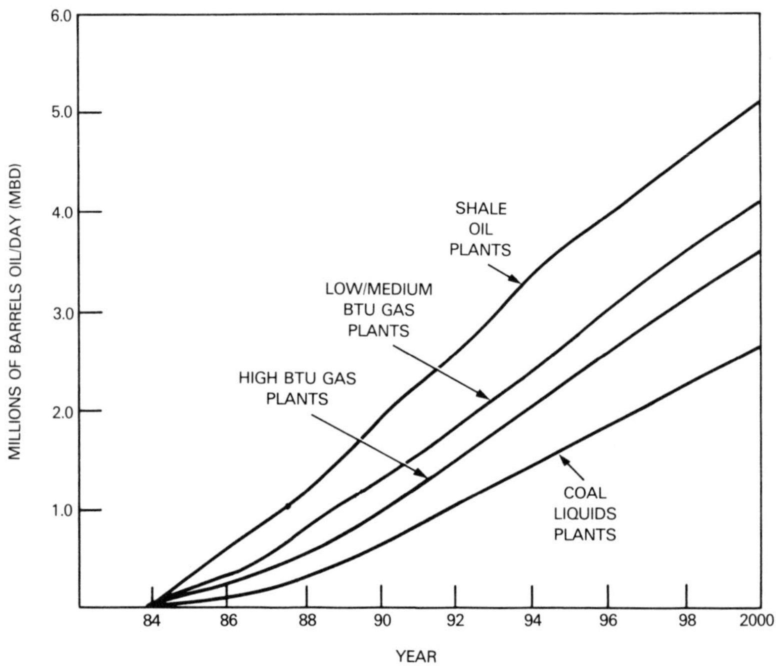

Fig. 7-4. Synfuels industry buildup for the accelerated production scenario.

The three scenarios describing possible synfuel industry buildup profiles provide a basis for projecting the market penetration of synfuel products in the near future. As these products enter the market, potential environmental impacts related to synfuels utilization must be considered.

Potential Market for Synfuel Products and By-products

The major synfuel products could be broadly classified into five groups:

- Gaseous products
  - High-Btu gas
  - Medium-Btu gas
  - Low-Btu gas
  - Liquified petroleum gas (LPG)

- Light Distillates
  - Gasoline
  - Naphtha
- Middle Distillates
  - Jet fuel
  - Kerosenes
  - Diesel oil
- Residue
  - Heavy fuel oil
  - Lubricants
- Petrochemicals

GASEOUS PRODUCTS

The high- and medium-Btu gases are suitable for essentially all industrial fuel applications that can be serviced by coal, oil, or natural gas. In some cases equipment modifications or special controls will have to be implemented to retrofit existing plants for medium-Btu gas, whereas this problem may not exist for high-Btu gas installation. However, there should be no difficulty in employing either high- or medium-Btu gas in new industrial installations. These products will be utilized by major energy-consuming industries such as food, textile, pulp and paper, chemicals, and steel. It appears that only chemical, petroleum, and steel industries will require sufficient fuel gas at a single location to economically justify the dedication of a single gasification plant. Other industrial plants will have to share the output distributed by pipeline from a central gasifier, or tap into the existing natural-gas-pipeline system for their need. Preliminary economic studies indicate that it is not economical to transport medium-Btu gas through pipelines for more than 200 miles. Medium-Btu gas can also be utilized by the petrochemical industries as chemical feedstock for the production of ammonia, methanol, and formaldehyde. Currently most of this requirement is met by reforming natural gas. The use of medium-Btu gasification appears especially attractive when integrated with new combined plants for utility applications.

The major characteristics of low-Btu gas are its high nitrogen content, low carbon monoxide and hydrogen content, and resulting heating value typically below 150 Btu/scf. Its flame temperature is also about 13% lower than that of natural gas. Because of these characteristics low-Btu gas is limited to on-site use and to industrial processes requiring temperatures below 2800–3000°F (1540–1650°C) and is generally unsuitable for use as a chemical feedstock. Furthermore, because of its low energy density it requires significant equipment modifications for retrofit applications. Today there are operating and planned low-Btu gasifiers in the United States for kiln firing of bricks, iron ore pelletizing, chemical furnace, and small boilers.

Liquified petroleum gas (LPG) has applications for industrial, domestic, and transportation uses. In domestic applications LPG is used mainly as a fuel for cooking and for water and space heating. In industry, LPG finds a large number of diverse outlets. Apart from use as a fuel in processes that require careful temperature control (glass and ceramics, electronics) or clean combustion gases (drying of milk, coffee, etc.), LPG is also used in the metallurgical industry to produce protective atmospheres for metal cutting and other uses. The chemical industry, particularly on the U.S. Gulf Coast, uses petroleum gases for cracking to ethylene and propylene as well as for the manufacture of synthesis gas. Small portions of LPG are also used to fuel automotive vehicles. Another use of LPG is to enrich lean gas made from other raw materials to establish proper heating value levels. On a volume basis, production of LPG in the United States exceeds that of kerosene and approaches that of diesel fuel. About 40% of LPG production is used by the chemical industry, another 40% is for domestic use, 10% for automotive use, and the remaining distributed among other industrial and agricultural fuel uses. Currently LPG is supplied primarily from refineries handling petroleum crudes. With the anticipated shortfall in the supply of these crudes, the resulting shortages of LPG will be met to some extent by LPG from synfuel plants.

LIGHT DISTILLATES

Gasoline, which is a major light distillate, is generally defined as a fuel designed for use in reciprocating, spark-ignition internal-combustion engines. Other uses for gasoline are of small volume. Primarily it is used as fuel for automotive ground vehicles of all types, reciprocating aircraft engines, marine engines, tractors, and lawn mowers. Other small-scale uses include fuel in appliances such as field stoves, heating and lighting units, and blow torches. By far the primary use of gasoline produced from coal will be for transportation applications. Currently we consume nearly 6.8 MMBPD of petroleum-derived gasoline and this corresponds to about 40% of the total petroleum consumption.

Naphthas have a wide variety of properties and serve many industrial and domestic uses. Their primary market is the petrochemical industry where they can be used for the manufacture of solvents, varnish, turpentines, rust-proofing compounds, pharmaceuticals, pesticides, herbicides, and fungicides. However, preliminary analysis indicates that there will be a relatively small amount of coal-derived naphthas entering the market.

MIDDLE DISTILLATES

The market for middle distillates, which essentially are jet fuel, kerosene, diesel oil, and light fuel oil, are jet aircraft, gas turbines, and diesel engines used for transportation and stationary applications, and residential and commercial heating.

RESIDUES

The market for residues, consisting mainly of fuel oil, is primarily for industrial, utility, and marine fuel use. Other applications for residues include preparation of industrial and automotive lubricants, metallurgical oils, roofing coatings, and wood preservative oils. Coke is another likely useful product from residue.

PETROCHEMICAL

Many synfuel products, in addition to their primary use as fuel, are likely to be used by the petrochemical industry for the production of several other by-products. Currently over 3000 petrochemical by-products are derived from petroleum and natural gas sources; these by-products include items like synthetic rubbers, plastics, synthetic fibers, detergents, solvents, sulfur, ammonia and ammonia fertilizers, and carbon black.

Petrochemicals from synfuels will generally fall under three broad groups based on their chemical composition and structure: aliphatic, aromatic, and inorganic. An *aliphatic* petrochemical is an organic compound that has an open chain of carbon atoms. Important petrochemicals in this group include acetic acid, acetic anhydride, acetone, ethyl alcohol, and methyl alcohol. Most aliphatic petrochemicals are currently made from methane, ethane, propane, or butane. Aliphatics currently represent over 60% of all petrochemicals and are the most important group economically.

An *aromatic* petrochemical is also an organic compound but one that contains or is derived from a basic benzene ring. Important in this group are benzene, toluene, and xylene, commonly known as the BTX group. Benzene is widely used in reactions with other petrochemicals; with ethylene it gives ethyl benzene which is converted to styrene, an important synthetic-rubber component. As a raw material it can be used to make phenol. Another use is in the manufacture of adipic acid for nylon. Toluene is largely used as a solvent in the manufacture of trinitrotoluene (TNT) for explosives. Xylene is used as a source of material for polyester fibers and isophthalic acid, among other petrochemicals.

An *inorganic* petrochemical is one which does not contain carbon atoms. Typical here are sulfur and ammonia, and its derivatives such as nitric acid, ammonium nitrate, and ammonium sulfate.

The different end-use applications of major synfuels products are summarized in Table 7-7. We see from this discussion that coal-derived synfuel products are likely to be used not only as a fuel, but also in the manufacture of a number of other by-products that will be used in multitudes of other applications.

Synfuels Market Penetration

The time frame over which issues regarding the use of various synfuel products, market development and penetration of these products must be anticipated is illustrated in Figs. 7-5, 7-6, and 7-7, over 1985–1987, 1988–1990, and 1991–2000 time frames.

SYNFUEL PRODUCT UTILIZATION EMPHASIS, 1985–1987

Oil-shale-derived synfuels will be introduced into the petroleum product markets about 1985, and, based on Scenario I, as much as 0.2 MMBPD of shale oil can enter the market by 1987. The first stage of synfuel market infrastructure development will be oriented toward transportation fuels (reference Fig. 7-5), because oil shale that is

Table 7-7. Major End-Use Applications of Synfuel Products

Major Synfuel Products	Likely Major End-Use Applications
High- and medium-Btu gas	Food, textile, pulp and paper, chemicals, iron and steel industries; residential/commercial heating
Low-Btu gas	Small boilers, kilns, pelletizing
LPG	Glass, electronics, chemical industries; domestic cooking and heating; automotive
Gasoline	Transportation
Naphtha	Petrochemical industry; solvents; varnish; turpentines
Middle distillates (kerosene, diesel, light fuel oil)	Transportation, gas turbines, residential and commercial heating
Residues	Industrial, utility, and marine fuel; metallurgical oils; roof coatings; wood preservatives, lubricants

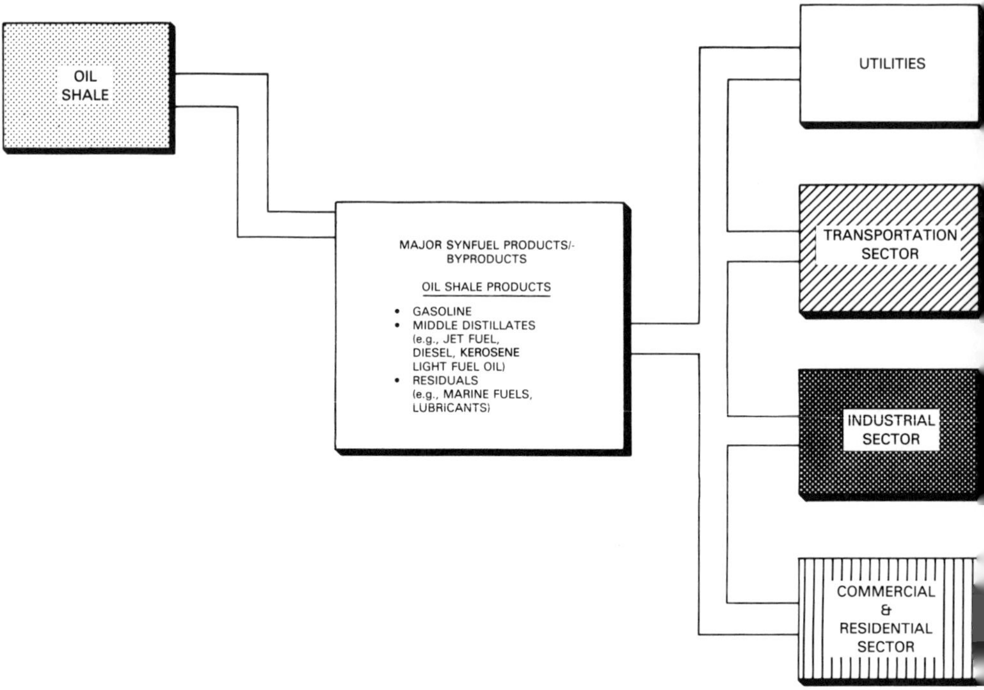

Fig. 7-5. Synfuel utilization during 1985–1987.

hydrotreated can be refined in existing refineries to such products as gasoline, jet fuel, diesel fuel, and marine fuel. The bulk of this supply will be in the form of middle distillates comprised of jet fuel and diesel oil. The demand for transportation during the late 1980s is expected to be around 10 MMBPD. Of this, about 5% is likely to be consumed by the military sector. It is conceivable, therefore, that the bulk of the shale oil products could be utilized by the military, possibly with a government synfuel purchase guarantee program.

It is anticipated that the oil shale industry will continue to grow, producing as much as 0.45 MMBPD by year 2000 as per Scenarios I and II and as much as 0.9 MMBPD as per Scenario III. The bulk of this production is anticipated for the transportation sector.

SYNFUEL PRODUCT UTILIZATION ADDITIONS, 1988–1990

Subsequent buildup of the synfuels industry during the 1988–1990 time period (reference Fig. 7-6) is expected to come from commercial-size, high-Btu gasifiers. As per Scenario I, the output from these high-Btu gasifiers may be as high as 0.4 MMBPD of oil equivalent by 1990; however, the conservative estimate based on Scenario II is that only around 0.17 MMBPD of oil equivalent is likely to be produced by that time. The high-Btu gasification will serve some of the energy needs of both the industrial and residential/commercial sectors as direct gas sales or through electric power generation by utilities. Some of the major industrial users of high-Btu gas are likely to be textile, food, steel, and chemical industries. Initially, following the current use pattern, it will be used not only as an industrial fuel but also as a chemical feedstock. It is expected that the existing natural gas pipeline network, with the exception of a few connecting pipelines, will be utilized for

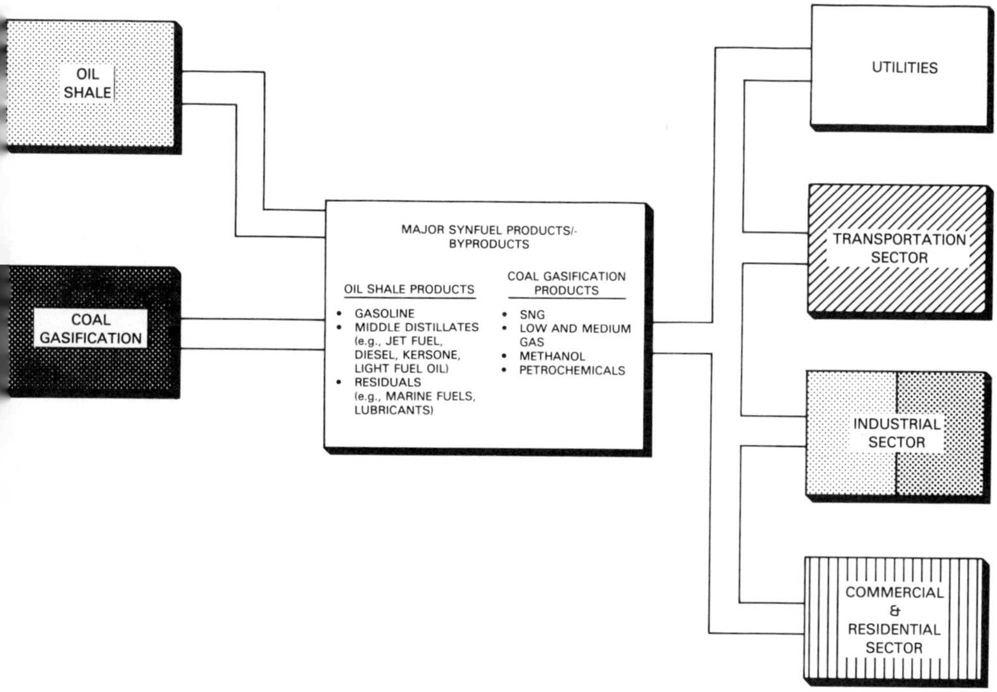

Fig. 7-6. Synfuel utilization during 1988–1990.

the distribution of high-Btu gas, and, therefore, introduction of high-Btu gas is not likely to cause major problems concerning distribution for end-use applications. During this time period low- and medium-Btu gasification plants will also be used by industries in a captive mode to supply some of their fuel and chemical feedstock needs. This may amount to as much as 0.3 to 0.4 MMBPD of oil equivalent based on the first two scenarios. The medium-Btu gas could be used as a synthesis gas for the production of different chemical products such as ammonia, which in turn could be used for the manufacture of such products as fertilizers, fiber and plastic intermediates, and explosives. Currently, the petrochemical industry derives its synthesis gas by reforming natural gas or naphtha. During this time period, it is likely that one to three small plants possibly producing methanol from medium-Btu gas may come on line. These are likely to be owned by industries primarily to supply internal needs, for example, the production of formaldehyde, a product with a number of end-use applications. It is unlikely that products from these plants will be entering the open market directly, on a large scale, for public consumption. During this time period the use of low-Btu gas will be limited to an industrial fuel in such applications as kilns, chemical furnaces, and small boilers. However, the use of low-Btu gasification by utilities on one or two demonstration units for combined-cycle applications cannot be ruled out.

During this time frame the shale oil output will continue to grow, reaching as much as 0.4 MMBPD in accordance with the National Goal Scenario. As a result it is anticipated that increasing amounts of shale oil products will be entering the transportation sector, with limited entry into the industrial sector for use as fuel.

SYNFUEL PRODUCT UTILIZATION ADDITIONS, 1991–2000

During the 1991–2000 time frame (reference Fig. 7-7), central coal liquefaction plants will introduce into the market a spectrum of products and by-products that will be consumed by the transportation, industrial, and residential/commercial sectors. Based on the nominal and accelerated scenarios, by the year 2000, 1.5–2.5 MMBPD of coal liquid products will be entering the market. Under these conditions, a significant segment of the transportation fleet could be running on synthetic fuel. Coal-derived liquids will be utilized not only by industry as a fuel source and chemical feedstock, but also by the residential and commercial sectors for space heating, hot water supply, and other domestic uses. Furthermore, many of the oil-fired utility plants given exemption from converting to coal in the interim will be burning coal-derived fuel oil. SRC-II plants will be the likely candidate to supply the bulk of this fuel. It is also expected that methanol from indirect coal liquefaction could be entering the market for use as turbine fuel for the production of electricity during this time period. In addition, SNG output from liquefaction plants could be as high as 20% of the total useful output from these plants in terms of heating value. LPG and naphtha produced from direct and indirect coal liquefaction processes and oil shale are likely to be used primarily by the petrochemical industries. For example, LPG may be used by the petrochemical industry as a raw material for the production of alcohols, organic acids, detergents, plastics, and synthetic rubber components. Naphthas may be utilized for the manufacture of such items as solvents, adhesives, pesticides, and chemical intermediates. Currently, the petrochemical industry uses about 11% of our crude oil supply for the production of various petrochemicals. During the 1990–2000 time frame, it is possible that the same percentage of available synfuels will be utilized by

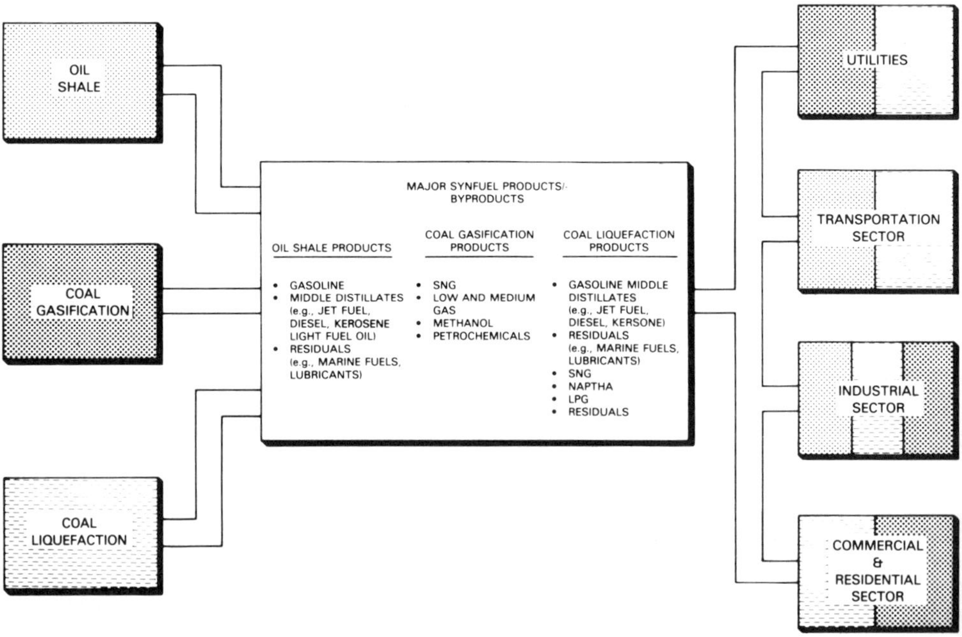

Fig. 7-7. Synfuel utilization during 1991–2000.

the petrochemical industry for the production of hundreds of petrochemical products. A major use of residuals from coal liquefaction processes and oil shale is likely to be the manufacture of different types of lubricants. These could be for such applications as lubrication of engines and general machinery, steam turbine bearings and reduction gears, compressors, insulating oils, and metalworking and -cutting oils.

So we see that synfuel products and by-products are likely to enter all the end-use sectors, in the course of time. The potential for exposure and for environmental impacts must be carefully considered.

Potential Environmental Exposures due to Synfuel Utilization

A major concern of the emerging synfuel industry is the potential environmental, health, and safety impacts associated with the use of synfuels. The potential exposure of the public to synfuels will depend on the rate of development of the synfuel industry's specific end-use markets. Since the market may cover a wide range of products and end uses, a significant portion of the population may be exposed. The products will enter the markets in varying quantities over the coming years. To illustrate the important environmental concerns, synfuel product production rates based on the National Goal Scenario (Scenario I) are considered and three time periods are examined for potential environmental exposure: 1985–1987, 1988–1990, and 1991–2000.

POTENTIAL EXPOSURE: 1985–1987

During this period, synfuels entering the market will be limited mostly to shale oil products. Approximately 0.2 MMBPD of products by 1987 is projected by Scenario I. Crude shale oil will most likely be transported to refineries in either the Gulf Coast or Midwest and is expected to be distributed by existing pipelines. Product quantities will be limited. The hazards of transporting and storing crude shale oil and shale oil products are expected to be minimal. Shale oil products will be used primarily as transportation fuels such as gasoline, diesel oil, and jet fuel and will be distributed by railroads, tankers, trucks, and barges. During this period the quantities handled are estimated to amount to 0.04 MMBPD of gasoline and a combined total of 0.16 MMBPD for the middle distillates. The major exposure to these products occurs at storage terminal unloading operations and service station storage tank loading operations, both of which have high spill potential. The end user (a passenger car, truck, or other vehicle) also poses a potential spill problem owing to the rapid expansion of self-service stations. Combustion of the fuels may expose a large segment of the population since most automobile traffic is generated in central business districts and their suburbs. By-products from shale oil refining such as lubricating oils and greases will be shipped from refinery bulk packing plants in secure containers, minimizing the likelihood of exposure.

Products from shale oil production could reach approximately 0.3–0.4 MMBPD during this period, as suggested by the accelerated rate scenario, with the potential exposure reaching twice the level suggested by the National Goal Scenario.

POTENTIAL EXPOSURE: 1988–1990

During this period, in addition to increased shale oil production, SNG low- and medium-Btu gas and some indirect liquefaction products will also be entering the market, which increases the complexity of the synfuel distribution network and increases the potential for public exposure to the products. It is a time period by which the EPA must have identified potential problems and have developed a plan for meeting the synfuel challenge.

Shale oil production during this period is projected to be 0.3–0.4 MMBPD under the National Goal Scenario, but could range from 0.2 MMBPD (nominal production rate scenario) to 0.8 MMBPD (accelerated production rate scenario) in 1990. The exposure potential to the products will increase proportionally during this time period compared to the previous period.

The SNG entering the market is projected to amount to an oil equivalent of 0.4 MMBPD by 1990, and will be transported by existing pipeline to the various markets. Although pipelines transporting SNG or crude shale oil present a low accident potential, pipelines either transect or terminate in densely populated areas, providing some degree of exposure potential to these products. First-generation coal gasification technology (Lurgi) buildup will occur near western U.S. coal deposits, the Northern Great Plains/Rocky Mountains area. The SNG from this area will enter the northern tier pipeline network and will be distributed across the upper Midwest. Medium- and low-Btu gases will also be in the market during this period, although they will probably be used for internal plant needs. This will minimize the exposure potential since these gaseous products will not require any transportation.

Some synthetic gases have different compositions than natural gas and may cause internal corrosion and stress-corrosion cracking in pipelines. Effects of impurities on the long-term degradation of some pipeline materials are unknown. Synthetic gaseous fuels also have different flammability and explosion limits that may require new techniques in the management of pipeline leaks. Gases with a high CO concentration are toxic and could present significant exposure problems.

In addition to their use in transportation and boiler applications, synfuel products will be used as feedstocks for industrial processes. These applications, although limited during this time period, present another avenue of exposure for which EPA must be prepared. The population exposed could include industrial plant personnel as well as the end users of the synthesis gas for the production of methanol and ammonia, each of which can be utilized as a finished product.

Although this period will be characterized by the emergence of many synfuel products, the main population potential will occur from crude shale oil transport by pipelines, product storage, and the combustion of these products.

A basic environmental concern with the transportation of liquid synfuels is the possibility of an accidental spill. A recent (1979) Department of Transportation analysis shows that of all the accidents resulting from pipelines carrying liquid petroleum products, the largest spillage occurs from LPG (58.6%) followed by crude oil (25.3%), with fuel oil (6.1%) and gasoline (4.5%) being the other major contributors.

As an example to illustrate the relative exposure of transporting petroleum products by pipeline in order to provide an awareness of the potential exposure in transporting shale oil, Table 7-8 presents a listing of oil pipeline accidents. Since existing pipelines will be used during this time period for transporting crude shale oil, these potential exposures and risks in each component of the carrier system must be considered by EPA.

POTENTIAL EXPOSURE: 1991–2000

This period is characterized by the large-scale entry into the market of direct and indirect liquefaction products and by-products for use primarily by the transportation, industrial, and utility sectors. Based on the National Goal Scenario, 1.0 MMBPD of coal

Table 7-8. Number of Oil Pipeline Carrier System Accidents

	79	78	77	76	75	74	73	72	71
Line pipe	207	194	177	169	185	203	215	238	264
Pumping station	20	30	32	11	24	13	23	31	14
Delivery point	6	4	4	4	5	5	5	3	2
Tank farm	5	15	12	14	30	22	21	24	11
Other	11	13	12	14	10	13	9	10	19
Total accidents	249	256	237	212	254	256	273	306	310

Source: Department of Transportation

liquids will be in the market by 2000, but may range up to 2.5 MMBPD. Utility and industrial boiler fuels produced by coal liquefaction processes will be most in demand in the Gulf Coast, Northeast, and Southern California regions, as shown in Fig. 7-8. These regions contain a significant portion of the U.S. population. The use of these fuels will also have some beneficial effects in areas that are sensitive to particulates and sulfur dioxide since these fuels have lower ash and sulfur contents. As more liquefaction capacity develops in the Appalachian and Interior regions, liquid fuels will more readily be used in the industrial areas of Indiana, Illinois, Ohio, and the upper Northwest. Shale oil products during the period may reach a level of 0.4 MMBPD under the National Goal Scenario and could reach as high as 0.9 MMBPD under the accelerated rate scenario. High-Btu gas under these two scenarios is estimated at 0.5 MMBPD and 1.0 MMBPD, respectively, by 2000. As coal gasification technology develops, it is likely that a key area for gasification will eventually be Appalachia, with SNG entering the existing pipelines and being distributed along the east coast to both industrial and residential users.

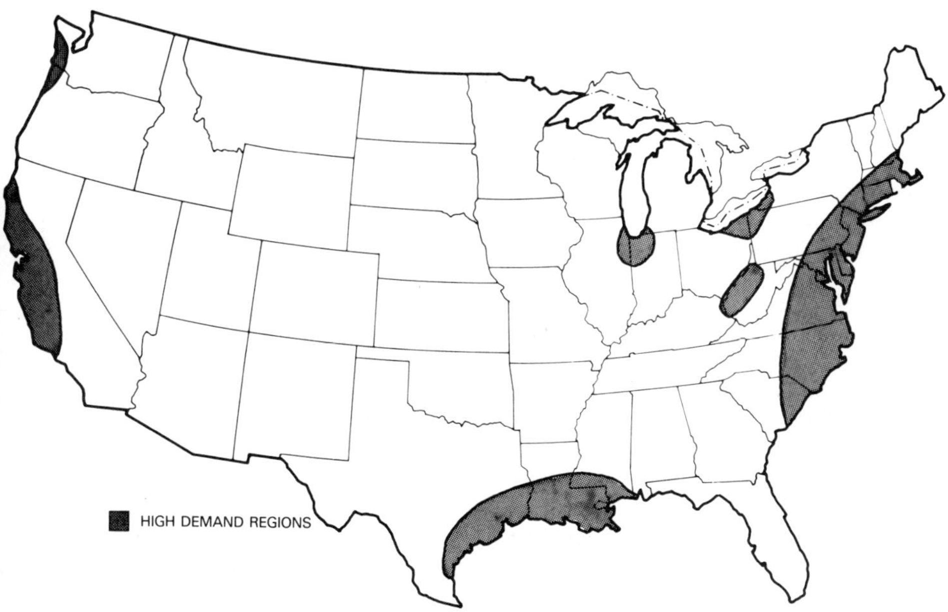

Fig. 7-8. Potential regions for synfuel demand for industrial fuel applications.

This period is also characterized by increased use of coal liquid products for chemical feedstocks and in the housing and commercial sectors for space heating and hot water supply. Naphthas produced by liquefaction processes are likely to be used by the petrochemical industries for manufacturing solvents, pesticides, and chemical intermediates. Residues from coal liquefaction processes may be used to manufacture several types of lubricants with a wide variety of applications. This market penetration significantly increases exposure potential as there is virtually no segment of the population that would be excluded from the use of synfuel products and by-products.

As the synfuel industry develops during this period, transportation modes other than pipelines will be utilized. Although there are associated risks, pipelines are considered to present less risk than other modes such as railroads, trucks, and tankers. As these modes are currently used for a wide variety of petrochemical products, it is expected that they will also be used as synfuels penetrate the market, thereby presenting another environmental concern.

Table 7-9 presents an estimate of the range of synfuel products to be shipped by various transportation modes beginning in the 1990s. Nearer to the year 2000, the relative amounts of products transported between the modes may vary. The majority of the synfuel products as well as crude shale oil will be transported by pipelines, which presents the least amount of exposure potential. On the other hand, railroads, which have a high accident potential, will transport the least amount of products. In order to supply the high demand regions (reference Fig. 7-8) the transport distribution networks may develop as illustrated in Fig. 7-9. The distribution system indicates that the crude shale oil, refined products, SNG, and coal liquids will each be transported across areas of high population density and industrial concentration, mostly in the eastern United States. A market for 2.2 MMBPD of synfuel products and by-products by 1982 under the National Goal Scenario indicates the magnitude of the problem for which EPA must prepare.

The transportation modes that will be utilized by the synfuel industry and which pose a greater accident potential than pipeline transport are railroads, trucks, and tankers. Railroads will be used primarily for the transportation of naphthas, which in 1992 are estimated to range from 0.04 to 0.07 MMBPD. This mode of transportation presents a high degree of accident risk owing to the poor condition of the nation's rail system. Derailments, grade crossing accidents, and collisions between trains pose potential risks to the transportation of any hazardous or toxic substance. Tank car accidents with hazardous materials are shown in Table 7-10, providing another example of potential risks associated with transporting synfuel products.

The use of tanker trucks will be extensive in transporting coal liquids and refined shale oil products. In 1992 under the National Goal Scenario, approximately 0.4 MMBPD of gasoline from coal liquefaction may be transported by truck, and up to 0.5 MMBPD under the accelerated rate scenario. Other products using this mode are middle distillates and naphthas. Potential exposure to the general population is high with tanker trucks since much of these products will be delivered to urban areas where trucks will face the normal amount of traffic accidents in congested areas. In addition, exposures to the products will occur in loading and unloading of trucks at storage terminals and service stations. Owing to vapor recovery requirements mandated by state implementation plans, evaporative emissions of volatile compounds are gradually being controlled, but pollution control systems must be improved to further reduce emissions. Accidents or defective emission control systems provide the chief potential for release of synfuel products by truck transport.

Table 7-9. Range of Synfuels Distributed by Mode of Transportation in 1990s

Synfuel Product	Pipeline	Rail	Truck	Tanker or Barge
High-Btu gas (MMSCFD)[a]	1900 – 4800	0	0	0
Medium-Btu gas (MMSCFD)	3800 – 5600	0	0	0
Liquefaction products (MMBPD)[b]				
Heavy fuel oils and middle distillates	0.025 – 0.080	0	0.014 – 0.044	0.007 – 0.022
Gasoline	0	0	0.072 – 0.520	0.018 – 0.130
Naphtha	0	0.042 – 0.067	0.005 – 0.008	0.005 – 0.008
LPG	0	0	0.0 – 0.018	0.0 – 0.005
Crude shale oil (MMBPD)	0.389 – 0.750	0	0	0
Refined shale (MMBPD)				
Gasoline	0.040 – 0.076	0	0	0.026 – 0.051
Jet fuel	0.047 – 0.090	0	0	0.031 – 0.061
Diesel oil	0.084 – 0.162	0	0.084 – 0.162	0.042 – 0.162
Residual oil	0.025 – 0.047	0	0.004 – 0.007	0.007 – 0.014
Accident risk	Low	High	High	Moderate

[a]MMSCFD = million standard cubic feet per day.
[b]MMBPD = million barrels per day.

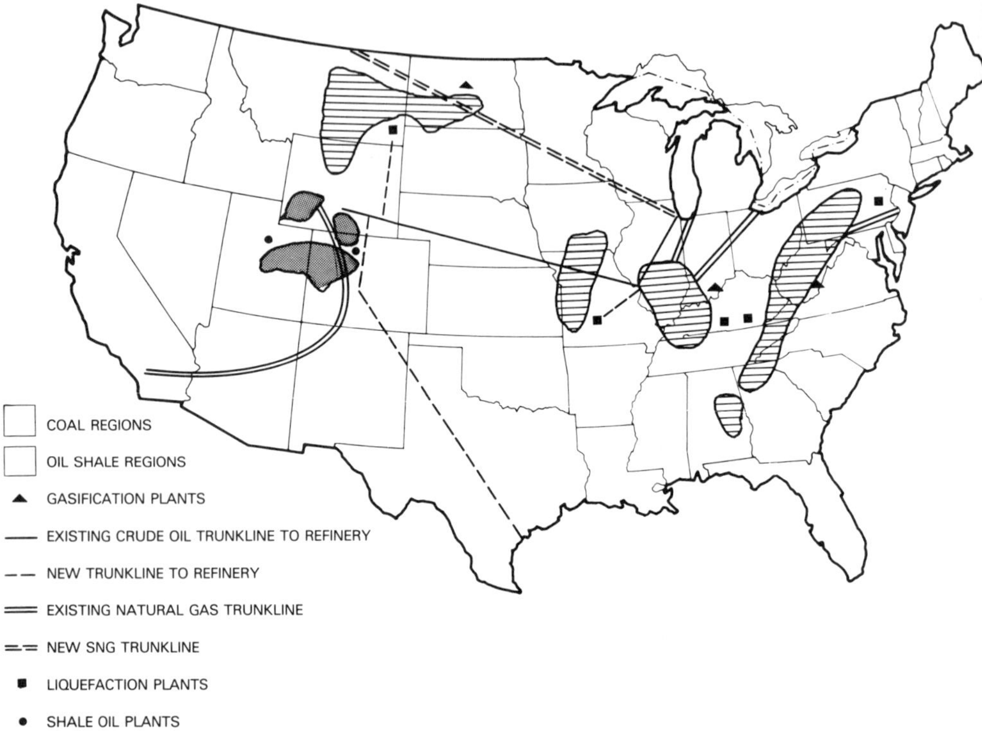

Fig. 7-9. Synfuel resources and distribution system in the 1990s.

Tankers and barges will also be used for the transportation of the refined shale oil and coal liquid products and could be used extensively if the markets are accessible to the Gulf Coast. Under the National Goal Scenario, 0.1 MMBPD each of gasoline and diesel oil may be transported by these modes in 1992. Other products to a lesser extent are naphtha, LPG, jet fuel, and residual oil. A significant amount of petrochemical products currently move along the Mississippi River to northern markets. The major emission source for this operation involves loading and unloading; however, the accident rate is less than that of surface transportation mode. As with truck loading, increased emission controls are being initiated for ship and barge loading, which will significantly decrease evaporative emissions by the time the synfuel industry is developed. Improvements are also being made to reduce spills of petrochemical products into waterways. Reduction of

Table 7-10. Railroad Tank Car Accidents with Hazardous Materials

	1979	1978
Total Accidents	937	1014
Accidents involving atmospheric release	165	228

Source: Federal Railway Administration

accidental spills and prevention of intentional releases are currently under regulation by the Coast Guard and EPA.

In addition to transportation and handling, the storage of synfuel products and by-products may pose potential environmental problems. These problems may occur primarily with refined shale oil and coal liquids. As with other petroleum products they will be stored at bulk storage terminals until used. By 1992, a total of 1.4 MMBPD of synthetic liquids will be produced under the National Goal Scenario, and ranging up to 1.7 MMBPD under the accelerated rate scenario. Exposures to these products at the terminals may occur during the loading and unloading operations, as well as breathing losses from the tanks during product storage. The potential for exposure depends on the volatility of the products and the frequency of loading operations. Since storage facilities are located at refineries, utility and industrial plants, airports, and numerous other facilities, exposure potential is significant. Concern over the uncertainties of the constituents of synfuels may lead to storage procedures for these products that are more rigid, and new storage vessels or containers for liquids may be required under stringent specifications. Some emissions may also occur from low-level leakage.

As with other major control requirements for loading and unloading petroleum products, vapor recovery techniques for bulk storage facilities are being improved, primarily by the use of floating roof tanks. Synfuels such as SRC-II liquids have pour pressures similar to No. 6 fuel oil, which has very low evaporative emissions and working losses compared to gasoline. Fugitive emissions of synfuels will always be present as they are with other petroleum products. However, there will be new control systems developed and emissions will be reduced over the next several years through improvements in emissions control procedures in transportation, handling, and storage operations. Only after thorough toxicity testing of synfuel products and by-products can an assessment be made of whether synfuel transportation, handling, and storage will pose environmental, health, and safety problems greater than those experienced in the petroleum refining and chemical manufacturing industries.

Chapter 8

Environmental Considerations

Natural Resources for Synthetic Fuels

This chapter provides basic information on the natural resources required for synthetic fuel production and briefly describes some of the possible geologic, hydrologic, and environmental considerations. No attempt is made to address the social, economic, and secondary environmental effects that would result from the large-scale production of synthetic fuels. In addition to the natural resource environmental information, the balance of this chapter describes the generic technical process environmental considerations.

Synthetic fuels can be made from coal, oil shale, tar sands, and various types of biomass such as wood, grains, and garbage. However, this chapter focuses on the potential production of synthetic fuels from coal and oil shale.

Mineral chemical composition is an important factor in determining the suitability of different coals for making synthetic fuels. In general, the lower-quality lignitic, subbituminous, and bituminous coals are more easily and efficiently converted to synthetic fuels than anthracites. Furthermore, for a given type of coal, the conversion efficiency may vary considerably, depending on the detailed mineralogy.

Oil shale is rich in kerogen, an organic substance similar to the materials which have produced naturally occurring oil and gas. Oil shale must be heated to high temperatures to convert its kerogen to syncrude (synthetic crude) oil.

COAL RESOURCES FOR SYNTHETIC FUELS

Figure 8-1 shows the location of the nation's coal resources that might be considered for synthetic fuel development. These resources include coal deposits to a depth of 3000 ft, in beds thicker than 28 inches for anthracite and bituminous coal and thicker than 5 ft for subbituminous coal and lignite. On this basis, the potential coal synthetic fuel resource of the nation is 1.1 trillion tons. Most (about 90%) of these resources are less than 1000 ft below the surface.

Figure 8-2 shows the state-by-state distribution of the 1.1-trillion-ton resource, 60% of which is concentrated in Montana, Illinois, Wyoming, Alaska, and North Dakota, and an additional 23% of which is located in Colorado, West Virginia, Pennsylvania, and Kentucky.

Of the total tonnage, 1% is anthracite, 48% is bituminous, 34% is subbituminous, and 17% is lignite. Over 80% of the anthracite and bituminous coal is in the eastern United States, whereas 99% of the subbituminous coal and lignite occurs in the western states. The heat content differs significantly for these four kinds of coal. This must be considered

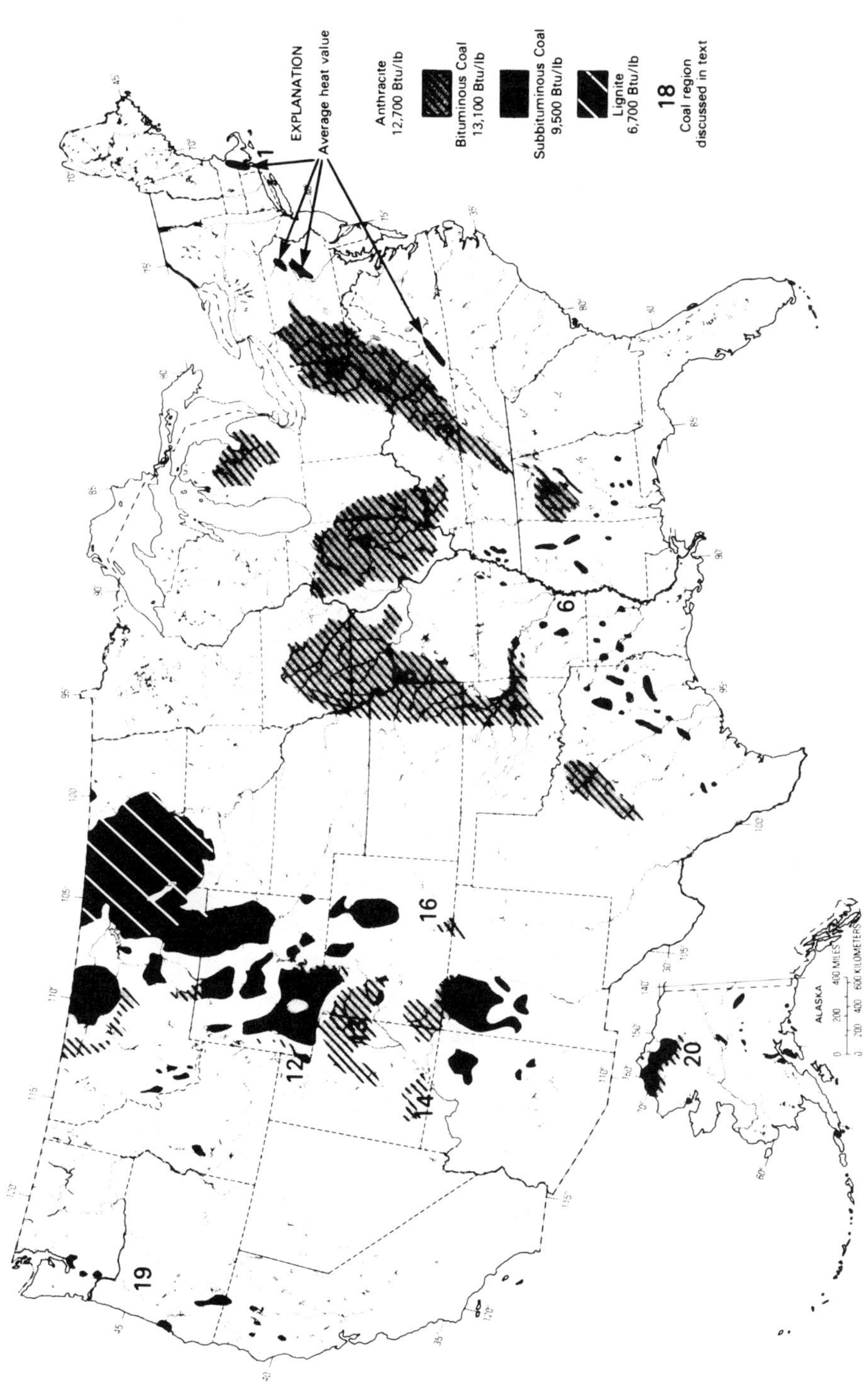

Fig. 8-1. Coal fields of the United States.

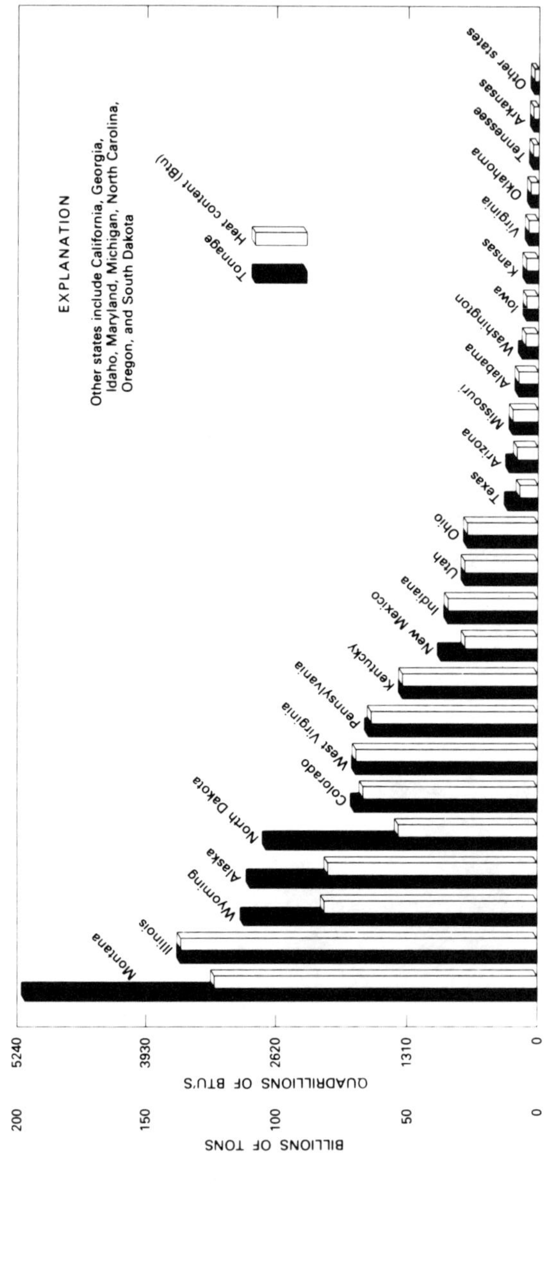

Fig. 8-2. Tonnage and heat content of coal resources having potential for synthetic fuel development.

when evaluating existing coal tonnages. For example, the total Montana resource of 198 billion tons is 45% larger than the 137 billion in Illinois. However, the total Btu content for the subbituminous coal and lignite in Montana is almost 10% less than that of the Illinois bituminous coal resource. Similarly, North Dakota lignite ranks fifth in tonnage but ninth in Btu content. Nearly 2 tons of North Dakota lignite are required to produce the Btu's available from 1 ton of Kentucky bituminous coal.

Fifty-four percent of the U.S. coal reserves lie west of the Mississippi River. In aggregate, as shown in Fig. 8-3, ownership of this western coal is 27% federal under federal lands, 34% federal under privately owned surface, 26% privately owned, 8% Indian owned, and 5% state owned. Complex patterns of coal ownership occur in many areas of the West. Such patterns have resulted from state and railway land grant procedures, coupled with laws which provided for severance of the surface ownership rights while reserving to the federal government the coal and other mineral rights. State ownership of coal typically consists of two or four sections per township.

In the East, most coal is privately owned. A few states, most notably Alabama, contain a relatively large percentage of federal coal under privately owned land. In general, federal coal ownership in the East consists of small tracts, from 40 to 320 acres, interspersed within private holdings.

GEOLOGIC CONSTRAINTS

Large-scale use of coal to make synthetic fuels will require expansion of coal production, including development of new deposits. This expansion faces a number of potential geologic constraints that differ from region to region. The constraints for mining coal for synthetic fuel production are very similar to those for ordinary coal mining. Figure 8-4 illustrates the potential geologic constraints to synthetic fuel development in principal coal regions. The key constraints are summarized below:

- In adequate coal bodies—a projected standard-sized synthetic fuel plant will be designed to produce 50,000 barrels of oil or 250 million standard cubic feet of gas per day. Depending on the type of coal and the type of synthetic fuel conversion process, such a plant will require about 20,000–40,000 tons of coal per day. Under average conditions of conventional mining and operation, this means that about 12 million–24 million tons of coal reserves will be needed to supply each plant for 1 year. Over the estimated 30-year life span of a plant, between 360 million and 720 million tons of coal reserves will be required. Such large tonnages in single blocks are unavailable in certain coal regions.
- Thin or discontinuous coal beds—In some coal regions, a substantial part of identified resources is in beds too thin to meet the present requirements of large-scale mechanized mining. This includes layers of bituminous coal and anthracite less than 28 in. thick and subbituminous coal and lignite less than 5 ft thick.
- Excessive depth of coal—Resources in many coal regions extend to depths greater than 3000 ft. These resources are beyond the range of current domestic mining practices.
- Complex structure of coal-bearing rocks—In many coal regions, the coal-bearing beds have been subjected locally to severe folding and faulting. The mining of highly deformed coal deposits will require detailed geologic investigations and, perhaps, special mining techniques to avoid high costs and delayed

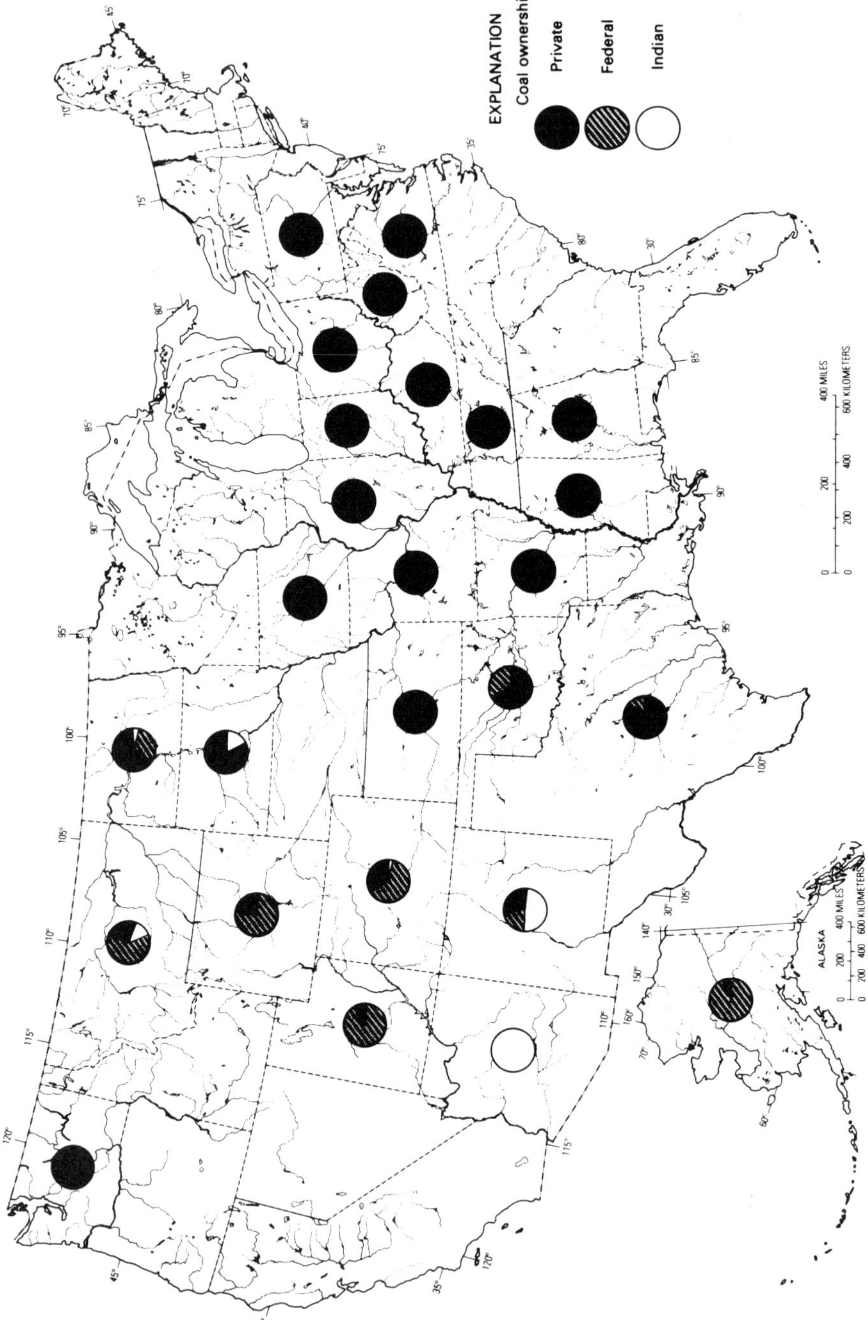

Fig. 8-3. Principal ownership of major coal resources.

EXPLANATION

- Regional constraints (solid black)
- Major local constraints (hatched)
- Insufficient data (dotted)

	Naragansett Basin (1)	Northern Appalachian (2)	Southern Appalachian (3)	Eastern (4)	Western (5)	Texas-Mississippi (6)	Fort Union (7)	Powder River (8)	North Central (9)	Bighorn Basin (10)	Wind River (11)	Hams Fork (12)	Uinta (13)	Southwestern Utah (14)	San Juan River (15)	Raton Mesa (16)	Denver (17)	Green River (18)	Pacific Coast (19)	Alaska (20)
PHYSICAL CONSTRAINTS TO MINING																				
Inadequate coal bodies	Regional				Major local	Major local			Regional	Major local	Regional						Regional		Regional	
Thin or discontinuous coal beds	Insufficient data		Major local		Major local				Major local	Major local	Regional		Major local		Major local	Major local	Major local		Major local	
Excessive depth of coal	Insufficient data	Major local	Major local			Major local		Major local		Major local	Major local	Major local	Major local		Major local	Major local		Major local	Major local	Major local
Complex structure of coal-bearing rocks	Regional		Major local						Major local	Major local	Major local	Major local	Major local		Major local	Major local	Major local	Major local	Major local	Major local
Thick glacial deposits or permafrost	Regional	Major local		Major local	Major local		Major local	Major local	Major local	Insufficient data										Regional
PHYSICAL AND CHEMICAL IMPACTS OF MINING																				
Aquifer disruption	Major local	Major local	Major local	Major local	Insufficient data	Insufficient data	Major local	Regional	Insufficient data	Insufficient data	Insufficient data	Insufficient data	Major local	Insufficient data	Insufficient data		Major local	Insufficient data	Insufficient data	
Underground mining hazards	Insufficient data	Major local	Major local	Major local	Major local		Regional	Regional	Major local	Insufficient data	Insufficient data		Major local	Major local	Major local	Major local	Major local	Major local	Major local	Major local
Post-mining hazards	Major local	Major local	Major local	Major local	Major local		Major local	Major local	Major local	Insufficient data	Insufficient data		Major local	Major local	Major local	Major local	Major local	Major local	Major local	Major local
Acid mine drainage	Insufficient data	Regional		Regional	Insufficient data	Insufficient data			Insufficient data	Insufficient data	Insufficient data	Insufficient data	Insufficient data	Insufficient data	Insufficient data	Insufficient data	Insufficient data	Insufficient data	Insufficient data	Insufficient data

Fig. 8-4. Potential geologic constraints to synthetic fuel development in principal coal regions.

production schedules that would adversely affect large scale synthetic fuel production.

- Thick glacial deposits or permafrost—In many regions, coal beds are partly covered by thick glacial deposits or glacially derived silts. Such deposits interfere with surface mining, because of increased overburden, and with underground mining, because of their commonly unconsolidated permeable character. Development of coal resources will be locally constrained by such deposits in parts of the Eastern (4),[1] Western (5), Northern Appalachian (2), Fort Union (7), Powder River (8), and North Central (9) regions. Moreover, development of coal deposits throughout the Alaskan (20) region could well be constrained by overlying thick permafrost.
- Aquifer disruption—Coal beds are valuable aquifers in several regions. Mining would remove part of the aquifer and could disrupt some supplies and degrade water quality.
- Underground mining hazards—Underground mining hazards are common in many regions because of the poor physical characteristics of rocks surrounding the coal and the occurrence of exceptionally thick (20−150 ft) coal beds. Such hazards may present regional problems for underground extraction of coal in poorly consolidated rocks of the Fort Union (7) and Powder River (8) regions. In fact, many mining experts are of the opinion that the thick beds of these regions cannot be deep mined with high recovery rates unless new, innovative mining techniques are developed. However, thick coal beds in the Fort Union (7) and Powder River (8) regions may be suitable for *in situ* gasification at depths below about 500 ft where fractures and joints are minimal.
- Postmining hazards—The principal postmining hazard associated with coal mining is the subsidence of beds that overlie mines. Subsidence is common locally in many coal areas, but is currently a major regional problem only in the Pennsylvania anthracite fields of the Northern Appalachian region (2). However, it is likely that widespread subsidence will occur in any coal region if thick beds are extracted at high recovery rates.
- Acid mine drainage—Thousands of miles of streams are degraded by acid mine water in the Northern Appalachian (2) and Eastern coal regions (4). This pollution results whenever iron sulfide minerals in coal beds and spoil areas are exposed to oxygen in flowing water. The minerals are oxidized, forming sulfuric acid. Almost all coal beds contain iron sulfide minerals. However, the contents differ from bed to bed, as does the groundwater flow. Therefore, different regions have different potentials for acid mine drainage.
- Coal quality—The quality of coal differs from deposit to deposit and from bed to bed. It is known that certain coal-quality parameters greatly affect the utility of coal for various synthetic processes. The important parameters include: content of moisture and ash, coking tendencies, heat content, the amounts and forms of sulfur, and the major trace element chemistry of the coal and coal ash. Knowledge of such properties in coal deposits is needed both to prevent adverse impacts and to facilitate synthetic fuel production. However, at the present time

[1] Refer to Fig. 8-4.

there are insufficient data on coal-quality parameters for the coal in the regions to warrant their inclusion in the matrix. Figure 8-4 shows which of the nine major constraints to synthetic fuel production occur in the nation's 20 principal coal regions. Most constraints are local in extent and will not greatly affect development of the resources. However, coal production could be seriously affected where the constraints are regional in scope or where several local constraints occur in combination. Lack of large regional blocks of coal or the presence of region-wide permafrost will both likely preclude synthetic fuel development.

Based on evaluation of geologic constraints, Fig. 8-5 shows 10 coal regions that have the best potential for synthetic fuel development. The following briefly describes the development potential of each region without attempting to rank their relative importance. It is also important to note that some coal regions excluded from this listing might sometime be deemed useful resources for synthetic fuel production. As prices of petroleum products increase, changes occur in the definition of what constitutes a usable coal resource.

Northern Appalachian Region

The coal resources of the Northern Appalachian region (2) are adequate to support several synthetic fuel plants. The major consideration is the difficulty in controlling acid mine drainage in an area already seriously affected. Additional constraints are: (1) Much of the better coal is of a grade suitable for metallurgical use and is already committed to the steel or power industries; and (2) the coal necessary to support a 30-year life for each synthetic fuel facility will require a large number of surface mines or several large underground mines.

Southern Appalachian Region

All statements given above for region 2 also apply to the Southern Appalachian region (3), except the reference to acid mine drainage.

Eastern Region

The Eastern region (4) contains adequate resources for supporting several synthetic facilities. Acid mine drainage, however, is already a serious problem in many places. Another constraint is that the coal areas of the region include some of the best farmland in the nation. Furthermore, much coal is already committed to the metallurgical and power industries. Most coal extraction for synthetic fuels probably would be from deep mines.

Fort Union and Powder River Regions

The Fort Union (7) and Powder River (8) regions contain large resources of subbituminous coal and lignite in thick beds that are sufficient to support several synthetic facilities. Aquifer disruption could be an important constraint in the Powder River Basin and a local one in the Fort Union region. Thick coal beds in both regions would be difficult to mine underground. Because of large coal resources, both regions are attractive for synthetic fuel development.

Hams Fork Region

Although small in area, the Hams Fork region (12) contains sufficient resources for potential synthetic fuel development. The principal drawback is that coal resources are largely committed. Coal can be both surface and underground mined.

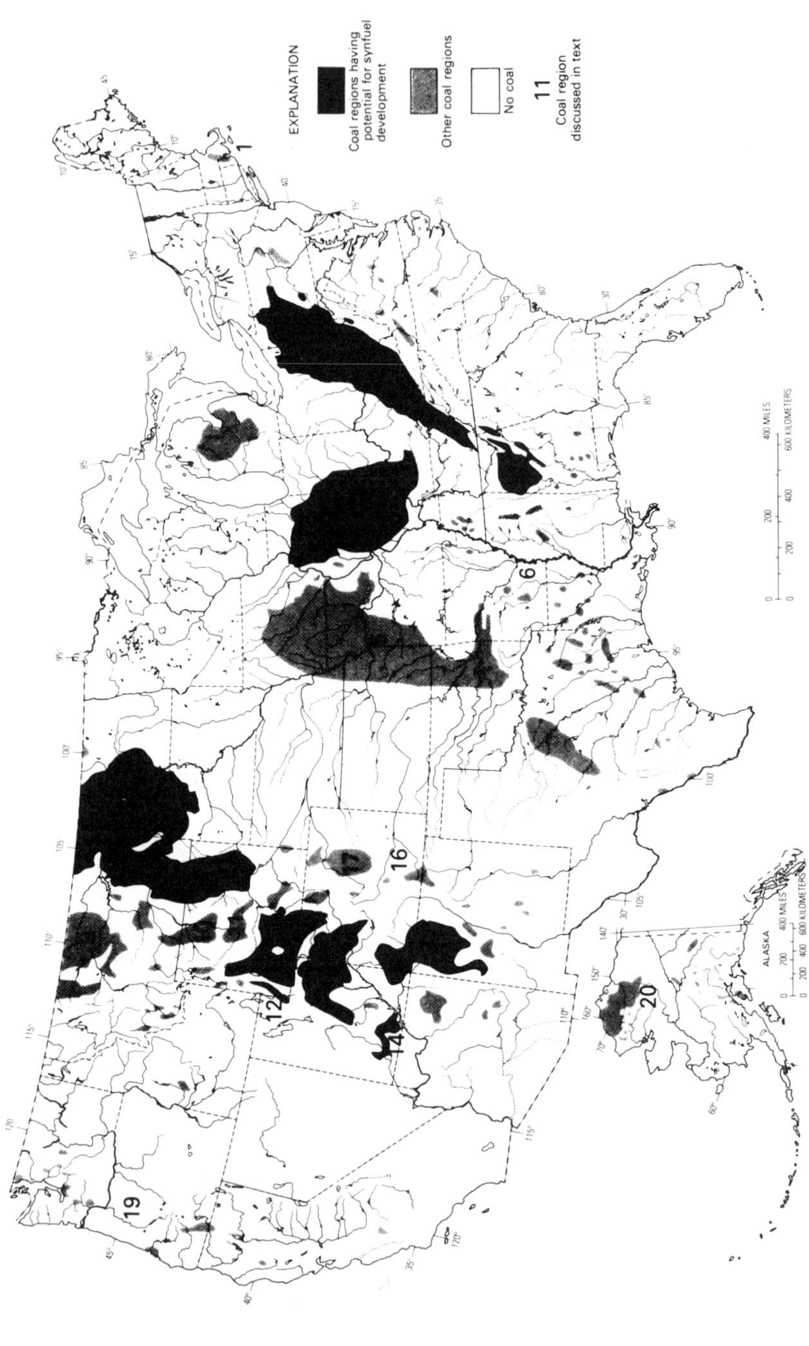

Fig. 8-5. Coal resources most likely to be used for synthetic fuel production.

Uinta, Southwestern Utah, and Green River Regions

The large Uinta (13), Southwestern Utah (14), and Green River (18) regions contain sufficient coal resources for a number of synthetic fuel facilities, but some of the coal is already committed. A principal constraint to the rapid siting of synthetic fuel facilities is a lack of site-specific geologic knowledge. The coal can be both surface and underground mined.

San Juan River Region

The San Juan River region (15) contains sufficient coal resources to supply several synthetic fuel facilities. Part of the coal is already committed to powerplants. The coal can be both surface or underground mined.

This assessment highlights the regions where adequate coal resources exist and can be efficiently mined. The first requisite of siting a synthetic fuel plant is that it must be within economic reach of a hydrocarbon resource. However, many additional factors will affect the siting of synthetic plants. For example, adequate water resources must be obtainable. In addition, numerous federal, state, and local guidelines must be met.

OIL SHALE RESOURCES

Large areas of the United States contain oil shale deposits. The richest deposits, however, are found in the Green River formation in Colorado, Utah, and Wyoming.

Colorado's Piceance Creek Basin contains 85% of this western high-grade deposit that is, in reality, more similar in nature to limestone than to shale. The kerogen-rich shale of this basin contains an equivalent of about 10 times more oil than has been consumed nationally to date. Moreover, in 1974, total oil and gas consumption in the United States was equivalent in Btu content to 12½ billion barrels of oil. At this rate of use, and assuming 100% recovery of all the shale oil, the Piceance Creek Basin deposits would supply the needs of the United States for 100 years.

The comparisons are presented for purposes of illustrating the magnitude of energy resources available in the Green River oil shale. However, it is important to note that under today's technology only part of the total oil resource is economically recoverable.

Figure 8-6 shows (1) that part of the Green River Formation containing oil shale deposits; (2) that part of the deposits, 15 ft or thicker, which will yield 15 gallons of oil or more per ton of rock; and (3) that part, 10 ft or thicker, which will yield 25 or more gallons per ton.

Deposits in the 15 gallon or more category underlie 10,000 square miles and contain an estimated 2 trillion barrels (1 barrel equals 42 gallons) of oil. The richest deposits (25 gallons or more) have an extent of about 5000 square miles and account for 730 billion of the 2 trillion barrels.

In the heart of the Piceance Creek Basin, the shale occurs in seven rich zones that total 2000 ft in thickness. The Mahogany Zone is the most uniformly rich and extensive area. The Mahogany Zone also extends into Utah and, in total, underlies more than 6000 square miles and contains about 250 billion barrels of oil.

The federal government holds the mineral rights to most of the western oil shale and also owns about 70% of the associated land surface, as shown in Fig. 8-7. The remaining surface ownership is mostly private, accounting for 21% of the oil shale land in Colorado, 9% in Utah, and 24% in Wyoming. In Utah, 8% of the oil shale land and mineral rights are Indian owned, whereas in both Wyoming and Utah about 6% is state land. It must be

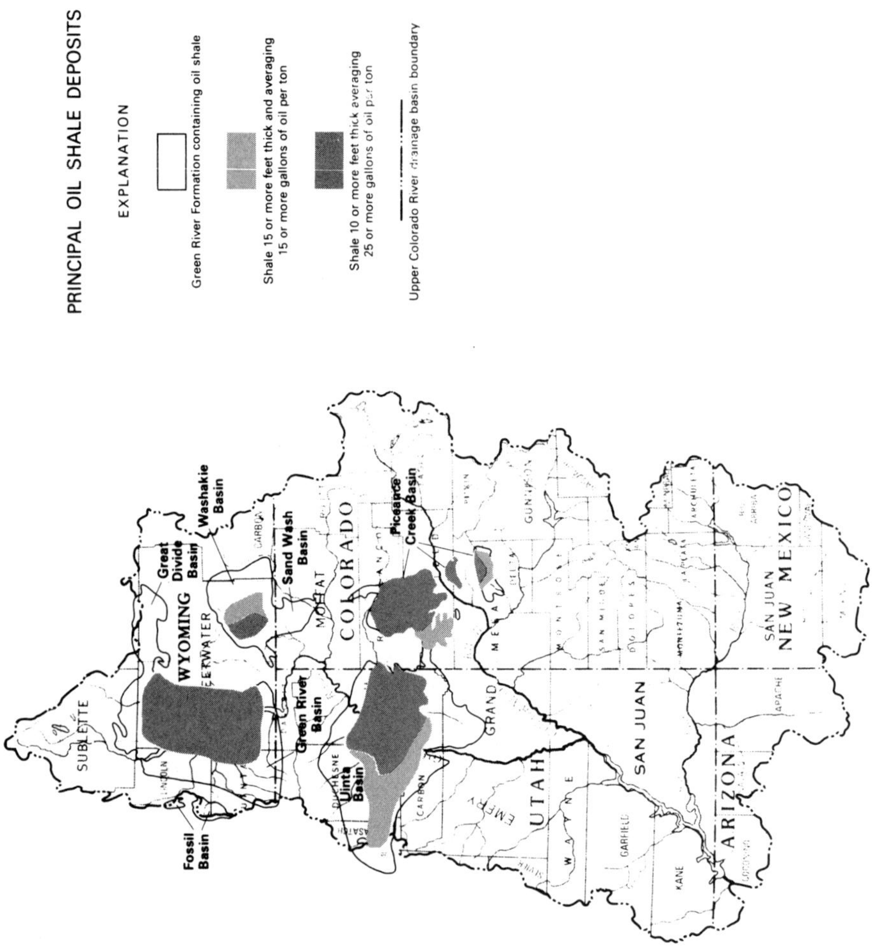

Fig. 8-6. Principal oil shale deposits.

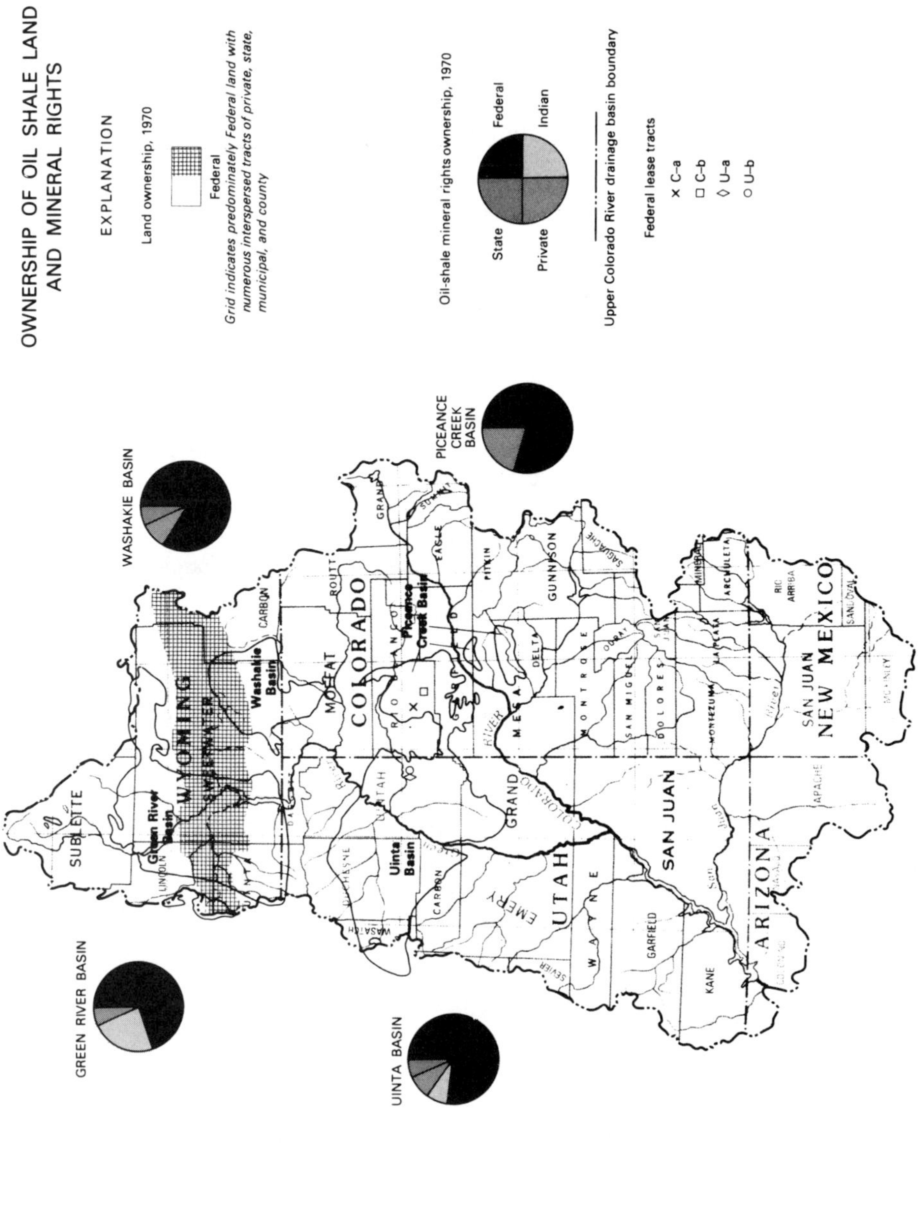

Fig. 8-7. Ownership of oil shale land and mineral rights.

noted, however, that Utah has laid claim to 150,000 acres of federal oil shale land including the area comprising oil shale tracts U-a and U-b.

To date, four federal tracts have been leased, two each in Utah and Colorado. In addition, several experimental or demonstration projects are in progress on private and state land. Also, about 20 other oil shale development projects are now being contemplated, mostly on private land.

GEOLOGIC CONSTRAINTS FOR OIL SHALE

Several geologic conditions may constrain development of a shale oil industry. Some of these conditions are local, whereas others occur over entire regions. A general problem with all the deposits is the limestone nature of the rock. This property of the rock increases the amount of energy required for high-temperature retorting to recover the oil.

Initial development of oil shale will probably be limited to areas having large amounts of the high-grade resources. This immediately restricts the prime locations for development to far less than the area of the Green River Formation that contains oil shale.

Another constraint is that the central areas of the oil shale basins have a thick rock mantle overlying the resource deposits. In some areas, this may prevent development. In other areas, it will add considerably to the expense of extracting the oil and will severely decrease the recoverable resource.

Saline minerals occur in the oil shale in the depositional centers of the Green River Basin, Wyoming; the Piceance Creek Basin, Colorado; and to a lesser extent, the Uinta Basin, Utah. In Wyoming and Colorado, thick beds of sodium chloride and highly soluble sodium carbonate minerals such as nahcolite and trona are interspersed in hundreds of feet of oil shale. In addition, analcime, a sodium aluminum silicate, occurs in all three states; and dawsonite, a sodium aluminum carbonate, occurs in the Colorado shale. At a given location, the presence of any of these associated minerals adds to the energy requirements for retorting the oil shale at high temperatures.

In the Piceance Creek Basin, groundwater has leached much of the soluble saline minerals from the beds throughout several hundred square miles. Locally, entire beds have been leached and a honey-combed framework of permeable material has formed. The leached beds now constitute an aquifer hundreds of feet thick in the middle of the basin. This aquifer is estimated to contain millions of acre-feet of water, much of which is fresh. In some areas, however, the water is saline, and in certain locations, it contains appreciable concentrations of fluoride, boron, and hydrogen sulfide. If such waters are not disposed of properly, the constituents could be harmful to the environment.

Two prominent fracture systems and several of lesser magnitude occur throughout the oil shale area. Faults locally displace the richer oil shale in both the Uinta and Piceance Creek Basins. The fractures and faults could cause problems in underground mining operations and could also interfere with *in situ* recovery of oil.

The initial development of the oil shale resource will probably be limited to areas having amounts of high-grade shale sufficient for sustaining production of 50,000 barrels of oil per day for 30 years. Based on present concepts of technology, the suitable shale deposits will be at least 10 ft thick and will average at least 25 gallons of oil per ton. The location of such deposits is shown in Fig. 8-8. It is important to note that federal leases C-a and C-b in the Piceance Creek Basin, Colorado, as well as leases U-a and U-b in Utah, are within such areas.

Within the areas identified in Fig. 8-8, the ideal deposits for synthetic fuel development would (1) be above the water table, (2) have thin overburden, (3) be unfractured,

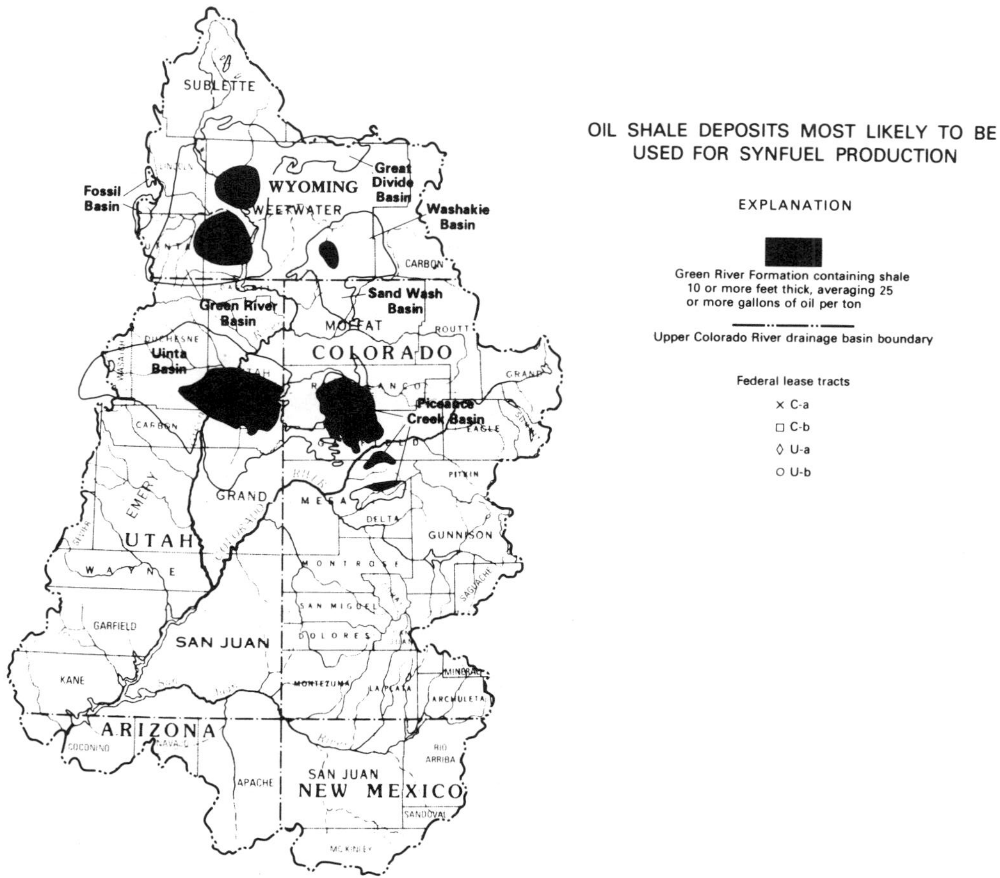

Fig. 8-8. Oil shale deposits most likely to be used for synthetic fuel production.

and (4) be lacking in unwanted accessory minerals and harmful trace elements. No such areas exist; therefore, evaluation and trade-offs will be necessary. The following describes what the trade-offs might be in two areas.

(1) In the Piceance Creek Basin, much of the area with the thickest and richest shale also contains a thick aquifer, the thickest zone of saline minerals, and the greatest amount of overburden. The thick overburden is definitely a considerable constraint. However, certain of the saline minerals might be marketable. Nahcolite, the naturally occurring sodium bicarbonate mineral, has potential value in scrubber technology for industrial stack gases. In addition, the dawsonite may have by-product value as a source of soda ash and elemental aluminum.

The associated groundwater will have to be pumped during either mining or *in situ* retorting of the shale. The groundwater pumping may be a mixed blessing. Over much of the Piceance Creek Basin the quality of the groundwater is quite good, with low quantities of dissolved solids and low levels of potentially harmful substances. Such water represents a resource for processing the shale, especially for cooling water and for moisturizing spent shale. However, in certain areas, the groundwater is highly saline, with dissolved solids in excess of 10,000 parts per million (about one-third that of sea water). In these areas and where potentially toxic substances occur, the water will be a major treatment or waste disposal problem rather than a resource. However, such water could possibly be disposed of by injecting it back into the aquifer.

(2) Local faulted areas in the Piceance Creek Basin would present problems for utilizing conventional mining and modified *in situ* recovery techniques. However, in these areas, the oil shale is moderately thick and rich, and the overburden has maximum depths of a few hundred feet. Such deposits are suitable for surface mining because of the fairly thin overburden coupled with a ratio of overburden depth to ore thickness less than 1:1. A surface-mining approach would essentially eliminate the likely problems associated with underground mining and *in situ* conversion. It also would minimize mining problems associated with groundwater. An added bonus would be a 90% or more recovery of the oil resource in contrast to much lower recoveries attainable with other extraction methods.

OTHER SYNTHETIC FUEL RESOURCES

Other potential synthetic fuel resources, shown in Fig. 8-9, include black Devonian shales, tar sands, and oil shale of the Phosphoria Formation in Montana.

Devonian black and dark brown shales rich in organic detritus are known by many names in the Appalachian, Illinois, and Michigan Basins. In contrast to western oil shales, the eastern Devonian shales contain more natural gas, but will yield less oil per unit volume of rock. It is important to note that the gas in all Devonian shale is locked tightly in the rock, and, consequently, only small amounts are economically recoverable by present technologies.

About 160,000 square miles of the Appalachian Plateau are underlain by Devonian black shales ranging in thickness from less than 10 ft in southern Tennessee to an aggregate of more than 1400 ft in northeastern Pennsylvania. The volume of the shales exceeds 12,600 cubic miles. The oil content of the shales is highest, 15–20 gallons per ton, in central Kentucky and Tennessee. It decreases to the northeast and is less than 1 gallon per ton in outcrops along the east side of the Allegheny Plateau of central Pennsylvania. Estimates of the gas in the Appalachian Devonian shale are as great as 3 million billion standard cubic feet. In the Michigan Basin, the Devonian shale ranges in thickness from 250 to 600 ft, and the oil content ranges from 1 to 15 gallons per ton. Scant

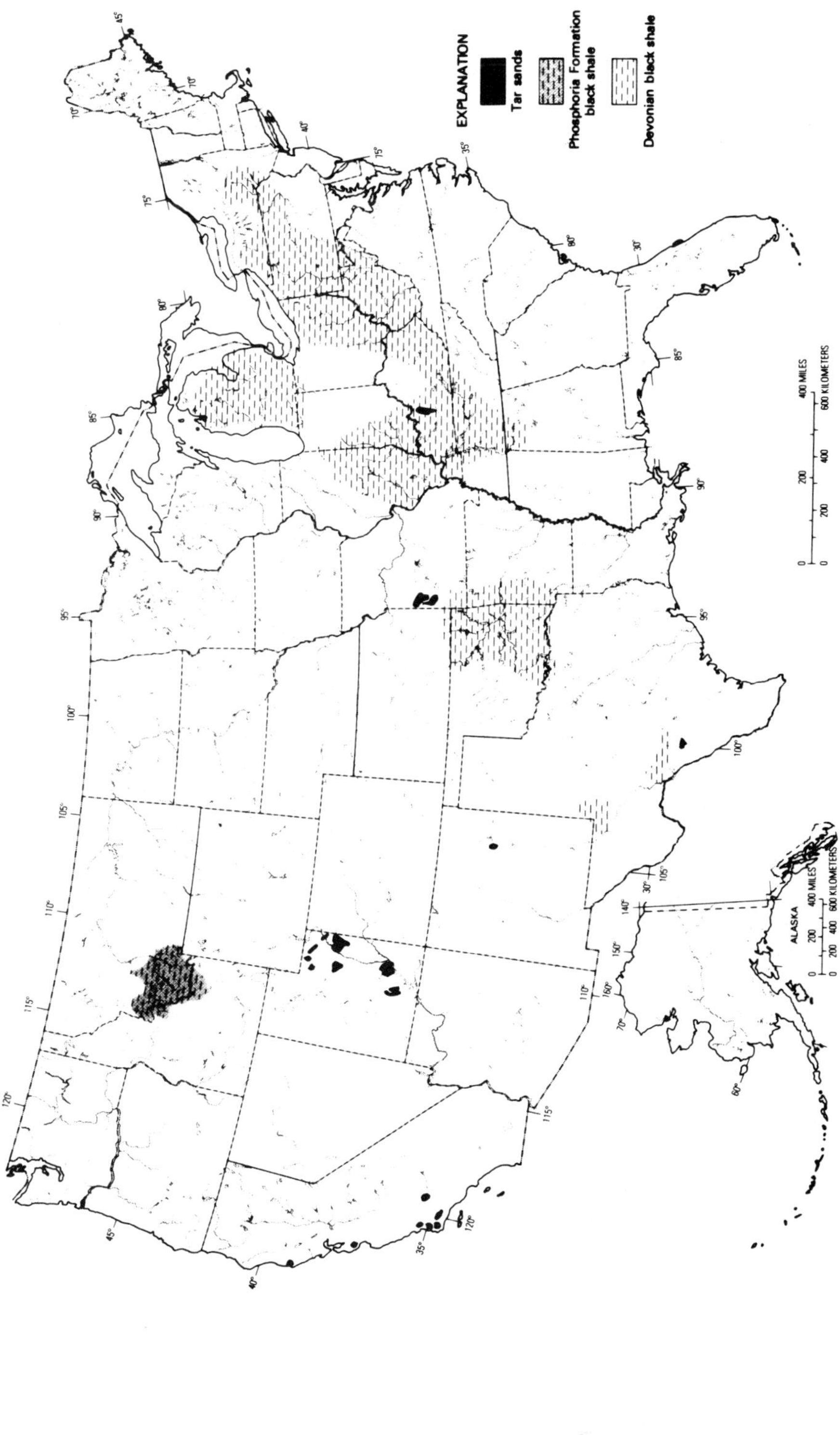

Fig. 8-9. Other potential synthetic fuel resources.

data suggest that the gas per unit volume of shale may closely approximate the value for the Appalachian Basin. In the Illinois Basin, the Devonian shales range in thickness from 150 to 450 ft, and the oil content is as much as 20 gallons per ton.

Tar sands are sandstones with pore spaces containing viscous to solid petroleum, which cannot be recovered by conventional methods. World tar sand resources are estimated to contain more than 2 trillion barrels of oil, most of it in Canada and Venezuela. The United States has an estimated resource of 29 billion barrels of tar sand oil, most of it in Utah. At present there is no commercial production of tar sand oil in the United States, although oil is produced commercially from similar deposits in the Saskatchewan Province of Canada.

Another significant potential source of oil from shale is the organic-rich Phosphoria Formation in Idaho and western Montana. In southwest Montana, as much as 24 gallons of oil have been extracted from a ton of shale. However, the oil shale occurs in structurally complex folded and faulted mountains, a situation that makes the mining or processing of large quantities very difficult.

WATER FOR SYNTHETIC FUEL PRODUCTION

The water requirements of different synthetic fuel production processes are an important consideration in the development of synthetic fuel plants. Even for a particular process, the water requirements will vary depending on the options selected for process application and cooling. Savings in water use usually entail additional costs and sometimes a decrease in the overall efficiency of the conversion process. Figure 8-10 shows estimated ranges of water requirements for major synthetic fuel processes and selected other uses of coal. The basis of comparison is the coal input required to produce 50,000 barrels of synthetic oil, or the equivalent input of coal for the other uses.

Coal Gasification

Published values for water use in coal gasification vary widely. Key factors affecting water requirements are:

- Coal Composition. Consumption of water for waste-gas cleaning increases with sulfur content of coal and generally decreases with increasing carbon content.
- Cooling methods. Nearly two-thirds of the water used is consumed as cooling water. Therefore, alterations in cooling methods can greatly change the net water requirements. For example, the low estimate in Fig. 8-10 for coal gasification reflects the case where 80% of the waste heat is removed by dry cooling. This includes direct dissipation of heat from hot-process streams to the surrounding atmosphere. In contrast, the high estimate for coal gasification corresponds to 50% dry cooling. Dry cooling systems in general are more capital intensive and expensive than wet cooling systems. However, as the cost of water increases, dry cooling may become more economical, and water consumption may decrease.
- Product gas heating value. The conversion of energy in coal to energy in gas becomes less efficient as the Btu content of the product gas increases. The less efficient the conversion, the greater the generation of waste heat and, therefore, the greater the volume of cooling water required for a given input of coal.

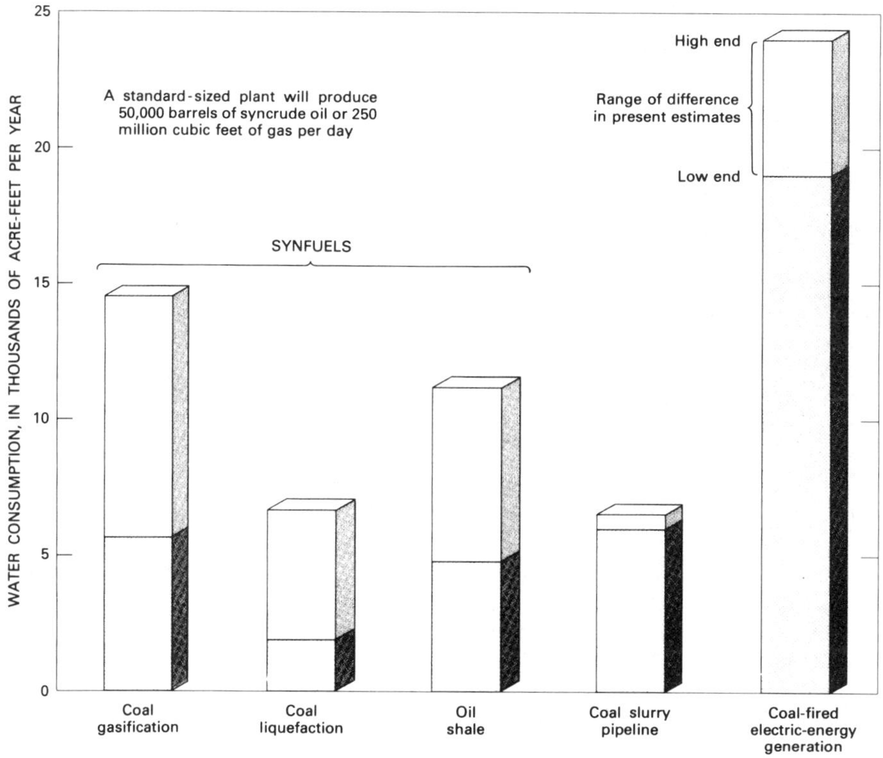

Fig. 8-10. Estimated water consumption for producing usable synthetic fuel products in standard-sized plants, and for slurry pipeline and electric-energy generation having coal input equivalent to that for a standard coal-conversion plant.

Coal Liquefaction

Water use in coal liquefaction is primarily influenced by the same three factors that influence water use in coal gasification. However, as shown on the graph, the water requirements for coal liquefaction are generally smaller.

Oil Shale Conversion

- Cooling methods. The water required for dry and wet cooling methods is much the same as described for coal gasification.
- Spent-shale disposal. In aboveground processing, water use for spent-shale disposal accounts for nearly 50% of the total water used. Underground *(in situ)* oil-shale processing eliminates this potential demand and, therefore, requires much less water.

Figure 8-10 illustrates that estimated water consumption in a coal gasification plant is about twice that for a coal liquefaction plant using the same input of coal. Water consumption in the aboveground production of shale oil is generally estimated to be between these values. However, developments of *in situ* technology may soon lower the requirements for shale oil production.

From a resource development viewpoint, it is interesting to compare the water requirements for producing synthetic fuels from coal with those for processing equivalent inputs of coal in other ways. For example, in comparison to standard-sized coal conversion plants, there is a low water requirement for transporting coal by slurry pipeline. In contrast, the same coal input would supply a 2000-MW thermal electric plant and create a water requirement considerably greater than those estimated for producing synthetic fuels.

There is abundant surface water in the nation, but the resources are distributed unevenly over both space and time.

Figure 8-11 shows the variations in average annual runoff of surface water in the United States. In the eastern half of the nation, annual runoff, which is a measure of stream-water outflow, varies, but is fairly uniform over large areas. In the West, however, particularly in the mountain regions, annual runoff differs greatly from place to place. For example, in Colorado and western Wyoming, average annual runoff can range from less than 1 in. to more than 20 in. within a distance of less than 50 miles. This difference results primarily from topographic influences; the greater depth of runoff per unit area occurs in the mountains.

Streamflow variations with time can be short term, seasonal, or long term. Seasonal variations can be marked. For example, Fig. 8-12 shows the June runoff in Clarks Fork of the Yellowstone River is about 18 times the monthly runoff from October to March. This results because precipitation is stored as snow during winter and then melts and runs off rapidly during the subsequent spring thaw. Such a variation illustrates that variability of streamflow is an important consideration in the planning of energy-resource development.

There are means of alleviating problems arising from low average runoff and highly variable streamflow. Where reservoir space can be made available (at the surface or underground), water can be stored during periods of excessive runoff for later release and use. Moreover, in some situations, water can be imported into local, water-deficient areas.

Groundwater is stored in aquifers, which are bodies of rock permeable enough to carry or yield water in useful quantities. Aquifers differ in thickness, extent, and depth of occurrence. The nation's groundwater resource is enormous—it is our largest freshwater resource in terms of volume in storage (see Fig. 8-13). Underlying the conterminous United States are about 200 billion acre-feet (65 quadrillion gallons) of groundwater within a few thousand feet of the land surface.

About one-third of the nation's annual streamflow is supplied by groundwater that emerges as natural springs and seeps to stream channels. In turn, surface water can be a major source of groundwater recharge. For many streams, much of the flow during the dry months comes from groundwater. In years of below-normal precipitation, all of the dry season streamflow may come from groundwater. Groundwater, therefore, is important to the continuity of streamflow.

About 48% of the total population and 95% of the rural population of the United States are dependent on groundwater to supply their domestic, agricultural, or industrial water needs. Currently, about 92 million acre-feet per year (82 billion gallons per day) are withdrawn from U.S. groundwater reservoirs. However, this rate, which represents about 20% of the nation's total water use, constitutes only a fraction of the potential for groundwater development.

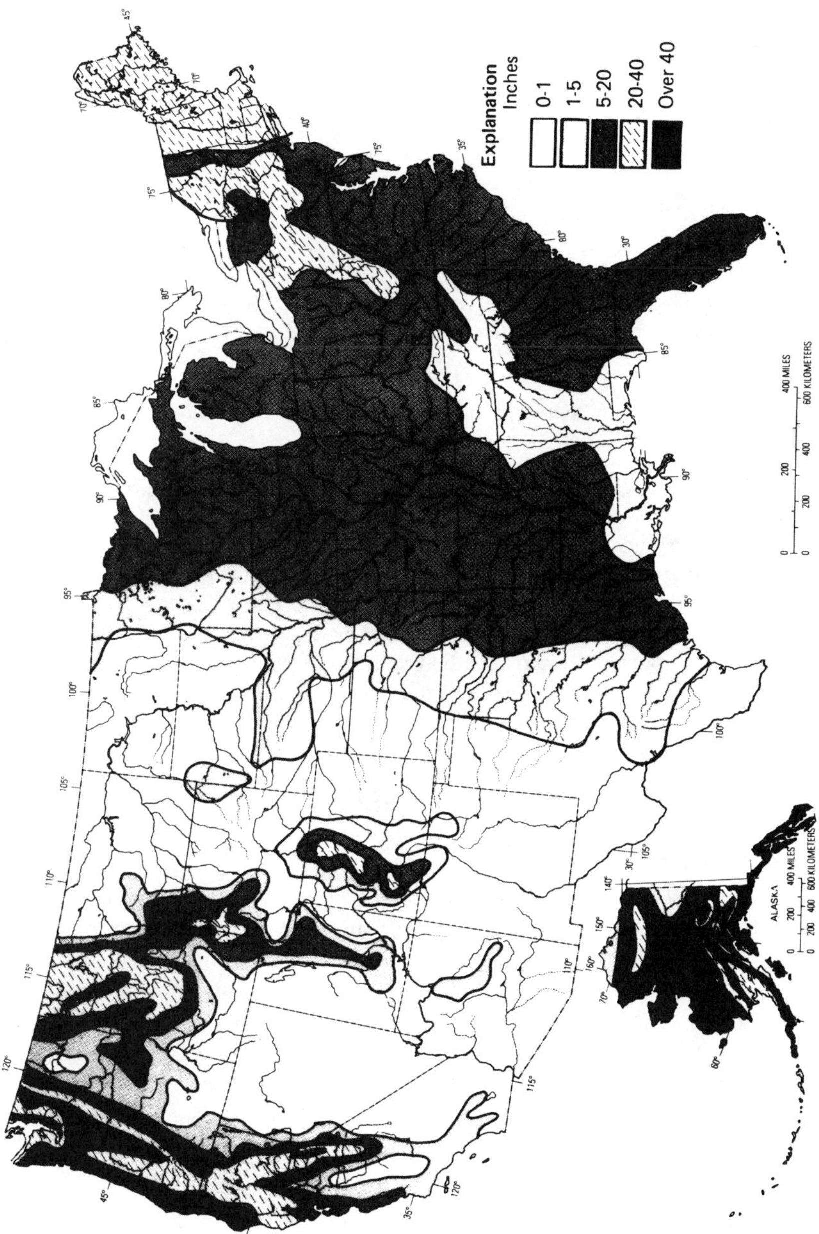

Fig. 8-11. Average annual runoff.

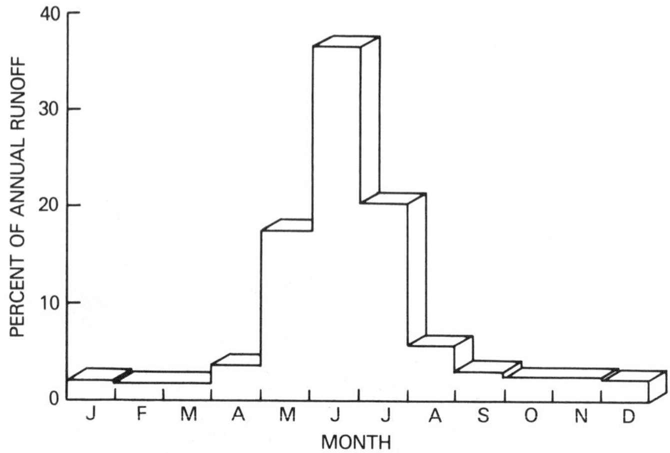

Fig. 8-12. Long-term average monthly runoff, Clarks Fork of the Yellowstone River near Belfrey, Montana.

The nation's systems of groundwater reservoirs range from those that are drained and refilled naturally on an annual cycle, to those in which the annual replenishment is small. In the arid West, many groundwater reservoirs receive small replenishment relative to the total volume in storage. In contrast, in the East, substantial yearly replenishment is more common. In general, the more widespread and productive aquifers are found in the humid East, whereas aquifers in the Far West tend to be less continuous and yield smaller quantities of water. However, because of the relative scarcity of surface water in the Far West, groundwater withdrawals are greater than in the East as shown in Fig. 8-14.

In a given region, the supply of water for use in synthetic fuel production is influenced by the following factors:

- The natural supply of water, which is influenced by precipitation, natural evapotranspiration, and aquifer recharge.
- Water imported into the region.
- Depletions of supply which include water evaporated as a result of industrial, agricultural, domestic, and commercial use; water evaporated from man-made reservoirs; and water exported from the region (alone or with some product).
- Regulation of flows (by reservoirs), which determines the availability of water in dry seasons and dry years. Without reservoir storage (either surface or subsurface), only a small fraction of the average stream supply can be put to continuous use in an industrial process such as synthetic fuel production.
- The possibility of groundwater mining (pumping from aquifers in excess of recharge rates), which can contribute to short-term water supplies for years, decades, or longer.

Figure 8-15 shows the first three factors are evaluated using data from the Water Resource Council's Second National Assessment. The map patterns illustrate the estimated percentage of water depleted in each subregion of the United States (total depletion/total supply) and thus indicate the relative magnitude of human impact on the water cycle. Estimated depletion is generally less than 50% of supply east of the 95th

Fig. 8-13. Groundwater resources.

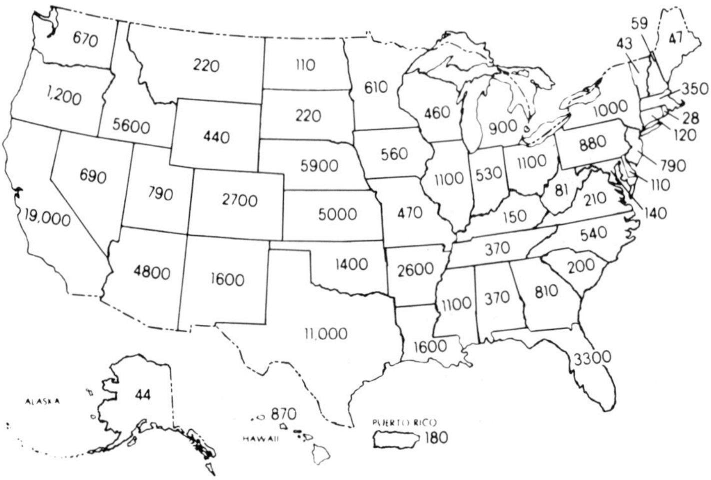

Fig. 8-14. Groundwater withdrawals, in million gallons per day, 1975.

meridian and in basins of the Northern Great Plains, the Rocky Mountains, the Pacific Intermountains, and the Pacific Northwest. Depletion exceeds 75% of supply primarily in the Southwest, where depletion exceeding the calculated supply occurs in some basins because of groundwater mining.

The number shown for each subregion is the estimated average annual outflow of water in billions of gallons per day. This average outflow is not necessarily a surplus; commonly part is required to satisfy instream-flow needs (as for fisheries, navigation, and hydropower) or downstream compact requirements.

A comparison of synthetic fuel resource areas with the water-depletion patterns indicates that estimated depletion does not exceed 50% of the available water supplies, except for small areas in southwest Utah, northwest New Mexico, and south-central Wyoming.

In certain areas there are geologic formations which, for one reason or another, have not been used as aquifers for traditional water uses but are capable of yielding significant quantities of water for energy development.

The Madison limestone in Montana, Wyoming, and the Dakota's shown in Fig. 8-16 is an example. In the past, a number of factors have combined to limit the pumping of groundwater from the Madison. Over much of the area, the formation occurs at great depth, so that high drilling costs have discouraged development. In some localities, the water has moderate to high concentrations of dissolved solids, making it unsuitable for irrigation or direct human consumption. Moreover, in some areas, the rock is relatively low in permeability, so that adequate well yields either cannot be obtained or require high pumping costs. Despite these limitations, it might be desirable to obtain groundwater from the Madison for energy-related purposes. This possibility is partly a reflection of economic considerations; water that is too expensive to develop for irrigation is often within economic reach for energy production. In addition, water-quality requirements for some energy-related applications are less stringent than those for agricultural use. Thus, some water from the Madison limestone unfit for irrigation or other uses might still be used for energy production.

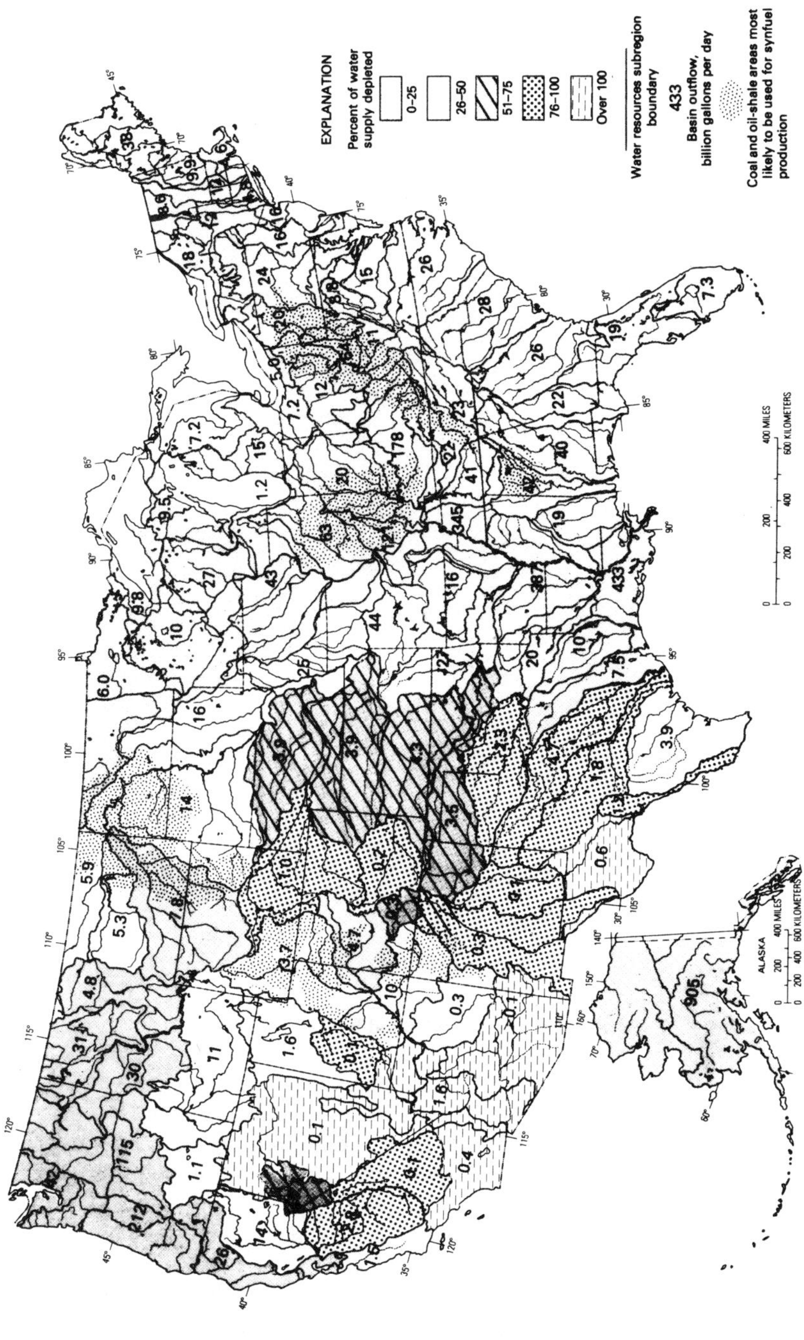

Fig. 8-15. Relative water depletion.

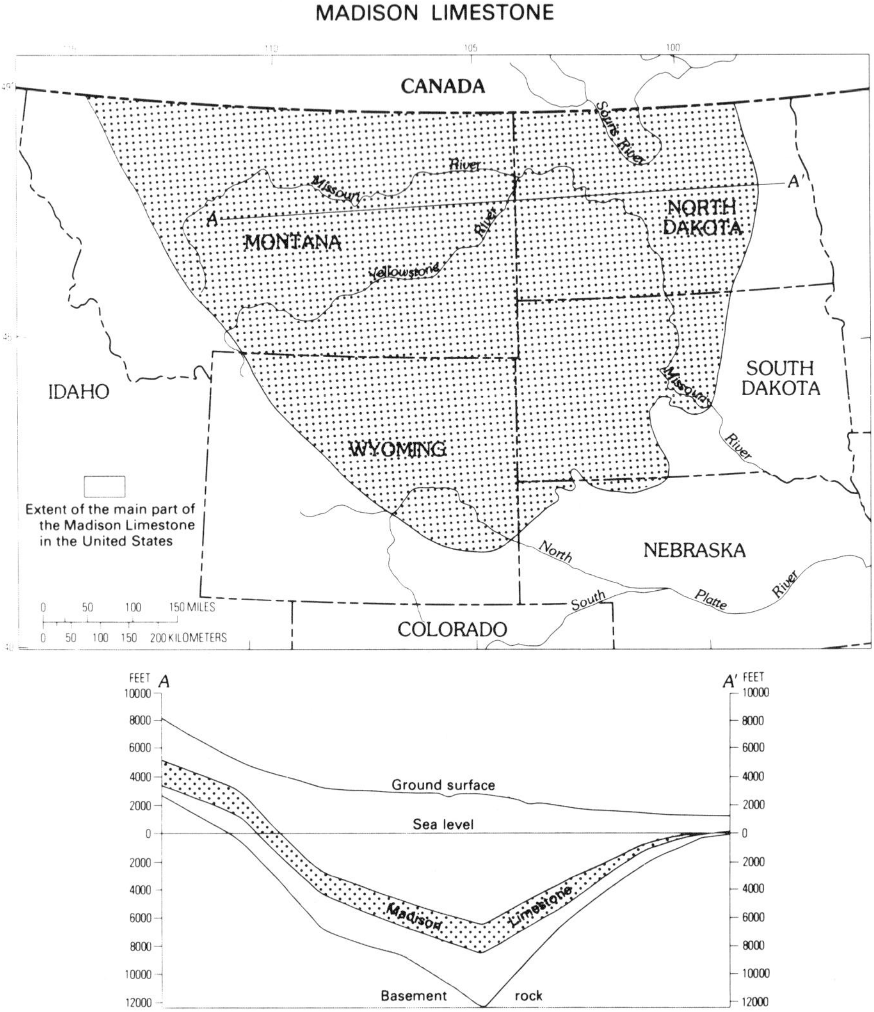

Fig. 8-16. Extent of Madison limestone in the United States.

There are several other examples of deep-lying and little-used aquifers in areas of potential synthetic fuel resources. For example, the Leadville limestone underlies much of the oil shale region of Colorado, as do several thick sandstone aquifers. These aquifers contain large quantities of stored groundwater, some of which might serve the development of oil shale.

An important consideration in the potential use of deep, little-used aquifers is that the water is largely unappropriated. Thus, depending on state laws, there may be few legal impediments to development of such water resources for energy purposes and, therefore, minimal need for trade-offs on present water use.

The availability of existing water resources for synthetic fuel development is affected by

- The extent of appropriation of existing water, including groundwater.
- The willingness on the part of existing water users to sell their water rights.
- The legislative, judicial, and administrative decisions regarding the transfer of water rights and changes in water uses.
- Uncertainties with respect to the extent of instream-flow requirements for fisheries, navigation, hydropower, and others, and unsettled claims for Indian and federal water rights.

Water allocation is primarily the responsibility of the states, subject to interstate compacts and international treaties. In much of the East and some areas of the West, unallocated water may be available for development as shown in Fig. 8-17. In the West, however, much of the physically available water may be subject to legal challenge owing to other claims, such as Indian water rights which have not been adjudicated. In some western states, water rights may be sold. Within those states, where waters are fully committed, some sales have been made by agricultural holders of water rights to energy-development interests. However, in North Dakota, sale of water rights for a change of use is prohibited. Moreover, in Montana, water law forbids the transfer of water rights from agricultural to industrial use if the right allows for diversion at a rate exceeding 15 cubic feet per second.

As previously stated, deep groundwater exists and may be available for synthetic fuel development in some areas of the West. In addition, in certain locations, dewatering of the oil shale or a coal seam might provide a significant part of the water required for its development.

Based on the availability of existing local water resources, synthetic fuel resource areas may be divided into three categories: (1) areas where unappropriated water resources occur in amounts sufficient for significant synthetic fuel development; (2) areas where water resources are sufficient for synthetic development, but are already appropriated or committed to other uses; and (3) areas where water resources are insufficient for synthetic fuel development. Delineation of these categories on the accompanying map is based on (1) surface reservoir storage that exists or is under construction; (2) economically usable groundwater resources; (3) existing interstate compacts; and (4) an assumed water-transportation radius of 50 miles. Changes in these assumptions or improved knowledge of the groundwater resources could change the distribution.

The map in Fig. 8-17 shows that most of the potential synthetic fuel areas of the nation have unappropriated local water that might be available for synthetic fuel development. In only a few areas do local water supplies appear to be insufficient for synthetic fuel development. However, there are areas of considerable size where local water would be sufficient for synthetic fuel development but is already fully appropriated.

GEOHYDROLOGIC PROBLEMS

There are a number of potential geohydrologic problems that could result from development of a large-scale synthetic fuel industry. However, if the problems are fully recognized, steps can be taken to prevent or minimize the potential impacts. This section

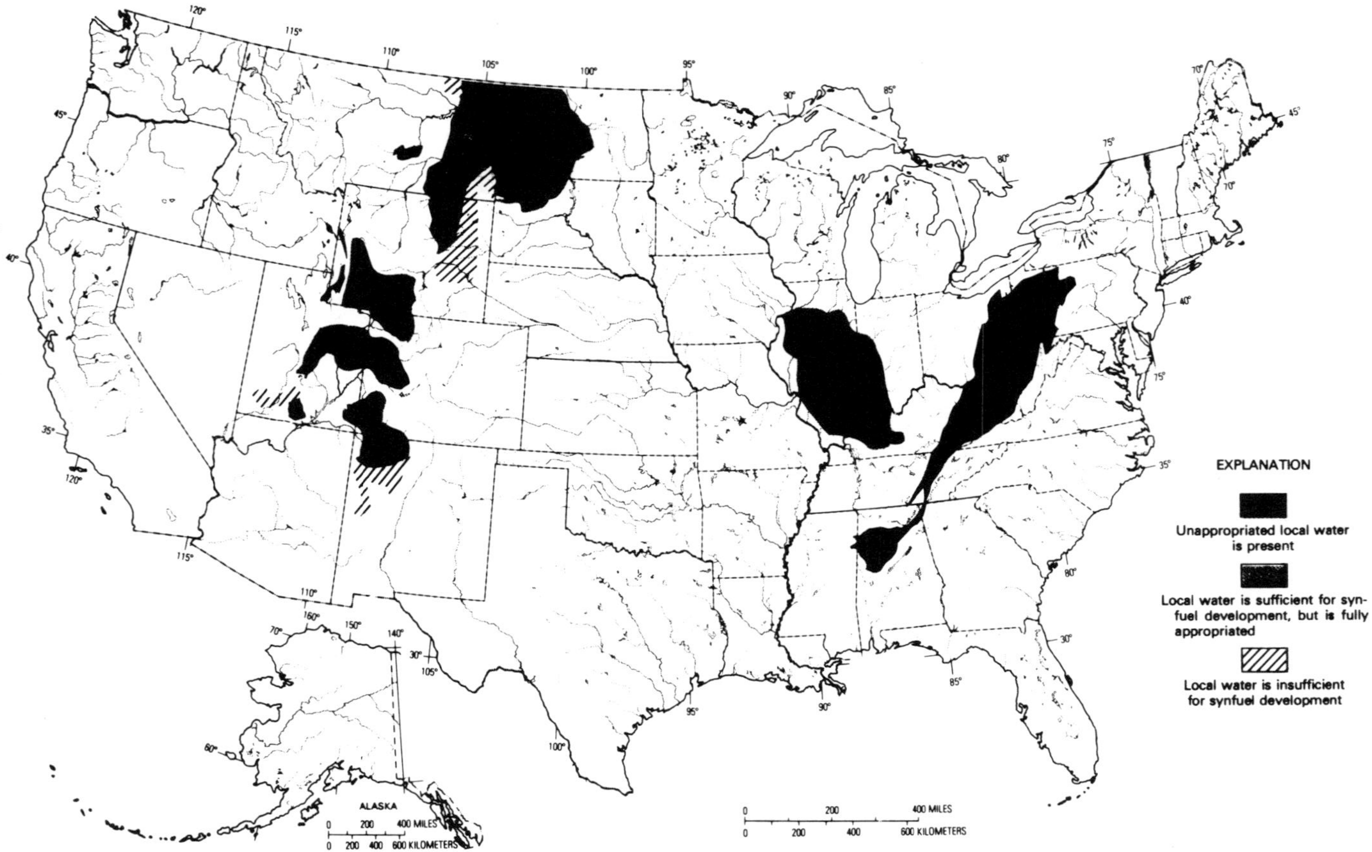

[1]Based on a canvass of State and Federal officials in the Western States by I. C. James II, September 1979

Fig. 8-17. Water resources available for additional use in areas of potential synthetic fuel development.

describes the types of problems that could arise from the mining of coal and oil shale, the conversion of the resources to synthetic fuel products, and the disposal of resultant waste products.

Landslides

Landslides are a great hazard in coal fields of the United States where only traditional mining methods are in use. Much of the northern and southern Appalachian coal regions are in rugged areas of great relief where the slope materials are inherently unstable. Most of the Rocky Mountain and Northern Great Plains Coal Provinces are underlain by earth materials that can fail under natural conditions or under conditions altered even slightly from the natural state. In addition, several areas in the low-lying Interior Coal Province are subject to slope failure.

Landslides always involve the failure of earth materials under shear stress. Factors contributing to slope failure are illustrated in Fig. 8-18. These include (A) accumulation of unconsolidated materials on the upper part of the slope; (B) addition of water, which adds weight to and lubricates the slope materials; and (C,D) removal of support from the bottom of the slope by mining or erosion. In most cases, several factors occur simultaneously and only await a triggering event such as vibration from heavy equipment or earthquakes.

Landslide damage resulting from development of synthetic fuel resources can be minimized through a program of geologic investigations of slide and slide-prone areas. Such investigations should include the study of factors contributing to sliding, and the examination of physical properties of earth materials in mined and unmined areas. Findings of such investigations can be used to design a program for minimizing slope stability problems in each synthetic fuel development area.

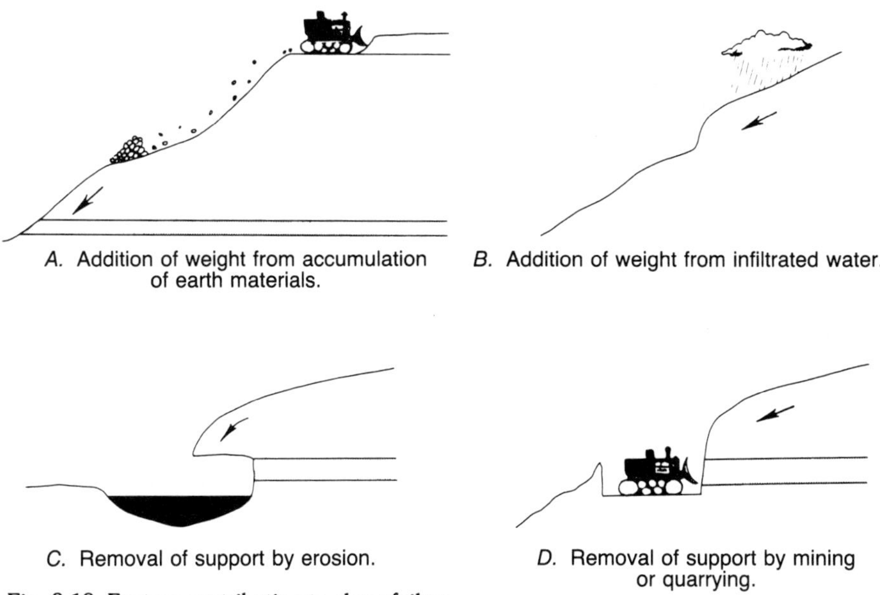

A. Addition of weight from accumulation of earth materials.

B. Addition of weight from infiltrated water.

C. Removal of support by erosion.

D. Removal of support by mining or quarrying.

Fig. 8-18. Factors contributing to slope failure.

Erosion and Sedimentation

Erosion and loading of streams with sediment is considered by many to be the most serious water-quality problem resulting from energy resource mining. Erosion rates from the active surface mining of coal have been estimated to be 10 times greater than from the same relative area of cropland. The impacts of erosion and sedimentation associated with mining include

- Flooding of communities due to filling of stream channels by sediment.
- Filling of reservoirs with sediment, resulting in reduced flood control, irrigation-water storage, power generation, and recreational facilities.
- Destruction or reduction of fish and other aquatic life.
- Increasing costs for maintaining navigation channels and for treating of industrial and domestic water.

Erosion most commonly occurs on poorly vegetated, steep, cut-and-fill slopes in areas having high rates of surface runoff. The problem is considered most acute in the more humid Eastern Coal Province or Appalachian Region.

In most cases, a high yield of sediment from mining operations reflects deficiencies in mining and reclamation practices. Sound practices include

- Preplanning to fit the mining operation to site conditions of topography, soils, and climate.
- Limiting the exposure of denuded land to the smallest possible area for the shortest time.
- Exercising soil-erosion and sediment-control practices concurrent with mining operations.
- Exercising maintenance of reclaimed mining areas.
- Properly disposing of solid wastes.

Figure 8-19 shows the areas where sediment from mining is a significant problem, but one that can be solved through application of sound practices and available engineering measures. Shown also are problem areas where adequate reclamation may not be entirely feasible. These are arid regions of the West where lack of rainfall may make effective revegetation extremely difficult. Careful assessment of each potential mining site will be called for in these areas. Some of these areas eventually may be considered unsuitable for mining because of the likelihood that erosion and sediment problems will be unmanageable.

The Surface Mining Control and Reclamation Act requires the issuing of permits for all coal-mining operations, together with strong measures to control erosion and sedimentation. No such national law exists to cover mining operations of oil shale. However, there are state and local regulations to cover erosion and sediment problems associated with oil shale mining. Through proper planning, engineering, and regulation, the mining of coal and oil shale for synthetic fuel production can be handled with minimum sediment-related impacts in most areas of the country.

Subsidence

Subsidence of the earth's surface follows removal of underground support by either natural or man-influenced causes. Figure 8-20 illustrates that subsidence can be caused

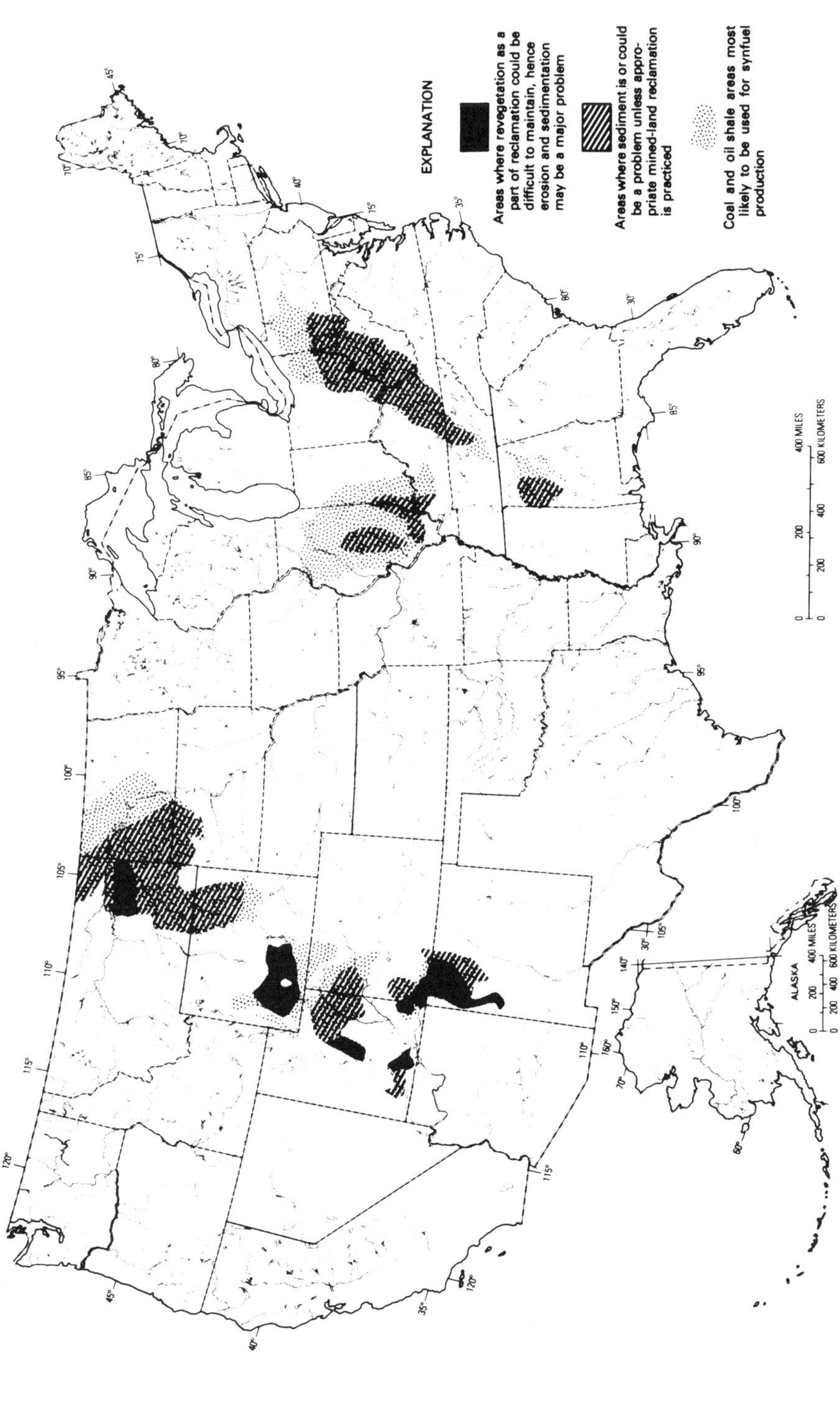

Fig. 8-19. Areas where waterborne sediment from mining could be a problem in synthetic fuel development.

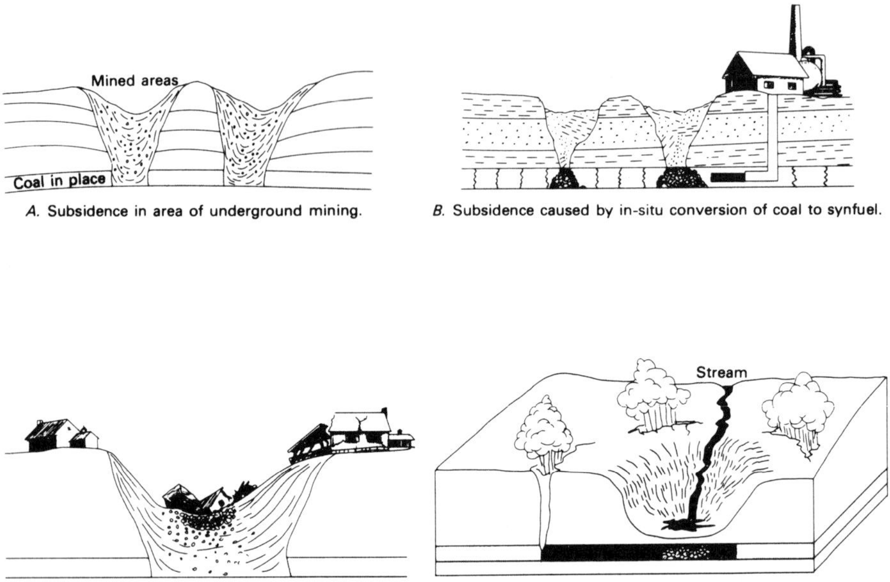

A. Subsidence in area of underground mining.

B. Subsidence caused by in-situ conversion of coal to synfuel.

C. Urban areas over mining districts are subject to severe disturbance.

D. Underground fire in coal bed causes cracks, results in fumes, explosions, diversion of stream.

Fig. 8-20. Causes of subsidence.

by (A) underground mining and (B) *in situ* conversion processes involving a loss of volume to subsurface rocks. Subsidence can be sudden or slow; it can occur soon after mining or after many years and can take many forms, ranging from cracks in the ground to gaping holes. In regions where coal is subject to spontaneous combustion, subsidence commonly allows entry of air into coal beds, increasing the possibility of fire.

Subsidence in urban areas that overlie underground coal workings has caused severe damage to property (C). In Scranton, Pennsylvania, and Sheridan, Wyoming, for example, burning mines abandoned many years ago undergo continuing unpredictable subsidence, spreading of mine fires, and steam explosions that discharge noxious gases (D).

Subsidence also can reduce developable coal reserves, decrease mine safety, and disrupt the local hydrologic system. In western coal fields, several mineable coal beds often occur one above the other. Subsidence caused by mining one or two such beds can create stress conditions in surrounding rocks that severely restrict or prohibit mining of the remaining coal. Cracks caused by subsidence can intercept and divert groundwater or methane gas associated with the coal and may even divert surface water drainage to underground workings. In some western coal fields, the entry of oxygen and water through subsidence cracks contributes to underground combustion.

In areas of *in situ* conversion, subsidence could cause all or some of the same problems as in mined out areas. This, in turn, would strongly affect recoverability of the total energy resources and future land use.

Subsidence can be avoided by backfilling voids with mining spoil or by leaving large amounts of coal or oil shale in place to support the overburden. Problems caused by subsidence can be decreased by avoiding the use of surface areas over underground mines. These approaches, however, do not provide easy or necessarily desirable results. The first is very expensive and technologically difficult; the second is wasteful of scarce resources; and the third involves a type of land-use planning that is unacceptable to many western landowners. Many of the geological hazards related to subsurface mining can be avoided by surface mining and aboveground conversion of coal and oil shale to synthetic fuels coupled with sound reclamation of the disturbed land.

Disruption of Aquifers

Some coal and oil shale deposits are aquifers. Although such aquifers are found nationwide, they are commonly regarded as undesirable for domestic or industrial water supplies for two reasons: (1) adequate capacities to supply sufficient quantities of water and (2) poor chemical quality of the water. In the eastern coal-producing areas, the latter problem is frequently one of excess dissolved metals, which are toxic to humans and animal life. Where the quality of the water is acceptable, the shallow coal aquifers are commonly tapped for private use of farms and ranches.

Most of the private wells which tap coal or oil shale aquifers are located in the Rocky Mountains and Northern Great Plains, particularly in Colorado, Wyoming, Montana, and North Dakota. In these areas, the water quality is considered acceptable, although the amounts of dissolved solids often exceed desirable health limits. Although the supply is small, so is the demand in this region of generally low population. A noteworthy exception is the lignite deposits of North Dakota which are a principal source of domestic water. This lignite is now considered to be a prime candidate for coal gasification.

Figure 8-21 shows how the mining of coal may impact the future use of an aquifer. Illustration (A) shows a water-saturated coal seam under premining conditions. Water flows from the recharge area through the coal seam to supply water to a shallow well and a stream. In (B), mining has interrupted the normal flow of water through the aquifer, and a poorly permeable backfill has caused flow in the aquifer to change direction. The well has gone dry and water from the stream now flows into the aquifer.

Illustration (C) depicts an idealized aquifer restoration in keeping with sound post-mining reclamation practices. The mined area has been filled with porous materials, allowing the surface to act as a recharge zone and permitting water to flow in a pattern similar to premining conditions. Use of the permeable backfill may improve recharge to the aquifer relative to natural premining conditions and, thereby, increase the supply of water to the well and stream. However, the quality of water might be somewhat impaired owing to solution of minerals which might occur in the fill.

The mining of coal and oil shale for synthetic fuel production will seldom impact a principal water supply aquifer. As noted, the most important exception may be the lignite beds of North Dakota. Before these deposits are mined, the area must be carefully evaluated and, if necessary, alternative sources of water identified. Premining planning will also be needed in certain local areas of the country where synthetic fuel production has the potential to disrupt small supply aquifers. Options for minimizing problems include the drilling of deeper wells and obtaining other supplies where impacts cannot be avoided.

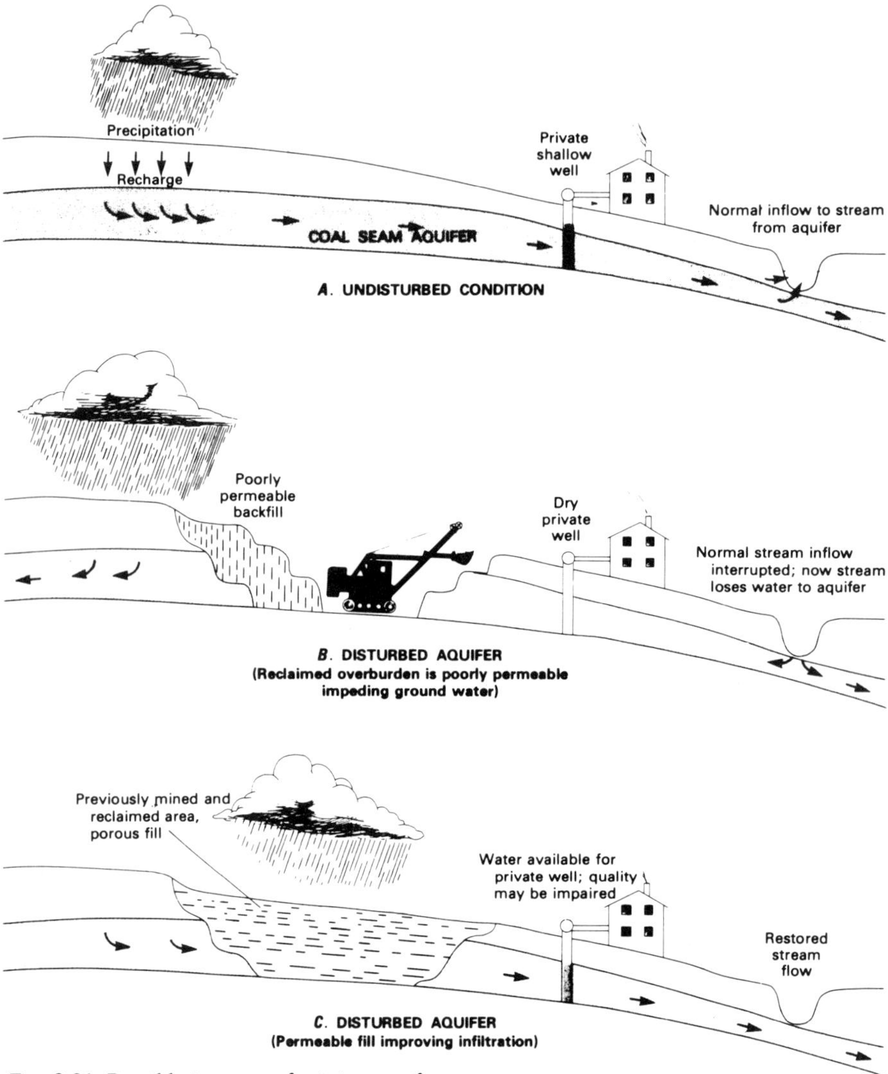

Fig. 8-21. Possible impacts of mining aquifers.

Acid Mine Drainage

Acid mine drainage refers to acidic water discharge from active or abandoned mines. The problem occurs whenever pyritic (iron sulfide) minerals associated with coal are exposed to oxygen in flowing water. The minerals are oxidized, releasing sulfuric acid. This may occur in mines, in unreclaimed spoil banks, and on reclaimed land. However, 7–80% of the acidic water originates from underground coal mines, most of them inactive.

Acid mine drainage affects about one-half of the stream mileage in the coal areas of eight Appalachian States, but is particularly prevalent in Pennsylvania, Maryland, Ohio,

Kentucky, and West Virginia. Acid mine drainage is not a significant problem in the Western Interior, Northern Great Plains, and Rocky Mountain Coal Provinces because pyritic materials are less abundant, water is less available, and neutralization readily occurs in the naturally alkaline soils and water.

The factors causing acid mine drainage and measures to control it are known and can be readily applied. The cost of control is considered small relative to the other costs of producing and converting coals to more usable energy forms.

The potential coal resources for synthetic fuel production are primarily located in regions having low potentials for acid mine drainage. However, even if synthetic fuel resources are mined in potential problem areas, acid mine drainage can be prevented or minimized through sound application of principles outlined in the Surface Mining and Reclamation Act of 1977.

Leaching of Waste Materials

Aboveground processes for converting coal to synthetic fuel generally leave a 10–30% residue of ash, slag, tar, and other products. Moreover, residues from the surface retorting of oil shale consist of rock particles and up to 75% of the original organic matter in the form of char, tar, and other materials. These residues from both coal and oil shale conversion contain numerous leachable organic and inorganic substances which have the potential for seriously degrading the quality of the water resources. Some of the leachable substances are toxic to stream organisms and to higher animals where present in sufficiently high concentrations. Certain of the other substances are carcinogenic.

The schematic shown in Fig. 8-22 shows that water infiltrating a spoil pile could leach contaminants into nearby streams or into the groundwater system. The quantity and character of leachate would depend mainly on the type and solubility of the waste materials, the amount and intensity of precipitation, and the infiltration capacity of the spoils (and any material overlying them).

Dissolved contaminants that enter groundwater may be reduced in concentration by a number of processes, including attachment to mineral surfaces, biological degradation, and dispersion. Once in the groundwater, contaminants move in the general direction of groundwater flow. Movement of groundwater is usually very slow compared to stream flow. Years, or even decades, might pass before contaminants in groundwater reach a stream, whereas those in surface runoff might enter a stream in a period of hours, days, or weeks. Moreover, contaminants in groundwater are correspondingly more persistent than in surface water. Even after contamination from the land surface has stopped, very

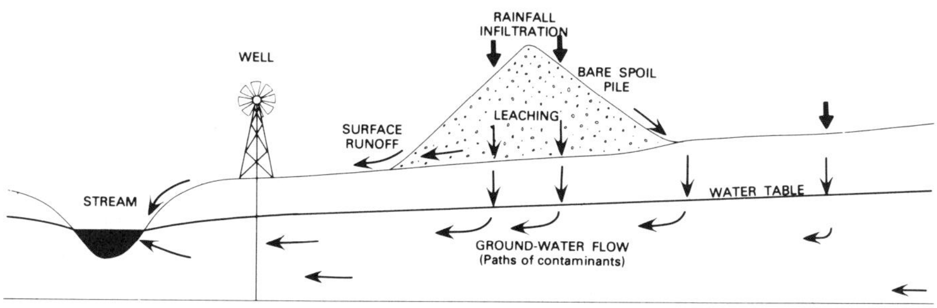

Fig. 8-22. Toxic substances leaching from spoils.

long periods of time may pass before contaminants in groundwater have declined to undetectable levels.

Depending on regional climate, the effects of spoil residues may be mitigated in several ways:

- Dispose of spoils in areas of groundwater discharge rather than areas of recharge; especially avoid recharge areas of extensive freshwater aquifers.
- Reduce the infiltration into spoils by covering them with a relatively impervious material. One method is to spread the spoils on ridge tops, cover them with gravel to reduce erosion, and to top this with a relatively impermeable soil material.
- Contaminated groundwater might be prevented from reaching well fields or streams by intercepting the water at pumped wells located upgradient from the area to be protected. Contaminated water pumped from such wells would need to be treated or disposed of in an acceptable manner. One possible method in certain areas would be reinjection through wells into selected subsurface zones that have little or no foreseeable value (such as brine-filled aquifers).

The extent to which mitigating measures will be successful depends largely on the climate and local geohydrologic conditions. These must be carefully evaluated before reliable waste-management plans can be made for a specific development.

IN SITU CONVERSION EFFECTS

Much research and development work has recently focused on methods to extract synthetic fuels from coal and oil shale in their natural underground *(in situ)* environment. Most schemes for the *in situ* processing of coal or oil shale involve the underground burning of part of the energy-bearing rock. The heat generated by the burn converts the solid organic material to gaseous and liquid products that can readily flow through pores and fractures in the rock, and, hence, be recovered through a production well. Only a relatively small fraction of the organic content (generally 10−20%) of the rock is consumed to sustain the burn.

In situ conversion appears to offer several advantages over conventional mining and subsequent processing in surface facilities. For example, disturbance of the land surface, hazards associated with mining, air pollution from retorting plants, and disposal of solid wastes would all be minimized. But there would be a price to pay. Precise engineering control of several factors that affect conversion efficiency, such as particle size, retorting temperature, and moisture content, may not be possible with an *in situ* process. In addition, most coal and oil shale beds are saturated with groundwater. Dewatering may be required prior to processing oil shale by the modified *in situ* method. In contrast, owing to differences in rock characteristics and methodology, dewatering may not be required prior to *in situ* conversion of coal. However, the underground conversion of both coal and oil shale will necessarily consume and may contaminate a part of the local groundwater resource.

In a given location, numerous factors must be considered to determine whether *in situ* or surface processing is the more acceptable. The ultimate acceptability of *in situ* processing will depend greatly on aquifer properties and on the existing or potential groundwater uses.

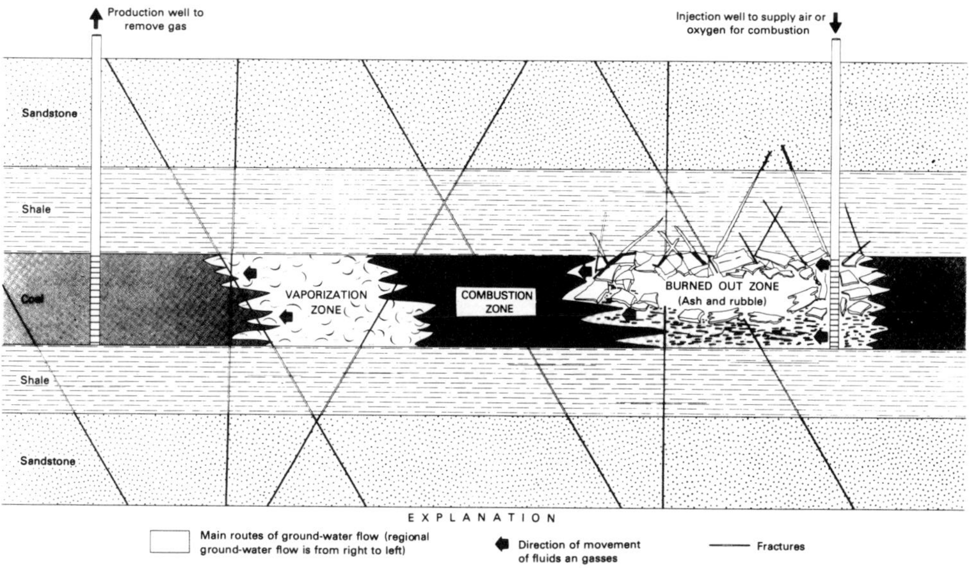

Fig. 8-23. Schematic cross section of an *in situ* coal gasification process.

A number of different operational designs have been proposed for underground conversion of both coal and oil shale. Figure 8-23 is a general representation of an *in situ* coal gasification process. The coal is initially ignited at an access point provided by a well. The well also allows the air or oxygen necessary to sustain combustion to be injected into the formation. Heat generated by the burn causes a vaporization zone to form in front of the advancing combustion zone. Within this area, water is vaporized and combustible gases are produced by chemical reactions among carbon, oxygen, and steam. Water is consumed both by steam production and chemical reactions, and groundwater adjacent to the vaporization zone (or condensed out of this zone) may become severely contaminated with organic and inorganic compounds.

As the combustion zone advances through the rock, it leaves behind a zone of ash and rubble. Some of the rubble forms by collapse of the overburden into the spaces that result from combustion or from subsequent leaching of soluble residues. If natural fractures or fractures caused by the *in situ* process extend into adjacent aquifers, groundwater may leak into the ash and rubble zone and become polluted by dissolving combustion by-products or inorganic salts.

Thus, groundwater may become contaminated either in front of or behind the advancing zone of combustion. If contaminated groundwater is removed by the production wells, it will present a disposal problem, and will probably require treatment prior to being safely discharged aboveground. The contaminated groundwater not removed by wells will naturally migrate through the rock formations and may eventually reach a well, spring, or stream.

Potential subsurface movement of some contaminants can be predicted after an adequate hydrogeologic and geochemical study of a specific site has been completed. The inclusion of such studies in the planning process provides a means of ensuring that *in situ* conversion will be located only where (1) subsequent downgradient migration of contaminants in the subsurface will have a negligible impact on man and the surface environment or (2) potential contamination can be controlled through engineering practices.

Process Environmental Considerations

This section describes the environmental considerations for generic synthetic fuel technologies, such as oil shale, coal gasification, and coal liquefaction.

OIL SHALE

The environmental considerations for oil shale are divided into two process categories, which include (1) surface oil shale and (2) *in situ* oil shale. The Environmental Protection Agency, the Department of Energy, and other governmental agencies and groups are studying the environmental aspects of oil shale development. Currently, there remain a number of unanswered environmental questions. It may not be possible to provide a meaningful environmental determination until experience with one or possibly more operating plants is acquired. The technologies may be deemed too new, the effected ecologies are not well understood, and the scale of operation is too massive to be able to predict (with a reasonable degree of confidence) the effects of "scale-up" for an oil shale industry. In addition, envisioned environmental controls for the oil shale industry are subject to large uncertainties.

The most significant problems and uncertainties are associated with impacts on air and water quality, waste management, and occupational health and safety aspects.

Surface and *in situ* Oil Shale—Identified Pollutants

Air Emissions: Atmospheric emissions can arise from several activities or operations during surface and *in situ* oil shale processing. The major source of SO_2, NO_x, and CO is fuel combustion for process heat; SO_2 is also emitted in the tail gases of sulfur-recovery operations. The use of fuel oils in mobile equipment and in explosives will result in emissions of CO and NO_x. Hydrocarbons are present in both combustion emissions and product storage tank vapors. Emissions of particulate matter can result from blasting, raw and spent shale handling and disposal, raw and spent shale dust in process gas stream, fuel combustion, and site activities which generate fugitive dust.

Particulate emissions from fuel combustion and fugitive dust from spent shale handling and disposal can contain polycyclic organic material (POM) and certain trace metals. Gaseous ammonia, hydrogen sulfide, and volatile organics may be released during moisturizing and subsequent cooling of retorted shale. Catalyst materials may release particulate matter containing trace metals to the atmosphere during regeneration, handling, or final disposal.

Actual SO_2 emission associated with individual retorting processes will depend on the degree of sulfur removal accomplished for in-plant fuels, the extent of on-site shale oil processing, and the degree of control applied to sulfur recovery tail gases. Combustion of any hydrocarbon fuel will produce oxides of nitrogen when air containing nitrogen is used as the source of oxygen. In addition, organic nitrogen contained in the fuel can be partially oxidized to NO and NO_2. In general, those processes which require small-sized shale feed will have more uncontrolled particulate emission during crushing and raw shale operations than processes which require large feed.

Site use activities which may generate fugitive dust generally are not process specific. The use of open pit versus underground mining will be the largest factor determining total fugitive emissions associated with extraction of oil shale. Overall fugitive dust emissions may present more of a problem for the Tosco II process than for some other processes. Ore storage and handling and disposal of the fine Tosco II retorted shale are potential fugitive sources.

The largest source of CO in an oil shale operation is mobile equipment used for mining and transport. The quantity of such emissions is a function of mining method and haul distances rather than retorting process.

The pyrolysis of essentially any type of organic material produces a certain amount of POM, and oil shale kerogen is no exception. Generally, POM compounds have a low volatility and will be associated with high-boiling-point liquid or solid products or particulate matter. And, although POM is known to be present in carbonaceous retorted shales, the biological availability and potential hazard of such material is not accurately known at present.

Release of POM to the atmosphere during oil shale processing can occur via three major pathways:

1. Handling and disposal of retorted shale—fugitive particulates and possible volatilization of hydrocarbons.
2. Combustion of shale derived oils containing POM.
3. Flue gases containing entrained retorted shale particulates, along with retort gas or spent shale coke combustion products.

Oil shale contains trace amounts of many elements; however, for elements other than Si, Fe, Al, Ca, Mg, and K, the concentrations in oil shale are less than generally found in coal. In addition, conditions during retorting are not severe enough to volatilize most metallic and heavy elements. With notable exceptions such as arsenic (As) and possibly antimony (Sb), most trace elements [e.g., nickel (Ni), vanadium (V), molybdenum (Mo)] remain with the spent shale, or are found as components of raw and spent shale solids entrained in retort gases and in raw shale oil. Arsenic in raw shale apparently forms a range of volatile oil-soluble compounds (perhaps organic arsines) during retorting, and appears in raw shale oil and all condensible oil fractions. If not removed during upgrading, arsenic will be present in shale oil combustion products.

Metals and their compounds are used as catalysts (Ni, Co, Mo, Cr, Fe, Zn) for hydrotreating dearsenating, and sulfur recovery, and trace metals can occur either during on-site regeneration or during handling and disposal. Catalyst use is not unique to shale oil processing, and much information and experience in preventing hazardous emissions can be borrowed from the petroleum and related industries.

Solid and Liquid Effluents: Construction, mining, and site use activities may potentially result in increased sediment and dissolved solids loading in surface run-off and receiving streams. This indirect source of potential water pollution is not unique to oil shale extraction and processing but may require careful control owing to the magnitude of site activities. Collection and impoundment of run-off may be necessary.

The need to process 1–3 tons of shale per barrel of oil results in a major solid waste disposal problem regardless of whether surface or *in situ* retorting is employed. A principal impact will be the necessary storage of overburden or refuse from open pit and room-

and-pillar mining. Large volumes of waste material must be disposed of in a satisfactory manner. If the overburden and refuse have properties toxic to the existing surroundings, vegetation, and ecosystem, then the technology must include control of these effects.

Aqueous wastes from oil shale processing can be broadly categorized as originating from direct or indirect sources. Direct sources are wastewaters generated from unit operations and/or processes, including wastewater from retorting operations; wastewater from upgrading operations; water from air emission control and gas cleaning systems; cooling water and boiler water blowdowns; water treatment systems; mine dewatering wastewater; and sanitary wastewaters. Indirect sources include leachate from retorted shale disposal areas; run-off and erosion resulting from construction and site use activities; and run-off from mining and transport activities.

Water is a direct product of oil shale retorting, resulting from the step in the release of free and inorganically bound water from raw shale, and combustion of organic material in shale. From 1 to 8 gallons of water are commonly produced per ton of input shale feed to a surface retort, depending on the retorting process and the composition of the shale processes.

This water can separate partially from crude shale oil during storage and/or can appear in aqueous waste streams of shale oil upgrading operations. Water remaining in retort gases after oil separation can be condensed during cooling or gas cleaning operations, or can appear in the flue gas stream from retort gas combustion. Water separated from crude shale oil contains mainly ammonia, carbonate and bicarbonate, sodium, sulfate, chloride, and dissolved or suspended organic compounds (phenolics, amines, organic acids, hydrocarbons, mercaptans). Smaller quantities of calcium, magnesium sulfides, and trace elements may also be present, along with suspended shale fines. Water condensed from retort gases contains primarily ammonia and carbonates, with traces of organic substances and sulfur-containing compounds.

The quality of wastewaters from an upgrading operation varies with the level of on-site upgrading or refining utilized. In general, a full-scale regaining operation may include any of the following wastewater streams: oily cooling water, process water, and wash water.

Wastewaters are also collected during retort gas cleaning, tail-gas cleanup, and foul water stripping. Major constituents in such waters are shale dust particulates, hydrocarbons, H_2S, NH_3, phenols, organic acids, and amines. Other constituents such as thiosulfate and thiocyanate may also be present.

Approximately 45–50% of the water required for an oil shale plant is expected to be used for moisturizing of retorted shale. Much of this water requirement will be supplied by minewater and process wastewaters. Because of the large quantities of water utilized and the exposure of retorted shale to rain and snowfall, a source of indirect water pollution may occur via leaching or run-off from retorted shale piles. However, the bulk of the water applied to retorted shale is expected to be held in capillarity or to be bound as simple hydrates. The suspended and dissolved constituents of wastewaters applied to retorted shale are expected to be partially immobilized by physical adsorption and/or chemical reaction with retorted shale. Leaching experiments in the laboratory and with small plots indicate that inorganic salts—Na, Mg, SO_4, Cl—may be leached from retorted shales. Small quantities of organic substances and trace elements are also water soluble.

In Situ Oil Shale—Identified Pollutants

Air Emissions: Atmospheric emissions can arise from several activities or operations during *in situ* oil shale processing. The major source of SO_2, NO_x, and CO is fuel combustion for process heat; SO_2 is also emitted in the tail gases of sulfur-recovery operations. The use of fuel oils in mobile equipment and in explosives will result in emissions of CO and NO_x. Hydrocarbons are present in both combustion emissions and in product storage tank vapors. Emissions of particulate matter can result from blasting, raw and spent shale handling and disposal, raw and spent shale dust in process gas stream, fuel combustion, and site activities which generate fugitive dust.

Emissions of potentially hazardous substances may occur during the extraction and processing of oil shale. Silica (quartz) may be present in dust derived from oil shale and associated rocks and in fugitive dust. Particulate emissions from fuel combustion and fugitive dust from spent shale handling and disposal can contain polycyclic organic material (POM) and certain trace metals. Gaseous ammonia, hydrogen sulfide, and volatile organics may be released during moisturizing and subsequent cooling of retorted shale. Catalyst materials may release particulate matters containing trace metals to the atmosphere during regeneration, handling, or final disposal.

Generally, the retorting operation itself does not involve atmospheric emissions; gaseous, liquid, and solid streams leaving the retort are handled by downstream systems before reaching an atmospheric interface. However, certain features inherent in the retorting method influence the nature and magnitude of emissions from other sources in the associated shale oil plant.

Sulfur in raw oil shale amounts to about 0.7% by weight, approximately one-third associated with the organic fraction and two-thirds as pyrite (Fe_2S). During kerogen pyrolysis, about 40% of the organic sulfur in shale appears as H_2S in the produced gases, and the other 60% as heavier sulfur compounds in raw shale oil and in the spent shale carbonaceous residue. Pyritic shale sulfur does not decompose under nonoxidizing retorting conditions.

Actual SO_2 emissions associated with individual retorting processes will depend on the degree of sulfur removal accomplished for in-plant fuels, the extent of on-site shale oil processing, and the degree of control applied to sulfur-recovery tail gases. Combustion of any hydrocarbon fuel will produce oxides of nitrogen when air containing nitrogen is used as the source of oxygen. In addition, organic nitrogen contained in fuel can be partially oxidized to NO and NO_2.

The feed to a surface retorting plant always presents a particulate control problem. Run-of-mine raw shale commonly contains about 5 weight percent of ore of less than ½-in. size. A sizable percentage of this segment will become minus 100-micron particulates as a result of primary crushing.

Emissions of hydrocarbons and CO occur during incomplete combustion of fuels in process heaters and in mobile equipment. Hydrocarbons may also be vaporized during product storage. Equipment use and evaporative hydrocarbon emissions are not expected to be process specific.

The largest source of CO in an oil shale operation is mobile equipment used for mining and transport. The quantity of such emissions is a function of mining method and haul distances rather than retorting process.

The pyrolysis of essentially any type of organic material produces a certain amount of POM, and oil shale kerogen is no exception. Generally, POM compounds have a low volatility and will be associated with high-boiling-point liquid or solid products of particulate matter. It should be noted that, although POM is known to be present in carbonaceous retorted shales, the biological availability and potential hazard of such material is not accurately known at present.

Release of POM to the atmosphere during oil shale processing can occur via three major pathways:

1. Handling and disposal of retorted shale, fugitive particulates, and possible volatilization of hydrocarbons.
2. Combustion of shale derived oils containing POM.
3. Fuel gases containing entrained retorted shale particulates, along with retort gas or spent shale coke combustion products.

Oil shale contains trace amounts of many elements. However, for elements other than Si, Fe, Al, Ca, Mg, Na, and K, the concentrations in oil shale are less than generally found in coal. In addition, conditions during retorting are not severe enough to volatilize most metallic and heavy elements. With notable exceptions such as arsenic (As) and possibly antimony (Sb), most trace elements [e.g., nickel (Ni), vanadium (V), molybdenum (Mo)] remain with the spent shale, or are found as components of raw and spent shale solids entrained in retort gases and in raw shale oil. Arsenic in raw shale apparently forms a range of volatile oil soluble compounds (perhaps organic arsines) during retorting, and appears in raw shale oil and all condensible oil fractions. If not removed during upgrading, arsenic will be present in shale oil combustion products.

Actual emissions of nonvolatile trace elements are anticipated in approximate proportion to raw and retorted shale particulate emissions for an oil shale extraction and retorting operation. Such emissions may not be different in nature or magnitude from those associated with the extraction and processing of other fuel and nonfuel minerals (coal, limestone, phosphate rock, etc.). Furthermore, the dolomitic and/or alkaline nature of shale immobilized many elements as relatively inert oxide, carbonate, or silicate salts.

Metals (Ni, Co, Mo, Cr, Fe, Zn) and their compounds are used as catalysts for hydrotreating, dearsenating, sulfur recovery, and trace sulfur removal. Emissions of particulate matters containing catalyst metals can occur either during on-site regeneration or during handling and disposal. Catalyst use is not unique to shale oil processing, and much information and experience in preventing hazardous emissions can be borrowed from the petroleum and related industries.

Solid and Liquid Effluents: The surface retorted shale from a modified *in situ* process can be a major problem. Surface retorted shale occupies a greater volume than the original shale and contains varying quantities of organic and inorganic residuals. The nature of these are dependent on many factors including process, site, and climatic variables.

Aqueous wastes from oil shale processing can be broadly categorized as originating from direct or indirect sources. Direct sources are wastewaters generated from unit operations and/or processes, including wastewater from retorting operations; wastewater from upgrading operations; water from air emission control and gas cleaning systems; cooling water and boiler water blowdowns; water treatment systems; mine dewatering wastewater; and sanitary wastewaters. Indirect sources include leachate from retorted shale disposal areas; run-off and erosion resulting from construction and site use activities; and run-off from mining and transport activities.

Water is a direct product of oil shale retorting, resulting from the release of free and inorganically bound water from raw shale, and combustion of organic material in shale. From 1 to 8 gallons of water are commonly produced per ton of input shale feed to a surface retort, depending on the retorting process and the composition of the shale processes. *In situ* process demonstrations have reportedly produced even greater amounts of water. Some water condenses with crude shale oil during separation of the oil from retort gases. This water can separate partially from crude shale oil during storage, or can appear in aqueous waste streams of shale oil upgrading operations. Water remaining in retort gases after oil separation can be condensed during cooling or gas cleaning operations, or can appear in the flue gas stream from retort gas combustion. Water separated from crude shale oil contains mainly ammonia, carbonate and bicarbonate, sodium, sulfate, chloride, and dissolved or suspended organic compounds (phenolics, amines, organic acids, hydrocarbons, mercaptans). Smaller quantities of calcium, magnesium sulfides, and trace elements may also be present, along with suspended shale fines. Water condensed from retort gases contains primarily ammonia and carbonates, with traces of organic substances and sulfur-containing compounds. In particular, it should be noted that *in situ* and modified *in situ* retorting will result in various liquid, solid, and gaseous products being left in the retort zone. Their effect on aquifers due to disruption and leaching is not known.

The quality of wastewater from an upgrading operation varies with the level of *in situ* upgrading or refining utilized. In general, a full-scale refining operation may include any of the following wastewater streams: oily cooling water, process water, and wash water.

Wastewaters are also collected during retort gas cleaning, tail-gas cleanup, and foul water stripping. Major constituents in such waters are shale dust particulates, hydrocarbons, H_2S, NH_3, phenols, organic acids, and amines. Other constituents such as thiosulfate and thiocyanate may also be present.

Cooling water is used in retorting and oil upgrading to absorb heat which cannot be economically recovered for use in the complex or absorbed by air fan coolers. Cooling water is generally circulated through a wet cooling tower system to release this heat to the atmosphere. Because of evaporative losses, there is a constant buildup of dissolved solids which require a portion of this recirculated water to be discharged as a blowdown from the cooling water system. Similarly, a fraction of the boiler water must be discharged as blowdown to minimize scaling of boilers. Both the cooling water and the boiler blowdown waters contain a high concentration of dissolved solids, and substances such as hexavalent chromium used for corrosion control.

Good quality water is needed to supply processing, cooling tower, steam generation, and other miscellaneous process uses. Wastes from water treatment systems generally consist of chemical sludges, backwash water from filtration system, and blowdown from zeolite-softening systems. The largest quantity of waste is lime sludge, which is characterized by high hardness and dissolved salts content.

An appreciable amount of the water required for an oil shale plant is used for moisturizing of retorted shale. Much of this water requirement will be supplied by minewater and process wastewaters. Because of the large quantities of water utilized and the exposure of retorted shale to rain and snowfall, a source of indirect water pollution may occur via leaching or run-off from retorted shale piles.

Construction, mining, and site use activities may potentially result in increased sediment and dissolved solids loading in surface run-off and receiving streams. This indirect source of potential water pollution is not unique to oil shale extraction and processing but may require careful control due to the magnitude of site activities.

COAL GASIFICATION

The environmental considerations for coal gasification are divided into two process categories which include (1) high-Btu gasification and (2) low/medium-Btu gasification.

As previously indicated, data applicable for environmental assessments of the overall gasification cycle are very limited. Any environmental assessments must be recognized as estimates until one can obtain data from on-line operating systems. Data from pilot plants may not be representative of effluents that would be produced by a commercial plant.

High-Btu Gasification—Identified Pollutants

Approximately 10% of the coal input to a gasification plant is used to generate steam. Combustion of this coal creates the same environmental emissions generally associated with steam generation.

Other wastes (e.g., tar oils, phenols, solids) would be of different composition than wastes from conventional coal-fired boiler steam generators. It must be emphasized that the handling of waste from a high-Btu coal gasification plant is an area where we have very limited information. The following material must be considered accordingly.

This discussion addresses the environmental aspects associated with the gasification process and does not cover coal extraction and transportation. The database for evaluating environmental, health, and safety aspects is very limited and reported information is frequently contradictory. Adverse health effects are particularly difficult to estimate, since no large-scale plants are operational. The only available data on emissions and effluents are based on limited information collected at pilot plant operations.

Air Emissions: The type and sources of potential air pollutants from coal conversion are shown in Table 8-1.

Sulfur dioxide is emitted principally from the tail-gas stream of the sulfur-recovery plant and from stack gases of auxiliary systems requiring fuel oxidation. These include plant boilerhouse and miscellaneous fossil-fuel-fired process heaters.

Particulate matter can be released as a fugitive dust and as a process or combustion-based stack emission. Fugitive emissions have a potential for occurring at receiving, handling, and storage areas for coal, solid waste, and from leakage from process equipment elements. Process stack emissions would include the exhaust of pollution control equipment (e.g., scrubbers and precipitators). Fuel combustion would provide the potential source of particulate matter.

Table 8-1. High-Btu-Gasification Air Pollutants

Pollutant	Process Generated	Combustion Generated
Particulate matter	x	x
Sulfur oxides	x	x
Reduced sulfur compounds	x	
Nitrogen oxides		x
Hydrocarbons	x	x
Carbon monoxide	x	x
Trace metals	x	x
Odors	x	
Other gases (including NH_3, HCN, HCl)	x	

Nitrogen oxide emissions would result from fossil fuel firing of boilers. Hydrocarbon emissions could occur from liquid storage areas, from system leaks, and from the evaporation of hydrocarbon liquids dissolved in cooling systems. Reduced sulfur compounds occur in the initial product stream of virtually all coal conversion processes.

Trace element emissions of such substances as mercury, beryllium, arsenic, and other heavy metals which are contained in coal in small amounts are expected in view of experience from coal-fired boilers. In addition, other gaseous emissions, especially hydrogen cyanides and ammonia (as well as hydrogen chloride and gaseous odorants) may also be associated with coal conversion plants.

Liquid Effluents: Wastewaters from coal conversion processes can originate from a number of sources. These include water of constitution, water added for stoichiometric process requirements, and water induced for gas scrubbing and by-product recovery. Such process waters come into contact with contaminants in coal and are likely to be a principal source of pollution. Table 8-2 indicates expected composition of wastewaters associated with one conversion approach (i.e., Synthane).

Table 8-2. Composition of Synthane By-Product Water

Pollutant	By-Product Water (mg/liter)
pH	7.9 – 9.3
COD	1,700 – 43,000
Ammonia	2,500 – 11,000
Cyanide	0.1 – 0.6
Thiocyanate	21 – 200
Phenols	200 – 6,000
Sulfide	N/D[a]
Alkalinity (as $CaCO_3$)	N/D
Specific conductance (as mhos/cm)	N/D

[a]N/D = not determined.

Solid Waste: Past analyses have indicated the expectation that in a commerical plant, residue from the gasifier would be burned along with tars to raise steam for the overall process. Table 8-3 provides representative analyses of coals and associated chars from the Synthane process. The chars would contain some trace elements. It should be noted that problems may exist with burning tars owing to residence time (in flame) and with SO_x stack gas cleaning when burning tar and char. In the past, consideration was given to solid waste disposal by means of burial in a mine. These waste materials would consist of ash from the boiler plant and gasifiers, coal wash plant waste (i.e., if wash plant is used), process sludge, and other waste. Undoubtedly, this whole disposal area requires additional effort. It should be noted that additional information for decision making is required so as to permit compliance with applicable environmental, health, and safety regulations.

Low/Medium-Btu Gasification—Identified Pollutants

Air Emissions: The type and sources of potential air pollutants from coal conversion are shown in Table 8-4.

Sulfur dioxide is emitted principally from the tail-gas stream of the sulfur-recovery plant and from stack gases of auxiliary systems requiring fuel oxidation. These include plant boilerhouse and miscellaneous fossil-fuel-fired process heaters.

Table 8-3. Representative Proximate and Ultimate Analyses of Coals and Chars, Weight Percent (Synthane Process)

	Illinois No. 6 Coal	Western Kentucky Coal	Wyoming Subbituminous Coal	North Dakota Lignite	Pittsburgh Seam Coal
Coals:					
Moisture	8.3	4.3	18.1	20.6	2.5
Volatile matter	37.5	34.6	31.9	32.9	30.9
Fixed carbon	43.0	44.5	32.0	38.2	51.5
Ash	11.2	16.6	18.0	8.3	15.1
Hydrogen	5.3	4.7	5.4	5.7	4.7
Oxygen	15.9	10.9	30.3	32.6	9.3
Carbon	63.0	62.7	45.2	51.5	68.4
Nitrogen	1.1	1.2	0.6	0.7	1.2
Sulfur	3.5	3.9	0.5	1.2	1.3
Chars (from above coals):					
Moisture	0.8	1.2	0.5	1.2	1.4
Volatile matter	4.0	4.8	5.1	10.0	1.6
Fixed carbon	69.9	63.3	38.1	50.2	69.3
Ash	25.3	30.7	56.3	38.6	27.7
Hydrogen	1.0	1.0	1.0	0.9	1.0
Oxygen	1.3	1.1	1.2	0.0	1.7
Carbon	70.4	64.5	40.6	58.9	68.9
Nitrogen	0.6	0.7	0.4	0.2	0.5
Sulfur	1.4	2.0	0.5	2.0	0.2

Table 8-4. Low/Medium-Btu-Gasification Air Pollutants

Pollutant	Process Generated	Combustion Generated
Particulate matter	x	x
Sulfur oxides	x	x
Reduced sulfur compounds	x	
Nitrogen oxides		x
Hydrocarbons	x	x
Carbon monoxide	x	x
Trace metals	x	x
Odors	x	
Other gases (including NH_3, HCN, HCl)	x	

Particulate matter can be released as a fugitive dust and as a process or combustion-based stack emission. Fugitive emissions have a potential for occurring at receiving, handling, and storage areas for coal and solid waste, and from leakage from process equipment elements. Process stack emissions would include the exhaust of pollution control equipment (e.g., scrubbers and precipitators). Fuel combustion would provide the potential source of particulate matter.

Nitrogen oxide emissions would result from fossil fuel firing of boilers. Hydrocarbon emissions could occur from liquid storage areas, from systems leaks, and from the evaporation of hydrocarbon liquids dissolved in cooling systems. Reduced sulfur compounds occur in the initial product stream of virtually all coal conversion processes.

Trace element emissions of such substances as mercury, beryllium, arsenic, and other heavy metals which are contained in coal in small amounts are expected in view of experience from coal-fired boilers. In addition, other gaseous emissions, especially hydrogen cyanide and ammonia (as well as hydrogen chloride and gaseous odorants) may also be associated with coal conversion plants.

Liquid Effluents: Wastewaters from coal conversion processes can originate from a number of sources. These include water of constitution, water added for stoichiometric process requirements, and water induced for gas scrubbing and by-product recovery. Such process waters come into contact with contaminants in coal and are likely to be a principal source of potential pollution.

There are potential sources of water effluents from boiler blowdown, the raw gas cooling system, and overfill of water clarifiers. It is expected that blowdown and raw gas cooling waters will be recycled via a clarifier and filter system for reuse. It has been estimated (i.e., for the Koppers–Totzek process) that 1.3 million gallons will be produced for every 10^{12} Btu of coal input to the gasifiers. In addition, the clarifiers will require 80 gallons per minute in makeup water because of evaporation losses in quenching the ash from the gasifier.

Solid Wastes: The solid waste generated by low-Btu gasification ranges from 3500 to 8500 tons for each 10^{12} Btu of coal input. The value is dependent on the heat and ash values of the coal. If the sulfur recovered in the process cannot be sold, it also will require disposal. The solid waste from the gasifier will resemble waste from coal cleaning and boiler plant operations. This waste (from a gasifier) contains small quantities of radioactive isotopes. Analyses for an agglomerating gasifier produced estimated levels of 0.00076 curie of radium-226 and 0.0128 curie of radium-228 and thorium-228 and -230 for each 10^{12} Btu coal input to the gasifiers.

COAL LIQUEFACTION

The environmental considerations for coal liquefaction are divided into two process categories which include (1) direct coal liquefaction and (2) indirect coal liquefaction.

Although many of the environmental issues associated with conventional fossil fuel utilization are common to coal conversion processes, liquefaction technology presents some unique problems, which include the identification of materials with carcinogenic, mutagenic, and related effects; characterization and treatment of wastes, fugitive emissions, and effluents; and disposal of sludges and solid wastes. These problems are generally common to all liquefaction technologies; however, particular processes may have to be evaluated individually. Liquefaction does have the inherent advantage of separating the processing of the coal from the ultimate utilization. Since impurities can be removed from the coal during liquefaction, a "clean" fuel can be delivered to the utilization site (possibly an urban area), and, thus, the power generation facility will not have to cope with the impurities. A major environmental consideration in direct coal liquefaction is product toxicity due to the presence of high-molecular-weight organic compounds.

Direct Coal Liquefaction—Identified Pollutants

Air Emissions: Development and commercialization of a coal liquefaction industry creates a concern with regard to the introduction of air pollutants into the atmosphere.

The typical materials produced in a coal liquefaction facility which could have a detrimental impact on air quality include hydrogen sulfide, ammonia, particulate matter (e.g., coal dust and process fines), hydrocarbons, sulfur dioxide, hydrogen cyanide, and small amounts of nitrogen dioxide, polycyclic hydrocarbons, and heavy metals. These emissions result from such activities as fuel combustion, coal preparation, sulfur recovery, ammonia storage, petroleum storage, and miscellaneous hydrocarbon losses.

The major air emissions from liquefaction facilities are generally known, and conventional control techniques may be applicable. The Dravo Corporation, in a 1976 handbook produced for the U.S. Government, provides information on a number of industrial sulfur removal systems (*Handbook of Gasifiers and Gas Treatment Systems*, FE-1772-11, February 1976). The majority of the proprietary systems described are for removing H_2S from industrial gases. Some systems in addition to removing H_2S also remove other gaseous effluents (e.g., CO_2, NH_2, HCN). Almost all of the addressed systems have been in existence for many years with significant industrial usage. Such systems include the Selexol and the Stretford processes that have been used for selectively cleaning up gaseous impurities from processes used to convert oil or coal to other fuel form(s). Liquefaction air emission streams may contain impurities which could reduce the capabilities of commercially available control technologies.

However, in some instances, advanced controls may have to be developed before such plants are constructed on a commercial scale. In addition, airborne pollutants will be transported into the general environment and possibly transformed into other compounds after emission from coal liquefaction facilities. Conventional models are generally adequate to predict atmospheric dispersion for a high stack and flat land scenario. However, more detailed atmospheric transformation and dispersion information will be required to fully utilize these models.

The 1977 Clean Air Act amendments mandate that fossil energy facilities, including coal conversion plants, utilize the "best available technology" to control pollutants. Coal liquefaction (and other process facilities) constructed in nonattainment areas will be subject to emission trade-off policies. The energy and cost penalties of applicable air pollution controls must be characterized as well as the secondary pollutants which may be emitted by the controls.

Liquid Effluents: Coal liquefaction processes may produce waste effluents which have broad temperature and pH ranges and may contain a variety of materials such as suspended particles, ammonia, toxic trace metals, phenols, aromatic hydrocarbons, thiophenes, aromatic amines, and other organic compounds. Conventional control and wastewater treatment techniques possibly could be applied to some of these materials. However, particularly troublesome areas requiring more study include phenols, trace metals, and the final disposal of the effluents.

Water quality may also be affected by gaseous streams, fugitive effluents, and air emissions, which may settle or be washed into water bodies by rain. Improper handling or disposal of solid wastes may also release dissolved and suspended solids and organics. Control and treatment options compatible with water discharge standards should be identified and their effects evaluated. Ultimate discharges (after treatment) can generally be projected for each coal liquefaction process. The general effects of these discharges at specific locations on indigenous aquatic organisms and communities can also be predicted.

Effluent constituents may accumulate and/or be transformed in the water column and biotic sediment or aquatic ecosystems. Current methods for predicting the movement of waste contaminants through surface and groundwater systems must be evaluated for locations where liquefaction facilities may be located.

Solid Waste: Solid wastes generated by coal liquefaction processes consist primarily of ash and refuse removed from the coal and sludges and solids recovered from waste-treatment processes. The major solid waste streams, as well as minor ones such as spent catalyst, must be characterized and appropriate disposal techniques determined. Where appropriate, new treatment and disposal techniques may need to be developed.

Conventional disposal of solid wastes (especially ash) in offsite landfills will require transport and handling equipment and relatively large areas of land. The handling, transportation, and disposal of wastes must be controlled to prevent fugitive dust emissions and accidental discharges. Groundwater leaching is another concern which must be evaluated if landfills are used as disposal areas for coal liquefaction wastes. Physical and chemical reactions involved, effects of various methods of disposal upon leachability, effective control and containment techniques, and compliance with new state hazardous waste disposal regulations should all be evaluated.

DOE has estimated that the total solid wastes to be disposed of by a large-scale facility would be about 1200–2800 tons per day for a coal liquefaction plant, and 1000–2500 tons per day for an SRC-I plant. Most of these wastes will be in the form of ash. Disposal of these wastes would cover approximately 300–700 acres to a depth of 10 ft over a 20-year period. Approximately 250–525 acres would be needed to dispose of wastes from an SRC plant.

Indirect Coal Liquefaction—Identified Pollutants

Air Emissions: Development and commercialization of a coal liquefaction industry creates a concern with regard to the introduction of air pollutants into the atmosphere. The typical materials produced in a coal liquefaction facility which could have a detrimental impact on air quality include hydrogen sulfide, ammonia, particulate matter (e.g., coal dust and process fines), hydrocarbons, sulfur dioxide, hydrogen cyanide, small amounts of nitrogen dioxide, polycyclic hydrocarbons, and heavy metals. These emissions result from such activities as coal handling and preparation, fuel combustion, coal gasification, raw gas and liquid product cleanup, sulfur recovery, catalyst regeneration, and product upgrading and storage.

The major air emissions from liquefaction facilities are generally known and conventional control techniques possibly could be effectively applied. The majority of the proprietary systems are for removing H_2S from industrial gases. Some systems in addition to removing H_2S also remove other gaseous effluents (e.g., CO_2, NH_2, HCN). Almost all of the addressed systems have been in existence for many years with significant industrial usage. Such systems include the Rectisol and the Stretford processes that have been used for selectively cleaning up gaseous impurities from processes used to convert oil or coal to other fuel form(s). Liquefaction air emission streams may contain impurities which could reduce the capabilities of commercially available control technologies.

However, in some instances, advanced controls may have to be developed before coal conversion plants are constructed on a commercial scale. In addition, airborne pollutants will be transported into the general environment and possibly transformed into other compounds after emission from coal liquefaction facilities.

The 1977 Clean Air Act amendments mandate that fossil energy facilities, including coal conversion plants, utilize the best available technology to control pollutants. Coal liquefaction (and other process facilities) constructed in nonattainment areas will be subject to emission trade-off policies. The energy and cost penalties of applicable air pollution controls must be characterized as well as the secondary pollutants which may be emitted by the controls.

Liquid Effluents: Coal liquefaction processes may produce waste effluents which have broad temperature and pH ranges and may contain a variety of materials such as suspended particles, ammonia, toxic trace metals, phenols, aromatic hydrocarbons, thiophenes, aromatic amines, and other organic compounds. Conventional control and wastewater treatment techniques can be applied to most of these materials. However, particularly troublesome areas requiring more study include phenols, trace metals, and the final disposal of the effluents.

Water quality may also be affected by gaseous streams, fugitive effluents, and air emissions, which may settle or be washed into water bodies by rain. Improper handling or disposal of solid wastes may also release dissolved and suspended solids and organics. Control and treatment options compatible with water discharge standards should be identified and their effects evaluated.

Effluent constituents may accumulate and/or be transformed in the water column and biotic sediment or aquatic ecosystems. Current methods for predicting the movement of waste contaminants through surface and groundwater systems must be evaluated for locations where liquefaction facilities may be located.

Solid Waste: Solid wastes generated by coal liquefaction processes consist primarily of (gasifier) ash and refuse removed from the coal and sludges and solids recovered from waste-treatment processes. The major solid waste streams, as well as minor ones such as spent catalyst, must be characterized and appropriate disposal techniques determined. Where appropriate, new treatment and disposal techniques may need to be developed.

Conventional disposal of solid wastes (especially ash) in offsite landfills will require transport and handling equipment and relatively large areas of land. The handling, transportation, and disposal of wastes must be controlled to prevent fugitive dust emissions and accidental discharges. Groundwater leaching is another concern which must be evaluated if landfills are used as disposal areas for coal liquefaction wastes. Physical and chemical reactions involved, effects of various methods of disposal upon leachability, effective control and containment techniques, and compliance with new state hazardous waste disposal regulations must all be evaluated.

DOE has estimated that the total solid wastes to be disposed of by a large-scale Fischer–Tropsch facility would be about 1000 to 4500 tons per day. Most of these wastes will be in the form of ash. Disposal of these solid wastes (from a Fischer–Tropsch plant) would cover approximately 250 to 1125 acres to a depth of 10 ft over a 20-year period.

Environmental Control Technology Needed for Synfuel Utilization

The utilization of synthetic fuel products and by-products will require improvements in existing environmental control technology and the development of new technologies. In order to assess the control technology requirements, it is necessary to first understand the hazards associated with synthetic fuel utilization. This may be accomplished by determining the constituents of synthetic fuel products and by-products, their transformation upon use, and ultimate fate in the environment. These data in turn must be tied closely to the product buildup rate described in each of the scenarios, since these impact the types of products produced and their rate of penetration into the market. Once these factors are understood, then control technology options may be evaluated.

EXISTING DATA REGARDING HAZARDS OF SYNTHETIC FUEL PRODUCTS IS SPARSE

At the current time there is a lack of sufficient data available to properly assess the potential risks associated with the utilization of synthetic fuel products and by-products. The development of these data will require significant efforts on the part of the government, industry, and the academic community to generate sound, reliable information to ensure minimum risks to the health and welfare of the nation as synfuels are introduced into the market. This synthetic fuel database must contain not only accurate and representative information about the physical properties, chemical composition, and biological activities of synthetic fuels, but must also contain equally comprehensive data on the end uses of the products and by-products.

The DOE and EPA are presently conducting research efforts on synthetic fuel product characterization. The results of some of the shale oil and coal liquid products are becoming available. An example of the preliminary analysis of these two products compared with petroleum crude is presented in Table 8-5. There are some similarities in the diaromatic content between shale oil and petroleum crude, with coal liquids having the highest content. This factor may be significant if a spill of these products occurred, as impacts on water pollution would be less than from coal liquids. A comparison between coal- and petroleum-derived gasolines indicates significant variations in aromatics and unidentified compounds. Owing to the high aromatic content of the coal-derived gasoline, potential adverse health effects may occur from widespread use of this fuel in automotive applications. Some of the synthetic fuel products and by-products may be classified as toxic chemicals under the Toxic Substances Control Act (TSCA).

Table 8-5. Diaromatic Content of Synthetic Crudes and Crude Oils

	Concentration (mg/g)[a]		
Constituent	Typical Shale Oil	Coal Syncrude	Petroleum Crude
Naphthalene	1.39	1.68	0.87
2-Methylnaphthalene	0.91	3.47	1.04
1-Methylnaphthalene	0.68	1.11	0.75
Biphenyl	0.06	0.44	T
2,6-Dimethylnaphthalene	0.10	0.81	0.08
1,3/1,6-Dimethylnaphthalene	1.63	3.01	1.48
2,3-Dimethylnaphthalene	0.28	1.53	0.51
1,5-Dimethylnaphthalene	0.03	0.67	0.08
1,2-Dimethylnaphthalene	0.19	0.23	0.31
Acenaphthalene	0.26	2.19	0.30
Acenaphthene	T	0.30	ND
Total	5.53	15.44	5.42

[a]T = trace, ND = not detected.
Source: Oak Ridge National Laboratory

Preliminary health effects studies have indicated that coal liquids have industrial toxicity ratings similar to those of benzoic acid, phosphoric acid, sodium tartrate, and polychlorinated biphenyls (PCB). Coal liquids have also been found to be less toxic than pesticides such as dieldrin and chlordane, and more toxic than crude petroleum and shale oil. Historical epidemiological and animal studies have established that coal tars and pitches from coal coking, gasification, and combustion possess a carcinogenic nature. Although these studies are not all directly comparable, it would appear that some high-boiling-point products from direct liquefaction processes or from coal pyrolytic processes may possess a high degree of carcinogenicity.

It is apparent that although work has started in the right direction to assess synthetic fuel hazards, much work still needs to be conducted. As the physical, chemical, and biological results are analyzed and potential risks are evaluated, decisions can start to be made as to the various pollution control technologies that can be most effectively applied in the utilization of synthetic fuels.

Product Buildup Rates Will Determine Magnitude of Environmental Impacts

Once the hazards of synthetic fuel products and by-products are known, their relative impacts on the environment will depend on the product buildup rate and market penetration as described for each of the scenarios. All media—air, water, and land—must be considered

Air pollution impacts will occur primarily from the combustion of synthetic fuels in stationary and mobile sources, with some impacts from fugitive emissions occurring during transportation, storage, and handling operations. Under the National Goal Scenario (noted in Chapter 7), coal liquids and shale oil products will contribute the greatest percentage of products. A level of 1.4 MMBPD of these fuels will be produced in 1992 and continue through 2000. Coal liquids will most likely be used in all sectors of the market including utilities, transportation, industry, and commerce. As most of the products will be used in stationary sources, the air pollution impacts are expected to be less than from shale oil products, all of which will be used by transportation sources. The individual mobile sources do not lend themselves to as effective emission controls as centralized stationary sources. Owing to the moderate amount of petroleum product use that is expected to be replaced by synthetic liquids, the air pollution impacts are expected to be moderate.

Under the nominal rate scenario (Scenario 2, in Chapter 7), only 0.5 MMBPD of liquid fuels will be produced in 1992, and the 1.3 MMBPD level will not be reached until 1998. This will provide relatively lower air pollution impacts from liquids combustion than the National Goal Scenario. The use of low- and medium-Btu gas is projected to be higher under Scenario 2 than Scenario 1, although air pollution impacts are not considered to be significant since these products will most likely be used for in-plant and feedstock applications.

The greatest relative impact would occur under the accelerated rate scenario, as the quantities of each product are higher than for each of the other two scenarios. By 1992, shale oil and coal liquid production reach a level of 1.8 MMBPD and as much as 3.5 MMBPD by 2000. Shale oil in this period is in excess of 0.9 MMBPD, all of which is used in transportation sources. As shale oil products can be used virtually anywhere in the United

States, there is very little of the population that may not be exposed to the combustion products. If the majority of the products are used by the military sector, the geographic area of use may be better defined. Significant market penetration under this scenario will also be made by SNG, which may be used in all sectors with the exception of transportation. As this product has widespread application, its combustion must be accurately defined to determine if combustion will produce air pollution impacts different from use of natural gas.

Water pollution will occur primarily from spills associated with the transportation of synthetic liquids. As the production of these is greatest under the accelerated rate scenario, it provides the greatest potential for these impacts. The crude and refined shale oils, as well as coal liquids, will be transported over long distances by pipelines, and then to the markets by various modes of transportation. The loaning of tankers and barges and transportation of the products by waterways provide a moderate degree of spill potential.

Solid wastes will be generated primarily by the pollution control systems used during synfuel utilization. These systems will be limited to stationary source applications where the coal liquids and gases are used in utilities and for industrial processes. As the quantities of solid wastes produced will be dependent on the amount of these fuels used, it will have the greatest impact under the accelerated rate scenario. Oil shale products will not contribute to these impacts since they will be used in transportation sources. By 1992 under this scenario, coal-derived fuels will be produced at a level of 1.8 MMBPD and 4.1 MMBPD by 2000. The majority of these products will be used in stationary sources with emission control systems producing solid wastes. Under Scenario 2, only 1.7 MMBPD of coal-derived fuels will be produced by 2000, and 1.8 MMBPD under the National Goal Scenario. As another example of the need to determine synthetic fuel composition, the solid wastes generated by control systems may contain toxic or hazardous components which upon disposal may leach into groundwaters at waste disposal sites.

On the basis of the information presented, significant data need to be developed to assess control technology options. The optimal method of control, if achievable, would be to upgrade the products to remove as much of the pollutant source content as possible rather than rely on downstream pollution controls. This would have significant benefits on pollution impacts that may occur prior to product utilization. As an example, fugitive emissions into the atmosphere or spills into waterways would not be expected to be severe if the majority of pollutants were removed during the manufacture of the product.

Once the products are ready for combustion, emission controls will be necessary if product combustion is not fully successful. Recent small-scale tests of synthetic fuel combustion have provided encouraging results from an environmental perspective. Several combustion tests of SRC liquids and solids, EDS and H-Coal liquids, shale oil, and coal-derived gases have been conducted. For test purposes, some of the combustion devices were not equipped with high-efficiency pollution control devices. Once the products are used in commerce, best available control technology (BACT) will be required.

Significant Environmental Issues for Synfuel Utilization

- Although a few synfuel products have been included in the toxic substances inventory, most synfuels may be designated as new products under TSCA. EPA will have to identify potential risks associated with the transport and use of synfuel products and by-products, as well as their end uses. Risk and exposure concerns depend on the market infrastructure and likely end use of the variety of products that will result. More diverse end uses and methods of handling, storage, and distribution will increase the exposure potential.

- In addition to TSCA, stipulations of the Clean Air Act will also impact the synfuels market. Atmospheric emissions from fugitive sources are potentially an environmental concern, as well as end-use combustion emissions. These emissions must be characterized so that BACT determinations can be made. Similarities and differences with related petroleum products need to be evaluated.
- Potential atmospheric emissions are much more diverse than the limited set of criteria pollutants which constitute the majority of air pollution concerns today. A critical issue is not so much that hydrocarbons may be an emission, but rather an assessment is needed of the types of other organic emissions and the associated risks.
- The potential of accidental spills in the transport and storage of synfuel products and by-products is one of the most critical concerns for protection of groundwater quality and dependent drinking water sources, as stipulated by the Clean Water Act. Additional contamination of receiving waters could be caused by area washdown and stormwater runoff at facilities where minor leakage occurs.
- RCRA requirements will include an integrated solid and hazardous waste management program. Waste oils, storage tank sludges, disposable materials (seals, packing, etc.), and ash residues can all be anticipated from synfuel usage, in addition to waste by-products.
- There is a high probability that synfuels will be blended with petroleum products, either as refinery and petrochemical feeds or as products at end-use locations. EPA will have to judge the applicability of existing regulations covering petroleum product transport and use when the product characterizations are related to blend ratios. Furthermore, synfuel materials that will be used as chemical feedstocks will require environmental assessments regarding their physical, chemical, and biological acceptability.
- The eventual complexity and diversity of the synfuels market infrastructure will represent a challenge to traditional environmental monitoring and inspection procedures, as well as control technology assessment.
- Some of the control approaches will be equipment and operations oriented. This characteristic will require a close EPA interface with other regulatory agencies (such as DOT, ICC, and Coast Guard) regarding transportation operations which are both safe and environmentally acceptable.
- The feasibility of segregating the handling and end-use of potentially hazardous synfuels will certainly have to be evaluated. Proper assessment of environmental risks from synfuel product end-use will be needed to establish exposure estimates.

Permitting and Progress

EPA's regulatory role in an emerging synfuels market will involve permitting for the production, storage, transportation, and end use of the products. Permitting procedures will have to be streamlined to eliminate unnecessary delays in the long-range national goal of reducing petroleum imports. TSCA requirements will be particularly critical in this emerging industry. Plans have been announced by some industries to begin construction of plants to supply SNG and chemical feedstocks. Synfuel projects scheduled for the mid-1980s include shale oil development in Colorado and Utah. With typical engineering and design efforts requiring 2 years, and construction another 2–3 years, it is essential that all permitting be complete with 1 year to keep these critical developments on schedule.

Chapter 9

Economic Comparisons

This chapter provides brief process descriptions and economic comparisons for synthetic fuel technologies. The specific processes addressed in this chapter are listed below:

Indirect Liquefaction

- Methanol (Lurgi Gasification)
- Fischer–Tropsch (Lurgi Gasification)
- Mobil M-Gasoline
- Mobil M-Gasoline (Gasoline Production Only)
- Mobil M-Gasoline (Advanced)

Direct Liquefaction

- Exxon Donor Solvent
- H-Coal
- SRC-II

Gasification

- Lurgi (Dual Feed)
- BiGas
- Synthane
- Hygas, Steam–Oxygen
- Hygas, Steam–Iron
- CO_2 Acceptor

Oil Shale

- Tosco II Aboveground Retorting
- Occidental Modified *In Situ*

The economic comparisons were adapted from the TRW *Synthetic Fuel Cost Estimating Notebook* and conform with the economic comparison bases described in Table 9-1. The precision of the resulting product selling prices is estimated to be −20–+40%.

Table 9-1. Economic Comparison Basis

- 20-Year Production Life
- 16-Year Depreciation—Sum-of-Years Digits
- 20% Investment Tax Credit
- DCF Rates—3%, 6.5%, 10%, 16%, 18%
- 5-Year Construction Schedule—2%/10%/30%/47%/11%
- Costs Updated to Mid-1979 via Nelson Refinery Index
- Working Capital Recovered after 20 Years
- 50% Production in First Year
- Coal Prices; Eastern Bituminous—\$1.20/MMBtu (\$25 – 27/ton)
  Western Subbituminous—\$0.55/MMBtu (\$9.30 – 10.00/ton)
- Tax Rate = 48.16% (Composite Federal and State)
- Advalorum Taxes @ 2.75% Investment
- 15% Contingency
- Full Tax Loss Pass Through to Parent Corporation
- Composite Pricing on a per MMBtu Basis

The product selling price ranges for the different process categories are summarized in Table 9-2. The selling prices for indirect liquefaction are about \$15/BCOE higher than for direct liquefaction. However, the indirect liquefaction processes produce transportation-grade fuels, while the products of the direct liquefaction processes must be upgraded by further processing to be of transportation fuel quality.

Table 9-2. Transportation Fuel Selling Price Summary

Process Category	Major Products	Selling Price Range (\$/BCOE, 16% DCF)
Indirect Liquefaction	SNG, Methanol, Naphtha Gasoline	48 – 78
Direct Liquefaction	SNG, Naphtha, Fuel Oil Syncrude	33 – 39
Gasification	SNG	36 – 45

In all the process descriptions and economic comparisons, the following apply:

1. Power at 3415 Btu/kWh.
2. BCOE at 5.8 MMBtu/lb.
3. $EFF = \frac{\text{HHV of Products + By-Products}}{\text{HHV of Coal + Power}}$, not including S or NH_3.
4. $\text{Coal to Product EFF} = \frac{\text{HHV Product}}{\text{HHV Coal}}$.

Indirect Liquefaction

METHANOL (LURGI GASIFICATION)

In this process, sized coal is gasified at pressure with oxygen and steam in a Lurgi gasifier, shown in Fig. 9-1. The raw synthesis gas is then shifted to yield an H to CO ratio of about 2:1, and the shifted gases are quenched with water and fed to an acid gas removal

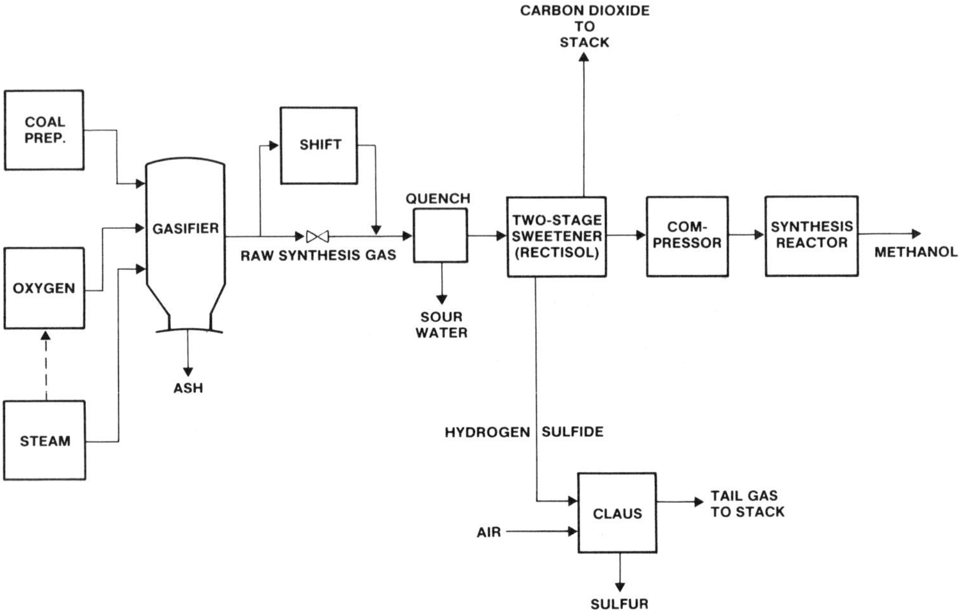

Fig. 9-1. Block diagram for methanol (Lurgi gasification) process.

unit to eliminate CO_2 and sulfur compounds. The purified synthesis gas is then fed to the methanol synthetic reactor. This reactor operates at about 1500 psi with a copper catalyst containing zinc and aluminum additives. The raw methanol from the synthesis reactor is dried by distillation.

FUEL CHARACTERIZATION
TYPE: Wyoming Subbituminous
HIGHER HEATING VALUE (Btu/lb): 8,509

FUEL/POWER INPUTS

	TPSD	*MMTPY*		*10^12 Btu/yr*
FUEL	27,334	8,979		152.81
POWER[1]	(6.55MW)			0.18
			TOTAL	152.99

PRODUCT/BY-PRODUCT CHARACTERIZATION/OUTPUT

PRODUCT/CHARACTERIZATION		*BCOE[2]/SD*	*MMBCOE/yr*	*10^12 Btu/yr*
SNG 975 Btu/scf	146.4 MMSCFD	24,610	8.08	46.89
MeOH, 9724 Btu/lb, 7080 TPD	50,840 BPSD	23,740	7.80	45.23
Naphtha 5.2 MMBtu/bbl	1,315 BPSD	1,179	0.39	2.25
	TOTAL	49,529	16.27	94.37

BY-PRODUCT CHARACTERIZATION

Sulfur	61 TPSD
Ammonia	103 TPSD

PROCESS EFFICIENCY

EFFICIENCY[3]: 61.7% COAL TO PRODUCT EFFICIENCY[4]: 61.8%

CAPITAL REQUIREMENT

ITEM	*$MM (mid-1979)*
PLANT INSTALLED COST	1,533.4
CONTINGENCY (15%)	230
OWNER COST	26
ROYALTIES	19
INITIAL CATALYST and CHEMICALS	6
START-UP COSTS	—
TOTAL PLANT INVESTMENT	1,805
WORKING CAPITAL	66
TOTAL CAPITAL REQUIREMENT	1,871

ANNUAL COST

ITEM	*$MM (mid-1979)*
FUEL ($0.55/MMBtu)	84.0
OPERATION AND MAINTENANCE	139.2
SUBTOTAL	223.2
BY-PRODUCT CREDITS	11.0
NET ANNUAL COST	212.2

COMPOSITE PRODUCT PRICE $/BCOE ($/MMBtu)

DCF %	*3*	*6.5*	*10*
COAL	5.16(0.89)	5.16(0.89)	5.16(0.89)
O&M	7.88(1.36)	7.88(1.36)	7.88(1.36)
CAPITAL	6.62(1.14)	11.42(1.97)	17.55(3.02)
TOTAL	19.66(3.39)	24.46(4.22)	30.59(5.27)

DCF %	*10 (No Tax)*	*16*	*18*
COAL	5.16(0.89)	5.16(0.89)	5.16(0.89)
O&M	7.88(1.36)	7.88(1.36)	7.88(1.36)
CAPITAL	16.24(2.80)	31.23(5.38)	36.69(6.33)
TOTAL	29.29(5.05)	44.28(7.63)	49.74(8.58)

ECONOMIC CONSIDERATIONS

STREAM FACTOR: 90%

ORIGINAL ESTIMATE DATE: 3Q 77

ESCALATION CORRECTIONS: CAPITAL 1.13 O&M 1.12

ALLOCATION OF PRODUCT PRICE

Required revenues = $1.515 MM/SD @ 10% DCF and $2.193 MM/SD @ 16% DCF. With SNG @ $3.50/MMBtu and naphtha @ $26/bbl the required selling price of methanol is $19.29/bbl (46¢/gal or $7.15/MMBtu) @ 10% DCF or $32.65/bbl (78¢/gal or $12.10/MMBtu) at 16% DCF.

UNIT OPERATION CONTRIBUTION TO COMPOSITE PRODUCT PRICE (CAPITAL INVESTMENT EFFECTS ONLY)

UNIT OPERATION	%
Coal Handling	6.3
Gasification	20.1
Shift	1.3
Gas Cooling	1.9

Gas Purification	7.2
Sulfur Recovery	5.9
Gas Liquor Separation	1.9
Phenol, NH_3	3.3
Methanation	1.8
CO_2 Removal, Drying	0.7
Methanol Synthesis	5.4
Methanol Distillation	0.7
Naphtha Hydrotreating and H_2 Recovery	0.6
Miscellaneous	2.1
Oxygen Plant	11.0
Boiler Facilities	11.6
Off-sites	13.6
General Facilities	4.6
	100.0

Reference

Research Guidance Studies to Assess Gasoline from Coal by Methanol-to-Gasoline and SASOL-Type Fischer-Tropsch Technologies, Mobil R&D Corp., FE-2247-13, February 1978.

FISCHER–TROPSCH (LURGI GASIFICATION)

The Fischer–Tropsch process, shown in Fig. 9-2, is a catalytic conversion system that produces hydrocarbon liquids from coal-derived intermediate-Btu synthesis gas. The process is in commercial operation in South Africa where a Lurgi gasifier is used to produce synthesis gas. The gas from the Lurgi gasifier is cleaned of hydrogen sulfide, carbon dioxide, and impurities, then shift-converted before it enters a catalytic reactor

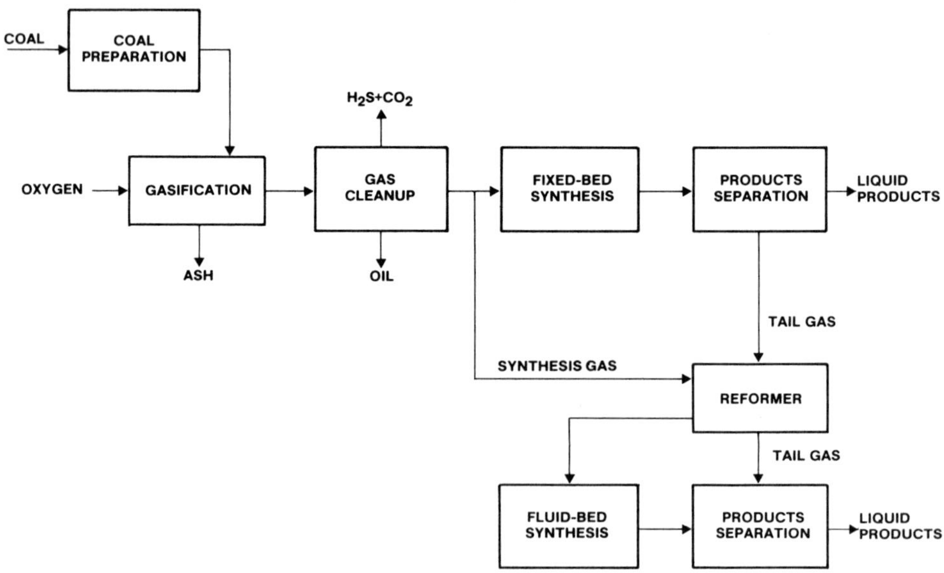

Fig. 9-2. Block diagram for Fischer–Tropsch (Lurgi gasification) process.

which produces hydrocarbon liquids. In this plant, two reactors, using different catalysts and temperatures, process gases with different carbon-to-hydrogen ratios into different products. Reactor products include gasoline, diesel/other fuel oils, waxes, alcohols, and ketones.

FUEL CHARACTERIZATION

TYPE: Wyoming Subbituminous

HIGHER HEATING VALUE (Btu/lb): 8,509

FUEL/POWER INPUTS

	TPSD	*MMTPY*		*10^{12} Btu/yr*
FUEL	27,698	9,099		154.84
POWER[1]	—	—		—
			TOTAL	154.84

PRODUCT/BY-PRODUCT CHARACTERIZATION/OUTPUT

PRODUCT/CHARACTERIZATION		*BCOE[2]/SD*	*MMBCOE/yr*	*10^{12} Btu/yr*
SNG, 1003 Btu/scf	173.3 MMSCFD	29,969	9.84	57.10
Gasoline, 87 Octane, 5.023 MMBtu/bbl	13,580 BPSD	11,761	3.86	22.41
Diesel Oil, 56°API, 5.345 MMBtu/bbl	2,307 BPSD	2,126	0.70	4.05
Heavy Fuel Oil, 41°API, 5.682 MMBtu/bbl	622 BPSD	609	0.20	1.16
Mixed Alcohols, 3.425 MMBtu/bbl	2,025 BPSD	1,196	0.39	2.28
Propane LPG, 3.815 MMBtu/bbl	1,107 BPSD	728	0.24	1.39
Butane LPG, 4.321 MMBtu/bbl	146 BPSD	109	0.04	0.21
	TOTAL	46,498	15.27	88.60

BY-PRODUCT CHARACTERIZATION		*BCOE/SD*	*MMBCOE/yr*	*10^{12} Btu/yr*
By-products	61 TPSD			
Sulfur	103 TPSD			
Power	7.01 MW	99	0.03	0.19
TOTAL OUTPUT			15.30	88.79

PROCESS EFFICIENCY

EFFICIENCY[3]: 57.4% COAL TO PRODUCT EFFICIENCY[4]: 57.2%

CAPITAL REQUIREMENT

ITEM	*$MM (mid-1979)*
PLANT INSTALLED COST	1,817
CONTINGENCY (15%)	273
OWNER COST	26
ROYALTIES	15
INITIAL CATALYST AND CHEMICALS	3
START-UP COSTS	—
TOTAL PLANT INVESTMENT	2,134
WORKING CAPITAL	68
TOTAL CAPITAL REQUIREMENT	2,202

ANNUAL COST

ITEM	*$MM (mid-1979)*
FUEL ($0.55/MMBtu)	85.2
OPERATION AND MAINTENANCE	170
SUBTOTAL	255.2
BY-PRODUCT CREDITS	8
NET ANNUAL COST	247.2

COMPOSITE PRODUCT PRICE $/BCOE ($/MMBtu)

DCF %	*3*	*6.5*	*10*
COAL	5.58(0.96)	5.58(0.96)	5.58(0.96)
O&M	10.61(1.83)	10.61(1.83)	10.61(1.83)
CAPITAL	8.30(1.43)	14.30(2.47)	21.98(3.79)
TOTAL	24.48(4.22)	30.49(5.26)	38.17(6.58)

DCF %	*10 (No Tax)*	*16*	*18*
COAL	5.58(0.96)	5.58(0.96)	5.58(0.96)
O&M	10.61(1.83)	10.61(1.83)	10.61(1.83)
CAPITAL	20.40(3.52)	39.15(6.75)	46.00(7.93)
TOTAL	36.59(6.31)	55.34(9.54)	62.19(10.72)

ECONOMIC CONSIDERATIONS

STREAM FACTOR: 90%

ORIGINAL ESTIMATE DATE: 3Q 77

ESCALATION CORRECTIONS: CAPITAL 1.13 O&M 1.12

ALLOCATION OF PRODUCT PRICE

Required revenues = $1.775 MM/SD @ 10% DCF and $2.573 MM/SD @ 16% DCF. With SNG @ $3.50/MMBtu, C_3 and C_4 LPG @ $15.00/bbl and $16.50/bbl, diesel oil @ $22.50/bbl, fuel oil (#1) @ $24.00/bbl, and mixed alcohols at $25/bbl, the required selling price for gasoline is $87.66/BCOE ($75.90/bbl or $1.81/gal) at 10% DCF and $155.50/BCOE ($134.68/bbl or $3.21/gal) at 16% DCF.

UNIT OPERATION CONTRIBUTION TO COMPOSITE PRODUCT PRICE (CAPITAL INVESTMENT EFFECTS ONLY)

UNIT OPERATION	%
Coal Handling	5.5
Gasification	16.9
Shift	1.1
Gas Cooling	1.6
Purification	6.0
Sulfur Recovery	5.0
H_2, HC Recovery	2.4
Methanation	1.0
CO_2 Removal, Drying	1.0
Fischer–Tropsch Synthesis	6.4
Hydrotreating, Fractionation Hydrogenation	2.4
Reforming, Isomerization, Polymerization, Alkylation	1.8
Alcohol Recovery, H_2 Purification	1.2
Miscellaneous	2.2
O_2 Production	9.3
Steam Boilers	12.4

Waste Water Treatment	2.2
Refrigeration	2.3
FT Catalyst Prep.	2.3
Off-sites	13.1
General Facilities	3.9
	100.0

Reference

Research Guidance Studies to Assess Gasoline from Coal by Methanol-to-Gasoline and SASOL-Type Fischer–Tropsch Technologies, Mobil R&D Corp., FE-2247-13, February 1978.

MOBIL M-GASOLINE

The Mobil M-Gasoline process, shown in Fig. 9-3, takes crude methanol produced from coal-derived synthesis gas and preheats it to 600°F (316°C) at 335 psig. It is then converted in a fixed-bed catalytic reactor to an equilibrium mixture of methanol, dimethyl ether, and water, releasing about 20% of the total heat of reaction. These products are diluted with cooled recycled gas composed mainly of process-derived light hydrocarbons, and fed to a second catalytic reactor for conversion to gasoline. The inlet temperature of the second reactor is about 680°F (360°C), and the adiabatic temperature rise, which is controlled by the recycle ratio, is about 200°F (112°C). The gaseous phase is recycled, and the condensed phase, raw gasoline, is separated from the water by-product and sent to storage.

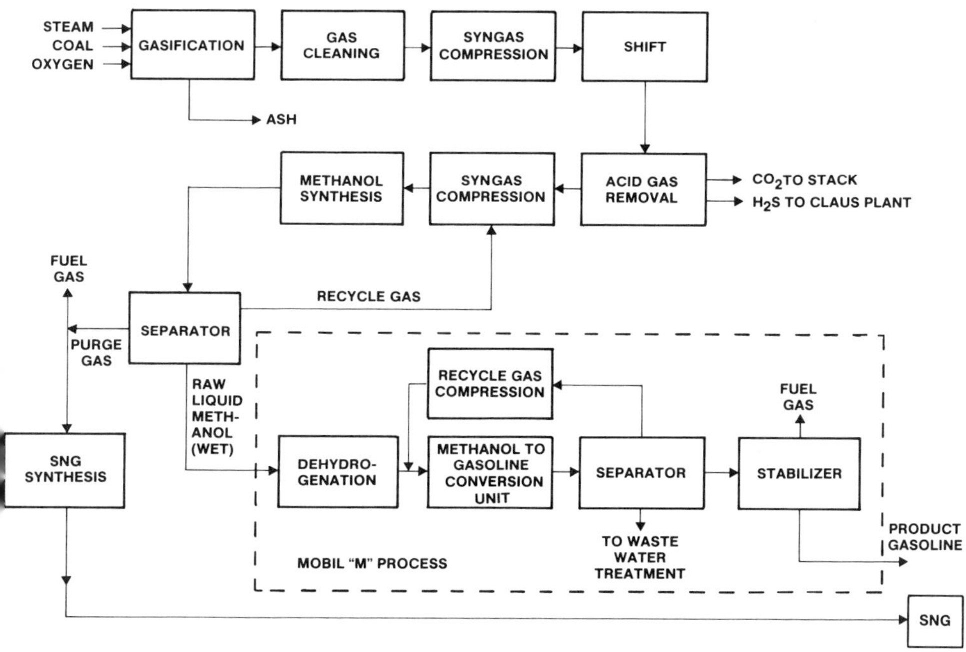

Fig. 9-3. Block diagram of Mobil M-Gasoline process.

FUEL CHARACTERIZATION
TYPE: Wyoming Subbituminous
HIGHER HEATING VALUE (Btu/lb): 8,509

FUEL/POWER INPUTS

	TPSD	*MMTPY*		*10^{12} Btu/yr*
FUEL	27,334	8,979		152.81
POWER[1]	—	—		—
			TOTAL	152.81

PRODUCT/BY-PRODUCT CHARACTERIZATION/OUTPUT

PRODUCT/CHARACTERIZATION		*BCOE[2]/SD*	*MMBCOE/yr*	*10^{12} Btu/yr*
SNG, 980.4 Btu/scf	148.52 MMSCFD	25,150	8.25	47.83
Gasoline, 87 Octane, 61.4°API, 5.105 MMBtu/bbl	22,045 BPSD	19,403	6.37	36.97
Propane LPG, 3.816 MMBtu/bbl	1,555 BPSD	1,023	0.34	1.95
Butane LPG, 4.19 MMBtu/bbl	2,205 BPSD	1,593	0.52	3.04
	TOTAL	47,124	15.48	89.79

BY-PRODUCT CHARACTERIZATION		*BCOE/SD*	*MMBCOE/yr*	*10^{12} Btu/yr*
Sulfur	61 TPSD			
Ammonia	103 TPSD			
Power	5.31 MW	75	0.02	0.14
Coal Fines (8509 Btu/lb)	1,230 TPD	3,609	1.19	6.88
	TOTAL BY-PRODUCTS	3,684	1.21	7.02
	TOTAL OUTPUT	50,808	16.69	96.81

PROCESS EFFICIENCY
EFFICIENCY[3]: 63.4% COAL TO PRODUCT EFFICIENCY[4]: 58.8%

CAPITAL REQUIREMENT

ITEM	*$MM (mid-1979)*
PLANT INSTALLED COST	1,662
CONTINGENCY (15%)	249
OWNER COST	26
ROYALTIES	19
INITIAL CATALYST AND CHEMICALS	6
START-UP COSTS	—
TOTAL PLANT INVESTMENT	1,952
WORKING CAPITAL	66
TOTAL CAPITAL REQUIREMENT	2,018

ANNUAL COST

ITEM	*$MM (mid-1979)*
FUEL ($0.55/MMBtu)	84.0
OPERATION AND MAINTENANCE	147.5
SUBTOTAL	231.5
BY-PRODUCT CREDITS	10.5
NET ANNUAL COST	221.0

COMPOSITE PRODUCT PRICE \$/BCOE (\$/MMBtu)

DCF %	*3*	*6.5*	*10*
COAL	5.43(0.94)	5.43(0.94)	5.43(0.94)
O&M	8.85(1.53)	8.85(1.53)	8.85(1.53)
CAPITAL	7.50(1.29)	12.93(2.23)	19.88(3.42)
TOTAL	21.78(3.75)	27.21(4.69)	34.15(5.89)

DCF %	*10 (No Tax)*	*16*	*18*
COAL	5.43(0.94)	5.43(0.94)	5.43(0.94)
O&M	8.85(1.53)	8.85(1.53)	8.85(1.53)
CAPITAL	18.43(3.18)	35.40(6.10)	41.60(7.17)
TOTAL	32.71(5.64)	49.68(8.57)	55.87(9.63)

ECONOMIC CONSIDERATIONS

STREAM FACTOR: 90%

ORIGINAL ESTIMATE DATE: 3Q 77

ESCALATION CORRECTIONS: CAPITAL 1.13 O&M 1.12

ALLOCATION OF PRODUCT PRICE

Required revenues = \$1.609 MM/SD @ 10% DCF and \$2.341 MM/SD @ 16% DCF. With SNG @ \$3.50/MMBtu, C_3 LPG @ \$15/bbl, and C_4 LPG @ \$16.50/bbl, the required selling price for gasoline is \$47.17/bbl (\$1.12/gal) @ 10% DCF and \$80.38/bbl (\$1.92/gal) @ 16% DCF. This compares to current spot prices of 60–69¢/gal for unleaded gasoline.

UNIT OPERATION CONTRIBUTION TO COMPOSITE PRODUCT PRICE (CAPITAL INVESTMENT EFFECTS ONLY)

UNIT OPERATION	%
Coal Handling	5.8
Gasification	18.5
Gas Cooling	1.8
Shift	1.2
Acid Gas Removal	6.6
Sulfur Recovery	5.4
Gas Liquor Separation	1.7
Phenol, NH_3 Recovery	3.0
Methanol Synthesis	5.0
H_2 Recovery and Methanation	1.9
CO_2 Removal	0.6
MeOH Conversion	5.5
Hydrotreating, Fractionation—Alkylation	1.0
Miscellaneous	1.9
O_2 Production	10.2
Steam Boiler, Precipitator	11.3
Wastewater Treatment	2.3
Product Storage	2.1
Off-sites	9.9
General Facilities	4.3
	100.0

Reference

Research Guidance Studies to Assess Gasoline from Coal by Methanol-to-Gasoline and Fischer–Tropsch Technologies, Mobil R&D Corp., FE-2447-13, February 1978.

MOBIL M-GASOLINE (GASOLINE PRODUCTION ONLY)

In this version of the Mobil M-Gasoline process, only gasoline is produced; the purge gas (containing some CH_4) is reformed to produce more synthesis gas. The methanol is then used as feed for the Mobil-M fuel process as shown in Fig. 9-3.

FUEL CHARACTERIZATION
TYPE: Wyoming Subbituminous
HIGHER HEATING VALUE (Btu/lb): 8,509

FUEL/POWER INPUTS

	TPSD	*MMTPY*		*10^{12} Btu/yr*
FUEL	27,439	9,014		153.40
POWER[1]	—	—		—
			TOTAL	153.40

PRODUCT/BY-PRODUCT CHARACTERIZATION/OUTPUT

PRODUCT/CHARACTERIZATION		*BCOE[2]/SD*	*MMBCOE/yr*	*10^{12} Btu/yr*
Gasoline, 5.105 MMBtu/bbl	37,430 BPSD	32,945	10.82	62.71
Propane LPG, 3.816 MMBtu/bbl	2,665 BPSD	1,753	0.58	3.34
Butane LPG, 4.191 MMBtu/bbl	3,800 BPSD	2,746	0.90	5.23
	TOTAL	37,444	12.30	71.28

BY-PRODUCT CHARACTERIZATION		*BCOE/SD*	*MMBCOE/yr*	*10^{12} Btu/yr*
Sulfur	61 TPSD			
Ammonia	103 TPSD			
Power	1.38 MW	19.5	0.006	0.04
	TOTAL	19.5	0.006	0.04

PROCESS EFFICIENCY
EFFICIENCY[3]: 46.5% COAL TO PRODUCT EFFICIENCY[4]: 46.5%

CAPITAL REQUIREMENT

ITEM	*$MM (mid-1979)*
PLANT INSTALLED COST	1,905
CONTINGENCY (15%)	286
OWNER COST	26
ROYALTIES	22
INITIAL CATALYST AND CHEMICALS	9
START-UP COSTS	—
TOTAL PLANT INVESTMENT	2,248
WORKING CAPITAL	86
TOTAL CAPITAL REQUIREMENT	2,334

ANNUAL COST

ITEM	*$MM (mid-1979)*
FUEL ($0.55/MMBtu)	84.4
OPERATION AND MAINTENANCE	167.1
SUBTOTAL	251.8
BY-PRODUCT CREDITS	6.5
NET ANNUAL COST	245.3

COMPOSITE PRODUCT PRICE $/BCOE ($/MMBtu)

DCF %	*3*	*6.5*	*10*
COAL	6.86(1.18)	6.86(1.18)	6.86(1.18)
O&M	13.06(2.25)	13.06(2.25)	13.06(2.25)
CAPITAL	10.92(1.88)	18.84(3.25)	28.95(4.99)
TOTAL	30.83(5.32)	38.76(6.68)	48.87(8.43)

DCF %	*10 (No Tax)*	*16*	*18*
COAL	6.86(1.18)	6.86(1.18)	6.86 (1.18)
O&M	13.06(2.25)	13.06(2.25)	13.06 (2.25)
CAPITAL	26.78(4.62)	51.52(8.88)	60.52(10.44)
TOTAL	46.70(8.05)	71.44(12.32)	80.44(13.87)

ECONOMIC CONSIDERATIONS
STREAM FACTOR: 90%
ORIGINAL ESTIMATE DATE: 3Q 77
ESCALATION CORRECTIONS: CAPITAL 1.13 O&M 1.12

ALLOCATION OF PRODUCT PRICE
Required revenues = $1.83 MM/SD @ 10% DCF and $2.675 MM/SD @ 16% DCF. With minor contributions from LPG sales the required selling price of gasoline is $52.42/BCOE ($45.14/bbl or $1.09/gal) @ 10% DCF and $78.07/BCOE ($68.72/bbl or $1.64/gal) @ 16% DCF. This compares with current spot prices of 60–69¢/gal for unleaded gasoline.

UNIT OPERATION CONTRIBUTION TO COMPOSITE PRODUCT PRICE (CAPITAL INVESTMENT EFFECTS ONLY)

UNIT OPERATION	%
Gasification	30.6
Gasoline Production	13.5
Oxygen Plant	9.5
Steam Plant	9.7
Water Facilities	4.7
Other Off-sites	10.4
General Facilities	3.0
Other Project Costs	18.6
	100.0

Reference

Research Guidance Studies to Assess Gasoline from Coal by Methanol-to-Gasoline and Fischer–Tropsch Technologies, Mobil R&D Corp., FE-2447-13, February 1978.

MOBIL M-GASOLINE (ADVANCED)

Coal-derived synthesis gas, cleansed of particulates, CO_2, and sulfur compounds, is fed directly to a catalytic reactor for conversion to gasoline. The reaction products, containing unreacted H_2 and CO, inerts, methane, and conversion products, are sent to a hydrocarbon recovery unit where the liquid hydrocarbons are separated from the gases. The liquids are fractionated and alkylated to produce gasoline and LPG, while the gases are used for SNG production.

FUEL CHARACTERIZATION
TYPE: Wyoming Subbituminous
HIGHER HEATING VALUE (Btu/lb): 8,509

FUEL/POWER INPUTS

	TPSD	*MMTPY*		*10^{12} Btu/yr*
FUEL	27,334	8,979		152.81
POWER[1]	—	—		0.03
			TOTAL	152.84

PRODUCT/BY-PRODUCT CHARACTERIZATION/OUTPUT

PRODUCT/CHARACTERIZATION		*BCOE[2]/SD*	*MMBCOE/yr*	*10^{12} Btu/yr*
SNG, 1002 Btu/scf	177.9 MMSCFD	30,734	10.10	58.56
Gasoline, 5.105 MMBtu/bbl 87.5 Octane	17,485 BPSD	15,390	5.06	29.32
Propane LPG, 3.822 MMBtu/bbl	1,675 BPSD	1,104	0.36	2.10
	TOTAL	47,228	15.52	89.98

BY-PRODUCT CHARACTERIZATION		*BCOE/SD*	*MMBCOE/yr*	*10^{12} Btu/yr*
Sulfur	61 TPSD			
Ammonia	103 TPSD			
Coal Fines	336 TPSD	986	0.32	1.88
	TOTAL	986	0.32	1.88

PROCESS EFFICIENCY

EFFICIENCY[3]: 60.1% COAL TO PRODUCT EFFICIENCY[4]: 58.9%

CAPITAL REQUIREMENT

ITEM	*$MM (mid-1979)*
PLANT INSTALLED COST	1,665
CONTINGENCY (15%)	250
OWNER COST	26
ROYALTIES	15
INITIAL CATALYST AND CHEMICALS	4
START-UP COSTS	—
TOTAL PLANT INVESTMENT	1,960
WORKING CAPITAL	59
TOTAL CAPITAL REQUIREMENT	2,019

ANNUAL COST

ITEM	*$MM (mid-1979)*
FUEL ($0.55/MMBtu)	84.0
OPERATION AND MAINTENANCE	149.8
SUBTOTAL	233.8
BY-PRODUCT CREDITS	6.9
NET ANNUAL COST	226.9

COMPOSITE PRODUCT PRICE $/BCOE ($/MMBtu)

DCF %	*3*	*6.5*	*10*
COAL	5.41(0.93)	5.41(0.93)	5.41(0.93)
O&M	9.21(1.59)	0.21(1.59)	9.21(1.59)
CAPITAL	7.48(1.29)	12.89(2.22)	19.81(3.41)
TOTAL	22.10(3.81)	27.51(4.74)	34.43(5.94)

DCF %	*10 (No Tax)*	*16*	*18*
COAL	5.41(0.93)	5.41(0.93)	5.41(0.93)
O&M	9.21(1.59)	9.21(1.59)	9.21(1.59)
CAPITAL	18.41(3.17)	35.28(6.08)	41.46(7.15)
TOTAL	33.02(5.69)	49.90(8.60)	56.08(9.67)

ECONOMIC CONSIDERATIONS
STREAM FACTOR: 90%
ORIGINAL ESTIMATE DATE: 3Q 77
ESCALATION CORRECTIONS: CAPITAL 1.13 O&M 1.12

ALLOCATION OF PRODUCT PRICE
Required revenues = \$1.626 MM/SD @ 10% DCF and \$2.357 MM/SD @ 16% DCF. With SNG @ \$3.50/MMBtu and C_3 LPG @ \$15.00/bbl, the required selling price is \$63.48/BCOE (\$55.88/bbl or \$1.33/gal) at 10% DCF and \$111/BCOE (\$97.68/bbl or \$2.33/gal) @ 16% DCF. This compares to current spot prices of 60–69¢/gal for unleaded gasoline.

UNIT OPERATION CONTRIBUTION TO COMPOSITE PRODUCT PRICE (CAPITAL INVESTMENT EFFECTS ONLY)

UNIT OPERATION	%
Gasification	32.2
Gasoline Production	8.9
SNG Production	1.9
Oxygen Plant	8.2
Steam Plant	10.4
Water Facilities	5.5
Other Off-Sites	10.7
General Facilities	3.5
Other Project Costs	18.6
	100.0

Reference

Research Guidance Studies to Assess Gasoline from Coal by Methanol-to-Gasoline and Fischer–Tropsch Technologies, Mobil R&D Corp., FE-2447-13, February 1978.

Direct Liquefaction

EXXON DONOR SOLVENT (EDS)

The EDS process, shown in Fig. 9-4, is a noncatalytic process that liquefies coal by the use of a hydrogen donor solvent obtained from coal-derived distillate. The donor solvent transfers hydrogen to the coal, thus promoting liquefaction of the coal.

Coal is ground and slurried with the recycle donor solvent. The slurry is heated by a fired heater, and preheated gaseous hydrogen is added. The reaction is carried out in a tubular reactor with no internals. Products from the liquefaction reactor are sent to several stages of separation units for recovery of gas, naphtha, middle distillate, and bottoms comprised primarily of unreacted coal and mineral matter. Distillation is the means for solid–liquid product separation.

The heavy bottoms from distillation are sent to a Flexicoker to produce additional liquids and low-Btu gas for in-plant fuel use. Hydrogen for in-plant use is provided by

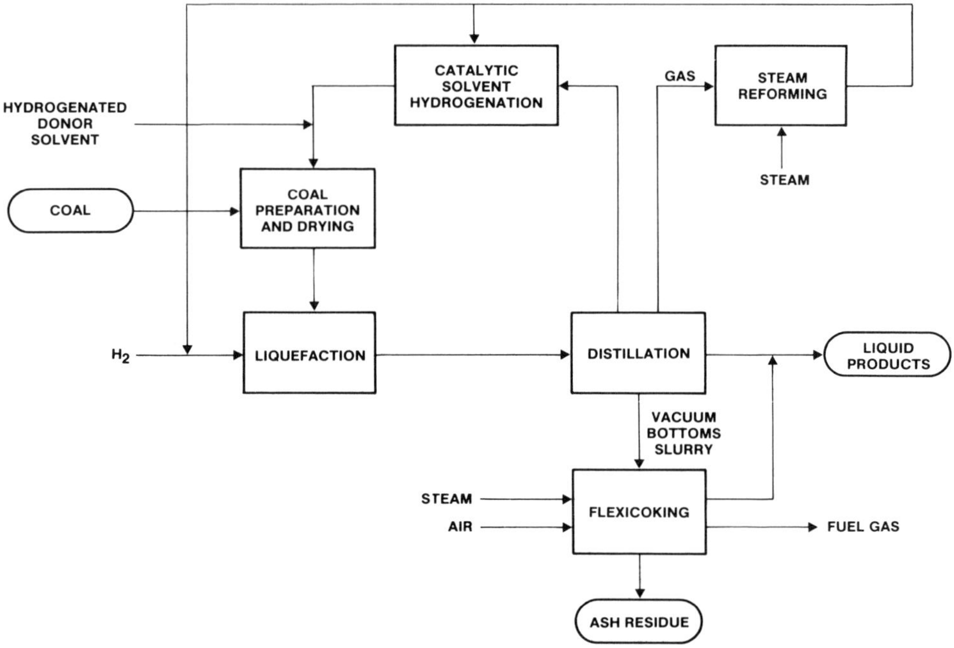

Fig. 9-4. Block diagram of Exxon Donor Solvent process.

steam reforming of C_1-C_2 gases produced in the process. The hydrogen is recycled to the liquefaction and solvent hydrogenation sections.

A portion of the middle distillate product is sent to the solvent hydrogenation step, using a catalytic fixed-bed reactor to produce donor solvent to be recycled to the slurry preparation step. Depending on the ultimate product utilization, the primary liquid products may be further refined.

The plant is balanced in that it is self-sufficient in both process fuel and H_2 requirements. The process gives high yields of low-sulfur liquids from either bituminous or subbituminous coals or lignites. For Illinois bituminous coal, the liquid yield is determined to be 2.6 barrels of C_4/1000°F (538°C) liquids per ton of dry coal feed.

FUEL CHARACTERIZATION

TYPE: Illinois No. 6
HIGHER HEATING VALUE (Btu/lb): 10,574

FUEL/POWER INPUTS

	TPSD	*MMTPY*		*10^{12} Btu/yr*
FUEL	28,743	199.68		199.68
POWER[1]	—	—		4.71
			TOTAL	204.39

PRODUCT/BY-PRODUCT CHARACTERIZATION/OUTPUT

PRODUCT/CHARACTERIZATION		*BCOE[2]/SD*	*MMBCOE/yr*	*10^{12} Btu/yr*
Fuel Oil, 3.4°API, 6.22 MMBtu/bbl	29,866 BPSD	32,029	10.52	61.02

Naphtha, 42.9°API, 5.2 MMBtu/bbl	21,984 BPSD	19,710	6.47	37.55
Propane LPG, 146.3° API, 3.86 MMBtu/bbl	4,089 BPSD	2,721	0.89	5.18
Butane LPG, 111.7°API, 4.33 MMBtu/bbl	3,889 BPSD	2,903	0.95	5.53
	TOTAL	57,363	18.83	109.28

BY-PRODUCT CHARACTERIZATION

Sulfur	784 TPSD
Ammonia	124 TPSD

PROCESS EFFICIENCY

EFFICIENCY[3]: 53.5% COAL TO PRODUCT EFFICIENCY[4]: 54.7%

CAPITAL REQUIREMENT

ITEM	*$MM (mid-1979)*
PLANT INSTALLED COST	1,294
CONTINGENCY (15%)	194
OWNER COST	26
ROYALTIES	—
INITIAL CATALYST AND CHEMICALS	3
START-UP COSTS	104
TOTAL PLANT INVESTMENT	1,621
WORKING CAPITAL	95
TOTAL CAPITAL REQUIREMENT	1,716

ANNUAL COST

ITEM	*$MM (mid-1979)*
FUEL ($1.2/MMBtu)($25.40/ton)	239.8
OPERATION AND MAINTENANCE	153.0
SUBTOTAL	392.8
BY-PRODUCT CREDITS	13.4
NET ANNUAL COST	379.4

COMPOSITE PRODUCT PRICE $/BCOE ($/MMBtu)

DCF %	*3*	*6.5*	*10*
COAL	12.73(2.20)	12.73(2.20)	12.73(2.20)
O&M	7.41(1.28)	7.41(1.28)	7.41(1.28)
CAPITAL	5.22(0.90)	9.03(1.56)	13.85(2.39)
TOTAL	25.37(4.37)	29.18(5.03)	34.00(5.86)

DCF %	*10 (No Tax)*	*16*	*18*
COAL	12.73(2.20)	12.73(2.20)	12.73(2.20)
O&M	7.41(1.28)	7.41(1.28)	7.41(1.28)
CAPITAL	12.71(2.19)	24.51(4.23)	28.73(4.95)
TOTAL	32.86(5.67)	44.66(7.70)	48.88(8.43)

ECONOMIC CONSIDERATIONS

STREAM FACTOR: 90%

ORIGINAL ESTIMATE DATE: 3Q 75

ESCALATION CORRECTIONS: CAPITAL 1.3 O&M 1.26

ALLOCATION OF PRODUCT PRICE

Required revenues = $1.95 MM/SD @ 10% DCF and $2.562 MM/SD @ 16% DCF. With naphtha @ $25/bbl, C_3 LPG @ $15/bbl and C_4 LPG @ $16.50/bbl, the required selling price of fuel oil is $42.70/bbl @ 10% DCF and $63.20/bbl @ 16% DCF. This compares to current spot prices @ $19.50/bbl for lighter (14°–16°API) No. 6 heating oil.

UNIT OPERATION CONTRIBUTION TO COMPOSITE PRODUCT PRICE (CAPITAL INVESTMENT EFFECTS ONLY)

UNIT OPERATION	%
Liquefaction	23.4
Solvent Hydrogenation	7.9
Flexicoker	15.6
Hydrogen Recovery and Generation	23.4
Gas and Water Treatment	4.7
Product Recovery	0.9
Off-sites (includes coal storage and preparation)	24.1
	100.0

OTHER INFORMATION

ANNUAL COSTS		*$MM (mid-79)*
Power (175 MW, 35 mill/kWh)		48.3
Water (6000 gpm, 60¢/Mgal)		1.7
Catalyst and Chemicals		6.3
Labor, Operation, and Maintenance		38.1
Repair Material, Advalorum		58.6
GROSS ANNUAL COST		153.0
BY-PRODUCTS		
Sulfur ($26/ton)	6.7	
Ammonia ($165/ton)	6.7	13.4

Reference

EDS Coal Liquefaction Process Development, Phase IIIA, Exxon Research & Engineering Company, FE-2353-13, January 1978.

H-COAL (SYNCRUDE MODE)

The H-Coal process, shown in Fig. 9-5, is a catalytic hydroliquefaction process that converts high-sulfur coal to either a boiler fuel that will meet sulfur-emission regulations or to a refinery syncrude. Coal is dried and crushed to minus 40 mesh, then slurried with recycled oil and pumped to a pressure of 200 atm. Compressed hydrogen is added to the slurry, and the mixture is preheated and charged continuously to the bottom of the ebullated-bed catalytic reactor. Upward passage of the internally recycled reaction mixture maintains the catalyst in a fluidized state (catalyst activity is maintained by the semicontinuous addition of fresh catalyst and the withdrawal of spent catalyst). The temperature of the ebullated-bed catalytic reactor is controlled by adjusting the temperature of reactants entering from the preheater. Typically, temperature of the mixture entering the reactor is 650– 700°F (343–371°C). Control of the ebullated bed of catalyst is a function of many process parameters. The interrelationships of these parameters are the subject of ongoing research.

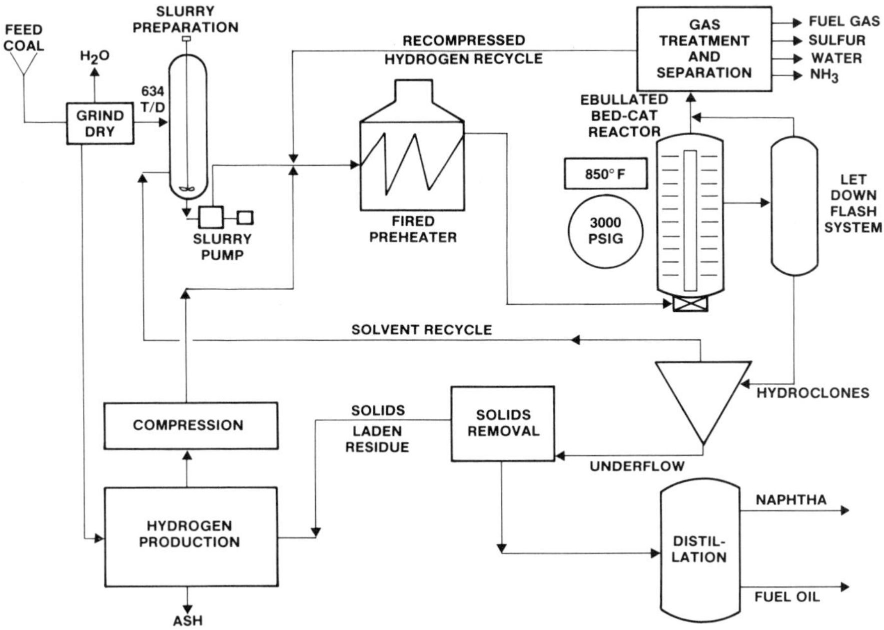

Fig. 9-5. Block diagram of H-Coal (Syncrude Mode) process.

The vapor product leaving the top of the reactor is cooled to condense the heavier components as a liquid. Light hydrocarbons, ammonia, and hydrogen sulfide are absorbed and separated from the remaining gas, leaving a hydrogen-rich gas which is recompressed and recycled to be combined with the input slurry. The liquid–solid product, containing unconverted coal, ash, and oil, is fed into a flash separator. The material that boils off is passed to an atmospheric distillation unit. The bottoms product containing solids and heavy oil is further separated with a hydroclone, a liquid–solid separation, and a vacuum still.

The gas and liquid products (hydrocarbon gas, hydrogen sulfide, ammonia, light distillate, heavy distillate, and residual fuel) may be further refined as necessary. Heavy distillate is recycled as the slurry medium. The stream containing unreacted carbon and some liquid will eventually be processed on-site to produce hydrogen needed for the process.

FUEL CHARACTERIZATION

TYPE: Illinois No. 6

HIGHER HEATING VALUE (Btu/lb): 11,900

FUEL/POWER INPUTS

	TPSD	*MMTPY*		*10^{12} Btu/yr*
FUEL	27,836	9,276		220.77
POWER[1]	—	—		—
			TOTAL	220.77

PRODUCT/BY-PRODUCT CHARACTERIZATION/OUTPUT

PRODUCT/CHARACTERIZATION	*BPSD*	*BCOE²/SD*	*MMBCOE/yr*	*10^{12} Btu/yr*
Syncrude, 13.70°API, 6.35 MMBtu/bbl	36,452	39,909	13.30	77.14
Naphtha, 40.5°API, 5.75 MMBtu/bbl	24,686	24,473	8.16	47.30
TOTAL	61,138	64,382	21.46	124.44

BY-PRODUCT CHARACTERIZATION		*BCOE/SD*	*MMBCOE/yr*	*10^{12} Btu/yr*
Sulfur	1,215 TPSD			
Ammonia	234.6 TPSD			
Phenols	23.7 TPSD			
Power	81.6 MW	1,153	0.38	2.23

PROCESS EFFICIENCY

EFFICIENCY[3]: 57.4% COAL TO PRODUCT EFFICIENCY[4]: 56.4%

CAPITAL REQUIREMENT

ITEM	*$MM (mid-1979)*
PLANT INSTALLED COST	1,495
CONTINGENCY (15%)	224
OWNER COST	20
ROYALTIES	27
INITIAL CATALYST AND CHEMICALS	10
START-UP COSTS	52
TOTAL PLANT INVESTMENT	1,828
WORKING CAPITAL	90
TOTAL CAPITAL REQUIREMENT	1,918

ANNUAL COST

ITEM	*$MM (mid-1979)*
FUEL ($1.2/MMBtu)	264.9
OPERATION AND MAINTENANCE	140.5
SUBTOTAL	405.4
BY-PRODUCT CREDITS	47.6
NET ANNUAL COST	357.8

COMPOSITE PRODUCT PRICE $/BCOE ($/MMBtu)

DCF %	*3*	*6.5*	*10*
COAL	12.34(2.13)	12.34(2.13)	12.34(2.13)
O&M	4.33(0.75)	4.33(0.75)	4.33(0.75)
CAPITAL	5.13(0.88)	8.86(1.53)	13.60(2.32)
TOTAL	21.80(3.76)	25.54(4.41)	30.27(5.19)

DCF %	*10 (No Tax)*	*16*	*18*
COAL	12.34(2.13)	12.34(2.13)	12.34(2.13)
O&M	4.33(0.75)	4.33(0.75)	4.33(0.75)
CAPITAL	12.53(2.16)	24.12(4.16)	28.30(4.88)
TOTAL	29.20(5.03)	40.79(7.03))	44.97(7.75)

ECONOMIC CONSIDERATIONS

STREAM FACTOR: 91.3% (8000 hr)

ORIGINAL ESTIMATE DATE: Mid-1975

ESCALATION CORRECTIONS: CAPITAL 1.3 O&M 1.26

ALLOCATION OF PRODUCT PRICE

Required revenues = $1.332 MM/SD @ 10% DCF and $2.009 MM/SD @ 16% DCF. With naphtha @ $25/bbl, the required selling price of syncrude is $36.50/bbl @ 10% DCF and $55.11/bbl @ 16% DCF.

UNIT OPERATION CONTRIBUTION TO COMPOSITE PRODUCT PRICE (CAPITAL INVESTMENT EFFECTS ONLY)

UNIT OPERATION	%
Coal Preparation	7.1
Liquefaction	18.8
Product Recovery	4.4
Gas Recovery	2.7
Sulfur Plant	1.8
Oxygen Plant	12.5
Hydrogen Plant	19.7
Sour Water Treatment	2.7
Storage Loading	3.6
Off-Sites and Utilities	26.7
	100.0

OTHER INFORMATION

ANNUAL COSTS	*$MM/yr*
Water (12,900 gpm, 60¢/Mgal)	3.72
Catalyst & Chemicals	16.59
Labor	18.37
Maintenance	44.00
Supplies	2.52
Overhead	5.04
Property Tax & Insurance	47.27
Waste Disposal	3.01
GROSS ANNUAL COSTS	140.52
BY-PRODUCTS	
Sulfur ($26/ton)	10.53
Ammonia ($165/ton)	12.90
Phenols ($170/ton)	1.34
Power (35 mill/kWh)	22.84
NET ANNUAL COST	47.61

Reference

H-Coal Commercial Evaluation, Fluor Engineers/Constructors, Inc., FE-2002-12, March 1976.

SRC-II

In the SRC process, coal is pulverized and mixed with a process-derived solvent. The slurry, at 450°F (232°C) and 2000 psig, is combined with hydrogen and is pumped to a preheater to increase the temperature to 700–750°F (371–399°C) and then fed into a dissolver as shown in Fig. 9-6.

The slurry stream from the dissolver is split into two portions, one of which is recycled to provide solvent for coal slurry mixing, while the other is fractionated to recover the

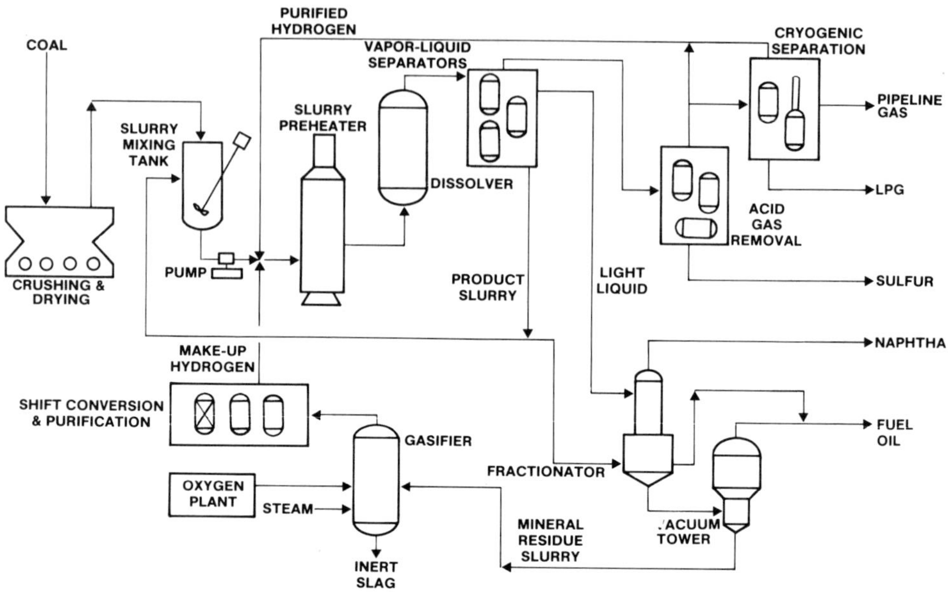

Fig. 9-6. Block diagram of SRC-II process.

primary products of the process. The light hydrocarbon liquid stream from the dissolver is also fractionated. The products from the fractionation system are naphtha, low-sulfur fuel oil, and a vacuum residue consisting of heavy oil, ash, and undissolved organic material from the coal. The gases from the dissolver are subsequently treated to remove hydrogen sulfide and carbon dioxide. A cryogenic separation unit yields liquid petroleum gases and pipeline gas for sale. Unreacted hydrogen is recovered and recycled to the process.

FUEL CHARACTERIZATION

TYPE: Illinois No. 6 High Volatile C Bituminous
HIGHER HEATING VALUE (Btu/lb): 11,500 (ROM)

FUEL/POWER INPUTS

	TPSD	*MMTPY*		*10^{12} Btu/yr*
FUEL	37,600 (ROM)	12,352		284.09
POWER[1]	—	—		—
			TOTAL	284.09

PRODUCT/BY-PRODUCT CHARACTERIZATION/OUTPUT

PRODUCT/CHARACTERIZATION		*BCOE[2]/SD*	*MMBCOE/yr*	*10^{12} Btu/yr*
Naphtha, 50°API, 5.425 MMBtu/bbl	9,379 BPSD	8,773	2.88	16.71
Fuel Oil, -8.2°API, 6.904 MMBtu/bbl	56,348 BPSD	67,074	22.03	127.80
SNG, 1,050 Btu/scf	165.9 MMSCFD	30,034	9.87	57.22
Propane LPG, 4.199 lb/gal, 3.814 MMBtu/bbl	6,030 BPSD	3,965	1.30	7.56

Butane LPG, 4.733 lb/gal, 4.238 MMBtu/bbl	4,082 BPSD	2,983	0.98	5.68
	TOTAL	112,829	37.06	214.97

BY-PRODUCT CHARACTERIZATION

Sulfur	1231.7 TPSD
Ammonia	90.1 TPSD

PROCESS EFFICIENCY

75.7%

CAPITAL REQUIREMENT

ITEM	*$MM (mid-1979)*
PLANT INSTALLED COST	1,900.2
CONTINGENCY (15%)	285.1
OWNER COST	26.0
ROYALTIES	10.4
INITIAL CATALYST AND CHEMICALS	94.2
START-UP COSTS	—
TOTAL PLANT INVESTMENT	2,315.9
WORKING CAPITAL	79.3
TOTAL CAPITAL REQUIREMENT	2,395.2

ANNUAL COST

ITEM	*$MM (mid-1979)*
FUEL ($1.2/MMBtu)	340.9
OPERATION AND MAINTENANCE	145.5
SUBTOTAL	486.4
BY-PRODUCT CREDITS	15.4
NET ANNUAL COST	471.0

COMPOSITE PRODUCT PRICE $/BCOE ($/MMBtu)

DCF %	*3*	*6.5*	*10*
COAL	9.20(1.59)	9.20(1.59)	9.20(1.59)
O&M	3.51(0.61)	3.51(0.61)	3.51(0.61)
CAPITAL	3.71(0.64)	6.38(1.10)	9.79(1.69)
TOTAL	16.42(2.83)	19.09(3.29)	22.50(3.88)

DCF %	*10 (No Tax)*	*16*	*18*
COAL	9.20(1.59)	9.20(1.59)	9.20(1.59)
O&M	3.51(0.61)	3.51(0.61)	3.51(0.61)
CAPITAL	9.10(1.57)	17.40(3.00)	20.43(3.52)
TOTAL	21.80(3.76)	30.02(5.19)	33.14(5.71)

ECONOMIC CONSIDERATIONS

STREAM FACTOR: 90%

ORIGINAL ESTIMATE DATE: 4Q 75

ESCALATION CORRECTIONS: CAPITAL 1.3 O&M 1.26

ALLOCATION OF PRODUCT PRICE

Required revenues = $2.538 MM/SD @ 10% DCF and $3.397 MM/SD @ 16% DCF. With SNG @ $3.50/MMBtu, naphtha @ $26/bbl, propane LPG @ $15/bbl and butane LPG @ $16.50/MMBtu, the required selling price of fuel oil is $27.10/bbl @ 10% DCF and $42.35/bbl

@ 16% DCF. This compares to current spot prices of \$19.50/bbl for lighter (14°–16°API) No. 6 fuel oil.

UNIT OPERATION CONTRIBUTION TO COMPOSITE PRODUCT PRICE (CAPITAL INVESTMENT EFFECTS ONLY)

UNIT OPERATION	%
Coal Crushing and Drying	1.6
Slurrying and Dissolving	23.3
Filtration, Solvent Recovery	4.5
Fractionation	3.4
Naphtha Hydrotreating	1.0
Dissolver Acid Gas Removal	2.2
SNG, LPG Production	5.2
Methanation	0.1
Process Gasifier	4.9
Shift	6.4
Acid Gas Removal	5.1
Oxygen Plant	9.7
Fuel Gas Generation	9.6
Sour Water Treatment	0.6
Sulfur Plant	1.7
Off-sites	20.8
	100.0

OTHER INFORMATION

Original design calls for 35.070 TPD (12.125 Btu/lb). This coal has been dried to 27% moisture. With ROM coal at 7.7% moisture, the coal requirements are 37,600 TPD (11,500 Btu/lb) (ROM).

OPERATING COSTS	*\$MM/Yr*
Operating Supplies	0.6
Maintenance Materials and Contract Labor	33.0
Catalyst and Chemicals	19.2
Operating Labor and Supervision	7.9
Maintenance Labor and Supervision	8.4
Payroll Burden	5.7
Overhead	10.6
Property Tax and Insurance	60.1
GROSS ANNUAL COST	\$145.5MM
BY-PRODUCTS	
Sulfur (\$26/ton)	10.5
Ammonia (\$165/ton)	1.9
NET ANNUAL COST	15.4

Reference

Oil/Gas Complex Conceptual Design/Economic Analysis: Oil and SNG Production, Ralph M. Parsons Co., FE-1775-8, March 1977.

Gasification

LURGI (DUAL FEED)

The Lurgi process, shown in Fig. 9-7, takes noncaking coal and crushes it. It is then screened to remove fines, and fed to a bank of Lurgi fixed-bed gasifiers through a lock hopper system. The coal is distributed by rotating arms and travels downward in the gasifiers where it is dried, devolatilized, and gasified at 300−500 psi with steam and oxygen. Ash is removed by rotating grates through an ash lock at the bottoms of the gasifiers. Raw gases leaving the top of the gasifiers are quenched and passed through a shift converter. Carbon dioxide and H_2S are removed in the purification system, and the product gas is methanated and dehydrated to produce pipeline gas.

FUEL CHARACTERIZATION
TYPE: Montana Subbituminous
HIGHER HEATING VALUE (Btu/lb): 8,800

FUEL/POWER INPUTS

	TPSD	MMTPY		10^{12} Btu/yr
FUEL	29,760	9,776		172.06
POWER[1]	—	—		—
			TOTAL	172.06

PRODUCT/BY-PRODUCT CHARACTERIZATION/OUTPUT

PRODUCT/CHARACTERIZATION		*BCOE[2]/SD*	*MMBCOE/yr*	*10^{12} Btu/yr*
SNG 902.6 Btu/scf	281.6 MMscf/SD 254.2 MMBtu/SD	43,829	14.40	83.51

BY-PRODUCT CHARACTERIZATION		*BCOE/SD*	*MMBCOE/yr*	*10^{12} Btu/yr*
Sulfur	96 TPSD			
Ammonia	140.4 TPSD			
Phenols	905 BPSD			
Naphtha	2,719 BPSD	8,739	2.87	16.65
Tar	2,098 BPSD			
Light Oil	3,689 BPSD			
Coal Fines	5,568 TPSD	16,894	5.55	32.19
	TOTAL OUTPUT	69,462	22.82	132.35

PROCESS EFFICIENCY
EFFICIENCY[3]: 76.9% COAL TO PRODUCT EFFICIENCY[4]: 48.5%

CAPITAL REQUIREMENT

ITEM	*$MM (mid-1979)*
PLANT INSTALLED COST	1,127.3
CONTINGENCY (15%)	169.1
OWNER COST	—
ROYALTIES	1.2
INITIAL CATALYST AND CHEMICALS	25.4
START-UP COSTS	40.9
TOTAL PLANT INVESTMENT	1,363.7
WORKING CAPITAL	32.0
TOTAL CAPITAL REQUIREMENT	1,395.9

+ BCOE @ 5.8 MMBtu/bbl

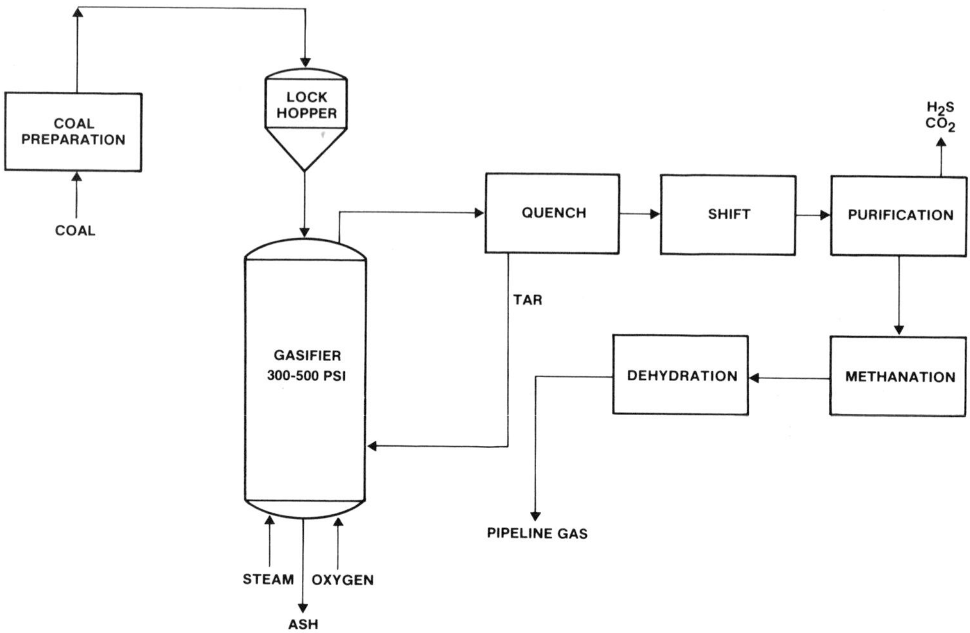

Fig. 9-7. Block diagram of Lurgi (dual feed) process.

ANNUAL COST

ITEM	*$MM (mid-1979)*
FUEL ($0.55/MMBtu)	94.63
OPERATION AND MAINTENANCE	114.20
SUBTOTAL	208.83
BY-PRODUCT CREDITS	65.54
NET ANNUAL COST	143.29

COMPOSITE PRODUCT PRICE $/BCOE ($/MMBtu)

DCF %	*3*	*6.5*	*10*
COAL	6.57(1.13)	6.57(1.13)	6.57(1.13)
O&M	3.38(0.58)	3.38(0.58)	3.38(0.58)
CAPITAL	5.56(0.96)	9.55(1.65)	14.65(2.53)
TOTAL	15.51(2.67)	19.50(3.36)	24.61(4.24)

DCF %	*10 (No Tax)*	*16*	*18*
COAL	6.57(1.13)	6.57(1.13)	6.57(1.13)
O&M	3.38(0.58)	3.38(0.58)	3.38(0.58)
CAPITAL	13.69(2.36)	26.08(4.50)	30.64(5.28)
TOTAL	23.64(4.08)	36.03(6.21)	40.59(7.00)

ECONOMIC CONSIDERATIONS

ORIGINAL ESTIMATE DATE: 1/76

ESCALATION CORRECTIONS: CAPITAL 1.22 O&M 1.19

ALLOCATION OF PRODUCT PRICE

Single product (SNG), no allocation necessary.

UNIT OPERATION CONTRIBUTION TO COMPOSITE PRODUCT PRICE (CAPITAL INVESTMENT EFFECTS ONLY)

UNIT OPERATION	%
Coal Storage	1.7
Coal Preparation	3.1
Coal Feed Gasification Raw Gas Quench Shift Conversion	23.0
Acid Gas Removal	14.1
Methanation	7.0
Sour Water Treatment	2.3
Sulfur Recovery	6.5
Solids Disposal	0.4
Steam and Utility Systems	21.4
Plant Water	2.6
Oxygen Plant	7.8
General Facilities	10.0
	100.0

OTHER INFORMATION

ANNUAL COSTS		*$MM/yr*
Catalysts and Chemicals		10.77
Water (60¢/Mgal)		0.69
Labor		31.80
Administration and Overhead		19.08
Supplies		16.21
Local Taxes and Insurance		35.65
GROSS ANNUAL COSTS		114.20
BY-PRODUCTS		
Sulfur ($26/ton)	0.82	
NH ($165/ton)	7.61	
Oil, Naphtha, Tar	43.79	
Fines ($0.41/MMBtu)	13.32	(65.54)
NET ANNUAL COSTS		48.66

Reference

Factored Estimates for Western Coal Commercial Concepts, C. F. Braun, FE-2240-5, October 1976.

BIGAS

The BiGas reactor, shown in Fig. 9-8, is a two-stage gasifier of the pressurized, entrained-flow, slagging type. The gasifier has three sections separated by formed restrictions: stage two (topmost) receives dense phase coal in gas-entrained flow through

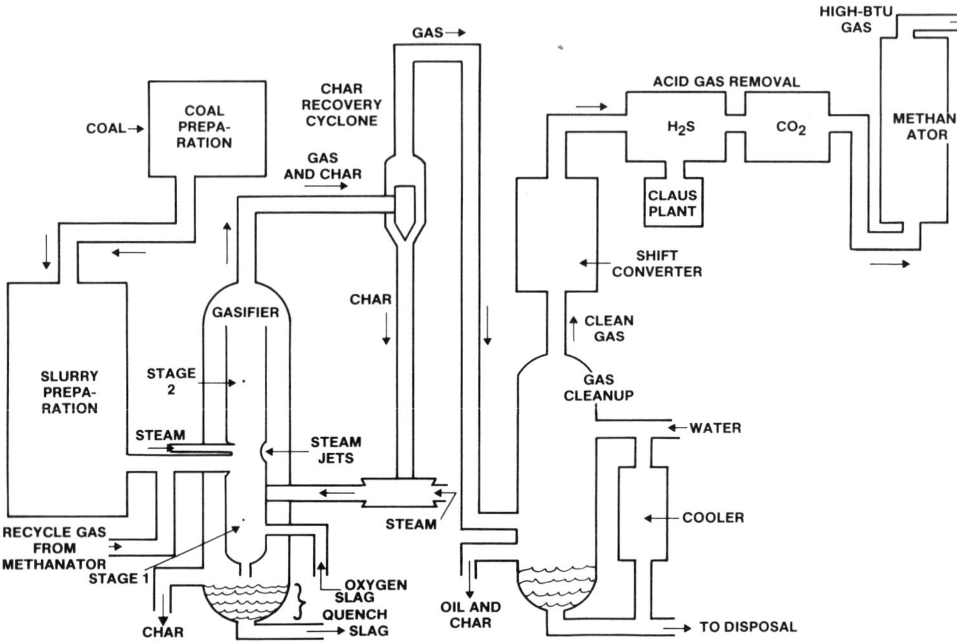

Fig. 9-8. Block diagram of BiGas process.

injectors located in a throat between stages two and one. Steam is also injected there. Process gas and char flow from the top of stage two of the gasifier to a cyclone. The char is recycled from the cyclone and fed to a steam nozzle and then, with oxygen and steam, to the bottom of stage one. The products of char gasification in stage one are hot gas feeding into stage two and slag dropping into stage three, the slag quench section. Molten slag runs down the walls of stage one of the pressurized gasifier.

Raw gas leaves the char recovery cyclone, passes through a water wash column, and flows to the shift reactor where the H_2 to CO ratio is adjusted to the desired value. H_2S and part of the CO_2 are removed in the Selexol unit upstream of the methanator. H_2S gas goes to a Claus unit. The product gas from the methanator goes through final CO_2 removal in the Selexol unit. The methanator is a fluidized bed with embedded water-cooled tubes, and it offers great potential for commercial development.

The BiGas reactor can handle all types of coal without the need for a pretreatment step, and because of its high operating temperature of over 2500°F (1370°C), there is no tar in the product gas, thus simplifying the gas cleanup step. High methane yields are achieved, and coals can be gasified without prior treatment, since the process uses an entrained rather than a fixed- or fluidized-bed system.

FUEL CHARACTERIZATION

TYPE: Eastern Bituminous (Pittsburgh Seam No. 8)
HIGHER HEATING VALUE (Btu/lb): 12,400

FUEL/POWER INPUTS

	TPSD	*MMTPY*		*10^{12} Btu/yr*
FUEL	14,960	4,914		121.88
POWER[1]	—	—		—
			TOTAL	121.88

PRODUCT/BY-PRODUCT CHARACTERIZATION/OUTPUT

PRODUCT/CHARACTERIZATION		*BCOE[2]/SD*	*MMBCOE/yr*	*10^{12} Btu/yr*
SNG, 894.8 Btu/scf	282.8 MMscf/SD (253 MMBtu/SD)	43,630	14.33	83.13

BY-PRODUCT CHARACTERIZATION

Sulfur	627.6 TPSD
	110.6 TPSD

PROCESS EFFICIENCY

EFFICIENCY[3]: 68.2%

CAPITAL REQUIREMENT

ITEM	*\$MM (mid-1979)*
PLANT INSTALLED COST	929.6
CONTINGENCY (15%)	139.4
OWNER COST	—
ROYALTIES	6.7
INITIAL CATALYST AND CHEMICALS	14.2
START-UP COSTS	43.1
TOTAL PLANT INVESTMENT	1,133.0
WORKING CAPITAL	31.8
TOTAL CAPITAL REQUIREMENT	1,164.8

ANNUAL COST

ITEM	*\$MM (mid-1979)*
FUEL (\$1.2/MMBtu)(\$29.76/ton)	146.26
OPERATION AND MAINTENANCE	78.27
SUBTOTAL	224.53
BY-PRODUCT CREDITS	11.35
NET ANNUAL COST	213.18

COMPOSITE PRODUCT PRICE \$/BCOE (\$/MMBtu)

DCF %	*3*	*6.5*	*10*
COAL	10.20(1.76)	10.20(1.76)	10.20(1.76)
O&M	4.67(0.80)	4.67(0.80)	4.67(0.80)
CAPITAL	4.66(0.80)	8.01(1.38)	12.29(2.12)
TOTAL	19.53(3.37)	22.88(3.95)	27.16(4.68)

DCF %	*10 (No Tax)*	*16*	*18*
COAL	10.20(1.76)	10.20(1.76)	10.20(1.76)
O&M	4.67(0.80)	4.67(0.80)	4.67(0.80)
CAPITAL	11.45(1.97)	21.84(3.77)	25.65(4.42)
TOTAL	26.33(4.54)	36.72(6.33)	40.53(6.99)

ECONOMIC CONSIDERATIONS

STREAM FACTOR: 90%

ORIGINAL ESTIMATE DATE: 1/76
ESCALATION CORRECTIONS: CAPITAL 1.22 O&M 1.19

ALLOCATION OF PRODUCT PRICE
Single product (SNG), no allocation necessary.

UNIT OPERATION CONTRIBUTION TO COMPOSITE PRODUCT PRICE (CAPITAL INVESTMENT EFFECTS ONLY)

UNIT OPERATION	%
Coal Storage and Reclaiming	1.8
Coal Preparation	4.4
Coal Feeding	2.0
Gasification	7.2
Raw Gas Quench	4.5
Shift	4.2
Acid Gas Removal	14.3
Methanation	4.1
Effluent Treatment	4.2
Sulfur Recovery	13.6
Solids Disposal	1.3
Steam and Power	16.6
Plant Water	4.5
Oxygen Plant	9.8
General Facilities	10.1
	100.0

OTHER INFORMATION

ANNUAL COSTS		*$MM/yr*
Catalyst and Chemicals		6.37
Water (60¢/Mgal)		2.10
Oper. Labor		3.97
Administration and Overhead		2.38
Oper. Supplies		0.99
Maintenance (Labor & Material)		32.06
Local Taxes & Insurance		29.40
Ash Disposal		1.00
GROSS ANNUAL COSTS		78.27
BY-PRODUCTS		
Sulfur ($26/ton)	5.36	
NH_3 ($165/ton)	5.99	(11.35)
NET ANNUAL COST		66.92

Reference

Factored Estimates for Eastern Coal Commercial Concepts, C. F. Braun and Company, FE-2240-31, September 1978.

FUEL CHARACTERIZATION
TYPE: Montana Subbituminous
HIGHER HEATING VALUE (Btu/lb): 8,800

FUEL/POWER INPUTS

	TPSD	*MMTPY*		*10^{12} Btu/yr*
FUEL	22,672	7,488		131.08
POWER[1]	—	—		—
			TOTAL	131.08

PRODUCT/BY-PRODUCT CHARACTERIZATION/OUTPUT

PRODUCT/CHARACTERIZATION		*BCOE[2]/SD*	*MMBCOE/Yr*	*10^{12} Btu/yr*
SNG 907.6 Btu/scf	276.7 MMscf/SD (25.1 MMBtu/SD)	43,295	14.22	82.49

BY-PRODUCT CHARACTERIZATION	*BCOE/SD*
Sulfur	108 TPSD
Ammonia	92.4 TPSD

PROCESS EFFICIENCY
EFFICIENCY[3]: 62.9%

CAPITAL REQUIREMENT

ITEM	*\$MM (mid-1979)*
PLANT INSTALLED COST	1,098
CONTINGENCY (15%)	167.7
OWNER COST	—
ROYALTIES	2.1
INITIAL CATALYST AND CHEMICALS	12.7
START-UP COSTS	34.6
TOTAL PLANT INVESTMENT	1,312.1
WORKING CAPITAL	29.1
TOTAL CAPITAL REQUIREMENT	1,341.2

ANNUAL COST

ITEM	*\$MM (mid-1979)*
FUEL (\$0.55/MMBtu)	72.09
OPERATION AND MAINTENANCE	103.94
SUBTOTAL	176.03
BY-PRODUCT CREDITS	5.92
NET ANNUAL COST	170.11

COMPOSITE PRODUCT PRICE \$/BCOE (\$/MMBtu)

DCF %	*3*	*6.5*	*10*
COAL	5.07(0.87)	5.07(0.87)	5.07(0.87)
O&M	6.89(1.19)	6.89(1.19)	6.89(1.19)
CAPITAL	5.42(0.93)	9.30(1.60)	14.28(2.46)
TOTAL	17.38(3.00)	21.26(3.67)	26.24(4.52)

DCF %	*10 (No Tax)*	*16*	*18*
COAL	5.07(0.87)	5.07(0.87)	5.07(0.87)
O&M	6.89(1.19)	6.89(1.19)	6.89(1.19)
CAPITAL	13.34(2.30)	25.44(4.39)	29.89(5.15)
TOTAL	25.30(4.36)	37.40(6.45)	41.85(7.22)

ECONOMIC CONSIDERATIONS
STREAM FACTOR: 90%

ORIGINAL ESTIMATE DATE: 1/76
ESCALATION CORRECTIONS: CAPITAL 1.22 O&M 1.19

ALLOCATION OF PRODUCT PRICE
Single product (SNG), no allocation necessary.

UNIT OPERATION CONTRIBUTION TO COMPOSITE PRODUCT PRICE (CAPITAL INVESTMENT EFFECTS ONLY)

UNIT OPERATION	%
Coal Storage and Reclaiming	1.6
Coal Preparation	4.2
Coal Feed	2.1
Gasification and Power Recovery	5.4
Raw Gas Quench	6.2
Shift Conversion	3.9
Acid Gas Removal	14.1
Methanation	3.6
Sour Water Treatment	4.6
Sulfur Recovery	7.3
Solids Disposal	1.1
Steam and Utility	22.7
Plant Water	3.1
Oxygen Plant	10.0
General Facilities	10.0
	100.0

OTHER INFORMATION

ANNUAL COSTS		*$MM/yr*
Catalysts and Chemicals		5.51
Water (60¢/Mgal)		0.62
Labor		29.83
Administration and Overhead		17.90
Supplies		15.36
Local Taxes and Insurance		34.72
GROSS ANNUAL COSTS		103.94
BY-PRODUCTS		
Sulfur ($26/ton)	0.92	
NH_3 ($165/ton)	5.00	(5.92)
NET ANNUAL COSTS		98.02

Reference

Factored Estimates for Western Coal Commercial Concepts, C. F. Braun & Company, FE-2240-5, October 1976.

SYNTHANE (EASTERN COAL)

The synthane process, shown in Fig. 9-9, takes coal, crushes it to −20 mesh. The coal is then dried, pressurized to approximately 70 atm, and transferred into the fluidized-bed pretreater by means of high pressure steam and oxygen. Pretreatment provides a mild oxidation of the coal particle surface so that caking coals will not agglomerate in the

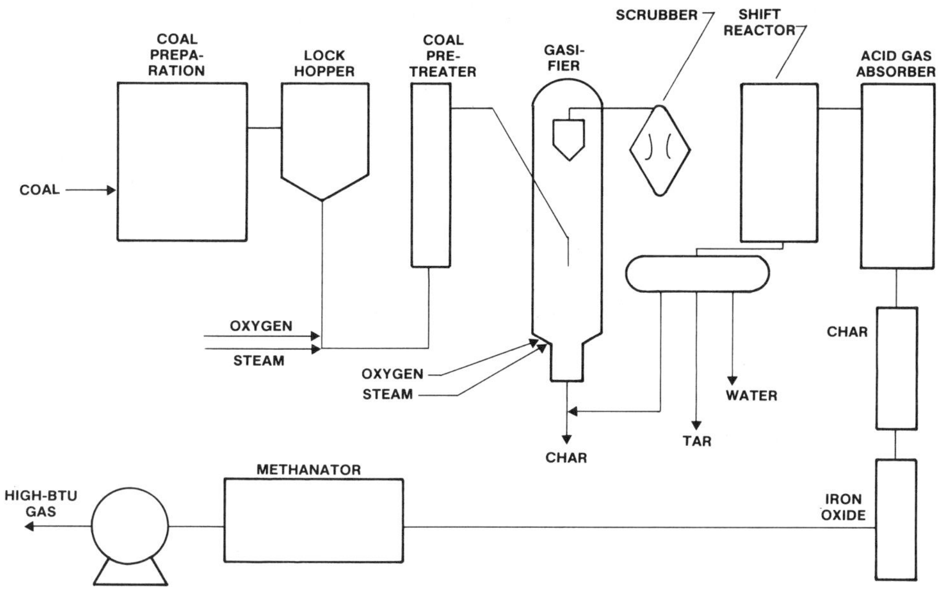

Fig. 9-9. Block diagram of Synthane process.

gasifier. The coal overflows from the pretreater into the top of the fluidized-bed gasifier about 10 ft above the bed level, falls through hot gases rising from the fluidized bed, and is devolatilized. This devolatilization contributes significantly to the methane yield. Steam and oxygen enter the gasifier just below the fluidizing gas distributor. The gasification reaction occurs within the fluidized bed. Unreacted char flows downward into a bed fluidized with steam and water sprays and is removed through lock hoppers. The product gas, containing methane, hydrogen, carbon monoxide, carbon dioxide, ethane, and impurities, is passed through a venturi scrubber and a water scrubber to remove carryover ash, char, and tars. The concentration of hydrogen and carbon monoxide in the gas is then adjusted to a three-to-one ratio in a shift converter. The acid gases are absorbed in a hot-potassium-carbonate scrubber. The purified gas is then sent to a methanator for final upgrading.

Two methanation systems have been proposed: a tube wall reactor, in which the catalyst is sprayed onto tube walls of an exchanger; and a hot gas recycle methanator, in which the catalyst is sprayed onto metal plates. Coal can be fed to the gasifier through a lock hopper system as shown or as a coal–water slurry.

FUEL CHARACTERIZATION

TYPE: Eastern Bituminous (Pittsburgh Seam No. 8)
HIGHER HEATING VALUE (Btu/lb): 12,400

FUEL/POWER INPUTS

	TPSD	*MMTPY*		*10^{12} Btu/yr*
FUEL	18,140	5,959		147.78
POWER[1]	—	—		—
			TOTAL	147.78

PRODUCT/BY-PRODUCT CHARACTERIZATION/OUTPUT

PRODUCT/CHARACTERIZATION		*BCOE[2]/SD*	*MMBCOE/yr*	*10^{12} Btu/yr*
SNG, 908 Btu/scf	275.8 MMscf/SD (250 MMBtu/SD)	43,104	14.16	82.13

BY-PRODUCT CHARACTERIZATION		*BCOE/SD*	*MMBCOE/yr*	*10^{12} Btu/yr*
Sulfur	700.8 TPSD			
Ammonia	160.2 TPSD			
Light Oil	1,474 BPSD	3,430	1.13	6.54
Tar	2,983 BPSD			
Char and Coal Fines (MAF) (13,950 Btu/lb)	678.5 TPSD	3,265	1.07	6.22
	TOTAL OUTPUT	44,799	16.36	94.89

PROCESS EFFICIENCY

EFFICIENCY[3]: 64.2% COAL TO PRODUCT EFFICIENCY[4]: 55.6%

CAPITAL REQUIREMENT

ITEM	*$MM (mid-1979)*
PLANT INSTALLED COST	1,067.5
CONTINGENCY (15%)	160.1
OWNER COST	—
ROYALTIES	7.7
INITIAL CATALYST AND CHEMICALS	12.2
START-UP COSTS	50.3
TOTAL PLANT INVESTMENT	1,297.8
WORKING CAPITAL	37.1
TOTAL CAPITAL REQUIREMENT	1,334.9

ANNUAL COST

ITEM	*$MM (mid-1979)*
FUEL ($1.2/MMBtu) ($29.76/ton)	177.34
OPERATION AND MAINTENANCE	84.21
SUBTOTAL	261.55
BY-PRODUCT CREDITS	39.79
NET ANNUAL COST	221.76

COMPOSITE PRODUCT PRICE $/BCOE ($/MMBtu)

DCF %	*3*	*6.5*	*10*
COAL	12.52(2.16)	12.52(2.16)	12.52(2.16)
O&M	3.14(0.54)	3.14(0.54)	3.14(0.54)
CAPITAL	5.41(0.93)	9.30(1.60)	14.26(2.46)
TOTAL	21.07(3.63)	24.96(4.30)	29.92(5.16)

DCF %	*10 (No Tax)*	*16*	*18*
COAL	12.52(2.16)	12.52(2.16)	12.52(2.16)
O&M	3.14(0.54)	3.14(0.54)	3.14(0.54)
CAPITAL	13.29(2.29)	25.36(4.37)	29.78(5.13)
TOTAL	28.95(4.99)	41.02(7.07)	45.44(7.84)

ECONOMIC CONSIDERATIONS

STREAM FACTOR: 90%

ORIGINAL ESTIMATE DATE: 1/76

ESCALATION CORRECTIONS: CAPITAL 1.22 O&M 1.19

ALLOCATION OF PRODUCT PRICE

Single product (SNG), no allocation necessary.

UNIT OPERATION CONTRIBUTION TO COMPOSITE PRODUCT PRICE (CAPITAL INVESTMENT EFFECTS ONLY)

UNIT OPERATION	%
Coal Storage and Reclaiming	1.5
Coal Preparation	3.4
Coal Feeding	1.4
Gasification	7.5
Raw Gas Quench	4.1
Shift	3.9
Acid Gas Removal	10.9
Methanation	3.7
Effluent Treatment	9.2
Sulfur Recovery	10.0
Solids Disposal	0.6
Steam and Power	21.7
Plant Water	4.1
Oxygen Plant	8.0
General Facilities	10.0
	100.0

OTHER INFORMATION

ANNUAL COSTS		*$MM/yr*
Catalyst and Chemicals		5.15
Water (60¢/Mgal)		2.22
Oper. Labor		4.05
Admin. and Overhead		2.38
Oper. Supplies		0.99
Maintenance (Labor & Material)		34.88
Local Taxes and Insurance		33.76
Ash Disposal		0.78
GROSS ANNUAL COST		84.21
BY-PRODUCTS		
Sulfur ($26/ton)	5.98	
NH_3 ($165/ton)	8.68	
Oil	19.53	
Fines ($0.90/MMBtu)	5.60	(39.79)
NET ANNUAL COST		44.42

Reference

Factored Estimates for Eastern Coal Commercial Concepts, C. F. Braun & Company, FE-2240-31, September 1978.

SYNTHANE (WESTERN COAL)

When the Synthane plant is fed with western coal, the coal pretreater is eliminated from the process train. Western coal is noncaking and does not require pretreatment prior to entering the fluidized bed. Otherwise, refer to the previous Synthane description.

FUEL CHARACTERIZATION
TYPE: Montana Subbituminous
HIGHER HEATING VALUE (Btu/lb): 8,800

FUEL/POWER INPUTS

	TPSD	*MMTPY*		*10^{12} Btu/yr*
FUEL	27,863	9,153		161.09
POWER[1]	—	—		—
			TOTAL	161.09

PRODUCT/BY-PRODUCT CHARACTERIZATION/OUTPUT

PRODUCT/CHARACTERIZATION		*BCOE[2]/SD*	*MMBCOE/yr*	*10^{12} Btu/yr*
SNG, 902.2 Btu/scf	277.2 MMscf/SD (2.50 MMBtu/SD)	43,113	14.16	82.125
BY-PRODUCT CHARACTERIZATION		*BCOE/SD*	*MMBCOE/yr*	*10^{12} Btu/yr*
Sulfur	118.8 TPSD			
Ammonia	86.4 TPSD			
Light Oil	1,440 BPSD	4,129	1.36	7.87
Tar	2,846 BPSD			
Char and Fines	1,840 TPSD	5,583	1.83	10.64
	TOTAL OUTPUT	52,825	17.35	100.635

PROCESS EFFICIENCY
EFFICIENCY[3]: 62.5% COAL TO PRODUCT EFFICIENCY[4]: 51.0%

CAPITAL REQUIREMENT

ITEM	*\$MM (mid-1979)*
PLANT INSTALLED COST	1,223.7
CONTINGENCY (15%)	183.5
OWNER COST	—
ROYALTIES	1.8
INITIAL CATALYST AND CHEMICALS	9.9
START-UP COSTS	38.5
TOTAL PLANT INVESTMENT	1,457.4
WORKING CAPITAL	32.8
TOTAL CAPITAL REQUIREMENT	1,490.2

ANNUAL COST

ITEM	*\$MM (mid-1979)*
FUEL (\$0.55/MMBtu)	88.60
OPERATION AND MAINTENANCE	108.06
SUBTOTAL	196.66
BY-PRODUCT CREDITS	25.81
NET ANNUAL COST	170.85

COMPOSITE PRODUCT PRICE \$/BCOE (\$/MMBtu)

DCF %	*3*	*6.5*	*10*
COAL	6.26(1.08)	6.26(1.08)	6.26(1.08)
O&M	5.81(1.00)	5.81(1.00)	5.81(1.00)
CAPITAL	6.05(1.04)	10.39(1.79)	15.95(2.75)
TOTAL	18.11(3.12)	22.45(3.87)	28.02(4.83)

DCF %	*10 (No Tax)*	*16*	*18*
COAL	6.26(1.08)	6.26(1.08)	6.26(1.08)
O&M	5.81(1.00)	5.81(1.00)	5.81(1.00)
CAPITAL	14.90(2.57)	28.41(4.90)	33.39(5.76)
TOTAL	26.96(4.65)	40.48(6.98)	45.46(7.84)

ECONOMIC CONSIDERATIONS
STREAM FACTOR: 90%
ORIGINAL ESTIMATE DATE: 1/76
ESCALATION CORRECTIONS: CAPITAL 1.22 O&M 1.19

ALLOCATION OF PRODUCT PRICE
Single product (SNG), no allocation necessary.

UNIT OPERATION CONTRIBUTION TO COMPOSITE PRODUCT PRICE (CAPITAL INVESTMENT EFFECTS ONLY)

UNIT OPERATION	%
Coal Storage and Reclaiming	1.6
Coal Preparation	2.4
Coal Feed	1.8
Gasification and Power Recovery	6.0
Raw Gas Quench	4.2
Shift	5.8
Acid Gas Removal	12.8
Methanation	3.7
Sour Water Treatment	4.7
Sulfur Recovery	6.8
Solids Disposal	1.0
Steam and Utilities	28.2
Plant Water	2.2
Oxygen Plant	8.8
General Facilities	10.0
	100.0

OTHER INFORMATION

ANNUAL COSTS		*$MM/Yr*
Catalysts and Chemicals		3.78
Water (60¢/Mgal)		0.62
Labor		30.70
Administration and Overhead		18.42
Supplies		15.84
Local Taxes and Insurance		38.70
GROSS ANNUAL COSTS		108.06
BY-PRODUCTS		
Sulfur ($26/ton)	1.01	
NH_3 ($165/ton)	4.68	
Oil and Tars	16.00	
Char ($6.80/ton)	4.12	(25.81)
NET ANNUAL COST		82.25

Reference

Factored Estimates for Western Coal Commercial Concepts, C. F. Braun & Company, FE-2240-5, October 1976.

HYGAS (STEAM–OXYGEN) EASTERN COAL

In the Hygas process, shown in Fig. 9-10, coal is reacted with hydrogen-rich gases to produce methane directly in a hydrogasifier. It is a high-pressure process for all types of coals. The hydrogasifier consists of a slurry drier and three high-pressure fluidized-bed reactors. Coal is dried, crushed, slurried with a process-derived oil, and compressed to 1500 psi. The slurry is then moved into the slurry drier section of the hydrogasifier, where it is sprayed onto a fluidized bed. The oil is vaporized and removed with the hot process gases rising through the bed from the lower stages of the gasifier. Vaporized oil is recovered by quenching and is reused. The dried coal is routed to the first stage of the gasifier, which is located below the drier and above the second stage. The coal enters the gasifier at midheight and is heated to about 1200°F (650°C) by the hot reaction gas from the second stage. The first stage [1200°F (650°C); 100 psi] favors the direct production of methane from the coal. Solids are channeled to the second stage, where rising gases contact the char for further gasification at about 1700°F (925°C). Unreacted char and ash drop into the steam–oxygen gasifier for final gasification, producing a hydrogen-rich gas and steam mixture, at about 1850°F (1010°C).

The raw gas leaves at the top of the slurry drying section through a quench tower which removes dusts, the slurry oil, and excess steam. Gas purification occurs in an acid removal system, which consists of an amine–water scrub followed by a caustic wash and a water wash. CO_2 and H_2S removal are essentially complete after the amine scrub. A

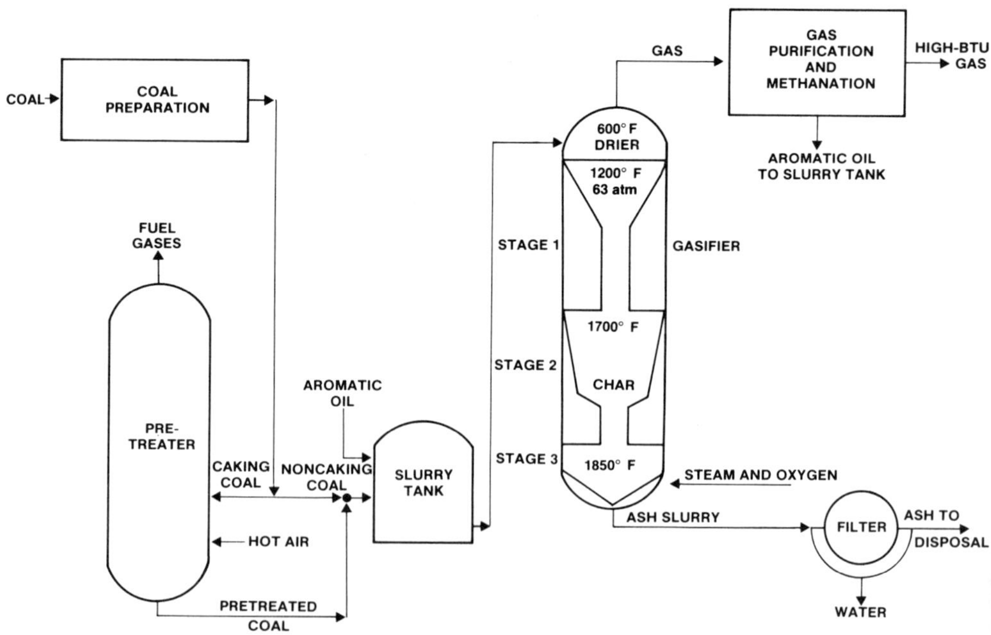

Fig. 9-10. Block diagram of Hygas (steam–oxygen) process utilizing eastern coal.

Claus plant converts the sulfur products to elemental sulfur. After shift reaction, methanation takes place in two packed-bed down flow reactors. Heat release control is by the Institute of Gas Technology (IGT) split-flow recycle method.

FUEL CHARACTERIZATION
TYPE: Eastern Bituminous (Pittsburgh Seam No. 8)
HIGHER HEATING VALUE (Btu/lb): 12,400

FUEL/POWER INPUTS

	TPSD	*MMTPY*		*10^{12} Btu/yr*
FUEL	15,864	5,211		129.24
POWER[1]	—	—		—
			TOTAL	129.24

PRODUCT/BY-PRODUCT CHARACTERIZATION/OUTPUT

PRODUCT/CHARACTERIZATION		*BCOE[2]/SD*	*MMBCOE/yr*	*10^{12} Btu/yr*
SNG, 929 Btu/scf	269.8 MMscf/SD (250.6 MMBtu/SD)	43,217	14.20	82.34
BY-PRODUCT CHARACTERIZATION		*BCOE/SD*	*MMBCOE/yr*	*10^{12} Btu/yr*
Sulfur	610.6 TPSD			
Ammonia	93 TPSD			
Light Oil (35°API, 5.4 MMBtu/bbl)	2,794 BPSD	2,603	0.855	4.96
	TOTAL OUTPUT	45,820	15.055	87.3

PROCESS EFFICIENCY
EFFICIENCY[3]: 67.5% COAL TO PRODUCT EFFICIENCY[4]: 63.7%

CAPITAL REQUIREMENT

ITEM	*$MM (mid-1979)*
PLANT INSTALLED COST	989.4
CONTINGENCY (15%)	148.4
OWNER COST	—
ROYALTIES	6.4
INITIAL CATALYST AND CHEMICALS	11.7
START-UP COSTS	45.8
TOTAL PLANT INVESTMENT	1,201.7
WORKING CAPITAL	33.6
TOTAL CAPITAL REQUIREMENT	1,235.3

ANNUAL COST

ITEM	*$MM (mid-1979)*
FUEL ($1.2/MMBtu)($29.76/ton)	155.09
OPERATION AND MAINTENANCE	83.31
SUBTOTAL	238.4
BY-PRODUCT CREDITS	28.61
NET ANNUAL COST	209.79

COMPOSITE PRODUCT PRICE $/BCOE ($/MMBtu)

DCF %	*3*	*6.5*	*10*
COAL	10.92(1.88)	10.92(1.88)	10.92(1.88)
O&M	3.85(0.66)	3.85(0.66)	3.85(0.66)
CAPITAL	4.99(0.86)	8.58(1.48)	13.16(2.27)
TOTAL	19.77(3.41)	23.36(4.03)	27.94(4.82)

DCF %	*10 (No Tax)*	*16*	*18*
COAL	10.92(1.88)	10.92(1.88)	10.92(1.88)
O&M	3.85(0.66)	3.85(0.66)	3.85(0.66)
CAPITAL	12.27(2.12)	23.41(4.04)	27.49(4.74)
TOTAL	27.05(4.66)	38.18(6.58)	42.27(7.29)

ECONOMIC CONSIDERATIONS

STREAM FACTOR: 90%
ORIGINAL ESTIMATE DATE: 1/76
ESCALATION CORRECTIONS: CAPITAL 1.22 O&M 1.19

ALLOCATION OF PRODUCT PRICE

Single product (SNG), no allocation necessary.

UNIT OPERATION CONTRIBUTION TO COMPOSITE PRODUCT PRICE (CAPITAL INVESTMENT EFFECTS ONLY)

UNIT OPERATION	%
Coal Storage and Reclaiming	1.6
Coal Preparation	2.7
Coal Pretreatment	4.7
Coal Feeding	2.5
Gasification	7.2
Raw Gas Quench	7.2
Shift	3.7
Acid Gas Removal	13.8
Methanation	3.6
Effluent Treatment	5.6
Sulfur Recovery	10.4
Solids Disposal	1.4
Steam and Power	15.3
Plant Water	4.7
Oxygen Plant	5.6
General Facilities	10.0
	100.0

OTHER INFORMATION

ANNUAL COSTS		*$MM/yr*
Catalyst and Chemicals		4.53
Water (60¢/Mgal)		2.34
Labor (Operation and Super)		4.21
Administration and Overhead		2.52
Operational Supplies		1.06
Maintenance (Material & Labor)		36.16
Local Taxes and Insurance		31.29
Ash Disposal		1.20
GROSS ANNUAL COST		83.31
BY-PRODUCTS		
Sulfur ($261/ton)	5.22	
NH_3 ($165/ton)	5.04	
Oil	18.35	(28.61)
NET ANNUAL COST		54.70

Reference

Factored Estimates for Eastern Coal Commercial Concepts, C. F. Braun, FE-2240-31, September 1978.

HYGAS (STEAM–OXYGEN) WESTERN COAL

When the Hygas plant is fed with western coal, the coal pretreater, shown in Fig. 9-10, is eliminated from the process train. Western coal is noncaking and does not require pretreatment prior to entering the fluidized bed. Otherwise, refer to the previous Hygas description.

FUEL CHARACTERIZATION
TYPE: Montana Subbituminous
HIGHER HEATING VALUE (Btu/lb): 8,800

FUEL/POWER INPUTS

	TPSD	*MMTPY*		*10^{12} Btu/yr*
FUEL	20,472	6,725		118.36
POWER[1]	—	—		—
			TOTAL	118.36

PRODUCT/BY-PRODUCT CHARACTERIZATION/OUTPUT

PRODUCT/CHARACTERIZATION		*BCOE[2]/SD*	*MMBCOE/yr*	*10^{12} Btu/yr*
SNG, 913.9 Btu/scf	273.5 MMscf/SD (250 MMBtu/SD)	43,100	14.16	82.125
BY-PRODUCT CHARACTERIZATION		*BCOE/SD*	*MMBCOE/yr*	*10^{12} Btu/yr*
Sulfur	99.6 TPSD			
Ammonia	88.8 TPSD			
Light Oil, 48.5°API, 5.59 MMBtu/bbl	4,594 BPSD or 631.2 TPSD or 25.7 MMBtu/SD	4,431	1.46	8.44
	TOTAL OUTPUT	47,531	15.62	90.565

PROCESS EFFICIENCY
EFFICIENCY[3]: 76.5% COAL TO PRODUCT EFFICIENCY[4]: 69.4%

CAPITAL REQUIREMENT

ITEM	*$MM (mid-1979)*
PLANT INSTALLED COST	923.5
CONTINGENCY (15%)	138.5
OWNER COST	—
ROYALTIES	1.9
INITIAL CATALYST AND CHEMICALS	10.6
START-UP COSTS	30.1
TOTAL PLANT INVESTMENT	1,104.7
WORKING CAPITAL	25.0
TOTAL CAPITAL REQUIREMENT	1,129.7

ANNUAL COST

ITEM	*$MM (mid-1979)*
FUEL ($0.55/MMBtu)($9.68/ton)	65.10

OPERATION AND MAINTENANCE	88.45
SUBTOTAL	153.55
BY-PRODUCT CREDITS	34.33
NET ANNUAL COST	119.22

COMPOSITE PRODUCT PRICE $/BCOE ($/MMBtu)

DCF %	*3*	*6.5*	*10*
COAL	4.60(0.79)	4.60(0.79)	4.60(0.79)
O&M	3.82(0.66)	3.82(0.66)	3.82(0.66)
CAPITAL	4.58(0.79)	7.87(1.36)	12.08(2.08)
TOTAL	13.00(2.24)	16.29(2.81)	20.50(3.53)

DCF %	*10 (No Tax)*	*16*	*18*
COAL	4.60(0.79)	4.60(0.79)	4.60(0.79)
O&M	3.82(0.66)	3.82(0.66)	3.82(0.66)
CAPITAL	11.29(1.95)	21.52(3.71)	25.28(4.36)
TOTAL	19.71(3.40)	29.94(5.16)	33.71(5.81)

ECONOMIC CONSIDERATIONS
STREAM FACTOR: 90%
ORIGINAL ESTIMATE DATE: 1/76
ESCALATION CORRECTIONS: CAPITAL 1.22 O&M 1.19

ALLOCATION OF PRODUCT PRICE
Single product (SNG), no allocation necessary.

UNIT OPERATION CONTRIBUTION TO COMPOSITE PRODUCT PRICE (CAPITAL INVESTMENT EFFECTS ONLY)

UNIT OPERATION	%
Coal Storage and Reclaiming	1.9
Coal Preparation	2.5
Coal Feed	3.8
Gasification and Power Recovery	6.6
Raw Gas Quench	2.8
Shift	4.5
Acid Gas Removal	15.8
Methanation	4.3
Sour Water Treatment	8.5
Sulfur Recovery	8.8
Solids Disposal	1.3
Steam and Utilities	20.0
Plant Water	2.5
Oxygen Plant	6.6
General Facilities	9.9
	100.0

OTHER INFORMATION

ANNUAL COSTS	*$MM/yr*
Catalysts and Chemicals	4.38
Water (60¢/Mgal)	0.54
Labor	25.76

Administrative and Overhead		15.46
Supplies		13.10
Local Tax and Insurance		29.21
GROSS ANNUAL COSTS		88.45
BY-PRODUCTS		
Sulfur ($26/ton)	0.85	
NH_3 ($165/ton)	4.81	
Light Oil	28.67	(34.33)
NET ANNUAL COSTS		54.12

Reference

Factored Estimates for Western Coal Commercial Concepts, C. F. Braun, FE-2240-5, October 1978.

HYGAS (STEAM–IRON)

The main units of this process, shown in Fig. 9-11, are a two-stage fluidized-bed hydrogasifier and a steam–iron process unit producing hydrogen for hydrogasification. Caking coal is first made nonagglomerating by pretreatment with hot air in a fluidized bed at 1 atm and 750°F (399°C) and is then mixed with light oil to form a slurry. This slurry is pumped into a fluidized drying bed, operating at 600°F (316°C) and 1000–1500 psi. The light oil evaporates, and coal from the drying bed passes successively through the first stage of the gasifier, where devolatilization and partial noncatalytic methanation occur at 1300–1500°F (704–816°C) in the presence of hydrogen-rich gas, and then as char into

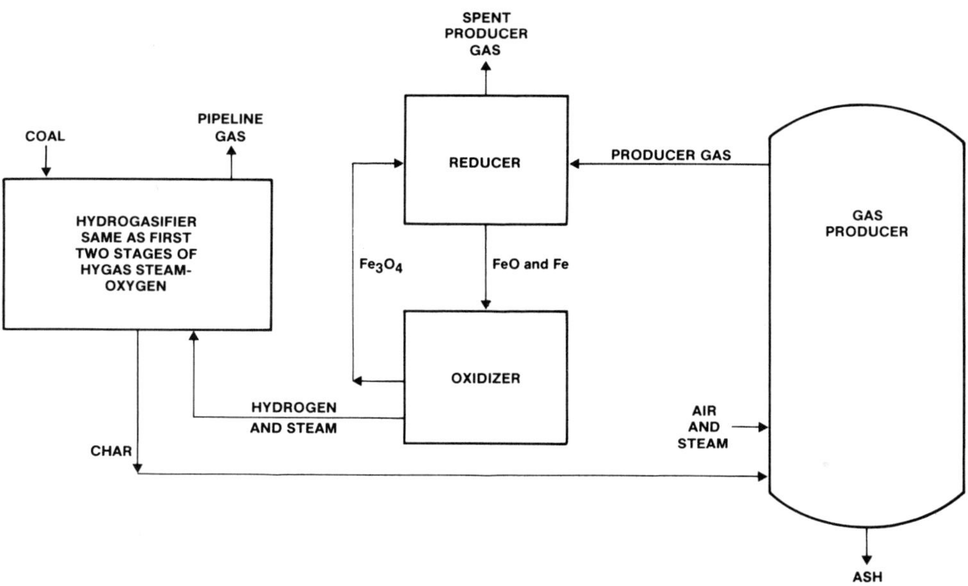

Fig. 9-11. Block diagram of Hygas (steam–iron) process.

the second stage, where partial gasification at 1700–1800°F (927–982°C) occurs by reaction with the steam and hydrogen-rich gas from the steam–iron process. The product gases from the drying bed flow to the purification and catalytic methanation systems.

Char from the second stage of the hydrogasifier is fed into the producer vessel where it reacts with air and steam to generate a gas capable of reducing magnetite (Fe_3O_4) to iron. The reaction takes place at 1900–2000°F (1038–1094°C) and 1100 psi. The hot producer gas is fed to the reducer vessel in which it contacts a recirculating stream of magnetite to produce iron. The reaction is at 1400°F (760°C) and 1100 psi. The iron is then fed to the oxidizer in which it contacts steam and is reoxidized to magnetite at the same temperature and pressure, producing a mixture of hydrogen and steam that is fed to the second stage of hydrogasifier. The magnetite from the oxidizer is recycled to the reducer.

FUEL CHARACTERIZATION
TYPE: Eastern Bituminous (Pittsburgh Seam No. 8)
HIGHER HEATING VALUE (Btu/lb): 12,400

FUEL/POWER INPUTS

	TPSD	*MMTPY*		*10^{12} Btu/yr*
FUEL	21,440	7.04		174.67
POWER[1]	—	—		—
			TOTAL	174.67

PRODUCT/BY-PRODUCT CHARACTERIZATION/OUTPUT

PRODUCT/CHARACTERIZATION		*BCOE[2]/SD*	*MMBCOE/yr*	*10^{12} Btu/yr*
SNG, 920.9 Btu/scf	272.2 MMscf/SD (250.7 MMBtu/SD)	43,331	14.20	82.35

BY-PRODUCT CHARACTERIZATION		*BCOE/SD*	*MMBCOE/yr*	*10^{12} Btu/yr*
Sulfur	640.8 TPSD			
Ammonia	138.7 TPSD			
Naphtha and Light Oil (4.98 MMBtu/bbl)	2,229 BPSD	1,916	0.63	3.65
Power	645.4 MW	9,120	2.99	17.38
	TOTAL OUTPUT	54,257	17.82	103.38

PROCESS EFFICIENCY
EFFICIENCY[3]: 59.2% COAL TO PRODUCT EFFICIENCY[4]: 47.1%

CAPITAL REQUIREMENT

ITEM	*$MM (mid-1979)*
PLANT INSTALLED COST	1,333.5
CONTINGENCY (15%)	200.0
OWNER COST	—
ROYALTIES	16.7
INITIAL CATALYST AND CHEMICALS	9.6
START-UP COSTS	61.0
TOTAL PLANT INVESTMENT	1,620.8
WORKING CAPITAL	45.4
TOTAL CAPITAL REQUIREMENT	1,666.2

ANNUAL COST

ITEM	*$MM (mid-1979)*
FUEL ($1.2/MMBtu)($29.76/ton)	209.60
OPERATION AND MAINTENANCE	108.08
SUBTOTAL	317.68
BY-PRODUCT CREDITS	204.99
NET ANNUAL COST	112.69

COMPOSITE PRODUCT PRICE $/BCOE ($/MMBtu)

DCF %	*3*	*6.5*	*10*
COAL	14.76(2.55)	14.76(2.55)	14.76(2.55)
O&M	−6.83(−1.18)	−6.83(−1.18)	−6.83(−1.18)
CAPITAL	6.73(1.16)	11.57(1.99)	17.75(3.06)
TOTAL	14.67(2.53)	19.51(3.36)	25.69(4.43)

DCF %	*10 (No Tax)*	*16*	*18*
COAL	14.76(2.55)	14.76(2.55)	14.76(2.55)
O&M	−6.83(−1.18)	−6.83(−1.18)	−6.83(−1.18)
CAPITAL	16.55(2.85)	31.56(5.44)	37.07(6.38)
TOTAL	24.48(4.22)	39.50(6.81)	45.01(7.76)

ECONOMIC CONSIDERATIONS

STREAM FACTOR: 90%
ORIGINAL ESTIMATE DATE: 1/76
ESCALATION CORRECTIONS: CAPITAL 1.22 O&M 1.19

ALLOCATION OF PRODUCT PRICE

Single product (SNG), no allocation necessary.

UNIT OPERATION CONTRIBUTION TO COMPOSITE PRODUCT PRICE (CAPITAL INVESTMENT EFFECTS ONLY)

UNIT OPERATION	%
Coal Storage and Reclaiming	1.2
Coal Preparation	2.4
Coal Pretreatment	4.1
Coal Feeding	2.0
Gasification	10.7
Power Recovery	33.9
Raw Gas Quench	5.4
Acid Gas Removal	7.9
Methanation	2.1
Effluent Treatment	4.0
Sulfur Recovery	9.5
Solids Disposal	0.9
Steam and Power	1.3
Plant Water	4.4
General Facilities	10.0
	100.0

OTHER INFORMATION

ANNUAL COSTS	*$MM/yr*
Raw Material	0.15

Catalysts and Chemicals		2.25
Water (60¢/Mgal)		3.33
Operations and Labor		4.21
Administration and Overhead		2.52
Operating Supplies		1.06
Maintenance (Labor and Material)		50.44
Local Taxes and Insurance		42.17
Ash Disposal		1.95
GROSS ANNUAL COST		108.08
BY-PRODUCTS		
Sulfur ($26/ton)	5.47	
NH_3 ($165/ton)	7.52	
Oil	13.90	
Power (3.5¢/kWh)	178.10	(204.99)
NET ANNUAL COST		(96.91)

Reference

Factored Estimates for Eastern Coal Commercial Concepts, C. F. Braun, FE-2240-33, September 1978.

HYGAS (STEAM–IRON) WESTERN COAL

When the Hygas plant is fed with western coal, the coal pretreater, shown in Fig. 9-11, is eliminated from the process train. Western coal is noncaking and does not require pretreatment prior to entering the fluidized bed. Otherwise, refer to the previous Hygas description.

FUEL CHARACTERIZATION
TYPE: Montana Subbituminous
HIGHER HEATING VALUE (Btu/lb): 8,800

FUEL/POWER INPUTS

	TPSD	*MMTPY*		*10^{12} Btu/yr*
FUEL	22,829	7,499		131.99
POWER[1]	—	—		—
			TOTAL	131.99

PRODUCT/BY-PRODUCT CHARACTERIZATION/OUTPUT

PRODUCT/CHARACTERIZATION		*BCOE[2]/SD*	*MMBCOE/yr*	*10^{12} Btu/yr*
SNG, 904 Btu/scf	276.6 MMscf/SD (250 MMBtu/SD)	43,100	14.16	82.125
BY-PRODUCT CHARACTERIZATION		*BCOE/SD*	*MMBCOE/yr*	*10^{12} Btu/yr*
Sulfur	90.0 TPSD			
Ammonia	105.6 TPSD			
Light Oil, 34.7°API, 5.4 MMBtu/bbl	2,379 BPSD	2,222	0.73	4.22
Power	193.26 MW	2,728	0.90	5.20
	TOTAL OUTPUT	48,050	15.79	91.545

PROCESS EFFICIENCY
EFFICIENCY[3]: 69.4% COAL TO PRODUCT EFFICIENCY[4]: 62.2%

CAPITAL REQUIREMENT

ITEM	*$MM (mid-1979)*
PLANT INSTALLED COST	1,250
CONTINGENCY (15%)	187.5
OWNER COST	—
ROYALTIES	9.2
INITIAL CATALYST AND CHEMICALS	8.9
START-UP COSTS	42.5
TOTAL PLANT INVESTMENT	1,498.1
WORKING CAPITAL	35.4
TOTAL CAPITAL REQUIREMENT	1,533.5

ANNUAL COST

ITEM	*$MM (mid-1979)*
FUEL ($0.55/MMBtu)($9.68/ton)	72.59
OPERATION AND MAINTENANCE	138.96
SUBTOTAL	211.55
BY-PRODUCT CREDITS	74.67
NET ANNUAL COST	136.88

COMPOSITE PRODUCT PRICE $/BCOE ($/MMBtu)

DCF %	*3*	*6.5*	*10*
COAL	5.13(0.88)	5.13(0.88)	5.13(0.88)
O&M	4.54(0.78)	4.54(0.78)	4.54(0.78)
CAPITAL	6.22(1.07)	10.68(1.84)	16.40(2.83)
TOTAL	15.89(2.74)	20.35(3.51)	26.07(4.49)

DCF %	*10 (No Tax)*	*16*	*18*
COAL	5.13(0.88)	5.13(0.88)	5.13(0.88)
O&M	4.54(0.78)	4.54(0.78)	4.54(0.78)
CAPITAL	15.31(2.64)	29.20(5.03)	34.31(5.92)
TOTAL	24.98(4.31)	38.87(6.70)	43.98(7.58)

ECONOMIC CONSIDERATIONS
STREAM FACTOR: 90%
ORIGINAL ESTIMATE DATE: 1/76
ESCALATION CORRECTIONS: CAPITAL 1.22 O&M 1.19

ALLOCATION OF PRODUCT PRICE
Single product (SNG), no allocation necessary.

UNIT OPERATION CONTRIBUTION TO COMPOSITE PRODUCT PRICE (CAPITAL INVESTMENT EFFECTS ONLY)

UNIT OPERATION	%
Coal Storage and Reclaiming	1.3
Coal Preparation	2.5
Coal Feed	2.8
Gasification and Power Recovery	45.7
Raw Gas Quench	8.0
Acid Gas Removal	7.5
Methanation	2.6
Sour Water Treatment	6.8
Sulfur Recovery	7.6
Solids Disposal	0.8

Steam and Utility		1.2
Plant Water		3.4
General Facilities		10.0
		100.0

OTHER INFORMATION

ANNUAL COSTS		*$MM/yr*
Raw Materials		0.25
Catalyst and Chemicals		5.95
Water (60¢/Mgal)		0.98
Labor		43.36
Administration and Overhead		26.01
Supplies		22.88
Local Taxes and Insurance		39.53
GROSS ANNUAL COSTS		138.96
BY-PRODUCT		
Sulfur ($26/ton)	0.87	
NH_3 ($165/ton)	5.72	
Light Oil	14.85	
Power (3.5¢/kWh)	53.33	(74.67)
NET ANNUAL COSTS		64.29

Reference

Factored Estimates for Western Coal Commercial Concepts, C.F. Braun, FE-2240-5, October 1976.

CO_2 ACCEPTOR

In the CO_2 Acceptor process, shown in Fig. 9-12, lignite is devolatilized at 140 psia in the presence of steam, carbon monoxide, hydrogen, and dolomitic calcine ($MgO \cdot CaO$) in a fluidized-bed devolatilizer kept at 1500°F (816°C) by addition of calcine at 1870°F (1021°C); char from the latter is fed to a gasifier bed containing calcine and operating at 1520°F (827°C) and 150 psia in a fluidized-bed regenerator which receives separate streams of partially carbonated calcine ($MgO \cdot CaCO_3$) from the devolatilizer and gasifier, returns regenerated calcine to the same units, and sends waste gas to an energy recovery system. The circulating solid material, introduced as dolomite ($MgCO_3 \cdot CaCO_3$), evolves carbon dioxide with absorption of sensible and chemical energy in the regenerator, and accepts carbon dioxide and releases both sensible and chemical energy in the devolatilizer and gasifier. Gas from the gasifier, rich in hydrogen and carbon monoxide, is fed with steam to the devolatilizer, the gas from which is purified, catalytically methanated, and then compressed.

The process is also designed to operate at about 300 psia, in which case temperatures in the regenerator and gasifier change to 1940°F (1060°C) and 1575°F (847°C), and the gasifier operates with a recycle stream.

FUEL CHARACTERIZATION
TYPE: Montana Subbituminous
HIGHER HEATING VALUE (Btu/lb): 8,800

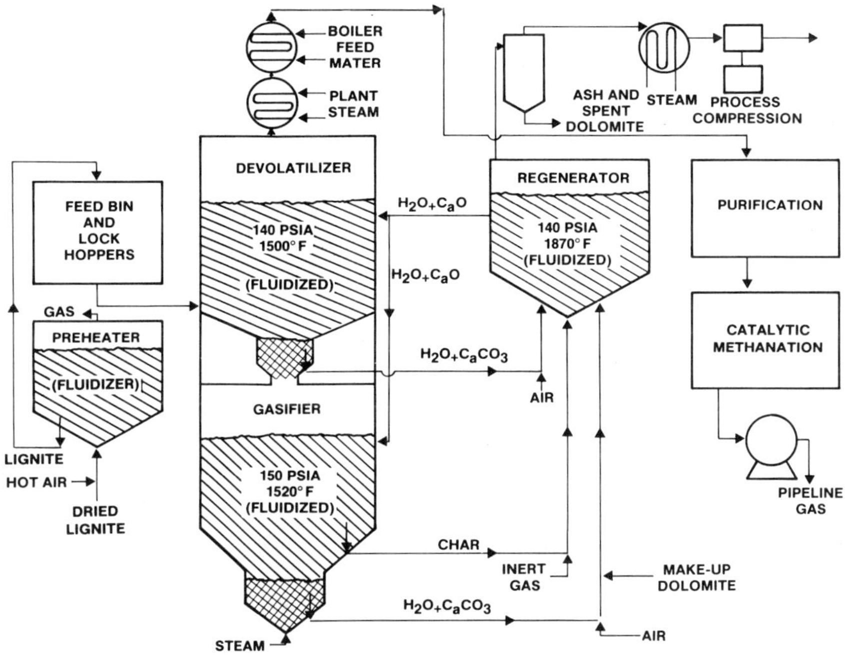

Fig. 9-12. Block diagram of CO_2 Acceptor process.

FUEL/POWER INPUTS

	TPSD	*MMTPY*		*10^{12} Btu/yr*
FUEL	21,892	7,192		126.57
POWER[1]	—	—		—
			TOTAL	126.57

PRODUCT/BY-PRODUCT CHARACTERIZATION/OUTPUT

PRODUCT/CHARACTERIZATION		*BCOE[2]/SD*	*MMBCOE/yr*	*10^{12} Btu/yr*
SNG, 889.9 Btu/scf	292.5 MMscf/SD (251.4 MMBtu/SD)	43,350	14.24	82.59

BY-PRODUCT CHARACTERIZATION		*BCOE/SD*	*MMBCOE/yr*	*10^{12} Btu/yr*
Sulfur	100.8 TPSD			
Ammonia	69.6 TPSD			
Power	60.5 MW	855	0.28	1.63
Coal Fines	1,308 TPSD	3,969	1.30	7.56
	TOTAL OUTPUT	48,174	15.82	91.78

PROCESS EFFICIENCY

EFFICIENCY[3]: 72.5% COAL TO PRODUCT EFFICIENCY[4]: 65.3%

CAPITAL REQUIREMENT

ITEM	*\$MM (mid-1979)*
PLANT INSTALLED COST	940.6
CONTINGENCY (15%)	141.1

OWNER COST	—
ROYALTIES	5.1
INITIAL CATALYST AND CHEMICALS	7.7
START-UP COSTS	37.8
TOTAL PLANT INVESTMENT	1,132.3
WORKING CAPITAL	28.0
TOTAL CAPITAL REQUIREMENT	1,160.3

ANNUAL COST

ITEM	*$MM (mid-1979)*
FUEL ($0.55/MMBtu)	69.61
OPERATION AND MAINTENANCE	121.48
SUBTOTAL	191.09
BY-PRODUCT CREDITS	24.96
NET ANNUAL COST	166.13

COMPOSITE PRODUCT PRICE $/BCOE ($/MMBtu)

DCF %	*3*	*6.5*	*10*
COAL	4.89(0.84)	4.89(0.84)	4.89(0.84)
O&M	6.78(1.17)	6.78(1.17)	6.78(1.17)
CAPITAL	4.68(0.81)	8.03(1.39)	12.33(2.13)
TOTAL	16.34(2.82)	19.70(3.40)	24.00(4.14)
DCF %	*10 (No Tax)*	*16*	*18*
COAL	4.89(0.84)	4.89(0.84)	4.89(0.84)
O&M	6.78(1.17)	6.78(1.17)	6.78(1.17)
CAPITAL	11.51(1.98)	21.95(3.78)	25.79(4.45)
TOTAL	23.18(4.00)	33.62(5.80)	37.46(6.46)

ECONOMIC CONSIDERATIONS

STREAM FACTOR: 90%
ORIGINAL ESTIMATE DATE: 1/76
ESCALATION CORRECTIONS: CAPITAL 1.22 O&M 1.19

ALLOCATION OF PRODUCT PRICE

Single product (SNG), no allocation necessary.

UNIT OPERATION CONTRIBUTION TO COMPOSITE PRODUCT PRICE (CAPITAL INVESTMENT EFFECTS ONLY)

UNIT OPERATION	%
Coal Storage and Reclaiming	1.9
Coal Preparation	8.5
Coal Feed	3.9
Gasification and Power Recovery	27.8
Raw Gas Quench	9.2
Acid Gas Removal	5.2
Methanation	5.2
Sour Water Treatment	2.6
Sulfur Recovery	7.5
Solids Disposal	1.2
Steam and Utility	14.4
Plant Water Systems	2.6
General Facilities	9.9
	100.0

OTHER INFORMATION

ANNUAL COSTS		*$MM/yr*
Raw Materials		20.23
Catalyst and Chemicals		5.38
Water (60¢/Mgal)		0.65
Labor		30.94
Administration and Overhead		18.56
Supplies		15.97
Local Taxes and Insurance		29.75
GROSS ANNUAL COSTS		121.48
BY-PRODUCTS		
Sulfur ($26/ton)	0.86	
NH_3 ($165/ton)	3.77	
Power (3.5¢/kWh)	16.69	
Fines	3.64	(24.96)
NET ANNUAL COSTS		96.52

Reference

Factored Estimates for Western Coal Commercial Concepts, C. F. Braun and Company, FE-2240-5, October 1978.

Oil Shale

TOSCO II ABOVEGROUND RETORTING

The Tosco II retort is an externally heated reactor, shown in Fig. 9-13, that uses hot ceramic balls to heat the shale to pyrolysis temperature. The shale is crushed and fed to a fluidized bed where it is preheated. After preheating, the shale is mixed in a reactor with the heated ceramic balls. Oil, steam, and gas are produced, the oil is sent to a fractionator, and the balls are recovered from the ash using a screen. Since no combustion takes place in the reactor vessel, the resulting gas has a higher energy content and the oil has a lower viscosity than that obtained from an internally heated retort.

FUEL CHARACTERIZATION
TYPE: Oil Shale, 34.8 gal/ton

FUEL/POWER INPUTS

	TPSD	*MMTPY*
FUEL	66,000	21.681
POWER[1]	99.6 MW	

PRODUCT/BY-PRODUCT CHARACTERIZATION/OUTPUT

PRODUCT/CHARACTERIZATION		*BCOE[2]/SD*	*MMBCOE/yr*	*10^{12} Btu/yr*
Hydrotreated Shale Oil, 40°API, 5.67 MMBtu/bbl	44,400 BPSD	43,405	14.26	82.70
LPG, 3.79 MMBtu/bbl	3,500 BPSD	2,287	0.75	4.36
TOTAL	47,900 BPSD	45,692	15.01	87.06

BY-PRODUCT CHARACTERIZATION		*BCOE/SD*	*MMBCOE/yr*	*10^{12} Btu/yr*
Sulfur	198.2 TPSD			

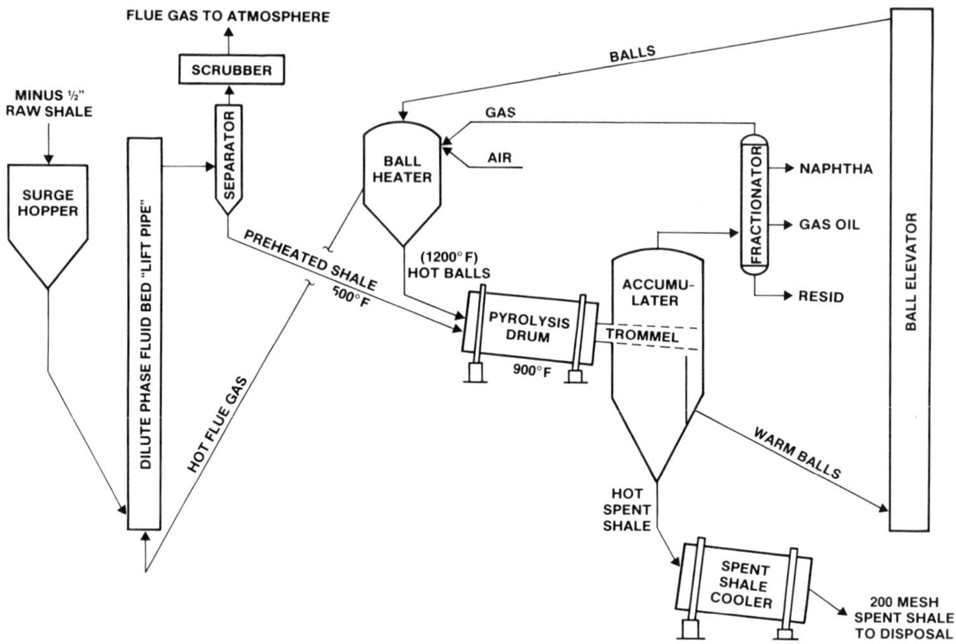

Fig. 9-13. Block diagram of Tosco II aboveground retorting process.

Ammonia	150	TPSD
Coke	836	TPSD

CAPITAL REQUIREMENT

ITEM	*$MM (mid-1979)*
PLANT INSTALLED COST	811
RESERVE COST	90
OWNER COST: CAPITALIZED	50
OWNER COST: EXPENSED	99
INITIAL CATALYST AND CHEMICALS	18
START-UP COSTS	36
TOTAL PLANT INVESTMENT	1,104
WORKING CAPITAL	25
TOTAL CAPITAL REQUIREMENT	1,129
DEFERRED INVESTMENT = $6.1 MM/yr	

ANNUAL COST

ITEM	*$MM (mid-1979)*
MINING, CRUSHING, DISPOSAL	39.9
OPERATION AND MAINTENANCE	76.9
SUBTOTAL	116.8
BY-PRODUCT CREDITS	12.8
NET ANNUAL COST	104.0

COMPOSITE PRODUCT PRICE $/BCOE ($/MMBtu)

DCF %	*3*	*6.5*	*10*
COAL	—	—	—
O&M	6.93(1.19)	6.93(1.19)	6.93(1.19)
CAPITAL	5.00(0.86)	8.14(1.40)	12.17(2.10)
TOTAL	11.93(2.06)	15.07(2.60)	19.10(3.29)

DCF %	*10 (No Tax)*	*16*	*18*
COAL	—	—	—
O&M	—	6.93(1.19)	6.93(1.19)
CAPITAL	—	21.28(3.67)	24.95(4.30)
TOTAL		28.21(4.86)	31.88(5.50)

ECONOMIC CONSIDERATIONS

STREAM FACTOR: 90%
ORIGINAL ESTIMATE DATE: 3Q 77
ESCALATION CORRECTIONS: CAPITAL 1.13 O&M 1.12

ALLOCATION OF PRODUCT PRICE

Required revenues = $0.872 MM/SD @ 10% DCF and $1.289 MM/SD @ 10% DCF. With LPG @ $15.00/bbl, the required selling price of oil is $18.47/bbl @ 10% DCF and $27.85/bbl @ 16% DCF.

UNIT OPERATION CONTRIBUTION TO COMPOSITE PRODUCT PRICE (CAPITAL INVESTMENT EFFECTS ONLY)

UNIT OPERATION	%
Mining, Crushing, Spent Shale Disposal	11.6
Retort and Oil Recovery	27.3
Oil Upgrading and H_2 Plant	10.4
By-Product Recovery	6.1
Utilities and Facilities	18.1
Reserve Costs	8.1
Capitalized Owner Cost	
Mobil Equipment	2.9
Catalyst and Chemicals	1.6
Spare Parts	0.5
Process Licenses	0.2
Miscellaneous	1.0
Expensed Owners Cost	
Project Management	3.1
Taxes, Insur. during Construction	0.7
Community Assistance Cost	3.1
Precommitment Cost	1.0
Mine Predevelopment Cost	1.1
Start-Up	3.3
	100.0

OTHER INFORMATION

Additional Economic Assumptions:
- Royalties at $0.17 ton of shale
- 15% depletion allowance, based on raw oil output from retorts and priced at 80% of hydrotreated oil

4% severance tax on revenues less O&M less straight-line depreciation
It is presumed that some contingency allowance was included in the plant installed cost.
By-products at $26/ton sulfur, $165/ton ammonia, $11/ton coke.
Quality and heating value of coke are not characterized.

References

R.G. Vawter, "The TOSCO II Process Ready for Commercialization," *International Symposium on Oil Shale Chemistry and Technology*, Jerusalem, Israel, October 1978.
J.F. Nutter and C.S. Waitman, "Oil Shale Economics Update," 14th Annual Technical Meeting, AICHE, So. California Section, April 18, 1978.

OCCIDENTAL MODIFIED IN SITU

The modified *in situ* shale oil recovery process, shown in Fig. 9-14, developed by Occidental (Oxy) involves the mining of approximately 20% of the shale from an underground chamber or retort, and then explosively rubblizing the remaining shale in the retort to fill up the mined out area. Thus, the oil shale remains underground during retorting, resulting in a process which Oxy feels is more economical than competitive processes. When the shale is retorted, a small amount of supplemental outside fuel, such as shale oil, is used to heat the top of the rubble pile to the required temperature of 900°F (482°C). After a predetermined amount of rubblized shale has been heated, the supplemental fuel burners are operated; this causes the shale to burn. The retorted shale oil flows down the retort by gravity, ahead of the burning shale, and is collected in a sump at the bottom of the retort. From there, pipelines carry the oil to storage at the surface.

FUEL CHARACTERIZATION
TYPE: Unknown

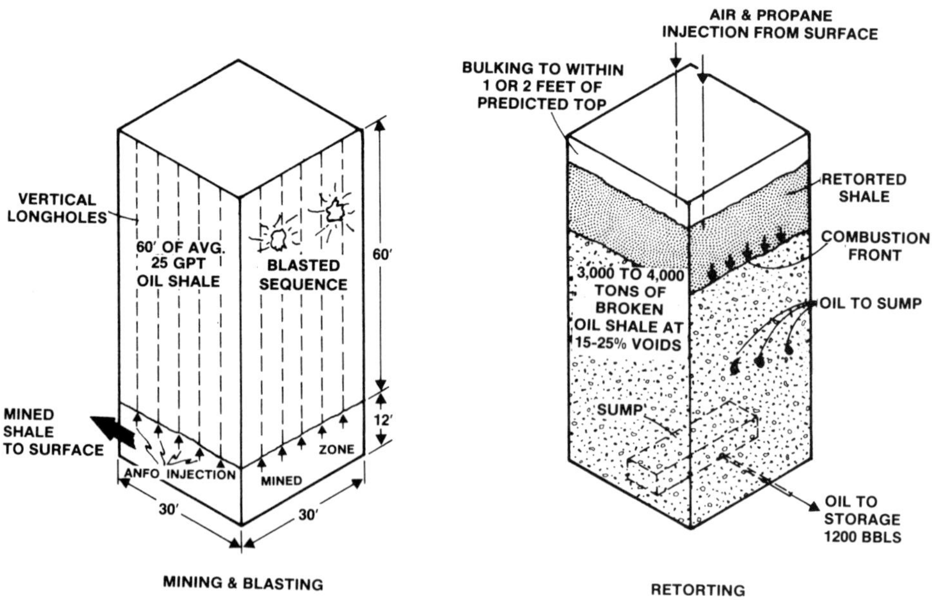

Fig. 9-14. Block diagram of Occidental modified *in situ* process.

FUEL/POWER INPUTS

	TPSD	*MMTPY*	10^{12} *Btu/yr*
SHALE	Unknown		
POWER[1]	Unknown		
		TOTAL	

PRODUCT/BY-PRODUCT CHARACTERIZATION/OUTPUT

PRODUCT/CHARACTERIZATION	*BCOE[2]/SD*	*MMBCOE/yr*	10^{12} *Btu/yr*
Raw Shale Oil, 22.5°API 30,500 BPD	Unknown		
BY-PRODUCT CHARACTERIZATION	*BCOE/SD*	*MMBCOE/yr*	10^{12} *Btu/yr*
Coke	Unknown		
Sulfur	Unknown		

CAPITAL REQUIREMENT

ITEM	*$MM (mid-1979)*
PLANT INSTALLED COST	366
RESERVE COST	60
OWNER COST: CAPITALIZED	33
OWNER COST: EXPENSED	66
INITIAL CATALYST AND CHEMICALS	—
START-UP COSTS	24
TOTAL PLANT INVESTMENT	549
WORKING CAPITAL	25
TOTAL CAPITAL REQUIREMENT	574

DEFERRED INVESTMENT = $9.4 MM/Yr

ANNUAL COST

ITEM	*$MM (mid-1979)*
FUEL ($ /MMBtu)	
OPERATION AND MAINTENANCE	
SUBTOTAL	
BY-PRODUCT CREDITS	
NET ANNUAL COST	71.4 MM

COMPOSITE PRODUCT PRICE $/BCOE ($/MMBtu)

DCF %	*3*	*6.5*	*10*
COAL	—	—	—
O&M	7.13	7.13	7.13
CAPITAL	4.80	7.23	10.38
TOTAL	11.93	14.36	17.51

DCF %	*10 (No Tax)*	*16*	*18*
COAL	—	—	—
O&M	—	7.13	7.13
CAPITAL	—	17.46	20.32
TOTAL	—	24.59	27.45

ECONOMIC CONSIDERATIONS

STREAM FACTOR: 90%

ORIGINAL ESTIMATE DATE: 1976

ESCALATION CORRECTIONS: CAPITAL 1.22 O&M 1.19

OTHER INFORMATION

Reserve costs, owner costs, etc., were prorated from the Tosco estimate on a per barrel basis. Definitive economic and process design data are currently unavailable. As such, only a nominal attempt was made to normalize the economics to the agreed set of guidelines.
To date, five retort tests have been completed with widely different results. As yet, there is no consistent database on which to definitize the process engineering and economic analysis.

Reference

TRW, *Oil Shale Data Book*, DOE Contract No. DE-AC01-78RA32012, June 1979.

Appendix A

Bibliography

Coal Gasification

Anastasia, L.J., and W.G. Bair, "The HYGAS Process," presented at Second CFFC Symposium, sponsored by IGT, Chicago, Illinois (June 23–27, 1975).

Bair, W.G., "Status of the HYGAS Program," presented at Ninth SPG Symposium, sponsored by AGA, Chicago, Illinois (October 31–November 2, 1977).

Banchik, I.N., "The Winkler Process for the Production of Low Btu Gas From Coal," paper presented at the Symposium on Clean Fuels from Coal, sponsored by the Institute of Gas Technology, Chicago, Illinois (1973).

Banchik, I.N., "The Winkler Process for the Production of Low-Btu Gas from Coal," presented at Second Clean Fuels from Coal Symposium, sponsored by IGT, Chicago, Illinois (June 23–27, 1975).

Bardos, R., "Gasifiers in Industry Program," presented at Fourth Annual International Conference on Coal Gasification, Liquefaction, and Conversion to Electricity, sponsored by University of Pittsburgh, Pittsburgh, Pennsylvania (August 2–4, 1977).

Barnett, F.M., *et al.*, "Medium Btu Gas Fits Refiner's Needs," *Hydrocarbon Processing*, June 1978, p. 131.

Blackwood, J.D., and D.J. McCarthy, "The Mechanism of Hydrogenation of Coal to Methane," *Australian Journal of Chemistry*, **19** (1966), 797–813.

Bodle, W.W., and K.C. Vyas, "Clean Fuels from Coal—Introduction to Modern Processes," presented at First CFFC Symposium, sponsored by IGT, Chicago, Illinois (September 10–14, 1973).

Bodle, W.W., *et al.*, "Clean Fuels from Coal—Technical and Historical Background and Principles of Modern Technology," presented at Second CFFC Symposium, sponsored by IGT, Chicago, Illinois (June 23–27, 1975).

Burt, U.H., and J.R. Kalafut, *Advanced Anthracite Technology and Research*, University of Scranton. ERDA Report No. FE-1774-T1 (January 1975), pp. 60–62.

Considine, D.M., *et al.*, *Energy Technology Handbook*. New York: McGraw-Hill, 1977.

Dirksen, H.A., and H.R. Linden, *Pipeline Gas from Coal by Methanation of Synthesis Gas*, Research Bulletin No. 31, Institute of Gas Technology, Chicago, Illinois (July 1963).

Dravo Corporation, *Handbook of Gasifiers and Gas Treatment Systems*, Energy Research and Development Administration, FE-1772-11 (February 1976).

Eastland, D.H., "Methanol from Coal with Winkler Synthesis Gas," presented at Second Clean Fuels from Coal Symposium, sponsored by IGT, Chicago, Illinois (June 23–27, 1975).

Eddinger, R.T., and R. Bloom, "Cogas Process—A Promise for the Future," presented at Fifth Annual International Conference on CGLCE, sponsored by University of Pittsburgh, Pittsburgh, Pennsylvania (August 1–3, 1978).

Energy Fact Book—1976, Tetra-Tech, Inc., Arlington, Virginia (July 1976).

Evaluation of Coal Gasification Technology, Part I, Pipeline Quality Gas, National Academy of Engineering (GPO), 1974, pp. 25–26.

Farnsworth, J.F., *et al.*, "Production of Gas from Coal by the Koppers–Totzek Process," Paper presented at the Symposium on Clean Fuels from Coal, sponsored by the Institute of Gas Technology, Chicago, Illinois (1973).

Fink, C.E., "The CO_2 Acceptor Process," paper presented at the Symposium on Clean Fuels from Coal, sponsored by the Institute of Gas Technology, Chicago, Illinois (1973).

Forman, H.R., "The Gasification of Low Volatile Fuels," *Journal of the Institute of Fuel*, **17** (April 1944), pp. 109–115.

Fusemann, G.T., *et al.*, "The Saarberg–Otto Gasifier," presented at Fifth Annual International Conference on Coal Gasification, Liquefaction, and Conversion to Electricity, sponsored by University of Pittsburgh, Pittsburgh, Pennsylvania (August 1–3, 1978).

Gilbert/Commonwealth, *Fixed Bed Coal Gasification for Production of Industrial Fuel Gas*, Department of Energy, FE-2220-26 (October 1977).

Grace, R., "Development of the Bi-Gas Process," paper presented at the Symposium on Clean Fuels from Coal, presented by the Institute of Gas Technology, Chicago, Illinois (1973).

Gray, R., "U-Gas, Memphis Light, Gas and Water Division," presented at Fifth Annual International Conference on CGLCE, sponsored by University of Pittsburgh, Pittsburgh, Pennsylvania (August 1–3, 1978).

Harrison, J.S., "Research on Coal-Gas Conversion in the U.K.," *Coal Processing Technology*, Vol. 2 (New York: American Institute of Chemical Engineers, 1975), pp. 70–71.

Haynes, W.P., and A.J. Forney, "The Synthane Process," presented at Second CFFC Symposium, sponsored by IGT, Chicago, Illinois (June 23–27, 1975).

Hoogendoorn, J.C., "New Applications of the Fischer–Tropsch Process," presented at Second Clean Fuels from Coal Symposium, sponsored by IGT, Chicago, Illinois (June 23–27, 1975).

Hoogendoorn, J.C., *Gas from Coal for Synthesis of Hydrocarbons, Status of SASOL II*, Ninth AGA Synthetic Pipeline Gas Symposium (November 1977).

Hottel, H.C., and J.B. Howard, *New Energy Technology—Some Facts and Assessments*, Cambridge, Massachusetts: The MIT Press, 1971.

Howard-Smith, I., and G.J. Werner, *Coal Conversion Technology*, Noyes Data Corporation, Park Ridge, New Jersey (1976).

Huebbler, J., "Coal Gasification: State of the Art," *Heating/Piping/Air Conditioning*, **45** (1), 1973, pp. 149–155.

Jones, F., "Commercial Gasification in North Dakota–Lurgi Process," presented at Fifth Annual International Conference on Coal Gasification, Liquefaction, and Conversion to Electricity, sponsored by University of Pittsburgh, Pittsburgh, Pennsylvania (August 1–3, 1978).

Jortberg, R., "Powerton Gasifier," presented at Fifth Annual International Conference on CGLCE, sponsored by University of Pittsburgh, Pittsburgh, Pennsylvania (August 1–3, 1978).

Kasturiranzan, V.N., *et al.*, "Indian Government Sponsored Study of Commercial Coal Gasification Processes—Koppers–Totzek, Lurgi and Winkler," reproduced by the Koppers Company, Inc., Pittsburgh, Pennsylvania (1973).

Krieb, K.H., "Combined Gas- and Steam-Turbine Process with Lurgi Coal Pressure Gasification," presented at Second Clean Fuels from Coal Symposium, sponsored by IGT, Chicago, Illinois (June 23–27, 1975).

Landers, J.E., "Coal Methanation Demonstration in Scotland," *Coal Processing Technology*, Vol. 2 (New York: American Institute of Chemical Engineers, 1975), pp. 48–50.

Lewis, R., "Update of Synthane Pilot Plant Status," presented at Ninth Synthetic Pipeline Gas Symposium, Chicago, Illinois (October 31–November 2, 1977).

A.D. Little, Inc., *A Current Appraisal of Underground Coal Gasification*, U.S. Department of the Interior, Bureau of Mines, Washington, D.C., PB 209 274, NTIS, Springfield, Virginia (April 1972).

Loeding, J.W., and C.L. Tsaros, "IGT U-Gas (Clean Utility Gas) Process," paper presented at the Symposium on Clean Fuels from Coal, sponsored by the Institute of Gas Technology, Chicago, Illinois (1973).

Lom, W.L., and A.F. Williams, *Substitute Natural Gas*. New York: John Wiley, 1976.

McGurl, G.J., "Gasifiers in Industry—Experiences on the First Few Projects," presented at Fifth

Annual International Conference on Coal Gasification, Liquefaction, and Conversion to Electricity, sponsored by University of Pittsburgh, Pittsburgh, Pennsylvania (August 1–3, 1978).

McWhorter, R.J., *et al.*, "Large Scale Clean Low-Btu Gas from Coal via the Kiln Gas System," 40th American Power Conference (April 1978).

Mark, H.F., J.J. McKetta, Jr., and D.F. Othmer, *Kirk–Othmer Encyclopedia of Chemical Technology*, Vol. 10. New York: Interscience Publishers, 1965, 2nd ed.

Massey, M.J., and D.V. Nakles, "ERDA's Coal Gasification Environmental Assessment Program—A Status Report," presented at Ninth Synthetic Pipeline Gas Symposium, sponsored by AGA, Chicago, Illinois (October 31–November 2, 1977).

Middleton, A.J. and C.A. Stokes, "The Manufacture of Industrial Fuel Gas from Coal on a Small Scale," paper delivered at the University of Pittsburgh, Pennsylvania, for the 4th Annual International Conference on Coal Gasification and Conversion to Electricity (August 2–4, 1977).

Mission, A., *Report on the Petroleum and Synthetic Oil Industry of Germany*, Ministry of Fuel and Power, London, England (1947), pp. 3–46.

National Research Council, Division of Engineering, *Evaluation of Coal Gasification Technology, Part I, Pipeline Quality Gas*, R&D Report No. 74, Interim Report No. 1, U.S. Department of the Interior, Office of Coal Research, Washington, D.C.: USGPO, 1973.

Nelson, S.G., "Applications for Industrial Fuel Gas, USA and International," presented at Fifth Annual International Conference on Coal Gasification, Liquefaction, and Conversion to Electricity, sponsored by University of Pittsburgh, Pittsburgh, Pennsylvania (August 1–3, 1978).

Newman, L.L., "Oxygen in the Production of Hydrogen or Synthesis Gas," *Industrial and Engineering Chemistry*, **40** (4), 1948, pp. 559–582.

The Northwestern Gas Board, *The Stretford Process*. London, England: Bradley & Son, Ltd., 1967.

Partridge, L.J., "Coal-Based Ammonia Plant Operations," *Chemical Engineering Progress* (August 1976), pp. 57–61.

Patel, J.G., and J.W. Loeding, "IGT-U-Gas Process," presented at Second CFFC Symposium, sponsored by IGT, Chicago, Illinois (June 23–27, 1975).

Percival, G., *Coal Gasification Developments at Westfield*, Fifty-Fifth Annual Convention, Gas Producers Assn. (March 1976).

Peters, W., "Coal Gasification Using Nuclear Energy," *Coal Processing Technology*, Vol. 2 (New York: American Institute of Chemical Engineers, 1975), pp. 72–75.

Project Independence, Synthetic Fuels from Coal, Department of Interior, FEA (1974), pp. 112–123.

Rudolph, P.F.H., and H.H. Bierbach, "Fuel Gas from Coal," presented at Second Clean Fuels from Coal Symposium, sponsored by IGT, Chicago, Illinois (June 23–27, 1975).

Sarma, K.R.K., "Ranking Coal Gasification Systems," presented at Fifth Annual International Conference on CGLCE, sponsored by University of Pittsburgh, Pittsburgh, Pennsylvania (August 1–3, 1978).

Segeler, C.G., *et al.*, *Gas Engineers Handbook*. New York: Industrial Press, 1965.

Seifert, G., "GDR Experience in Fixed Bed Gasification of Brown Coal," presented at Ninth SPG Symposium, sponsored by AGA, Chicago, Illinois (October 31–November 2, 1977).

Seifert, G., *et al.*, "State and Prospects of Pressure Gasification of Coal in the German Democratic Republic," presented at Ninth Synthetic Pipeline Gas Symposium, sponsored by AGA, Chicago, Illinois (November 1977).

Sudbury, J.D., *et al.*, "Gasification of Caking Coals in the British Gas Corporation/Lurgi Slagging Gasifier—A Status Report," presented at Fifth Annual International Conference on Coal Gasification, Liquefaction, and Conversion to Electricity, sponsored by University of Pittsburgh, Pittsburgh, Pennsylvania (August 1–3, 1978).

Synthetic Gas Coal Task Force, *The Supply-Technical Advisory Task Force—Synthetic Gas-Coal*, Final Report, Federal Power Commission, National Gas Survey, Supply-Technical Advisory Committee, Washington, D.C. (April 1973).

Von Fredersdorff, C.G., and M.A. Elliott, "Coal Gasification," *Chemistry of Coal Utilization* (Editor:

H.H. Lowry), Supplementary Volume, New York: Wiley, 1963, pp. 892–1022.

Walker, K.E., "Status of the BI-GAS Pilot Plant Program," presented at Ninth SPG Symposium, sponsored by AGA, Chicago, Illinois (October 31–November 2, 1977).

Wintrell, R., "The Koppers–Totzek Gasification Process," *Coal Processing Technology*, Vol. 2 (New York: American Institute of Chemical Engineers, 1975), pp. 121–123.

Coal Liquefaction

Badische, A., and A.G. Soda Fabrick, *High-Pressure Hydrogenation at Ludwigshafen-Heidelberg*, Volume IV B, Chemical Section, FIAT Final Report No. 1320, ATI No. 88, 725, Central Air Documents Office, Dayton, Ohio (March 1951).

Battelle Columbus Laboratories, *Liquefaction and Chemical Refining of Coal* (July 1974).

Bodle, W.W., and K.C. Vyas, "Clean Fuels from Coal—Introduction to Modern Processes," presented at First CFFC Symposium, sponsored by IGT, Chicago, Illinois (September 10–14, 1973).

Bodle, W.W., *et al.*, "Clean Fuels from Coal—Technical and Historical Background and Principles of Modern Technology," presented at Second CFFC Symposium, sponsored by IGT, Chicago, Illinois (June 23–27, 1975).

Carlson, F.B., *et al.*, "The Toscoal Process for Low-Temperature Pyrolysis of Coal," paper presented at the 65th Annual Meeting of the American Institute of Chemical Engineers, New York, New York (November 1972).

Conoco Coal Development Co., *Zinc Halide Hydrocracking Process for Distillate Fuels from Coal*, Quarterly Technical Progress Report, FE-1743-37 (February 1–April 30, 1977).

Donath, E.E., "Hydrogenation of Coal and Tar," *Chemistry of Coal Utilization*, Supplementary Volume, H.H. Lowery, New York: Wiley, 1968.

Downs, W., *et al.*, *Preparation and Burning of Solvent Refined Coal*, 39th American Power Conference (April 1977).

Eastland, D.H., "Methanol from Coal with Winkler Synthesis Gas," presented at Second CFFC Symposium, sponsored by IGT, Chicago, Illinois (June 23–27, 1975).

Edwards, R.G., *et al.*, *Social, Economic, and Environmental Impacts of Coal Gasification and Liquefaction Plants*, Institute for Mining and Minerals Research, Contract #201-65-8HO50-S3343 (April 1976).

Energy Fact Book–1976, Arlington, Virginia: Tetra-Tech, Inc., July 1976.

Farley, R., and D.J. Ray, "The Design and Operation of a Pilot Scale Plant for Hydrocarbon Synthesis in the Slurry Phase," *Journal of the Institute of Petroleum*, **50** (1964), pp. 27–46.

Fieldners, A.C., H.H. Storch, and L.L. Hirst, *Bureau of Mines Research on the Hydrogenation and Liquefaction of Coal and Lignite*, U.S. Bureau of Mines Technical Paper No. 666 (1944).

Fluor Engineerings and Constructors, Inc., *Coal Liquefaction Design Practices Manual*, EPRI-AF-199 (July 1976).

Friedman, S., *et al.*, "The Synthoil Process," presented at Second CFFC Symposium, sponsored by IGT, Chicago, Illinois (June 23–27, 1975).

Green, N.W., "Synthetic Fuels from Coal—The Garrett Process," presented at Second CFFC Symposium, sponsored by IGT, Chicago, Illinois (June 23–27, 1975).

Harrison, J.S., "Coal Liquefaction in the U.K.," *Coal Processing Technology*, Vol. 2. New York: American Institute of Chemical Engineers, 1975, pp. 20–24.

Higginson, G., "Chemical Refining of Coal-SRC," presented at Fifth Annual International Conference of CGLCE, sponsored by University of Pittsburgh, Pittsburgh, Pennsylvania (August 1–3, 1978).

Howard-Smith, I., and G.J. Werner, *Coal Conversion Technology*, Park Ridge, New Jersey: Noyes Data Corporation, 1976.

Jahnig, C.E., *Evaluation of Pollution Control in Fossil Fuel Conversion Processes, Liquefaction: Section 3, H-Coal Process*, Exxon Research & Engineering Company, EPA-650/2-74-0090M (October 1975).

Johnson, C.A., *et al.*, "Present Status of the H-Coal Process," presented at Second CFFC Symposium, sponsored by IGT, Chicago, Illinois (June 23–27, 1975).

Jones, J.F., "Project COED," presented at Second CFFC Symposium, sponsored by IGT, Chicago, Illinois (June 23–27, 1975).

The M.W. Kellogg Co., *A SASOL Type Process for Gasoline, Methanol, SNG, and Low-Btu Gas from Coal*, EPA-650/2-47-072 (July 1974).

Kloepper, D.L., *et al.*, *Solvent Processing of Coal to Produce a De-Ashed Product*, R&D Report No. 9, U.S. Department of the Interior, Office of Coal Research, Washington, D.C. (1965).

Libowitz, H.E., "Clean Liquids and Solid Fuels from Coal," presented at Fifth Annual International Conference on CGCLE, sponsored by University of Pittsburgh, Pittsburgh, Pennsylvania (August 1–3, 1978).

Lowry, H.H., ed., *Chemistry of Coal Utilization*, Vol. 1. New York: Wiley, 1945.

Mobil Research & Development Corp., *Development Studies on Selected Conversion of Synthesis Gas from Coal to High Octane Gasoline*, Quarterly Report, FE-2276-9 (March 1977).

Mobil Research & Development Corp., *Fluid Bed Process Studies on Selective Conversion to Methanol to High Octane Gasoline*, FE-2490-13 (November 1977).

Occidental Research Corp., *Flash Pyrolysis Coal Liquefaction Process Development*, FE-2244-11 (January–March 1977).

R.M. Parsons Co., *Coal Liquefaction Process Research, Mobil M-Gasoline Process*, R&D Interim Report No. 1, ORNL-Sub-718617 (June 24, 1977).

Phinney, J.A., "Clean Fuels Via the CSF Process," presented at Second CFFC Symposium, sponsored by IGT, Chicago, Illinois (June 23–27, 1975).

Pittsburgh & Midway Coal Mining Co., *Solvent Refined Coal (SRC) Process: Development of a Process for Producing an Ashless, Low Sulfur Fuel from Coal*, Vol. 3, Part 3, ERDA FE-496-T13 (June 1977).

Pittsburgh & Midway Coal Mining Co., *Solvent Refined Coal (SRC) Process*, FE-496-129 (April 1977), FE-496-126 (February 1977), FE-496-131 (May 1977).

Ralph M. Parsons Co., *Coal Conversion and Utilization-Liquefaction; Clean Coke Process*, FE-1234-7 (November 1976).

Rockwell International Corp., *Partial Liquefaction of Coal by Direct Hydrogenation*, FE-2044-22 (April–June 1977).

Salmon, R., *et al.*, *Evaluation of the Synthoil Process*, Oak Ridge National Laboratory, ORNL-5209 (June 1977).

Salmon, R., *et al.*, *Evaluation of the Synthoil Process*, Oak Ridge National Laboratory, ORNL-5210 (June 1977).

Sass, A., "The Production of Liquid Fuels from Coal," *Minerals Science and Engineering*, **4** (4), 1972, pp. 18–22.

Schmid, B.K., and D.M. Jackson, "Commercial Scale Development of the SRC-II Process," presented at Fifth Annual International Conference on CGLCE, sponsored by University of Pittsburgh, Pittsburgh, Pennsylvania (August 1–3, 1978).

Sherwood, P.W., *High-Pressure Hydrogenation in Germany, I, the Liquid Phase*, FIAT Final Report No. 952, Joint Intelligence Objectives Agency, Washington, D.C. (May 1947).

Sinatt, F.S., *The Action of Solvent on Coals*, Department of Science and Industry, Fuel Research Technical Paper No. 37, Great Britain (1933).

Snell, G.T., and R.H. Long, *Developmental Research Program for Clean Industrial and Transportation Fuels from Coal*, Monthly Technical Report, The Lummus Company, FE-2514-8 (June 1977).

Stanford Research Institute, *Evaluation of Processes for the Liquefaction and Gasification of Solid Fossil Fuels, Phase III*, Vol. 1, SRI Project EDD-2896 (January 1977).

Storch, H.H., N. Columbic, and R.B. Anderson, *The Fischer–Tropsch and Related Syntheses*. New York: Wiley, 1951.

Tramm, H., "25 Years of Fischer–Tropsch Synthesis with Fixed-Bed Catalysts, Present Status and Future Possibilities," paper 27, Proceedings of the 5th World Petroleum Congress held in New York, New York (1959).

U.S. Bureau of Mines, *Liquid Fuel from Coal Plant-Synthoil Process*, Morgantown, W. Va., BM 74–27 (April 1975).

U.S. Bureau of Mines, *50,000 Barrel Per Day Liquid Fuels Plant Synthoil–Wyodak Coal: An Economic Analysis*, Morgantown, W. Va., BM 75-32 (June 1975).

U.S. Department of Energy, *Gasification of Residual Materials from Coal Liquefaction*, Quarterly Report (July–September 1977).

U.S. Department of Energy, *Conceptual Design of a Coal to Methanol Commercial Plant,* Interim Final Report for the Period July 16, 1976–February 15, 1978, Executive Summary, Badger Plants, Inc., FE-2416-24 (February 1978).

U.S. ERDA, *Fossil Energy Technical Program Summary Book FY 1977* (February 15, 1977).

U.S. ERDA, *Program Plan for Coal Liquefaction* (September 30, 1977).

U.S.S., Engineers and Consultants, Inc., *Clean Coke Process—Process Development Studies*, FE-1220-28 (August 1977).

Van Krevelen, D.W., "Solvent Extraction of Coals," Coal—Typology, Chemistry, Physics, and Constitution. New York: Elsevier Publishing Co., 1962.

Vestal, M.L., and W.H. Johnston, "Chemistry and Kinetics of the Hydrosulfurization of Coal," American Chemical Society, Division of Fuel Chemistry, Preprints **14** (1), 1970, pp. 1–11.

Voltz, S.E., and J.J. Wise, *Development Studies on Conversion of Methanol and Related Oxygenates to Gasoline*, Final Report, Mobil Research and Development Corp., FE-1773-25 (November 1976).

Wise, J.J., and A.J. Silvestor, "Mobil Process Efficiently Converts Methanol to Gasoline," *Oil and Gas Journal* (November 22, 1977), pp. 140–142.

Wiser, W.H., in *Proceedings of the EPRI Conference on Coal Catalysis*, Santa Monica, California (September 24–25, 1973), pp. 3–53.

Wu, W.R.K., and H.H. Storch, *Hydrogenation of Coal and Tar*, U.S. Bureau of Mines Bulletin No. 633, U.S. Department of the Interior, Bureau of Mines, Pittsburgh Coal Research Center, Pittsburgh, Pennsylvania (1968).

Coal Mining, Preparation, and Transportation

Anonymous Individual Author, *Technical and Cost Comparisons for Chemical Coal Cleaning Processes*, Bechtel Corporation (May 1977).

AP Release, "Coal Slurry Pipelines Opposed," *Washington Post* (24 January 1978).

A Technology Assessment of Coal Slurry Pipelines, Office of Technology (March 1978).

Bituminous Coal Research Staff, *An Analysis of Coal Gasification, Liquefaction, Fluidized-Bed Combustion and Coal Cleaning*, National Coal Association (June 1977).

Campbell, T.C., and S. Katell, *Long-Distance Coal Transport: Unit Trains or Slurry Pipelines*, U.S. Bureau of Mines Information Circular 18 8690 (1975).

Cassidy, S.M., *Elements of Practical Coal Mining*, Coal Division of SME, AIMM & PE (1973).

Coal Data, National Coal Association (1975 and 1976).

Cummins, A.B., and I.A. Given, *SME Mining Engineering Handbook*, Volumes I and II: SME Division of AIMM & PE (1973).

Davenport, C. (Chairman of Task Force), *Transporting the Nation's Coal*, The Coal Transportation Task Force, U.S. Department of Transportation (January 1978).

Dawson, P. "Coal-Slurry Pipelines Have Strong Foes," *The Billings Gazette* (22 January 1978).

Duerbrouck, A.W., "United States Coal Preparation Practices," *World Coal*, **3**, No. 5 (May 1977), pp. 24–26.

Goode, C.A., "Underground Coal Mine Innovations," Seventieth Annual AIChE Meeting (1977).

"Hydraulic Mining—A Refined and Growing Technological Art," *World Coal*, **3**, No. 7 (July 1977), p. 55.

HYGAS 1964 to 1972 Pipeline Gas from Coal-Hydrogenation (IGT Hydrogasification Process), Volume 3, Part VI, FE-381-T9-P3 (July 1975).

Leonard, J.W., and D.R. Mitchell, *Coal Preparation*, AIMM & PE, Third Edition (1968).

"Multi-Year Program Plan of Solid Fuels Mining and Preparation," prepared for the Division of Solid

Fuels Mining and Preparation, U.S. Department of Energy, by Systems Consultants, Inc., Arlington, Virginia (May 10, 1978).

Olson, J.J., and S. Tandanand, *Mechanized Longwall Mining, A Review Emphasizing Foreign Technology*, Bureau of Mines Information Circular 8740 (1977).

Paul, B. "German Mines Uproot Thousands as Nation Exploits Coal Reserves," *Wall Street Journal* (2 February 1978).

Pfleider, E.P., *Surface Mining*, The American Institute of Mining, Metallurgical and Petroleum Engineers, Inc., New York, New York.

Phillips, P.J., *et al.*, *Coal Preparation for Combustion and Conversion*, Gibbs & Hill, Inc., Project 466-1, for EPRI (1976).

"Productivity and the UMW," *Coal Age* (July 1975).

Report of the Panel on Civilian Technology on Coal Slurry Pipelines, U.S. Department of the Interior (1962).

Rieber, M., and S.L. Soo, *Comparative Coal Transportation Costs: An Economic and Engineering Analysis of Truck, Belt, Barge and Coal Slurry and Pneumatic Pipelines*, Center for Advanced Computation, University of Illinois, Urbana, Illinois (August 1977), BuMines ORF 146 (1-8-77) (CAC Doc. #223).

Weiss, A., and N. Fusmura, *World Mining and Metals Technology*, Volume I: MMIJ (Japan) and AIMM & PE (1976).

Flue-Gas Desulfurization for Conventional Combustion

Addison, G., *et al.*, "Shell FGD Process—Recent Evaluation of Non-Polluting Pollution Control," University of Pittsburgh (August 1, 1978).

Ando, J., "Status of SO_2 and NO_x Removal Systems in Japan," EPA FGD Symposium (November 1977).

Ando, J., *et al.*, "SO_2 Abatement for Stationary Sources in Japan," Report for EPA/IERL done by Pedco, Environmental Specialists (January 1976).

Ando, J., *et al.*, "NO_x Abatement for Stationary Sources," EPA-600/7-77-103B, PB 276948 (1977).

Arneson, A., F. Nooy, and J. Pohlenz, "The Shell FGD Process Pilot Plant Experience at Tampa Electric," EPA FGD Symposium (November 1977).

Clasen, D.D. and H. Idemura, "Limestone/Gypsum Jet Bubbling Scrubbing System," EPA FGD Symposium (November 1977).

Coal Technology, Casey Publications Co., Washington, Vol. 1, #4 (August 1978), p. 4.

Esche, M., "Saarberg–Holter FGD Process: Second Generation Line Based FGD System," EPA FGD Symposium (November 1977).

Felsvang, K., *et al.*, "SO_2 Spray Absorption with Dry Wastes: Niro Atomizer," company report.

Herlihy, J., "FGD in Power Plants-Status Report," EPA Office of Enforcement (April 1977).

Hirai, M., *et al.*, "MHI Flue Gas Desulfurization Systems Applied to Several Emission Sources," Proceedings: Symposium on Flue Gas Desulfurization (March 1976).

Holighaus, R., "Desulfurization Technology in the Federal Republic of Germany," EPA FGD Symposium (November 1977).

The M.W. Kellogg Co., "A SASOL Type Process for Gasoline, Methanol, SNG, and Low-Btu Gas from Coal," EPA-650/2-27-072 (July 1974).

Knoblauch, K., "BF-STEAG Demonstration Unit Operational Experience and Performance," Proceedings: Symposium on Flue Gas Desulfurization (March 1976).

Laseke, B. and T. Devitt, "Status of Flue Gas Desulfurization Systems in the United States," EPA FGD Symposium (November 1977).

McGlamery, G.G., *et al.*, 6th EPA FGD Symposium (March 1976).

"November 1977 Survey of FGD Facilities in the U.S.," report for EPA by Pedco (January 25, 1978); and "January 1978 Survey," March 1978; "March Survey," June 1978; "May Survey," September 1978.

Quackenbush, V., *et al.*, "Ammonia Scrubbing Pilot Activity at Calvert City," EPA FGD Symposium (November 1977).

Rush, R., and R. Edwards, "Operational Experience with Three 20 MW Prototype Flue Gas Desulfurization Processes at Gulf Power Company's Scholz Electric Generating Station," EPA FGD Symposium (November 1977).

Start, W., *et al.*, "The Spring–Nobel Hoechst Process for Sulfur Dioxide Recovery from Stack Gases," Proceedings: Symposium on Flue Gas Desulfurization (March 1976).

"Sulfur Oxide Control Technology: Visits in Japan—March 1974," Report by Interagency Technical Committee (EPA and TVA).

Tougeno, H., T. Takashi, and T. Itoh, "Operating Experiences with Kawasaki Magnesium–Gypsum Flue Gas Desulfurization Process," EPA FGD Symposium (November 1977).

Fluidized-Bed Coal Combustion

Abbi, Y., "Development of FBC Boilers for Indian Coals," presentation at the AIChE Meeting (Chicago, Illinois, November 1976).

"AEP Designing PFB-GT Plant," *Gas Turbine World* (March 1977), p. 24.

Babcock & Wilcox, Ltd., "Studies of Advanced Electric Power Generation Techniques and Coal Gasification" prepared for British Columbia Hydro & Power Authority, Volumes I–III, No. 267 (1976).

Berglund, B.E.A., *et al.*, "Method of Burning Sulphur Containing Fuels Such As Shale," United States Patent Office, Patent No. 2,928,718 (March 15, 1960).

Bush, J., *Development of HTHP Electrostatic Precipitator*, EPA Report 600/7-77-132 (November 1977).

"Coal Burning Fluidized Bed Combustion System," Coal Research of CSIRO, No. 34 (April 1968), pp. 2–5.

"Coal Research in Australia 1977," Department of Natural Resources for International Committee for Coal Research, Louisville (October 17, 1977).

Covell, R., *et al.*, *Proceedings of the FBC Technology Exchange Workshop Conference*, 770447, pp. 1, 205.

Demmy, R., 2nd Hueston Woods Conference (1970), p. IV.3.1.

Demonstration Program for Coal-Oil Mixture Combustion in Electric Utility Boilers, DOE Report HCP/T2564-01 (June 1978).

5th International FBC Conference, 1977, Washington, D.C.

"Fluid Bed Coal Combustors Commercialized," *Chemical and Engineering News* (February 6, 1978), p. 24.

Fossil Energy R&D Program of the U.S. DOE, DOE/ET-0013(78), March 1978.

4th International FBC Conference, 1975, McLean, Virginia, pp. I.5, I.6, II.7, VII.5, VII.6.

Fuel Economy Consultants, Inc., "Survey and Analysis of FBC Work Being Conducted in Europe," report to ERDA (April 1976).

Godel, A., and P. Cosar, *AIChE Symposium Series*, No. 116, **67** (1971), p. 210.

Grant, A., "Fluidized Coal Combustion: What Can Be Done Now?," presentation at the University of Pittsburgh (Pittsburgh, Pennsylvania, August 3, 1976).

Grant, A., "Impact of FBC on U.S. Coal Mining and Use," NCA Coal Conference, Louisville, October 18, 1977.

Harboe, H., "Coal for Peak Power" by Stal-Laval Ltd., ACS Chicago, August 1973, Division of Fuel Chemistry Preprints, 18, (4).

Hoke, R., *et al.*, *Miniplant Studies of Pressurized FBC*, EPA Report 600/7-78-069 (April 1978).

Industrial and Utility Application, Coal-Oil Mixture Data Index, DOE Report HCP/T2637-01 (May 1978).

International Fluidized Combustion Conference, London, England (September 1975).

Journal of Fuel and Heat Technology, CSIRO, 15, No. 5 (1968), pp. 11–12.

1977 Keystone Coal Industry Manual. New York: McGraw-Hill, 1977, pp. 720–733.

KFA Julich, "Jahresbericht 1977 über das Programm Energieforschung und Energietechnologien 1977–1980" (1978).

Lurgi Brochure, *Fluid Bed Technology and Environmental Protection* (1977).

McKenzie, E., "Fluidized Bed Firing Boilers," CEA Conference, Eastbourne (1976).

Mesko, J., Trip Report to CKD Dukla, Czechoslovakia (1976).

Mesko, J., Trip Report to B&W, Ltd., Renfrew, Scotland (February 1977).

The Mitre Corporation, "International Activities in Fluidized-Bed Combustion," Working Paper No. 12326 (April 1977).

Murthy, K.J., *Air Pollution Control Association*, **28**, 213 (1978).

National Fluidized-Bed Combustion Program, The Mitre Corporation, Report No. M-75-58, Volume 3, by Battelle Columbus Laboratories (December 1974).

Novotny, P., "Fluid Bed Combustion of High Ash Fuels," *International Technology Digest (Prague)*, **2**, No. 12 (1968), pp. 12–15.

Pope, Evans, & Robbins, "Annual Report for FY 75," Report FE-1237-38.

Pradt, L.A., Zimpro Inc., *Wet Oxidation Boiler-Incinerator*, ASME Conference (1977).

Prior, M., *Control of Sulphur Oxides Emitted in Coal Combustion*, IEA Economic Assessment Service Report BI/77 (December 1977).

Proceedings of Coal-Mix Combustion Workshop, CONF-761019 by ERDA (October 29, 1976).

Reh, L., "Today's Possibilities in FBC of Fossil Fuels and Energy Bearing Minerals," presented LKAB Symposium, Hjo, Sweden (September 22, 1977).

"Research into the FBC Process Carried Out in the FRI, Bechovice," Sbornik prednasek k. 50 Vyroci Ustavu pro Byzkum a Vyuziti Paliv (1972), pp. 104–110 (Czechoslovakia).

Roberts, A.G., *et al.*, "Fluidized Combustion of Fossil Fuels," *Coal Processing Technology*, Vol. 2; (New York: American Institue of Chemical Engineers, 1975), pp. 34–39.

Sharan, H., "RDD&C Programs for Better Utilization of Coal," B.H.E.L., New Delhi (September 1977).

Skinner, D., *Fluidized Combustion of Coal.* London: Mills and Boon, Ltd., 1971.

Steam, Babcock and Wilcox Co., 38th ed. (1975).

Svoboda, J., 2nd Hueston Woods Conference (1970), p. I.6.1.

Szpindler, G., *et al.*, "FBC as a Solution to the Environmental Problem of Coal Mining Waste," National Chem. Engr. Conference, July 10, 1974, pp. 391–403.

Takaki, S., Japan Coal Association, personal communication to M. Card, DOE (March 29, 1978).

Walker, D., *et al.*, *FBC Technology for Future Industrial Boilers*, 40th American Power Conference (1978).

Wilson, J., Proceedings of the FBC Technology Exchange Workshop Conference, 770447, pp. 2, 3.

General Literature

Coal—Bridge to the Future. Report of the World Coal Study. Cambridge, Mass.: Ballinger Publishing Company, 1980.

Energy—Global Prospects 1985–2000. Workshop on Alternative Energy Strategies. New York: McGraw-Hill Book Company, 1977.

Energy in America's Future. Baltimore, Maryland: The Johns Hopkins University Press, 1979.

Energy—The Next Twenty Years. Cambridge, Mass.: Ballinger Publishing Co., 1979.

Hunt, V. Daniel. *Handbook of Conservation and Solar Energy.* New York: Van Nostrand Reinhold, 1982.

Hunt, V. Daniel. *The Gasohol Handbook.* New York: The Industrial Press, 1981.

Hunt, V. Daniel. *Handbook of Energy Technology—Trends and Perspectives.* New York: Van Nostrand, 1982.

Yergin, Daniel, *et al.* (eds.). *Energy Future—Report of the Energy Project at the Harvard Business School.* New York: Random House, 1979.

Appendix B

Glossary

absorption. An imprecise term suggesting the taking up of one substance by another by either a physical process or a chemical combination.

acceptor. Calcined carbonate that absorbs carbon dioxide evolved during gasification, liberating heat.

acid gas removal. The process of selectively removing hydrogen sulfide and carbon dioxide from a gas stream.

activated carbon. Carbon obtained by carbonization in the absence of air, preferably in a vacuum; it has the property of absorbing large quantities of gases and vapors; used also for clarifying liquids.

adiabatic. Any process where heat is neither given off nor absorbed.

adsorption. The process by which the surface of a solid or liquid attracts and holds any atom, molecule, or ion from a solution or gas with which it is in contact.

agglomerate. Assemblage of ash particles rigidly joined together, as by partial fusion (sintering).

air blown gasifier. A gasifier using air as a combustion source.

aliphatic. Referring to a class of hydrocarbons in which the carbon atoms form open chains, e.g., propane, hexane.

anthracite coal. Anthracite, or hard coal, is the highest rank of coal, containing 86–98% fixed carbon and small percentages of volatile materials and ash. It is hard, jet black, with a high luster and is used for generating electricity and space heating. Its ignition temperature is about 900°F (482°C). Anthracite is mined in northeastern Pennsylvania.

API gravity. A scale adopted by the API for measuring the density of oils:

$$°\text{API} = \frac{141.5}{\text{specific gravity, } 60°\text{F}} - 131.5$$

area mining. Area mining is practiced on flat terrain and recovers coals by mining long strips successively; the material excavated from the strip being mined is deposited in the strip previously mined.

aromatic. Referring to a class of hydrocarbons in which the carbon atoms form closed rings with conjugated double bonds, e.g., benzene, toluene.

aromatic hydrocarbon. A cyclic hydrocarbon containing one or more six-carbon (benzene) rings.

ash. Solid residue remaining after the combustion of coal.

ASTM. American Society for Testing Materials.

auger machine. An auger machine is a large horizontal drill, generally 3 ft or more in diameter and up to about 1000 ft long.

auger mine. An auger mine recovers coal through the use of a large-diameter drill driven into a coal bed in a hillside. It usually follows contour surface mining, particularly when the overburden is too costly to excavate.

autoclave. A vessel, constructed of thick-walled steel, for carrying out chemical reactions under high pressures and temperatures.

bench-scale unit. A small-scale laboratory unit for testing process concepts and operating parameters as a first step in the evaluation of a process.

beneficiation. A generic term referring to any process for improving the chemical or physical properties of mined coal prior to sale or use.

Best Available Control Technology (BACT). Proposed EPA rules (*Federal Register*, September 19, 1978, p. 42154), requires that BACT be determined on a case-by-case basis for each pollutant, by the permit issuing ("permitting") authority. It is based on the maximum degree of reduction achievable at a specific facility in the view of the permitting authority. Energy, environmental, and economic impacts are to be considered in designating BACT.

binder. Carbon products, tars, etc., used to impart cohesion to the body to be formed; a coal extract may be used to prepare formed coke pellets from noncaking coals.

biogas. Gas generated by anaerobic digestion technologies, which contains about 50–60% methane, with the remainder primarily carbon dioxide.

biomass. Organic matter, such as trees, crops, manure, and aquatic plants, that is available on a renewable basis.

bituminous coal. Bituminous coal, or soft coal, is the most common coal. It is dense, black, often with well-defined bands of bright and dull material, and is used for generating electricity, making coke, and space heating. The ignition temperature is 700–800°F (371–427°C). Bituminous coal is mined chiefly in the Appalachian and Interior coal fields.

blow down. Periodic or continuous removal of water from a boiler to prevent accumulation of solids.

blowdown water. Water effluent containing relatively high concentrations of dissolved salts and other residues, which is periodically purged from steam boilers or wet cooling towers.

boiler efficiency. Defined by ASME Power Test Code No. 4.1. The output of the boiler in terms of Btu per hour in the steam divided by the fuel input gross heat value expressed in Btu per hour.

bottoming cycle. The lower-temperature thermodynamic power cycle of a combined-cycle system.

breeze. A term referring to the dust or fines from coal, coke, or charcoal.

Btu. A measure of energy, it is the amount of heat needed to raise the temperature of 1 pound of water by 1 degree Fahrenheit. A Btu is a convenient measure by which to compare the energy content of various fuels.

bulldozer. A bulldozer is a tractor with a movable steel blade mounted on the front.

caking. The softening and agglomeration of coal as a result of the application of heat.

caking coal. A property of certain coals that manifests itself during rapid heating. The crushed or powdered coal swells and the surface becomes "tacky," preventing it from remaining free-flowing and causing problems in transporting or processing.

calcination. The process of heating a solid to a high temperature to cause the decomposition of hydrates and carbonates.

calorific value. The quantity of heat obtained by the complete combustion of a unit mass of a fuel under prescribed conditions.

carbon fiber. Fine filaments of carbon about 8 microns (312 microinches) in diameter, which are used in composite material, being bound with resins.

carbonization. Destructive heating of carbonaceous substances with the production of a solid, porous residue, or coke, and the evolution of a number of volatile products. For coal, there are two principal classes of carbonization, high-temperature coking [about 1650°F (900°C)] and low-temperature carbonization [about 1290°F (700°C)].

captive coal. This refers to coal produced and consumed by the mine operator, a subsidiary of the parent company (for example, steel companies and electric utilities). It is not sold in the open market.

carrier gas. An inert gas used to transport feed material through a conversion process.

carryall scraper. A carryall scraper is a self-loading machine, usually self-propelled, with a scraper-like retractable bottom. It is used to excavate and haul overburden.

catalyst. A substance that accelerates the rate of a chemical reaction without itself undergoing a permanent chemical change.

centrifuge. An apparatus rotating at high speed that utilizes the centrifugal force generated to separate materials of different densities, e.g., undissolved residue from coal solution in the SRC process.

char. The solid residue remaining after the removal of moisture and volatile matter from coal.

Claus plant. A Claus plant processes gaseous emission streams containing 10% or more hydrogen sulfide and oxidizes the hydrogen sulfide in the presence of a solid catalyst, producing elemental sulfur of high purity.

cleaning. A mechanical form of coal beneficiation which separates mine trash, e.g., rock gangue, and blasting wire, from the coal.

coal rank. This classification of coal is based on fixed carbon, volatile matter, and heating value. It is an indication of the progressive alternation, or coalification, from lignite to anthracite.

- *Lignite*, the lowest rank of coal, is brownish black and has a high moisture content; it is used mainly to generate electricity. It is mined in North Dakota, Montana, and Texas.
- *Subbituminous coal*, or black lignite, is dull black. It is used for generating electricity and space heating. This rank of coal is mined in the western coal fields.
- *Bituminous coal*, or soft coal, is the most common coal. It is dense black, often with well-defined bands of bright and dull material, and is used for generating electricity, making coke, and space heating. The ignition temperature is 700–800°F (371–427°C). Bituminous coal is mined chiefly in the Appalachian and Interior coal fields.
- *Anthracite*, or hard coal, is the highest rank of coal. It is hard, jet black, with a luster, and is used for generating electricity and space heating. Its ignition temperature is about 900°F (482°C). Anthracite is mined in northeastern Pennsylvania.

coal type. This classification is based on physical characteristics or microscopic constituents (for example, bright coal, splint coal, cannel coal).

coalification. Metamorphosis of vegetable debris into coal.

cocurrent. A mode of operation in which the materials, for example, gases, liquid, or solids, all flow in the same direction (in parallel).

cogeneration. A mode of industrial or institutional boiler operation in which steam is raised at pressures exceeding the in-plant usage requirements, and backpressure turbines are used to extract shaft power (producing electricity) from the steam prior to its conventional process use. When plant electricity/steam demand profiles coincide, high coal-to-electricity conversion efficiencies are achieved and the thermal discharge (waste heat) of a conventional utility generating plant is avoided. Cogeneration is also applicable to gas turbine power plants with waste heat boilers.

coke. In general, coke is made from bituminous coal (or blends of bituminous coal) from which the volatile constituents have been driven off by heat, so that the fixed carbon and ash are fused together. Coke is hard and porous.

coke breeze. The fine screenings from crushed coke usually passing a ½- or ¾-in. screen opening.

combined cycle. Two sequential thermodynamic power conversion systems operating at different temperatures.

combustion gas. Gas formed by the combustion of coal, e.g., burning.

combustor. A vessel in which combustion takes place.

commercial impact. A plant or process of sufficient size to be considered commercial by the industry for which it is intended, and with demonstrated continuous operation of 1–2 years.

continuous mining machine. A continuous mining machine cuts or rips coal from the coal face and loads it into shuttle cars or conveyors in one operation. It can mine coal at the rate of 8–15 tons per minute.

contour mining. Contour mining is practiced when coal is mined on hillsides. The mining follows the contour of the hillside until the overburden becomes uneconomical to remove. This method creates a shelf, or bench, on the hillside.

countercurrent. A mode of operation in which two streams of material flow in opposite directions.

coupon. A polished metal strip used to measure the rate of corrosion of the metal in a specific gaseous or liquid environment.

cracking. The partial decomposition of high-molecular-weight organic compounds into lower-molecular-weight compounds, generally as a result of high temperature.

criteria pollutant. EPA has determined that six air pollutants are dangerous to human health: SO_2, NO_x, CO, total suspended particulate matter (TSP), photochemical oxidants, and nonmethane hydrocarbons. It is EPA's intent to add others to the list following completion of health effects studies or of developments of reliable ambient source sampling and analytical techniques.

crude gas. Impure gas produced in a gasifier.

culm. A term referring to fine waste, particularly a mixture of coal dust and dirt.

cyclone. A mechanical separation device which imparts a rotational flow to the fluid stream, using the resulting centrifugal forces to cause separation of denser particles suspended in the fluid.

cyclone separator. Essentially a settling chamber to separate solid particles from a gas, in which gravitational acceleration is replaced by centrifugal acceleration.

degasification. A process for removing naturally occurring methane from coal seams.

delayed coking. A process wherein coal is subject to a long period of carbonization at moderate temperatures to form coke.

demineralization. Removal of mineral matter (ash) from coal by solvent extraction, usually under hydrogen atmosphere.

dense media. A process for upgrading mined coal through removal of trash rock (gangue). The process incorporates a dense fluid, e.g., a suspension of finely powdered iron ore in water, which allows the heavy gangue to settle out. The cleaned coal is then separated from the dense fluid by washing.

dense (heavy) medium process. Dense (heavy) medium processes use a thick solution, usually a mixture of magnetite and water, to separate coal from impurities. They work by gravity separation.

depolymerization. The change of a large molecule into simpler molecules usually accompanied by the substitution of hydrogen for oxygen in the molecular structure.

destructive distillation. The distillation of coal accompanied by its thermal decomposition.

desulfurization. The removal of sulfur from hydrocarbonaceous substances by chemical reactions.

devolatilization. The removal of a portion of the volatile matter from medium- and high-volatile coals.

diatomaceous earth. A yellow, white, or light gray, siliceous porous deposit made up of opaline sheets or diatoms; used as a filter aid, paint filler, adsorbent, abrasive, and thermal insulator. Also known as kieselguhr.

dissolution. The taking up of a substance by a liquid with the formation of a homogeneous solution.

distillation. A process of vaporizing a liquid and condensing the vapor by cooling; used for separating liquids into various fractions according to their boiling points or boiling ranges.

dolomite. A carbonate of calcium and magnesium having the chemical formula $CaMg(CO_3)_2$.

downdraft gasifier. A vertical shaft gasifier in which the oxidation gas (air or oxygen) flows cocurrent with the feedstock flow.

Downtherm. Trademark for a series of eutectic mixtures of diphenyl oxide and diphenyl used as high-temperature heat-transfer fluids.

dragline excavator. A dragline excavator removes overburden to expose the coal by means of a scoop bucket that is suspended from a long boom. The dragline digs by pulling the bucket toward the machine by means of a wire rope.

drift mine. A drift mine is a mine driven horizontally into coal exposed or accessible in a hillside.

doubly promoted. A term applied to a catalyst in which two materials are added—in relatively small amounts—to enhance or otherwise direct the activity of the catalyst.

dragline. An excavating machine in which a bucket is attached by cables and operates by being drawn toward the machine.

ebullated bed. Gas, containing a relatively small proportion of suspended solids, bubbles through a higher-density fluidized phase with the result that the system takes on the appearance of a boiling liquid.

economizer. Heat exchanging mechanism for recovering heat from flue gases.

effluent gas. Gas given off from a process vessel.

elemental analysis. An assay to determine the relative percentage of the elements (C, N, H, O, etc.) in a substance.

elutriation. The preferential removal of the small constituents of a mixture of solid particles by a stream of high-velocity gas.

endothermic. Adjective for a reaction that requires heat input to proceed.

endothermic reaction. A process in which heat is absorbed.

enhanced landfill. A sanitary landfill in which solid waste and sewage sludge are placed in such a manner that the anaerobic digestion of the organic material can be controlled (nutrient, buffer, and moisture additions) to maximize methane gas recovery potential.

enthalpy. A thermodynamic quantity giving the heat content (H) of a body.

enthalpy change. The increase or decrease in heat content of a substance or system, which accompanies its change from one state to another under constant pressure.

entrained bed (flow). A bed in which solid particles are suspended in a moving fluid and are continuously carried over in the effluent stream.

entrained flow. A generic term referring to a process for placing a solid and gas in contact in which the gas velocity is so high that the solid particles (usually very fine) are swept along with the gas.

eutectic. That combination of two or more components that produces the lowest melting temperature.

exothermic. Indicating liberation or escape of heat; a chemical reaction that produces heat.

exothermic reaction. A process in which heat is liberated.

extraction. A method of separation in which a solid or solution is contacted with a liquid solvent (the two being essentially mutually insoluble) to transfer components into the solvent.

extractive coking. Similar to a delayed coking process, with the emphasis on high tar yields to produce liquids.

filter aid. Finely divided solids used to increase the efficiency of filtering.

filter cake. The moist residue remaining from the filtration of a slurry to produce a clean filtrate.

filtrate. A liquid free of solid matter after having passed through a filter.

filtration. The separation of solids from liquids by passing the mixture through a suitable medium, e.g., cloth, paper, or diatomaceous earth.

Fischer assay. Method for determining the tar and light oil yields from coal or oil shale; conducted in a retort under an inert atmosphere with a prescribed increase in temperature to 932°F (500°C).

Fischer – Tropsch. A well-established catalytic process for reacting hydrogen and carbon monoxide synthesis gas to form mixtures of hydrocarbons, e.g., motor fuel.

Fischer – Tropsch catalyst. Catalysts developed for the catalytic synthesis of liquid fuels from coal-derived synthesis gas; these catalysts contain, principally, iron, cobalt, nickel, or ruthenium.

Fischer – Tropsch process. Method of hydrogenating mixtures of carbon monoxide and hydrogen produced from coal, lignite, or natural gas by means of steam, at 1–10 atm and 360–410°F (182–210°C) to yield liquid and gaseous fuels, and a wide spectrum of industrial chemicals.

fixed bed. A generic term referring to any process for contacting a solid and fluid (gas or liquid) in which the solid material remains at rest, each particle resting on its neighbor. In the terminology of the gasification industry, the term fixed bed signifies a fuel bed supported at the bottom by a grate and maintained at a constant depth during gasification by the addition of fuel at the top. The top and bottom extremes are fixed in space but, within the bed, fuel moves slowly from top to bottom, and the ungasified residue is discharged through the bottom grate. Strictly speaking this constitutes a moving bed, but the gasification idiom will be used.

fixed carbon. The solid residue, other than ash; obtained by destructive distillation; determined by definite prescribed methods.

flash carbonization. A carbonization process characterized by short residence times of coal in the reactor to optimize tar yields.

flotation process. Flotation processes treat fine-sized coal with an oil-based reagent that attracts air bubbles in a liquid medium; the coal floats to the surface as a froth, leaving the refuse below.

flue gas. Gaseous combustion products.

fluid (fluidized) bed. A generic term referring to any process for placing a solid and gas in contact in which the gas is used to cause a boiling action of the solid particles with a great deal of relative motion between solid particles as they are propelled about in the process vessel. In general, the gas passes through the fluidized bed while the solids are retained.

fluidization (dense phase). The turbulent motion of solids particles in a fluid stream; the particles are close enough so as to interact and give the appearance of a boiling liquid.

fluidization (entrained). Gas–solid contacting process in which a bed of finely divided solid particles is lifted and agitated by rising streams of gas.

fluidized bed. Assemblage of small solid particles maintained in balanced suspension against gravity by the upward motion of a gas.

fly ash. A fine ash from the pulverized coal burned in power station boilers, or entrained ash carried over from a gasifier.

fractionation. Distillation process for the separation of the various components of liquid mixtures.

freeboard. The space in a fluidized-bed reactor between the top of the bed and the top of the reactor.

free swelling index. A measure of the caking tendency of a specific coal, with higher values reflecting a stronger tendency to cake.

Friedel – Crafts reaction. A substitution reaction, catalyzed by aluminum chloride in which an alkyl (*R*—) or acyl (*R*CO—) group replaces a hydrogen atom of an aromatic nucleus to produce a hydrocarbon or a ketone.

front-end loader. A front-end loader is a tractor with a digging bucket mounted and operated on the front.

froth flotation. A process to separate solid mixtures.

fuel cell. A galvanic cell in which the chemical energy of a conventional fuel is utilized to produce electricity.

fuel gas. Low-heating-value (150–350 Btu/scf) product generally utilized on site for power generation or industrial use.

gasification. The process of converting a solid fuel into gas-phase molecules, e.g., N_2, O_2, CO, CO_2, H_2.

gasification of coal. The conversion of solid coal into a gaseous form by various chemical reactions with steam.

gasifier. A vessel in which gasification occurs, usually utilizing fluidized-bed, fixed-bed, or entrained-bed units.

gob area. An area of collapsed overburden or broken rock rubble such as that produced by longwall mining.

grade. A means of classifying coal by the quantity of impurities.

grade of coal. In general, this refers to coal quality with regard to use. Metallurgical coal (or coking coal) meets the requirements for making coke. Steam coal is best suited for generating steam to produce electricity or for other purposes. Low-sulfur coal generally contains 1% or less sulfur by weight. Cleaned coal or prepared coal has been processed to reduce the amount of impurities present and improve the burning characteristics.

head. A term which indicates discharge pressures of pumps and compressors.

heat capacity. Quantity of heat required to raise the temperature of 1 pound of a substance 1 degree Fahrenheit.

heating value of coal. Determined in terms of Btu, both on an "as-received" basis (including moisture) and on a dry basis.

high-Btu gas. A gas having a heating value of 900–1000 Btu/scf, which approaches the value for natural gas.

higher heating value (HHV). The heat liberated during a combustion process in which the product water vapor is condensed to a liquid and the heat of condensation is recovered.

hydraulic processes. Hydraulic processes use currents of waste to separate coal from impurities.

hydroclone. A small cyclone extractor for removal of suspended solids from a flowing liquid by means of the centrifugal forces set up when the liquid is made to flow through a tight conical vortex.

hydrocoking. Coking of tars, SRC, etc., under hydrogenating conditions to form liquid products.

hydrocracking. The combination of cracking and hydrogenation of organic compounds.

hydrogenation. Chemical reactions involving the addition of gaseous hydrogen to a substance in the presence of a catalyst under high temperatures and pressures.

hydrogen-donor solvent. A solvent, such as anthracene oil, tetralin (tetrahydronaphthalene), decalin, etc., which transfers hydrogen to coal constituents causing depolymerization and consequent conversion to liquid products of a lower-boiling-point range that are then dissolved by the solvent.

hydroliquefaction. A process for the manufacture of hydrocarbon liquids from coal by reaction with hydrogen (or a hydrogen proxy such as hydrogen-donor solvent or carbon monoxide and steam).

hydrolysis. Decomposition of a compound by reaction with water.

hydrotreating. A process to catalytically stabilize petroleum or other liquid hydrocarbon products and/or remove objectionable elements from products or feedstocks by reacting them with hydrogen.

ideal gas. Any gas whose equation of state is expressed by the ideal gas law, namely, $PV = nRT$, where P is the pressure, V is the volume, R is the gas constant, T is the absolute temperature, and n is number of moles.

ignition temperature. The minimum temperature necessary to initiate self-sustained combustion of a substance.

inerts. Constituents of a coal which decrease its efficiency in use, e.g., mineral matter (ash) and moisture in fuel for combustion.

in situ. In its original place, e.g., underground gasification of a coal seam.

intermediate-Btu gas. A mixture of hydrogen, carbon monoxide, and methane. When burned it yields about 40% of the heat that could be produced by burning a like volume of natural gas.

landfill gas. Biogas generated by the decomposition of wastes in a sanitary landfill.

life cycle costs. A method of economic analysis that not only examines the first cost (capital cost) of a project but all costs over the lifetime of the project (e.g., operating and maintenance costs) in determining the least cost alternative.

lignite. The lowest rank of coal; it is brownish black, has a high moisture content, and is used mainly to generate electricity. It is mined in North Dakota, Montana, and Texas.

limestone. Sedimentary rock containing 50% carbonate (CO_3) of lime or magnesia. Chemical formula (for calcite limestone) is $CaCO_3$.

liquefaction. Conversion of a solid to a liquid; with coal, this appears to involve the thermal fracture of carbon–carbon and carbon–oxygen bonds, forming free radicals. These radicals abstract hydrogen atoms yielding low-molecular-weight gaseous and condensed aromatic liquids.

liquefied petroleum gas (LPG). Those hydrocarbons that have a vapor pressure [at 70°F (21°C)] slightly above atmospheric (such as propane and butane); kept in liquid form under a pressure higher than 1 atm.

lock hopper. A mechanical device that permits the introduction of a solid into an environment of different pressure.

low-Btu gas. A gas having a heating value up to 350 Btu/scf.

lower heating value (LHV). The heat liberated by a combustion process assuming that none of the water vapor resulting from the process is condensed, so that its latent heat is not available.

longwall mining machine. A longwall mining machine shears coal from a long straight coal face (up to about 600 ft) by working back and forth across the face. The broken coal is transported by conveyor.

longwall mining system. In a longwall mining system, large sections of coal are removed and no pillars are left to support the mined-out areas; the caged area (gob) compacts and, after initial subsidence, supports the overlying strata.

MAF. Moisture and ash-free; a term that relates to the organic fraction in coal.

major facility. EPA defines major facilities as those emitting over 100 tons/year of any criteria pollutant.

mesh. Measure of fineness of a screen, e.g., a 400-mesh sieve has 400 openings per linear inch.

methanation. A process by which the heating value of gases containing hydrogen and carbon monoxide may be increased. It entails a catalytic reaction to form methane, i.e., $CO + 3H_2 \rightarrow CH_4 + H_2O$, and with the proper synthesis gas composition, can produce a gas having a heating value approaching that of natural gas (1000 Btu/scf).

methane (CH_4). A colorless, odorless, and tasteless gas, lighter than air; it is the chief component of natural gas.

micron (μm). A unit of length equal to one millionth of a meter; 10^{-6} m.

middlings. A generic term for a mixture of coal and gangue encountered in cleaning processes which do not make a sharp cut between coal and gangue. It is not sufficiently clean for use as a high-quality product and contains too much valuable coal for disposal. It can be recycled or used for raising steam.

molecular weight. Molecular weight equals the total of atomic weights for all elements comprising a single typical molecule.

mountaintop mining. Mountaintop mining refers to the mining of a coal bed that underlies the top of a mountain. The overburden removed is deposited in the mined-out area.

moving bed. A generic term referring to a process for contacting a solid and fluid (gas or liquid) in which the solid material moves through the bed *en masse*, with no "boiling" or relative motion between solid particles. The flow of fluid is usually, but not always, countercurrent to the movement of the solids.

multiple-effect evaporator. A series of evaporators in which the vapors removed from each unit are used to supply heat to the next unit in the series.

municipal solid waste (MSW). Combined residential and commercial wastes generated within a municipal area and consisting of any materials, including food wastes, that are discarded or rejected as spent, worthless, useless, or excess, and which are not wet enough to be free flowing.

natural gas. Naturally occurring gas extracted from sedimentary structures; it consists mainly of methane and has a higher-heating value of approximately 1050 Btu/scf.

nonattainment. An air quality control region (AQCR) or area with air quality worse than National Ambient Air Quality Standards (NAAQS) (see *criteria pollutant*) is in nonattainment status. Permits to construct facilities cannot be issued unless emissions in the region from existing sources are in some way reduced (offset) in an amount more than sufficient to offset the new plant's emissions, in order to produce net air quality benefits. Also, the plant must restrict emissions to the Lowest Achievable Emission Rate (LAER), and all of applicant's existing facilities in the AQCR must be in compliance with source emission regulations.

noncoking coal. A coal that does not form coke under normal coking conditions.

olefinic hydrocarbon. A class of unsaturated hydrocarbons containing one or more double bonds and having the general chemical formula C_nH_{2n}.

open cycle. A thermodynamic power cycle in which the working fluid passes through the system only once and is then exhausted to the atmosphere.

open market coal. This refers to coal sold on the open market.

open-pit coal mining. Open-pit coal mining is essentially a combination of contour and area mining methods and is used to mine beds that are thick or steeply inclined. The overburden is removed by both power shovels and trucks.

organic sulfur. Sulfur that is chemically bonded to coal hydrocarbons rather than to the inorganic ash constituents of coal.

overburden. Any material, consolidated or unconsolidated, that overlies a coal deposit. The overburden ratio refers to the amount of overburden that must be removed to excavate a given quantity of coal.

overburden ratio. A measure of the quantity of overburden which must be removed to extract a given quantity of coal, usually expressed as a ratio (e.g., tons of overburden per ton of coal or cubic yards of overburden per ton of coal).

overhead. The relatively low-boiling-point liquids removed from the top of a distillation unit.

partings. The waste material within a seam that separates the coal into horizontal layers.

peat. An unconsolidated, hydrophilic, yellowish-brown to brownish-black, carbonaceous sediment, formed by accumulation of partially fragmented and decomposed plant remains in swamps and marshes which retains more than 75% inherent moisture and less than 12% mineral matter in saturated natural deposits.

petrochemicals. Those derived from crude oil or natural gas, or their coal-derived substitutes; they include light hydrocarbons such as butylene, ethylene, and propylene, the raw materials for the production of plastics by polymerization.

phenols. A group of aromatic compounds having the hydroxyl (OH) group directly attached to the benzene ring.

pilot plant. A chemical process plant containing all the processes of a commercial unit, but on a smaller scale, for the purpose of studying the technical and economic feasibility of the process.

pipeline gas. A methane-rich gas that conforms to certain standards and has a higher heating value between 950 and 1050 Btu/scf.

plenum chamber. An enclosed space through which air is forced for slow distribution through ducts.

pneumatic processes. Pneumatic processes use currents of air to separate coal from impurities.

power shovel. A power shovel removes overburden and loads coal by means of a digging bucket mounted at the end of an arm suspended from a boom. The shovel digs by pushing the bucket forward and upward. It does not dig below the level at which it stands.

precipitated iron catalyst. Ferric hydroxide, a gelatinous amorphous material, is dehydrated and reduced to particles of elemental iron with high specific surface area. The tubular catalytic reactor containing a fixed bed of this catalyst also contains boiler water to remove the heat of reaction. Commercial reactors using several thousands of 2-in. diameter tubes give little further potential for scale-up. Catalyst cost is one major disadvantage of the precipitated iron ARGE system.

precoat. Layer of suitable filtering medium, e.g., diatomaceous earth, laid down on a rotary filter cloth prior to operation.

prevention of significant deterioration. Many AQCR's have (for a specific criteria pollutant) better air quality than required by NAAQS. In the case of SO_2 and TSP, such areas (regions or subdivisions) are designated PSD area. Three classes of PSD areas exist. In each class, maximum allowable increases in SO_2 and TSP pollutant levels are set by regulation. Most recent information is published in *Federal Register*, November 3, 1977, p. 57459.

Class I areas are international parks, national wilderness areas, and national memorial parks over 6000 acres. The PSD increment is 3% or less of the NAAQS (no single facility may use up all of the increment).

Class II areas are national monuments, primitive areas, recreation areas, wild and scenic rivers, wild life refuges, lakeshores and seashores, future national parks and wilderness areas. The PSD increment is less than 30% of the NAAQS.

Class III areas may be designated by state governments after a hearing and approval process. PSD increments are 50% or less of the NAAQS. New or modified major facilities in PSD areas must be BACT and comply with other permitting PSD increments and monitoring requirements.

prilling tower. A tower that produces small solid agglomerates by spraying a liquid solution in the top and blowing air up from the bottom.

process development unit (PDU). A process research facility of intermediate size between bench scale and pilot plant.

producer gas. A low-heating-value fuel gas (130–170 Btu/scf) generated from the reaction of air and steam with a solid fuel. The gas contains a mixture of hydrogen, carbon monoxide, and nitrogen and was generated on-site for industrial heating in the latter 19th century and early 20th century.

proximate analysis. Determine on an "as-received basis," the moisture content, volatile matter (gases released when coal is heated), fixed carbon (solid fuel left after the volatile matter is driven off), and ash (impurities consisting of silica, iron, alumina, and other non-combustible matter). The moisture content affects the ease with which coal can be handled and burned. The amount of volatile matter and fixed carbon provide guidelines for determining the intensity of the heat produced. Ash increases the weight of coal, adds to the cost of handling, and can affect the burning characteristics.

purification. Removal of a wide range of impurities present in gases from coal gasification.

pyritic sulfur. Sulfur in the mineral form of iron disulfide, FeS_2, some of which can be removed from coal by conventional beneficiation techniques.

pyrolysis. Thermal decomposition of organic compounds in the absence of oxygen.

quad. A unit of energy equivalent to 10^{15} (quadrillion) Btu.

quenching. Cooling by immersion in oil, water bath, or water spray.

Raney nickel catalyst. Specially prepared nickel catalyst used in the hydrogenation of organic materials and the methanation of synthesis gas to methane.

rank. A means of classifying coal by the degree of carbonization, according to the natural series of lignite, subbituminous, bituminous, and anthracite.

raw gas. The product gas directly from a coal gasification reactor; frequently requiring cleaners to remove ash, hydrogen sulfide, ammonia, tars, and untreated coal fines or further processing to adjust composition for SNG manufacture, Fischer–Tropsch synthesis, etc.

reactivity. Susceptibility to chemical change; for example, in coal liquefaction, the reactivity of the coal for conversion to liquid products is a function of the coal rank, among other things.

reactor. Vessel in which coal-conversion reactions take place.

rectisol process. A process for the purification of coal-gasification gas based on the capability of cold methanol to absorb all gas impurities in a single step; gas naphtha, unsaturated hydrocarbons, sulfur compounds, hydrogen cyanide, and carbon dioxide are removed from the stream by the methanol at temperatures below 32°F (0°C).

reducing gas. A gas which, at high temperatures, lowers the state of oxidation of other chemicals.

reforming processes. A group of proprietary processes in which low-grade or low-molecular-weight hydrocarbons are catalytically converted to higher-grade or higher-molecular-weight materials; also applies to the endothermic reforming of methane, for the production of hydrogen, by the reaction of methane and steam in the presence of nickel catalysts.

refractory. A material capable of withstanding extremely high temperatures and having relatively low thermal conductivity.

residence time. Time spent by a typical particle in a particular zone.

room-and-pillar mining system. In a room-and-pillar mining system, the most common method, the mine roof is supported by coal pillars left at regular intervals. Rooms are places where the coal is mined; pillars are support columns of coal left between the rooms.

rotary breaker. A machine to reduce coal particle size.

run-of-mine (ROM). A generic term referring to coal as it comes from the mine without screening or beneficiation.

saturated hydrocarbon. A carbon–hydrogen compound with all carbon bonds filled; that is, there are no double or triple bonds as in olefins and acetylenes.

scalp. To remove oversize chucks from broken coal using a coarse screen or grid.

scrubber. Apparatus in which a gas stream is freed of tar, ammonia, and hydrogen sulfide.

seam coal. Coal which is intermediate in rank between bituminous coal and anthracite; contains 8–22% volatile matter and from 91 to 93% carbon.

semiwater gas. A mixture of carbon monoxide, carbon dioxide, hydrogen, and nitrogen, obtained by passing an air–steam mixture through a hot bed of coke; it has a higher-heating value of about 120 Btu/scf.

sensible heat. That heat which results in only the elevation of the temperature of a substance with no phase changes.

sewage gas. Biogas generated by the decomposition of sewage sludge at wastewater treatment plant digesters.

sewage sludge. Solid, semisolid, or liquid residues generated from municipal wastewater treatment plants.

shaft mine. A shaft mine is driven vertically into the coal deposit.

shift. A catalytic process for increasing the hydrogen content of a CO-containing process stream, e.g., from a coal gasifier, to meet the needs of a subsequent processing step or end use. The reaction is $CO + H_2O \rightarrow CO_2 + H_2$ and it may be used to increase the hydrogen content of the stream or to adjust the hydrogen/carbon monoxide ratio as required for methanation or synthesis.

shift conversion. Process for the production of gas with a desired carbon monoxide content from crude gases derived from coal gasification; carbon-monoxide-rich gas is saturated with steam and passed through a catalytic reactor where the carbon monoxide reacts with steam to produce hydrogen and carbon dioxide, the latter being subsequently removed in a scrubber employing a suitable sorbent.

shortwall mining machine. A shortwall mining machine is, in general, a continuous mining machine used with a powered roof support system. It shears coal from a short coal face (up to about 150 ft long). The broken coal is hauled by shuttle areas to a conveyor belt.

shortwall mining system. A shortwall mining system, in general, combines room-and-pillar operations with small-sized longwall operations.

silviculture energy farms. Intensely managed acreage on which fast-growing tree species are raised for the purpose of fuel production.

sintering. The agglomeration of solids at temperatures below their melting point, usually as a consequence of heat and pressure.

slag (slagging). Slag is molten ash and slagging refers to the operation of a gasification unit such that the ash melting temperature is exceeded, causing the ash to be discharged as a viscous melt or an agglomeration of tacky particles rather than a free-flowing powder.

slagging gasifier. A gasifier that reaches temperatures effecting fusion of the fuel's ash content.

slope mine. A slope mine is driven at an angle to reach the coal deposit.

slurry. A mixture of liquid with solid particles, usually for the purpose of washing or for enabling the solid to "flow."

solvation. The association or combination of molecules of solvent with solute ions or molecules.

solvent. That component of a solution which is present in excess; liquid used to dissolve a substance.

solvent extraction. Selective solution of coal constituents from finely divided coal particles into a suitable solvent after intimate mixing, usually at high temperatures and pressures in the presence of hydrogen, with or without a catalyst, followed by phase separation.

solvent-refined coal (SRC). A coal product derived by solvent extraction; a brittle, vitreous solid [melting point, 300–400°F (149–204°C)] containing about 0.1% ash and about 10% of the sulfur in the original coal feedstock; calorific value is about 16,000 Btu per pound; may be used as a clean fuel for power generation by combustion; utilized for the production of high-grade metallurgical coke, anode carbon, and activated carbon by coking, or hydrogenated to produce synthetic crude oil.

space velocity. Volume of a gas (measured at standard temperature and pressure) or liquid passing through a given volume of catalyst in a unit time.

specific gravity. Ratio of the weight of any volume of a substance to the weight of an equal volume of water at 4°C (39°F).

specific heat. Heat capacity of a substance as compared with the heat capacity of an equal weight of water.

standard cubic foot (scf). The volume of a gas at standard conditions of temperature and pressure. The American Gas Association uses moisture-free gas at 60°F (16°C) and 30 in. of mercury (1.0037 atm) as its standard conditions. The pressure standard is not universal in the gas industry: 14.7 psia (1.000 atm) and 14.4 psia (0.980 atm) are also used. The scientific community uses 32°F (0°C) and 1 atm as standard conditions.

State Implementation Plan. Under 1977 CAA amendments, each state had to submit to the

Environmental Protection Agency prior to December 1, 1978, a State Implementation Plan (SIP) revision stating how the state will comply with provisions of the amendments with regard to both primary and secondary ambient standards. Federal review of the SIP is a lengthy process; both mobile and stationary sources are covered; litigation is possible and, in some instances, the old SIP resulting from the 1970 Clean Air Act amendments have not yet been approved. After June 30, 1979, each state has to have an approved SIP revision that provides for attainment of primary ambient air quality standards by December 31, 1982, before a major new source may locate in a nonattainment area. Once new sources must be included in the SIP, it is to the advantage of the facility operator to communicate development plans to the state permitting authority as early as possible.

steam-catalyzed gasifier. A gasifier in which an alkali carbonate (catalyst) is mixed with input fuel and subjected to high-pressure, high-temperature steam.

steam rate. Industrial boilers are rated in pounds-per-hour (pph) steam output. Some smaller or older units are rated in horsepower. Considering boiler efficiency and feedwater heating factors, about 1300 Btu in coal are used to make 1 pound of industrial steam. Utility electrical output and boiler steam output rates do not precisely interconvert due to steam/condensate cycle differences between various plants. For a 1000 MW plant, steaming rate may be between 6,500,000 and 7,500,000 pph. For existing plants to make 1 kWh of electricity, about 10,500 Btu in coal are required.

steam reformation. A process of reacting hydrocarbon fuel (methane or naphtha) with water at high temperature and pressure, forming carbon monoxide and hydrogen as major products.

stoichiometry. The definite proportions in which molecules react chemically to form new molecules.

stripping. The removal of the more volatile components from a liquid mixture of compounds.

subbituminous coal. Subbituminous coal, or black lignite, is dull black. It is used for generating electricity and space heating. This rank of coal is mined in the western coal fields.

subsidence. A sinking of the surface of the earth because of underground excavation.

substitute natural gas (SNG). Any methane-rich gas mixture, derived from either coal or petroleum, which may be substituted directly for natural gas.

sulfonation. To treat with sulfur dioxide.

superficial velocity. The linear velocity of a fluid flowing through a bed of solid particles calculated as though the particles were not present.

superheater. A heat exchanger which adds heat to the saturated steam leaving a boiler.

syncrude. Synthetic crude oil; oil produced by the hydrogenation of coal, coal extracts, oil sands, or oil shale, which is similar to petroleum crude.

syngas. A synthetic gas, chemically similar to natural gas, that is obtained through the gasification or anaerobic digestion of biomass.

synthesis gas (syngas). A generic term for mixtures of hydrogen and carbon monoxide with minor diluents (nitrogen and carbon dioxide) which are to be used as feedstocks for subsequent processing steps. Principal syngas preparation methods include

- $H_2O + CH_4 \rightarrow CO + 3H_2$
- Partial oxidation of naphtha
  $(CH_2)_n + \frac{1}{2} O_2 \rightarrow CO + H_2$
- Steam–oxygen gasification of coal
  $2C + H_2O + O_2 \rightarrow CO + CO_2 + H_2$

The principal uses of syngas include

- Ammonia synthesis (preceded by CO/CO_2 removal)
  $3H_2 + N_2 \rightarrow 2NH_3$
- Methanol synthesis
  $CO + 2H_2 \rightarrow CH_3(OH)$
- Hydrocarbon synthesis
  $CO + 2H_2 \rightarrow H_2O + (CH_2)_n$

synthetic gasoline. A liquid hydrocarbon that is chemically similar to gasoline but that has been derived from biomass.

tail gas. A gas issuing from a gas-treatment unit which may be recycled to the process or exhausted.

tar (coal). A dark brown or black, viscous, combustible liquid formed by the destructive distillation of coal.

therm. A unit of heat used as a basis for the sale of natural gas; equal to 100,000 Btu.

tipping fee. The price charged to dump or tip solid wastes at a disposal facility; the price covers disposal costs.

topping cycle. The higher-temperature thermodynamic power cycle of a combined-cycle system.

town gas. A generic term for a class of gas mixtures which were in common use by utilities prior to World War II. Produced from coal, it consisted of a mixture of hydrogen, carbon monoxide, and nitrogen to which lesser amounts of methane and/or propane were added to adjust the heating value.

turndown ratio. The minimum ratio of actual flow rate to design flow rate at which a process unit can be operated.

ultimate analysis. Ultimate analysis determines the amount of carbon, hydrogen, oxygen, nitrogen, and sulfur in coal.

updraft gasifier. A gasifier in which the fuel is fed in at the top of a vertical shaft and the oxidation gas (air or oxygen) is fed below the solid fuel; the streams are countercurrent.

venturi scrubber. A gas cleaning device which involves the injection of water into a stream of dust-laden gas flowing at a high velocity through a contracted portion of a duct, thus transferring the dust particles to the water droplets which are subsequently removed.

volatile matter. Those constituents of coal, exclusive of moisture, that are liberated from a sample when heated to 1750°F (955°C) for 7 minutes in the absence of oxygen.

water gas. Gas produced by the reaction of carbon (in coal or coke) and steam to yield mixtures of carbon monoxide and hydrogen; similar to synthesis gas.

water gas shift. The reaction between water vapor and carbon monoxide to produce hydrogen and carbon dioxide or the reverse: $CO + H_2O \rightleftarrows H_2 + CO_2$.

working fluid. A gas stream which directly does work, e.g., powering a gas turbine.

Appendix C

Acronyms and Abbreviations

A

ac	alternating current
ACFM	actual cubic feet per minute
ACP	aqueous carbonate process
A&E	architect and engineer
AE&CI	African Explosives and Chemical Industries
AECT	Advanced Environmental Control Technology
AEDC	Arnold Engineering Development Center
AEP	American Electric Power Company
AFB	atmospheric fluidized-bed
AFBC	atmospheric fluidized-bed combustion
AFB/CTIU	atmospheric fluidized-bed component test and integration unit
AF/Y	acre-feet per year
AGA	American Gas Association
AHF	advanced hydraulic fracturing
AL	Ames Laboratory
AMOCO	American Oil Company
ANL	Argonne National Laboratory
AOSTRA	Alberta Oil Sands Technology and Research Authority
APCI	Air Products and Chemicals, Inc.
API	American Petroleum Institute
APL	Applied Physics Laboratory, Johns Hopkins University
APS	advanced power systems
APT	advanced process technology
AQCR	Air Quality Control Region
Ar	argon
ARCO	Atlantic Richfield Company
ARTD	Advanced Research and Technology Development
ASFE	Assistant Secretary for Fossil Energy
ASPEN	Advanced System for Process Engineering
ASRA	Assistant Secretary for Resource Applications
ASTM	American Society for Testing and Materials
atm	atmosphere (of pressure)

B

BACT	best available control technology
BASF	Badische Anilin und Soda Fabrik
bbl	barrel
bbl/day	barrels per day
bcf	billion cubic feet
BCR	Bituminous Coal Research, Inc.
BCURA	British Coal Utilization Research Association
BETC	Bartlesville Energy Technology Center
Bi	quantity of product per time
BITE	built-in test equipment
BMP	best management practices
BNL	Brookhaven National Laboratory
BOD	biochemical oxygen demand
BOM	Bureau of Mines
BPCT	best practical control technology
BSB	blind shaft borer
Btu	British thermal unit
BTX	benzene, toluene, and xylene
B&W	Babcock & Wilcox

C

C	carbon
CAA	Clean Air Act
CAFB	chemically active fluidized-bed
CaS	calcium sulfide
CATEC	combustor augmented with TEC
CBC	carbon burnup cell
CC	combined cycle
CCC	coal combustion for cogeneration
CCGT	closed cycle gas turbine
CCU	coal conversion and utilization
CDIF	component development and integration facility
CDTF	component development test facility
CE	Combustion Engineering, Inc.
CEF	chemical explosive fracturing
CEMS	central energy management systems
CERCHAR	French government equivalent of the U.S. Bureau of Mines, and the National Bureau of Standards
CFFF	coal-fired flow facility
CFR	continuous flow reactor
CG	coal gasifiers
CM	continuous miner

CMTC	Carbondale Mining Technology Center
CO	carbon monoxide
COED	coal–oil energy development
COGAS	COGAS Development Company
COM	coal–oil mixture
CONOCO	Continental Oil Company
CO_2	carbon dioxide
COS	carbonyl sulfide
CPS	central power systems
CRE	Coal Research Establishment
CSD	Critical Solvent De-Ashing Process
CSIRO	Commonwealth Scientific and Industrial Research Organization
CSL	Combustion Systems Limited
CTAS	Cogeneration Technology Alternatives Study
CTIU	component test and integration unit
CURL	Coal Utilization Research Laboratory
CVD	chemical vapor deposition
CW	coal–water (mixture)
CY	calendar year

D

dc	direct current
DISC	direct-injected stratified charge
DOC	Department of Commerce
DOD	Department of Defense
DOE	Department of Energy
DOI	Department of the Interior
DPS	dispersed power systems

E

EA	environmental assessment
EAS	Economic Assessment Service
EAST	Energy Applied Systems Test facility
EB	evaluation board
EC	executive committee
ECAS	Energy Conversion Alternatives Study
ECLP	Exxon Coal Liquefaction Project
ECTP	Environmental Control Technology
EDIMS	Environmental Data Information Management System

EDP	environmental developmental plan
EDS	Exxon donor solvent process
EGR	enhanced gas recovery
EGSP	Eastern Gas Shales Project
EH&S	Environmental Health & Safety
EIA	Energy Information Administration
EIA	environmental impact assessment
EIS	environmental impact statement
EMAC	Energy Materials Advisory Committee
EOR	enhanced oil recovery
EPA	Environmental Protection Agency
EPRI	Electric Power Research Institute
ERA	Energy Reorganization Act
ERCO	Energy Resources Company, Inc.
ERDA	Energy Research and Development Administration
ER&E	Exxon Research and Engineering Company
ESCOE	Engineering Societies Commission on Energy
ESECA	Energy Supply and Environmental Coordination Act
ESEERCO	Empire State Electric Energy Research Corporation
ESP	electrostatic precipitator
ETC	energy technology center
ETEC	Engineering Test Evaluation Center
ETF	engineering test facility

F

°F	degree Fahrenheit
FBC	fluidized-bed combustion
FBHX	fluidized-bed heat exchanger
FE	fossil energy
FERC	Federal Energy Regulatory Commission
FFE	fossil fuel extraction
FFU	Division of Fossil Fuel Utilization
FGC	fuel gas cleanup
FGD	flue gas desulfurization
FGR	Federal Republic of Germany
f_i	product value factor for product *i*
FOB	free on board
FRI	Fuel Research Institute
FT	Fischer–Tropsch
ft^3	cubic feet (cu ft)

FWPCA	Federal Water Pollution Control Act
FY	fiscal year

G

g	gram
gal	gallon
GBF	granular-bed filter
GCC	gasification combined cycle
GDP	gaseous diffusion plant
GFETC	Grand Forks Energy Technology Center
GISP	geotechnically instrumented seafloor probe
GPGA	Great Plains Gasification Associates
gpm	gallons per minute
GPP	general plant projects
GRI	Gas Research Institute

H

H	hydrogen
H_2	hydrogen gas
H/C	mass ratio of hydrogen to carbon
H-Coal	hydrogen-fed coal
HGMS	high-gradient magnetic separation
HHV	higher heating value
HMF	high mass flux
HMIS	horizontal modified *in situ*
H_2O	water
hp	horsepower
HPDE	high-performance demonstration experiment
HRI	Hydrocarbon Research, Incorporated
HRSR	heat recovery and seed recovery
H_2S	hydrogen sulfide
HTHP	high temperature, high pressure
HTTT	high-temperature turbine technology
HTW	High-Temperature Winkler Gasifier
HUD	Department of Housing and Urban Development

I

IBG	intermediate-Btu gas
IC	internal combustion
I&C	instrumentation and control
ICI	Imperial Chemical International
ICE	increased conversion efficiency

ICGG	Illinois Coal Gasification Group
IEA	International Energy Agency
IES	integrated energy systems
IGCC	Integrated Gasifier Combined Cycle
IGT	Institute of Gas Technology
IITRI	Illinois Institute of Technology Research Institute
INEL	Idaho National Energy Laboratory
IU/liter/hr	international units/liter/hour

J

JPL	Jet Propulsion Laboratory

K

kcal	kilocalorie
kg	kilogram
kJ	kilojoule
KOH	potassium hydroxide
kW	kilowatt
kWh	kilowatt hour
KT	Koppers–Totzek

L

LAER	lowest achievable emission rate
LASL	Los Alamos Scientific Laboratory
lb	pound
LBG	low-Btu gas
lb/hr	pound(s) per hour
LBL	Lawrence Berkeley Laboratory
LETC	Laramie Energy Technology Center
LFFC	liquid fossil fuel cycle
LG	longwall generator
LHV	lower heating value
LLL	Lawrence Livermore Laboratory
LNG	liquified natural gas
LPG	liquefied petroleum gas
LSCF	large scale cold flow
LTAS	Liquefaction Technology Assessment Study
LVW	linked vertical well

M

m^3	cubic meter
M	thousand
MAF	moisture- and ash-free
Mag-Ox	Magnesium Oxide Process
MBG	medium-Btu gas
MCFC	molten carbonate fuel cell

MCSD	Molten Carbonate Systems Development
md	millidarcy (unit of porous permeability)
METC	Morgantown Energy Technology Center
MFB	multicell fluidized-bed boiler
mg	milligram
MHD	magnetohydrodynamics
MHF	massive hydraulic fracturing
Minnegasco	Minnesota Gas Company
MIS	modified *in situ*
MIT	Massachusetts Institute of Technology
MIUS	modular integrated utility systems
ML	Mound Laboratory
MLGW	Memphis Light, Gas and Water Division
MM	million
mm	millimeter
MMbbl/d	millions of barrels per day
MMF	mineral-matter-free
MMscf	millions of standard cubic feet
MR&D	Mining Research and Development
MRCP	Methane Recovery from Coalbeds Project
MSA	major systems acquisition
msec	millisecond
MSW	municipal solid waste
MT	metric ton
MTBE	methyl tertiary-butyl ether
MTC	mining technology center
MTCH	Mining Technology Clearing House
MW	megawatt
MWe	megawatt electrical
MWt	megawatt thermal

N

N	nitrogen
NA	nonattainment area
NAAQS	National Ambient Air Quality Standards
NaCl	sodium chloride (table salt)
NASA	National Aeronautics and Space Administration
NBS	National Bureau of Standards
NCB	National Coal Board (British)
NEA	National Energy Act
NEP	National Energy Plan
NEPA	National Environmental Policy Act

NEPSCO New England Power Services Company
NGPA Natural Gas Policy Act
NIOSH National Institute of Occupational Safety and Health
NL national laboratory
NO_x nitrogen oxides
NOAA National Oceanographic and Atmospheric Administration
NRL Naval Research Laboratory
NSF National Science Foundation
NSPS New Source Performance Standards
NTIS National Technical Information Service
NYU New York University

O

O oxygen
OART Office of Advanced Research and Technology
OCGT open-cycle gas turbine
OCM Office of Coal Mining
OCR Office of Coal Research
OCS outer continental shelf
O&M operations and maintenance
OMB Office of Management and Budget, Executive Office of the President
ORC Occidental Research Corporation
ORC organic Rankine cycle
ORNL Oak Ridge National Laboratory
OSHA Occupational Safety and Health Administration
OS/IES onsite/integrated energy systems

P

PA performance assurance
PACE power at combined efficiencies
PAF principal alternate fuels
PAS Performance Assurance System
PC&U product characterization and utilization
PDU process development unit
PEF process evaluation facility
Penn/GE University of Pennsylvania/General Electric Company
PETC Pittsburgh Energy Technology Center
PFB pressurized fluidized-bed
PFBC pressurized fluidized-bed combustion

PFB/CTIU	pressurized fluidized-bed/ component test and integration unit
PG&E	Pacific Gas & Electric Company
P.L.	Public Law
P&M	Pittsburgh & Midway Coal Mining Company
PMTC	Pittsburgh Mining Technology Center
PNL	Pacific Northwest Laboratory
PON	program opportunity notice
PPBS	Policy, Programming and Budget System
pph	pounds per hour
ppm	parts per million
PRDA	program research and development announcement
PSD	prevention of serious deterioration
psi	pounds per square inch
psia	pounds per square inch, absolute
psig	pounds per square inch, gauge

Q

Q_e	quads of electrical energy
Q_f	quads of fossil fuel consumption
Q_r	quads of heat rejection
quad	10^{15} (quadrillion) Btu

R

RCRA	Resource Conservation and Recovery Act
R&D	research and development
RD&D	research, development, and demonstration
REAP	Residual Energy Applications Program
RefCOM	refuse conversion to methane system
RESOX	proprietary process of Foster–Wheeler Corporation for conversion of sulfur dioxide to elemental sulfur
RFP	request for proposal
RI	Rockwell International
ROM	run of mine (coal)
RW	residue fuels

S

S	sulfur
SARS	Safety Analysis and Review System

SASOL	South African Coal, Oil and Gas Corporation, Ltd.
SBA	Small Business Administration
SBM	shaft boring machine
SCE	Southern California Edison
scf	standard cubic feet
scf/day	standard cubic feet per day
scf/h	standard cubic feet per hour
SCS	Southern Company Services, Inc.
SDB	steeply dipping bed
SDC	Systems Development Corporation
SEB	source evaluation board
sec	second
SECV	State Electricity Commission of Victoria
SEMS	seafloor earthquake measurement system
SERI	Solar Energy Research Institute
SFC	Synthetic Fuels Corporation
SGT	small gas turbine
SiC	silicon carbide
SIC	Standard Industrial Code
SIP	State Implementation Plans
SL	Sandia Laboratory
SNG	substitute natural gas
SOFC	solid oxide electrolyte fuel cell
SORU	shale oil refining and utilization
SO_2	sulfur dioxide
SO_x	oxides of sulfur
SRC	solvent-refined coal
SRL	solvent-refined lignite
SRT	short residence time
SSF	simultaneous saccharification and fermentation
STAG	steam turbine and gas (turbine)
$stdft^3$	standard cubic foot
STEAG	a power company in the Federal Republic of Germany
STP	standard temperature and pressure
Syngas	synthesis gas

T

TAS	Technology Alternatives Study
TBD	to be determined
TBM	tunnel boring machine
TC	technical committee
tcf	trillion cubic feet
t/d	tons per day
TEC	total estimated cost

TETAS	Total Energy Technology Assessment Study
TG	transport gas
t/h	tons per hour
TIS	Technical Information Service
TPD	tons per day
TPO	Texaco Information Service
TQ	transition quarter
TSL	two-stage liquefaction
TSP	total suspended particles
TTU	turbine test unit
TVA	Tennessee Valley Authority

U

UCG	underground coal gasification
UGR	unconventional gas recovery
U.K.	United Kingdom
UOP	Universal Oil Products
USGS	U.S. Geological Survey
UTC	United Technologies Corporation
UTRC	United Technology Research Center

V

V	volts
VMIS	vertical modified *in situ*

W

WAO	wet air oxidation
WGSP	Western Gas Sands Project
WVU	West Virginia University

Z

$ZnCl_2$	zinc chloride

Appendix D

Sources of Additional Information

- U.S. Synthetic Fuels Corporation
  1900 L Street, N.W.
  Washington, D.C. 20586
  (202) 653-4400
- U.S. Department of Energy
  Energy Information Administration
  Clearinghouse
  1726 M Street, N.W.
  Washington, D.C. 20461

  Office of Public Inquiries
  1000 Independence Avenue, S.W.
  Washington, D.C. 20585

  Technical Information Center
  P.O. Box 62
  Oak Ridge, Tennessee 37830
- Mine Safety and Health Administration
  U.S. Department of Labor
  4015 Wilson Blvd.
  Arlington, Virginia 22203
- National Coal Association
  1130 17th Street, N.W.
  Washington, D.C. 20036
- National Technical Information Service
  U.S. Department of Commerce
  Springfield, Virginia 22161
- U.S. Department of the Interior
  18th & C Streets, N.W.
  Washington, D.C. 20240
- Pennsylvania Anthracite Institute
  240 North Third Street, Suite 508
  Harrisburg, Pennsylvania 17101
- U.S. Environmental Protection Agency
  401 M Street, S.W.
  Washington, D.C. 20460
- U.S. Geological Survey
  The National Center
  12201 Sunrise Valley Drive
  Reston, Virginia 22092

Note: Many publications about synfuels and coal-related subjects are sold by the Superintendent of Documents, U.S. Government Printing Office, Washington, D.C. 20402. Reference copies of many of these publications are available to the public in federal depository libraries located nationwide. A list of depository libraries is available free upon request from: Library, Public Documents Department, U.S. Government Printing Office, Washington, D.C. 20402.

Index

D

E

F

G

H

I

J

K

L

M

R

S

T

U

W